Ionic Equilibria in Analytical Chemistry

Jean-Louis Burgot

Ionic Equilibria in Analytical Chemistry

Springer

Prof. Jean-Louis Burgot
Prof. Honoraire des Universités
France

ISBN 978-1-4419-8381-7 e-ISBN 978-1-4419-8382-4
DOI 10.1007/978-1-4419-8382-4
Springer New York Dordrecht Heidelberg London

Library of Congress Control Number: 2012932899

Printed on acid-free paper

Springer is part of Springer Science+Business Media (www.springer.com)

Foreword

"*Ionic Equilibria in Analytical Chemistry*" Draft English translation of "*Chimie analytique et équilibres ionique*" by Jean-Louis Burgot, professeur honoraire de chimie analytique.

When I started writing *Ionic Equilibrium* in the early 1960s [Addison-Wesley, 1964], I was trying to clarify in my mind a confusing part of the analytical chemistry curriculum that I was teaching. I found the key in mass and charge balances, which were already in use in Scandinavia through the efforts of Lars Gunnar Sillén and colleagues (see L. G Sillén, P. W. Lange, and C. O Gabrielson, *Problems in Physical Chemistry* (translated from the Swedish), New York: Prentice Hall, 1952).

Much later, after my *Carbon Dioxide Equilibria* was published, I was contacted by a group of medical scientists who were teaching mass and charge balances in treating the equilibria of physiological fluids. I also learned that the methods I had been teaching were used in quantifying the dissolution of carbon dioxide in the oceans, an important sink for the greenhouse gas produced by industrial activity. Still more recently, the acidification of the oceans by carbon dioxide has been of concern because of its potential dissolution of calcium carbonate skeletons of marine organisms.

Jean-Louis Burgot presented an elegant version in French of the methods by which equilibria in aqueous solutions are calculated, but it had yet to be brought to an English-speaking audience. His special sections on various analytical methods, such as those employing ceric ion, iodine, permanganate ion, and dichromate ion, are valuable. He amplifies his discussion of complex formation with examples using mercury, ethylene diamine tetra-acetic acid (EDTA), and other reagents in analysis of inorganic and organic materials. Following a discussion of the solubility product, he presents detailed precipitation titrations using silver ion and some gravimetric methods based on the same kind of reactions.

This book presents the "arithmetic" of quantitative analytical chemistry, and the more students are aware of this background, the more they will understand the methods and limitations of more automated methods, as well as the various computer methods for calculating solution equilibria. (See Chap. 12 by David R. Cogley in *Ionic Equilibrium* by James N. Butler, 1998 edition, John Wiley & Sons, New York.)

I am pleased to offer my support to a work that carries on the basic theory of solution chemistry. Even as simple exercises for musical instruments can lead to symphonic works, I hope that students who learn the elements of solution chemistry will be better equipped to use it in many practical applications.

James N. Butler
Professor emeritus,
Gordon Mc Kay,
Professor of applied chemistry,
Harvard school of engineering and applied sciences

Preface

A lot of operations in chemical analysis are carried out in solution, in particular in aqueous solutions. Because water is both a dissociating and ionizing solvent, the chemical reactions occurring within it are generally ionic reactions. From another standpoint, in order to obtain valid analytical conclusions, the chemical reactions proposed to perform the analysis must be carried out until their term, that is, until their equilibrium has been reached.

One of the great interests in the analytical chemistry practiced in aqueous solutions lies in the fact that it can be quasi-systematically described by mathematical equations, which can be grouped themselves in mathematical systems, systems being, in principle, always determined. Because they are not so easily soluble, they induce the use of informatics or the adoption of pertinent simplifications. Without any doubt, adopting pertinent simplifications is a difficult task for anybody who has not thoroughly mastered the discipline. Therefore, in my opinion, it is imperative, systematically, to write first all the equations describing the phenomena in solution that must be satisfied. Next, one or several simplifications may be made according to the conditions of the experiment. The results of the calculations issuing from the preceding simplifications can then be challenged experimentally in the third stage. This methodology constitutes the heart of this book.

Briefly, it is the fortunate balance existing between the theoretical aspects resulting from the mathematical equations governing the phenomena and the purely experimental aspects that confer the great academic interest on analytical chemistry in aqueous solutions. Therefore, analytical chemistry appears to be as close to pure physics as to chemistry. Furthermore, the IUPAC classifies analytical chemistry in the realm of physical chemistry. It is one of the goals of this book to provide an example of such an assertion.

Analytical chemistry consists of studying the physical and chemical phenomena that are applied in chemical analysis. Hence, the two disciplines should not be confused. However, in this book, numerous examples coming, of course, from the realm of chemical analysis are given in order to illustrate the principles of analytical chemistry that are studied. These examples are taken from the fields of inorganic and organic chemistries and even from that of biochemistry. Finally, a large place is given to the analysis of pharmaceutically active ingredients.

I particularly want to thank Professor Maurice Bernard, honorary dean of the faculty of sciences of Caen, for having introduced me to and helped me into this kind of chemistry. I do not forget the very fascinating time during which I worked with him. I also thank my wife, Gwenola Burgot, professor of analytical chemistry in the faculty of pharmacy of the University of Rennes I, who, of course, immediately understood the interest in such a book and who has always encouraged me to continue the great task that was its writing. I still thank Gabriel Gorre, holder of the superior chair of physics in the Lycee Joliot-Curie of Rennes, for his thorough reading of the manuscript.

My great thanks also go to André Le Goff, who was one of my English teachers in school. He has considerably helped me in the translation of the French version of this book into the English one. My English needed his help!

Finally, I also thank Annick Simon-Malard for the diligency and devotion she exhibited during the preparation of the different versions of this book.

Last but not least, I dedicate this book to James Newton Butler, professor of chemistry in the division of applied sciences at Harvard University, for the book he wrote in 1964 entitled *Ionic Equilibrium, a Mathematical Approach*. My book has only the pretense of being a continuation of his masterpiece in analytical chemistry.

Rennes, France
May 2010

Prof. Jean-Louis Burgot

Contents

Part I
General Considerations

Chapter 1
Solvents—Composition of Solutions

Numerous operations involved in chemical analysis are carried out in solutions. For this reason, it is interesting to recall some of their properties. In particular, we will recall the expressions of their composition.

1.1 Definitions

A *solution* is a liquid phase containing at least two different substances. Here we will consider only liquid solutions. One of the components is called the *solvent*, the other one the *solute*. We arbitrarily determine which is the solvent and which is the solute. Nevertheless, if one of the components is liquid, it is usually called the solvent. In this case, the liquid will dissolve the solute. Most of the time, the solvent is the major component of the solution. Let's anticipate what we will determine later by saying that in terms of the activities of components in dilute solutions, the distinction between solvent and solutes is of major importance.

1.2 Molecular Solvents

Molecular solvents are the ones most commonly used in chemical analysis. Their principal characteristic is that they are composed of molecules that more or less self-associate. They conduct electric current very weakly and can be organic or mineral. Among organic molecular solvents are alcohols, ketones, carboxylic acids, amines, and others, and among mineral molecular solvents are nitric and sulfuric acids, liquid ammonia, and others . The most striking example of a molecular solvent is water.

The solvating properties of molecular liquids are related to some of their physical properties, such as their relative permittivity, their dipole moment, and their ability to form hydrogen bonds (H bonds) (see Sect. 1.4).

J.-L. Burgot, *Ionic Equilibria in Analytical Chemistry*,
DOI 10.1007/978-1-4419-8382-4_1, © Springer Science+Business Media, LLC 2012

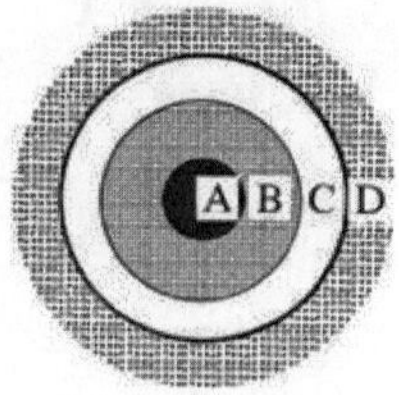

Fig. 1.1 Model of a solvated ion in water

1.3 Solvation of Solutes in a Molecular Solvent

The phenomenon of dissolution can be thought as being the reduction of two phases (the initial ones) into only one, the solution. It involves physicochemical properties that depend on both the solute and the solvent. The solvent action is called *solvolysis*. Dissolution results in the integration of the substance among solvent molecules. As a consequence, the substance has become the solute. This integration is the consequence of solute–solvent interactions called *solvation phenomena*. As a result, the solute is spread among solvent molecules in the state of either solvated molecules or solvated ions. Compounds giving rise to solvated ions after dissolution are known as *electrolytes*. We can distinguish them into two groups:

- the *ionophores*, which are already ions in the solid state (e.g., sodium chloride),
- the *potential electrolytes*, which are molecules in the pure state giving rise, at least in part, to ions after dissolution and solvolysis.

It is worth noting that there exist not only solute-solvent interactions but also solute-solute interactions. These latter are at the origin of the behavior of weakly diluted solutions. This leads to the concept of activity (see Chap. 3).

We already know a lot about the state of the ions in aqueous solutions, but we can explore them further. An often cited model is given in Fig. 1.1.

Ions are enveloped by a first solvation sheath in which the solvent molecules are in direct interaction with the ion. This appears in region A in Fig. 1.1. Beyond region A exists a second solvation sheath (region B). Although solvent molecules are no longer in direct interaction with the ion, they are still under the influence of its electrical field. In sheath D, the solvent molecules are sufficiently far from the ion not to be influenced by it. Sheath D is the bulk water. Between sheaths B and D, water molecules are more or less influenced by the electrical field. This is sheath C. The latter sheath seems to exist only in solvents that are well structured by H bonds, such as water and alcohols.

The numbers of molecules of water in sheaths A and B are not well known yet, and their determination remains a challenge. Several methods have been used, including nuclear magnetic resonance (NMR) spectroscopy and measurements of transport properties, such as the ionic conductivities, mobilities, and viscosities of solutions. One major difficulty encountered is the residence-time interval during which the solvent molecule is in interaction at the ion's surface. It is generally very brief. It can be much less than 10^{-4} s. This is the time interval required in NMR for the nucleus to be under a peak distinct from that given by the solvent.

The primary solvation number most frequently encountered in water seems to be 6. In the block of the sp elements, this is the case with Mg^{2+}, Zn^{2+}, Al^{3+}, Ga^{3+}, In^{3+}, and Sc^{3+}. Only Be^{2+} seems to exhibit number 4. Among transition ions, V^{2+}, Mn^{2+}, Fe^{2+}, Co^{2+}, Ni^{2+}, Ti^{3+}, V^{3+}, Cr^{3+}, and Fe^{3+} exhibit the number 6. For Pd^{2+} and Pt^{2+}, the number is 2, and for thorium (actinid), it is 9.

1.4 Water as Solvent

Water is a solvent of the utmost importance from several points of view. This is the reason why the chemistry in aqueous solutions is the most well known of all the "in-solution chemistries."

Let us recall, without describing all its physicochemical properties, that liquid water consists of networks of associated water molecules through the formation of H bonds in equilibrium with free water molecules. Its solvating properties result from

- its ability to form H bonds,
- its dipolar moment high value,
- its dissociating power in relation to its high relative permittivity.

1.4.1 Ability to Give H Bonds

Its ability to give H bonds permits water to dissolve covalent compounds, which possess hydroxyl and amino groups, that is, groups with electronegative atoms of a weak volume.

1.4.2 High Value of Its Dipolar Moment

The dipolar moment of water is high: $\mu = 1.87\,\text{D}$ (D: debye) ($1\,\text{D} = 3.336 \cdot 10^{-3}$ cm). In a water molecule, the oxygen atom is located near the center of a tetrahedron, with the two OH bonds directed toward two corners (angle $\cong 180°$). The oxygen atom also brings two lone pairs of electrons. Their orbitals are directed toward the other two corners. Due to the respective electronegativities of oxygen and hydrogen, the center of gravity of the negative charge does not coincide with that of the positive charge (Fig. 1.2).

Sometimes the water molecule is simply represented by the dipole vector. Hence, the water molecule can be assimilated to an electrical dipole from the electrical standpoint.

The electrical quadrupolar model is a more accurate representation of the water molecule. It consists of two negative electrical charges located near the oxygen atom and of one positive electrical charge located near each of the two hydrogen atoms (Fig. 1.3).

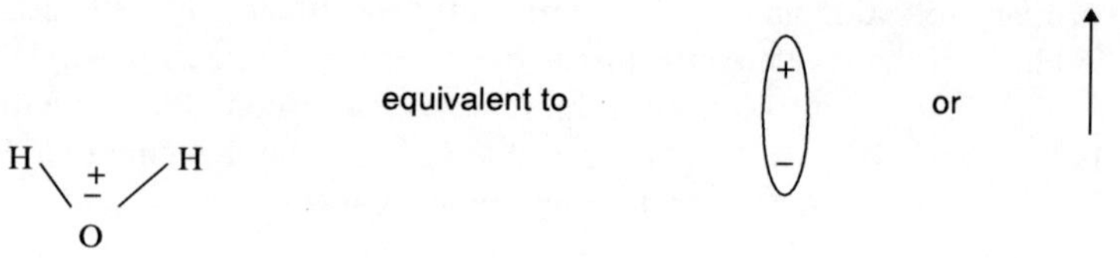

Fig. 1.2 Dipolar character of water molecule and simplified representations: + gravity center of positive charges; – gravity center of negative charges

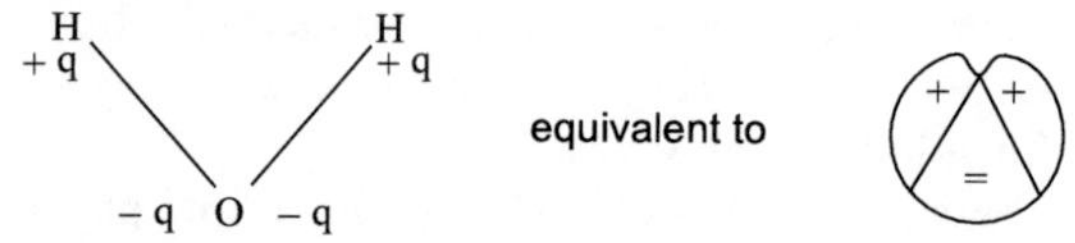

Fig. 1.3 The quadrupolar model of the water molecule

The dipolar character of water and, even more so, its quadrupolar one are interesting for three reasons:

1. If the substance to solubilize is an ionophore, the dipole exerts an attractive force on the electrolyte's bonds,. inducing its dissolution. Figure 1.4 represents this process in the case of sodium chloride.

   The whole process can be symbolized by the equation

   $$A^+B^-_{(s)} + (p+q)H_2O \rightarrow A^+, pH_2O + B^-, qH_2O,$$

   where p and q are the numbers of water molecules solvating the ions A^+ and B^-, and p and q are not yet exactly known because of the occurrence of several solvation sheaths,
2. its dipolar property permits the solvation of covalent molecules by water through the development of Van der Waals, Keesom, Debye, and London forces. This solvation process is an electrostatic one, as in the preceding case. Both are physical processes.

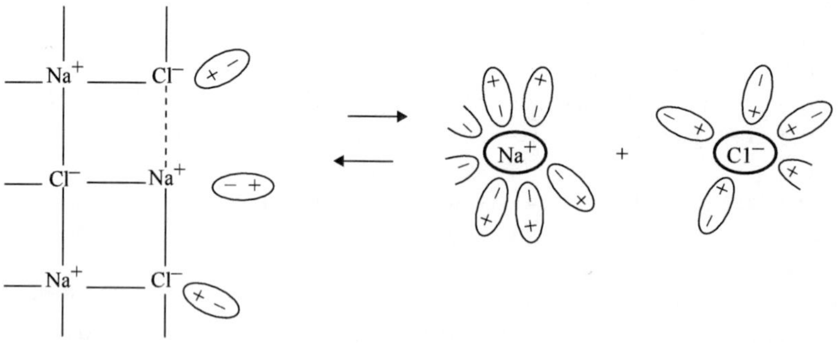

Fig. 1.4 Electrostatic attraction between the ions of the ionophore in the solid state and the water dipoles

3. it can induce the ionization of some molecules, thereby revealing their character of potential electrolytes. In this process, not only is the dipolar character of water operant, but its amphoteric character is operant as well (see Chap. 4). This process of solvolysis by water is a chemical one since there is a break of a covalent bond. It is helped, of course, by the dipolar character of water.

The quadrupolar model of water is of great theoretical importance. Indeed, it justifies some extrapolating procedures that have permitted the determination of some properties of individual ions. Prior to these procedures, this determination was quite a challenge. In fact, it must be emphasized that a single ion alone cannot be introduced into a solvent. It must be accompanied by a counter ion of the opposite electrical charge with respect to the electroneutrality principle. As a result, only the properties of the whole electrolyte can be determined in the realm of classical thermodynamics.

1.4.3 Dissociating Power of Water

Water is endowed with a high value of its relative permittivity (dielectric constant) ε_r. Let us recall that if two ions of opposite electrical charges are forming in a solution, they will attract each other according to Coulomb's law:

$$f = z_1 \cdot z_2 e^2 / d^2 \varepsilon_r,$$

where f is the intensity of the electrostatic force, $z_1 e$ and $z_2 e$ the charges of the ions (e is the elementary charge 1.602 18.10.10^{-19} C, and z_1 and z_2 their charge numbers), and d the distance between both ions. The water dielectric constant is high ($\varepsilon_r = 80$ at 25°C); as a result, the attractive force is weak. This is the reason why the formation of ion pairs is negligible in water except in the case of polycharged species in high concentrations. The thermal agitation in water is sufficient to separate ions. In fact, an ion pair can be defined as a chemical entity in which both ions are trapped in each other's coulombic field. Unlike free ions, both ions in the ion pairs can no longer continue their independent movement.

1.5 Definition of the Solution Composition

The *solution composition* is the relative proportion of the solute and of the solvent. Until now, we have referred to it as the solution concentration. This definition is no longer valid since the concept of concentration has now been endowed with a more limited meaning (see ahead).

1.6 Quantity of a Substance

The quantity of a substance is the number of moles of a substance in a sample. Its symbol is n. Its SI unit is the mole, abbreviated as "mol." For example, a sample chlorine quantity is written as $n(Cl_2)$ mol or n_{Cl_2} mol.

1.7 Different Expressions of the Composition

We mention here only the most frequently used expressions.

1.7.1 *Composition Expressed in Quantity of a Substance: The Molar Composition*

The composition in the quantity of a substance is the solute mole number (quantity of substance) contained in a given solution volume. If n_i is the quantity of substance i contained in the volume V, the concentration of i is

$$C_i = n_i/V.$$

The symbol is C. Its SI unit is mol/m^3. For practical reasons, mol/dm^3 and mol/L are used in chemical analysis[1]. A 1 mol/L solution is often called a *molar solution*, written 1 M. The word *molarity* is often incorrectly used instead of the word *concentration*. The symbol [i] is very often encountered instead of C_i.

Exercise 192.6 g of potassium nitrate is dissolved in water in order to obtain 1 L of solution. Calculate the solution concentration (the molar weight of potassium nitrate is $101 \cdot 10^{-3}$ kg/mol):

$$C(KNO_3) = 192.6/101 \text{ mol/L},$$
$$C(KNO_3) = 1.906 \text{ mol/L}.$$

It is worth noting that V is the volume of the whole solution once the solution is at equilibrium. It is not the volume of the pure solvent used to prepare the solution. In other words, V is the sum of the volumes occupied by the solvent and by the solute in the final solution. That is to say, V is the sum of their two partial molar volumes (see Chap. 2).

1.7.2 *Molality*

The *molality* of a solution is the mole number of the solute for one kilogram of pure solvent. Its symbol is m. Its SI unit is mol/kg. If the quantity of solute i dissolved into the mass m_0 of pure solvent is n_i, the molality m_i[2] is

$$m_i = n_i/m_0.$$

---

[1] Since the General Conference on Weights and Measures in 1964, the liter has been a special name of the cubic decimeter.

[2] Unfortunately, m has two meanings here: that of a mass for the solvent (m_0) and that of a molality for the solute (m_i). The context often provides an easy answer to the ambiguity.

The molality is of little use in analytical chemistry but is used very often in physical chemistry. Its great interest is that it is not dependent on temperature. This is not the case for the concentration. For diluted aqueous solutions, the molality value of a solution is close to that of its molarity. The more diluted the solution is, the closer both values are. For the example given before,

$$M_{KNO_3} = 2.004 \text{ mol/kg } (1.906 \text{ mol/L}) 25°C.$$

The numerical values are close despite the fact that the solution is not very diluted in the analytical meaning.

1.7.3 Molar Fraction

The *molar fraction* is the ratio of the mole number of the solute and the overall number of the whole solution. Its symbol is x. It is a dimensionless number. Let us consider a binary solution. If n_i and n_0 are, respectively, the mole numbers of the solute and of the solvent, the molar fractions are

$$x_i = n_i/(n_0 + n_i).$$

and

$$x_0 = n_0/(n_0 + n_i).$$

For the example given above, we find

$$x(KNO_3) = 0.0348, \quad x(H_2O) = 0.9652.$$

The numerical value $x(KNO_3)$ differs significantly from those of $C(KNO_3)$ and $m(KNO_3)$. Other composition expressions exist as well. Let us mention the normality, which is still frequently used. However, the International Union for Pure and Applied Chemistry (IUPAC) does not recommend its use. The normality and its intimately associated concept of equivalent are encountered in titrimetric analysis. They will be introduced in the Chap. 7 devoted to general considerations concerning titrations.

1.8 Calculation of the Molality and the Molarity of a Solution from Its Molar Fraction

Calculating the molality and molarity of a solution from its molar fraction is not trivial. Such calculations involve the solution's density. They emphasize the fact that in the molarity, it is the solution volume that is considered, whereas in the molality, it is the mass of the pure solvent.

Let's suppose that the solution contains n_0 solvent moles, n_A, n_B moles of solutes A and B. Let us consider solute A:

By definition:

$$x_A = n_A/\Sigma n_i, \quad \text{with} \quad \Sigma n_i = n_0 + n_A + n_B + \ldots \tag{1.1}$$

and

$$C_A = n_A/V, \tag{1.2}$$

where V is the solution volume. We want to find a relationship between C_A and x_A. The solution mass $\Sigma n_i M_i$ (grams) is

$$\Sigma n_i M_i = n_0 M_0 + n_A M_A + n_B M_B + \ldots,$$

and so forth, where M_0, M_A, and M_B are the molar masses (g/mol) of the solvent and of the solutes A, B, and so forth. The solution volume V is

$$V = \Sigma n_i M_i/\rho \cdot 1000, \tag{1.3}$$

where ρ is the volumic mass of the solution (g/cm^3). The factor 1,000 permits us to express the volume in liters. We deduce immediately from Eqs. (1.1), (1.2), and (1.3) that

$$C_A = (1000\rho/\Sigma n_i M_i)n_A, \tag{1.4}$$

$$C_A = (1000\rho/\Sigma n_i/\Sigma n_i M_i)x_A. \tag{1.5}$$

From a strict mathematical standpoint, C_A and x_A are not in a linear relationship since Σn_i and $\Sigma n_i M_i$ do change with x_A. However, the linear relationship appears when the solution is sufficiently diluted. In this case, indeed,

$$\Sigma n_i \cong n_0$$

and

$$\Sigma n_i M_i \cong n_0 M_0.$$

Thereby we have

$$C_A = (1000 \cdot \rho/M_0)x_A, \tag{1.6}$$

where ρ is the volumic mass of the pure solvent since the solution is diluted. In particular, for water at 25°C,

$$\rho \cong 1,$$

$$C_A = (1000/M_0)x_A \quad \text{(room-temperature water)}. \tag{1.7}$$

Then, by definition:

$$m_A = 1000 \cdot n_A/n_0 M_0. \tag{1.8}$$

The factor 1,000 is introduced since M_0 is expressed in grams while m_A is in mol/kg[1]. By introducing (1.1) into (1.8), we find

$$m_A = (\Sigma n_i \cdot 1000/n_0 M_0)x_A. \tag{1.9}$$

Again, m_A and x_A are not in a linear relationship. However, in diluted solutions,

$$\Sigma n_i \cong n_0,$$

$$m_A = (1000/M_0)x_A. \tag{1.10}$$

Relations (1.5) and (1.9) permit the calculation of C_A and m_A from x_A. The relationship between C_A and m_A is obtained after eliminating x_A in (1.5) and (1.9). We find

$$m_A/C_A = \Sigma n_i M_i/\rho n_0 M_0. \tag{1.11}$$

Relation (1.11) clearly shows that in a diluted aqueous solution,

$$\Sigma n_i M_i \cong n_0 M_0,$$

$$m_A \cong C_A$$

since $\rho \to 1$ and

In brief, it is only in sufficiently diluted aqueous solutions that numerical values of the molarity and the molality of a solute are equal and that, in these conditions, the molarity and the molality are proportional to their molar fraction.

Chapter 2
Thermodynamics and Equilibrium

The concept of the equilibrium state of a system is of utmost importance in analytical chemistry. To illustrate this assertion briefly, recall that numerous chemical reactions are performed with analytical goals in mind. The conclusions they provide can be easily reached if and only if the reactions proceed through to their natural completion, that is, reach their equilibrium state.

The equilibrium concept cannot be separated from that of the free enthalpy change during the process of concern. In turn, the free enthalpy concept comes directly from the second principle of thermodynamics. In actuality, the free enthalpy concept is nothing more than the second principle expressed in a form that is particularly easy for chemists to handle, at least in some well-defined experimental conditions. These are the reasons why in this chapter we recall some general conclusions stemming from the thermodynamics framework.

The link between thermodynamics and analytical chemistry is strikingly apparent when we consider the work performed by electrochemical cells. This is done with the help of Nernst's law. Due to its importance, we will review it in detail. It will be particularly useful for the study of redox phenomena.

2.1 Chemical Potential

The *chemical potential* μ_i of a substance i is expressed in term of its activity a_i. The relationship between them is as follows:

$$\mu_i = \mu_i^\circ + RT \ln a_i . \tag{2.1}$$

μ_i° is the chemical potential of i in its standard state. More simply, it is its standard chemical potential. The definition of the standard state is given below. Of importance now is the fact that μ_i° is a constant characteristic of i for a given temperature. (In the case of solutions, it depends only weakly on pressure.) R is the ideal gas constant and T the absolute temperature of the system. a_i is the activity of species i in the experimental conditions, more clearly in the thermodynamic state of the system (see below). For this brief recall, it is sufficient to visualize an activity as a kind of

J.-L. Burgot, *Ionic Equilibria in Analytical Chemistry*,
DOI 10.1007/978-1-4419-8382-4_2, © Springer Science+Business Media, LLC 2012

concentration. We must also know that in the case of solutes, the numerical value of a_i tends toward its concentration value C_i when the solution becomes more and more diluted. The differences between the numerical values of the activities and those of the corresponding concentrations are more pronounced for ions than for uncharged compounds. Finally, to terminate temporarily with the activity concept (see Chap. 3), let's note that an activity is quantified by a dimensionless number. Coming back to relation (2.1), we notice that the chemical potential of a substance is equal to its standard potential when its activity is unity (see Chap. 3).

A chemical potential is an intensive partial molar quantity. Its value does not depend on the sample mass since, by definition, a molar quantity concerns one mole. The chemical potential is also called the *partial molar free enthalpy* G_i:

$$\mu_i = G_i.$$

In order to grasp the very deep meaning of the chemical potential, consider that G. N. Lewis[1] described it as the "escaping tendency" of the species. The escaping tendency is the tendency of the substance to leave its thermodynamic state by either a physical or chemical process. A simple example taken in the realm of physical processes involves the partitioning of a solute i between two immiscible phases α and β. In the initial state (which is not the equilibrium state of the system), the solute is supposed to be wholly present in phase α. Its chemical potential in this phase is $\mu_{i\alpha}$, while it is $\mu_{i\beta} = 0$ in the other:

$$\mu_{i\alpha} \neq \mu_{i\beta}$$

and

$$\mu_{i\alpha} > \mu_{i\beta}.$$

By the mechanical stirring of both phases, the system is equilibrated. A fraction of the solute goes into phase β. The state of equilibrium partitioning is attained when the solute concentrations in both phases no longer change with time. Then the chemical potentials of the solute in both phases are equal:

$$\mu_{i\alpha} = \mu_{i\beta}.$$

This is the condition of *equilibrium partitioning*.

Partitioning occurred because the chemical potentials in both phases were different in the initial state. Therefore, it appears that the chemical and electrical potentials play analogous parts. There is a displacement of electrons because of the electrical potential difference between two points of a circuit. Likewise, a chemical reaction takes place because, at a given moment, a linear combination of the chemical potentials of the reactants and products differs from zero (see below).

---

[1] Gilbert Newton Lewis: American physicist and chemist (1875–1946), especially well known for his works concerning covalency and one definition of acids and bases.

- For a solution of n components including the solvent, the Gibbs free energy of the whole chemical system (solvent plus solutes) is given by the relationship

$$G_{\text{syst}} = n_1\mu_1 + n_2\mu_2 + \ldots ;\ n_i\mu_i + \ldots + n_n\mu_n \tag{2.2}$$

or

$$G_{\text{syst}} = \sum n_i\mu_i,$$

where the n_i are the numbers of moles of each component. In a binary solution composed of the solvent and the solute only, the Gibbs free energy is given by

$$G_{\text{syst}} = n_0\mu_0 + n_1\mu_1,$$

where the subscript 0 is usually related to the solvent. Contrarily to the case of chemical potentials, the Gibbs free enthalpy G is an extensive property since, quite evidently, it depends on the components' mole numbers.

- The chemical potentials or partial molar free enthalpies are defined by the following partial derivative:

$$\mu_i = G_i = (\partial G_{\text{syst}}/\partial_{ni})_{T,P,nj}(i \neq j). \tag{2.3}$$

The chemical potential is the partial derivative of the free enthalpy of the whole system with respect to the mole number of i; the other variable values (temperature, pressure, and other species mole number n_j) remain constant. It is also the change in the system free enthalpy when the mole number n_i is varied by an infinitely small amount, while other variables defining the thermodynamic state remain constant. Most of the time, partial molar quantities do vary with the system composition. This is the case with chemical potentials. A familiar example is given by methanol/water mixtures. The total volume V_{syst} of the mixture is given by the relationship

$$V_{\text{syst}} = n_W V_W + n_m V_m,$$

where V_w and V_m are the partial molar volumes of water and methanol in the solution. They are defined by the relationships

$$V_w = (\partial V_{\text{syst}}/\partial n_w)_{T,P,nm} \quad \text{and} \quad V_m = (\partial V_{\text{syst}}/\partial n_m)_{T,P,nw}.$$

It is a well known fact that V_w and V_m change with the mixture composition. Hence, if the percentage in mass of methanol is 40%, $V_m = 39.0\,\text{cm}^3/\text{mol}$ and $V_w = 17.5\,\text{cm}^3/\text{mol}$. If it is 60%, $V_m = 39.8\,\text{cm}^3/\text{mol}$ and $V_w = 16.8\,\text{cm}^3/\text{mol}$. The molar volumes of the pure components are $V_m = 40\,\text{cm}^3/\text{mol}$, $V_w = 18\,\text{cm}^3/\text{mol}$ ($T = 298.15$ K and $P = 1$ bar).

In some exceptional cases, the partial molar quantities do not change with the composition. Their values are then purely and simply equal to those of the molar quantities of the components in their pure state.

The absolute values of some partial molar quantities, such as the volume, can be determined experimentally. Other absolute values, however, cannot be; these include partial molar enthalpies, free enthalpies, and free entropies. However, their changes are measurable.

2.2 Gibbs Free Energy Change ΔG_{syst} and Useful Work Available from the Process

Let's consider a closed system in which the studied process takes place and examine its surroundings. According to the definition of a closed system, the system exchanges work and heat with its surroundings but does not exchange matter. If the initial and final temperatures are equal and also equal to the constant temperature of surroundings (heat bath) and the surroundings are displaced at constant pressure, the maximum work W' that can be done by the system of interest is equal to its Gibbs free energy decrease:

$$W'_{\text{max}} = \Delta G_{\text{syst}} \quad \text{(reversible process)}.$$

W' does not include the work of displacement. W' is the useful work. It can be, for example, an electrical work. This is the maximum work obtained from a galvanic cell (see Sect. 2.8). The preceding relationship holds only in the case of a reversible process. Let's examine the concept of reversibility in some depth. The reversibility conditions are approached when the process takes place very slowly. In practice, a galvanic cell is an interesting device to obtain quasi-reversibility. When the process is irreversible, the following inequality holds:

$$|W'| < |\Delta G_{\text{syst}}| \quad \text{(irreversible process)}.$$

In this case, the work done is less than the decrease of the Gibbs free energy of the system.

All the preceding considerations are quite general provided that only closed systems are under consideration. The processes can be physical, chemical, or biological, and they can be applied to the case of chemical reactions. In the usual experimental conditions in which the latter are performed, the variables that define the thermodynamic state of the system are its temperature, pressure, and composition. The thermodynamic system itself can be defined by the reactants and the products of the reaction. Its Gibbs free energy change during the process is given by

$$dG_{\text{syst}} = -S\,dT + V\,dP + \sum_{i=1}^{n} \mu_i dn_i. \tag{2.4}$$

This relationship comes from general considerations. S and V are the entropy and the volume of the system, respectively, while n_i and μ_i are, respectively, the mole numbers and the chemical potentials of the species. Let's consider the following reaction:

$$\nu_A A + \nu_B B \rightarrow \nu_M M + \nu_N N, \tag{2.5}$$

which is often written as

$$\sum \nu_J R_J = 0.$$

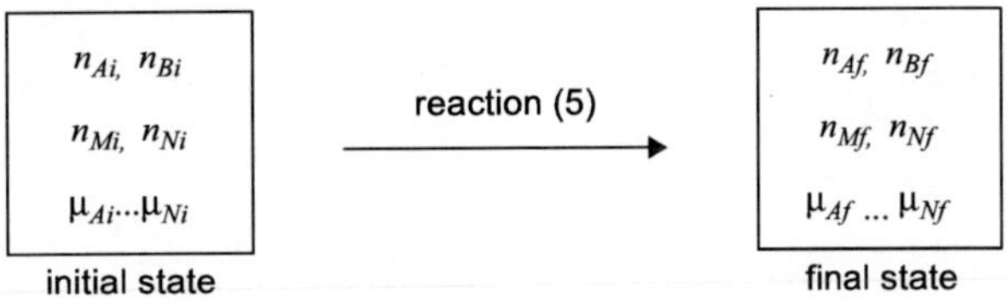

Fig. 2.1 Definition of the initial and final states of reaction (2.5)

For the reactants (A, B), the stoichiometric coefficients are preceded by the minus sign because of the decrease in their concentrations during the reaction.

At the beginning of the process, the system consists of species A and B and possibly of species M and N, which can already be present in the reactor. Suppose that the reaction takes place at a constant pressure and temperature. The mole numbers of ions facing each other are n_{Ai}, n_{Bi}, n_{Mi}, and n_{Ni} and their chemical potentials are μ_{Ai}, μ_{Bi}, μ_{Mi}, and μ_{Ni} (i: initial).The initial state of the system is fully defined since the composition, temperature, and pressure are known. In the final state, the mole numbers of species have changed. They have become n_{Af}, n_{Bf}, n_{Mf}, and n_{Nf} (f: final), and the corresponding chemical potentials are μ_{Af}, μ_{Bf}, μ_{Mf}, and μ_{Nf}. The latter have changed quite evidently since the composition has changed (Fig. 2.1). (Incidentally, it is interesting to note that the final state can be chosen arbitrarily. For example, it can be realized experimentally by abruptly stopping the reaction at a given moment. Of course, the final state may also be the state in which equilibrium is achieved.)

The Gibbs free energy change during the process is

$$\Delta G_{\text{syst}} = \left[n_{Af}\mu_{Af} + n_{Bf}\mu_{Bf} + N_{Mf}\mu_{Mf} + n_{Nf}\mu_{Nf}\right] - \left[n_{Ai}\mu_{Ai} + n_{Bi}\mu_{Bi} + n_{Mi}\mu_{Mi} + n_{Ni}\mu_{Ni}\right]. \tag{2.6}$$

Equation (2.6) gives the maximum useful work available from the chemical reaction.

An inconsistency between Eqs. (2.4) and (2.6) appears at first sight. In fact, let's write Eq. (2.6) in its compact form:

$$\Delta G_{\text{syst}} = \sum n_j \mu_j$$

and differentiate it. We find

$$dG_{\text{syst}} = \sum n_j d\mu_j + \sum \mu_j dn_j.$$

By differentiation, Eq. (2.4) gives

$$dG_{\text{syst}} = \sum \mu_j dn_j;$$

both relationships were obtained at a constant pressure and temperature. The inconsistency comes from the fact that in the last equation, the term $\sum n_j d\mu_j$ is missing. This inconsistency is merely apparent.

The missing term, $\sum n_j d\mu_j$, indeed, is null; that is, the chemical potentials of a mixture's components cannot change independently of each other at constant pressure and temperature. The relationship

$$\sum n_i d\mu_i = 0 \quad (dT = 0,\ dP = 0)$$

is called Gibbs–Duhem's[2] equation. It originates in the mathematical nature of free enthalpy. The free enthalpy is a first-degree homogeneous function with respect to the mole number of each constituent.[3]

2.3 Molar Reaction Gibbs Function

The solute mole numbers change during a chemical reaction. They are linked by the equalities

$$(dn_A/\nu_A) = (dn_B/\nu_B) = (dn_M/\nu_M) = (dn_N/\nu_N) = d\xi.$$

ξ is called the *extent reaction.* In these equations, the changes dn_A and dn_B are negative because A and B disappear during the reaction, dn_M and dn_N being positive. When taking these equalities into account, the free enthalpy change accompanying reaction (2.5), given by Eq. (2.4) at constant temperature and pressure, becomes

$$dG_{\text{syst}} = \left(-\nu_A\mu_A - \nu_B\mu_B + \nu_M\mu_M + \nu_N\mu_N\right) d\xi \quad (dT = 0,\ dP = 0). \tag{2.7}$$

The entire term within parentheses in Eq. (2.7) is called the *molar reaction Gibbs function*, $\Delta_r G$. More precisely, it is the molar free enthalpy change accompanying the reaction at a given instant of its evolution, that is, for an extent ξ for which the instantaneous chemical potentials are μ_i:

$$\Delta_r G = -\nu_A\mu_A - \nu_B\mu_B + \nu_M\mu_M + \nu_N\mu_N. \tag{2.8}$$

The molar reaction Gibbs function varies with respect to the extent reaction. After Eq. (2.7), it is the partial derivative of the system free enthalpy with respect to ξ at constant pressure and temperature.

We must not confuse the free enthalpy change ΔG_{syst} accompanying a chemical reaction and the molar reaction Gibbs function change $\Delta_r G$. They do not have the same mathematical status. The former change is a difference, the latter a partial derivative. They no longer have the same physical status. The former is an extensive quantity, the latter an intensive one.

---

[2] Pierre Duhem: French mathematician, physicist, and philosopher (1861–1916).

[3] See Euler's theorem in mathematics.

The change dG_{syst} is negative when the process is spontaneous. As a result, at constant pressure and temperature, the molar reaction Gibbs function $\Delta_r G$ must be negative in the case of a spontaneous process:

$$\Delta_r G < 0 \quad \text{(spontaneous process).}$$

In the case of reaction (2.5), we can write

$$\nu_A \mu_A + \nu_B \mu_B > \nu_M \mu_M + \nu_N \mu_N \quad (dT = 0,\ dP = 0).$$

The molar reaction Gibbs function $\Delta_r G$ quantity, whose algebraic sign has been inversed, is called the *chemical affinity* (de Donder's definition). It is positive for a spontaneous process:

$$A = -\Delta_r G,$$

where A is the chemical affinity.

2.4 Evolving Reactions and Equilibrium Conditions

According to the meaning of the free enthalpy function, the equilibrium state is obtained when

$$dG_{\text{syst}} = 0 \quad \text{(equilibrium).}$$

Then the system free enthalpy G_{syst} exhibits its minimum value. When the evolving system consists of reactants together with their reaction products, the equilibrium state is obtained for the extent ξ' such as

$$\left(\partial G_{\text{syst}}/\partial \xi\right)_{T,P,\xi=\xi'} = 0 \quad (\text{equilibrium};\ dT = 0,\ dP = 0).$$

Then Eq. (2.7) shows that

$$\nu_A \mu_A + \nu_B \mu_B = \nu_M \mu_M + \nu_N \mu_N.$$

or

$$\Delta_r G = 0 \text{ or } A = 0.$$

As a result, for a spontaneous chemical change starting from the sole reactants A and B, the molar reaction Gibbs function $\Delta_r G$ becomes increasingly less negative when the extent of the reaction increases. Likewise, the reaction course starting from the sole reaction products is such that the molar reaction Gibbs function is increasingly less positive. Each point of the G/ξ curve can represent either the initial or the final instant of the reaction. Except for the farthest points, each curve point represents a state where the reactants and products A, B, M, and N are all present in the mixture.

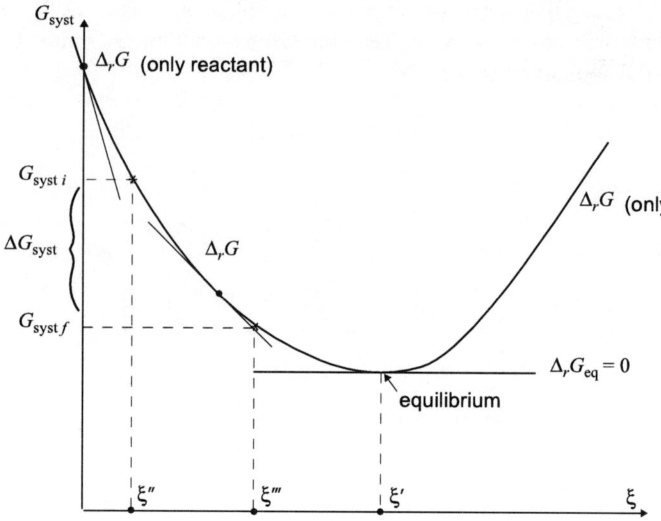

Fig. 2.2 Difference between the molar reaction Gibbs function and the free enthalpy change during a chemical reaction

Their chemical potentials are related to each other through Eq. (2.8). For example, the reaction can be initiated for $\xi = \xi''$ and stopped for $\xi = \xi'''$.

The extreme points represent the cases in which only reactants and only product are present in the mixture.

The system free enthalpy change is then (f: final, i: initial)

$$\Delta G_{syst} = G_{syst\,f} - G_{syst\,i}.$$

Figure 2.2 shows that $\Delta_r G$ also changes all along the process. At $\xi = \xi'''$, the reaction equilibrium is not obtained since the slope $\Delta_r G$ is not null. Slopes are maximal when the sole reactants or products are present in the reaction mixture (extreme points).

A question unavoidably comes to mind: What the equilibrium conditionsmust be respected when several chemical reactions simultaneously evolve? This is a problem encountered in chemical analysis when, for example, we titrate a mixture of two compounds and when they can react simultaneously with the common titrant. The answer to the question comes from the general criterion $dG_{syst} = 0$. It is inferred from this condition that the molar reaction Gibbs functions $\Delta_r G$ of each reaction must be simultaneously null provided that the different reactions are independent of each other. For example, in a system containing the compounds CO, H_2, H_2O, CH_3OH, and C_2H_6, all in gaseous state, only two independent reactions exist:

$$CO + 2H_2 \rightarrow CH_3OH,$$

$$2CO + 5H_2 \rightarrow C_2H_6 + 2H_2O.$$

The general equilibrium conditions results in the simultaneous occurrence of the two equalities

$$\mu_{CH_3OH} - \mu_{CO} - 2\mu_{H_2} = 0,$$
$$\mu_{C_2H_6} + 2\mu_{H_2O} - 2\mu_{CO} - 5\mu_{H_2O} = 0 \quad \text{(equilibrium)}.$$

The problem is to recognize the independent reactions.

2.5 Equilibrium Conditions and Mass Law

The expression of the molar reaction Gibbs function $\Delta_r G$ given by Eq. (2.8) is valid for each extent value ξ. After introducing the chemical potentials expressions (2.1) into it, we find

$$\Delta_r G = \left(\nu_M \mu_M^\circ + \nu_N \mu_N^\circ - \nu_A \mu_A^\circ - \nu_B \mu_B^\circ\right) + RT \ln\left\{\left[a_M^{\nu M} \cdot a_N^{\nu N}\right] / \left[a_A^{\nu A} \cdot a_B^{\nu B}\right]\right\}. \tag{2.9}$$

The activities in the logarithm argument can take any value. They depend only on the initial concentrations and on the reaction extent. The expression that constitutes the logarithm argument is called the *quotient reaction, Q*:

$$Q = \left[a_M^{\nu M} \cdot a_N^{\nu N}\right] / \left[a_A^{\nu A} \cdot a_B^{\nu B}\right],$$

or

$$Q = \Pi_i a_i^{\nu i}.$$

Equation (2.9) can be written as

$$\Delta_r G = \left(\nu_M \mu_M^\circ + \nu_N \mu_N^\circ - \nu_A \mu_A^\circ - \nu_B \mu_B^\circ\right) + RT \ln Q, \tag{2.10}$$

or, with the introduction of the chemical affinity, as

$$A^\circ = \left(\nu_A \mu_A^\circ + \nu_B \mu_B^\circ - \nu_M \mu_M^\circ - \nu_N \mu_N^\circ\right),$$
$$\Delta_r G = -A^\circ + RT \ln Q. \tag{2.11}$$

The terms enclosed within parentheses in Eqs. (2.10) and (2.11) are respectively called the standard molar reaction Gibbs function change $\Delta_r G^\circ$[4] and the standard chemical affinity A°. By setting up

$$\Delta_r G^\circ = \left(\nu_M \mu_M^\circ + \nu_N \mu_N^\circ - \nu_A \mu_A^\circ - \nu_B \mu_B^\circ\right),$$

it is quite evident that

$$A^\circ = -\Delta_r G^\circ.$$

---

[4] Or, more quickly, standard molar reaction free enthalpy. This denomination is misleading because it is a measurable quantity, whereas a free enthalpy cannot be measured.

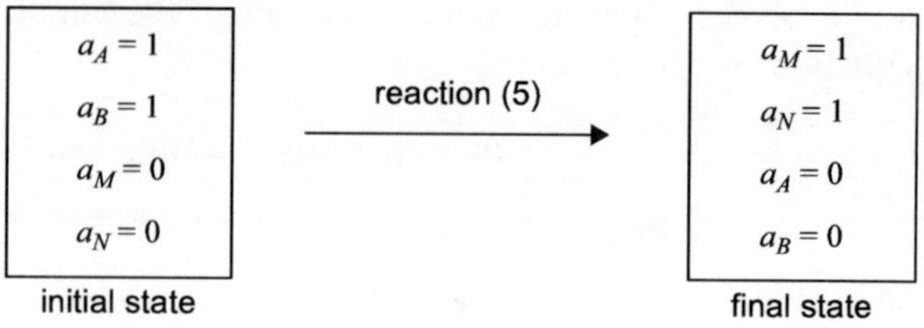

Fig. 2.3 Initial and final states of reaction (2.5) defining $\Delta_r G°$ and $A°$

Equation (2.9) then becomes

$$\Delta_r G = \Delta_r G° + RT \ln\left[a_M^{\nu M} \cdot a_N^{\nu N}\right] / \left[a_A^{\nu A} \cdot a_B^{\nu B}\right]. \tag{2.12}$$

$A°$ and $\Delta_r G°$ are both constant quantities at a constant temperature since the standard chemical potentials are constant in these conditions (see Sect. 2.1). Both quantities can be determined. They are equal (in absolute values) to the *useful work*, or the maximum work available. Such is the case when the chemical reaction takes place reversibly and when the initial and final states are the standard ones. They are described in Fig. 2.3.

By introducing these activities of unit value (which, in fact, characterize the standard states of species A, B, M, and N—see Sect. 2.1) in Eqs. (2.9)–(2.11), we find $Q = 1$ and

$$\Delta_r G = \Delta_r G° \quad (A = A°).$$

It is interesting to note that the standard molar reaction Gibbs function $\Delta_r G°$ does not have the same mathematical status as $\Delta_r G$ at each extent ξ value. $\Delta_r G°$, indeed, is not a derivative, unlike $\Delta_r G$. $\Delta_r G$ is simply the system free enthalpy change ΔG_{syst} between two particular states, which happen to be the standard ones.

The fact that the standard molar reaction Gibbs function change $\Delta_r G°$ and the standard chemical affinity $A°$ are constant at a given temperature is of utmost importance. Its major outcome is the mass law. Indeed, the equilibrium condition $\Delta_r G = 0$ introduced into Eq. (2.9) gives

$$-RT \ln \left\{\left[a_{M\ \text{eq}}^{\nu M} \cdot a_{N\ \text{eq}}^{\nu N}\right] / \left[a_{A\ \text{eq}}^{\nu A} \cdot a_{B\ \text{eq}}^{\nu B}\right]\right\} = \Delta_r G°.$$

Since $\Delta_r G°$ is constant, the term enclosed within brackets is also a constant. By definition, it is the thermodynamic constant $K°$ of reaction (2.5):

$$K° = \left[a_{M\ \text{eq}}^{\nu M} \cdot a_{N\ \text{eq}}^{\nu N}\right] / \left[a_{A\ \text{eq}}^{\nu A} \cdot a_{B\ \text{eq}}^{\nu B}\right], \tag{2.13}$$

$$-RT \ln K° = \Delta_r G°. \tag{2.14}$$

The adjective "thermodynamic" obligatorily implicates that the mass law is expressed in activity terms. Other kinds of equilibrium constants are also used (see Chap. 3). The thermodynamic equilibrium constant is a dimensionless number since all activities are dimensionless.

It is important to note that the activities in Eq. (2.13) cannot have any value. They can have only the values at equilibrium. They are related to each other by Eq. (2.13).

Equilibrium constants are systematically used in general analytical chemistry. They are of huge importance. Their values permit us to calculate the concentrations of the reaction species, once the equilibria have been attained. Some equilibrium constants bear a particular name according to the kind of chemical reaction they quantify. For example,

- In the case of acid–base phenomena, the dissociation constant K_a is introduced. It quantifies the strength of the acid. An acid HA dissociates in aqueous solution and gives a hydrated proton H^+ and the conjugated base A^- (see the Part II) according to

$$HA \rightarrow H^+{}_{\mathrm{w}} + A^-{}_{\mathrm{w}}.$$

K_a is defined by the equation

$$K_a = (a_{H+\mathrm{eq}} \cdot a_{A-\mathrm{eq}})/a_{HA\,\mathrm{eq}}.$$

The molar reaction Gibbs function of acid ionization is given by the equation

$$\Delta_r G^\circ = -RT \ln K_a.$$

- For complexation phenomena, stability constants are used, in particular the overall stability constants β_n (see Part IV). In brief, a complexation reaction can be defined as resulting from the action of a ligand (an electrically neutral or negatively charged species) with a metallic ion, according to the equations

$$M + mL \rightarrow ML_m,$$

$$\beta_m = a_{MLm\mathrm{eq}}/(a_{M\mathrm{eq}} \cdot a_{L\mathrm{eq}}{}^m),$$

where L is the ligand and M the metallic ion (electrical charges are missing for the sakes of both generality and simplicity). The molar reaction Gibbs function of the overall complexation is given by the equation

$$\Delta_r G^\circ = -RT \ln \beta_m.$$

- For precipitation phenomena, equilibrium considerations introduce the concept of the solubility product K_s. This constant concerns the heterogeneous equilibrium between a precipitated compound and its constituent ions in solution (see Part V). For example, for silver chloride, one of the satisfied equilibria is

$$\mathrm{AgCl_s} \rightarrow \mathrm{Ag^+}{}_w + \mathrm{Cl^-}{}_w,$$

$$K_s = [a_{\mathrm{Ag+w}} \cdot a_{\mathrm{Cl-w}}]/a_{\mathrm{AgCls}},$$

which, due to the conventions upon activities (see Chap. 3), reduces to

$$K_s = a_{\mathrm{Ag+w}} \cdot a_{\mathrm{Cl-w}}.$$

The molar reaction Gibbs function of dissolution is

$$\Delta_r G^\circ = -RT \ln K_s.$$

2.6 Chemical Potentials and Standard States

Relation (2.1) involves the standard chemical potential μ° of a species. Definitively, it cannot be determined experimentally, nor is its chemical potential μ_i. However, we noticed when we were on the subject of mass law that a judicious linear combination of standard chemical potentials is experimentally accessible. In truth, it is impossible to know the absolute value of a chemical potential because it is a free enthalpy and because the absolute value of any free enthalpy itself cannot be known. However, the fact that an absolute value of a quantity cannot be determined is not rare. An example that is more common than those of enthalpies, free enthalpies, and free entropies is provided by an altitude point. The difference in altitude between two points is measurable. It is not the case for their absolute altitude. Numerous other quantities also follow suit.

In order to determine the standard molar reaction Gibbs function change $\Delta_r G^\circ$, the standard states of the reaction reactants and products must be defined. They must be chosen or, at a minimum, they must correspond to physical states, such as the physical property differences between them being endowed with an unambiguous physical meaning. As an example, a possible standard state of a solute is the state in which its concentration is 1 mol/L and in which the solution it forms with the solvent is an ideal one. Standard states are chosen conventionally for practical reasons. Fortunately, the conventions are universally agreed upon.

The fact that standard states are conventional may be somewhat troublesome. Indeed, the question immediately arises about equilibrium constant values with different arbitrarily chosen standard states. Quite evidently, when standard states other than the usual ones are chosen, the value of the standard molar reaction Gibbs function change $\Delta_r G^\circ$ is different; according to Eq. (2.14), it is also the case with K°. However, this is not the case with $\Delta_r G$ and ΔG_{syst}, which remain constant for a given process regardless of the adopted conventions. Actually, in Eq. (2.12),

$$\Delta_r G = \Delta_r G^\circ + RT \ln Q.$$

If $\Delta_r G^\circ$ possesses different numerical values (due to the different choices of standard states), the reaction quotient Q is endowed with a different value, but the $\Delta_r G$ value remains the same as before. There exists a very subtle compensation between $\Delta_r G^\circ$ and the numerical values of the activities that depend on the chosen standard states. $\Delta_r G$ and ΔG_{syst} are the true invariants of the thermodynamic process. This subtle compensation takes its roots in the depth of the thermodynamics framework.

Let us give, incidentally, the important methods of $\Delta_r G$ and more precisely of $\Delta_r G°$ determinations. They are grounded on the following bases:

- equilibrium constants' measurements. They themselves involve the determination of reactants and products concentrations at equilibrium;
- the use of thermodynamic tables. The literature contains numerous data that are the standard molar enthalpies and entropies of formation $\Delta fH°$ and $\Delta fS°$ of compounds and ions. They permit, in a first step, the calculation of the standard enthalpy and entropy of the reaction of concern via the following equations:

$$\Delta_r H° = \sum \nu_p \Delta_f H°_p - \sum \nu_r \Delta_f H°_p,$$

$$\Delta_r S° = \sum \nu_p S°_p - \sum \nu_r S°_r,$$

where the indices r and p are related to the reactants and products. In the second step, the following relationship is used to calculate $\Delta_r G°$:

$$\Delta_r G° = \Delta_r H° - T\Delta_r S°.$$

(The data mentioned in the tables are not absolute. They are relative but self-coherent. As a result, the calculations of changes in thermodynamic functions are accurate.);

- calculations of statistical thermodynamics. They necessitate the knowledge of the molecular energy levels' differences, which can be accessible by spectroscopy;
- the use of the third law of thermodynamics;
- the determination of the electromotive force (EMF.) of galvanic cells. The methodology is detailed in the next section.

The relationship between the equilibrium constant of a redox reaction and the EMF of the corresponding galvanic cell (potential cell) is the subject of Nernst's law. It is of tremendous interest in analytical chemistry from both theoretical and practical points of view.

2.7 Redox Reaction: Redox Couples

The preceding considerations are quite general. As a result, they apply to redox reactions. The relationship between the equilibrium constant of a redox reaction and the EMF of the corresponding electrochemical cell mentioned just above necessitates the brief introduction of redox reactions (they will be studied further in Part III).

Let's consider the spontaneous chemical reaction

$$Zn_s + Cu^{2+}{}_w \rightarrow Zn^{2+}{}_w + Cu_s. \tag{2.15}$$

It involves an electron exchange between Zn_s and Cu^{2+}. We say that zinc is oxidized by cupric copper while this last one is reduced by zinc in copper. Zn_s and $Zn^{2+}{}_w$ form a redox couple. Both species are interrelated by the half redox reaction:

$$Zn^{2+}{}_w + 2e^- \rightarrow Zn_s.$$

It is the same state of affairs for the couple Cu^{2+}/Cu_s, the two elements of which are interrelated by the redox half-reaction

$$Cu^{2+}{}_w + 2e^- \rightarrow Cu_s.$$

Reaction (2.15), which results from the exchange of electrons between the two pairs $Zn^{2+}{}_w/Zn_s$ and $Cu^{2+}{}_w/Cu_s$, is a redox reaction. Some redox reactions are more complicated than (2.15). Let's take a look, for example, at the following reaction:

$$a\mathrm{Ox}_1 + b\mathrm{Red}_2 \rightarrow a\mathrm{Red}_1 + b\mathrm{Ox}_2, \tag{2.16}$$

where the different stoichiometric coefficient a and b values result from the different electron numbers exchanged in each redox half-reaction $\mathrm{Ox}_1/\mathrm{Red}_1$ and $\mathrm{Ox}_2/\mathrm{Red}_2$.

The molar reaction Gibbs functions of reactions (2.15) and (2.16) are given by the following equations:

$$\Delta_r G = -\mu_{\mathrm{Zn}} - \mu_{\mathrm{Cu}^{2+}} + \mu_{\mathrm{Zn}^{2+}} + \mu_{\mathrm{Cu}}, \tag{2.17}$$

$$\Delta_r G = -a\mu\mathrm{Ox}_1 - b\mu\mathrm{Red}_2 + a\mu\mathrm{Red}_1 + b\mu\mathrm{Ox}_2. \tag{2.18}$$

As stated previously, the value of these reactions does change with the extent reaction value ξ. In the case of reaction (2.16), Eq. (2.12) becomes

$$\Delta_r G = \Delta_r G^\circ + RT \ln \left\{\left[a_{\mathrm{Red1}}{}^a \cdot a_{\mathrm{Ox2}}{}^b\right]/\left[a_{\mathrm{Ox1}}{}^a \cdot a_{\mathrm{Red2}}{}^b\right]\right\}. \tag{2.19}$$

2.8 Brief Description of an Electrochemical Cell: Daniell's Galvanic Cell

Let's consider the electrochemical cell represented in Fig. 2.4. It contains two compartments. One consists of a zinc strip dipping into a zinc sulfate solution, and the other is a copper strip dipping into cupric copper sulfate. The two metallic conductors are called the *electrodes*. Both electrode compartments are linked by a conducting *bridge*. An *electrolyte solution* completes the full circuit. The bridge precludes the mixing of the two electrolyte solutions. Moreover, the two electrodes dip into two electrolyte solutions. Finally, the two electrodes are in mutual electrical contact through a metallic thread, for example, a platinum thread. One galvanometer and possibly an electrical motor are inserted into the circuit. A millivoltmeter is connected in parallel between the two platinum threads. When the circuit is closed, a spontaneous current passes through it until equilibrium has been attained. Two kinds of electrical current exist in the cell. In the external circuitry, the electrical current is nothing more than the electrons' displacement. It exists through the electrodes and the metallic threads. In the electrolytic solutions, the current is due to the ions' displacements. It is an ionic current. The same current also exists in the conducting bridge. What happens at the interface of both kinds of current, that is, on the electrode's surfaces, is particularly interesting. According to Kirchhoff's law, the current

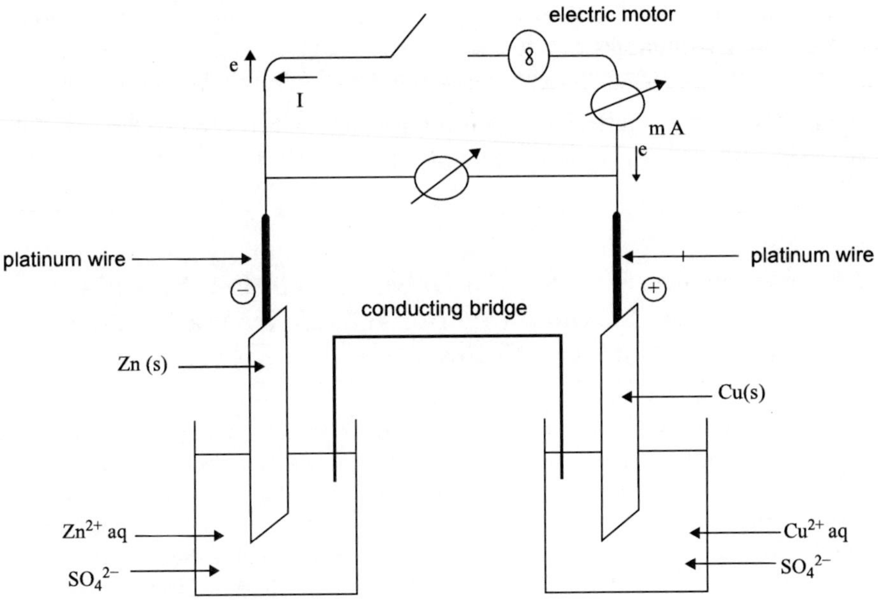

Fig. 2.4 Daniell's galvanic cell

continuity must exist. It can be realized only if the two following reactions take place simultaneously and separately:

$$Zn_s \rightarrow Zn^{2+}{}_w + 2e^-(\text{metal}), \tag{2.20}$$

$$Cu^{2+}{}_w + 2e^-(\text{metal}) \rightarrow Cu_s. \tag{2.21}$$

The zinc strip loses electrons. The formed Zn^{2+} ions at the surface of the electrode go into the solution to give the hydrated $Zn^{2+}{}_w$ ions. Thus, they contribute to the ionic current. The electrons liberated simultaneously (at the same electrode) go into the electric conductor, where they circulate toward the copper strip. At the surface of the copper electrode, the cupric $Cu^{2+}{}_w$ ions capture these electrons brought by the metallic conductor. They also contribute, as do the Zn^{2+} and $SO_4{}^{2-}$ ions, to the ionic current. The full circuitry is then ensured.

Reactions (2.20) and (2.21), which take place at the two interfaces, are called *electrochemical reactions*. The chemical reaction, which formally corresponds to the sum of the two preceding electrochemical reactions, is called the *cell reaction.* It is reaction (2.15) in the chosen example. When the cell reaction is equivalent to a spontaneous chemical reaction (under the same conditions of composition, temperature, and pressure), the electrochemical cell is called a *galvanic cell.* It is an energy-producing device.

In Daniell's galvanic cell, electrons spontaneously go from the zinc strip to the copper one. This indicates that, spontaneously, the copper strip is more positive than the zinc strip. The electrode signs indicated in Fig. 2.4 follow this conclusion.

Remark A potential difference can be measured only between two threads of the same chemical nature in order to be endowed with a physical meaning (see the two platinum threads in Fig. 2.4).

2.9 Electromotive Force of a Galvanic Cell, Cell Potential Difference, Maximum Work Available from a Chemical Reaction, and Nernst's Equation[5]

It is an experimental fact that reaction (2.15) performed by purely chemical means leads to the same final state as the same reaction performed by electrochemical means through the two half-reactions (2.20) and (2.21). This is the case, of course, provided that the initial states and the temperature and pressure are the same in both experiences. It is also the case if, moreover, no work is recuperated in the electrochemical experience. The galvanic cell must be short-circuited. More precisely, initial identical states mean that concentrations $[Cu^{2+}{}_w]i$ and $[Zn^{2+}{}_w]i$ (i: initial) are identical in the chemical reactor and in both electrode compartments of the cell together with identical temperature and pressure. This must be the same state of affairs for the final states. When the galvanic cell is definitely used, its concentrations $[Cu^{2+}{}_w]f$ and $[Zn^{2+}{}_w]f$ (f: final) are identical to those existing in the equivalent chemical reactor at the end of the chemical reaction.

Since the final and initial states are identical in both experiences, the free enthalpy change ΔG_{syst} is the same in the two processes. This result is general and applies to all redox reactions. A galvanic cell discharge is equivalent to a chemical process that can be decomposed into two electrochemical half-reactions that take place simultaneously but separately, the thermodynamic evaluations of both processes being identical.

Nernst's equation results from the equality of the work furnished by the electrochemical system under conditions of reversibility and free enthalpy decrease. The work performed in the cell is that of the transport of the $Cu^{2+}{}_w$ and $Zn^{2+}{}_w$ ions in both compartments against the electrical field since the copper electrode is positively charged. The charge q brought by dn mol/L of the $Cu^{2+}{}_w$ and $Zn^{2+}{}_w$ ions of valence two crossing the cell during a very brief time interval dt from the left to the right is

$$dq = 2F\ dn,$$

where F is the faraday. Since the interval dt is very brief, the potential difference between both electrodes can be considered constant. If its value is E, the work done

[5] Hermann Walther Nernst: a German physicochemist (1864–1941); received Nobel Prize in chemistry (1920). Nernst's equation was published in 1889 in terms of concentrations. Some authors consider Nernst as the father of modern electrochemistry.

by the system for the displacement of positive ions is in absolute value:

$$|dw'| = |E\ dq|$$

or

$$|dw'| = |2\ FE\ dn|.$$

This results from law of electrostatics. Since it is evolved by the system, we find $|dw'| = -2FE\ dn$. From that, we have

$$dG_{\text{syst}} = -2FE\ dn. \quad (2.22)$$

The free enthalpy change accompanying reaction (2.15) is

$$dG_{\text{syst}} = (-\mu_{\text{Cu}^{2+}} - \mu_{\text{Zn}} + \mu_{\text{Cu}} + \mu_{\text{Zn}^{2+}})d\xi;$$

by comparison with (2.22), we obtain

$$\mu_{\text{Cu}} + \mu_{\text{Zn}^{2+}} - \mu_{\text{Cu}^{2+}} - \mu_{\text{Zn}} = -2FE \quad (2.23)$$

since $d\xi = dn$ with $\nu_{\text{Cu}}{}^{2+} = \nu_{\text{Zn}}{}^{2+}$. The left-hand term in Eq. (2.23) is the molar reaction Gibbs function change of the chemical reaction equivalent to the reaction cell. Relation (2.23) can be generalized. For reaction (2.16) performed via electrochemical means, we can write

$$\Delta_r G = -nFE, \quad (2.24)$$

$$(-a\mu_{\text{Ox1}} - b\mu_{\text{Red2}} + a\mu_{\text{Red1}} + b\mu_{\text{Ox2}}) = -nFE, \quad (2.25)$$

where $n = ab$ is the number of electrons exchanged by the two pairs Ox_1/Red_1 and Ox_2/Red_2.

Relations (2.24) and (2.25) are the basis of Nernst's law. They are of enormous interest in physical and analytical chemistries. Among other features, they permit us to predict the redox reactions, at least from the thermodynamic standpoint, by introducing the concept of the standard electrode potential.

We must emphasize two points. The first one has already been mentioned. The molar reaction Gibbs function change is endowed with an instantaneous value. It depends on the reaction extent and, of course, on the initial reactants' and products' concentrations. As a result, the potential difference E is also instantaneous. This is the reason why the above reasoning involved differential quantities.

This is also the reason why the reversible potential difference of a cell is mentioned with the system composition. Formerly, the reversible potential difference was named the "electromotive force" of the cell. (For the meaning of the word "reversible," see immediately below). For example, the following galvanic cell (see Chap. 13 for the electrochemical cell representations)

$$\text{Pt}|\text{H}_{2(\text{g})}(P = 92.1\ \text{kPa})|\text{HCl}_{(\text{g})}(P = 1.01\ \text{kPa})|\text{Hg}_2\text{Cl}_{2(\text{s})}|\text{Hg}_{(1)}$$

exhibits the electromotive force $E = 0.011$ V ($T = 298$ K). The chemical process is

$$H_{2(g)}(P_{H_2} = 91.2\ \text{kPa} + Hg_2Cl_{2(s)} \rightarrow 2HCl_{(g)}(P_{HCl} = 1.01\ \text{kPa}) + 2Hg_{(l)}.$$

The molar reaction Gibbs function change is

$$\Delta_r G = -\mu_{H_2}(g : P = 91.2\ \text{kPa}) - \mu_{Hg_2Cl_2(s)} + 2\mu_{HCl}(g\text{: } P = 1.01\ \text{kPa}) + 2\mu_{Hg(1)}.$$

The second point to emphasize is that the electromotive forces mentioned above are measured at null current. They are the zero cell potentials. This is a reversibility condition. In Daniell's galvanic cell experiment, electrical current can drive an electrical motor, for example. There is then an electrical energy recovery from the free enthalpy initially contained in the chemical system. Due to the current's occurrence, there are thus a release of heat due to the Joule effect as well as an entropy creation. The process is irreversible. However, the current can be very weak with the use of an opposition montage. In this condition, by avoiding any other irreversibility source, the free enthalpy decrease can be fully converted into electrical energy. Then the process becomes reversible. (The opposition montage even permits the reaction course to be inverted in some cases. The cell is then called a reversible one; see Chap. 13.)

An important particular case is that in which the system is in standard conditions. Then

$$\Delta_r G^\circ = -nFE^\circ, \tag{2.26}$$

where E° is the electromotive force of the galvanic cell in standard conditions or the standard potential cell. Equation (2.19) becomes

$$E = E^\circ - (RT/nF)\ \ln\left[\left(a_{\text{Red1}}{}^a \cdot a_{\text{Ox2}}{}^b\right)/\left(a_{\text{Ox1}}{}^a \cdot a_{\text{Red2}}{}^b\right)\right]. \tag{2.27}$$

This equation expresses Nernst's law written in activity terms.

2.10 Electrode Potentials

Relation (2.27) can be written equivalently as

$$E = E^\circ - (RT/nF)\ \ln\left[(a_{\text{Red1}}{}^a)/(a_{\text{Ox1}}{}^a)\right] + (RT/nF)\ \ln\left[(a_{\text{Red2}}{}^b)/(a_{\text{Ox2}}{}^b)\right].$$

It is possible to set the standard potential E°, which is a constant equal to a difference of two constants, temporarily called $E^\circ{}_1$ and $E^\circ{}_2$; that is,

$$E^\circ = E^\circ{}_1 - E^\circ{}_2.$$

Next, two submembers appear on the right-hand side of the preceding equation, written in brackets:

$$\begin{aligned} E = &\left\{E^\circ{}_1 - (RT/nF)\ \ln\left[(a_{\text{Red1}}{}^a)/(a_{\text{Ox1}}{}^a)\right]\right\} \\ &- \left\{E^\circ{}_2 - (RT/nF)\ \ln\left[(a_{\text{Red2}}{}^b)/(a_{\text{Ox2}}{}^b)\right]\right\}. \end{aligned}$$

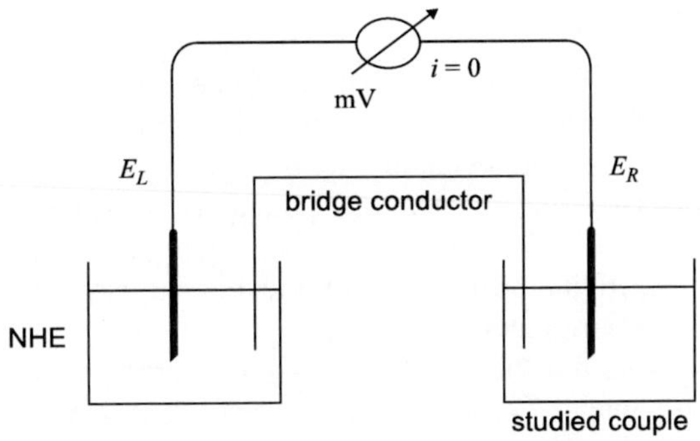

Fig. 2.5 Schematic representation of a cell permitting the determination of standard potentials of redox couples

Since the measured cell potential difference is actually the potential difference between two electrodes, it immediately comes to mind to assimilate each of the bracketed terms into the potential of each of the electrodes. They are called *electrode potentials.* $E°_1$ and $E°_2$,,which are in the two subgroups, exhibit characteristic values of both couples Ox_1/Red_1 and Ox_2/Red_2. These constants are called *standard potentials* of both couples and are symbolized $E°(Ox_1/Red_1)$ and $E°(Ox_2/Red_2)$. Assigning numerical values to $E°_1$ and $E°_2$ has been a problem since the experimental determination of absolute electrode potentials; hence, assigning those to standard electrode potentials is impossible (see the electrochemistry part). It was solved by assigning relative values to them. The strategy was based on the fact that if absolute electrode potentials are not measurable, the difference between them can be. Thus, an electrode standard potential has been chosen conventionally for the couple $H^+{}_w/H_{2(g)}$ (hydrogen electrode). Its standard electrode has been set definitively to the value 0.0000 V at every temperature:

$$E°(H^+{}_w/H_{2(g)}) = 0.0000\ V$$

(the standard conditions for the members of both couples being $P_{H2} = 1$ bar and $a(H^+{}_w) = 1$: this is the *normal hydrogen electrode*, or *NHE*). The electromotive force of a cell that consists of the couple under study and by the couple $H_{2(g)}/H^+{}_w$ in these conditions is, by definition, the standard potential electrode of the first couple. Standard electrode potentials have been established according to this principle (see Chap. 13). Note that cells that permit such determinations must be built so that the hydrogen electrode is on the left and that studied on the right (Fig. 2.5).

The accepted value $E°(Ox/Red)$ is

$$E°(Ox/Red) = E_R - E_L$$

(*R*: right and *L*: left). It can be positive or negative. In the first case, the Ox form of the studied couple is more oxidant than $H^+{}_w$. In the second case, it is the opposite: $E° = E_R - E_L$. Hence, $E°(Ox/Red)$ values are determined under reversibility conditions, that is, for a null intensity.

Remarks

- There are other kinds of electrodes in addition to those described above. However, the preceding considerations can be applied to all kinds of electrodes.
- It is not obligatory to be under standard conditions in order to determine the standard electrode potentials with the help of the preceding general cell (Fig. 2.5). Successive measurements in experimental conditions such as the activity coefficients of the couple members approaching unity permit their determination.
- From a theoretical standpoint, there is another way to introduce the electrode potential concept. This can be done by the use of electrochemical potentials (see a general course on electrochemistry).

Finally, let us point out that the absolute standard electrode potential value of the couple $H^+{}_w/H_{2(g)}$ is actually about 4.5 V. This value cannot be verified since we cannot measure an absolute potential. It was obtained by using thermodynamic cycles, taking into account some thermodynamic data such as the proton hydration enthalpy and entropy. These last ones have been approached by considering the quadrupole model of water (see Chap. 1). It is quite evident that the value of 4.5 V differs considerably from the conventional one (0.00 V). However, it does not change the redox phenomena provision since only the standard electrode potential differences are taken into account.

The reason why it is not possible to measure electrode potentials is of an operational nature. An absolute electrode potential is the potential difference settled at equilibrium between the electrode metal and the ionic solution where it dips. The measurement necessitates introducing another electrode into the solution. As a result, a supplementary electrode potential exists. The measurement with the help of a voltmeter cannot give another value than that of an electrode potential difference.

2.11 Addition of Free Enthalpies and Calculation of Standard Electrode Potentials from Other Standard Electrode Potentials

Free enthalpy is a state function. Its change does not depend on the stages through which the process takes place. It is equal to the sum of the free enthalpies of each stage. It is the same for the standard free enthalpy. This fundamental property is interesting because of the equivalence $\Delta_r G°/E°$, which permits the calculation of standard potentials of some redox couples that are difficult to obtain (see examples below). The principle of the calculation of the free enthalpy change of a redox process is to sum the already known free enthalpies of each stage. We will give some examples here.

The standard potential of the couple $Br_{2(l)}/Br^-$ symbolized by $E°(Br_{2(l)}/Br^-)$. The problem is to calculate the standard potential of the couple $Br_{2(w)}/Br^-$ from the former one. The half-reaction under study is

$$Br_{2(w)} + 2e^- \rightarrow 2Br^- \Delta G°_3.$$

It can be considered as resulting from the sum of the following two processes:

$$Br_{2(w)} \rightarrow Br_{2(l)} \Delta G°_1,$$
$$Br_{2(l)} + 2e^- \rightarrow 2Br^- \Delta G°_2.$$

We can write

$$\Delta G°_3 = \Delta G°_1 + \Delta G°_2.$$

$\Delta G°_1$ is given by the equation

$$\Delta G°_1 = RT \ln (s/C°),$$

where s is the dibrome's solubility in water and C° its solution concentration in its standard state. Otherwise,

$$\Delta G°_2 = -2FE°(Br_{2(l)}/Br^-).$$

Finally, by applying Nernst's law to the system $Br_{2(w)}/Br^-$, we find

$$\Delta G°_3 = -2FE°(Br_{2(w)}/Br^-).$$

Remark The above equation giving $\Delta G°_1$ comes from the equality of the dibrome's chemical potentials in a liquid state and in its saturated solution (see Part V):

$$\mu°_{Br_2(l)} = \mu°_{Br_2(w)} + RT \ln (s/C°).$$

Now, let's consider the examples given by the pairs $I_{2(w)}/I^-$ and $I_{2(s)}/I^-$. Both are of analytical interest. How do we calculate the standard potential of one using the other's?

$$I_{2(w)} + 2e^- \rightarrow 2I^-$$
$$\Delta G°_1 = -2FE°(I_{2(w)}/I^-);$$
$$I_{2(s)} \rightarrow I_{2(w)}$$
$$\Delta G°_2 = -RT \ln(s/C°);$$
$$I_{2(s)} + 2e^- \rightarrow 2I^-$$
$$\Delta G°_3 = \Delta G°_1 + \Delta G°_2$$
$$\Delta G°_3 = -2FE°(I_{2(s)}/I^-).$$

Now, consider now the redox pairs $HClO/Cl^-$ and ClO^-/Cl^-:

$$HClO + H^+ + 2e^- \rightarrow 2Cl^- + H_2O,$$
$$\Delta G°_1 = -2FE°(HClO/Cl^-),$$

$$ClO^- + H^+ \rightarrow HClO$$
$$\Delta G°_2 = -RT \ln\{1/[K_a(HClO, ClO^-)]\}$$
$$ClO^- + 2H^+ + 2e^- \rightarrow Cl^- + H_2O$$
$$\Delta G°_3 = \Delta G°_1 + \Delta G°_2$$
$$\Delta G°_3 = -2FE°(ClO^-/Cl^-).$$

Remark $K_a(HClO, ClO^-)$ is the acid dissociation constant of hypochlorous acid:

$$K_a(HClO, ClO^-) = (a_{H+w} \cdot a_{ClO^-})/a_{HClO^-}.$$

Redox pairs BrO_3^-/Br^-in acidic medium and BrO_3^-/Br^- in basic medium:

$$BrO_{3^-} + 6H^+ + 6e^- \rightarrow Br^- + 3H_2O$$
$$\Delta G°_1 = -6FE°(BrO_{3^-}/Br^-; H^+);$$
$$6H_2O \rightarrow 6H^+ + 6OH^-$$
$$\Delta G°_1 = -6RT \ln k_w;$$
$$BrO_{3^-} + 2H_2O + 6e^- \rightarrow Br^- + 6OH^-$$
$$\Delta G°_3 = -\Delta G°_1 + \Delta G°_2,$$
$$\Delta G°_3 = -6FE°(BrO_{3^-}/Br^-; OH^-).$$

Remark K_w is the ionic product of water: $K_w = (H^+)(OH^-)$.

Redox couples $Au^{3+}/Au_{(s)}$ and $AuCl_4^-/Au_{(s)}$:

$$Au^{3+} + 3e^- \rightarrow Au_{(s)}$$
$$\Delta G°_1 = -3FE°(Au^{3+}/Au_{(s)});$$
$$AuCl_{4^-} \rightarrow Au^{3+} + 4Cl^-$$
$$\Delta G°_2 = -RT \ln(1/\beta_4);$$
$$AuCl_{4^-} + 3e^- \rightarrow Au_{(s)} + 4Cl^-$$
$$\Delta G°_3 = -\Delta G°_1 + \Delta G°_2$$
$$\Delta G°_3 = -3FE°(AuCl_{4^-}/Au_{(s)}).$$

Remark β_4 is the overall stability constant of the ultimate complex [$AuCl_4^-$], (see the Part IV):

$$\beta_4 = (AuCl_4^-)/\left[(Au^{3+})(Cl^-)^4\right].$$

Redox couples $Ag^+/Ag_{(s)}$ and $AgCl_{(s)}/Ag_{(s)}$:

$$Ag^+ + 1e^- \rightarrow Ag_{(s)}$$
$$\Delta G°_1 = -1FE°(Ag^+/Ag_{(s)});$$

$$AgCl_{(s)} \rightarrow Ag^{+} + Cl^{-}$$
$$\Delta G°_{2} = -RT \ln K_{s};$$
$$AgCl_{(s)} + 1e^{-} \rightarrow Ag_{(s)} + Cl^{-}$$
$$\Delta G°_{3} = \Delta G°_{1} + \Delta G°_{2}$$
$$\Delta G°_{3} = -1FE°(AgCl_{(s)}/Ag_{(s)}).$$

Remark K_s is the solubility product of the silver chloride: $K_s = (Ag^{+})(Cl^{-})$ (see Part V).

Pairs $Zn^{2+}{}_{(w)}/Zn_{(s)}$ and $Zn(OH)_{2(s)}/Zn_{(s)}$:

$$Zn^{2+}{}_{(w)} + 2e^{-} \rightarrow Zn_{(s)}$$
$$\Delta G°_{1} = -2FE°(Zn^{2+}/Zn_{(s)});$$
$$Zn(OH)_{2(s)} \rightarrow Zn^{2+}_{(w)} + 2OH^{-}$$
$$\Delta G°_{2} = -RT \ln K_{s};$$
$$2OH^{-}{}_{(w)} + 2H^{+}{}_{(w)} \rightarrow 2H_{2}O$$
$$\Delta G°_{3} = -RT \ln (1/K_{w});$$
$$Zn(OH)_{2(s)} + 2e^{-} + 2H^{+} \rightarrow Zn_{(s)} + 2H_{2}O$$
$$\Delta G°_{4} = \Delta G°_{1} + \Delta G°_{2} + \Delta G°_{3}$$
$$\Delta G°_{4} = -2FE°(Zn(OH)_{2(s)}/Zn_{(s)}; H^{+}).$$

This is an example in which the standard potential $E°$ can be determined only by calculation. $Zn(OH)_2$, indeed, cannot exist at a pH equal to 0.

The so-called Latimer[6]–Luther's rule, which is frequently used in these calculations, is a particular case of the free enthalpy addition principle. It applies to the calculation of the standard potential of every redox pair once the potentials of other pairs where the same element is involved are already known. Let's consider, for example, the iron pairs Fe^{3+}/Fe^{2+} ($E° = 0.77$ V) and $Fe^{2+}/Fe_{(s)}$ ($E° = -0.44$ V). What is the standard potential of the pair $Fe^{3+}/Fe_{(s)}$? The free enthalpy addition principle leads to the following relationships:

$$Fe^{3+}{}_{(w)} + 1e^{-} \rightarrow Fe^{2+}{}_{(w)}$$
$$\Delta G°_{1} = -1FE°(Fe^{3+}/Fe^{2+});$$
$$Fe^{2+}{}_{(w)} + 2e^{-} \rightarrow Fe_{(s)}$$
$$\Delta G°_{2} = -2FE°(Fe^{2+}/Fe_{(s)});$$
$$Fe^{3+}{}_{(w)} + 3e^{-} \rightarrow Fe_{(s)}$$

[6] Wandell Mitchell Latimer: American physicochemist, at the—University of California at Berkeley (1893–1955).

$$\Delta G^{\circ}{}_3 = -3FE^{\circ}(Fe^{3+}/Fe_{(s)});$$
$$-\,3FE^{\circ}(Fe^{3+}/Fe_{(s)}) = -1FE^{\circ}\left(Fe^{3+}/Fe^{2+}\right) - 2FE^{\circ}(Fe^{2+}/Fe_{(s)}).$$

From the above, we have

$$E^{\circ}(Fe^{3+}/Fe_{(s)}) = \left[1E^{\circ}(Fe^{3+}/Fe^{2+}) + 2E^{\circ}(Fe^{2+}/Fe_{(s)})\right]/3,$$
$$E^{\circ}(Fe^{3+}/Fe_{(s)}) = -0.037\mathrm{V}.$$

This mathematical relationship expresses Latimer–Luther's rule. We infer from it the fact that it is unnecessary to calculate the standard molar free enthalpy changes, but this result is not general. The rule is valid only for processes that involve only redox phenomena. This was not the case in the examples given before.

Chapter 3
Activities and Activity Coefficients

The activity concept was introduced in 1907 by G. N. Lewis in order to study, from a thermodynamic viewpoint, nonideal solutions with the same mathematical formalism as that used for ideal solutions. A typical case of nonideal solutions is given by ionic solutions, which are rarely ideal except when they are very diluted. The activity concept has turned out to be very fruitful. It is one of the most important concepts in physical and analytical chemistries.

3.1 Chemical Equilibrium. Mass Law and Species Activities

Let us consider the following equilibrated chemical reaction:

$$aA + bB + \ldots \rightarrow mM + nN + \ldots. \tag{3.1}$$

At equilibrium, the species activities are such that the following relationship is satisfied:

$$K^\circ = \left[a_{\mathrm{M}}{}^{\mathrm{m}}{}_{\mathrm{eq}} \cdot a_{\mathrm{N}}{}^{\mathrm{n}}{}_{\mathrm{eq}}\right] / \left[a_A{}^a{}_{\mathrm{eq}} \cdot a_B{}^b{}_{\mathrm{eq}}\right]. \tag{3.2}$$

K° is the thermodynamic equilibrium constant. Its numerical value depends only on the temperature and on the conventions chosen to define activities (for condensed phases, it depends very little on the system pressure). Equation (3.2) expresses the mass law. As already demonstrated in Chap. 2, it results from thermodynamic considerations.

3.2 On the Physical Meaning of An Activity

A species activity can be thought of as being its concentration fraction that actually participates in the process. It can be a chemical reaction or a physical transformation (for example, it may be a distribution between two solvents).

J.-L. Burgot, *Ionic Equilibria in Analytical Chemistry*,
DOI 10.1007/978-1-4419-8382-4_3, © Springer Science+Business Media, LLC 2012

In very diluted solutions, there are no electrostatic interactions among the solute molecules or ions. Each solute ion or molecule has its own random movement in solution. Hence, each solute ion or molecule does fully participate in its reactive potential, that is, in its chemical potential. These very diluted solutions are said to exhibit an *ideal behavior*. However, when the solution is less diluted, the electrostatic interactions between solute ions and molecules become nonnegligible and are noticeable. In this condition, their movements are no longer independent of each other. Then only one fraction of the solute concentration participates in its chemical potential. This fraction is called its *activity*.

It is easily conceivable that the more diluted the solutions are, the weaker the probability of electrostatic interactions among solute ions and molecules is. In fact, the interactions' occurrence requires weak distances between the interacting species. The more concentrated the solution is, the smaller these distances are. Hence, in sufficiently diluted solutions, the solute activities become equal to their concentrations. From these general considerations, we immediately become aware of a difference in the behavior of ions and molecules. According to Coulomb's law, the electrostatic force between ion-ion interactions is $1/r^2$, where r is the distance between the two ion species. The molecule-ion interactions occur at $1/r^3$, and the dipole-dipole ones (between molecules) at $1/r^6$. The ion-ion interactions are hence more operative at higher distances than others. Of course, the presence in solution of other ions than those that participate to the process also generates deviations from the solution's ideality.

To summarize, the activity concept was introduced to overcome the problem of the electrostatic interactions between solute ions and molecules in solution by using the same thermodynamic formalism as that applied to ideal solutions.

Despite the fact that activities can be considered effective concentrations, they are defined in thermodynamics in such a way that they are dimensionless quantities (see Sect. 3.4). As a result, the equilibrium thermodynamic constants $K°$ are dimensionless.

3.3 Ionic Strength of a Solution

After the preceding considerations, it is quite understandable that the total ionic concentration of a solution is of major importance. Depending on its value, it may govern its thermodynamic behavior. Lewis was the first to note, on purely experimental grounds, that it is the ionic strength of a solution that is important in this respect. The ionic strength of a solution is defined as

$$I_c = \frac{1}{2}\sum_i c_i z_i^2. \quad (3.3)$$

This means that the concentration C_i of each ion i is multiplied by the square of its charge z_i, with all the terms for the various ions in solution totaled. I_c is expressed in mol/L, thus explaining the c subscript. Some years later, Debye-Hückel's law (discussed ahead) theoretically confirmed the importance of ionic strength. This parameter indeed is naturally included in calculations. When ionic "concentrations" are expressed in molalities, the equation defining the ionic strength becomes

$$I_m = \frac{1}{2}\sum_i m_i z_i^2. \tag{3.4}$$

Exercise 1 What is the ionic strength of a 10^2 mol/L sodium chloride solution?

Solution chloride is a strong electrolyte. It is fully ionized in aqueous solution. Then

$$[Na^+{}_w] = [Cl^-{}_w] = 10^{-2}\ \text{mol/L}$$

(this relation expresses the mass balance—see Chap. 5);

$$I_c = \frac{1}{2}\left[0.01 \cdot 1^2 + 0.01(-1)^2\right],$$

$$I_c = 10^{-2}\ \text{mol/L}.$$

In this example, the ionic strength is equal to the solute concentration.

Remark The ions $H_3O^+{}_w$ and $OH^-{}_w$ issuing from water dissociation must also be taken into account. However, sodium and chloride ions are not endowed, respectively, with acidic and basic properties in water. As a result, the solution is neutral, that is,

$$[H_3O^+{}_w] = [OH^-{}_w].$$

It is easy to check that their contribution to the solution's ionic strength is negligible, since their concentration is very low: 10^{-7} mol/L.

Exercise 2 What is the ionic strength of a 10^{-2} mol/L sodium sulfate solution?

Sodium sulfate can be considered to be a strong electrolyte.[1] Both ions are not endowed with acidic and basic properties in aqueous solutions. Two moles of sodium ions are present in solution for one mole of sulfate ions:

$$[Na^+{}_w] = 2 \cdot 10^{-2}\ \text{mol/L},$$

$$[SO_4{}^{2-}{}_w] = 10^{-2}\ \text{mol/L},$$

$$I_c = \frac{1}{2}(3 \cdot 10^{-2})\ \text{mol/L}.$$

The ionic strength is higher than the concentration. It very quickly becomes high when polycharged ions are present in the solution.

[1] pKa (HSO_4^-, SO_4^{2-}) = 1.90. See Chap. 5.

3.4 Link Between Activities and Concentrations: The Activity Coefficients

The concept of *activity coefficient* was introduced together with that of activity. In fact, both are indissociable. Activity coefficients are symbolized by γ if they concern the solute or by f if they concern the solvent.

An activity coefficient can temporarily be thought of (see below) as being the ratio between the species activity and its concentration.

For a solute i and according to the scale of "concentrations" used,

$$\gamma_{c,i} = a_{c,i}/C_i, \tag{3.5}$$

$$\gamma_{m,i} = a_{m,i}/m_i, \tag{3.6}$$

$$\gamma_{x,i} = a_{x,i}/x_i. \tag{3.7}$$

$\gamma_{c,i}$, $\gamma_{m,i}$, and $\gamma_{x,i}$ are the activity coefficients related to concentrations expressed in molarities, molalities, and mole fractions, respectively.

For the solvent, the activity coefficient f_o is defined by

$$f_o = a_o/x_o. \tag{3.8}$$

(We give the activity coefficient here only when its "concentration" is expressed in a mole fraction. It is a quasi-universal convention to express the solvent "concentration" in a mole fraction.)

Despite their apparent mathematical definition, Eqs. (3.5)–(3.8) are not linear between activities and "concentrations" because the activity coefficient value changes with concentrations.

From the viewpoint of dimensions, activity coefficients are dimensionless. This assertion is apparently in contradistinction with Eqs. (3.5) and (3.6) since C_i and m_i are endowed with dimensions, whereas activities are dimensionless. Equations (3.5) and (3.6) have been simplified. The simplification results from the choice of some standard states to define activities (see Sect. 2.5 of Chap. 2).

3.5 Standard States and Activity Coefficients

By definition, the definition of an activity requires choosing a standard state. For example, for a solute i, the standard state can be chosen as being the state in which its concentration is $C°_i$ (or its molality $m°_i$ or its mole fraction is $x°_i$), the temperature is T, and the solution is ideal (recall that pressure exerts a very weak influence on the behavior of condensed phases). At concentration $C°_i$ (which is that in the standard state), the solute chemical potential is its standard chemical potential $\mu°_i$. Hence, when the solution is ideal, the solute chemical potential μ_i at concentration C_i or at molality m_i is given by the expressions

$$\mu_i = \mu^\circ{}_{i,c} + RT \ln (c_i/c^\circ{}_i) \quad \text{ideal solution,}$$

$$\mu_i = \mu^\circ{}_{i,m} + RT \ln (m_i/m^\circ{}_i) \quad \text{ideal solution.}$$

We can see that, effectively, when $C_i = C^\circ{}_i$ or $m_i = m^\circ{}_i$, $\mu_i = \mu_i^\circ$.

In the case of real (nonideal) solutions, we introduce activities and activity coefficients defined by

$$\gamma_{c,i} = a_{c,i}/(c_i/c^\circ{}_i), \tag{3.9}$$

$$\gamma_{m,i} = a_{m,i}/(m_i/m^\circ{}_i). \tag{3.10}$$

Activity coefficients are, indeed, dimensionless, as discussed earlier in this chapter. However, standard states, which are arbitrary, are chosen usually such that $C^\circ{}_i = 1$ mol/L and $m^\circ{}_i = 1$ mol/kg. The simplified equations (3.5) and (3.6) result from this choice. Note that on the scale of mole fractions, activities are necessarily dimensionless quantities.

For real solutions, the solute chemical potentials are given by

$$\mu_i = \mu^\circ{}_{c,i} + RT \ln a_{c,i},$$

$$\mu_i = \mu^\circ{}_{c,i} + RT \ln \ (c/c^\circ{}_i) + RT \ln \gamma_{c,i} \quad \text{(real solutions/concentrations scale),}$$

$$\mu_i = \mu^\circ{}_{m,i} + RT \ln a_{m,i},$$

$$\mu_i = \mu^\circ{}_{m,i} + RT \ln (m_i/m^\circ{}_i) + RT \ln \gamma_{m,i} \quad \text{(real solutions/molalities scale).}$$

These relationships are interesting and must be carefully considered. Clearly, it appears that the standard state is the one in which the solute is at the concentration 1 mol/L (or 1 mol/kg) and in which it is endowed with an ideal behavior. In this case, indeed, $\gamma_{c,i}$, $\gamma_{m,i}$ are equal to unity. It is rare that at 1 mol/L (or 1 mol/kg), a solution exhibits an ideal character. As a result, solute standard states are very often virtual ones. The preceding four relationships also show that the solute activity coefficient alone measures the chemical—potential change arising from interactions among the solute particles. Finally, they show that the logarithm arguments are dimensionless. This is necessary from a mathematical standpoint.

3.6 Different Ways to Write the Mass Law

Equation (3.2), which expresses the general mass law, can be written with the help of Eq. (3.5):

$$K^\circ = C_M{}^m C_N{}^n / C_A{}^a C_B{}^b \cdot \gamma C_M{}^m \cdot \gamma C_N{}^n / \gamma C_A{}^a \cdot \gamma C_N{}^b.$$

The constant related only to concentrations at equilibrium is given by

$$K_c = C_M{}^m \cdot C_N{}^n / C_A{}^a \cdot C_B{}^b$$

and is called the *apparent* or *formal concentration equilibrium constant*. It is not necessarily dimensionless, which can easily be verified. The important point is the fact that K_c does change with the ionic strength, while $K°$ does not. This is the reason why we find in the literature concentration equilibrium constant data together with the ionic strength values for which they have been determined. Of course, the temperatures of determination are also mentioned since both $K°$ and K_c depend on them.

Diluted solutions tend to ideality. Activities are then equal to their "concentrations" in numerical values, and activity coefficients tend to be unity. As a result,

$$K_c \rightarrow K° \quad \text{(very diluted solutions)}.$$

Likewise, when the "concentration" equilibrium constant is related to molalities, the same behavior is encountered:

$$K_m = [m_M{}^m \cdot m_N{}^n]/[m_A{}^a \cdot m_B{}^b],$$

$$K_m \rightarrow K° \quad \text{(very diluted solutions)}.$$

Using "concentration" equilibrium constants expressed in mole fractions is, by far, a rarer case than the two preceding ones, at least for solutes.

It is important to recall the fact that activity coefficient values increase with the solution's ionic strength as soon as the latter exceeds about 0.3 mol/L. Below this value, they decrease. As a result, activity coefficients exhibit a minimum value of about 0.3 mol/L. Activity coefficients get back the unit value for ionic strengths in the range 1–4 mol/L, depending on the specific chemical species. This finding, which is of an experimental origin, has led to the determination of equilibrium constants in this range, as the values found could not be very different from the thermodynamic ones.

The increase in activity coefficients originates from a solvation phenomenon. It is now a solute-solvent interaction that is added to the preceding solute-solute interaction. The higher the solute concentration is, the fewer sufficient solvent molecules required to achieve the solute solvation.

Remark Sometimes the term "conditional constant" appears in place of "formal" or "apparent concentration equilibrium constant." However, the term "conditional constant" must be used only when all the experimental conditions, and not only the ionic strength value, are known.

3.7 Usual Conventions for Activities

The numerical activity values tend toward those of the corresponding concentrations for very diluted solutions. That is to say, they tend toward those of molarities, molalities, or mole fractions of the species under study. As an immediate result, the numerical values of the thermodynamic constant (and also that of the formal one) are different depending on the chosen "concentration" scale. Recall, in this respect, that for diluted solutions, molalities differ little from molarities, but mole fractions vary greatly from the preceding ones. Usually, in tables of equilibrium constants, only one of their values is mentioned. Tables do not take into account the possibility of several values. The reason behind this is very simple: Conventions for the quantification of activities have been quasi-universally chosen. In other words, some particular standard states have been universally adopted and their choice is implicit. The conventions are as follows:

- for a solute (ion or molecule), the numerical value of its activity is systematically related to its molarity or molality. As a result, in the case of sufficiently dilute solutions, since $C^{o}{}_{i} = 1$ mol/L or $m^{\circ}{}_{i} = 1$ mol/kg, its numerical value tends toward that of its molarity or molality:

$$a_{i,c} \rightarrow c_i \quad \text{(numerical value of diluted solution)},$$

$$a_{i,m} \rightarrow m_i \quad \text{(numerical value of diluted solution)};$$

  since in these conditions, $c_i \approx m_i$, we have

$$a_{i,c} \approx a_{i,m} \quad \text{(diluted solution)};$$

- for the solvent, its activity value is related to its mole fraction. In sufficiently diluted solutions, the value of this quantity tends toward unity, as in

$$x_0 = n_0/(n_0 + n_i)$$

  since $n_i \ll n_0$ for diluted solutions.
  As a result,

$$a_o \rightarrow 1 \quad \text{(diluted solutions)}.$$

  It is the reason why, when the solvent also participates in the chemical reaction and when the solution is sufficiently diluted, the solvent activity is purely and simply omitted in the mass law;
- for every pure phase (for example, a pure liquid or a pure solid), the activity value is taken to be exactly equal to unity. This is why a pure metal electrode has an activity value of unity;
- for every liquid in a mixture of liquids, its activity value is chosen to be approximately equal to its mole fraction.

Due to the different conventions chosen for solutes and solvents, at this stage in the study of chemical thermodynamics, the importance of assigning the status of solvent or solute to the different solution components is evident.

3.8 Determination of Activities

3.8.1 *Uncharged Solutes*

It is possible to experimentally determine the uncharged solutes activities, for example, by measurements of the solvent's vapor pressure above the solution.

3.8.2 *Activity of An Ion: Activity of the Whole Electrolyte*

It is impossible to measure the activity of an ion experimentally. A solution is necessarily electroneutral; for this principle to be obeyed, an ion is necessarily accompanied by a counterion of the opposite electrical charge. However, the activity of the whole electrolyte can be determined experimentally. For example, in an aqueous solution of sodium chloride, only the activity a_{NaCl} can be reached, but a_{Na^+} and a_{Cl^-} cannot.

3.9 Calculation of Activity Coefficients and of Activities

3.9.1 *Activity Calculation of Uncharged Species*

Experimental measurements show that up to an ionic strength of 0.1 mol/L, the activity value of an uncharged species does not differ from its concentration value by an error higher than 1%. In the range of ionic strengths 0.1–5 mol/L, the following relationship is verified:

$$\log \gamma_c = kI_c.$$

In most cases, but not often, the activity coefficients of uncharged species are above unity.

3.9.2 *Calculation of Activity Coefficients and Activities of Ions*

Despite the fact that the activity coefficient of an ion cannot be determined experimentally, their values can be reasonably approached, in all probability,[2] by calculation, with the help of the Debye-Hückel theory.[3] They can be done for at least some ranges of ionic strengths.

---

[2] This reads "in all probability" because the calculation results cannot be directly verified experimentally (see ahead).

[3] Peter Debye: Dutch physicochemist, naturalized American (1884–1966); Erich Hückel: German physicochemist (1896–1980). He is also known for "Hückel method of molecular orbitals."

3.9.2.1 The Debye-Hückel Limiting Law

The first basic assumption of the theory is that ions behave like point charges. The next one is that they move in a continuous medium with a relative permittivity equal to that of the solvent. Then Debye and Hückel deduced the limiting law from the laws of electrostatics and thermodynamics. The limiting law is given by

$$\log \gamma_x = -Az^2\sqrt{I_c}, \tag{3.11}$$

where γ_x is the activity coefficient of the ion and z its charge. A is a constant for a given temperature and for a given solvent. More precisely, the nature of the solvent intervenes through its relative permittivity value. For water at 25°C, $A = 0.5115$ and

$$\log\gamma_x = -0.5115z^2\sqrt{I_c} \quad (I_c < 10^{-3}\ \text{mol/L}; 25°\text{C}). \tag{3.12}$$

This limiting law is obeyed at ionic strengths below 10^{-3} mol/L. At higher concentrations, the predicted activity coefficients are smaller than those measured (indirectly) experimentally.

Notice that when I_c tends toward zero, γ_x tends toward unity, in agreement with the above considerations. We also notice that all the ions, cations or anions, whichever they are, exhibit the same activity coefficient value at the same ionic strength, provided they possess the same electrical charge. Finally, in Eqs. (3.11) and (3.12), the ionic strength is expressed in mol/L, whereas the calculated activity coefficient γ_x is related to the solute mole fraction. Fortunately, we can demonstrate that, under the ionic strength conditions in which the Debye-Hückel limiting law is valid,

$$\gamma_c = \gamma_m = \gamma_x.$$

(this result is obtained by reasoning analogous to that used in Chap. 2). Hence, Eqs. (3.11) and (3.12) can also be written as

$$\log \gamma_c = -Az^2\sqrt{I_c}, \tag{3.13}$$

$$\log \gamma_c = -0.5115z^2\sqrt{I_c} \quad (\text{aqueous solutions} - 25°\text{C} - I_c < 10^{-3}\ \text{mol/L}). \tag{3.14}$$

Exercise Calculate the activity coefficients and the activities of the different ions present in solution after the dissolution of potassium chloride 10^{-4} mol/L and magnesium sulfate 10^{-4} mol/L in water. Both electrolytes are strong. Ions exhibit neither an acidic property nor a basic character.

The ionic strength of the solution is

$$I_c = \tfrac{1}{2}\left[1^2 \cdot 10^{-4} + (-1)^2 \cdot 10^{-4} + 2^2 \cdot 10^{-4} + (-2)^2 \cdot 10^{-4}\right],$$

$$I_c = 5 \cdot 10^{-4}\ \text{mol/L},$$

$$\log \gamma_c(\text{Cl}^-) = \log \gamma_c(\text{K}^+) = -0.5115\sqrt{5 \cdot 10^{-4}},$$

$$\gamma_c(\text{Cl}^-) = \gamma_c(\text{K}^+) = 0.974,$$

$$\log \gamma_c(Mg^{2+}) = \log \gamma_c(SO_4{}^{2-}) = -0.51155 \cdot 2^2\sqrt{5} \cdot 10^{-4},$$
$$\gamma_c(Mg^{2+}) = \gamma_c(SO_4{}^{2-}) = 0.900,$$
$$a(Cl^-) = a(K^+) = 9.74 \cdot 10^{-5},$$
$$a(Mg^{2+}) = a(SO_4{}^{2-}) = 9.00 \cdot 10^{-5}.$$

For ionic strengths greater than 10^{-3} mol/L, Eq. (3.11) gives activity coefficients that are too weak and, hence, activities that are also too weak (see Sect. 3.10).

3.9.2.2 Extended Debye-Hückel Law

If we take the same theory as the preceding one but now consider that an ion is endowed with a radius of finite size, we get the equation

$$\log \gamma_c = -\left[Az^2\sqrt{I_c}\right]\Big/\left[1 + Ba\sqrt{I_c}\right], \qquad (3.15)$$

called the *extended Debye-Hückel law*. Compared to the limiting law, a denominator with two new parameters appears. B is a constant depending only on the solvent (through its relative permittivity) and on the temperature. For water at 25°C, $B = 0.3291$. The parameter a, measured in angstroms (10^{-10} m), can be considered the radius of the hydrated ion. Its occurrence in Eq. (3.15) permits ion individualization through the radius value a. This was not possible with the limiting law, as we have noticed. In water at 25°C, the extended Debye-Hückel law is

$$\log \gamma_c = -\left[0.5115z^2\sqrt{I_c}\right]\Big/\left[1 + 0.3291a\sqrt{I_c}\right] \quad (\text{water} - 25°\text{C} - I_c < 0.1\ \text{mol/L}). \qquad (3.16)$$

It is only valid for ionic strengths lower than 0.1 mol/L. For higher ionic strengths, the activity coefficients and the activities are also too weak. It is interesting to note that for ionic strengths less than 0.1 mol/L, we can consider

$$I_c = I_m,$$
$$\text{and} \quad \gamma_m = \gamma_c = \gamma_x,$$

3.9.2.3 Other equations used for higher ionic strengths

In the 10^{-1}–8.10^{-1} mol/L interval, equations such as

$$\log \gamma_c = -\left[Az^2\sqrt{I_c}\right]\Big/\left[1 + Ba\sqrt{I_c}\right] + CI_c \qquad (3.17)$$

are used. According to the values given to the parameters a and C, they are called the *Davies* or *Guggenheim equations*. The new parameter C appears in Eq. (3.17). Its occurrence stemmed from a purely experimental origin. (The next section gives the experimental data that have permitted us to add C.) However, it has been theoretically but uneasily justified since its introduction. This has been done using reasoning in the realm of statistical thermodynamics by taking into account the solute-solvent interactions together with the solute-solute ones. This was not done in the Debye-Hückel theory.

3.10 Justification of Debye-Hückel's Theory

The question of the accuracy of Debye-Hückel's theory has arisen since an ion's activity cannot be measured. Hence, calculations cannot be compared with experimental values. Fortunately, the Debye-Hückel equations can be indirectly checked. This is achieved by comparing the calculated values of mean activity coefficients with those determined experimentally. Mean activity coefficients can indeed be determined experimentally, unlike the activity coefficients of the sole ions. Moreover, they can be calculated from the values found for the individual ions of the electrolyte by using the Debye-Hückel laws. As a result, we can see that it is possible to compare the experimental and calculated values of some activity coefficients.

The mean activity coefficients are defined in the following way. Let us consider a one-to-one electrolyte (example: sodium chloride). The mean activities $a_\pm$ of both ions Na^+ and Cl^- are given by the equation

$$a_\pm{}^2 = a_{Na^+} \cdot a_{Cl^-}. \tag{3.18}$$

Likewise, the mean activity coefficient $\gamma_\pm$ is defined by

$$a_\pm = \gamma_\pm C, \tag{3.19}$$

where C is the electrolyte concentration. Handling these two equations gives us

$$\gamma_\pm{}^2 C^2 = \gamma_{Na^+} \cdot C_{Na^+} \cdot \gamma_{Cl^-} C_{Cl^-}.$$

Since the electrolyte is strong, $C_{Na^+} = C_{Cl^-} = C$ and

$$\gamma_\pm{}^2 = \gamma_{Na^+} \cdot \gamma_{Cl^-}. \tag{3.20}$$

Otherwise, the intuitive assumption that the chemical potential of a strong electrolyte is equal to the sum of the chemical potentials of every ion component of the electrolyte is proven. The sum of the chemical potentials gives

$$a_{NaCl} = a_{Na^+} \cdot a_{Cl^-}. \tag{3.21}$$

From the above equation, it appears that the whole electrolyte activity is equal to the product of the activities of each ion. Equations (3.18) and (3.21) give

$$a_\pm{}^2 = a_{NaCl}. \tag{3.22}$$

As a result, since the activity a_{NaCl} of the whole electrolyte is measurable, the mean activity $a_{\pm}$ and the mean activity coefficient $\gamma_{\pm}$ are also measurable through Eq. (3.19).

From the viewpoint of calculations, now the mean activity coefficient $\gamma_{\pm}$ is calculated through Eq. (3.20), where, in succession, γ_{Na^+} and γ_{Cl^-} are calculated by Debye-Hückel equation (3.13) or (3.15). From Eq. (3.20), by taking the logarithms, and by noting that $z_+ = z_-$ in the case of a one-to-one electrolyte, we find

$$-\log \gamma_i = Az_+z_-\sqrt{I_c} \tag{3.23}$$

and

$$-\log \gamma_i = \left(Az_+z_-\sqrt{I_c}\right) \Big/ \left(1 + Ba\sqrt{I_c}\right). \tag{3.24}$$

Yet Eqs. (3.13) and (3.15) have now been transformed into Eq. (3.24), which can be directly compared with experimental data. (These considerations can be generalized to every type of strong electrolyte, such as one-to-two, one-to-three, two-to-three, etc.). Hence, the indirect checking of the Debye-Hückel laws is grounded on the hypothesis that if the product of activities a_{Na^+} and a_{Cl^-} is correctly predicted, it is very probable that it will be the same for the individual activities a_{Na^+} and a_{Cl^-}.

Actually, the Debye-Huckel laws predict accurate values for mean activity coefficients. Herein lies their great triumph. It has been found this way that the activity coefficients of ions with the same charge number are identical for weak ionic strengths, no matter what the ions (limiting law) are. It has also been found this way that for higher ionic strengths, the activity coefficients predicted by the limiting law were too weak and that they varied from one ion to another. Finally, in order to predict the correct mean activity coefficients, the parameter C was added for the higher ionic strengths (Davies' and Guggenheim's equations).

The Debye-Hückel theory has achieved enormous success. It is considered among the greatest discoveries of the 20th century in the realm of physical chemistry. However, it is not fully satisfactory. It leads to difficulties in some cases. For example, the parameter a can be endowed with a value that is not that of a hydrated ion radius. It can sometimes be negative! Debye himself said that the theory was awarded more success than it deserved. However, the Debye-Hückel laws are now irreplaceable. As just one example, they justify extrapolation procedures to obtain thermodynamic equilibrium constants to null ionic strength.

The activity concept is of utmost importance, particularly in the case of the activity concept of an ion. To convince ourselves of this importance, it is sufficient to know, for example, that a solution's pH is related to the activity of the hydrated proton and not to its concentration. The Debye-Hückel laws intervene here by permitting reasonable pH values to be assigned to some reference solutions.

Part II
Acids and Bases Equilibria—Analytical Applications

Chapter 4
Definitions of Acids and Bases: Strength of Acids and Bases

There are several definitions of acids and bases. We are especially interested here in the Brønsted–Lowry definition and its consequences. After reviewing definitions, we shall study the strength concept of acids and bases. It is important because it allows us to predict acid–base reactions.

4.1 Arrhenius Definition

The Arrhenius definition is considered to have been the first true theory of acids and bases. It results from Arrhenius and Ostwald's[1] idea concerning electrolytes' dissociation in aqueous solutions. Arrhenius theory is grounded on the ionization of water.

An acid HA is a species that ionizes in water to give one hydrated proton $H^+_{(w)}$[2] and one anion according to the more or less equilibrated reaction

$$HA \rightleftharpoons A^-{}_{(w)} + H^+{}_{(w)}. \tag{4.1}$$

A base BOH is a substance that ionizes in water to give one hydroxide anion $OH^-_{(w)}$ and one cation according to

$$BOH \rightleftharpoons B^+{}_{(w)} + OH^-{}_{(w)}. \tag{4.2}$$

There are also polyacids and polybases. *Polyacids* give several hydrated protons per molecule, while *polybases* give several hydroxide anions.

---

[1] Svante Arrhenius: swedish chemist (1859–1927), winner of Nobel Prize in Chemistry in 1903. His theory figured in his state thesis defended at Uppsala University when he was 25 years old. His thesis was judged too revolutionary and the examining board was doubtful!

Wilhelm Ostwald: german chemist (1853–1932). He was awarded the Nobel Prize in Chemistry in 1909. His work concerned chemical equilibria. He was one of the mentors of the famous Walther Nernst.

[2] In Sect. 4.4, we consider in more detail the structure of the hydrated proton in water, which, at the present time, we write as $H^+_{(w)}$.

J.-L. Burgot, *Ionic Equilibria in Analytical Chemistry*,
DOI 10.1007/978-1-4419-8382-4_4, © Springer Science+Business Media, LLC 2012

The strength of acids and bases is related to the extent of the ionization process, that is, to the equilibrium points of reactions (4.1) and (4.2).

The Arrhenius definition can only be applied to aqueous solutions. It is not sufficiently general. However, it is satisfactory in aqueous solutions. From a quantitative viewpoint, it permits us to correctly predict acid–base reactions.

4.2 Brønsted–Lowry Definition

The following definition was given independently in 1923 by the two chemists Brønsted and Lowry.[3] An acid is a proton donor according to the equation

$$HA \rightleftharpoons A^- + H^+. \tag{4.3}$$

A base is a proton acceptor according to the equation[4]

$$B + H^+ \rightleftharpoons BH^+ \tag{4.4}$$

In Eq. (4.3), the species A^- can accept a proton. According to the definition, it is a base. After Eq. (4.4), BH^+ is an acid. HA and A^-, on the one hand, and BH^+ and B, on the other hand, form an acid–base couple. HA and A^- are conjugate. It is the same for BH^+ and B. Further examples of conjugate acid–base couples are the ammonium ion $NH_4{}^+$ and ammoniac NH_3, hydrogen chloride HCl and the chloride ion Cl^-. So far, no particular solvent has been mentioned.

Equations (4.3) and (4.4) also show that the acidic or basic character of a substance is independent of its electrical charge. Ammoniac, a neutral substance, is a base. The chloride ion, which is negatively charged, is also a base. It is the same state of affairs for hydrogen chloride and ammonium chloride, which are both acids.

Polyacids donate two or more protons upon ionization. This is the case, for example, with sulfuric acid, which successively gives one proton and the hydrogen sulfate ion:

$$H_2SO_4 \rightarrow H^+ + HSO_4{}^-$$

and a second one by the ionization of the above-mentioned ion:

$$HSO_4{}^- \rightleftharpoons H^+ + SO_4{}^{2-}.$$

Polybases accept several protons. Hence, the orthophosphate ion successively accepts three protons to give hydrogen phosphate and dihydrogen phosphate ions, and orthophosphoric acid, respectively:

$$PO_4{}^{3-} + H^+ \rightleftharpoons HPO_4{}^{2-},$$

[3] J. N. Brønsted: danish chemist and biochemist (1879–1947). T. M. Lowry: english chemist (1874–1936).

[4] In this book, we symbolize bases by B or B^- depending on whether or not they are charged.

$$HPO_4^{2-} + H^+ \rightleftharpoons H_2PO_4^-,$$

$$H_2PO_4^- + H^+ \rightleftharpoons H_3PO_4.$$

Some compounds can donate and accept protons. They are called *amphoteric substances*, or *ampholytes*. The best-known example is water. It gives a proton according to Eq. (4.5) and accepts a proton according to (4.6):

$$H_2O \rightleftharpoons H^+ + OH^-, \tag{4.5}$$

$$H_2O + H^+ \rightleftharpoons H_3O^+. \tag{4.6}$$

In Eq. (4.6), the species H_3O^+ is called the hydroxonium ion or simply the oxonium ion. Another example of an ampholyte is the hydrogen sulfide ion HS^-, which is an acid according to the reaction

$$SH^- \rightleftharpoons S^{2-} + H^+$$

and a base according to

$$SH^- + H^+ \rightleftharpoons H_2S.$$

Both reactions are new examples of the independence of the acidic and basic characters from the electrical charge.

Exhibiting an acidic or basic character is not the prerogative of small molecules or that of mineral or organic ions, as the above examples would tend to prove. Mineral and organic polymers often exhibit an acidic or basic character. It is the same case for natural huge molecules such as polypeptides and proteins, among others.

4.3 Inexistence of the Proton in Solution

The proton is a particular species. Whereas all the ions possess one or several electrons around their nucleus, such is not the case for the proton, which is only one nucleus constituted by one particle: the proton. Whereas other ions exhibit radii on the order of the nanometer (10^{-9} m), the proton radius is on the order of the fermi (10^{-15} m). An intense electrical field exists on its surface. This enables it to approach a neighboring ion or atom or molecule much closer than any other ion or atom could. Hence, the naked proton does not exist in solution. As a result, reactions (4.3) and (4.4) are never seen. Only reaction (4.7) is seen; it is the result of the two previous ones:

$$HA \rightleftharpoons A^- + H^+ \tag{4.3}$$

$$B + H^+ \rightleftharpoons BH^+ \tag{4.4}$$

$$HA + B \rightleftharpoons A^- + BH^+ \tag{4.7}$$

Reaction (4.7) is a proton-exchange reaction between two acid–base couples. Equilibria (4.3) and (4.4), which represent the definition of acids and bases, are only virtual ones. It is the same thing for the equilibria described in the previous section.

The fact that acid–base reactions in solution are equivalent from the sole viewpoint of resulting from a proton exchange between two acid–base pairs by no means prejudges the true mechanism of the overall reaction. In the course of the reaction, there may actually be no exchange at all of hydrated protons between the two acid–base pairs considered. This is the reason why some authors speak of conventional protons H (0) in the preceding definition reactions. However, the half-reactions (acid–base) involving the exchange of true protons

$$B_{(g)} + H^{+}{}_{(g)} \rightleftharpoons BH^{+}{}_{(g)}$$

are observed in the gas phase by mass spectrometry.

As a result, the acidic or basic character of a species can appear in solution only when it is in the presence of an antagonist couple, which necessarily participates in the proton exchange. Speaking of an acidic or basic character in an absolute sense is devoid of any meaning. Although it is not obligatory, it is the solvent that usually plays the role of the antagonist couple. In this case, the solvent reveals the acidic or basic character of the couple under study.

These considerations are valid for any solvent that can play this role. Hence, the solvent must exhibit an acidic or basic character, or both.

4.4 Brønsted Acidity and Basicity in Water: Nature of the Hydrated Proton in Water

Ethanoic acid, more commonly called acetic acid, is an acid in water since the following reaction takes place:

$$CH_3COOH + H_2O \rightleftharpoons CH_3COO^{-}{}_{(w)} + H_3O^{+}{}_{(w)}; \qquad (4.8)$$

from which came its common name. Likewise, hydrogen chloride is an acid in water due to the reaction

$$HCl + H_2O \rightarrow Cl^{-}{}_{(w)} + H_3O^{+}{}_{(w)}, \qquad (4.9)$$

which justifies the name "hydrochloric acid."

Reaction (4.9) is not equilibrated. This fact must not be interpreted as being a consequence of a shortcoming in the Brønsted–Lowry theory. Actually, and simply, the chloride ion is too weak a base in water. Additionally, it is recognized as being a base in some other solvents.

As Eqs. (4.8) and (4.9) show, water simultaneously plays the roles of the reactant and the solvent. The latter is recalled in the formalism by the subscript (w).

An important point to recall further is the nature of the hydrated proton in water. So far, we have symbolized it successively by $H^{+}{}_{(w)}$ and $H_3O^{+}{}_{(w)}$. For reasons previously explained, the proton does not exist in water. A calculation due to Fajans gives as its concentration the inconceivable value of 10^{-130} mol/L! A body of thermodynamic

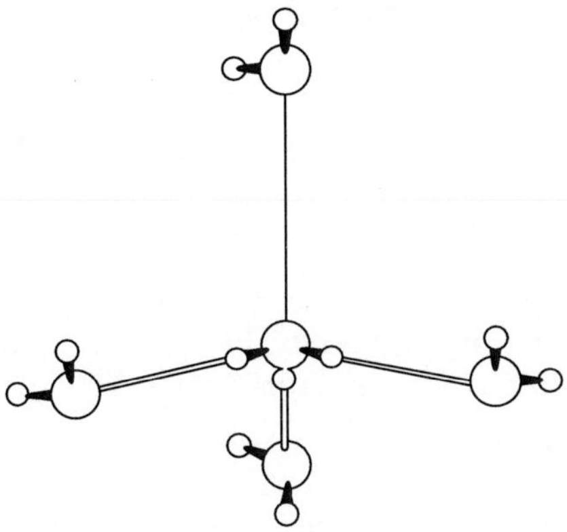

Fig. 4.1 Schematic representation of the hydroxonium ion in solution

and structural proofs shows that it settles down to a water molecule to give the oxonium ion, which is a defined chemical entity. It has a rather flat trigonal pyramidal structure, with the hydrogen at the corners of the pyramid and the oxygen in the middle, as shown in Fig. 4.1. Its structure resembles that of the ammoniac molecule. The oxonium ion itself is associated with three water molecules in its first solvation shell. The whole group is linked to an extra water molecule, which is electrostatically bound to it. This is the reason why we believe it judicious to symbolize the hydrated proton by $H_3O^+{}_{(w)}$. In this chapter, we use this symbolism, but in the rest of this book, we shall interchangeably use the symbols $H^+{}_{(w)}$, $H_3O^+{}_{(w)}$, and even H^+, with the understanding that the three symbols represent the same species: the hydrated proton.

4.5 Nomenclature

Some particular terms related to the Brønsted–Lowry are often used. The term *protolytes* refers to acids and bases in general, regardless of which are the solvents and which are the antagonist couples. We also speak of *protogenic* and *protophilic behaviors*. They have, respectively, acidic and basic behaviors. A solvent is called *protic* when it can give or accept a proton or when it can do both. Hence, it reveals the acidic and basic characters of protolytes. If it cannot exchange protons, it is called *aprotic* or *inert*.

4.6 About the Equivalence of the Arrhenius and Brønsted Theories in Aqueous Solutions

It is easy to find that both Arrhenius and Brønsted theories are very similar in aqueous solutions. When all is said and done, the acidic and basic characters, according to both theories, result in the appearance of the same species, that is, the hydroxonium cation and the hydroxide anion. Let us consider the acid HA ionization according to both theories:

$$HA \rightleftharpoons A^- + H^+{}_{(w)} \quad \text{Arrhenius,}$$

$$HA + H_2O \rightleftharpoons A^- + H_3O^+ \quad \text{Brønsted.}$$

They are equivalent if we identify the species $H^+{}_{(w)}$ and $H_3O^+{}_{(w)}$. This identification is a physical reality. Let's now consider the conjugated base A^-. It is, of course, a Brønsted base according to the following reaction:

$$A^-{}_{(w)} + H_2O \rightleftharpoons HA + OH^-{}_{(w)}.$$

But this reaction also shows that a hydroxide ion is liberated, as the Arrhenius theory points out. The only difference lies in the origin of this ion, which comes from the solvent according to this reaction. This is not the case in the Arrhenius theory, according to which it comes from the species. To sum up, if the origin of the hydroxide ions is not taken into account and if the species $H^+{}_{(w)}$ and $H_3O^+{}_{(w)}$ are intermingled, which must be done, both theories are identical at least concerning the species appearing in the process.

Actually, in aqueous solutions, the differences are theoretical rather than phenomenological. The amphoteric character of water provides an example of this assertion. In the Arrhenius theory, water is amphoteric because it simultaneously liberates one proton and one hydroxide ion per molecule:

$$H_2O \rightleftharpoons H^+{}_{(w)} + OH^-{}_{(w)}.$$

In the Brønsted theory, the amphoteric character requires the participation of two water molecules to be explained:

$$H_2O + H_2O \rightleftharpoons H_3O^+{}_{(w)} + OH^-{}_{(w)}.$$

The first molecule plays the part of a base, the second that of an acid. Hence, the amphoteric character of water results from a proton exchange between two different couples in which the water molecule is one member of each of them: the couples H_3O^+/H_2O and H_2O/OH^-. Again, both theories are different in principle, but in the end, the formed species are the same and the participation of two water molecules is not detectable.

It may happen, but exceptionally, that some species exhibit an acidic or basic character according to one of the two theories and do not exhibit such behavior according to the other. Let us consider a metallic hydroxide in the solid state. After its dissolution in water, hydroxide ions appear:

$$MOH(s) \rightleftharpoons M^+{}_{(w)} + OH^-{}_{(w)}.$$

MOH(s) is a base according to Arrhenius. This is not the case from the Brønsted point of view. The conjugate acid of the hydroxide ion is not the metallic hydroxide but quite obviously H_2O. Other examples could be given in this respect.

4.7 Other Theories of Acids and Bases

Some other theories of acids and bases exist. Let us very briefly quote those of Lewis and Werner. According to Lewis (1923), acids can accept an electron pair while bases can give an electron pair. According to Werner,[5] an acid is a species that accepts hydroxide ions. For example, orthoboric acid, H_3BO_3, is a Werner acid since it reacts with water according to the reaction

$$H_3BO_3 + 2H_2O \rightleftharpoons B(OH)_4{}^-{}_{(w)} + H_3O^+{}_{(w)}.$$

The most useful theory in analytical chemistry is that of Brønsted–Lowry.

4.8 Qualitative Considerations Concerning the Strength of Acids and Bases in Water

It is necessary to quantify the strength of acids and bases in order to predict the acid–base reactions. Let us consider the acid HA in water. Knowing the amount of the proton transfer from it to water provides us with important information about its strength. The acid is strong if the transfer is total. The following reaction occurs:

$$HA + H_2O \rightarrow H_3O^+{}_{(w)} + A^-{}_{(w)}. \tag{4.10}$$

The acid is weak if the transfer is incomplete. The corresponding reaction is

$$HA + H_2O \rightleftharpoons H_3O^+{}_{(w)} + A^-{}_{(w)}. \tag{4.11}$$

It is balanced. The inspection of these two reactions shows that the extent of the transfer depends on two independent factors. The acid strength depends

- first on its own ability to donate protons,
- on the solvent's ability (here water) to accept protons, that is, on its basic character. This second factor is often underestimated. For example, hydrogen chloride is strong in water. The proton transfer to water is total, hence its name: hydrochloric acid:

$$HCl + H_2O \rightarrow Cl^-{}_{(w)} + H_3O^+{}_{(w)}.$$

This is not the case with ethanoic acid when it is used as solvent, as the following equilibrated reaction shows:

$$HCl + CH_3COOH \rightleftharpoons Cl^-{}_{(solv)} + CH_3COOH_2{}^+{}_{(solv)}.$$

[5] Alfred Werner: swiss chemist (1866–1919), won the Nobel Prize (1913) in Chemistry. Author of several works concerning complexes (see Part IV).

Hence, in acetic acid playing the part of the solvent, hydrogen chloride is a weak acid. It is still the same acid species.

In this example, ethanoic acid is the solvent, but it also plays the part of a base since it accepts a proton from the hydrogen chloride to give the solvated proton $CH_3COOH_2^+$, which has sometimes been called the acetylium ion. Quite obviously, ethanoic acid is less basic than water since the proton transfer to it is equilibrated, whereas it is total to water. Hydrogen chloride is a weak acid in ethanoic acid and is a strong one in water. Conversely, these are the reasons why it is called hydrochloric acid and why ethanoic acid is called acetic acid in water. It is also interesting to notice that in this example the chloride ion exhibits a true basic character in acetic acid, in agreement with the Brønsted theory.

This example is somewhat disconcerting since acetic acid plays the solvent part and the base parts at the same time. This is unusual. The disconcerting factor is the usual name "acetic acid" given to ethanoic acid, which is a remnant of its behavior in water. There is a contradistinction between its usual name and its behavior in this example. In any case, it shows the relative character of the strength of acids and bases.

Remark in this book we shall call *strong* acids those that are fully dissociated and *weak* acids the incompletely dissociated ones (personal definition).

4.9 Quantitative Considerations Quantifying the Strengths of Acids and Bases: Dissociation Acid Constants K_a and pK_a

4.9.1 *Acids' Strength*

A quantitative insight into the strength of an acid in water is provided by the equilibrium constant value $K°$ of reaction (4.11):

$$HA + H_2O \rightleftharpoons H_3O^+{}_{(w)} + A^-{}_{(w)}, \qquad (4.11)$$

$$K° \approx \left[a_{H_3O^+} \cdot a_{A^-}\right] / a_{HA},$$

where the activities are those at equilibrium. Very often in analytical chemistry we are faced with diluted solutions. In these conditions, the mole fraction and hence the activity of the solvent are very near unity, and

$$K° \approx \left[a_{H_3O^+} \cdot a_{A^-}\right] / a_{HA}.$$

The constant $K°$ is called the *acid dissociation constant* of the couple HA/A^-. It is symbolized by K_a:

$$K_a = \left[a_{H_3O^+} \cdot a_{A^-}\right] / a_{HA}.$$

K_a does possess the meaning of the thermodynamic constant but only in diluted solutions.[6] K_a is a particularly well-suited constant to quantify the acid's strengths. The more ionized the acid, the higher the numerator and the lower the denominator. Usually, to simplify the writing, the decimal cologarithm of K_a is used:

$$pK_a = -\log_{10} K_a.$$

The stronger the acid is, the higher K_a is and the lower the pK_a values are. Like other constants expressed in activities, K_a is dimensionless. Since it is merged into $K°$ in diluted solutions, K_a only depends on temperature and, to a lesser extent, on the pressure. Equilibrium constants $K_{a,c}$ and $K_{a,w}$ related to concentrations and molalities:

$$K_{a,c} = [H_3O^+][A^-]/[HA],$$

$$K_{a,m} = (m_{H_3O^+} \cdot m_{A^-})/m_{HA}$$

are also used. They are, of course, endowed with a dimension. As we have already seen, $K_{a,c}$ and $K_{a,m}$ values also change with concentrations. When the ionic strengths are weak, their values together with that of K_a are very close to each other:

$$K_{am} \approx K_{ac} \approx K_a \qquad \text{(numerical values)}.$$

The question of how, for example, K_a changes with the solution's ionic strength has arisen. The thermodynamic constant can be written as

$$K_a = \{[A^-][H^+]/[HA]\} \cdot \{\gamma_{A^-}\gamma_{H^+}/\gamma_{HA}\},$$

$$K_a = K_{ac}(\gamma_{A^-}\gamma_{H^+}/\gamma_{HA}).$$

The undissociated acid HA has an activity coefficient very close to unity for dilutions for which K_a still possesses its thermodynamic meaning; therefore,

$$K_a = K_{ac}\gamma_{A^-} \cdot \gamma_{H^+}.$$

As a result, the higher the ionic strength, the lower the activity coefficients and the higher the concentration equilibrium constant $K_{a,c}$ since the thermodynamic one is constant. The ionic strength increase enhances the acid dissociation. This property is used in homogeneous catalysis overall when it involves hydroxonium ions. However, for ionic strength increases that are too large, activity coefficients also increase. Hence, the K_a value passes through a maximum. However, we must not forget that for ionic strengths that are too high, K_a loses its thermodynamic meaning, since then the water activity differs from unity markedly.

---

[6] Curiously, neither does an official definition (IUPAC) of K_a exist, nor is a name available for it. The definition and the nomenclature used here are those of R. G. Bates in his book, *Determination of pH: Theory and Practice*, John Wiley and Sons, New York, 1975. The constant K_a is also called the acid ionization constant.

Remark From now on, the subscript *c* will be omitted when no ambiguity is present.

4.9.2 Bases' Strength

Let us consider the base B. Its basic behavior is revealed by the reaction

$$B + H_2O \rightleftharpoons BH^+{}_{(w)} + OH^-{}_{(w)}.$$

The equilibrium constant is

$$K^\circ = (a_{BH^+} \cdot a_{OH^-}) / (a_B \cdot a_{H_2O}).$$

In diluted solutions, the water activity is very close to unity. Therefore, in these conditions, K° is called the *base dissociation constant* of the couple BH^+/B. It is symbolized by K_b:

$$K_b = a_{BH^+} \cdot a_{OH^-} / a_B. \tag{4.12}$$

The stronger the base, the higher the constant K_b. The pK_b is defined by

$$pK_b = -\log_{10} K_b.$$

The stronger the base, the lower the pK_b value. The constants $K_{b,c}$ and $K_{b,m}$ are also defined. They vary with the ionic strength. Actually, it is not necessary to define the constant K_b to characterize the strength of a base. Constants K_b or pK_b are rarely used.

4.10 Water Dissociation

Water, even *purissime*, is endowed with an electrical conductivity. In this case, it is very weak. It is the mark of its dissociation according to the equilibrium

$$H_2O + H_2O \rightleftharpoons H_3O^+{}_{(w)} + OH^-{}_{(w)}. \tag{4.13}$$

The mass law permits us to write

$$K^\circ = a_{H3O^+} \cdot a_{OH^-} / a_{H_2O}{}^2.$$

Conductivity measurements give

$$[H_3O^+] = 10^{-7} \text{mol/L} \qquad \text{(pure water } -25°\text{C)}.$$

The solution is very diluted in ions. The water molecule's activity can be safely taken to unity; hence:

$$K^\circ = a_{H_3O^+} \cdot a_{OH^-}.$$

The product of both activities a_{H_3O+} and a_{OH-} is named the *ion-product constant* for water. It is symbolized by K_w. Its numerical value is determined by measuring the concentrations of the $OH^-$and H_3O^+ ions. The result is

$$K_w = 10^{-14} \qquad \text{(at 25°C)}.$$

The water ion product varies with temperature. Dissociation increases with it.

$$\text{At 25°C, } K_w = 1.0 \cdot 10^{-14}.$$

$$\text{At 50°C, } K_w = 5.6 \cdot 10^{-14}.$$

$$\text{At 100°C, } K_w = 6.0 \cdot 10^{-13}.$$

In every aqueous solution diluted in ions, the water ion product is constant at a given temperature no matter what the OH^- and H_3O^+ ions concentrations may be, provided the solution remains diluted. We say that the water ion product is satisfied. It is the same thing for the acid dissociation constant K_a, which is constant whatever the HA and A^- concentrations may be, provided the solution remains diluted.

Solutions in which $[OH^-]=[H_3O^+]$ are said to be neutral. This is, of course, from the viewpoint of acido-basicity and not from that of electroneutrality. When an acid is dissolved in pure water, the $[H_3O^+]$ concentration increases, and conversely, the $[OH^-]$ concentration decreases for the ion product to be satisfied. It is the reverse when a base is added to the solution. The solution, of course, becomes basic.

4.11 Uselessness of the K_b Notion

Let us recall the K_b definition and multiply both the numerator and the denominator of Eq. (4.12) by the solution's proton activity. In the result obtained, both the expression of the water ion product and that of the constant K_a of the conjugate acid appear. Then, we immediately have

$$K_b = \left(a_{BH^+} \cdot a_{OH^-} \cdot a_{H_3O^+}\right) / \left(a_B \cdot a_{H_3O^+}\right);$$

that is, $K_b = K_w/K_a$ and $pK_w = pK_a + pK_b$. pK_w is the decimal cologarithm of the water ion product. The constant K_a, which appears just above, is the acid dissociation constant of the conjugate acid. This must be emphasized. The error to avoid in this respect is to handle constants K_a and constants K_b that do not belong to the same couple. For example, the hydroxide ion OH^- is not the conjugate base of the hydroxonium ion H_3O^+. It is that of the acid H_2O.

From these considerations, we see that the stronger an acid is, the weaker its conjugate base is. We also see that once the K_a value is known, so is the case of K_b. There is redundancy between the two strength concepts.

4.12 A Brief View of the Concept of pH

Let's reason in terms of concentrations and let the thermodynamic and formal equilibrium constants be intermingled for the sake of simplicity. Let's consider the relationship that defines the constant K_a:

$$K_a = [H_3O^+][A^-]/[HA].$$

We introduce the dissociation extent α, which is defined as the ratio of the concentration of the acid actually dissociated and its whole concentration, that is,

$$\alpha = [A^-]/([HA] + [A^-])$$

(from now on, we shall omit the subscript w because the remainder of this book is devoted to aqueous solutions). Handling the expressions of α and K_a immediately leads to the relation

$$\alpha = K_a/(K_a + [H_3O^+]).$$

The dissociation extent of an acid is function of the hydroxonium ion's concentration in the solution. Since the hydroxonium ion is the common species present in all acid solutions, Sørensen (a danish chemist, 1868–1939) proposed to identify the solution's acidity by its hydroxonium ion's concentration. He quantified it by the decimal cologarithm of this concentration. This is the quantity known as pH:

$$\text{pH} = -\log [H_3O^+].$$

A few years later, in 1918, Sørensen suggested replacing this first definition by the decimal cologarithm of the hydroxonium ion's activity:

$$\text{pH} = -\log a_{H_3O^+}$$

This is what today is known as the formal definition of pH.

Sørensen made this last suggestion because he had observed that potentiometric measurements realized with the help of a glass electrode (which responds to the hydroxonium ions) were not exactly matched with their $[H_3O^+]$ concentration. The notion of a potentiometric chain enters in the full definition of pH. The pH definition in terms of activities, which cannot escape it, brings great complications since the activity of an ion cannot be measured (see Chap. 3).

4.13 The Polyacid Case

Let's consider the polyacid H_nA. Each acidity is endowed with a thermodynamic equilibrium constant corresponding to each of the following successive equilibria:

$$H_nA + H_2O \rightleftharpoons H_{n-1}A^- + H_3O^+,$$

$$H_{n-1}A^- + H_2O \rightleftharpoons H_{n-2}A^{2-} + H_3O^+, \quad \text{etc.,}$$

with

$$K_{a1} = \left(H_{n-1}A^-\right)\left(H_3O^+\right)/(H_nA),$$

$$K_{a2} = \left(H_{n-2}A^{2-}\right)\left(H_3O^+\right)/\left(H_{n-1}A^-\right), \quad \text{etc.}$$

This is a simple generalization of the monoacid case. It is interesting to notice that the ionization of several acidities can take place simultaneously, that is, to the same pH value.

4.14 Distribution Diagrams

Distribution diagrams are very useful. Their inspection immediately provides information about the fractions of the acidic and basic forms to a given pH value. They are particularly interesting in the case of polyacids.

In the case of a monoacid, the ratio $[A^-]/[HA]$ of the conjugate forms is determined once the pH value is known by virtue of the relation

$$\left[A^-\right]/[HA] = K_a/\left[H^+\right]$$

(this is strictly exact only when activities and concentrations are intermingled). A distribution diagram is a graphical representation of this result.

If C is the acid analytical concentration, we have seen that the two following relations:

$$\left\{\left[A^-\right]\right\}\left[H^+\right]/[HA] = K_a,$$
$$[HA] + [A^-] = C$$

lead to the relation

$$\alpha_{A^-} = K_a/\left(K_a + \left[H^+\right]\right),$$

which expresses the dissociation extent defined by $\alpha_{A^-} = [A^-]/C$. The acid formation extent is defined by $\alpha_{HA} = [HA]/C$. It is expressed by the relation

$$\alpha_{HA} = \left[H^+\right]/\left(K_a + \left[H^+\right]\right).$$

The distribution diagram of an acid is a graph α_{A^-}/pH and α_{HA}/pH. The first one is called the *dissociation curve* and the second the *formation curve*. We give the two curves for acetic acid in Fig. 4.2.

With the previous considerations in mind, it is obvious that

$$\alpha_{HA} + \alpha_{A^-} = 1.$$

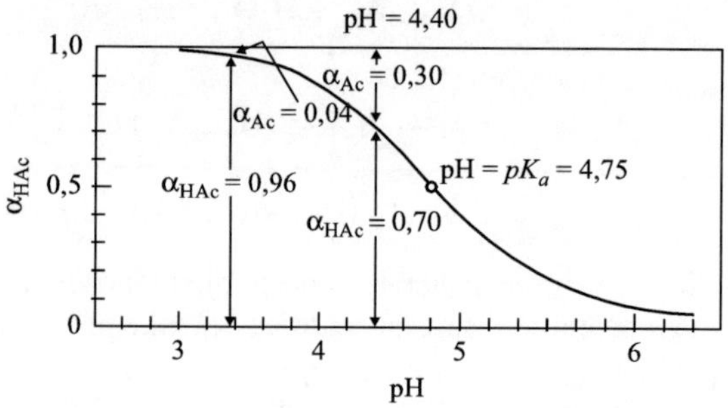

Fig. 4.2 Distribution diagram of acetic acid

The diagram is limited on the ordinate axis to unity. The vertical drawing for the solution's pH value immediately gives the percentages of the acidic and basic forms α_{HA} and α_{A^-}. Hence, at pH $= 4.4$, there exist 70% of undissociated acetic acid and 30% of acetate ion. A mathematical study of the formation curve shows that it always decreases whereas that of dissociation always increases when the pH is increasing. The inflection point is located for pH $= pK_a$.

In the case of a diacid H_2A, the handling of the system

$$[HA^-][H^+]/[H_2A] = K_{a1},$$

$$[A^{2-}][H^+]/[HA^-] = K_{a2},$$

$$[A^{2-}] + [HA^-] + [H_2A] = C$$

gives for coefficients $\alpha_{A^{2-}} = [A^{2-}]/C$, $\alpha_{HA^-} = [HA^-]/C$, $\alpha_{H_2A} = [H_2A]/C$, the expressions

$$\alpha_{A^{2-}} = K_{a1} \cdot K_{a2} / \left(K_{a1} K_{a2} + K_{a1} [H^+] + [H^+]^2 \right),$$

$$\alpha_{HA^-} = K_{a1}[H^+] / \left(K_{a1} K_{a2} + K_{a1} [H^+] + [H^+]^2 \right),$$

$$\alpha_{H_2A} = [H^+]^2 / \left(K_{a1} K_{a2} + K_{a1} [H^+] + [H^+]^2 \right).$$

The distribution diagram is given in Fig. 4.3 as an example for fumaric acid ($pK_{a1} = 3.05$ and $pK_{a2} = 4.49$).

The diagram shows, among other results, that there exists no pH range where the diacid is present only as the single species HA^-. A mathematical study of the α_{HA^-} fraction as a function of the constant values K_{a1} and K_{a2} for every pH value shows that the difference $\Delta pK_a = pK_{a2} - pK_{a1}$ must be higher than 4:

$$\Delta pK_a \geq 4,$$

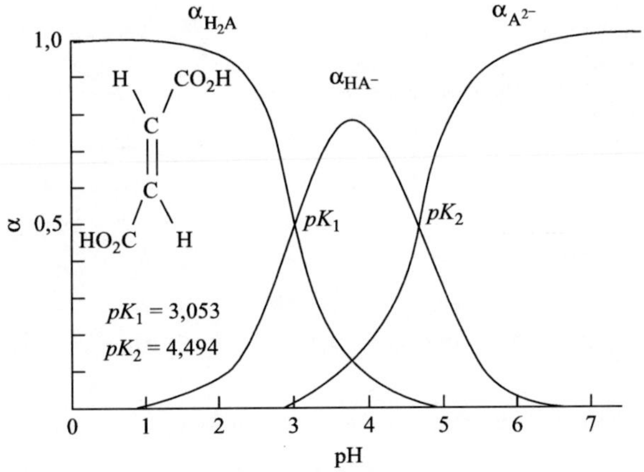

Fig. 4.3 Distribution diagram of fumaric acid

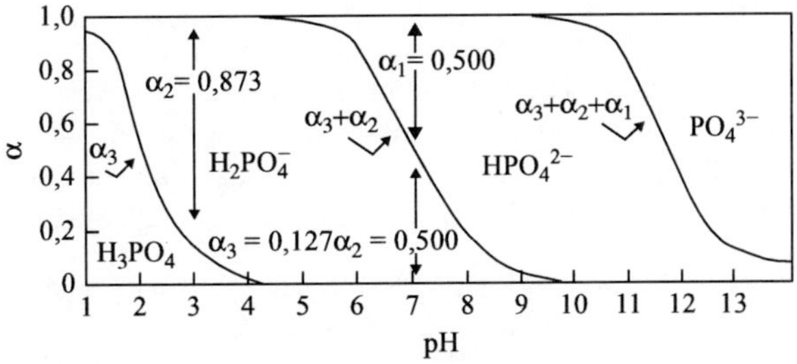

Fig. 4.4 Distribution diagram of orthophosphoric acid ($pK_{a1} = 2.10, pK_{a2} = 7.20, pK_{a3} = 12.35$)

and so the intermediate form should be considered as being the sole species in solution, even in a narrow pH range. This point will be considered again with the study of acid–base titration curves.

In the case of orthophosphoric acid, the mathematical expressions of the different species fractions are found by reasons similar to those followed previously. They are

$$\alpha_{H_3PO_4} = [H^+]^3 / \left([H^+]^3 + K_{a1}[H^+]^2 + K_{a1}K_{a2}[H^+] + K_{a1}K_{a2}K_{a3}\right),$$

$$\alpha_{H_2PO_4-} = K_{a1}[H^+]^2 / \left([H^+]^3 + K_{a1}[H^+]^2 + K_{a1}K_{a2}[H^+] + K_{a1}K_{a2}K_{a3}\right),$$

$$\alpha_{HPO_4{}^{2-}} = K_{a1}K_{a2}[H^+] / \left([H^+]^3 + K_{a1}[H^+]^2 + K_{a1}K_{a2}[H^+] + K_{a1}K_{a2}K_{a3}\right),$$

$$\alpha_{PO_4{}^{3-}} = K_{a1}K_{a2}K_{a3} / \left([H^+]^3 + K_{a1}[H^+]^2 + K_{a1}K_{a2}[H^+] + K_{a1}K_{a2}K_{a3}\right).$$

The distribution diagram is given in Fig. 4.4.

We find that for pH = 4.60 and for pH = 9.90, the only species present are, respectively, $H_2PO_4{}^-$ and $HPO_4{}^{2-}$. This is an illustration of the previous assertion stating that the pK_a difference must be higher than 4 so that intermediate species might be seen alone. This is the case with orthophosphoric acid.

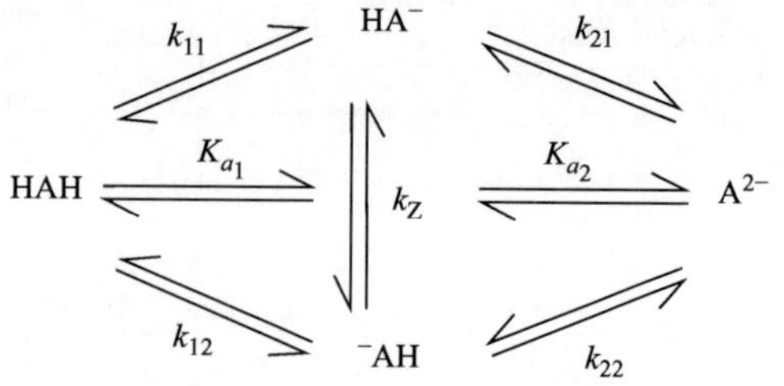

Fig. 4.5 Definitions and symbolism of microscopic constants. The second number in the subscript indicates the ionization

4.15 Macroscopic and Microscopic Equilibrium Constants

In the polyacid case, there are actually two sorts of acid equilibrium constants:

- the macroscopic constants,
- the microscopic ones.

Both are identical in the case of monoacids.

Let's consider, for example, the diacid HAH. It has two acid–base sites. Its ionization can follow the two ways shown in Fig. 4.5.

Like every acid dissociation constant, the microscopic ones are defined by the following relationships (mixing activities and concentrations):

$$k_{11} = [HA^-][H^+]/[HAH], \quad k_{21} = [A^{2-}][H^+]/[HA^-],$$
$$k_{12} = [^-AH][H^+]/[HAH], \quad k_{22} = [A^{2-}][H^+]/[^-AH].$$

The equilibrium constants K_a, which are determined experimentally, are the macroscopic ones. They quantify together the two ionization ways. They are overall acid-dissociation equilibrium constants. They are defined as follows:

$$K_{a1} = [H^+]\{[^-AH] + [HA^-]\}/[H_2A]$$

and

$$K_{a2} = [H^+][A^{2-}]/\{[^-AH] + [HA^-]\}.$$

The comparison of the expressions of macroscopic and microscopic constants leads to the following relationships that link them:

$$K_{a1} = k_{11} + k_{12}, \tag{4.14}$$

$$1/K_{a2} = 1/k_{21} + 1/k_{22}, \tag{4.15}$$

$$K_{a1}K_{a2} = k_{11}k_{21} = k_{12}\,k_{22}, \tag{4.16}$$

and

$$[HA^-]/[^-AH] = k_{11}/k_{12} \cdot k_{22}/k_{21} = k_z. \tag{4.17}$$

The macroscopic constants are the ones most frequently mentioned in the literature.

Equations (4.16) and (4.17) are not independent from each other in the mathematical sense. Equation (4.17) is interesting for its physical meaning. It indicates that the concentration ratio [$^-$AH]/[HA$^-$] is constant regardless of the pH value. Another important point to be emphasized is that Eqs. (4.14)–(4.17) form a system of three independent relations for four unknowns k_{11}, k_{12}, k_{21}, k_{22}, the macroscopic constants being accessible experimentally. Hence, microscopic constant values cannot be determined without any ambiguity. Additionally, the framework of classical thermodynamics does not include the concept of microscopic constants. However, they can be approached by starting with extrathermodynamic hypotheses, the ones that are not issued from classical thermodynamics. Hence, to solve the problem of determining the microscopic constants (but unavoidably with ambiguity), we can suggest the following hypotheses:

- a hypothesis about one microscopic constant value grounded on some structural analogies between the compound under study and some monoacid;
- a hypothesis about the absorbance of one microscopic form;
- a hypothesis about the proton chemical displacement (or of some other element) located in a particular position in the microscopic form.

When the HAH molecule is symmetric, the species HA$^-$ and $^-$AH are identical. As a result,

$$k_{11} = k_{12}$$

and

$$k_{21} = k_{22}.$$

We find from the above considerations that

$$K_{a1} = 2k_{11} \quad \text{or} \quad K_{a1} = 2k_{12}$$
$$K_{a2} = k_{21}/2 \quad \text{or} \quad K_{a2} = k_{22}/2.$$

Even in this particular case, the macroscopic constants are not equal to the microscopic constants. We can notice, however, that in this case the microscopic constants are accessible. This is an exception with regard to the previous assertion.

The previous considerations relative to diacids are generalizable to polyacids. However, the problem quickly becomes difficult because of the fact that the microscopic constant number to take into account increases far more quickly than the number of acido-basic sites.

The study of microscopic constants permits us to demonstrate that the second macroscopic constant, K_{a2}, is necessarily lower than the first one in the case of a diacid. Let's consider, for example, the case of a symmetric diacid HAH in which A is a polymethylenic chain, $(CH_2)_n$, in which n tends toward infinity. We can admit that the microscopic constants k_{11} and k_{21} (and evidently $k_{12} = k_{22}$) are equal. The electrical field due to the electrical negative charge of the intermediate species $^-$AH (identical to HA$^-$) tends to be null due to the distance between the groups A$^-$ and HA. In these conditions, the triple equality

$$K_{a1} = k_{21} = k_{12} = k_{22}$$

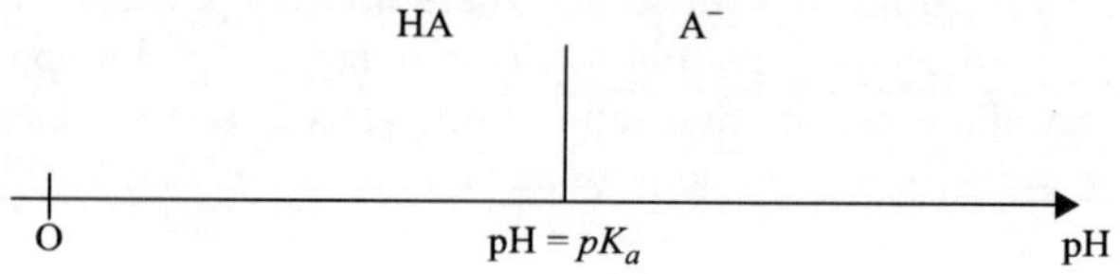

Fig. 4.6 Predominance area of HA and A^- species in relation with the pH values (the monoacid case)

gives the relation

$$K_{a1} = 4K_{a2}.$$

Hence,

$$pK_{a2} \geq pK_{a1} + 0.6.$$

4.16 Predominant Species Area

We have seen that the knowledge of the pK_a value of a monoacid and that of the pH of its solution permits us to know the ratio of its acidic and basic forms immediately. The relationship that defines the ionization constant K_a can be written as

$$\text{pH} - pK_a = \log(a_{\text{A}^-}/a_{\text{HA}})$$

and by intermingling activities and concentrations:

$$\text{pH} - pK_a = \log([\text{A}^-]/[\text{HA}]).$$

We see that when pH $= pK_a$, the acid is half-ionized. For values pH $= pK_a - 2$ units, $pK_a - 1$ unit, $pK_a + 1$ unit, $pK_a + 2$ units, the couple is at 99, 90, 10, and 1%, respectively, under the acidic form. The concentration decrease is very fast once we are far from the value pH $= pK_a$. This leads to the concept of a predominance area. Let's draw a vertical line for the value pH $= pK_a$ on a horizontal axis graduated in pH units (Fig. 4.6).

The basic form A^- predominates for pH $> pK_a$ and the acidic form HA for pH $< pK_a$. It is important to notice the very fast increases (according to exponentials) of the ratios $[A^-]/[HA]$ and $[HA]/[A^-]$, respectively, to the right and left of the vertical pH $= pK_a$.

The concept of predominance area also applies to the polyacids case. Let us consider that of a diacid H_2A. The ionization equilibria are

$$H_2A + H_2O \rightleftharpoons HA^- + H_3O^+,$$

$$HA^- + H_2O \rightleftharpoons A^{2-} + H_3O^+,$$

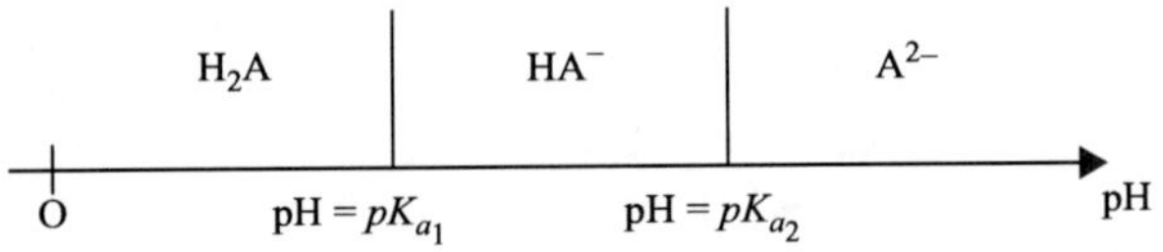

Fig. 4.7 Predominance area of species H_2A, HA^- and A^{2-} in relation to the pH value

and the acid dissociation constants are (after the intermingling of activities and concentrations)

$$K_{a1} = \left[HA^-\right]\left[H^+\right]/[H_2A] \quad \text{and} \quad K_{a2} = \left[A^{2-}\right]\left[H^+\right]\left[HA^-\right].$$

The concentration $[H^+]$ and hence the pH are the same whether they are calculated from K_{a1} or from K_{a2}:

$$pH = pK_{a1} + \log\left\{\left[HA^-\right]/[H_2A]\right\} = pK_{a2} + \log\left\{\left[A^{2-}\right]/\left[HA^-\right]\right\}.$$

Let's draw on the pH axis the verticals for $pH = pK_{a1}$ and for $pH = pK_{a2}$ (Fig. 4.7).

It is evident that HA^- is located on the right of the vertical $pH = pK_{a1}$. But HA^- is also the acid of the couple HA^-/A^{2-} (it is an ampholyte). Hence, its predominance area is limited on the right by the vertical $pH = pK_{a2}$. Beyond it is the A^{2-} predominance area.

After the previous considerations, the symbol HA^- in the middle area actually represents a mixture of the microscopic forms HA^- and ^-AH in a constant ratio. The middle area is the predominance region of the microscopic forms HA^- and ^-AH.

4.17 Prevision of Acid–Base Reactions: Equilibrium Constant of Acid–Base Reaction

Let's consider the possible following acid–base reaction:

$$\text{acid}_1 + \text{base}_2 \rightarrow \text{base}_1 + \text{acid}_2 \quad (K°). \tag{4.18}$$

Two questions can be asked:

- The first involves the direction of the spontaneous reaction: Does the reaction evolve spontaneously from the left to the right, or is it the reverse?
- The second is that of the reaction extent. We will examine this concept again later on. This last question arises, for example, when a satisfactory titration is desired. Let's anticipate some future considerations by saying that a titration reaction equilibrium must be displaced as much as possible toward the right so that the titration should be as accurate as possible.

From the qualitative viewpoint, the reaction evolves spontaneously from left to right when the equilibrium constant $K°$ is higher than unity (when it is lower than unity, it also evolves to the right but in a minute manner according to the mass law requirements). From the quantitative viewpoint, the knowledge of the equilibrium constant provides an answer to the question.

Equilibrium (4.18) results from the competition for the proton by the two couples $acid_1/base_1$ and $acid_2/base_2$ endowed with constants K_{a1} and K_{a2}. The constant $K°$ defined by

$$K° = a_{base1} \cdot a_{acid2} / a_{acid1} \cdot a_{base2}$$

can be expressed easily from the K_{a1} and K_{a2} constants. It is sufficient to multiply its numerator and denominator by the proton activity to find

$$K° = K_{a1}/K_{a2}$$

$$\text{or} \quad \log K = pK_{a2} - pK_{a1}.$$

To know the constant $K°$, it is sufficient to calculate the ratio of both constants' K_{a1} and K_{a2} values, which are tabulated once and for all. We find that reaction (4.18) evolving from left to right is all the more quantitative when the pK_{a2} value is high and the pK_{a1} value is low.

It is very important to realize that the reaction extent does not depend only on the equilibrium constant. It also depends on the initial concentrations of species, reactants, and products that participate in the reaction (see ahead).

From the mechanistic standpoint, the examination of reaction (4.18) suggests that it results from a direct proton exchange between $acid_1$ and $base_2$. However, the calculation of the constant $K°$ through the constants K_{a1} and K_{a2} suggests that the reaction results from its decomposition into two reactions in which water plays a role according to the following reaction:

$$\begin{aligned} acid_1 + H_2O &\rightleftharpoons base_2 + H_3O^+, \\ base_2 + H_3O^+ &\rightleftharpoons acid_2 + H_2O, \\ acid_1 + base_2 &\rightleftharpoons base_1 + acid_2, \end{aligned} \qquad (4.18)$$

whose result is precisely reaction (4.18). Thermodynamics make no difference between both processes, which are identical since the initial and final states are the same.

4.18 Acidity Scale in Water

The previous considerations can be presented with the help of the acidity scale in water. This is a filing of the different acid–base couples on a horizontal (or on a vertical) axis. The filing is based on the pK_a values. The acid–base couples are arranged by increasing pK_a values. The scale is limited conventionally to the values 0 and 14 (Fig. 4.8).

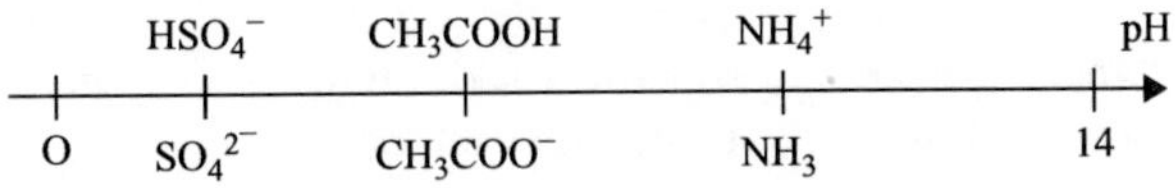

Fig. 4.8 Acidity scale in water

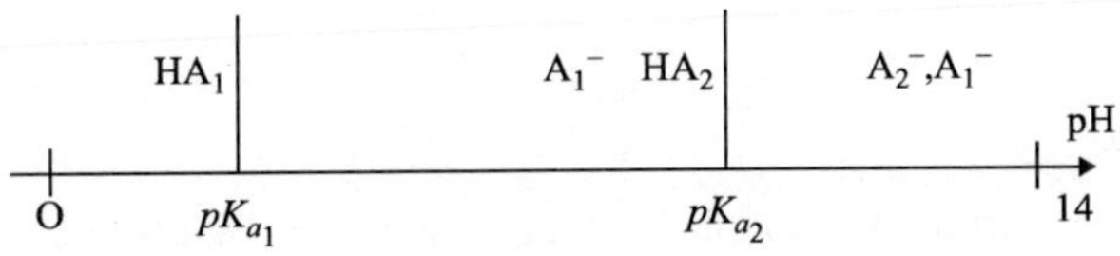

Fig. 4.9 Predominance area diagram of the species of the couples $HA_1/A_1{}^-$ and $HA_2/A_2{}^-$

The scale is oriented in such a way that the stronger acids are located on the left (above), and vice versa. The acidic form of a couple 1 reacts with every base of a couple 2 located on its right (under). The more quantitative the reaction is, the greater is the distance that separates them on the scale.

The concept of acidity scale provokes a discussion about the strength of the couples H_3O^+/H_2O and H_2O/OH^- in water. Since the constant K_a has the significance of the equilibrium constant $K°$ of reaction (4.11), we have

$$K_a = a_{H_3O^+} \cdot a_{A^-}/a_{H_2O} \cdot a_{HA} \quad \text{with} \quad a_{H_2O} = 1,$$

It appears that K_a quantifies the competition of the couples HA/A^- and H_3O^+/H_2O for the proton. It is the same with the reaction

$$B + H_2O \rightleftharpoons BH^+ + OH^-,$$

which characterizes the base B's strength. This reaction results from the competition between couples BH^+/B and H_2O/OH^- for the proton. Hence, the couples H_3O^+/H_2O and H_2O/OH^- can be considered as reference couples in order to characterize the acid and base strengths in water. It would be of great interest to locate them on the acidity scale in water. Some authors locate them respectively at values 0 and 14, some others at values −1.75 and 15.75.

It is the opinion of this author that this discussion is meaningless. Attributing a pK_a value to the couple H_3O^+/H_2O corresponds to determining the equilibrium constant of the reaction

$$H_3O^+ + H_2O \rightarrow H_2O + H_3O^+$$

since acidic pK_a values are defined after equilibria of this sort. However, this reaction does not exist. The initial state is identical to the final state. It is the same state of affairs for the couple H_2O/OH^- in water. These considerations do not mean that these couples are devoid of acid–base properties in water. These considerations simply deny any thermodynamic meaning to their pK_a definition. Their acid–base strength in water cannot be quantified from this concept the way ordinary couples can be.

The qualitative prediction of acid–base reactions can also be realized from predominance area diagrams. Let's consider the two couples $HA_1/A_1{}^-$ and $HA_2/A_2{}^-$ and suppose that $pK_{a1} < pK_{a2}$ (Fig. 4.9).

The acid HA_1 reacts with the base A_2^- because the predominance areas of these two species are not superimposed. They are separate. HA_2 and A_1^-, present in the same region, do not react with each other.

4.19 Leveling of Acids and Bases in Water

It is an experimental fact that some acids and some bases are wholly dissociated in water according to

$$HA + H_2O \rightarrow A^- + H_3O^+,$$

$$B^- + H_2O \rightarrow BH + OH^-.$$

This means that the acid HA and the base B^- are respectively stronger than H_3O^+ for the acid and stronger than OH^- for the base. The equilibrium constants of both reactions, based on the usual conventions concerning weak electrolyte activities, tend toward infinity. pK_a values of the acids HA and BH exhibit values that are certainly on the left and on the right of the conventional ones, 0 and 14. The important point to be emphasized is that it is impossible to distinguish two strong acids from each other in terms of acidity. Let's consider two strong acids HA_1 and HA_2. They are wholly transformed into the ion H_3O^+, the most acidic species that can exist in water. It is the same for the two strong bases B_1 and B_2. They are wholly transformed into the ion OH^-, the strongest base in water. Even if the two acids exhibit different acidity strengths in some other solvents, they have the same behavior in water from the standpoint of acido-basicity. The solution's acidity is due in both cases to the same ion: the hydroxonium ion. We say that these acids are *leveled* (at the acidity level of the hydroxonium ion). It is the same for the strong bases B_1 and B_2. For example, hydrochloric, perchloric, bromhydric, and iodhydric acids are leveled according to the reaction

$$HX + H_2O \rightarrow X^- + H_3O^+,$$

Likewise, the ions ethanolate $C_2H_5O^-$ and amidure NH_2^- are leveled according to

$$B^- + H_2O \rightarrow BH^+ + OH^-.$$

The pK_a determination of the leveled couples in water is impossible in this solvent. The reason lies in the fact that the lower pH value that can be reached in water is 0 (at about few percent) and the higher attainable value is 14 (at about few percent) regardless of the concentration of the strong acid or base. This is due to activity effects and also to the strong buffering capacities of the couples H_3O^+/H_2O and H_2O/OH^- (see Chap. 6). In fact, pK_a determination necessitates working with solutions whose pH values are close to that of the pK_a to be determined. This is the reason why this determination is impossible in the cases of strong acids and bases in water.

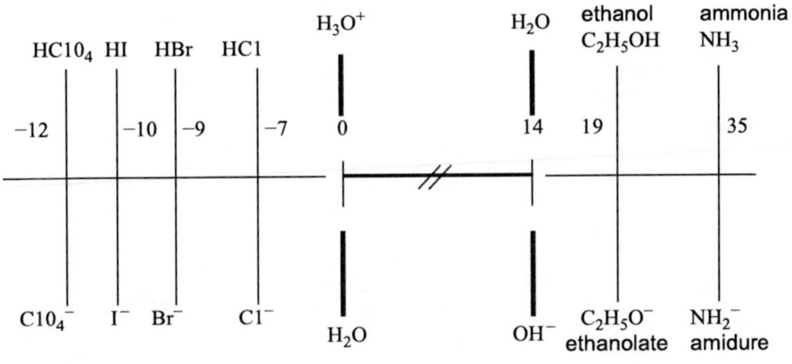

Fig. 4.10 Leveling in water

However, in some cases, one can estimate these values from the study of thermodynamic cycles. Hence, perchloric, iodhydric, bromhydric, and chlorhydric acids would exhibit very low pK_a values of −12, −10, −9, and −7 in water. These values, although markedly different when we compare them, do not induce different acido-basic behaviors of these acids in water since they are all leveled to the hydroxonium ion. Hence, the acidity scale in water can be extended beyond the conventional values but only virtually, that is, without any practical consequence (Fig. 4.10).

It must be noticed that, on this scale, the couples H_3O^+/H_2O and H_2O/OH^- are located conventionally about 0 and 14. They are located according to the same logic as that in Fig. 4.8. We recall that every acid located on the left and sufficiently far from the couple H_3O^+/H_2O is entirely transformed into the ion hydroxonium and likewise for couples located to the right of the couple H_2O/OH^-. Their bases are entirely transformed into the ion hydroxide.

The leveling phenomenon leads to an important analytical consequence. It precludes the possibility of a protometric sequential titration of strong acids and bases since only one acidic (or basic) species remains in solution: the hydroxonium (or the hydroxide) ion. The estimation of the pK_a of hydrochloric acid in water, for example, is based on the evaluation of the standard ionization free enthalpy $\Delta G°_i$. The pK_a is related to the latter by the relation (see Chap. 2)

$$\Delta G°_i = -RT\ln K_a.$$

The standard ionization free enthalpy is given by the relation

$$\Delta G°_i = \Delta H°_i - T\Delta S°_i,$$

where $\Delta H°_i$ and $\Delta S°_i$ are the standard ionization enthalpy and entropy, respectively, at temperature T. The standard ionization enthalpy is estimated from the thermodynamic cycle given in Fig. 4.11.

The difficulty in this cycle is the nonexistence of the undissociated hydrochloric acid in water. Hence, the measurement of the standard solution enthalpy

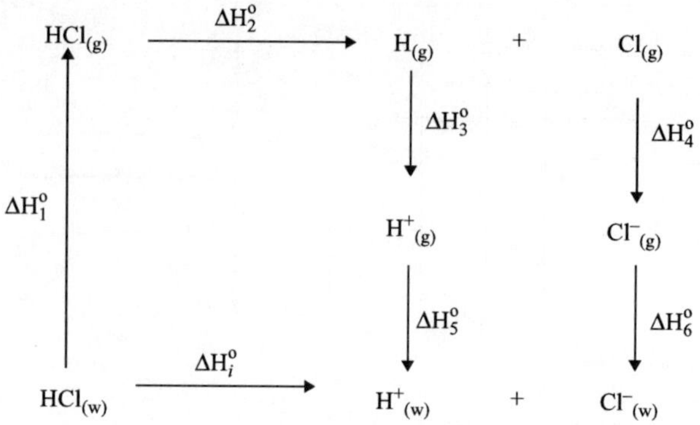

Fig. 4.11 Thermodynamic cycle permitting the estimation of the standard ionization enthalpy $\Delta H°_i$ of hydrochloric acid

$\Delta H°_i$ is impossible. However, an estimation of the enthalpy $\Delta H°_i$ of the process $HCl_{(w)} \rightarrow HCl_{(g)}$ is possible. $\Delta H°_1$ is the hydration enthalpy of hydrochloric acid (with its sign inverted) without any dissociation. It can be considered as little different from the mean of the hydration enthalpies of argon (−11.3 kJ/mol) and of methyl bromide (−23.8 kJ/mol). The argon atom is of the same size as the hydrochloric acid molecule and the methyl bromide molecule is of the same polarity. As a result,

$$\Delta H°_1 \approx 17.6\ \text{kJ/mol}.$$

$\Delta H°_2$ is the standard dissociation enthalpy of the molecule into its component atoms in a gaseous phase at 25°C. We have found its measurement to be

$$\Delta H°_2 = 431.4\ \text{kJ/mol}.$$

$\Delta H°_3$ is the ionization potential of the hydrogen atom at 25°C:

$$\Delta H°_3 = 1316.7\ \text{kJ/mol}.$$

$\Delta H°_4$ is the electronic affinity of the chlorine atom, with its sign inverted:

$$\Delta H°_4 = 364.9\ \text{kJ/mol}.$$

The hydration enthalpies of H^+ and Cl^- ions are also known:

$$\Delta H°_5 + \Delta H°_6 = -1458.8\ \text{kJ/mol}.$$

As a result of that calculation, we find

$$\Delta H°_i = -58.1\ \text{kJ/mol}.$$

The standard ionization entropy is calculated via the relation

$$\Delta S°_i = \left(S°_{H^+(w)} + S°_{X^-(w)}\right) - S°_{HX(w)}$$

The values $S^\circ_{H+(w)}$ (0.005 J/mol K) and $S^\circ_{X(w)}$ (55.05 J/mol K) are known. The standard entropy $S^\circ_{HX(w)}$ of the undissociated acid can be set equal to the sum of the standard entropy of the gas under the pressure of 1 bar (known: 186.4 J/mol K) and the standard entropy of hydration of the gas (under 1 bar) to give an undissociated solute at the molality 1 mol/kg. This last value is estimated to be −75.24 J/mol K. The estimation is based on the hydration entropy values of hydrocyanic acid (−71.06 J/mol K) and of hydrogen sulfide (−83.60 J/mol K), which are measurable since these compounds are not wholly dissociated in water. We find

$$\Delta S^\circ{}_i = -56.0 \text{ J/mol K}.$$

The standard ionization free enthalpy is

$$\Delta G^\circ{}_i = -58{,}100 - 298.15(-56) \text{ kJ/mol},$$

$$\Delta G^\circ{}_i = -41.4 \text{ kJ/mol},$$

and

$$K_a = 10^7, \qquad pK_a(\text{HCl}) = -7.$$

Chapter 5
Calculations of pH Values in Aqueous Solutions

It is important to know how to calculate the pH value of an aqueous solution. The calculation is carried with the help of equations that result from general reasoning often encountered in analytical chemistry.

Let's recall that the pH is expressed in terms of activities. In this chapter, activities and concentrations are mixed. This means that the solutions studied here are very diluted. Taking into account activities markedly complicates calculations, as we briefly show at the end of the chapter. The rate of acid–base reactions must be noted also. We can consider that acid–base equilibria are obtained immediately. Hence, there is no kinetic consideration in this chapter. This is quite justified since neutralization reactions in water are very fast (kinetic constant $k = 1.3 \times 10^{11}$ L/mol/s at 25°C) after Eigen's work.[1]

5.1 Analytical Concentration

The analytical concentration is the whole quantity of matter of the species under study contained in 1 L of solution, regardless of its form. For example, acetic acid is weak in water. It is partially dissociated in acetate ions. If 10^{-4} mol of acetic acid is weighed and dissolved in water at 25°C to give one solution liter, the acid and its conjugate base concentrations are

$$[CH_3COOH] = 0.67 \times 10^{-4}\ \text{mol/L},$$
$$[CH_3COO^-] = 0.33 \times 10^{-4}\ \text{mol/L}.$$

However, the analytical concentration of the acetic acid is 10^{-4} mol/L, and not 0.67×10^{-4} mol/L, which is the actual concentration. "Analytical concentration" and "total concentration" are synonymous terms.

[1] Max Eigen: A german chemist; awarded the Nobel Prize in Chemistry in 1967.

J.-L. Burgot, *Ionic Equilibria in Analytical Chemistry*,
DOI 10.1007/978-1-4419-8382-4_5,

5.2 pH of Pure Water

We know that in pure water

$$pH = 7 \quad \text{(pure water at 25°C)}.$$

In order to introduce the following reasoning and calculations, we'll recall how this result can be obtained from a general standpoint.

Pure water is very slightly dissociated according to

$$H_2O + H_2O \rightleftharpoons H_3O^+ + OH^-.$$

In every sufficiently diluted aqueous solution (see Chap. 4), the water ion product is satisfied, regardless of the species in the solution:

$$[H_3O^+][OH^-] = K_w \quad (10^{-4} \text{ at } 25°C). \tag{5.1}$$

Otherwise, a solution containing ions must be electrically neutral. The charge balance relation that results from this principle is obtained by counting the total number of positive charges per unit volume and setting it equal to the total number of negative charges per unit volume. In the case of pure water, only the hydroxonium and hydroxide ions are present. Therefore, the charge balance relation, which is also called the *electroneutrality relation*, is

$$[H_3O^+] = [OH^-]. \tag{5.2}$$

This system of the two simultaneous equations (5.1) and (5.2) in two unknowns $[H_3O^+]$ and $[OH^-]$ is immediately solved. It gives $pH = 7$ (at 25°C).

5.3 Calculation of pH in Solutions of Strong Acids

We call a *strong acid* an acid that is completely dissociated (personal definition) according to

$$HX + H_2O \rightarrow X^- + H_3O^+.$$

5.3.1 General Relation

The problem to be solved is to calculate the pH of the solution obtained by dissolving C moles of acid HX in pure water to form one solution liter. The relations that must be satisfied are

$$[H_3O^+][OH^-] = K_w, \tag{5.1}$$

$$[H_3O^+] = [OH^-] + X^-, \tag{5.3}$$

$$X^- = C. \tag{5.4}$$

Relations (5.3) and (5.4) are, respectively, the charge balance and mass balance relations. A mass balance relation states that the number of atoms of a given kind must remain constant throughout chemical reactions. If a group of atoms remains intact, it must be used for a mole balance. Equation (5.4) is the mass balance in acid. Equations (5.1), (5.3), and (5.4) are necessarily satisfied. This system in three unknowns can be reduced to the sole equation

$$[H_3O^+]^2 - C[H_3O^+] - K_W = 0 \quad \text{general equation.} \tag{5.5}$$

This is the general equation that permits the calculation of the pH value of a strongly acidic solution. It is legitimate provided the solution remains sufficiently diluted. The solution's pH value is given by the relation

$$\text{pH} = [C + \surd(C^2 + 4K_w)]/2 \quad \text{general equation.} \tag{5.6}$$

Remarks

- As we shall see for every kind of solution, pH calculations are, as a rule, always possible from a purely mathematical point of view. Indeed, the expressions of equilibria and those of charge and mass balances are always required to provide the same number of equations as the number of unknowns. However, the resulting polynomial equations may be difficult to solve because their order is higher than 2 in the existing unknown. Then numerical calculations may be made. Additionally, they can be performed with pocket calculators. Another way is to make approximations, which greatly simplify the solution of the problem (see below);
- in Eqs. (5.5) and (5.6), no parameter is characteristic of some particular acid, provided it is strong. The result is that every strongly acidic solution exhibits the same pH value for the same analytical concentration;
- in the above Eqs. (5.1), (5.3), and (5.4), we have not set up

$$[H_3O^+] = C, \tag{5.7}$$

as uggested by the dissociation reaction of a strong acid, which liberates one hydroxonium ion for one hydroxide ion. To be rigorous, Eq. (5.7) is not correct since the water ionization reaction is superimposed onto that of the acid dissociation, which also involves hydroxonium ions. This point is definitively proved by comparing Eq. (5.7) with the rigorous Eq. (5.3), which is the charge balance. The difference between them is the presence of the concentration $[OH^-]$ in the last one. However, we shall see that in some conditions, Eq. (5.7) can often be used.

5.3.2 Simplified Equations

Let's rewrite Eq. (5.5) under the form

$$C = [H_3O^+] - K_w/[H_3O^+]; \tag{5.8}$$

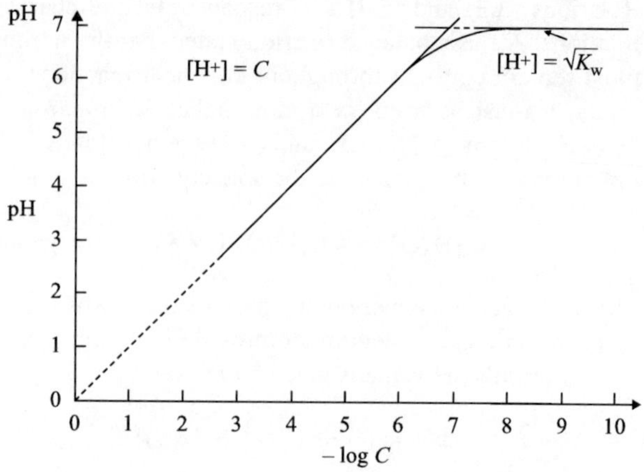

Fig. 5.1 Predominance area of species present in a strongly acidic solution and usable relations for pH calculations

- For high values of C, the term $K_w/[H_3O^+] = [OH^-]$ becomes negligible and Eq. (5.7) can be used. From Eq. (5.7), we have

$$pH = -\log C \text{ (strong acid sufficiently concentrated).} \tag{5.9}$$

  More precisely, this relation can be used for concentrations higher than about 5×10^{-6} mol/L, that is, for usual concentrations in analytical chemistry;
- for concentrations C becoming increasingly weak:

$$H_3O^+ \rightarrow K_w/H_3O^+$$
$$\text{and} \quad pH = 7.$$

  From a practical standpoint, the equality $pH = 7$ is satisfied for C less than 10^{-8} mol/L;
- in the intermediary region, no simplification is possible. The general Eq. (5.5) must be used.

5.3.3 Logarithmic Diagram

All the previous considerations can be shown in a logarithmic diagram. This is a plot of pH as a function of $-\log C$ (Fig. 5.1). The dotted curve recalls the activities problem when the concentration is higher than about 10^{-3} mol/L. Then, a_{H3O^+} differs markedly from H_3O^+.

5.4 pH in Solutions of Strong Bases

Let's consider the strong base B, which reacts quantitatively with water according to

$$B + H_2O \rightarrow BH^+ + OH^-.$$

The general relation permitting the pH calculation is found by expressing the water ion product and the charge and mass balances of the solution. If C is the analytical concentration of the base, the following system of equations holds:

$$[H_3O^+][OH^-] = K_w,$$
$$[BH^+] + [H_3O^+] = [OH^-],$$
$$[BH^+] = C.$$

It can be reduced to the general equation

$$[H_3O^+]^2 + C[H_3O^+] - K_w = 0, \quad (5.10)$$

which can be written equivalently as

$$C = [OH^-] - K_w/[OH^-]\,.$$

It becomes simplified when the concentration is higher than about 5×10^{-6} mol/L since H_3O^+ is then negligible. The pH value is given by the following reactions:

$$[H_3O^+] = K_w/C,$$
$$pH = -\log K_w + \log C,$$
$$pH = pK_w + \log C. \quad (5.11)$$

In very diluted solutions ($C < 10^{-8}$ mol/L), pH = 7. In the intermediate concentrations' region, no simplification is possible. The general equation must be used. Hence, the pH calculation of a strong base's solution is quite analogous to that of the pH of a strong acid's solutions. In particular for concentrations higher than 10^{-3} mol/L, Eq. (5.11) is only an approximation because of the ionic strength.

Remark The calculation could have been developed by choosing the hydroxide concentration as variable. Then the general relation would have been

$$C = [OH^-] - K_w/[OH^-],$$

which is perfectly analogous to Eq. (5.8). If we introduce the quantity $pOH = -\log[OH^-]$, the relation $pOH = -\log C$ would be found for concentrated solutions. An analogous process can also be done for weak bases. The formulas that permit us to calculate pOH are perfectly analogous to those permitting the pH calculations.

5.5 pH in Solutions of Salts of Strong Acids and Bases

A salt is a strong electrolyte. It is completely dissociated. Moreover, the anion and the cation that result from the dissociation are neither acidic nor basic in water; otherwise, the acid and the base would not be strong. For example, in the case of sodium chloride, these considerations can be summarized in the following way:

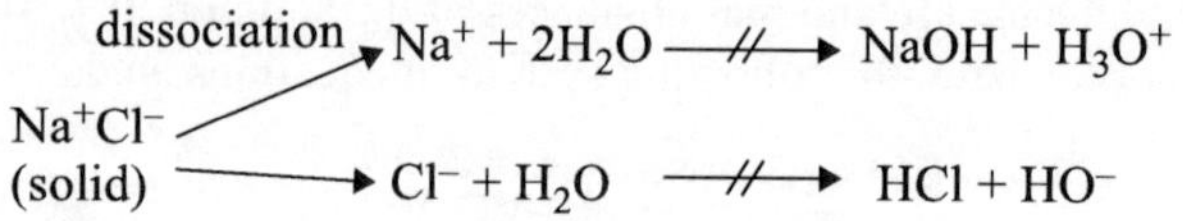

If 1 L of solution contains C moles of the salt, the mass balance equations are

$$[\mathrm{Cl}^-] = C,$$

$$[\mathrm{Na}^+] = C,$$

and the charge balance is

$$[\mathrm{Cl}^-] + [\mathrm{OH}^-] = [\mathrm{Na}^+] + [\mathrm{H_3O}^+].$$

It results in

$$[\mathrm{H_3O}^+] = [\mathrm{OH}^-].$$

The solution is neutral. At 25°C, pH = 7. The same result is obtained with a solution containing a mixture of a strong acid and a strong base in equal quantities. The equations that describe this case are, indeed, the same as above.

5.6 Ostwald's Dilution Law

In order to better grasp the significance of the obtained results when we calculate pH in weak acid and base solutions, we'll now introduce the subject by recalling Ostwald's dilution law. Let's consider a weak acid HA that dissociates according to

$$\mathrm{HA} + \mathrm{H_2O} \rightleftharpoons \mathrm{A}^- + \mathrm{H_3O}^+.$$

If C is its analytical concentration, the dissociated fraction at equilibrium is (see Chap. 4)

$$\alpha = [\mathrm{A}^-]/C,$$

$$[\mathrm{A}^-] = \alpha C \text{ and } [\mathrm{HA}] = C(1 - \alpha).$$

If we do not take water dissociation into account, we have

$$[\mathrm{H_3O}^+] = \alpha C,$$

and the dissociation constant is given by the relation

$$K_a = \alpha^2 C/(1 - \alpha).$$

As a result, the dissociation extent is a function of the equilibrium constant (as anticipated) and also of the acid's analytical concentration. To go further into this last point, suppose that the acid is weakly dissociated, that is, $\alpha < 1$. We immediately find

$$\alpha = \sqrt{(K_a/\mathrm{C})}.$$

The weaker the acid concentration is, the more dissociated it becomes. In the extreme case of a highly diluted solution, the weak acid behaves like a strong one. The study of the change in α as a function of C, when α is nonnegligible with respect to unity, gives the same result. This is Ostwald's law. Let's recall, however, that this law is not rigorous. It has been found by neglecting water dissociation (see above). Moreover, it does not take activities into account. Nevertheless, even after these approximations, it remains completely satisfactory, at least from a qualitative standpoint.

5.7 pH in Solutions of Weak Acids

5.7.1 General Equation Permitting the pH Calculation

If C is the acid's analytical concentration, the following relations are satisfied:

$$[\mathrm{H_3O^+}][\mathrm{OH^-}] = K_\mathrm{w},$$

$$[\mathrm{A^-}][\mathrm{H_3O^+}]/[\mathrm{HA}] = K_a,$$

$$[\mathrm{H_3O^+}] = [\mathrm{A^-}] + [\mathrm{OH^-}], \tag{5.12}$$

$$[\mathrm{HA}] + [\mathrm{A^-}] = C. \tag{5.13}$$

It is a system of four equations in four unknowns: $[\mathrm{H_3O^+}]$, [HA], $[\mathrm{A^-}]$, and $[\mathrm{OH^-}]$. It can be reduced to the third-order equation in $[\mathrm{H_3O^+}]$:

$$[\mathrm{H_3O^+}]^3 + K_a[\mathrm{H_3O^+}]^2 - (K_\mathrm{w} + K_a\mathrm{C})[\mathrm{H_3O^+}] - K_\mathrm{w}K_a = 0. \tag{5.14}$$

It has only one positive root because of the positive values of the constants K_a, K_w, and C.

Equation (5.14) shows that $[\mathrm{H_3O^+}]$ depends on C and also on its strength. It is through the K_a value that the acid's identity can be discerned. Contrary to strong acids, two weak acids at the same analytical concentration do not exhibit the same pH value apart from the accidental case in which both acids have the same pK_a. The pH values of solutions of hydrocyanic acid ($pK_a = 9.32$), acetic acid ($pK_a = 4.75$), and hydrofluoric acid ($pK_a = 3.17$) in relation with $-\log C$ are given in Fig. 5.2.

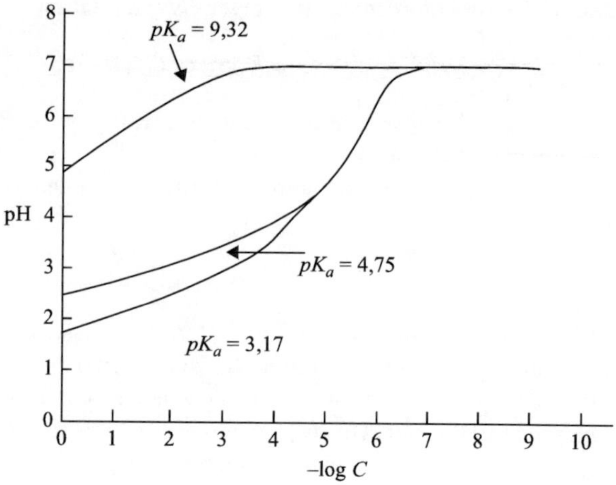

Fig. 5.2 pH of solutions of hydrocyanic, acetic, and hydrofluoric acids in relation with the co-logarithm of the analytical concentration

5.7.2 pH Calculations by Approximations

The general relation (5.14) is difficult to handle because it is third-order in $[H_3O^+]$. However, it can be simplified in some concentration ranges as a function of the pK_a value. Let's again consider the four relations that are reduced to the general Eq. (5.14). We'll make some successive approximations.

5.7.2.1 The Hydroxide Concentration Is Negligible in the Charge Balance

This is the first approximation coming to mind since it is an acid solution. From this simplification,

$$[A^-] = [H_3O^+].$$

As a result,

$$K_a = [H_3O^+]^2/(C - [H_3O^+]) \quad ([OH^-] \text{ negligible}). \tag{5.15}$$

There is only one positive root. This approximation is feasible only when the acid is sufficiently concentrated and strong for it to be dissociated. Then, in Eq. (5.12), $[A^-] \gg [OH^-]$. For example, with acetic and fluorhydric acids, Eq. (5.15) is usable for $C > 5 \times 10^{-6}$ mol/L. For hydrocyanic acid, which is far weaker than the previous ones, it is usable only for $C > 10^{-4}$ mol/L.

- When the acid is very weak, we can add a second approximation to the previous one. It consists of neglecting $[A^-]$ vs. [HA] in the mass balance (5.13). In these conditions, we find

$$[H_3O^+]^2/C = K_a,$$

$$pH = 1/2pK_a - 1/2\log C \quad ([OH^-] \text{ negligible and } [A^-] < [HA]). \tag{5.16}$$

It is the most well-known formula permitting the calculation of the pH of weakly acidic solutions. It must be emphasized that it results from two successive approximations that are not always legitimate. For hydrocyanic acid, it is usable for $C > 10^{-4}$ mol/L, for acetic acid for $C > 10^{-3}$ mol/L, and for hydrofluoric acid for $C > 10^{-2}$ mol/L. Again, we find Ostwald's law content: The stronger the acid is, the less weak its concentration must be in order not to be too dissociated.

- Another approximation is to neglect [HA] vs. $[A^-]$ while continuing to neglect $[OH^-]$ vs. $[A^-]$. Quite evidently this approximation excludes the previous one. Then

$$[A^-] = C,$$
$$[H_3O^+] = C,$$
$$pH = -\log C \quad ([HA] \ll [A^-] \quad \text{and} \quad [OH^-] \ll [A^-]).$$

This is the formula that gives the pH of strongly acidic solutions when their concentration is not too weak. The pH becomes independent of the pK_a. The acid must be sufficiently strong for this relation to be legitimate. This is never the case for hydrocyanic acid, regardless of its concentration. It is the case for acetic acid, for concentrations about 10^{-5} mol/L; and for fluorhydric acid, the concentration range goes until 10^{-4} mol/L. For less weak concentrations, acids are less dissociated and the condition $[HA] \ll [A^-]$ is no longer fulfilled.

5.7.2.2 The Concentration $[OH^-]$ Is No Longer Negligible

- When the acid's concentration is very weak ($C < 5 \times 10^{-6}$ mol/L), it is strongly dissociated. [HA] is negligible, but $[OH^-]$ is not with respect to $[A^-]$ in the charge balance relation, since C is very weak by hypothesis, even when $[A^-] \approx C$. In these conditions,

$$[H_3O^+] - K_w/[H_3O^+] = C \quad (10^{-8}\ \text{mol/L} < C < 5 \cdot 10^{-6}\ \text{mol/L}).$$

This relation is usable in the concentration range $10^{-8} < C < 5 \cdot 10^{-6}$ mol/L. It is the general relation found for strongly acidic solutions if they are sufficiently diluted.

When $C < 10^{-8}$ mol/L,

$$[H_3O^+] = [OH^-],$$

i.e.,

$$pH = 7, \quad C < 10^{-8}\ \text{mol/L},$$

a relation already found with strong acids. With these considerations in mind, it appears that some weak acids may exhibit a behavior exactly superimposable on that of strong acids provided they are sufficiently diluted. Figure 5.3 summarizes all these considerations in the case of acetic acid. Indeed, the pH solution can be superimposed onto that exhibited by a strong acid when the concentration is weak. The diagram pH/$-\log C$ shown in Fig. 5.3 is called Flood's diagram.

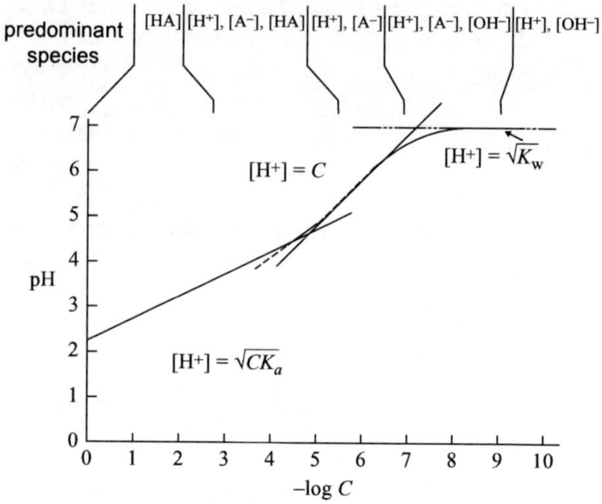

Fig. 5.3 Predominance area of the different species in an acetic acid solution as a function of its analytical concentration and formula permitting the pH calculations ($pK_a = 4.75$)

A good means to check the use of these approximations consists of introducing the values obtained from them into the exact relations, verifying if the other concentrations calculated from them are coherent, and especially verifying if the species that were neglected were effectively negligible.

5.7.3 Calculations with Hägg's Diagrams

Hägg's diagrams are diagrams that show the decimal logarithm of the different species concentrations[2] vs. the pH of the solution. Their being easily drawn permits us immediately to obtain approximate but often satisfactory solute concentrations values in a solution containing protolytes. The diagrams are drawn mixing activities and concentrations. Let's consider a 10^{-2} mol/L acetic acid solution. The pH definition immediately gives

$$\log |H_3O^+| = -\mathrm{pH}.$$

$\log |H_3O^+|$ is represented on the diagram by a straight line with slope -1. For $|OH^-|$, the relation

$$\log |OH^-| = \mathrm{pH} - pK_w$$

is satisfied. It is also represented by a straight line. Its slope is $+1$ and its intercept $-pK_w$. Combining the mass balance and acid dissociation constant relations gives

$$|CH_3COO^-| = C \cdot K_a/[K_a + |H_3O^+|],$$
$$|CH_3COOH| = C|H_3O^+|/[K_a + |H_3O^+|]\ .$$

---

[2] To have pure mathematical rigor, taking the logarithm of a quantity endowed with a unity is heresy.

The first one becomes simpler:

- at weak pH, namely, $|H_3O^+| \gg K_a$,

$$|CH_3COO^-| = CK_a/|H_3O^+|,$$

or for the diagram drawing,

$$\log|CH_3COO^-| = \log C - pK_a + pH.$$

This is a straight line with slope +1 and with intercept ($\log C - pK_a$).

- at high pH, namely, $|H_3O^+| \ll K_a$,

$$|CH_3COO^-| = C,$$
$$\log|CH_3COO^-| = \log C$$

This is a horizontal straight line. The two lines intersect each other for pH = pK_a and $\log C = -2$.This is the system point. Close to this point, these straight lines are no longer representative since $|H_3O^+|$ cannot be neglected with respect to K_a, and vice versa. $\log|CH_3COO^-|$ is then represented by a curve that joins the two previous straight lines.

The second of both equations above may also become simpler. At weak and high pH values, we find, respectively,

$$\log|CH_3COOH| = \log C,$$
$$\log|CH_3COOH| = \log C + pK_a - pH.$$

These two straight lines also intersect each other for the system point. Around this point, they are no longer representative. They are replaced by a curve. For pH = pK_a,

$$|CH_3COO^-| = |CH_3COOH| = 1/2 \times 10^{-2} \text{ mol/L}.$$

The corresponding point is located at 0.3 logarithmic units under the system point. Hägg's diagram is represented in Fig. 5.4.

The determination of the pH solution is immediate. Let's consider the charge balance:

$$|H_3O^+| = |CH_3COO^-| + |OH^-|,$$

and follow the straight line $\log|H_3O^+|$/pH until it intersects the $\log|CH_3COO^-|$/pH curve. At the intersecting point, the diagram shows that the concentration $|OH^-|$ is negligible. The intersecting point gives the value pH = 3.37. The concentrations of other species are found by drawing the vertical line for pH = 3.37 and noting the coordinates of its intersection with the different curves or straight lines of the diagram. The following values are found:

$$\log|H_3O^+| = -3.37 \quad \text{and} \quad |H_3O^+| = |CH_3COO^-| = 4.26 \times 10^{-4},$$
$$\log|CH_3COOH| = -2 \quad |CH_3COOH| = 1.00 \times 10^{-2},$$
$$\log|OH^-| < -9 \quad |OH^-| < 10^{-9} \text{ (mol/L)}.$$

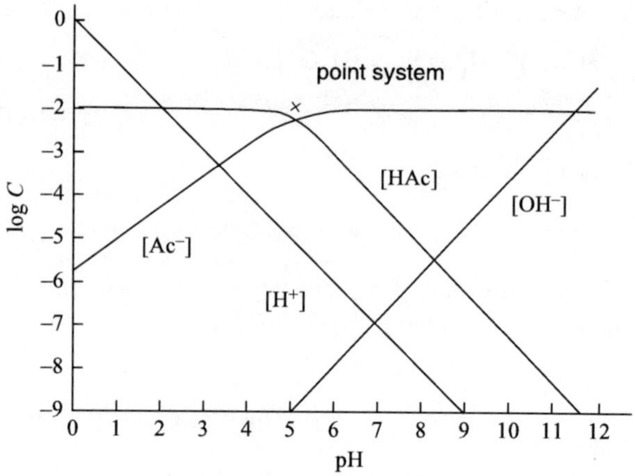

Fig. 5.4 Hägg's diagram for an acetic acid solution (10^{-2} mol/L)

To check these values, we'll add them. We find 1.04×10^{-2} mol/L for 1.00×10^{-2} mol/L. The result is correct at about 5%.

Hägg's diagrams can be built for bases, polyacids, polybases, and different systems. Notice that if the positions of the curves log |HA| and log $|A^-|$/pH do change with the acid's concentration and pK_a values, the straight lines representing $|H_3O^+|$ and $|OH^-|$ are permanent.

5.8 pH in a Weak Base Solution

Reasoning analogous to those followed in the case of weak acids can also be achieved in the case of weak bases. We find the following general third-order relations:

$$[OH^-]^3 + K_b[OH^-]^2 - (K_w + K_bC)[OH^-] - K_wK_b = 0,$$

$$[H_3O^+]^3 + (K_a + C)[H_3O^+]^2 - K_w[H_3O^+] - K_aK_w = 0.$$

They permit the calculation of the solution's pH. In some concentration conditions and depending on the base's strength, these relations may become considerably simpler.

5.8.1 The Base Concentration Is High

The concentration $[H_3O^+]$ can be neglected in the charge balance relation; thus,

$$[OH^-]^2 = K_b(C - [OH^-])$$

(the relations in $[OH^-]$ and K_b are given here to stress the perfect analogy with the case of acids).

- If, moreover, the base is very weak, it is weakly protonated:

$$[\mathrm{OH}^-]^2/C = K_b,$$
$$\mathrm{pH} = 1/2pK_a + 1/2pK_w + 1/2\log C,$$

  where pK_a is the dissociated constant of the conjugate acid.
- when the base is somewhat protonated, it can exhibit the behavior of a strong base if it is sufficiently diluted and

$$[\mathrm{OH}^-] = C,$$
$$\mathrm{pH} = \log C + pK_w.$$

5.8.2 *The Basic Solution Is Highly Diluted*

When the basic solution is highly diluted, the base behaves like a strong base and using the following general relation is legitimate:

$$K_w/[\mathrm{H_3O^+}] - [\mathrm{H_3O^+}] = C.$$

It is interesting to note that all the relations given in this section are found in either the Arrhenius or Brønsted theory.

Exercise 1 Calculate the pH of a sodium cyanide solution of analytical concentration C.

Sodium cyanide is a salt. It is completely dissociated in water. The obtained sodium ion is indifferent from an acido-basicity standpoint. The cyanide ion, however, exhibits a basic character due to its reaction with water, sometimes called a *hydrolysis reaction*:

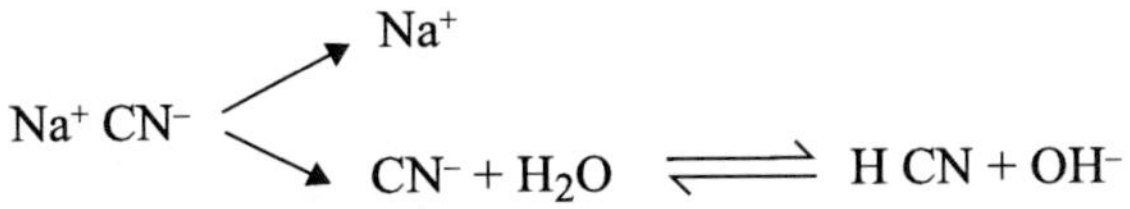

It is precisely because of this reaction that the cyanide ion is basic in water. The relations that must be satisfied are

$$[\mathrm{H_3O^+}][\mathrm{OH^-}] = K_{w,}$$
$$[\mathrm{H_3O^+}][\mathrm{CN^-}] = K_a[\mathrm{HCN}],$$
$$[\mathrm{H_3O^+}] + [\mathrm{Na^+}] = [\mathrm{CN^-}] + [\mathrm{OH^-}],$$
$$[\mathrm{CN^-}] + [\mathrm{HCN}] = C,$$
$$[\mathrm{Na^+}] = C.$$

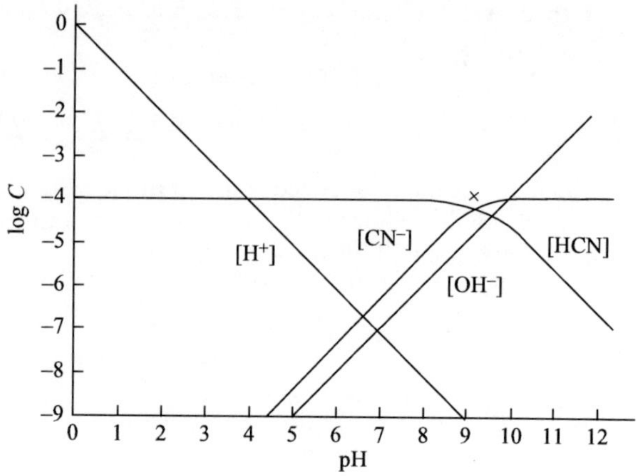

Fig. 5.5 Hägg's diagram of a sodium cyanide solution ($C = 10^{-2}$ mol/L)

By neglecting $[H_3O^+]$ in the charge balance, we get

$$C = [CN^-] + [OH^-],$$
$$[HCN] = C - [CN^-],$$
$$[HCN] = [OH^-].$$

As a result, we have

$$[H_3O^+](C - [OH^-]) = K_a[OH^-],$$

a relation permitting the pH calculation at the first level of simplification. Moreover, by setting up

$$[OH^-] \ll [CN^-], \quad \text{i.e.,}$$
$$[OH^-] \ll C,$$
$$[H_3O^+]C = K_a[OH^-].$$

From this, we have

$$\text{pH} = 1/2pK_w + 1/2pK_a + 1/2\log C,$$

a relation we found previously.

Exercise 2 Draw the Hägg's diagram representing a 10^{-2} mol/L solution of sodium cyanide.

The diagram is analogous to that obtained with acetic acid. It is given in Fig. 5.5. The combination of the following relations:

$$[H_3O^+] + [Na^+] = [OH^-] + [CN^-],$$
$$[CN^-] + [HCN] = 10^{-2} \text{ mol/L},$$
$$[Na^+] = 10^{-2} \text{ mol/L}$$

gives the relation

$$[OH^-] = [HCN]+[H_3O^+],$$

sometimes called the *proton condition*. The junction of the straight line log $[OH^-]$/pH with the curve log [HCN]/pH gives the value log [HCN] = −3.34. The pH value is 10.66. Indeed, $[H_3O^+]$ is negligible in the proton condition relation.

5.9 pH of a Mixture of Strong Acids

Let's consider a solution of two strong acids HX_1 and HX_2 of analytical concentrations C_1 and C_2. The following relations are satisfied:

$$[H_3O^+][OH^-] = K_w,$$
$$[H_3O^+] = [X_1{}^-] + [X_2{}^-] + [OH^-],$$
$$[X_1{}^-] = C_1,$$
$$[X_2{}^-] = C_2.$$

From the above equations, we have

$$[H_3O^+] - K_w/[H_3O^+] = C_1 + C_2.$$

This general equation must be compared to Eq. (5.8), which applies to the case of a solution of only one strong acid. The mixture of two strong acids exhibits the same behavior as the solution of only one strong acid. The analytical concentration C of this virtual sole acid would be equal to the sum of concentrations C_1 and C_2 of each of the mixture's acid components. In particular, if

$$C_1 + C_2 > 10^{-5}\ \text{mol/L},$$
$$\text{pH} = -\log(C_1 + C_2).$$

The fundamental origin of this result is the leveling of strong acids in water.

Remark The concentrations C_1 and C_2 are the actual concentrations once the mixture is prepared. These values are different from C'_1 and C'_2, which are the concentrations of the solutions that are mixed to prepare the solution under study. In order to stress the difference between the two types of concentrations, let's express the relationship between C_1 and C'_1. We'll mix V_1 cm^3 of the acid HX_1 solution of concentration C'_1 in V_2 cm^3 of the acid HX_2 solution of concentration C'_2. The number of moles of acid HX_1 in the final solution is

$n(HX_1) = V_1 C'_1 \times 10^{-3}$ mol (presence of the factor 10^{-3} since C'_1 is expressed in mol/L and V_1 in cm^3),

and its concentration C_1 is

$$C_1 = V_1 C'_1 \times 10^{-3} / \left\{(V_1 + V_2) \times 10^{-3}\right\} \text{mol/L}$$

or

$$C_1 = V_1 C_1^{'}/(V_1 + V_2)\ \text{mol/L}.$$

It is the same state of affairs for acid HX_2:

$$C_2 = V_2 C_2^{'}/(V_1 + V_2)\ \text{mol/L}.$$

5.10 pH of a Mixture of Strong Bases

A mixture of strong bases behaves like a strong monobase, whose concentration should be the sum of the concentrations of each component base.

5.11 pH of a Mixture of a Strong and a Weak Acid: Ionization Repression

Let's consider the weak acid HA. The dissociation equilibrium is

$$\text{HA} + \text{H}_2\text{O} \rightleftharpoons \text{H}_3\text{O}^+ + \text{A}^-.$$

The dissociation constant is

$$K_a = [\text{A}^-][\text{H}_3\text{O}^+]/[\text{HA}].$$

The addition of a strong acid HX to the previous solution, by definition completely dissociated, increases the concentration $[H_3O^+]$. However, the constant K_a invariability must be respected. Since $[H_3O^+]$ increases due to this addition, $[A^-]$ will decrease and [HA] will increase simultaneously. This can be obtained only by the reaction

$$\text{A}^- + \text{H}_3\text{O}^+ \rightarrow \text{HA} + \text{H}_2\text{O}.$$

The addition of the strong acid induces the protonation of the conjugate base of the weak acid. The weak acid is less ionized when the strong acid is present than when it is alone at the same concentration. This phenomenon is called *ionization repression* (of the weak acid). The weak acid dissociation is repressed. For example, in a solution of 10^{-4} mol/L of hydrochloric acid and 10^{-2} mol/L of acetic acid, calculations based on the principles recalled above give the result $[H_3O^+] = 4.71 \times 10^{-4}$ mol/L. In a solution of acetic acid alone at the same analytical concentration, $C = 1.0 \times 10^{-2}$ mol/L, calculations give 4.18×10^{-4} mol/L. In the mixture of both acids, the acidity due to acetic acid can be calculated by taking away the hydrochloric acid's acidity from the total acidity:

$$4.71 \times 10^{-4} - 1.0 \times 10^{-4} = 3.71 \times 10^{-4}\ \text{mol/L}$$

since the hydrochloric acid is completely dissociated. The ionization repression of acetic acid is quantified by the difference

$$4.18 \times 10^{-4} - 3.71 \times 10^{-4} = 0.47 \times 10^{-4} \text{ mol/L}.$$

It is about 10%.

5.12 pH of a Mixture of a Strong and a Weak Base

There is also an ionization repression of the weak base.

5.13 pH of an Equimolecular Mixture of a Weak Base and a Weak Acid

The solutions can be obtained by dissolution of the acid and the base. They can also be prepared by the dissolution of the salt of a weak acid with a weak base.

In the second case, we can conceive, from a thermodynamic point of view, two successive steps:

- the dissolution strictly speaking. This is a purely physical process. At the end of this step, the chemical equilibria are not attained. The salt, initially in the solid state, is completely dissociated into both ions A^- and BH^+. Their mole number is equal to that of the salt:

$$BH^+, A^-{}_{(s)} \rightarrow BH^+{}_{(w)} + A^-{}_{(w)} \quad \text{(dissolution);}$$

- setting up the chemical equilibria according to the proton exchange equilibria:

$$BH^+{}_{(w)} + H_2O \rightleftharpoons B + H_3O^+,$$

$$A^-{}_{(w)} + H_2O \rightleftharpoons HA + OH^-.$$

At the end of this second step, of course, the BH^+ and A^- concentrations are no longer equal to the initial concentration of the salt because of the achievement of the equilibria above.

The pH can be calculated rigorously by taking into account all the relations that must be satisfied (charge and mass balances, equilibrium dissociation constants, water ion product). The resulting equation, however, is difficult to solve. An approximation of the resulting equation consists of neglecting the concentrations $[H_3O^+]$ and $[OH^-]$ in the charge balance relation. It results then in

$$\text{pH} = 1/2(pK_{a1} + pK_{a2}) \quad \text{(approximate relation),}$$

where pK_{a1} and pK_{a2} are, respectively, the equilibrium ionization constants of the acids HA and BH^+. The pH value is independent of the salt concentration. This relation is legitimate when we have simultaneously

$$[H_3O^+] \ll [BH^+] \quad \text{and}$$
$$[OH^-] \ll [A^-].$$

This signifies that the pH value must not be located far from the neutrality. This hypothesis is reasonable due to the nature of the mixture. This also implies that the salt concentration is not too weak.

5.14 pH of Polyacid and Polybase Solutions

In this section, we will only study the polyacid case.

Let's consider the weak diacid H_2A, which gives rise to the following two ionization processes:

$$H_2A + H_2O \rightleftharpoons HA^- + H_3O^+,$$
$$HA^- + H_2O \rightleftharpoons A^{2-} + H_3O^+.$$

The following relations are satisfied:

$$[H_3O^+][OH^-] = K_w,$$
$$[HA^-][H_3O^+] = [H_2A]K_{a1},^3$$
$$[A^{2-}][H_3O^+] = [HA^-]K_{a2},$$
$$[H_2A] + [HA^-] + [A^{2-}] = C,$$
$$[HA^-] + 2[A^{2-}] + [OH^-][H_3O^+].$$

In the last equation (charge balance), the coefficient 2 is present. It simply recalls the fact that each A^{2-} ion brings two negative charges. The previous system can be reduced into a fourth-order equation that is difficult to solve. However, some particular conditions permit approximations.

The constant K_{a2} may be markedly weaker than K_{a1}. In this case, only the latter plays a part. We are again in the case of a monoacid. The only noticeable ionization is

$$HA_2 \rightleftharpoons HA^- + H^+ \quad (K_{a1}).$$

The pH value is calculated through one of the formulas already seen in the case of weak monoacids.

[3] The designations K_{a1} and K_{a2} are in agreement with that recommended by the IUPAC.

If the constants K_{a1} and K_{a2} values are close to each other, the fourth-order equation, evoked above, must be solved. However, if the two (close) acidities are not too weak and if the diacid's concentration is also not too weak, the hydroxide concentration is negligible. Therefore, all that remains to do is to manipulate the following third-degree equation:

$$[H_3O^+]^3 + K_{a1}[H_3O^+]^2 - K_{a1}C[H_3O^+] - 2CK_{a1}K_{a2} = 0,$$
$$(K_{a1} \approx K_{a2} \quad \text{and} \quad [OH^-] \text{ negligible}).$$

It would be too long a job to discuss the conditions that command the utilization of any of the three equations given, since the three parameters K_{a1}, K_{a2}, and C play a part.

5.15 pH of a Monosalt of a Diacid Solution—pH of an Ampholyte Solution

Let's calculate the pH value of the salt HANa resulting from the half-neutralization of the diacid H_2A. This case must not be mistaken with the previous one. If C is the analytical concentration of the salt and K_{a1} and K_{a2} the acid dissociation constants (with $K_{a1} > K_{a2}$), the following relations are satisfied:

$$[H_3O^+][OH^-] = K_w,$$
$$[HA^-][H_3O^+] = K_{a1}[H_2A],$$
$$[A^{2-}][H_3O^+] = K_{a2}[HA^-],$$
$$[H_2A] + [HA^-] + [A^{2-}] = C,$$
$$[HA^-] + 2[A^{2-}] + [OH^-] = [H_3O] + [Na^+],$$
$$[Na^+] = C.$$

Before solving this system, note that it differs from the previous one by the presence of sodium ions in the charge and mass balances. Solving this system is difficult. An approximation can be done under some conditions. It results from the fact that the intermediary species can be considered as resulting from the reaction of the diacid H_2A with the base A^{2-}, the two species being at the same concentration, according to

$$H_2A + A^{2-} \rightleftharpoons 2HA^-.$$

We are dealing with the case of the salt of a weak acid and a weak base. As a result, after what was said in Sect. 5.13:

$$pH = 1/2(pK_{a1} + pK_{a2}),$$

where pK_{a1} and pK_{a2} are the pK_a of the corresponding couples. Recall that this formula had been found by neglecting $[H_3O^+]$ and $[OH^-]$ in the charge balance.

This is the simplification that can also be done here. By neglecting $[H_3O^+]$ and $[OH^-]$ in the charge balance again, we find

$$[A^{2-}] = [H_2A]$$

and

$$[H_3O^+]^2 = K_{a1} \times K_{a2}, \quad pH = 1/2(pK_{a1} + pK_{a2}).$$

This approximation involves the facts that the solution must not be too diluted and that the *pKa* values are not too close to 0 and 14. It is interesting to note that the species HA^- is an ampholyte because it gives rise to the following two reactions with water:

$$HA^- + H_2O \rightleftharpoons A^{2-} + H_3O^+,$$

$$HA^- + H_2O \rightleftharpoons H_2A + OH^-.$$

There is a formation of two buffer couples: H_2A/HA^- and HA^-/A^{2-} (see Chap. 6).

5.16 pH of a Solution of an Amino-Acid

Amino acids are certainly among the most important organic molecules in biology. More precisely, in this section, we call amino acids the 2-aminocarboxylic acids or α–amino acids. They are molecules in which the amino and carboxylic groups are brought by the same carbon. (There are, of course, other sorts of amino acids. For example, there are the cases of molecules in which the two groups are not brought by the same carbon atom, of amino acids in which the acid function is a sulfonic one (taurine), of amino acids bringing a supplementary amino or carboxylic group, etc.)

The simplest example is that of glycine or glycocolle:

$$\underset{\text{glycin}}{NH_2 - CH_2 - COOH}.$$

We can already predict that amino acids are both acidic and basic since they possess a carboxylic function and an amino function. Hence, it is not surprising that the pH calculation of an amino acid solution is analogous to that of an ampholyte solution, but this assertion deserves further comments.

Several experimental facts indicate that in aqueous solutions, two glycine isomers do exist. They are the neutral form A° and the zwitterionic form $A^{(+-)}$:

$${}_2HN - CH_2 - COOH \qquad {}_3HN^+ - CH_2 - COO^-$$

neutral form — zwitterion form

It is the same state of affairs for all the amino acids of the same sort. They are also a mixture of an uncharged form A° and of the corresponding zwitterionic form $A^{(+-)}$:

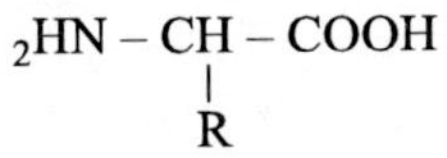

A°

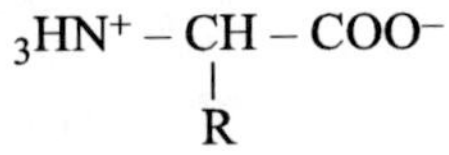

A±

Both forms A° and $A^{(+-)}$ have the property to give the same conjugate acid form A^+ by protonation:

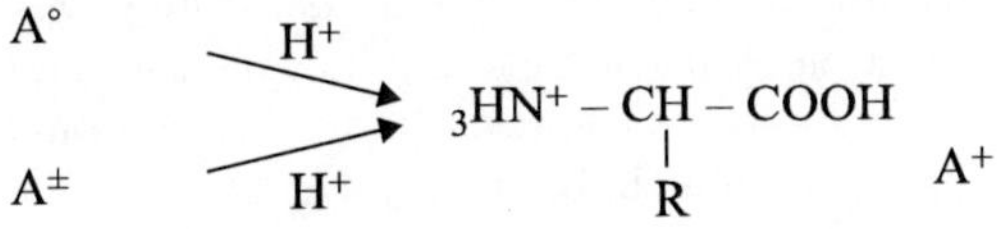

Hence, they exhibit their basic character. Likewise, both forms A° and $A^{(+-)}$ give the same conjugate basic form A^- after the loss of a proton:

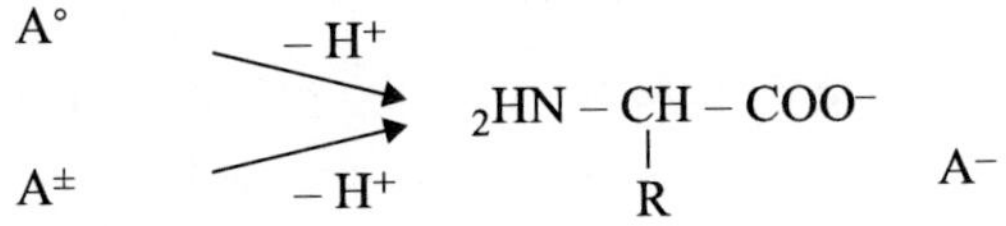

Then they behave like acids.

When we consider the two general ionization schemes

$$H_2A \underset{}{\overset{-H^+}{\rightleftharpoons}} HA^- \overset{-H^+}{\rightleftharpoons} A^{2-}$$

$$A^+ \overset{-H^+}{\rightleftharpoons} A^{\pm} + A^{\circ} \overset{-H^+}{\rightleftharpoons} A^-$$

we find that the analogous behavior of the monosalt of the diacid HA^- and that of the mixture of the uncharged and zwitterionic forms $(A^{\circ} + A^{(+-)})$, considered as a whole, is great. If K_{a1} and K_{a2} are the equilibrium constants of the two equilibria,

$$A^+ + H_2O \overset{Ka_1}{\rightleftharpoons} \{A^{\pm} + A^{\circ}\} + H_3O^+$$

or

$$\{A^{\pm} + A^{\circ}\} + H_2O \overset{K_{a_2}}{\rightleftharpoons} A^- + H_3O^+$$

the pH calculation of the amino acid solution is based on the following relations in five unknowns $[H_3O^+]$, $[OH^-]$, $[A^+]$, $[A^-]$, and $[(A^{\circ} + A^{(+-)})]$:

$$[H_3O^+][OH^-] = K_w, \tag{5.1}$$

$$[A^+] + [(A^{\circ} + A^{(+-)})] + [A^-] = C, \tag{5.17}$$

$$[A^+] + [H_3O^+] = [A^-] + [OH^-], \tag{5.18}$$

$$[(A^\circ + A^{(+-)})][H_3O^+] = K_{a1}[A^+], \tag{5.19}$$

$$[A^-][H_3O^+] = K_{a2}[(A^\circ + A^{(+-)})]. \tag{5.20}$$

K_{a1} and K_{a2} characterize the basic and acidic properties of the mixture $(A^\circ + A^{(+-)})$, which is regarded as being a sole species. The approximation to be achieved is evident. The pH value cannot be too far from the neutrality since the carboxylic and amine groups are both functions that are the vehicle of a weak acidity or basicity. Moreover, when equilibria are attained in solution, it may exist as concentrations approximately equal in undissociated carboxylic acid and amine functions, since their analytical concentrations are identical. In these conditions, the concentrations $[H_3O^+]$ and $[OH^-]$ are negligible in Eq. (5.18). Hence,

$$[A^+] = [A^-],$$
$$[H_3O^+]^2 = K_{a1} \times K_{a2},$$
$$pH = 1/2(pK_{a1} + pK_{a2}),$$

a relation already found in the case of the ampholyte HA^- solution.

Exercise 3 Calculate the pH value of a 10^{-2} mol/L solution of glycine ($K_{a1} = 4.5 \times 10^{-3}$ and $K_{a2} = 1.5 \times 10^{-10}$). Applying the previous formula gives pH = 6.09. The reverse calculation with this value gives $[A^+] = 1.6 \times 10^{-6}$ mol/L and $[A^-] = 2.3 \times 10^{-6}$ mol/L. These concentrations are really negligible with respect to 10^{-2} mol/L.

The pH value of an amino acid solution in pure water is called the *isoionic point*. Another particular pH value that is mentioned in the literature is the *isoelectric point*, pH_i. It is the pH value for which the concentrations of the diacid and dibasic species are equal, that is, for which

$$[A^+] = [A^-] \quad \text{(isoelectric point)}.$$

The isoelectric point is close to the isoionic point. It is obvious that the approximation used above in order to calculate the pH of an amino acid solution is equivalent to confuse these two points. Also, notice that the more diluted the amino acid solution is, the more the isoionic point tends toward the isoelectric point, as shown by Eq. (5.18).

The constants K_{a1} and K_{a2} are called *macroscopic constants* (see Chap. 4). The knowledge of their values and that of their analytical concentration is sufficient to calculate the solution's pH value. However, it is interesting to make a further study of the ionization constants of species A° and $A^{(+-)}$ taken separately. By definition, we can write

$$A^+ + H_2O \rightleftharpoons A^\circ + H_3O^+ \quad (k_{11}),$$
$$k_{11} = [A^\circ][H_3O^+]/[A^+] \tag{5.21}$$

and

$$A^+ + H_2O \rightleftharpoons A^{(+-)} + H_3O^+ \quad (k_{12}),$$
$$k_{12} = [A^{(+-)}][H_3O^+]/[A^+]. \tag{5.22}$$

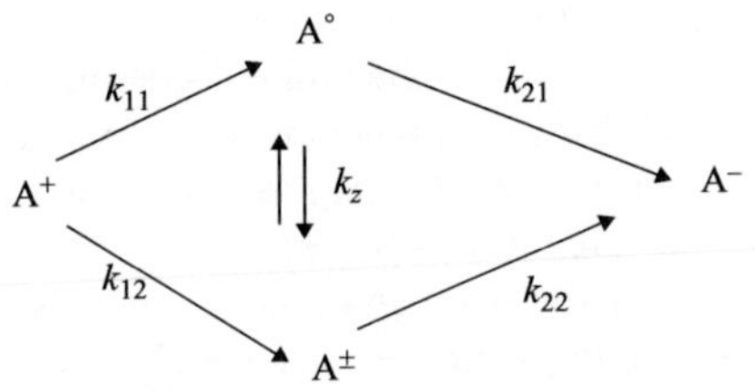

Fig. 5.6 Definition of ionization microscopic constants of an amino acid

k_{11} and k_{12} are the first ionization microscopic constants (index 1 = first ionization). Likewise,

$$A^\circ + H_2O \rightleftharpoons A^- + H_3O^+ \quad (k_{21}),$$
$$k_{21} = [A^-][H_3O^+]/[A^\circ], \tag{5.23}$$
$$A^{(+-)} + H_2O \rightleftharpoons A^- + H_3O^+ \quad (k_{22}),$$
$$k_{22} = [A^-][H_3O^+]/[A^{(+-)}]. \tag{5.24}$$

k_{21} and k_{22} are the second ionization microscopic constants. Figure 5.6 summarizes the symbolism used.

Recall (see Chap. 4) that comparing Eqs. (5.21), (5.22), and (5.19), on the one hand, and (5.23), (5.24), and (5.20), on the other, gives the relations

$$K_{a1} = k_{11} + k_{12},$$
$$1/K_{a2} = 1/k_{21} + 1/k_{22}, \tag{5.25}$$
$$k_z = k_{12}/k_{11} = k_{21}/k_{22}.$$

The ratio of the uncharged and zwitterionic forms' concentrations is constant and independent of the pH. Moreover,

$$k_{11}k_{21} = k_{12}k_{22} = K_{a1}K_{a2}.$$

The macroscopic constants K_{a1} and K_{a2} are accessible by experimental means.

The microscopic constants are not accessible, except through an extrathermodynamic hypothesis.

The macroscopic constant K_{a2} of glycocolle ($K_{a2} = 1.5 \times 10^{-10}$: $pK_{a2} = 9.82$) is of the same order as the pK_a values of aliphatic protonated aliphatic amines ($pK_a = 9.5$, a well-established statistical value). It results from Eq. (5.25) that

$$k_{22} \approx K_{a2}.$$

The chemical meaning of this result is that the amino acid is quasi-completely under the zwitterionic form in aqueous solution. It is, actually, this kind of reasoning that is followed in order to approach the microscopic constants' values.

Strictly speaking, there is no weakening of the carboxylic acid strength due to the presence of the amine function, as had been mentioned in older literature. The weakening phenomenon is simply due to the quasi-exclusive presence of the zwitterionic form in the solution. In other words, there exists no carboxylic acid function. They are totally ionized.

Finally, from an organic structural standpoint, we notice that, in the case of the example given here (glycocolle), the value $pK_a = 2.35$ is markedly weaker than that of normal carboxylic acids ($3.5 < pK_a < 4.8$). This signifies that the protonated amino group located close to the carboxylic acid function strengthens it very highly, probably by directly and indirectly withdrawing inductive effects. Anomalous pK_a values are also exhibited by other amino acids of this sort.

The very weak K_a value explains why it is not possible to titrate in water an amino acid by a strong base nor is it possible to titrate, in the same conditions, protonated aliphatic amines. This is the same problem as that encountered for the titration of the third acidity of phosphoric acid by sodium hydroxide (see Chap. 9).

5.17 pH of a Mixture of Two Weak Acids

Let's consider a mixture of two weak acids HA_1 and HA_2 of dissociation constants K_{a1} and K_{a2} and of analytical concentrations C_1 and C_2. Taking into account all the relations that are necessarily satisfied gives a fourth-order equation in $[H_3O^+]$ that is difficult to solve. However, neglecting the concentration $[OH^-]$ and formulating the hypothesis that both acids are weakly dissociated gives the resulting formula:

$$[H_3O^+]^2 \approx K_1C_1 + K_2C_2$$
$$\text{or} \quad [H_3O^+] = \sqrt{}(K_1C_1 + K_2C_2).$$

Notice that formulating the hypothesis of a weak dissociation of both acids implicates:

$$[HA_1] \approx C_1 \quad \text{and} \quad [HA_2] \approx C_2.$$

Let's also recall that neglecting $[A_1{}^-]$ and $[A_2{}^-]$ is, by far, not always legitimate. Finally, neglecting the hydroxide concentration implicates that concentrations C_1 and C_2 are not too weak.

5.18 pH of a Mixture of a Weak Acid and a Weak Base in Any Proportion: Interest in the Principal Reaction Concept

In this section, we give a methodology that permits us to obtain an approximate pH value of a mixture of a weak acid and a weak base in any proportion. It is based on the principal reaction concept. In some cases, we must consider several principal reactions occurring successively.

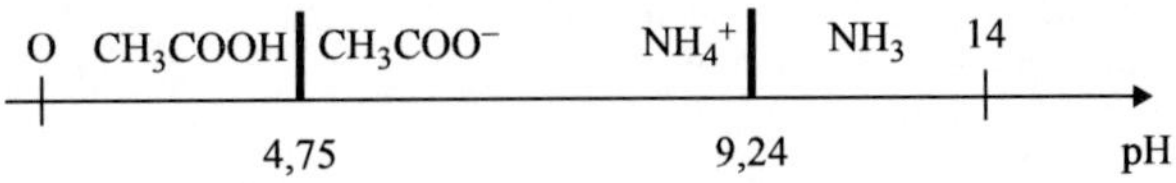

Fig. 5.7 Preponderance-area diagram (example of the mixture of ammonia and acetic acid)

Example 1 Let's consider a mixture of ammonia and acetic acid in aqueous solution, at the respective analytical concentrations 5×10^{-3} mol/L and 2×10^{-2} mol/L. The system of the necessarily satisfied relations (water ion product, charge and mass balances, equilibrium ionization constants) can be reduced to only one equation, which is of the fourth order in $[H_3O^+]$. It is difficult to solve it quickly. Calculations are made considerably easier by applying the concept under study. The first step is the search for the preponderance area of the different species. It is made by drawing the acidity scale (Fig. 5.7).

It is clear that ammonia and acetic acid belong to two disconnected regions. After what has been said about the prediction of acid–base reactions (see Chap. 4), the following reaction must occur. It is called the *principal reaction*:

$$CH_3COOH + NH_3 \rightarrow CH_3COO^- + NH_4{}^+.$$

According to the concept, it is assumed to be total. It must be emphasized that, in this assumption, this methodology is approximate. After the reaction has (supposedly) completely taken place, it is an easy task to calculate the different concentrations. They are $[\ CH_3COOH] = 1.5 \times 10^{-2}$ mol/L, $[CH_3COO^-] = 5 \times 10^{-3}$ mol/L, and $[NH_4^+] = 5 \times 10^{-3}$ mol/L. Indeed, according to the working hypothesis, the entire quantity of ammonia has completely disappeared together with the same quantity of acetic acid. Hence, what remains in solution is the buffer mixture acetic acid/acetate ion (see Chap. 6) and the weak acid $NH_4{}^+$. Introducing these calculated values of acetate ion and acetic acid concentrations into the expression of the acetic acid ionization constant,

$$[CH_3COO^-][H^+]/[CH_3COOH] = 10^{-4.75},$$

gives the value pH = 4.3. Since, by definition, a buffer mixture does not exhibit a great pH value change after a strong acid is added to its solution, it is clear that the presence of the weak acid $NH_4{}^+$ will not change this pH value appreciably.

Example 2 Another example is provided by the pH calculation of an ammonium phosphate solution of analytical concentration 10^{-2} mol/L. (This solution would have been prepared by dissolution of 10^{-2} mol/L of sodium orthophosphate PO_4Na_3 and 3×10^{-2} mol/L of ammonium chloride to obtain finally one liter of solution.) The preponderance area and pK_a values are given in Fig. 5.8.

The following reaction (assumed to be complete) takes place:

$$NH_4{}^+ + PO_4{}^{3-} \rightarrow HPO_4{}^{2-} + NH_3.$$

When the solution (supposedly complete) is terminated, the species concentrations are $HPO_4{}^{2-}$ ($C = 10^{-2}$ mol/L), NH_3 ($C = 10^{-2}$ mol/L), and $NH_4{}^+$ ($C = 2 \times 10^{-2}$

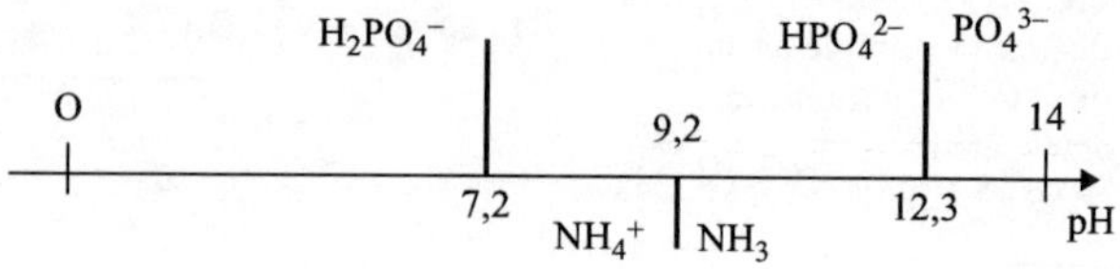

Fig. 5.8 Preponderance-area diagram for an ammonium orthophosphate solution

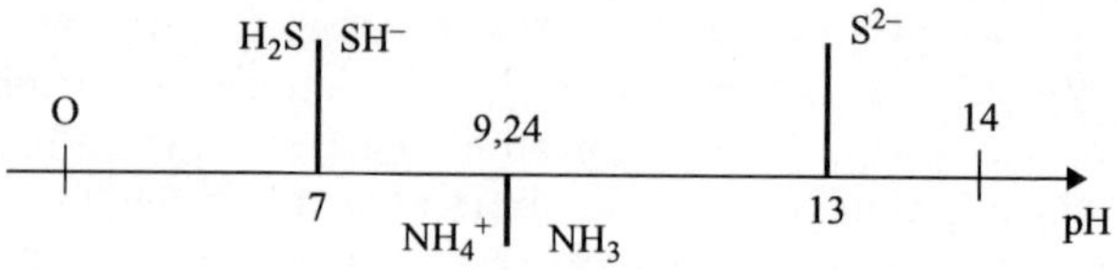

Fig. 5.9 Preponderance-area diagram of species from ammonium sulfide

mol/L). These three species exhibit a connected area. As a result, there is no further acid–base reaction. The pH of the solution must be very close to that of a buffer solution, $NH_4{}^+/NH_3$ with $[NH_4{}^+] = 2 \times 10^{-2}$ mol/L and $[NH_3] = 10^{-2}$ mol/L. The solution also contains the ampholyte $HPO_4{}^{2-}$ at the concentration 10^{-2} mol/L. Introducing the ammonia and ammonium ion concentrations into the expression of the ionization constant,

$$[NH_3][H^+]/[NH_4{}^+] = 10^{-9.34},$$

gives the value pH = 8.91.

Example 3 Let's study an ammonium sulfide $S(NH_4)_2$ solution at concentration 10^{-2} mol/L (this solution, which was prepared by dissolution of the salt, could also be prepared by mixing10^{-2} mol of sodium sulfide and 2×10^{-2} mol of ammonium chloride for 1 L of solution). The preponderance-area diagram is shown in Fig. 5.9.

The following reaction (supposedly complete) takes place:

$$NH_4{}^+ + S^{2-} \rightarrow NH_3 + SH^-.$$

After the reaction, NH_3 and $NH_4{}^+$ are in about equal concentrations and SH^- remains in solution. Their predominance areas are connected. The pH value must be near 9.24 (pK_a value of the ammonium ion). Let's notice in relation with this example that an ammonium hydrogen sulfide (NH_4 HS) solution would exhibit the value pH = 8.10. This would actually be the pH value of a solution of a salt of a weak acid and of a weak base $NH_4{}^+$ and SH^-.

Example 4 A more difficult problem to solve is that of the calculation of the pH value of a 10^{-2} mol/L quinine phosphate solution. It can be considered as resulting from the neutralization of quinine base by orthophosphoric acid. Quinine B is a dibase. The corresponding conjugate acids are symbolized by BH^+ and $BH_2{}^{2+}$. Hence, the quinine phosphate formula is symbolized by $(BH_2)_3(PO_4)_2$. The preponderance-area diagram of the different species and the corresponding pK_a values are given in Fig. 5.10.

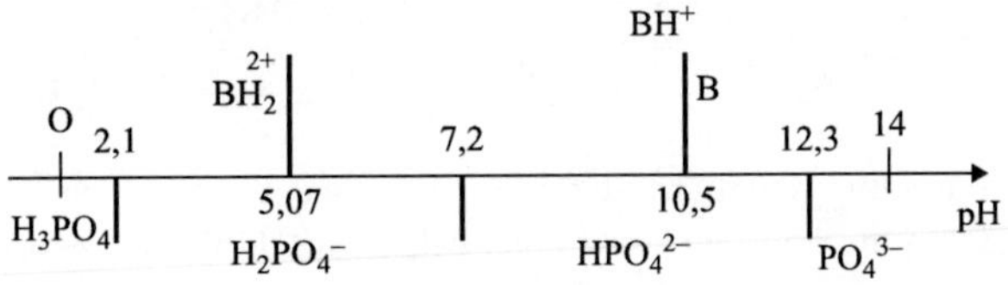

Fig. 5.10 Preponderance-area diagram of the species resulting from the dissolution of quinine phosphate

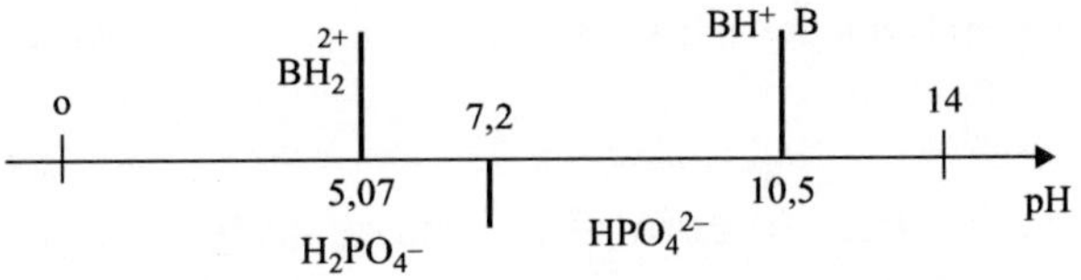

Fig. 5.11 Predominant species after the first principal reaction has taken place (case of quinine phosphate)

At the very beginning, before any (supposed) reaction, the concentrations are $[BH_2^{2+}] = 3C$ and $[PO_4^{3-}] = 2C$. The first reaction that must take place appears immediately after the study of the predominance-area diagram. It is

$$BH_2^{2+} + PO_4^{3-} \rightarrow BH^+ + HPO_4^{2-}.$$

For the time being, the species BH_2^{2+} ($[BH_2^{2+}] = 1C$), BH^+ ($[BH^+] = 2C$), and HPO_4^{2-} ($[HPO_4^{2-}] = 2C$) exist in solution. Due to this reaction, the diagram is modified (Fig. 5.11).

Then another principal reaction takes place since the BH_2^{2+} and HPO_4^{2-} predominance areas are disconnected:

$$BH_2^{2+} + HPO_4^{2-} \rightarrow BH^+ + H_2PO_4^-.$$

Considering this reaction as complete (recall that this is one of the cornerstones of the methodology), the following species remain: HPO_4^{2-} ($[HPO_4^{2-}] = 1C$), BH^+ ($[BH^+] = 3C$), $H_2PO_4^-$ ($[H_2PO_4^-] = 1C$) in the solution. Introducing HPO_4^{2-} and $H_2PO_4^-$ into the ionization constant expression,

$$[HPO_4^{2-}][H^+]/[H_2PO_4^-] = 10^{-7.2},$$

gives the approached value $pH = 7.2$.

This value is relatively far from the true one. The failure in the reasoning was to consider the second principal reaction as total. Actually, the pK_a values of the couples BH_2^{2+}/BH^+ and $H_2PO_4^-/HPO_4^{2-}$ are too close to each other for the reaction to be complete. Hence, we must consider the second principal reaction as being equilibrated and study the following system:

$$[HPO_4^{2-}][H^+]/[H_2PO_4^-] = 10^{-7.2},$$

$$[BH^+][H^+]/[BH_2^{2+}] = 10^{-5.07},$$

$$[H_2PO_4^-] + [HPO_4^{2-}] = 2C,$$

$$[BH^+] + [BH_2{}^{2+}] = 3C,$$

$$[BH^+] + 2[BH_2{}^{2+}] + [H^+] = [OH^-] + 2[HPO_4{}^{2-}] + [H_2PO_4{}^-].$$

It is solved easily by neglecting the concentrations $[H^+]$ and $[OH^-]$, as can be inferred from the previous erroneous value found, pH = 7.2. Finally, we find pH = 5.6.

In brief, the so-called principal reaction method, which is necessarily an approached one, permits us to know quickly what species can be neglected in the general equations. It simplifies the calculations considerably.

5.19 pH Calculations Taking Activities into Account

The previous relations and formulas have been found after mixing concentrations and activities. When the solutions are not sufficiently diluted, this is no longer possible. The ionic strength effects are not negligible. It is quite clear that in these conditions, the results found with the previous formula are only approached ones. We must work with activities.

We shall limit ourselves here to giving the principle of the pH calculation of a weak acid HA at the analytical concentration C, taking into account the part played by activities.

The relations that must be satisfied are

$$a_{H_3O^+} \times a_{OH^-} = K_w{}^\circ,$$

$$a_{H_3O^+} \times a_{A^-}/a_{HA} = K_a,$$

$$[HA] + [A^-] = C,$$

$$[A^-] + [OH^-] = [H_3O^+].$$

It is important to note that both charge and mass balances are expressed in concentration terms, whereas equilibrium constants are in activity terms since K_a and K_w are thermodynamic constants. Here is the problem: Solving it involves making the system homogeneous. The two methodologies consist, in a first step, of transforming the thermodynamic constants into apparent ones. In the second step, the system (which is now homogeneous in concentrations) is solved as before. The concentration $[H_3O^+]$ is hence calculated. The third step consists of going back from the concentration $[H_3O^+]$ to the activity a_{H3O+}.

According to the first strategy, the first step is often achieved by successive iterations. At the beginning of the iterative process, the system is solved by mixing activities and concentrations. A value $|H_3O^+|$ is obtained. It is not endowed with the significance of an activity or with that of a concentration. (This explains the symbolism used with the presence of vertical lines. This symbolism is personal.) This assertion is evident as the starting relations involve both activities and concentrations. However, this first $|H_3O^+|$ value permits us to calculate the pseudoconcentrations (or pseudoactivities) $|A^-|$, $|OH^-|$, and $|HA|$ and the pseudo ionic strength I_1, which permit us, in turn, to use one Debye–Hückel relation and, hence, a first set of activity coefficients for all species. As a result, it becomes possible to calculate a first set of apparent equilibrium constants through the following relations:

$$K_{a1} = K_a \gamma_{HA1}/\gamma_{A_1^-} \times \gamma_{H_3O_1^+} \quad \text{and} \quad K_{w1} = K_w{}^\circ/\gamma_{H_3O_1^+}$$

(usually $\gamma_{HA} = 1$; see Chap. 3).The first iteration is finished. At this point, the K_{a1} and K_{w1} constants already do possess a greater significance in apparent constants than the corresponding thermodynamic ones, of course, did possess. In the second iteration, the same process is started again but with the only difference that the initial equilibrium constants are now those found at the end of the last iteration. A new pseudo-ionic strength I_2 is then calculated, and so forth, until the pseudo-ionic strength I_n remains equal to I_{n-1}. (Usually, the convergence is obtained after four or five iterative processes.) At this point, the true ionic strength is known. This is because all the relations are homogeneous in terms of concentrations. The system is now self-consistent. Together with the true ionic strength, we also know the concentrations of all the species in the solution, including $[H_3O^+]$ and the activity coefficients. Hence, the second step is achieved simultaneously with the first. Finally, the third step is achieved easily, as the concentration $[H_3O^+]$, the ionic strength, and the activity coefficients are known. The following relation is used:

$$a_{H_3O^+} = [H_3O^+]_n \times \gamma_{H_3O^+n} \quad \text{(end of the iteration process- } n\text{th iteration).}$$

The second strategy consists of adding a supporting electrolyte that is inert from the chemical standpoint to the solution. Its concentration is intentionally far higher than that of the acid. Calculations show that the ionic strength is then exclusively fixed by it. Hence, the ionic strength is known and constant. The calculations of apparent constants and that of the concentration $[H_3O^+]$ are then easy. As at the end of the first process, the activity aH_3O^+ and the pH values are immediately calculated.

Not all solutions have been investigated in this chapter. An especially interesting case is provided by a solution of a mixture of an acid and its conjugate base. This is a buffer solution. Buffer solutions will be studied in the next chapter.

Chapter 6
Buffer Solutions

It is an experimental fact that solutions obtained by dissolving a weak acid and its conjugate base exhibit only a very weak pH change and even no change at all when a strong acid or base is added to them, at least in certain concentration conditions. These solutions are called *buffers*. We also say that they have acidity and basicity "in reserve." They are of considerable interest. It is sufficient to be convinced that the maintenance of a certain life form requires very narrow pH ranges. This must be related to the enzymatic systems, which actually can work only in very narrow pH regions.

6.1 pH of a Buffer Solution Before Addition of a Strong Acid or Base

Let's consider a solution of a weak acid HA at the analytical concentration C mol/L and of its conjugate base A^- at the concentration C mol/L dissolved as the sodium salt, for example. The pH value is calculated by solving the following system of simultaneous equations that must be satisfied:

$$[H_3O^+][OH^-] = K_w,$$

$$[A^-][H_3O^+]/[HA] = K_a,$$

$$[H_3O^+] + [Na^+] = [A^-] + [OH^-],$$

$$[HA] + [A^-] = C_{HA} + C_A,$$

$$[Na^+] = C_A,$$

where activities and concentrations are assumed to be equal. This system does not show the equalities $[HA] = C_{HA}$ and $[A^-] = C_A$. In fact, it is not correct to set them up since there are water dissociation and acid dissociation equilibria, which superimpose themselves.

J.-L. Burgot, *Ionic Equilibria in Analytical Chemistry*,
DOI 10.1007/978-1-4419-8382-4_6, © Springer Science+Business Media, LLC 2012

The system can be reduced to a third-order equation in $[H_3O^+]$, which would be difficult to solve. It is sometimes written in the literature in the form

$$[H_3O^+] = \{K_a(C_{HA} - [H_3O^+] + [OH^-])\}/(C_A + [H_3O^+] - [OH^-]).$$

The system becomes considerably simplified by assuming that the concentrations $[H_3O^+]$ and $[OH^-]$ can be neglected with respect to those of $[Na^+]$ and $[A^-]$ in the charge balance relation. With this hypothesis, the system can be reduced to the following relations:

$$[H_3O^+] = K_a C_{HA}/C_A,$$

$$pH = pK_a + \log(C_A/C_{HA}). \quad (6.1)$$

In the case of the couple BH^+/B, in which the acid would be dissolved under the hydrochloride form, for example, the relations found after the same simplifications as above would be perfectly analogous and would be written

$$pH = pK_a + \log(C_B/C_{BH}). \quad (6.2)$$

Equation (6.1) or (6.2) is called Henderson–Hasselbach's equation. From its examination, it appears that the pH value of a buffer solution only depends on the ratio of the concentrations of conjugate acid and basic forms and also on the nature of the acid through its pK_a value. It does not depend on the total concentration $C_{HA} + C_A$ or $C_{BH} + C_B$. In other words, the pH value of a buffer solution does not change with its dilution.

We emphasize the fact that the Henderson–Hasselbach equation is an approached one. Its use is valid if the concentrations $[OH^-]$ and $[H_3O^+]$ are negligible in the charge balance equation. This is true only if the pK_a value of thc couple is not too close to 0 or to 14. In this case, the concentrations $[H_3O^+]$ and $[OH^-]$ would no longer be negligible. In order for the Henderson–Hasselbach to be valid, the concentrations C_{HA} and C_A (or C_{BH} and C_B) must not be too weak. This is the reason why it is often stipulated that buffer solutions are rather concentrated solutions in the weak acid and in its conjugate base. When the buffer solution is too diluted, the Henderson–Hasselbach equation is no longer valid and the pH then depends on the total concentration. It increases with the buffer dilution (Fig. 6.1).

In the example of the acetic acid/acetate buffer ($C_{HA} = C_A$), the pH independence with the dilution is manifest only for $C > 10^{-3}$ mol/L. It is interesting to note that neglecting the concentrations $[H_3O^+]$ and $[OH^-]$ in the charge balance equation is equivalent to setting up $[HA] = C_{HA}$ and $[A^-] = C_A$, that is, to neglecting the water dissociation and the acid ionization.

Exercise 1 Calculate the pH value of a solution of 10^{-2} mol/L acetic acid and 10^{-2} mol/L sodium acetate ($pK_a = 4.75$).

The Henderson–Hasselbach equation gives the value pH = 4.75. Introducing this value into the rigorous system of equations shows that this approximate value is about 5% accurate.

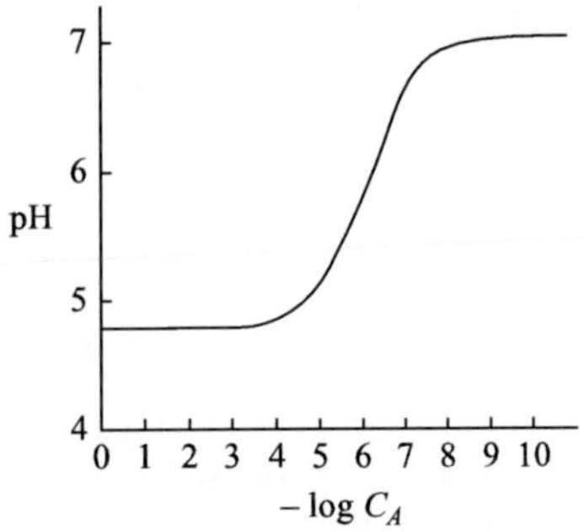

Fig. 6.1 pH of an acid acetic/sodium acetate buffer solution as a function of the concentration $C\ (C = C_{\mathrm{HA}} = C_{\mathrm{A}})$

6.2 pH of a Buffer Solution After a Proton Addition

Let's consider a solution of the buffer couple HA/A^- and of the strong acid HX. The analytical concentrations are C_{HA}, C_{A}, and C mol/L. In comparison with the preceding system of equations, the charge balance equation is modified and becomes

$$[H_3O^+] + [Na^+] = [OH^-] + [A^-] + [X^-].$$

We must also add the following relation to the preceding system:

$$[X^-] = C.$$

Neglecting the concentrations $[OH^-]$ and $[H_3O^+]$ in the charge balance equation gives the relations

$$[H_3O^+] = K_a(C_{\mathrm{HA}} + C)/(C_{\mathrm{A}} - C),$$

$$\mathrm{pH} = pK_a + \log[(C_{\mathrm{A}} - C)/(C_{\mathrm{HA}} + C)]. \tag{6.3}$$

Exercise 2 Calculate the pH value of the solution of the preceding example after adding hydrochloric acid at the concentration 10^{-4} mol/L. (We suppose that the addition of the hydrochloric acid does not change the solution's volume and hence the analytical concentrations $C_{\mathrm{HA}} = C_{\mathrm{A}} = 10^{-2}$ mol/L.)

Applying Eq. (6.3) gives pH = 4.67. The result is correct at about 5%.

The results found in these two exercises show that the pH value is changed very little after the addition of a strong acid (ΔpH = −0.08 unit). For the sake of comparison, we must recall that the preparation of a 10^{-4} mol/L hydrochloric solution starting from pure water gives a pH decrease of 3 pH units. A more realistic comparison is to consider the pH change when we add a strong acid to the solution of a strong acid already at pH = 4.75, so as to make the final concentration of the added acid equal to 10^{-4} mol/L. Calculations immediately give

$$[H_3O^+] = 10^{-4.75} + 10^{-4},$$

$$\mathrm{pH} = 3.93,$$

$$\Delta\mathrm{pH} = -0.82.$$

Whatever the comparison, it is clear that the addition of a strong acid to a buffer solution gives a quasi-null pH change of the latter. It is the same state of affairs when a strong base is added. The quasi-cancellation of the pH change is called the *buffer effect*.

6.3 Mechanism of the Buffer Effect

The mechanism of the buffer effect can be studied from two points of view: a chemical standpoint and a mathematical standpoint. We will examine both viewpoints here.

6.3.1 Chemical Standpoint

In the limits that prevail when setting up Henderson–Hasselbach's relation (see above), it has been found that after the addition of C mol/L of a strong acid, the concentrations of the base A^- and of the acid HA (in the case of a buffer of the kind HA/A^-) have become

$$[A^-] = C_A - C,$$
$$[HA] = C_{HA} + C.$$

These results are issued from a comparison of Eqs. (6.1) and (6.3). C mol/L of base A^- have disappeared and C mol/L of acid HA have appeared, whereas C mol/L of strong acid have also disappeared. The result is that the complete reaction

$$A^- + H_3O^+ \rightarrow HA + H_2O$$

has been induced by the addition of the strong acid.

The C moles (per solution liter) of hydroxonium ions added have been completely transformed into C moles of the weak acid HA. Briefly, due to the buffer effect, the addition of C mol/L of the strongest acid that can exist in water (H_3O^+) has been commuted to the addition of C mol/L of a weak acid that is very poorly dissociated. This is the chemical mechanism of the buffer effect.

6.3.2 Mathematical Standpoint

The weak pH change also originates in the mathematical properties of the function pH/C. The corresponding curve clearly shows the following property (see Fig. 6.2) between the asymptotes its slope is weak. This is not the case close to the asymptotes $C = \pm\, C_{HA} = C_A$. However, this last point is not of great significance since we must not calculate pH from Henderson–Hasselbach's equation, as it is not legitimate near the asymptotes.

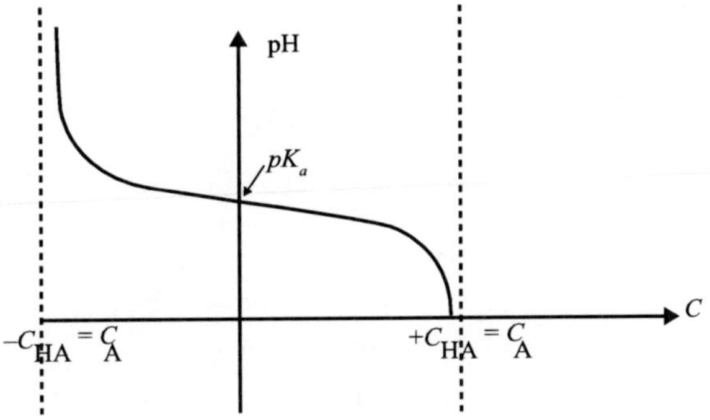

Fig. 6.2 pH/strong acid concentration C ($C_{HA} = C_A$ for the sake of simplification)

6.4 Buffer Capacity—Buffer Index

A buffer resists any change in pH resulting from the addition of a strong acid. A measure of the buffer capacity is the amount of strong acid or strong base required to change the pH by a given amount. The larger this quantity is, the better the buffer is. The buffer capacity is quantified by the *buffer index*. This latter is symbolized by β.

When a strong base is added, the buffer index is defined by the relation

$$\beta = dC_b/d\text{pH},$$

where dC_b is the number of moles of strong base added to 1 L of the buffer solution and dpH is the pH change due to the addition. In the case of the addition of a strong acid, the buffer index is defined by the relation

$$\beta = -dC_a/d\text{pH},$$

where dC_a is the number of moles of strong acid added to 1 L of the buffer solution and dpH the pH change. The minus sign is introduced in this case so that β remains positive. The buffer index is conveniently defined in differential terms since β changes with the initial solution pH.

6.5 Mathematical Expression of the Buffer Index

The buffer index can be easily expressed as a function of the pH value and of some parameters related to the buffer. From a general standpoint, let's consider a solution containing C_a mol/L of hydrochloric acid, C_b mol/L of sodium hydroxide, and the analytical concentration C mol/L of the weak acid HA. Before writing the relations that must be satisfied, it is important to notice that the simultaneous presence of sodium hydroxide and weak acid HA results in the formation of the weak base

A^- and, hence, in the formation of the buffer HA/A^-. The following relations are satisfied:

$$[H_3O^+][A^-]/[HA] = K_a,$$
$$[H_3O^+][OH^-] = K_w,$$
$$[Na^+] = C_b,$$
$$[Cl^-] = C_a,$$
$$[HA] + [A^-] = C,$$
$$[H_3O^+] + [Na^+] = [OH^-] + [Cl^-] + [A^-].$$

The system can be reduced to the sole equation

$$C_b = C_a + K_w/[H_3O^+] - [H_3O^+] + CK_a/(K_a + [H_3O^+]). \tag{6.4}$$

At first, we'll consider the case of the addition of a strong base to the preceding buffer solution. The strong acid does not exist in the solution ($C_a = 0$). By recalling the relations

$$dC_b/d\text{pH} = (dC_b/d[H_3O^+]) \times (d[H_3O^+]/d\text{pH})$$

and

$$d[H_3O^+]/d\text{pH} = -2.303[H_3O^+],$$
$$\beta = -2.303[H_3O^+](dC_b/d[H_3O^+]),$$

and by differentiating Eq. (6.4), we find

$$\beta = 2.303\left\{K_w/[H_3O^+] + [H_3O^+] + CK_a[H_3O^+]/(K_a + [H_3O^+])^2\right\}. \tag{6.5}$$

It is easy to verify that the same relation applies to express the buffer index when a strong acid is added to the buffer solution ($C_b = 0$). Several conclusions can be drawn from Eq. (6.5). In the first place, β is a function of the solution's pH. This justifies the use of differentials. In the second place, it is easy, through calculations involving derivatives $d\beta/d[H_3O^+]$ and $d^2\beta/(d[H_3O^+])^2$, to demonstrate that β is maximum when $[H_3O^+] = K_a$. This signifies that the buffer capacity is maximum when $C_{HA} = C_A$. It is minimal when HA or A^- is the only existing form of the buffer that exists in solution. When the solution does not contain the acid or the conjugate base of the buffer, the same calculation as that above gives the relation

$$\beta = 2.303\left\{K_w/[H_3O^+] + [H_3O^+]\right\}. \tag{6.6}$$

Evidently, Eq. (6.6) gives the buffer capacity of pure water. Moreover, Eq. (6.6) already existed in Eq. (6.5). Hence, we can say that the buffer capacity of the solution is the sum of two buffer capacities: one due to the buffer couple itself and the other

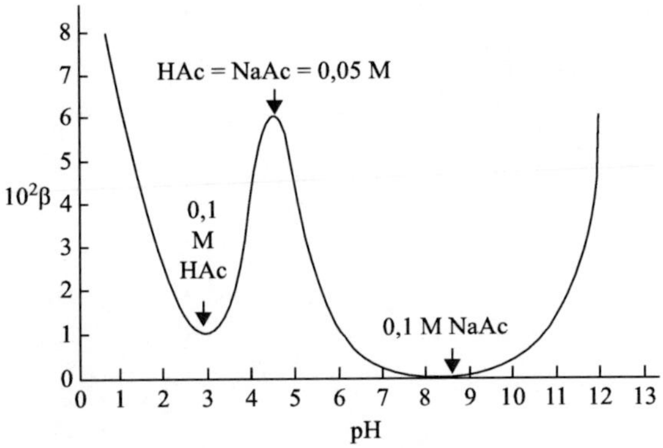

Fig. 6.3 Buffer index of an acetic acid solution ($C = 10^{-1}$ mol/L) in relation with the pH

due to pure water. A calculation involving differentials shows that the buffer capacity of the water couples is minimal for pH = 7. It is maximal for pH = 0 and pH = 14. For these values, it exhibits huge values due, respectively, to the couples H_3O^+/H_2O and H_2O/OH^-. It is precisely for this reason that in aqueous solutions, pH values never fall below 0 and never take values higher than 14 (apart from some hundredths for both values) even if more and more strong acid (or base) is added. Paradoxically, even, after an important addition of strong acid, the pH ends up increasing due to ionic strength and solvation effects. Indeed, when the solution becomes too concentrated in strong acid, there are not enough solvent molecules to solvate normally hydroxonium ions. It is the same state of affairs on the basic side of the acidity scale.

Figure 6.3 summarizes these considerations. It is the diagram of buffer index β/pH for an acetic acid solution ($C = 10^{-1}$ mol/L). In the region ($8 < \text{pH} < 10$), the buffer capacity is due to the acetate ion, the only species present in a noticeable quantity. The buffer index is very weak: $= 3.5 \times 10^{-5}$. At pH = 2.5, we could believe that the value would be about the same as immediately above because the only present form of the buffer couple is the acidic one. Actually, we find $= 8.3 \times 10^{-3}$. The value is far higher than anticipated. In fact, at this pH value, the buffer capacity of the couple H_3O^+/H_2O is already noticeable. Finally, Fig. 6.3 shows that the buffer index of the couple is maximum for $\text{pH} = pK_a$.

Remark The titration curve slope of an acid by a strong base is equal to $d\text{pH}/dC$, that is, to the inverse of the buffer index (see Chaps. 9 and 10).

6.6 Buffer Range

In general, we may state, on experimental grounds, that the buffering capacity is maintained for mixtures within the range of concentrations ratios 1 acid/10 salt and 10 acid/1 salt. Hence, the approximate pH range of a buffer is

$$pK_a - 1 < \text{pH} < pK_a + 1 \quad \text{pH units.}$$

It is called the *buffer range*.

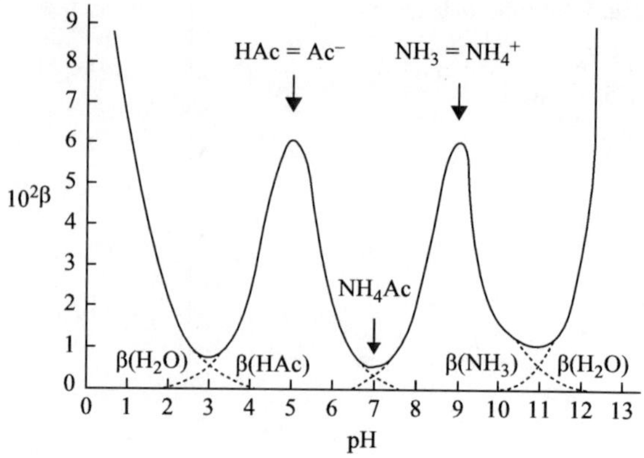

Fig. 6.4 Buffer index of a 0.1 mol/L ammonium acetate vs. the pH value

6.7 Mixtures of Several Buffers

The case of mixtures of several monoprotic acids and of their conjugate base mixtures is considered in this section. That of polyacids (and their conjugate bases) is studied in the following section.

It turns out that when several buffers are present, there exist as many buffer ranges as buffer couples. The buffer index can be calculated easily. In the case of a two-buffer mixture (HA_1/A_1^- and HA_2/A_2^-), the buffer index's expression is

$$\beta = 2.303\left\{([H_3O^+] + K_w/[H_3O^+]) + C_1K_{a1}[H_3O^+]/(K_{a1} + [H_3O^+])^2 + C_2K_{a2}[H_3O^+]/(K_{a2} + [H_3O^+])^2\right\}$$

where $C_1 = [HA_1] + [A_1^-]$,

$C_2 = [HA_2] + [A_2^-]$.

It is derived from the charge balance equation as in the case of one couple.

The buffer capacity is purely and simply the sum of the buffer components' mixture. This property is true regardless of the couple's number and the electrical charge of the acidic form. In order to illustrate this point, let's consider a 0.1 mol/L ammonium acetate solution [pK_a (acetic acid) = 4.75) and (pK_a (ammonium ion) = 9.24]. The preceding relation permits us to calculate β as a function of the pH value (Fig. 6.4). The latter is adjusted by adding a strong acid or a strong base.

It is evident that since the solution's acidity displaces the proton equilibria, one or both buffer couple(s) (CH_3COOH/CH_3COO^- and NH_4^+/NH_3) present in the solution may be operative according to the pH value. It is easy to demonstrate that β is maximum for $[CH_3COOH] = [CH_3COO^-]$ and $[NH_4^+] = [NH_3]$. In this example, the two pK_a values are far from each other and the result of this property is the occurrence of an intermediary pH value (pH about 7) for which the buffer index is quasi-null. This is the pH value for which only one form of each of the two

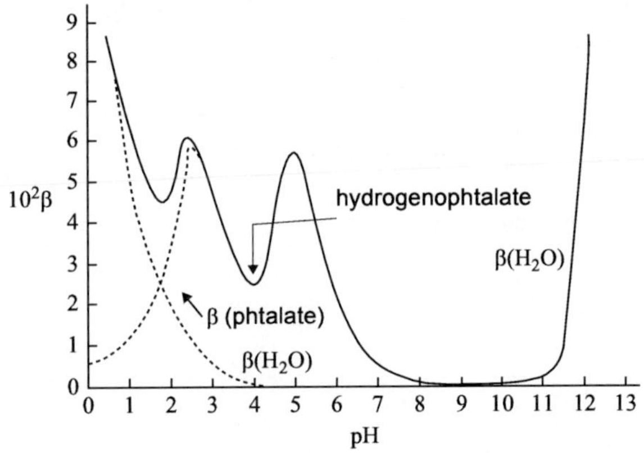

Fig. 6.5 Buffer index of a 0.1 mol/L potassium hydrogen phthalate solution as a function of the pH

buffers is present, namely, CH_3COO^- and NH_4^+. At pH values 3 and 11, the total buffer capacity does not vanish although the species CH_3COOH and NH_4^+ are alone, respectively. This is because, at these pH values, the buffer capacities of the couples H_3O^+/H_2O and H_2O/OH^- are already noticeable. From a more general standpoint, the shape of the β/pH curve depends, of course, on the respective concentrations and pK_a values of both buffers.

6.8 Buffer Capacity of a Polyacid

We'll consider the diacid H_2A. According to the pH value and its pK_a values, it can be in one or two (indeed three) of the three forms: H_2A, HA^-, and A^{2-}. Hence, if it is the case, the two buffer couples H_2A/HA^- and HA^-/A^{2-} exist in the solution. This consideration can be extended to the polyacid H_n/A case. If the pK_a values of the successive acidities are sufficiently far from each other, the curve β/pH exhibits as many maxima as acidities. They are located at the pK_a values, that is, for equimolecular concentrations of the corresponding conjugate acids and bases. For example, in the orthophosphoric acid case, the buffer capacity is maximum for $pH = pK_a = 7.20$, for which $[H_2PO_4^-] = [HPO_4^{2-}]$. It would present two other maxima for $pH = pK_a = 2.23$ and $pH = pK_a = 12.32$, but they are not visible because of the superimposition of the buffer capacities of the H_3O^+/H_2O and H_2O/OH^- couples to them. Another interesting case is that of potassium hydrogen phthalate, which is a primary standard (see below). Phthalic acid is a diacid ($pK_a = 2.95$ and $pK_a = 5.41$). (We notice, incidentally, that the pH of its 0.05 mol/L solution is 4.005 at 25°C. It must be emphasized that this value is assigned to this solution, in these conditions, once and for all. This is the basis of the pH definition.) The solution buffer capacity exhibits two maxima at the $pH = pK_a$ values (Fig. 6.5) as expected, but it does not vanish at the intermediary $pH = 4.18$. The explanation is that the pK_a values are sufficiently near each other so that, for that pH, the three species are present. As

Table 6.1 Structures and *pKa* values of some buffers (the only given structure is that of the acid form)

Name	Structure	pK_a
Phosphoric acid	$H_3\,PO_4$	2,15
Acetic acid	$CH_3\,COOH$	4,75
Imidazole (hydrochloride)	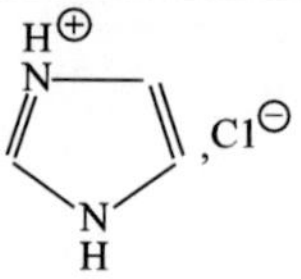	6,99
Dihydrogen phosphate	$H_2\,PO_4^-$	7,20
Trist(hydroxymethyl)amino-methane, hydrochloride	$(CH_2OH)_3-C-\overset{+}{N}H_3,Cl^-$	8,08
Boric acid	H_3BO_3	9,24
Monohydrogen phosphate	HPO_4^{2-}	12,35

a result, the two buffers are simultaneously operative and the buffer index remains high. This is one of the reasons why it has been chosen as a primary standard.

In the case of a diacid solution and concerning its buffer index calculation, it is not possible to consider it a simple equimolecular mixture of two monoacidic buffers as long as the ratio of the two ionization constants remains lower than 5%.

6.9 Some Buffers

Table 6.1 mentions some buffer couples together with their pK_a values.

There are some other organic buffers that derive from

- the piperazine cycle:

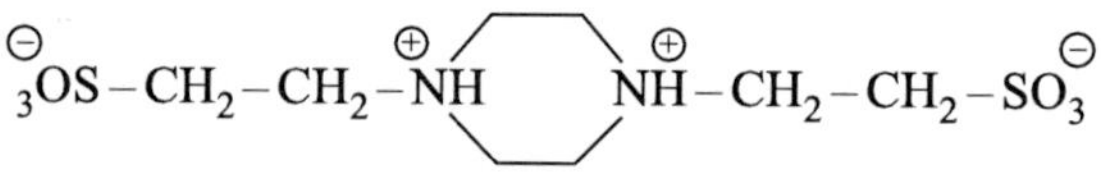

piperazine-N,N'bis (2-ethane sulfonic) acid. PIPES (pK_a = 6,80)

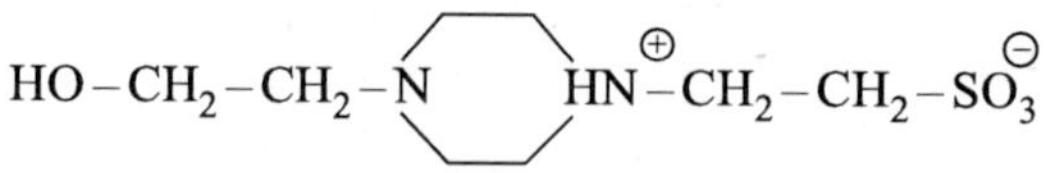

N-2-hydroxyethylpiperazine-N'-2-ethanesulfonic acid. HEPES (pK_a = 7,56)

N-2-hydroxyethylpiperazine-N-'3-propanesulfonic acid. HEPPS (pK_a = 8,00)

- the morpholine cycle:

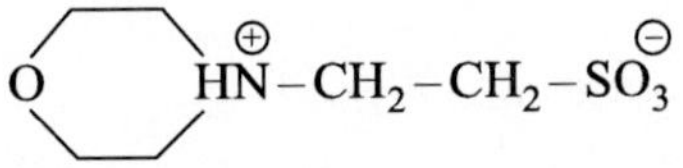

2-(N-morpholino)ethanosulfonic acid. MES (pK_a = 6,15)

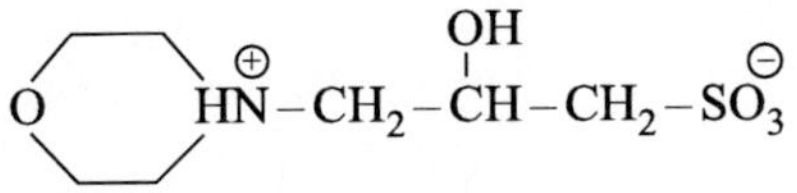

3-(N-morpholino)-2-hydroxypropanosulfonic acid. MOPSO (pK_a = 6,95)

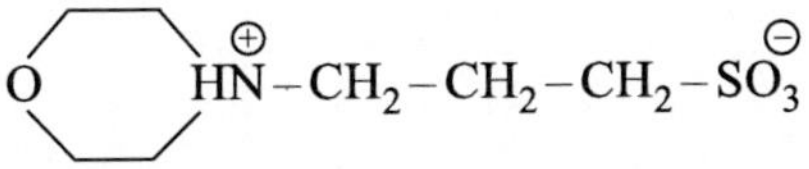

3-(N-morpholino)propanesulfonic acid. MOPS (pK_a = 7,20)

- cyclohexylamine:

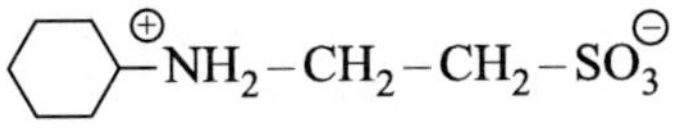

2-cyclohexylamineethanesulfonic acid. CHES (pK_a = 9,50)

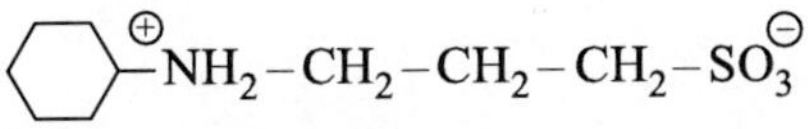

3-cyclohexylaminepropanesulfonic acid. CAPS (pK_a = 10,40)

There are also the derivatives of 2-amino-acetamide:

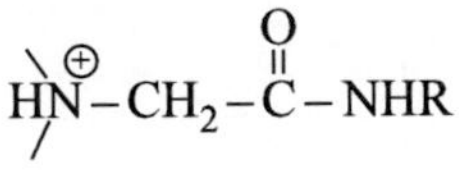

The values are given for the temperature of 25°C. After some preceding considerations, it is clear that pH buffer ranges change with temperature and the ionic strength. In the latter case, it takes place when the pK_a values are the apparent ones. Some particular buffers have been devised for some special applications. For example, the use of buffers that do not absorb in the wavelength range of the radiations used to perform a spectrophotometric analysis is required. This is often the case in the visible domain. Likewise, electrochemical measurements preclude the electroactive buffers' use.

Chapter 7
Some General Points Concerning Titrations

Titrimetric analysis is most likely the oldest method of quantitative analysis.[1] Its commonplace use has continued to date, due to its great simplicity of use and low cost. It is an absolute method to determine a quantity of matter. It is often used in official methods as well as in reference methods of analysis, but it has equal success as a routine quantitative method of analysis.

7.1 General Principle of Titrimetric Methods

Titrimetry permits us to determine a compound's concentration in a given solution by quantitatively measuring the quantity of reactant reacting with it. Let us titrate a compound A in a solution at concentration C_A. Increasing and known volumes of its reactant B at concentration C_B are added to an exactly known volume V_A of solution A. B is chosen to react quantitatively with A according to a stoichiometric reaction. Let us call V_{Bf} the added volume when the reaction has just completed. The quantity of B that has disappeared is $V_{Bf}\,C_B$. This is also the quantity of A that was initially present in the solution to analyze (after having taken into account the reaction stoichiometry). Then the determination of C_A becomes immediate. For example, if one molecule of A reacts with one molecule of B (a one-to-one reaction stoichiometry), we find

$$V_A C_A = V_{Bf} C_B,$$

$$C_A = V_{Bf} C_B / V_A. \tag{7.1}$$

V_{Bf} is accessible via any experimental means that indicates the end of the titration reaction; C_B and V_A are known.

---

[1] Joseph Gay-Lussac (1778–1850) seems to have been the first chemist to give the status of exact quantitative analysis method to titrimetry after his work in 1824 devoted to the determination of active chlorine, potassium hydroxide, and silver ion. Other chemists in this field must also be mentioned. We shall be content here with recalling Karl Friedrich Mohr (1800–1879) and Carl Remigius Fresenius (1818–1897).

J.-L. Burgot, *Ionic Equilibria in Analytical Chemistry*,
DOI 10.1007/978-1-4419-8382-4_7,

7.2 Terminology

The solution of A is called the *titrand solution*, and that of B is the *titrant solution*. In some countries (France, for example), a solution is called *titrated* if it contains a known amount of substance in a known volume. The *titer* of a solution is the reactant proportion (and thus the solvent proportion) in the solution. In the largest sense, it means its concentration (see Chap. 1). In a more restrictive sense, it is the solute concentration expressed in mol/L or its normality (see Sect. 7.3).

The titrant solution must be standardized before the titration. In other words, its titer must be precisely known. (In certain countries, a "standard solution" is purely and simply what we called a titrated solution above). The standardization of the titrant solution is achieved by its titration with a standard solution. This can be a primary or secondary (standard) solution.

A primary standard solution is prepared simply by weighing the reagent in the pure state (called for this goal a primary standard), by dissolving it into the solvent, and making up the solution to a known volume. A secondary standard is a substance that can also be used for standardizations. However, its solution content in the active substance has been found by comparison with a primary standard.

The point at which the titration reaction is just complete is called the *equivalence point* or the *theoretical* or *stoichiometric endpoint*. The volume added to the equivalence point is simply called the *added volume to the equivalence point*.

The completion of the titration reaction may be detectable with the help of a chemical indicator. After the titration reaction is complete, an auxiliary reagent dissolved previously in the titrand solution, called a *color indicator*, causes a color change in the solution being titrated. In this case, the indicator is called an internal indicator. An external one can also be used. A solution sample (a few drops) is taken and brought face to face with the indicator out of the titration vessel. In some cases there is no need for a color indicator since the titrant or titrand itself is colored. Then, at the equivalence point, a color may appear or disappear. A change in a physical property can also indicate this point. It can be detected with the help of an instrumental analysis method. The curve point at which the color changes or the physical change occurs is called the *endpoint* of the titration.

In an ideal titration, the endpoint and the equivalence point coincide. In practice, a difference occurs.

In the literature, the term "volumetric analysis" is encountered frequently instead of "titrimetric analysis" because there are volume measurements in titrimetric analysis. The term "volumetric analysis" must be avoided in this context and devoted solely to gas determinations.

7.3 Titration Error

Since the endpoint and the equivalence point do not coincide, an error occurs, called the *titration error*. More precisely, it is the titration error related to the indication of the equivalence point.

If V_f is the added volume at the endpoint (an experimental point) and V_{ep} is that added in principle at the equivalence point (a theoretical point), the absolute and relative errors are given by

$$\text{absolute titration error}: V_f - V_{ep}.$$

$$\text{relative titration error}: 100(V_f - V_{ep})/V_{ep}.$$

Naturally, the titration error must be minimized. It may have two origins:

- the choice of an indicator whose change does not occur sufficiently close to the equivalence point. (In the case of a protometric titration, for example, the change does not occur about the pH value of the equivalence point);
- a titration reaction that is not sufficiently quantitative. This case is investigated further in Sect. 7.7 and in several other chapters.

The titration error is a systematic error. Other systematic errors exist, among them

- errors related to poor glassware gauging,
- reading errors, in particular those due to parallax,
- errors due to the temperature differences between that at which the vessel was calibrated and that at which the present analysis is performed,
- errors due to wetting of the graduated flasks,
- errors due to not respecting the draining time,
- errors due to the delivery of any drop adhering to the outside of the pipette.

Random errors are added to the systematic ones. The precision of a titrimetric titration is about 0.2%.

In the case of protometric titrations, the most important error is usually due to the uncertainty related to the pH determination at the equivalence point.

7.4 Equivalents and Normal Solutions

The titration reaction is not necessarily of a one-to-one stoichiometry. From a general standpoint, it must be written as

$$\nu_A A + \nu_B B \rightarrow \text{products}.$$

As a result, the relation that permits us to calculate C_A from C_B and V_{Bf} is not always as simple as Eq. (7.1) since, quite evidently, the stoichiometric coefficients ν_A and ν_B must intervene into it. A development of profound importance has been the introduction of the concept of an *equivalent*. The titrant or titrand equivalent is its mole number, integer or fractional, which must be dissolved to prepare one liter of solution for Eq. (7.1) to be obeyed. The number of equivalents in one solution liter is called its *normality*. The normality is hence another expression of the quantity of matter per solution liter. Its symbol is N. After the introduction of the normalities of a solute, we can use the relationship

$$N_A V_A = N_B V_{Bf}. \tag{7.2}$$

in order to calculate the titer of the solution under study. A comparison of Eqs. (7.1) and (7.2) shows that the introduction of the concept of normality, as it is, transforms a reaction with a complicated stoichiometry into one of apparent one-to-one stoichiometry. This is done by modifying the expression of concentrations. The coefficient by which we must multiply the molar mass to obtain the corresponding equivalent is called the equivalent factor f_{eq}. It is a pure number that results from examining the titration reaction. We have

$$N = Mf_{eq}.$$

In the case of an acid–base reaction, f_{eq} is chosen in such a way that only one proton exchange is considered during the titration reaction. For example, hydrochloric acid can liberate only one proton per mole. Hence, its equivalent is equal to its molar mass, and its equivalent factor is equal to 1:

$$N(HCl) = M(HCl),$$

$$f_{eq}(HCl) = 1.$$

Sulfuric acid liberates two protons that can be titrated. We thus find

$$N(H_2SO_4) = M(H_2SO_4)/2,$$

$$f_{eq}(H_2SO_4) = \tfrac{1}{2}.$$

The equivalent is its half-molar mass. In the case of a redox reaction, the equivalent is related to the number of exchanged electrons, and the equivalent factor is chosen for one electron exchanged. The equivalent concept is also used for titrations by the formation of complexes or by precipitation. Its introduction has been of profound importance in the development of titrimetric analysis.

There is, however, an important point to emphasize: The equivalent and the equivalent factor are both ambiguous notions since their values depend entirely on the titration reaction of concern. Indeed, the same reactant may exhibit different values of these two quantities according to the considered reaction. We shall see some examples of this later in the book.

7.5 Some Titration Forms

Several titration forms exist, including the following:

- direct titrations: the titrant solution is added to the titrand solution.
- inverse titrations: the solution of concern is added to a known volume of the titrant solution.

- back titrations: in the first step, a known volume of the titrant solution in excess is added to the solution under study. In the second step, the excess is titrated with any standard solution. Back titrations are used when the titrand cannot be directly titrated for various reasons, such as the kinetics of the titration reaction which are too slow. In this case, a higher concentration of reactant increases the reaction speed. They can also be used for practical reasons, such as having on hand an easier detection method for the equivalence point;
- titrations by displacement or by substitution: the titrand is replaced mole by mole with another species that is easier to titrate. For example, to the titrand M solution is added another species NA, which reacts quantitatively with it according to the reaction

$$M + NA \rightarrow MA + N.$$

Since the reaction is quantitative, M is replaced mole by mole with N, which is then titrated directly. As explained just above, these titrations are practiced for both theoretical and practical reasons.

7.6 Types of Chemical Reactions Used in Titrations and Titration Designations

The chemical reactions used in titrations belong to the four great reaction groups that are usually recognized in the field of the chemical analysis in solution, namely,

- neutralization reactions based on protons exchanges,
- oxidation–reduction reactions based on electron exchanges,
- complex formation reactions,
- precipitation reactions.

Other titration designations have also been given in the literature. Some are named after the instrumental method used to detect the equivalence point. We can list, for example, conductometric, potentiometric, amperometric, spectrophotometric, and thermometric titrations. Others are named after the titrant nature. We speak then of iodometric, complexometric, and acidimetric titrations.

7.7 Conditions That the Titration Reaction Must Fulfill

For use in titrimetric analysis, the titration reactions must fulfill several conditions. Some are of a theoretical nature, while others are of a more practical nature. Examining Eqs. (7.1) and (7.2) already permits us to find two of them at first sight:

- for these relations to be satisfied, the titration reaction must be complete. In other words, it must not be equilibrated except if the equilibrium is very weak. Indeed, if the reaction is not sufficiently complete and when the equivalence point is reached (detected, for example, by a color change), too great a volume of titrant solution has been added. The phenomenon is complicated since titration curves get out of shape with the quantitative character (equilibrium constant) of the titration reaction. As a result, the endpoint differs too markedly from the equivalence point. Hence, it can be inferred from this result that the error inherent to a poor choice of an indicator (color change for a bad pH value) and that due to a lack of "quantitativity" of the titration reaction (even if the indicator is well chosen) are two complementary points of the titration error (see Chap. 9). Recall that all the reactions are equilibrated even if they appear complete (see Chap. 2). As a result, only the equilibrium constant knowledge permits us to know if the reaction will be satisfactory for a titration. The problem is determining the weakest value of the equilibrium constant that permits a titration to be satisfactory. The answer to this problem varies with the titration reaction types. Answers will be given for each of them;

- the stoichiometry of the titration reaction must be perfectly known; otherwise, Eqs. (7.1) and (7.2) can no longer be used. For the same reason, no other reaction involving the titrant or the titrand must take place simultaneously with the titration reaction. In this case, the species would be consumed another way.

From the practical standpoint, it may be impossible (in particular, in organic analysis) to summarize all the chemical phenomena taking place during the titration by one sole reaction that would be the titration reaction. As a result, quite evidently, there is no definite stoichiometry. Another case frequently encountered is that in which the quantities that have reacted do change in a complex manner with the titrand's concentration. Also, the involved reactions are often more sensitive to some factors, such as the temperature and pH, than is a unique one. These drawbacks are not necessarily redhibitory, however. Some satisfactory determinations can still be performed with an accuracy and precision sometimes as good as those found with a unique reaction.

In order to overcome these difficulties, an empirical calibration curve (value found by titrimetry: a volume y) as a function of the known concentrations of the solute x is drawn.[2] The same determination for the sample under study as those already done to construct the calibration curve is then achieved. Reporting the found value y' on the curve gives the sought after value x' (Fig. 7.1).

The crux of the success of this methodology is to perform the sample measurement absolutely under the same experimental conditions as those in which the calibration curve was prepared. The best-known example is probably that provided by using cuprimetry to determine reducing sugars according to Bertrand's method (see Part II of this book). The literature mentions correspondence tables: the ratio of the quantity of reducing sugar in the sample versus the added volume of potassium permanganate in well-described experimental conditions.

Despite their interest, using empirical calibration curves is not recommended. A doubt always persists regarding the possible occurrence of a singularity in the

[2] This methodology also applies to several other quantitative methods of analysis.

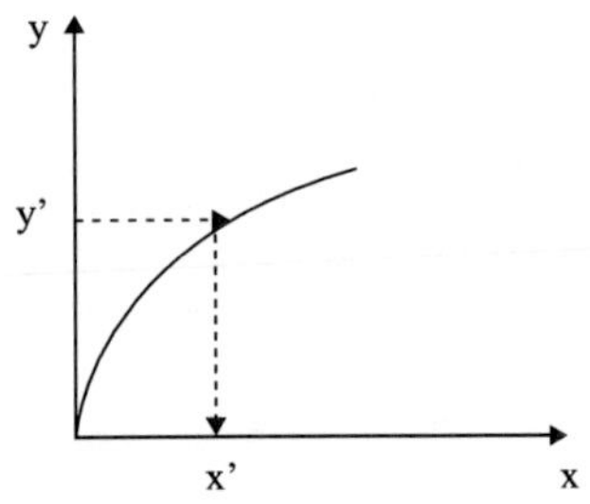

Fig. 7.1 Using a calibration curve when the titration reaction is not stoichiometric

calibration curve near the searched value. As a result, the necessary interpolation would be meaningless:

- the reaction must be practically instantaneous or at least must proceed with very great speed. In fact, some drawbacks may take place when the titration is too long. For example, the titrand's concentration may change by solvent evaporation;
- finally, from the experimental standpoint, an easily detectable change (regardless of which property) must occur at the equivalence point.

7.8 Glassware Used in Titrimetry

The glassware used in titration is essentially the graduated type, that is, graduated flasks, pipettes, and burettes. The glassware is standardized and normalized. It is purchased bearing appropriate certificates.

7.9 Titrations and Microinformatics: Current Trends

A very important point must be emphasized: A result obtained by titrimetry (that described immediately above) depends on the determination of only one titration curve point—the equivalence point. That is, it depends on one piece of experimental information. However, the titration curve is composed by plenty of other points, each of which brings a new piece of information about the unknown concentration. This is the reason why the current trend is to treat all points of the curve in order to benefit from all the available information and, hence, to reduce the uncertainties and inaccuracies resulting from the determination of a sole point. This is done by mathematical processes with the help of informatics programs (easy to write ourselves). The general mathematical treatment of titration curves is given at the end of Chap. 9.

Chapter 8
Neutralization or Acid-Base Indicators

Neutralization indicators, or acid-base indicators or pH indicators, are auxiliary reagents added to the titrand solution in order to detect the equivalence point in acid-base titrations. They can also be used for an accurate quantitative measure of the pH. Tournesol, a natural pigment extracted from some blue-green lichens, was the first pH indicator to be used (1850). Phenolphthalein and methyl orange were introduced somewhat later (1877 and 1878, respectively). Undeniably, the chief interests in the use of acid-base indicators are their low cost and ease of handling. However, they give rise to less precise and less accurate endpoints than some instrumental methods.

8.1 General Considerations on Neutralization Indicators

Neutralization indicators are organic substances that are acids or bases themselves. The color of their acidic form is different from that of the basic form. The ionization constant of the indicator couple is given by the equation

$$K_a = a_{H_3O^+} \times a_{In^-} / a_{HIn},$$

where HIn and In^- symbolize the conjugate forms of the indicator. By assuming activities equal to the concentrations, we find

$$[HIn]/[In^-] = [H_3O^+]/K_a,$$
$$\log([InH]/[In^-]) = pK_a - pH.$$

These relations show that the acid and base concentrations' ratio depends on the pH through the indicator pK_a value. The solution color depends on the pH value. Some indicators exhibit only one colored form; they are called *unicolor indicators*. For example, this is the case of phenolphthalein. The theory concerning them is the same as that concerning bicolor indicators except for the influence of their concentration on the color-change interval (see Sect. 8.4).

Recall that assuming activities equal to concentrations, as was the case above, is not always legitimate.

J.-L. Burgot, *Ionic Equilibria in Analytical Chemistry*,
DOI 10.1007/978-1-4419-8382-4_8,

8.2 Origin of the Color Change

The appearance of the color or its change is due to structural changes of the indicator after the capture or the loss of a proton. The new structure is endowed with an increase or a decrease in its resonance possibilities with respect to the initial one. This is the origin of the color change.

From the simple qualitative point of view, we have known for more than a century that colored molecules bring characteristic atom groups called *chromophores* (part responsible for bringing the color). They are unsaturated groups, such as

$$-N = N - \text{(azo)} \qquad - CH = N - \text{(imino)},$$
$$-N = O \text{ (nitroso)} \qquad > C = S \text{ (thiocarbonyl)},$$
$$-NO_2 \text{ (nitro)} \qquad > C = O \text{ (carbonyl)},$$
$$-N(O) = N - \text{(azoxy)} \qquad > C = C < \text{(ethenyl)}.$$

Sometimes the sole presence of one of these groups is not sufficient to confer a color to the molecule that brings it. In order to develop a color, it must possess a supplementary group called an *auxochrome* (increasing the color). In most cases, but not necessarily, the coloration change due to the medium's acidity is related to the appearance of a p-quinonic structure a or, more rarely, to that of an o-quinonic structure b:

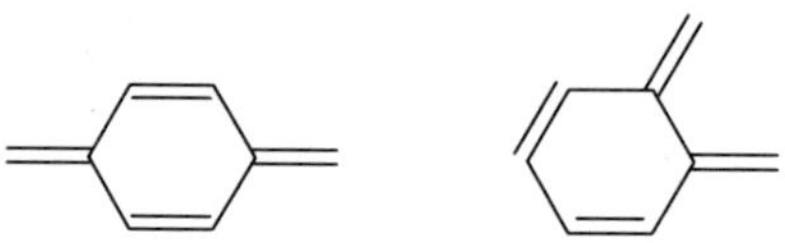

a b

A first example is that of methyl orange (helianthin), whose bicolor indicator is the sodium salt of the 4-dimethylaminoazobenzen-4′-sulfonic acid whose color-change interval is 4.4 < pH < 6.2. The structures that participate in the equilibrium

$$HIn \rightleftharpoons H^+ + In^-$$

are

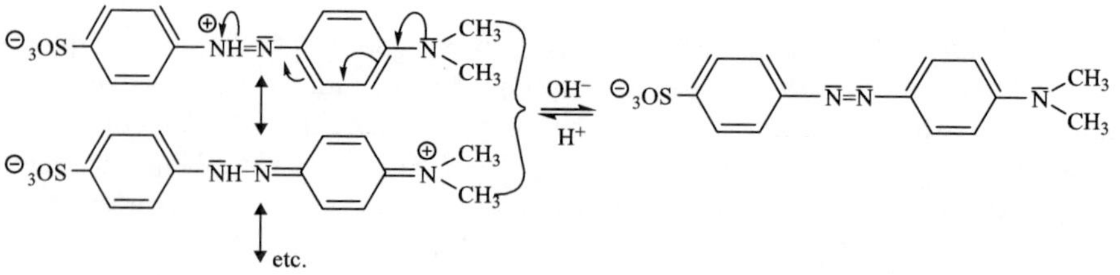

acidic form (red) *basic form (yellow)(azoic).*

The acidic form (red) exhibits two principal limit forms, one with an azo rest and the other one with a p-quinonediimine rest. The basic form (yellow) presents a principal limit form with an azo structure. Another example is provided by phenolphthalein, which is a unicolor indicator. It is a derivative of triphenylmethane. Its acidic form, symbolized by H_2In,

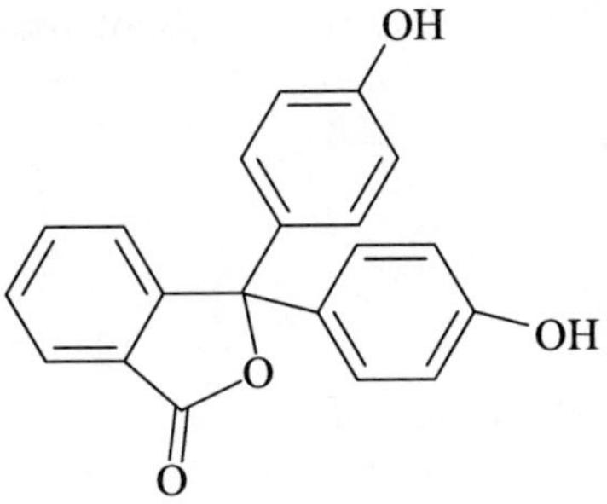

is colorless. It can lose two protons almost equally easily since the ionization constants of both phenol functions are very close to each other. The pink form In^{2-} is then obtained.

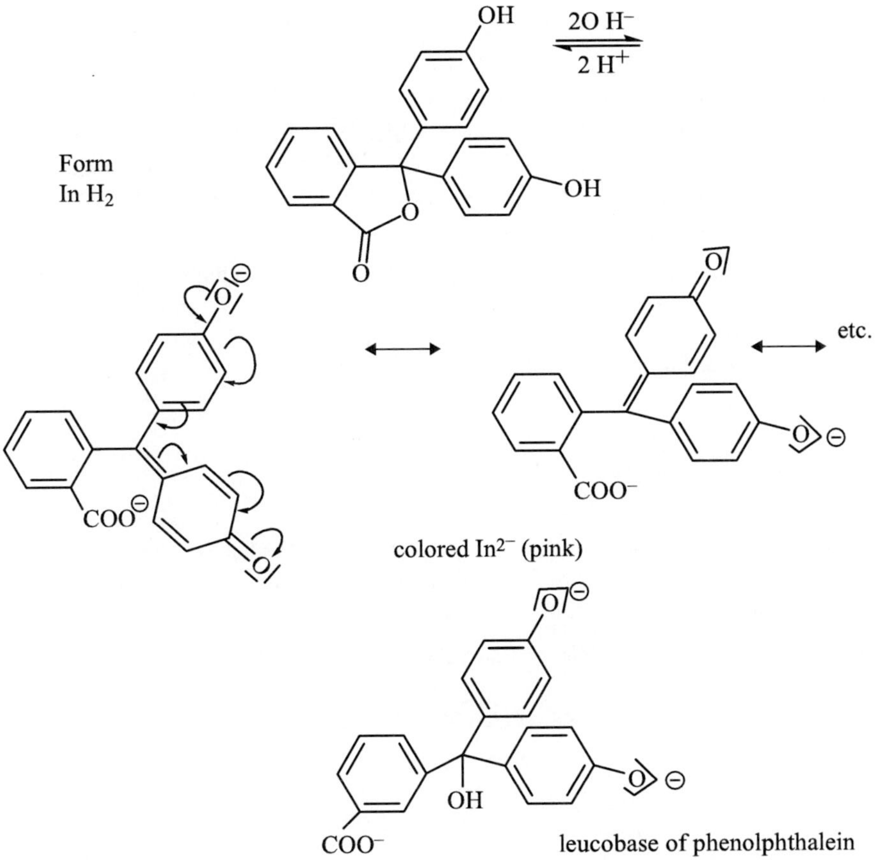

We note that both principal limit forms of In^{2-} are p-quinonic. In sufficiently concentrated alcoholic alkali, the pink form adds a hydroxide ion, yielding another colorless form, InOH.

8.3 Categories of Neutralization Indicators

We can distinguish among

- bicolor indicators (e.g., methyl orange).
- unicolor indicators (e.g., helianthin).
- mixed indicators, which are mixtures of two indicators. They can be
  - a mixture of two pH-sensitive indicators, for example, the mixture of phenolphthalein and 1-naphtolphtalein in a suitable proportion for the titration of phosphoric acid to the diprotic stage (equivalence point $pH = 8.7$).
  - a mixture of one pH-sensitive indicator and another one that is pH-insensitive. An example is provided by Tashiri's indicator used for the titration of ammonia solutions by sulfuric acid. It is a mixture of methyl red (pH-sensitive) and methyl blue (pH-insensitive). The color change is from violet (red + blue) for $pH < 4.2$ to green (yellow + blue) for $pH > 6.2$. At $pH = 5.5$, the color is pink.

  The mixed indicators are used when one wants to obtain a sharp color change over a narrow and selected range of pH.
- universal or multiple-range indicators, which are mixtures of several indicators. They permit color changes over a considerable portion of the pH range. They are suitable only for qualitative goals such as the approximate determination of a solution's pH.

Let's also discuss the indicators by fluorescence, which emit radiation by fluorescence in the visible region, and the indicators by turbidimetry, which induce turbidity in the solution for some pH ranges. The turbidity may be more visible if an adsorption indicator is added to the medium.

8.4 Some Indicators

We mention here just some indicators with their pH color-change intervals and their color changes by passing from the acidic to the basic form.

For more extreme pH ranges, let's mention, for example,

- methyl violet. Its color-change interval is $0.1 < pH < 1.5$. Its color change is from yellow to violet-blue;
- s-trinitrobenzoic acid, whose color-change interval is $12.0 < pH < 13.5$ (colorless to orange).

Let's keep in mind, incidentally, that the use of colored indicators may permit us to quantify the acidity or the basicity of very strong acidic or basic solutions. This is not the case with the simple measurement of the solution's pH. Indeed, it is intuitive that 5 mol/L of a strong acid and another at 2 mol/L are not endowed with the same level of absolute acidity. Paradoxically, however, they exhibit the same pH value: pH = 0. This is a striking example of the fact that the pH value is a bad indication of the acidity of strong acidic or basic solutions. However, in the presence of methyl violet, the above solutions are, respectively, yellow and green. Hence, using a color indicator enables discriminating between two strong acidity levels. The concept of Hammett acidity functions is based on this principle.

8.5 Conditions for Use of Color Indicators

8.5.1 Color-Change Interval

Henderson-Hasselbach's relation applied to the ionization of a pH indicator becomes

$$\mathrm{pH} = pK_a + \log\,[\mathrm{In}^-]/[\mathrm{InH}].$$

The proton addition induces the disappearance of the basic form In^- in favor of the form InH.

It is an experimental fact that the acidic form InH imposes its coloration when the two forms of concentration obey the rule

$$[\mathrm{InH}] \gg 10\left[\mathrm{In}^-\right];$$

the inverse is true when

$$[\mathrm{In}^-] \gg 10[\mathrm{InH}].$$

In logarithmic terms, these inequalities imply that the color-change interval is such that

$$pK_a - 1 < \mathrm{pH} < pK_a + 1.$$

Within this range, the indicator will appear to change from one color to the other. The color-change interval is hence

$$\mathrm{pH} = pK_a \pm 1.$$

Between these limits, the color change will, of course, be gradual since it depends on the ratio of the two colored forms. According to some authors, the color-change interval is narrower (see Table 8.1).

Table 8.1 Some usual indicators

Trivial name	Color-change interval	Color change (acidic to basic form)	
m-Cresol purple	1.2–2.8	Red	Yellow
Methyl orange	3.1–4.4	Red	Orange
Bromocresol green	3.8–5.4	Yellow	Blue
Methyl red	4.2–6.3	Red	Yellow
Bromothymol	6.0–7.6	Yellow	Blue
Neutral red	6.8–8.0	Red	Orange
Cresol red	7.2–8.8	Yellow	Red
Phenolphthalein	8.4–10	Colorless	Red
Thymolphthalein	9.4–10.5	Colorless	Blue

8.5.2 *Influence of the Indicator Concentration on the Color-Change Interval*

8.5.2.1 Bicolor Indicators

The Henderson relation does not take into account the total concentration of the indicator $C = [\mathrm{HIn}] + [\mathrm{In}^-]$. Only the ratio of these two forms plays a part. It is independent of the concentration C, since for every weak acid of this kind,

$$[\mathrm{HIn}] = \left[\mathrm{H_3O^+}\right] C/\left(\left[\mathrm{H_3O^+}\right] + K_a\right) \text{ and } \left[\mathrm{In^-}\right] = K_a C/\left(\left[\mathrm{H_3O^+}\right] + K_a\right).$$

The color-change interval of a bicolor pH indicator is independent of its total concentration. The assertion is correct only when the Henderson relation is legitimate.

8.5.2.2 Unicolor Indicators

Let us suppose that the basic form In^- is the colored one (phenolphthalein) and that its color is perceptible for a concentration at least higher than a mol/L. The Henderson relation gives

$$\mathrm{pH} = pK_a + \log\,[a/(c-a)].$$

The limiting concentration a is constant. Hence, the pH value for which the color is perceptible is a function of the indicator's concentration. In practice, this effect is of little importance because of the weak indicator concentrations used. For phenolphthalein, a is about 10^{-5} mol/L.

8.5.3 *pH Change of the Solution Under Study by Addition of the Indicator*

Since neutralization indicators are themselves acid-base couples, the question that immediately comes in mind is the following one: does their addition significantly change the pH value of the solution under study? Of course, the answer must be no. It is intuitive that the ratio of the solute and the indicator concentrations must play an important part in the answer to this problem.

In the case of the titration of strong acids by strong bases, and vice versa, there is no significant error due to the addition of the indicator, at least when the concentration of the latter is less than 10% that of the titrand. A calculation performed on the same grounds as those that permitted us to study the ionization repression (*recul d'ionisation*;- see Chap. 5) shows that a 10^{-4} mol/L of sodium hydroxide and 10^{-5} mol/L of methyl orange ($K_a = 1.6 \cdot 10^{-4}$) exhibits a pH value pH of 10.00. This is exactly the same value as that exhibited by sodium hydroxide alone at the same concentration.

In the case of the titration of a weak acid or base by a strong antagonistic protolyte, the effect of adding the indicator is less negligible than it was previously. The determining factors are the products K_aC for both the solute and the indicator. The approximate equation that permits the calculation of the pH of a mixture of two weak acids is

$$[H_3O^+] \approx (K_{a1}C_1 + K_{a2}C_2)^{1/2}.$$

It shows that the pH value does not differ significantly from that obtained with the solution of the solute alone when

$$K_{a2}C_2 \ll K_{a1}C_1,$$

where the subscript 2 designates the indicator.

8.5.4 *Ionic Strength of the Solution*

In a medium of high ionic strength, activities and concentrations can no longer be intermingled. Henderson's relation, taking into account ionic strength effects, is

$$\text{pH} = pK_a + \log\left([\text{In}^-]/[\text{HIn}]\right) + \log\left(\gamma_{\text{In}^-}/\gamma_{\text{HIn}}\right),$$

and it involves the indicator activities. However, the solution coloration is given by the ratio of concentrations and not by that of activities. The result is that two solutions exhibiting the same coloration may have a pH value that differs significantly from that of the other solution according to the ionic strengths of the solutions. The pH value obtained through the use of colored indicators depends on the ionic strength of the solution. This phenomenon is called the *salt effect*. For example, with helianthin, the pH value at the equivalence point decreases from 0.31 units when the solution ionic strength is, respectively, 0.1 and 1.0 mol/L. In both cases, the solution ionic strength was fixed by the addition of potassium nitrate and, of course, for the sake of comparison, the electrolyte concentrations were the same. Recall that nitrate and potassium ions are devoid of any acidic or basic property in water.

8.5.5 Nature of the Other Substances Present in Solution

Indicators may participate in other chemical equilibria than those involving proton exchanges. When this is the case, their indicator property may be modified and even suppressed. For example, they can enter into complexation equilibria with proteins in biological media. This is the case with albumin. Based on this property, the reactive papers to detect it are quickly operant.

8.5.6 Temperature

The color-change interval depends on temperature since its pK_a value, which is present in Henderson's relation, depends on it (as do all the equilibrium constants).

8.6 Uses of Neutralization Indicators

The neutralization indicators are used essentially for two finalities, whose practices are somewhat contradictory. They are

- the detection of the equivalence point in titrimetry. Actually, they only detect the endpoint. In this case, only indicators giving sharp and narrow color changes about the equivalence point are used;
- the pH determination of solutions and, by extension, the acidity of extreme solutions through the use of acidity functions. In this case, indicators giving a gradual color change over a wide pH range are preferred. The pH is then determined more accurately and precisely.

Chapter 9
Acid–Base Titration Curves

The theoretical study of titration curves is necessary because it allows us, among other things, to choose the optimal conditions for detecting the equivalence point in order to minimize the titration error.

In this chapter, we will be content with describing the curves, justifying them in an approached manner, and giving some rules that illuminate satisfactory titration conditions before we performs any experiments. We shall give a more detailed study of titration curves in the next chapter. Throughout this chapter, we shall assume activities are equal to concentrations although the quantity pH is defined in terms of activities. Let's also recall that the kinetics of proton exchange reactions are very fast and thus can be considered immediate. As a result, no kinetic consideration will be developed.

9.1 Terminology of Acid–Base Titrations

Acid–base titrations are also called *prototropic titrations* or *acidimetric* and *alkalimetric titrations*. According to the IUPAC, the term "acidimetric titration" is reserved for the titration of a base with a standard acid solution. It is the opposite of the term "alkalimetric titration." In France, the two terms' connotations completely oppose those recommended by the IUPAC.

9.2 General Considerations Concerning Acid–Base Titration Curves: Fraction Titrated

Let's consider the volume V_0 (cm^3) of the solution of an acid (or a base) of concentration C_0 mol/L that we want to titrate with a solution of a base (or of an acid) of concentration C. At each titration stage, the added volume is V(cm^3) (Fig. 9.1).

The acid–base titration curve or neutralization curve is the diagram given by the pH of the solution as a function of the volume V (Fig. 9.2).

J.-L. Burgot, *Ionic Equilibria in Analytical Chemistry*,
DOI 10.1007/978-1-4419-8382-4_9, © Springer Science+Business Media, LLC 2012

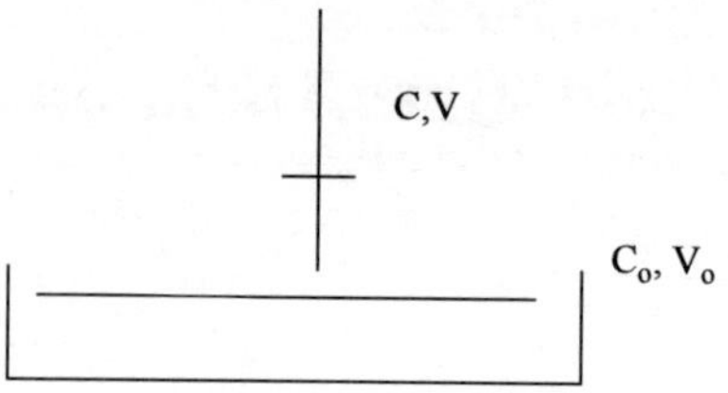

Fig. 9.1 Symbolism used to study titration curves

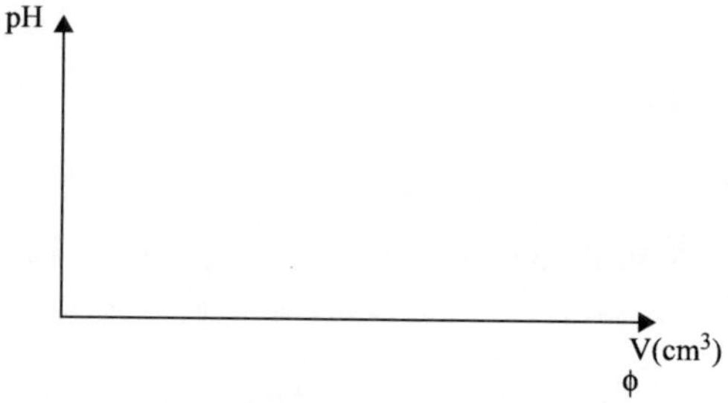

Fig. 9.2 Definition of an acid–base titration curve

Equivalently, neutralization curves are also the diagrams of pH/φ, where φ is the fraction titrated. It is the ratio of the number of moles of added titrant and the initial number of moles of the titrand:

$$\varphi = CV/C_0V_0.$$

By definition, for the equivalence point, we find

$$\varphi_{ep} = 1.$$

Handling the fraction titrated φ rather than the volume V allows us to use dimensionless numbers of low values on the abscissa coordinate. (The titrated fraction is a reduced coordinate.) Hence, the numbers applied to both coordinate axes are necessarily limited.

From the standpoint of one possible classification of the different kinds of titration, acid–base titration curves must be classified as logarithmic curves. Indeed, the measured quantity (the pH value) is directly related to the logarithm of the concentration (activity) of one species (here, H_3O^+).

9.3 Neutralization of a Strong Acid with a Strong Base and Vice Versa

9.3.1 Shape of the Titration Curve

Let's investigate the titration of 1 mol/L hydrochloric acid with 1 mol/L sodium hydroxide and two other titrations with concentrations of 10^{-1} and 10^{-2} mol/L. Table 9.1 lists the pH values for each volume V added; these values are obtained from

Table 9.1 Calculated pH values for each titration stage of a strong acid with a strong base at the same concentrations (298 K; activities and concentrations assumed to be equal)

Added NaOH cm^3	Concentrations of titrand and titrant solutions		
	1 mol/L	10^{-1} mol/L	10^{-2} mol/L
0	0.0	1.0	2.0
50	0.5	1.5	2.5
75	0.8	1.8	2.8
90	1.3	2.3	3.3
98	2.0	3.0	4.0
99	2.3	3.3	4.3
99.5	2.6	3.6	4.6
99.8	3.0	4.0	5.0
99.9	3.3	4.3	5.3
100.0	7.0	7.0	7.0
100.1	10.7	9.7	8.7
100.2	11.0	10.0	9.0
100.5	11.4	10.4	9.4
101.0	11.7	10.7	9.7
102.0	12.0	11.0	10.0
110.0	12.7	11.7	10.7
125.0	13.0	12.0	11.0
150.0	13.3	12.3	11.3
200.0	13.5	12.5	11.5

the rigorous relationships satisfied at every stage of the titration but after assuming activities are equal to concentrations (see Chap. 10). (Actually, the experimental pH values differ a little from those calculated here due to the activity problem, which is deliberately neglected.). The curves obtained are plotted in Fig. 9.3.

Some important points emerge from the study of the curve. They must be emphasized:

- at the equivalence point ($V = 100\,cm^3$, $\varphi = 1$), the pH value is exactly 7.0;
- for the fractions titrated $0 < \varphi < 0.9$ and $1.1 < \varphi < 2.0$, the curves exhibit weak pH changes. Their weak changes are due to the buffering action of the couples of water H_3O^+/H_2O and H_2O/OH^-, which are effective in the zones under consideration;
- in the same zones, the curves do not exhibit an inflection point. This result is emphasized by a comparison with titration curves given by weak acids or bases (see below);
- the curves exhibit an extremely large change in pH occurring in the vicinity of the equivalence point. In the case of 1 M solutions, the pH value changes from 3.3 to 10.7 when φ changes from 0.999 to 1.001. The strong pH change is due to the buffer capacities of the couples H_3O^+/H_2O and H_2O/OH^-, which are null at pH = 7. This explains the fact that it is impossible to stop such a titration for exactly pH = 7;
- The weaker the concentration is, the weaker the change at the equivalence point will be, too (Table 9.1).

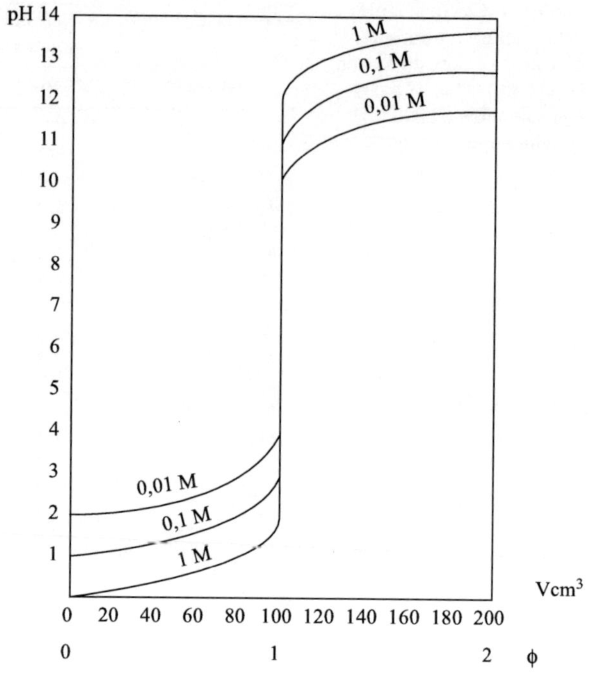

Fig. 9.3 Calculated neutralization curves of 100 cm³ of a hydrochloric acid solution with sodium hydroxide solutions of the same concentrations

In another respect, experiments show that the curves obtained are identical regardless of the strong acids and bases that are facing each other provided they are in the same concentration and volume conditions. This has been proven theoretically (see Chap. 10).

9.3.2 *Justifications*

Since it is the titration of a strong acid with a strong base that is in question, protolytes are completely dissociated. Hence, the titration reaction is merely the neutralization reaction:

$$H_3O^+ + OH^- \rightarrow 2H_2O.$$

In the following reasoning, the reaction is assumed to be complete. From this hypothesis, the following points result:

- by definition, before the equivalence point occurs, the H_3O^+ ions remain in excess in the titration vessel. Their concentration is calculated by taking into account the fact that for CV moles of added hydroxide ions, CV moles of hydroxonium ions disappear (since the reaction is considered to be complete). We find that

$$[H_3O^+] = (C_0V_0 - CV)/(V_0 + V_0). \tag{9.1}$$

Then the solution pH is calculated via

$$pH = -\log[(C_0V_0 - CV)/(V_0 + V)] \quad \text{(before the endpoint).} \tag{9.2}$$

Another expression of pH is

$$pH = -\log[C_0V_0(1 - \varphi)/(V_0 + V)]. \tag{9.3}$$

(V is kept in this expression only in order to lighten it. However, V can also be expressed as a function of φ.) Equations (9.2) and (9.3) can be markedly simplified if the added volume V of the titrant solution is negligible compared to V_0. In this case, they become

$$pH = -\log(C_0 - CV/V_0), \tag{9.2'}$$

$$pH = -\log[C_0(1 - \varphi)]; \tag{9.3'}$$

- at the equivalence point exactly, the solution obtained is that of a salt of a strong acid with a strong base. For example, in the case of the titration of hydrochloric acid by sodium hydroxide:

$$[Cl^-] = [Na^+]$$

since these ions are unconcerned with the acidobasicity (if they were concerned, hydrochloric acid and sodium hydroxide would not be strong protolytes!). As a result,

$$pH = 7 \quad (25°C) \quad \text{(equivalence point);}$$

after the equivalence point, there is an excess of hydroxide ions (the strongest base in water), whose concentration is

$$[OH^-] = (CV - C_0V_0)/(V_0 + V). \tag{9.4}$$

It results in

$$pH = pK_w + \log[(CV - C_0V_0)/(V_0 + V)] \tag{9.5}$$

or

$$pH = pK_w + \log[C_0V_0(\varphi - 1)/(V_0 + V)]. \tag{9.6}$$

These equations can be simplified if the titrant solution is very concentrated (V negligible with respect to V_0). They then become

$$pH = pK_w + \log[(CV/V_0) - C_0], \tag{9.5'}$$

$$pH = pK_w + \log[C_0(\varphi - 1)]. \tag{9.6'}$$

We already know that using Eqs. (9.2), (9.3), (9.2′), (9.3′) and (9.5), (9.6), (9.5′), (9.6′) is not fully legitimate in the domain $0.9 < \varphi < 1.1$. Their use is the least legitimate for φ close to 1. In this domain, the $[H_3O^+]$ and $[OH^-]$ concentrations are not negligible compared to the others in the charge balance. This is a serious problem because the equivalence point is located in this region.

9.3.3 *Practical Conclusion: Choice of the Indicator*

The study of the curves shown in Fig. 9.3 indicates that

- with the 1 mol/L solutions, any indicator whose color-change interval is located between pH 3.0–10.5 can be suitable. The color change is sharp since a great pH change occurs around the equivalence point. The titration error is negligible (see below). Methyl orange (color-change interval: 3.1–4.4), methyl red (4.2–6.1), bromothymol blue (6.0–7.5), phenol red (6.8–8.4), phenolphthalein (8.4–10.0), and thymolphthalein (9.5–10.5) are suitable;
- with the 0.1 mol/L solutions, any indicator whose color-change interval is located between 4.5–9.5 is suitable. Using methyl orange and phenolphthalein may present some difficulties. It induces a titration error of about 0.2%, which is approximately the same as that due to the graduation of glass flasks;
- with the 10^{-2} mol/L solutions, the color-change interval is located between 5.5–8.5. Methyl red, bromothymol blue, and phenol red are still suitable. Using helianthin induces an error of 1–2%.

9.3.4 *Titration Error*

Recall (see Chap. 6) that the titration error is the error resulting from the fact that the endpoint does not coincide with the equivalence point. If the volume added to the equivalence point is V_{ep} and that added to the final (or end) point is V_{fp}, the relative titration error, expressed in volumes, is

$$\text{relative titration error} = (V_{fp} - V_{ep})/V_{ep} \quad \text{(expressed in volumes)}$$

and that expressed in fraction titrated is

$$\text{relative titration error} = (\varphi_{pf} - 1)/1 \quad \text{(expressed in fraction titrated).}$$

The titration error is a systematic error. Several relationships permitting the quantification of titration error have been given in the literature. They are roughly equivalent. One of them is

$$\varphi_{pf} - 1 = [(C_0 + C)/C_0C]\left\{K_w/[H_3O^+{}_{pf}] - [H_3O^+{}_{pf}]\right\}, \tag{9.7}$$

where $\varphi_{fp} - 1$ is the relative error and $[H_3O^+{}_{pf}]$ is the concentration of hydroxonium ions at the endpoint. C_0 and C are the titrand and titrant concentrations, respectively. All the equations that permit us to calculate the titration error do possess the term $K_w/H_3O^+{}_{pf} - H_3O^+{}_{pf}$ in Eq. (9.7). The fact that it systematically takes place in all titration error equations will be justified in the next chapter.

Exercise 1 Determine the titration error for a titration of 10^{-1} M hydrochloric acid with a 10^{-1} M sodium hydroxide solution. We know, after comparing the color of the studied solution with that of another solution whose pH is exactly known by pH-metry, that at the endpoint of the titration under study, pH = 5.0.

The hydroxonium concentration at the endpoint is

$$[H_3O^+{}_{pf}] = 10^{-5} mol/L.$$

Applying Eq. (9.7) gives

$$\varphi_{pf} - 1 = -2 \cdot 10^{-4},$$

that is, the relative titration error = −0.02%.

The error is negligible compared to the precision of the graduated glassware.

9.3.5 *Titration of a Strong Base with a Strong Acid*

With C_0 and V_0 the initial concentration and volume of the solution of the strong base to titrate, and C and V the values of the corresponding quantities related to the titrant, a strong acid, it is easy to prove, via reasoning analogous to that followed in the previous case, that

- before the equivalence point, the solution in the titration vessel is that of a strong base whose concentration is

$$[OH^-] = (C_0V_0 - CV)/(V_0 + V); \qquad (9.8)$$

- at the equivalence point, pH = 7;
- after the equivalence point, the solution is that of a strong acid whose concentration is given by

$$\left[H_3O^+\right] = (CV - C_0V_0)/(V_0 + V). \qquad (9.9)$$

The titration curves obtained can be superimposed onto those obtained with the inverse titrations studied before. In the present case, however, we must read the titration curves from right to left. These considerations can be easily justified. Let's consider the two kinds of titrations that have molar solution. Equations (9.1) and (9.9), on the one hand, give

$$\left[H_3O^+\right] = C_0 - CV/V_0$$ beginning of the titration of a strong acid with a strong base

and

$$\left[H_3O^+\right] = CV/V_0 - C_0$$ end of the titration of a strong base with a strong acid.

Equations (9.4) and (9.8), on the other hand, give

$[OH^-] = CV/V_0 - C_0$ end of the titration of a strong acid with a strong base

and

$[OH^-] = C_0 - CV/V_0$ beginning of the titration of a strong base with a strong acid.

We observe that the curves can be quasi-superimposed provided that in the titration of strong bases, the increasing added volumes are noted from right to left on the abscissa axis. Actually, the curves cannot exactly be superimposed due to the presence of V in the denominator of Eqs. (9.1) and (9.9), which usually is not negligible with respect to V_0 at the end of the titration. These differences are very minor. In any case, the equivalence point is identical in both cases.

The color indicators that can be used are the same as those used for the previous titrations. The titration error can be calculated by

$$\varphi_{fp} - 1 = [(C_0 + C)/CC_0]\left\{\left[H_3O^+{}_{fp}\right] - K_w/\left[H_3O^+{}_{fp}\right]\right\}, \quad (9.10)$$

9.3.6 Concentration Conditions That Must Be Respected to Obtain Satisfactory Titrations of Strong Acids and Bases

The more diluted solutions are, the less satisfactory the titrations will be. With concentrations about 10^{-3} mol/L, titrations are still satisfactory since the random error resulting from the imprecision due to the pH measurements remains weaker than that inherent to the volumes measured with graduated glass flasks (error about 0.1%). Until concentrations of 10^{-4} mol/L, titrations are still reasonably good. For 10^{-5} mol/L solutions, the precision is about 1%, provided care is taken to eliminate carbonic gas.

9.4 Neutralization Titration Curve of a Weak Acid with a Strong Base

9.4.1 Shape of the Curve

Table 9.2 mentions the calculated pH values in the titration of 100 cm^3 of 10^{-1} mol/L solutions of acetic acid ($pK_a = 4.75$) and of a weak acid ($pK_a = 7.00$) with a 10^{-1} mol/L solution of sodium hydroxide. Figure 9.4 shows the curves obtained.

We see that

- each acid has its own curve. Since titration conditions are identical from one acid to the next, their different behavior is necessarily due to their difference in strength (pK_a);

Table 9.2 Calculated pH for the titration of two weak acids, $pK_a = 4.75$ and $pK_a = 7.00$ ($C_0 = 10^{-1}$ mol/L, $V = 100\ \text{cm}^3$) with 10^{-1} mol/L sodium hydroxide solution

Added NaOH cm^3	Acid $pK_a = 4.75$	Acid $pK_a = 7.00$
0	2.9	4.0
10	3.8	6.0
25	4.3	6.5
50	4.7	7.0
90	5.7	8.0
99	6.7	9.0
99.5	7.0	9.3
99.8	7.4	9.7
99.9	7.7	9.8
100.0	8.7	9.9
100.2	10.0	10.0
100.5	10.4	10.4
101.0	10.7	10.7
150	12.3	12.3
200	12.5	12.5

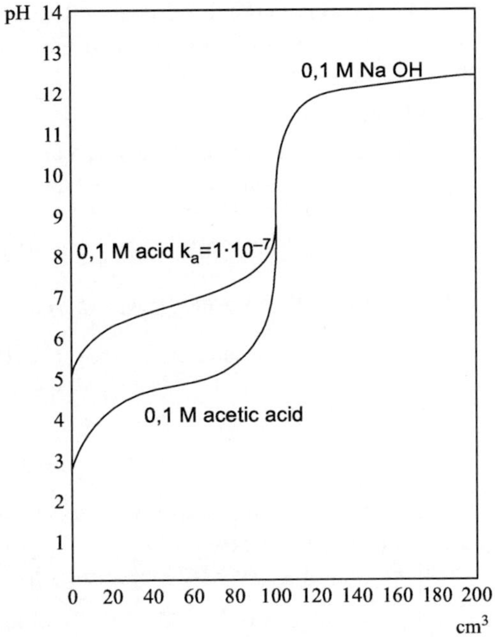

Fig. 9.4 Neutralization curves of two weak acids ($pK_a = 4.75$ and $pK_a = 7.00$ and $C_0 = 10^{-1}$ mol/L, $V_0 = 100$ ml) with 10^{-1} mol/L sodium hydroxide solution

- the pH-change intervals at the equivalence point are not identical. The change is weaker with the weaker acid;
- for both acids, the equivalence point is located in basic medium. Its accurate pH value depends on the acid. The weaker the acid is, the higher the pH value will be at the equivalence point;
- after the equivalence point, the two curves can be superimposed;

- the identical part of both curves, located after the equivalence point, can be superimposed onto that obtained in the titration of a strong acid by a strong base under the same conditions of concentration and volume;
- before the equivalence point, the two curves exhibit an S shape with an inflection point. This one is located at the half-neutralization, for which it appears that $pH = pK_a$.

9.4.2 Justifications

The justifications for the previous points are based on the hypothesis that the titration reaction is the following one and that it is complete:

$$HA + OH^- \rightarrow A^- + H_2O.$$

According to this hypothesis, when CV moles of the strong base are added, CV moles of the weak acid disappear and CV moles of the weak base A^- are formed. Therefore,

- before the equivalence point, the reaction vessel contains the buffer constituted by the mixture of the remaining acid HA and of its conjugate form A^- resulting from the titration reaction. The pH value can be calculated from Henderson's relation:

$$pH = pK_a + \log([A^-]/[HA]),$$

with $[A^-] = CV/(V_0 + V)$ and $[HA] = (C_0V_0 - CV)/(V_0 + V)$,

$$pH = pK_a + \log[CV/(C_0V_0 - CV)], \tag{9.11}$$

or, equivalently,

$$pH = pK_a + \log[\varphi/(1 - \varphi)].$$

Quite evidently, these relationships show that the pH value depends on the pK_a value. Hence, this is also the case for the curve's shape. These relationships also show that for $CV = C_0$ and $V_0/2$ ($\varphi = 0.5$), $pH = pK_a$, as we suspected it at first sight;

- at the equivalence point, all of the acid exists in its conjugate basic form A^-. Hence, we can write

$$[A^-] = C_0V_0/(V_0 + V_{ep}).$$

The pH value can be calculated by the simplified relation

$$pH_{ep} = 1/2\ pK_w + 1/2\ pK_a + 1/2\ \log[C_0V_0/(V_0 + V_{ep})],$$

and when the titrant solution is sufficiently concentrated, it is calculated by

$$pH_{ep} = 1/2\ pK_w + 1/2\ pK_a + 1/2\ \log(C_0).$$

The pH at the equivalence point is indeed located in basic medium. Its value depends on the pK_a value;

- after the equivalence point, two bases remain: OH^- and A^-. The latter, which is a weak base, is of little importance with respect to the former because of the repression of the ionization of the weak base by the strong base OH^-. The farther from the equivalence point we are, the truer it is.

9.4.3 Practical Conclusions: Choice of the Indicator

We must use indicators whose color-change interval is located in basic pH values. Their choice depends on the pK_a value of the acid. For example, for the titration of acetic acid in the above conditions, pH = 8.7 at the equivalence point. Using phenolphthalein, thymolphthalein, and thymol blue is satisfactory. For the second acid (pK_a 7.00), pH = 9.9 at the equivalence point, only thymolphthalein is satisfactory. (Recall that in this second example, the color-change interval is narrower than in the first one—see the curve's shape). For very weak acids ($pK_a > 7$), no simple indicator can be used. Only some mixtures of judiciously chosen indicators can be used. This is due to the fact that the pH change at the equivalence point is very weak. The very deep reason behind this fact is that the neutralization reaction can no longer be considered quantitative (see Chap. 10).

9.4.4 Titration Error

One relation giving the titration error is

$$\varphi_{\text{fp}} - 1 = [(C_0 + C)/CC_0]\left\{K_{\text{w}}/\left[\text{H}_3\text{O}^+{}_{\text{fp}}\right] - \left[\text{H}_3\text{O}^+{}_{\text{fp}}\right]\right\} - \left[\text{H}_3\text{O}^+{}_{\text{fp}}\right]/K_a. \tag{9.12}$$

It differs from that applying to the titration of strong acids by the term $-[H_3O^+{}_{pf}]/K_a$. This term is often preponderant.

Exercise 2 Calculate the titration error obtained for the titration of 50 cm^3 of a 10^{-1} mol/L acetic acid solution ($pK_a = 4.75$). The titrant is a solution of 10^{-1} mol/L sodium hydroxide. The endpoint is detected at pH = 8.00, whereas the equivalence point, theoretically, is located at pH = 8.72.

Applying Eq. (9.12) gives

$$\varphi_{\text{pf}} - 1 = 2.0 \cdot 10^{-5} - 5.7 \cdot 10^{-4},$$

$$\varphi_{\text{pf}} - 1 = -5.5 \cdot 10^{-4}.$$

The titration error is approximately $5 \cdot 10^{-2}$%, which is lower than that due to the graduation of the glassware. Note that, indeed, the term $[H_3O^+{}_{fp}]/K_a$ is preponderant.

9.4.5 Conditions That Must be Fulfilled for Satisfactory Titrations

The titration of a weak acid with a strong base will be satisfactory only in some concentration conditions, depending on the strength of the acid. The parameter that is conditional for the success of the titration is β, which, in the present case (see Chap. 10), is defined by

$$\beta = K_w/(K_a C_0).$$

We note that both parameters K_a and C_0 are linked in the expression of β. We accept that the titration is satisfactory when $\beta < 10^{-5.6}$. This value has been derived experimentally. It means that for higher values, the equivalence point can no longer be satisfactorily detected (see the following chapter). The parameter rationalizes some experimental facts. For example, it indicates that the titrations with a strong base of orthoboric acid ($pK_a = 9.10$) and the ammonium ion ($pK_a = 9.24$) (both 10^{-1} mol/L) cannot be satisfactory. At first sight, we can consider that the greater the difference is between the pK_a value and the acidity scale limit (14), the more satisfactory the titration of a weak acid by a strong base will be. In other words, the couple HA/A^- must be located as far as possible from the couple H_2O/OH^- on the acidity scale.

9.5 Titration of a Weak Base with a Strong Acid

A typical example is provided by the titration of aqueous ammonia by a hydrochloric solution. The curve obtained is given in Fig. 9.5.

The results are quite analogous to those obtained in the titration of a weak acid by a strong base. Starting from the hypothesis that the neutralization reaction

$$B + H_3O^+ \rightarrow BH^+ + H_2O$$

is complete, it is easy to justify that

- before the equivalence point, the buffer BH^+/B is present. This explains the S shape of the curve and the long flat portion about the half-neutralization;
- at exactly the equivalence point, the solution in the titration vessel is that of the weak acid BH^+. Therefore, the pH value can be calculated by the relation (see Chap. 5)

$$pH_{eq} = 1/2\ pK_a - 1/2 \log\left[C_0 V_0/\left(V_0 + V_{eq}\right)\right].$$

Its value depends on the strength of the base through the pK_a value of its conjugate acid;

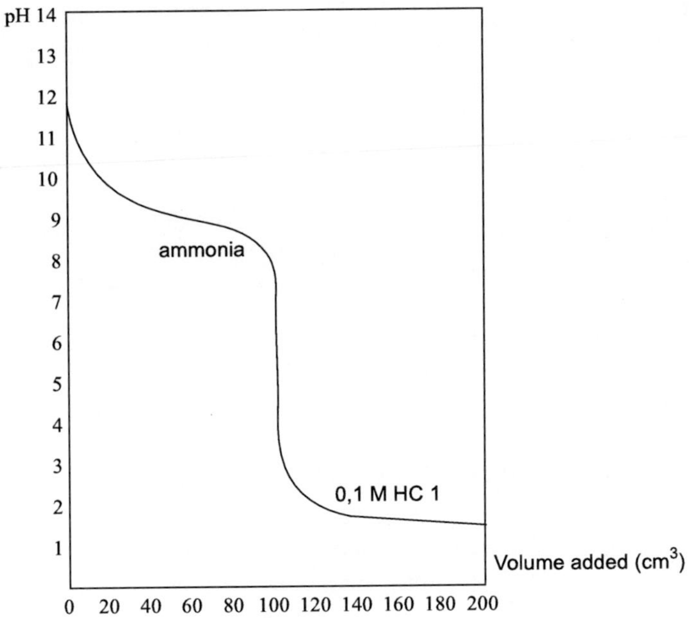

Fig. 9.5 Neutralization curve of aqueous ammonia [C_0 = 0.1 mol/L,—$pK_a(NH_4^+) = 9.24$] $V_0 = 100\,cm^3$] with a hydrochloric acid solution: ($C = 0.1$ mol/L)

- beyond the equivalence point, the solution contains an excess of hydroxonium ions and the weak acid BH^+. Hydroxonium ions predominate due to the ionization repression of the weak acid. The curve is not different from that of a strong base titrated with a strong acid.

From a practical standpoint, the neutralization indicators to be used must have their color-change interval located in the acid range. For example, for the above-mentioned titration of ammonia, methyl red and helianthin are suitable. This is also the case of bromophenol blue and bromocresol green.

The following relation:

$$\varphi_{fp} - 1 = [(C_0 + C)/CC_0]\left\{H_3O^+{}_{pf}\right\} - K_w/\left[H_3O^+{}_{pf}\right] - K_w/\left(K_b\left[H_3O^+{}_{pf}\right]\right)$$

permits us to calculate the titration error.

The success of such a titration depends on the value of the parameter $\beta = K_b C_0$. Actually, the inverse of this parameter must be lower than $10^{-5.6}$. This explains the inability to titrate acetate ions by a strong acid. At first sight, the farther from the couple H_3O^+/H_2O the base under study is located, the more satisfactory the titration of a weak base with a strong acid will be. The couple CH_3COOH/CH_3COO^- is too close to the couple H_3O^+/H_2O.

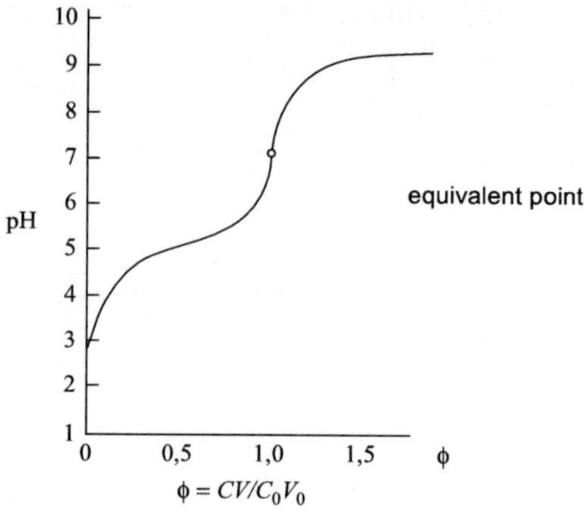

Fig. 9.6 Titration curve of 10^{-1} mol/L acetic acid with 10^{-1} mol/L ammonia

9.6 Titration of a Weak Acid with a Weak Base

From what has been said about weak protolytes, we can infer that the titration reaction is rather equilibrated. It can be symbolized by the equation

$$\text{acid}_1 + \text{base}_2 \rightleftharpoons \text{base}_1 + \text{acid}_2.$$

These titrations are seldom practiced because of the pH change near the equivalence point, which is very gradual. There is no sudden pH change (see Fig. 9.6). Because of this, the titration error may be high with neutralization indicators. They do not exhibit a sharp endpoint.

The gradual change near the equivalence point must be ascribed to the fact that the neutralization reaction is too equilibrated. This is the case when the couples HA_1/A_1^- and HA_2/A_2^- are too close to each other on the acidity scale. Reciprocally, when they are sufficiently far from each other, we can expect to achieve an accurate titration. The example of the titration of acetic acid ($pK_a = 4.75$) by ammonia [$pK_a(NH_4^+) = 9.21$] is interesting to consider since it is on the border of satisfactory and nonsatisfactory titrations. Indeed, with 10^{-1}mol/L—solutions and with a pH-metric detection of the equivalence point, the titration error is about 0.5%. It is higher than that due to the graduation of the glassware (0.2%). However, it remains acceptable. The lack of precision is higher, by far, when neutralization indicators are used.

In the present case, it is the ratio $\beta = K_{a1}/K_{a2}$ that determines the titration's success. It must be higher than 10^5.

9.7 Titration of a Mixture of Strong Acids with a Strong Base and Inversely

We have already seen that a mixture of strong acids exhibits the behavior of a solution of a sole strong acid whose concentration is the sum of those of the different acids constituting the mixture. As a result, the titration curve is strictly identical to that

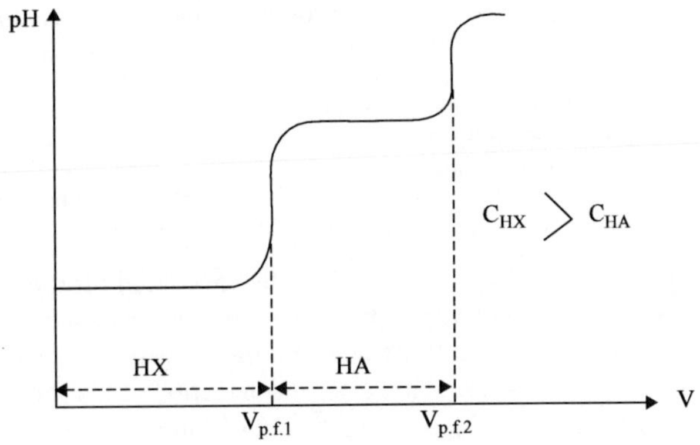

Fig. 9.7 Titration of a mixture of a strong acid and a weak acid with a strong base ($5 < pK_a$ of the weak acid < 10; $\beta < 10^{-5.6}$)

obtained with only one strong acid. The added volume at the equivalence point permits us to determine the total concentration of all the strong acids. So it is with the titration of a strong base's mixtures by a strong acid.

Hence, it is impossible to titrate such mixtures sequentially. In this case the adverb "sequentially" means that it is possible to visualize the successive sharp pH changes corresponding to the neutralization of each member of the mixture. These successive changes would enable us, of course, to titrate each component.

9.8 Titration of a Mixture of a Strong Acid and a Weak Acid with a Strong Base and Inversely

The question that arises is to know if it is possible to titrate a mixture of a strong acid and a weak acid with a strong base sequentially. The answer is yes. It is possible to perform such a titration successfully, but only under some conditions. The first condition to be fulfilled concerns the location of the couple of the weak acid HA on the acidity scale. It must not be too close to the couple H_3O^+/H_2O. The $pK_a(HA/A^-)$ value must be at least higher than about 5. Additionally, this limit may vary a little with the concentration. From another standpoint, the remarks concerning the titration of a weak acid by a strong base that were previously made remain correct and true. The $pK_a(HA/A^-)$ value must befar lower than 14. This is the second condition of the sequential titration. Both conditions oblige the weak acid to possess a pK_a value located near the middle of the acidity scale. If these conditions (both concerning the weak acid) are fulfilled, the sequential titration is possible. The titration curve is represented in Fig. 9.7. The first sharp pH change means the end of the neutralization of the strong acid, the second one the end of the neutralization of the weak acid.

The volume permitting the appreciation of the titer of the strong acid solution is V_{fp1}. The difference $V_{fp2} - V_{fp1}$ allows us to appreciate the titer of the weak acid solution. A point must be emphasized: Figure 9.7 represents the case in which $C_{oHX} > C_{oHA}$. The other case is possible, of course. Then the difference $V_{fp2} - V_{fp1}$

is greater than V_{fp1}. However, in any case, it is always the strong acid that is titrated first.

Using the word "first" is probably not judicious since it implicates a kinetic phenomenon. This is absolutely not the case in the occurrence. Only thermodynamics play a part here. According to calculations based on the relations necessarily satisfied in solutions, we can prove that the hydroxonium ions are indeed neutralized first and that after additional strong base has been added, it is the weak acid's turn to be neutralized.

It must be noted that even when the sequential titration is possible, the detection of the first equivalence point is generally not very precise. The first equivalence point corresponds to the strong acid. Further, it is less imprecise when $C_{\text{HX}} > C_{\text{HA}}$ than if it is the inverse case. However, the detection of the second equivalence point may be very precise provided $\beta < 10^{-5.6}$. For example, in the titration of a mixture of acetic acid and hydrochloric acid, the precision of determining the former remains higher than 1% regardless of the concentrations and the means of detection.

When the weak acid is too strong ($pK_a < 5$), the sequential titration is no longer possible. The titration curve does not exhibit the expected first sharp endpoint. Only one sharp endpoint is seen. It corresponds to the neutralization of both acids. (This was the second endpoint before.) Hence, it is possible to determine only the sum of the concentrations of both acids in this case.

Briefly, the success of the sequential titration of a mixture of a strong acid and a weak acid depends on the pK_a value of the weak acid and also on the different concentrations. All these considerations can be transposed to a mixture of a weak base and a strong base when they are titrated with a strong acid.

9.9 Titration of a Mixture of Weak Acids with a Strong Base

Now, let's discuss the possibility of the sequential titration of two weak acids HA_1 and HA_2 in a mixture. Let C_1 and C_2 be their analytical concentrations and K_{a1} and K_{a2} their ionization constants. Since one of these constants must necessarily be greater than the other, let's suppose that $K_{a1} > K_{a2}$.

It has been proven experimentally that the pK_a difference $\Delta pK_a = pK_{a2} - pK_{a1}$ must be greater than 5 for the sequential titration of two weak acids at about the same concentration to be successful. As described previously, the stronger acid is titrated first. Otherwise, as before, the parameter β_{HA2} must be smaller than $10^{-5.6}$ in order for the weaker acid HA_2 to be accurately titrated. This condition is not new. It was already encountered for the titration of a mixture of a strong acid and a weak acid.

The rule $\Delta pK_a > 5$ is necessary but not sufficient. In addition, the concentrations must be approximately equal, or else the stronger acid must be more concentrated. However, even if all these conditions are fulfilled, the titration of the first acid is poorly satisfactory. Its titration index (see Chap. 10) is weak.

If $\Delta pK_a < 5$, the titration curve no longer exhibits the sharp pH change corresponding to the end of the first neutralization, a fortiori, if $C_2 > C_1$. Only the sum of both concentrations is obtained. Figure 9.8 illustrates this matter.

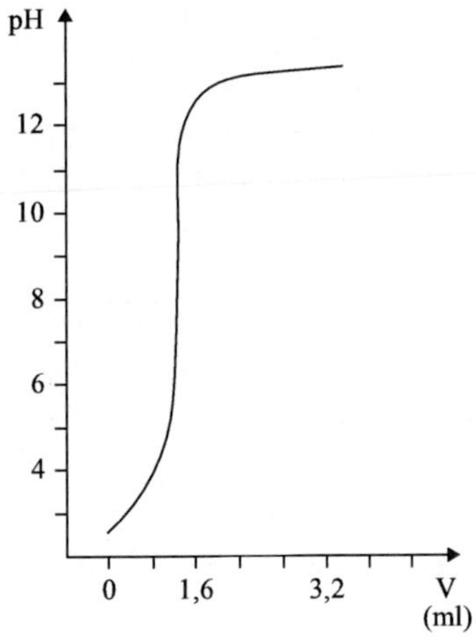

Fig. 9.8 Titration of a mixture of acetic acid ($1.10 \cdot 10^{-2}$ mol/L^{-}, $pK_a = 4.75$) and bromoacetic acid ($1.10 \cdot 10^{-2}$ mol/L^{-}, $pK_a = 2.90$) with a sodium hydroxide solution; only the global equivalence point is visible

It is difficult to study the titrations of mixtures easily any further because, from a mathematical standpoint, a sequential titration depends on several parameters. For a binary mixture, for example, the titration curve depends on the four parameters C_1, K_{a1}, C_2, and K_{a2}.

All these considerations can be transposed immediately, by analogy, to the titrations of a mixture of weak bases with a strong acid.

9.10 Titration of a Polyacid with a Strong Base

At first insight, the case of a polyacid can be assimilated to that of a mixture of several acids whose concentrations would be identical. With this reasoning in mind, the previous experimental rules can be applied to this case. Two striking examples are provided by the titrations of phosphoric acid and citric acid.

Phosphoric acid exhibits the values $pK_{a1} = 2.23$, $pK_{a2} = 7.21$, and $pK_{a3} = 12.32$. Its titration curve with potassium hydroxide is given in Fig. 9.9.

The differences between the pK_a values permit us to detect clearly and successively the ends of

- the neutralization of H_3PO_4,
- the neutralization of $H_2PO_4^-$,

according to the following reactions, which occur successively:

$$H_3PO_4 + OH^- \rightarrow H_2PO_4^- + H_2O,$$

$$H_2PO_4^- + OH^- \rightarrow HPO_4^{2-} + H_2O.$$

The fact that we do not see the neutralization of the third acidity (HPO_4^{2-}) is not due to too weak a pK_a difference ($\Delta pK_a = pK_{a3} - pK_{a2} = 5.10$). The origin of this

Fig. 9.9 Titration of 50 ml of 0.10 mol/L phosphoric acid with 0.10 mol/L potassium hydroxide

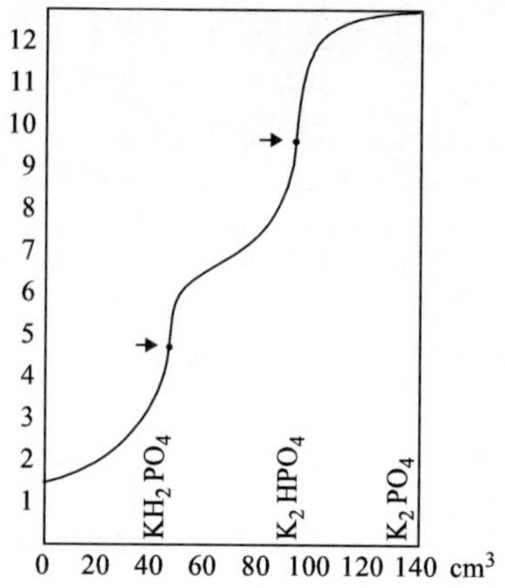

phenomenon lies in couple HPO_4^{2-}/PO_4^{3-} being too close to the couple H_2O/OH^- on the acidity scale for the titration reaction

$$HPO_4^{2-} + OH^- \rightleftharpoons PO_4^{3-} + H_2O$$

to be sufficiently quantitative. The value of the parameter

$$\beta = K_w/(K_{a3}C_0)$$

is too high.

Hence, the first pK_a difference $(pK_{a2} - pK_{a1})$ is high enough to predict the pH at any stage of the titration by using approximate reasoning and relations, at least for the first two thirds of the curve.

Therefore,

- concerning the first neutralization:

  at the first point ($V = 0\,\text{cm}^3$), we can consider that the solution present in the reaction vessel is that of a monoacid, with $pK_a = pK_{a1} = 2.23$. The reasoning behind this assertion is that the first pK_a difference is high enough for the second acidity not to be yet dissociated. Therefore, we can use the relation

$$\text{pH} \approx (1/2)\,2.23 - 1/2\log\left(10^{-1}\right) \quad \left(\text{assuming } C_0 = 10^{-1}\text{mol/L}\right),$$

$$\text{pH} = 1.60.$$

Likewise, before the first equivalence point, the exclusive neutralization reaction is

$$H_3PO_4 + OH^- \rightarrow H_2PO_4^- + H_2O.$$

Thus, the buffer $H_3PO_4/H_2PO_4^-$ appears. During the course of the titration and before the first equivalence point, the solution in the titration vessel is that of this buffer with successively different ratios of concentrations of its acid and conjugate base. Henderson's relation can be applied. In particular, we can posit that

$$pK_a = \text{pH} \quad \text{at the first half-neutralization.}$$

At the first equivalence point, keeping in mind the hypothesis that the second acidity is not yet dissociated, we can consider that the only present species is $H_2PO_4^-$, which is an ampholyte. It can be considered as having resulted from the following reaction:

$$H_3PO_4 + HPO_4^{2-} \rightleftharpoons 2H_2PO_4^-.$$

Hence, the pH at the first equivalence point is that of a salt of a weak acid with a weak base. It is approximately given by the relation

$$\text{pH} = 1/2(pK_{a2} + pK_{a1}),$$

with $pK_{a1}(H_3PO_4/H_2PO_4^-) = 2.2$,

$$pK_{a2}(H_2PO_4^-/HPO_4^{2-}) = 7.2,$$
$$\text{pH} = 4.7.$$

The study of the distribution diagram (see Fig. 4.4 in —Chap. 4) confirms that the species HPO_4^{2-} does not exist at the obtained value of pH = 4.7 ($\alpha H_2PO_4^- \approx 1$). The calculated and experimental results are self-consistent.

- concerning now the second neutralization:

for the same reason as the previous one (judicious pK_a difference), we can consider that the only evolving neutralization reaction is

$$H_2PO_4^- + OH^- \rightarrow HPO_4^{2-} + H_2O.$$

Moreover, it is complete due to the locations of the couples $H_2PO_4^-/HPO_4^{2-}$ and H_2O/OH^- on the acidity scale. Henderson—Hasselbach's equation can be used once again. In particular, we find

$$pK_{a2} = \text{pH} \quad \text{at the second half-neutralization.}$$

At the second equivalence point, the only species existing in the solution is the amphiprotic one: HPO_4^{2-}. It can be considered as having resulted from the reaction

$$H_2PO_4^- + PO_4^{3-} \rightleftharpoons 2HPO_4^{2-},$$

from which we have

$$\text{pH} = 1/2(pK_{a2} + pK_{a3}),$$
$$\text{pH} = 9.8.$$

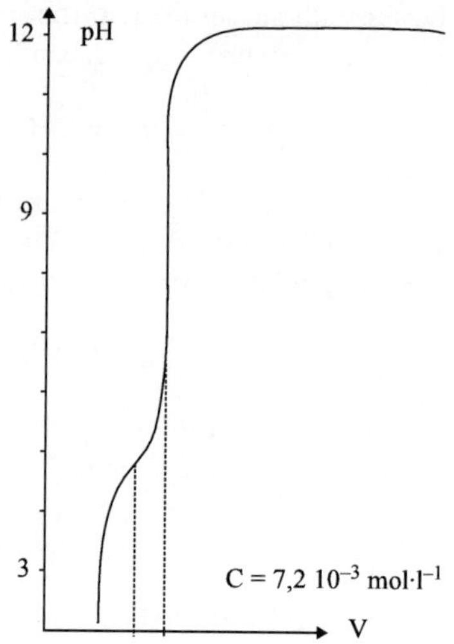

Fig. 9.10 Titration of $7.20 \cdot 10^{-3}$ mol/L citric acid solution with 1.90 mol/L sodium hydroxide

The distribution diagram mentioned above confirms that at pH = 9.8, only the species $HPO_4{}^{2-}$ exists (α $HPO_4{}^{2-} \approx 1$).

These calculations, although approximate, give fairly accurate pH values.

From a practical standpoint, we can infer from these considerations that there are two ways to titrate phosphoric acid: We base it on either the first or second sharp endpoint. In the first case, only the first acidity is neutralized and the color-change interval is located near pH = 4.8; congo red is suitable. In the second case, the first two acidities are neutralized and the color-change interval is located near pH = 9.8; thymolphthalein and phenolphthalein are suitable. The volumes at the equivalence point, which must be taken into account to calculate the titer, vary, of course, from simple to double according to the chosen strategy.

If the acidities of the polyacid are too close to each other, it is simply impossible to see them separately. With a triacid in this case, only the sum of the three acidities can be titrated. This is the case with citric acid ($pK_{a1} = 2.94$, $pK_{a2} = 4.14$, $pK_{a3} = 5.82$) (see Fig. 9.10).

It is no longer possible to calculate the titration curve at any stage by making simple approximations, as was done with phosphoric acid. The distribution diagram given in Fig. 9.11 shows that at the first two equivalence points (the two maxima of the diagram), the hydrogen and dihydrogen citrate ions AH^{2-} and $AH_2{}^{-}$ do not exist alone. This signifies that the two neutralization reactions

$$\text{citrate-H}_3 + OH^- \rightarrow \text{citrate-H}_2{}^- + H_2O,$$

$$\text{citrate-H}_2{}^- + OH^- \rightarrow \text{citrate-H}^{2-} + H_2O$$

are evolving more or less simultaneously in the neighborhood of the first equivalence point. It is the same state of affairs for the second equivalence point, the stage at which

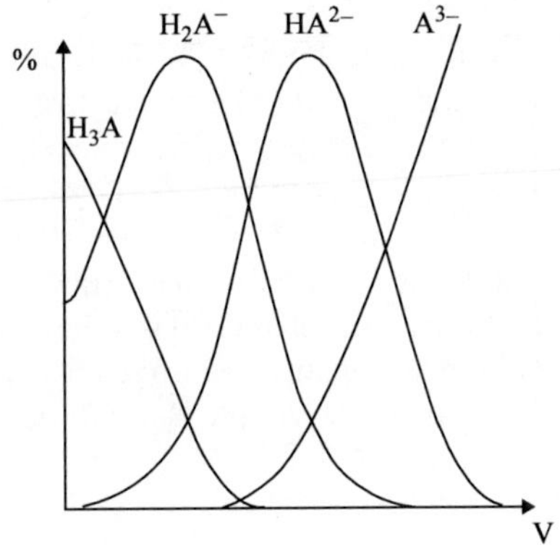

Fig. 9.11 Distribution diagram of citric acid ($C = 7.20 \cdot 10^{-3}$ mol/L)

the following two reactions occur more or less simultaneously:

$$\text{citrate-H}_2{}^- + OH^- \rightarrow \text{citrate-H}^{2-} + H_2O,$$

$$\text{citrate-H}^{2-} + OH^- \rightarrow \text{citrate}^{3-} + H_2O.$$

The exact calculation of the citric acid neutralization curve involves handling unwieldy equations in $[H_3O^+]$.

9.11 Titration of the Monosalt of a Diacid

Let's consider the monosalt HANa of the diacid H_2A. The species HA^- is amphiprotic according to the following reactions:

$$HA^- + H_3O^+ \rightleftharpoons H_2A + H_2O,$$

$$HA^- + H_2O \rightleftharpoons A^{2-} + H_3O^+.$$

As a result, two possibilities are conceivable for its titration. The first is to titrate it as an acid, that is, with sodium or potassium hydroxide according to

$$HA^- + OH^- \rightarrow A^{2-} + H_2O.$$

The titration is conditioned by the pK_{a2} value and by the analytical concentration of the acid. If the couple HA^-/A^{2-} is too close to the couple H_2O/OH^- on the acidity scale, the predicted titration reaction is too equilibrated to be interesting. In other words, the corresponding parameter β is not endowed with an adequate value. The second possibility is to titrate it as a base with a strong acid, according to the reaction

$$HA^- + H_3O^+ \rightarrow H_2A + H_2O.$$

The reaction is conditioned by the pK_{a1} value (couple H_2A/HA^-) and by the concentration of the base. If the species is located too close to the couple H_3O^+/H_2O, the titration cannot be satisfactory.

These considerations can also be developed for the case of salts of polyacids. We have already seen that the dihydrogen phosphate ion $H_2PO_4^-$ can be titrated satisfactorily with hydroxide ions, whereas the hydrogen phosphate ion HPO_4^{2-} cannot, due to its location on the acidity scale. Inversely, the orthophosphate PO_4^{3-} and hydrogen phosphate HPO_4^{2-} ions can be titrated without any problem with a strong acid, whereas the dihydrogen phosphate ion cannot. The couple $H_3PO_4/H_2PO_4^-$ is too close to the couple H_3O^+/H_2O on the acidity scale.

Chapter 10
Acid–Base Titrations: Further Theoretical Studies

In this chapter, we complete our preceding study of titration curves. In particular, we set up the exact equations of the titration curves as well as the relations giving the titration errors, and we introduce the useful concept of the sharpness index.

10.1 Exact Equation of the Titration Curve of a Strong Acid with a Strong Base and Conversely; Formula Giving the Titration Error

The relationships that are necessarily satisfied at any stage of the titration of hydrochloric acid with sodium hydroxide are,

$$[Na^+] = [CV]/[V_0 + V]$$

$$[Cl^-] = [C_0V_0]/[V_0 + V],$$

$$[Na^+] + [H_3O^+] = [OH^-] + [Cl^-].$$

The first two equations express the mass balances, and the third the charge balance. From them, we immediately deduce the exact equation of the titration curve:

$$[CV - C_0V_0]/[V + V_0] = K_w/[H_3O^+] - [H_3O^+]. \qquad (10.1)$$

In the preceding chapter, we used the approximate equation:

$$[H_3O^+] = [C_0V_0 - CV]/[V_0 + V]$$

in order to calculate the pH before the equivalence point. The approximate equation differs from the exact one (10.1) in the term $K_w/[H_3O^+] = [OH^-]$. It neglects the hydroxide ion concentration. This is legitimate until the fraction titrated $\varphi = 0.9$, but not for the range $0.9 < \varphi < 1.0$. This assertion is proved by rigorous calculations

J.-L. Burgot, *Ionic Equilibria in Analytical Chemistry*,
DOI 10.1007/978-1-4419-8382-4_10,

performed, in fact, with the help of the above exact equation. After the titrated fraction value reaches about 0.9, the pH increases sharply, as shown both by calculations and with experiments. The concentration $[OH^-]$ can, of course, no longer be neglected with respect to the hydroxonium ion concentration $[H_3O^+]$. This is the reason why the above-mentioned approximation is called the *acid approximation*. It is interesting to recall that in the range $0.9 < \varphi < 1.0$, the relation $pH = -\log C$ (which was used to describe the titration curve in the last chapter) cannot be used since the hydroxide ions cannot be neglected with respect to hydroxonium ions. Likewise, the relation

$$[OH^-] = [CV - C_0V_0] \,/\, [V_0 + V]$$

used in the preceding chapter for $\varphi > 1.0$ differs from the exact one by the absence of the term $[H_3O^+]$, whence the name *basic approximation*. It cannot be used in the range $1.0 < \varphi < 1.1$ since the hydroxonium ion concentration is not, in fact, negligible.

Writing Eq. (10.1) in terms of fractions titrated gives Eq. (10.2):

$$[C_0V_0(\varphi - 1)]/[V_0 + V] = K_w/[H_3O^+] - [H_3O^+]. \qquad (10.2)$$

This relation is interesting. Indeed, the term $(\varphi - 1)$, which intervenes in the definition of the titration error, is already present in it. However, the following distinction must be noticed: Equation (10.2) is the equation of the titration curve $[H_3O^+]/\varphi$, whereas the titration error corresponds to only one point of this curve. This is the point at which the color of the neutralization indicator changes sharply. More generally, it is the one at which a change in some physical or chemical property of the solution under the influence of the abrupt pH change is detected.

Since the endpoint is one point of the titration curve, its coordinates satisfy Eq. (10.2). At the endpoint, $\varphi = \varphi_{pf}$ and $[H_3O^+] = [H_3O^+{}_{pf}]$. Equation (10.2), applying to this point, becomes

$$[C_0V_0(\varphi_{pf} - 1)]/[V_0 + V_{pf}] = K_w/[H_3O^+{}_{pf}] - [H_3O^+{}_{pf}].$$

Hence, once we know $[H_3O^+{}_{pf}]$, its value can immediately be applied to calculate the titration error. The fact that the difference $V_{pf} - V_{pe}$ is weak simplifies the calculation of the titration error. It permits us to replace V_{pf} with V_{pe} in the last equation. We then find

$$(\varphi_{pf} - 1) = \{C_0C/[C_0 + C]\} \cdot \{K_w/[H_3O^+{}_{pf}] - [H_3O^+{}_{pf}]\}.$$

This expression was given in the previous chapter.

10.2 Exact Equation of the Titration Curve of a Weak Acid with a Strong Base and Conversely: Titration Error

The necessarily satisfied equations and relations governing this case are

$$[Na^+] = CV/[V_0 + V],$$
$$[HA] + [A^-] = C_0V_0/[V_0 + V],$$
$$\{[A^-][H_3O^+]\}/[HA] = K_a,$$
$$[H_3O^+] + [Na^+] = [A^-] + [OH^-].$$

This set of simultaneous equations in many unknowns can be reduced by substitution and elimination to the following rigorous equation:

$$CV/[V_0 + V] = K_a/\{K_a + [H_3O^+]\} \cdot (C_0V_0/[V_0 + V]) + K_w/[H_3O^+] - [H_3O^+]. \tag{10.3}$$

It is satisfied at any stage of the titration. This knowledge permits us to know how the following equation, which was used in the preceding chapter to describe the first part of the titration curve, was approximate:

$$pH = pK_a + \log[(CV/(C_0V_0 - CV)] .$$

Recall that it resulted from Henderson's equation, where we set

$$[A^-] = CV/(V_0 + V),$$

$$[HA] = (C_0V_0 - CV)/(V_0 + V)$$

starting from the hypothesis that the neutralization reaction

$$HA + OH^- \rightarrow A^- + H_2O$$

was complete.

If we recall that the mass balance in sodium ions is exactly

$$[A^-] = CV/(V_0 + V)$$

and we recall that in the approximate approach we set

$$[A^-] = [Na^+],$$

after a glance at the charge balance, we can deduce that the concentrations $[H_3O^+]$ and $[OH^-]$ become negligible according to this approximation. Depending on the titration stage under study, the strength and the analytical concentration of the acid, this approximation is not always legitimate. At first sight, we can say that the approximation is legitimate in the buffering zone of the titration curve and for solutions that are not too diluted.

Equation (10.3), set up above, permits us to calculate the titration error. After introducing the fraction titrated and by writing the term ($\varphi - 1$) explicitly, we find:

$$\varphi_{pf} - 1 = -\left([H_3O^+{}_{pf}]/(K_a + [H_3O^+{}_{pf}])\right.$$
$$\left. + [(V_{pf} + V_0)/C_0V_0]\{K_w/[H_3O^+{}_{pf}] - [H_3O^+{}_{pf}]\}\right).$$

At the endpoint, we can set $V_{pf} = V_{eq}$ for calculation purposes and, moreover, set

$$[H_3O^+] \ll K_a$$

since it is located in the basic range. As a result, we have

$$\varphi_{pf} - 1 = -[H_3O^+{}_{pf}]/K_a + \{(C_0 + C)/CC_0\}\{K_w/[H_3O^+{}_{pf}] - [H_3O^+{}_{pf}]\},$$

which has already been given.

10.3 Exact Equations of the Titration Curves of Mixtures of Acids, Bases, Polyacids, Polybases, etc.

The exact equations of the titration curves of mixtures of acids, bases, polyacids, polybases, and so on are easily established by starting from the relationships expressing the equilibria, the charge, and the mass balances in solution. The equations giving the corresponding titration errors are found in the same way as earlier. They are set up from the corresponding exact titration curves expressed in fractions titrated. One titration error exists, of course, for each equivalence point.

10.4 Precision of Acid–Base Titrations Related to the Sharpness Index

Recall that the precision quantifies random errors. Let's consider the titration curve especially about the equivalence point (Fig. 10.1). It appears that the lower the slope is at the equivalence point, the higher the propagation of the error due to the pH measurement is. This is sufficient to justify the introduction of the parameter η to quantify the phenomenon. η is named the *sharpness index*. It is defined as being the magnitude of the slope of the titration curve:

$$\eta = |d\text{pH}|/|d\varphi|,$$

whence

$$\Delta\varphi \approx \Delta\text{pH}/\eta.$$

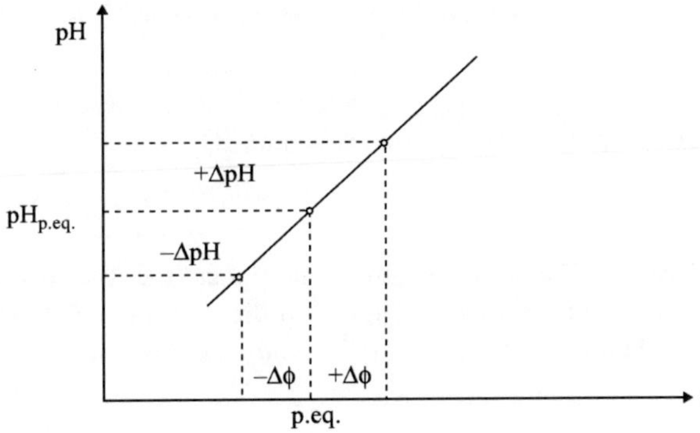

Fig. 10.1 Uncertainty about the determination of the equivalence point related to that governing the pH measurement

The larger the slope of the titration curve is, the smaller the titration error and the more accurate the titration for a given error in measuring pH will be.

For high values of the sharpness index, the titration precision is not limited by the uncertainty governing the pH measurement. It is limited by the precision of the graduated glassware. Conversely, if η is too low, the precision is determined by the detection of the equivalence point. One admits that if $\eta = 10^3$, the titration can be classified as being excellent even if the uncertainty affecting the pH measurement is as high as one unit pH! Then it is the precision affecting the graduated glassware (of the order of 0.1–0.2%) that limits the global precision. If the pH is measured with the help of a pH-meter, the uncertainty affecting its value is, at worst, around 0.1 unit. Then, for a value $\eta = 10^2$, the titration remains excellent. The value $\eta = 10$ gives a poor titration even with a pH-meter. Without any particular experimental condition, the uncertainty affecting the pH measurement with color indicators is located near 0.4 pH unit. The sharpness index must then be around $4 \cdot 10^2$ at least.

10.5 Expressions of the Sharpness Index

Knowledge of the mathematical relationship expressing the sharpness index as a function of the parameters that define the conditions of the titration permits us to extricate those that govern the success of the titration.

The strategy to derive the mathematical expression from the sharpness index is always the same, regardless of the type of titration. It consists of the following steps:

- first, expressing the derivative $d\text{pH}/d\varphi$ from the titration curve equation. It is found by applying the chain rule (see Appendix A);
- second, introducing the value $[H_3O^+{}_{ep}]$ in this derivative. The value $[H_3O^+{}_{ep}]$ is calculated for $\varphi = 1$ with the help of the titration curve equation.

The following expressions of the sharpness index are approximate, as will be seen in Appendices B, C, and D.

10.5.1 Titration of a Strong Acid with a Strong Base

For the titration of a strong acid with a strong base and conversely, the expression of the sharpness index is (see Appendix B)

$$(d\eta/d\varphi)_{\text{eq.p.}} = 0.217 \cdot 10^7 (C_0 C/(C + C_0)).$$

If we set $C = C_0$, for example, it appears that the titration is excellent until a concentration as low as 10^{-4} mol/L if the pH can be determined with an uncertainty of 0.1 pH unit, since $\eta_{\text{eq.p.}} = 10^2$ (see the last section).

10.5.2 Titration of a Weak Acid with a Strong Base

For the titration of a weak acid with a strong base, η at the equivalence point is given by the expressions (see Appendix C)

$$\eta_{\text{eq.p.}} = 0.434[C_0 C/(C_0 + C)][\mathrm{H_3O^+}_{\text{ep}}]/K_{\text{w}}$$

or

$$\eta_{\text{eq.p.}} = 0.434\{[CC_0 K_{\text{a}}]/[(C + C_0)K_{\text{w}}]\}^{1/2}.$$

We notice that the parameter $\beta = K_{\text{w}}/K_{\text{a}}C_0$ recognized above (Chap. 9) as conditioning this kind of titration occurs naturally here.

10.5.3 Titration of a Weak Base with a Strong Acid

For the titration of a weak base with a strong acid, the sharpness index at the equivalence point is (see Appendix C)

$$\eta_{\text{eq.p.}} = 0.434\{[CC_0 K_{\text{b}}]/[(C + C_0)K_{\text{w}}]\}^{1/2}.$$

Again, as just above, the parameter $\beta = K_{\text{b}}C_0$, already noticed as conditioning this kind of titration, appears naturally from the calculations.

10.5.4 Titration of a Weak Acid with a Weak Base

For the titration of a weak acid with a weak base, the sharpness index at the equivalence point is given by the expression

$$\eta_{\text{eq.p.}} = 0.217_{\text{eq.p.}}[1 + (\mathrm{K_a K_b/K_w})^{1/2}],$$

where K_{a} and K_{b} are the ionization constants of the acid and the base, respectively. This relationship is legitimate only when the concentrations $[\mathrm{H_3O^+}]$ and $[\mathrm{OH^-}]$ can be neglected with respect to [HA] and [B] at the equivalence point. With the same symbolism as that used in the Sect. 9.5 of the preceding chapter: K_a becomes K_{a1} and K_w/K_b becomes K_{a2}. The value of the ratio $\beta = K_{a1}/K_{a2}$, which has been already recognized as conditioning the titration, is indeed determinant in its success. For example, by investigating the titration of 0.1 M ammonia with 0.1 M acetic acid, we find $\eta = 20$ at the equivalence point. If the pH is measured with a pH-meter, the uncertainty on the equivalence point determination is about 0.5%. This value can be considered satisfactory. In this case, according to its mathematical expression, the sharpness index is apparently independent of the concentrations C and C_0. However, we must not forget that they must be sufficiently high for the relation to be legitimately applied (see Appendix D).

The concept of the sharpness index is quite general and can be applied to all kinds of titrations.

Remark: It is an easy matter to demonstrate that the sharpness index is related to the buffer index of the solution. When we consider the following expressions:

$$\eta = d\text{pH}/d\varphi$$

or

$$\eta = d\text{pH}/d[H_3O^+] \cdot d[H_3O^+]/d\varphi$$

and $\beta = dC_b/d\text{pH}$ or $\beta = dC_b/d[H_3O^+] \cdot d[H_3O^+]/d\text{pH}$ (see Chap. 6),

we immediately see that the link between β and η is the derivative $d[H_3O^+]/d\text{pH}$ or its inverse.

10.6 Extent of the Titration Reaction

We limit ourselves to the study of the case of the titration of a weak acid with a strong base. The titration reaction is

$$HA + OH^- \rightleftharpoons A^- + H_2O.$$

According to the working hypothesis, it is not complete. The equilibrium constant is (assuming activities equal to concentrations)

$$K = [A^-]/[HA][OH^-],$$

$$K = K_a/K_w.$$

The link between K and the sharpness index in this case (see Sect. 10.5.2) is

$$\eta_{\text{eq.p.}} = 0.434\{[CC_0/(C_0 + C)]^{1/2}(K)^{1/2}\}.$$

We see that if the constant K_a is too weak, the global constant K of the titration reaction is also too weak. In other words, the acid—base couple under study is located too close to the couple H_2O/OH^- on the acidity scale.

Let's continue with this case. Suppose that n^0(HA) is the amount of matter to titrate and C_0 its analytical concentration [$C_0 = n^0(\mathrm{HA})/V_0$]. Let n^0(OH) be the number of moles of hydroxide ions added at each titration stage. Because the titration reaction is equilibrated, the numbers of moles of hydroxide ions and of acid HA that have not reacted are

$$n(\mathrm{OH}) = n^0(\mathrm{OH}) - n_a,$$

$$n(\mathrm{HA}) = n^0(\mathrm{HA}) - n_a,$$

where n_a is the number of moles of base formed. Let's introduce the fraction α. It represents the fraction of the added titrant that has reacted, namely,

$$n_a = \alpha n^0(\mathrm{OH})$$

The parameter α enables us to quantify the quantitative character of the titration reaction. At every stage of the titration, the concentrations are

$$[\mathrm{A}^-] = \alpha n^0(\mathrm{OH})/(V_0 + V),$$

$$[\mathrm{OH}^-] = n^0(\mathrm{OH})(1 - \alpha)/(V_0 + V),$$

$$[\mathrm{HA}] = [n^0(\mathrm{HA}) - \alpha n^0(\mathrm{OH})]/(V_0 + V),$$

where V is the added volume at this stage. The mass law expression is

$$K_\mathrm{a}/[(V_0 + V)K_\mathrm{w}] = \alpha n^0(\mathrm{OH}) / \{ [n^0(\mathrm{HA}) - \alpha n^0(\mathrm{OH})][n^0(\mathrm{OH})(1 - \alpha)]\},$$

and by introducing the fraction titrated $\varphi = n^0(\mathrm{OH})/n^0(\mathrm{HA})$, it becomes

$$K_\mathrm{a}/[(V_0 + V)K\mathrm{w}] = \alpha/[(1 - \alpha\varphi)\,(1 - \alpha)].$$

If the titrant solution is markedly more concentrated than the solution to titrate, $V \ll V_0$ (this hypothesis does not alter the generality of the reasoning), and since $C_0 = n_0(\mathrm{OH}) / V_0$, the parameter $\beta = \mathrm{K_w}/K_\mathrm{a}\ C_0$ naturally enters the calculation. We find

$$\alpha^2\varphi - \alpha(\varphi + \beta + 1) + 1 = 0.$$

A good approximation of the root is

$$\alpha \approx 1/(\varphi + \beta + 1)$$

which is obtained after an expansion in series and a truncation; see Appendix E). The parameter β has already been recognized as conditioning the success of the titration. The fraction α of the titrant that has reacted is a function of the parameter β, as is the sharpness index. Another point to emphasize is that α is also a function of the fraction titrated φ. The reaction is less and less quantitative as φ increases. On the contrary, it is clear that the reaction can be total at the very beginning of the titration ($\varphi \rightarrow 0$) provided the parameter β is sufficiently weak.

This reasoning, which was applied to the titration of a weak acid with a strong base, can be generalized to all the other kinds of titrations. The sharpness index and the extent of a titration reaction are, therefore, intimately linked.

10.7 Gran's Diagram

In a titration, the goal is to detect the equivalence point with the highest precision as possible together with a titration error that is as low as possible.

Unfortunately, the equivalence point of a titration reaction is the least quantitative one, as we demonstrated in the preceding section. This is why logarithmic titrations are the most sensitive to the extent of the titration reaction. This is not the case for linear titrations. Let's first recall that linear titrations are those in which the dependent variable (the registered one) is directly proportional to the fraction titrated or to the concentration of the independent variable (in no case to the logarithm of its concentration or of its activity). Of course, the independent variable may be the activity rather than the concentration. Examples of linear titrations are spectrophotometric, amperometric, thermometric titrations, and so forth. This is the reason why Gran's method is interesting. Indeed, its principle involves the linearization of the logarithmic titration curves.

Let's take, for example, the titration of a weak acid with a strong base. Since we have seen that the titration reaction is the most quantitative at the very beginning of the titration, we can consider without any error of reasoning that it is, actually, complete in this range of the titration:

$$HA + OH^- \rightarrow A^- + H_2O \quad \text{(beginning of the titration: } \varphi \approx 0\text{)}.$$

As a result of this assertion and with the symbolism already used, we can write

$$[A^-] = CV/(V_0 + V),$$

$$[HA] = [C_0V_0 - CV]/(V_0 + V),$$

$$K_a = [H_3O^+]CV/[C_0V_0 - CV],$$

and after rearranging, we have

$$C_0V_0K_a/C - VK_a = [H_3O^+]V.$$

This is the equation of a straight line $[H_3O^+]V$ vs. V. Its slope is—K_a and its intercept is $C_0V_0K_a/C$. Considering this relation only, we immediately see that when the dependent variable is null, $[H_3O^+]V = 0$ (outside $V = 0$), the intercept on the x-axis gives the relation

$$CV = C_0V_0,$$

where the volume V was obtained in such a way that it is at the equivalence point.

However, an inconsistency remains: The ordinate is never null regardless of the titration stage, since $[H_3O^+]$ never tends toward zero! The reason for this is as follows: Gran's method is based on the expression of K_a only. It does not take into

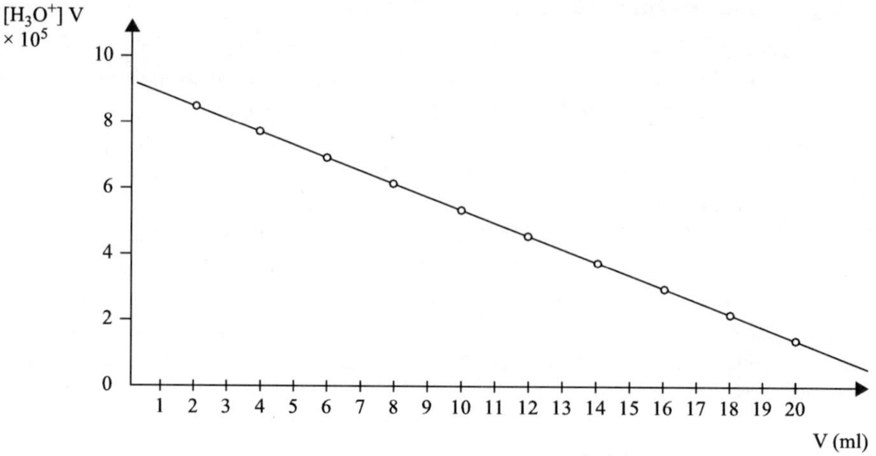

Fig. 10.2 Gran's plot corresponding to the titration of the weak acid HA with sodium hydroxide ($C_0 = 6 \cdot 10^{-2}$ mol/L, $V_0 = 40$ ml, $C = 0.1$ mol/L, $pK_a = 5.40$—see text)

account the charge-balance equation, which introduces the concentrations $[H_3O^+]$ and $[OH^-]$ at every stage of the titration. In other words, applying Gran's method is justified only when these concentrations are negligible vs. $[A^-]$ and $[Na^+]$. It is equivalent to say that the titration reaction is quite complete. The extrapolation of the straight line toward zero embodies this approximation and, therefore, overcomes the difficulty. Hence, obtaining the volume in such a way is accurate. In support of this reasoning, it is interesting to note that, actually, the experimental Gran diagram does not exhibit a linear part throughout the titration, in good agreement with the fact that the concentrations $[H_3O^+]$ and $[OH^-]$ are not always negligible.

The main advantage of Gran's method is to limit the titration error that might be due to the use of a neutralization reaction that is too equilibrated. Indeed, the straight line involved in Gran's method is drawn from the first points of the titration, for which, as we have demonstrated, the neutralization reaction is the most quantitative.

Actually, applying Gran's strategy is more complicated than stated, because the pH value is an activity value. Hence, the activities of species HA and A^- must be considered and K_a must be the true thermodynamic constant. As a result, activity coefficients must be used and Gran's equation must be written as follows:

$$K_a(C_0V_0/C) - VK_a = (\gamma_{A^-}/\gamma_{HA})(H_3O^+)V,$$

where (H_3O^+) is the activity of H_3O^+,

$$K_a(C_0V_0/C) - VK_a = (\gamma_{A^-}/\gamma_{HA})(10^{-pH})V.$$

We see that an accurate straight line can be obtained only if the ratio of the activity coefficients is constant. As γ_{HA} may be considered equal to unity for the concentrations involved, this implies that γ_{A^-} and, hence, the ionic strength of the solution

Table 10.1 Some data concerning the titration of the weak acid HA ($pk_a = 5.40$) with sodium hydroxide ($C_0 = 6 \cdot 10^{-2}$ mol/L, $V_0 = 40$ ml, $C = 0.1$ mol/L)

Added volume V (ml)	pH	(H_3O^+) V ($\times 10^5$)	I (mol/L)
0	3.31		
2.0	4.36	8.67	$4.8 \cdot 10^{-3}$
3.95	4.70	7.89	$9.0 \cdot 10^{-3}$
6.0	4.92	7.17	
8.0	5.10	6.36	
9.9	5.25	5.59	
12.0	5.40	4.78	
14.0	5.55	3.96	
16.0	5.70	3.19	
18.0	5.88	2.38	
20.0	6.10	1.59	
22.0	6.44	0.80	
22.9	6.71	0.45	$3.6 \cdot 10^{-2}$
24.2	10.41	$9.39 \cdot 10^{-5}$	$3.8 \cdot 10^{-2}$

remain rather constant all along the titration. We can also notice that for the titration of a strong acid with a strong base, the ionic strength does not significantly change since H^+ ions are replaced mole by mole with Na^+ ions.

Figure 10.2 represents the Gran plot corresponding to the titration of 40 ml of a $6 \cdot 10^{-2}$ mol/L solution of the acid HA with a 0.1 mol/L sodium hydroxide solution. It is built from data collected in Table 10.1. Table 10.1 also contains approximate values of the ionic strength I calculated by mixing activities and concentrations (see the end of Chap. 5).

We can check off that the last point cannot be retained to build Gran's plot because $[OH^-] = 2.57 \cdot 10^{-4}$ mol/L is not completely negligible vs. $[A^-] = 3.74 \cdot 10^{-2}$ mol/L. It is the same matter for the second point ($V = 2.0$ ml) because of the value $[H^+] = 4.37 \cdot 10^{-5}$, which is not completely negligible vs. $[Na^+] = 4.73 \cdot 10^{-3}$ mol/L.

We can also notice that the ionic strength of the solution changes by a factor of 10 between the beginning and the endpoint of the titration. With the concentrations used, it does not seem to matter markedly, since the slope obtained by mixing concentrations $[A^-]$ and [HA] and activities (H_3O^+) (with K_a normally defined in terms of activities) leads to the awaited endpoint and to the awaited K_a value.

Chapter 11
Acid–Base Reactions and Chemical Analysis

Because of their huge number, it is quite impossible to list all the analytical applications of the acid–base reactions. It is even impossible to delineate homogeneous headings that group them. In this chapter, we briefly review the concept of pH, the pH value as a parameter conditioning the success of several analytical applications, and the acidity of an aqueous solution as a purity index of a solute. Above all, we comment on several acid–base titrations.

11.1 The Concept of pH

The concept of pH is one of the great ideas of chemistry. It is endowed with an immense practical interest in addition to some very interesting theoretical considerations concerning it. The concept is solidly based on these considerations, which are thermodynamic in nature.

One of the great practical interests in the concept of pH lies in the fact that its value is a satisfactory measurement of the protonic activity of a solution, in other words, of its actual acidity. This point signifies that the operation of pH measurement does not alter the solution's acidity, that is, does not shift the acid–base equilibria of the solution. (This kind of measurement can be found in other branches of physical sciences. An example is provided by thermometry, which acknowledges that the use of a thermometer does not change the temperature of the system under study.) Let's incidentally note that it is not at all the same when we measure the acidity of a solution by a titration procedure in which the acid–base equilibria are completely shifted by the titration reaction. In such a case, we are speaking of the *total* acidity instead of the *true* acidity.

11.2 Analytical Operations and pH

The acidity of a solution is often a parameter that is easy to modify in order to improve the conditions of an analysis. Briefly, the pH is one of the, if not the first, modifiable parameters that must come to mind when we want to find the best conditions for an

J.-L. Burgot, *Ionic Equilibria in Analytical Chemistry*,
DOI 10.1007/978-1-4419-8382-4_11, © Springer Science+Business Media, LLC 2012

analysis. This is the reason why we briefly mention some kinds of equilibria that can be shifted in acidic and basic media.

These shifts can occur

- in a homogeneous medium. This is the case of the redox properties of some couples that are pH-dependent (see Part III of this book). This is also the case of the complexes' formation, which can be pH-dependent (see Part IV);
- in a heterogeneous medium. For example, a judicious choice of several successive pH values may permit the separation of the components of a solution by selective precipitation or by extraction of the initial aqueous solution with an immiscible solvent (see Part V).

Inducing equilibria displacements such as those mentioned above is the basis of the most fundamental kind of reasoning followed in the analysis of solutions.

The pH value is also an important parameter to take into account when instrumental methods of analysis are used. Among such methods are the chromatographic and electrochemical methods.

11.3 Acidity of a Medium as an Index of Its Purity

The (true) solution acidity may constitute a purity index of the solute dissolved into it. Pharmacopeias often call for this kind of assay. Quite evidently, an abnormal pH value found after dissolution of the drug indicates the presence of an acidic or basic impurity. The impurity may come from the synthesis or the degradation of the drug. Most of the time, the determination of the acidity is performed by pH-metry.

11.4 On the Choice of Examples of Acid–Base Titrations

In the remainder of this chapter, we shall choose our acid–base titration examples in some pharmacopeias, especially the european one. Indeed, most of the pharmacologically active chemical products (about 85%) do possess one or several acidic–basic sites due to the presence of some kinds of functional groups.

Most of them are organic compounds, and the practical problem that often arises is the precipitation in water of the basic or acidic form of the compound liberated during the titration. More simply, even the problem of the dissolution in water of the derivative to be titrated arises. In order to overcome this difficulty, an organic solvent miscible with water is often added. It may also be the case that the compound to titrate is soluble only in an organic solvent. In this case, if the adjuvant solvent is miscible in water, the solute is dissolved initially in the organic solvent in the titration vessel and then the purely aqueous solution of the titrant is added.

In this chapter, we only mention examples of titrations in strictly aqueous solutions and in some aqueous organic mixtures. According to a first approximation, we can consider that such mixtures exhibit the same behavior as strictly aqueous solutions. Hence, the principles of these titrations are the same as before. If necessary, we shall mention the changes due to the addition of the miscible organic solvent to the aqueous solution. Limiting ourselves to such comments is satisfactory if we consider the analytical applications only. The origin of the similarity of the above-evoked behaviors of strictly aqueous solutions and of the hydro-organic mixtures must be imputed to the fact that the latter retain some properties of aqueous solutions even when the fraction of water titrated in the mixture is low. The presence of the organic miscible solvent may, of course, considerably change the thermodynamic properties of the solution, in particular, the pK_a of the protolyte, the water ion product, and others. This is true also for the pH values and even, more fundamentally, for their significance.

We choose to comment on examples for which equivalence points are detected with the help of color indicators or of pH-metry. For now, it is sufficient to know that pH-metry is a method that indicates the equivalence point and that it is more precise than methods using neutralization indicators.

11.5 Direct Titrations of Acid Compounds

Acid compounds are usually titrated with sodium or potassium hydroxide solutions. Apart from some mineral compounds, the other compounds titrated are essentially carboxylic acids, phenols, imides, and quaternary ammoniums resulting from the protonation of amines.

Among the mineral compounds, we'd like to mention the titration of the sodium dihydrogen phosphate. It is performed in the presence of phenolphthalein according to the reaction:

$$H_2PO_4^- + OH^- \rightarrow HPO_4^{2-} + H_2O.$$

It behaves like a weak monoacid (see Chap. 9).

Numerous titrations of carboxylic acids have been proposed in the literature, in strictly aqueous solutions as well as in hydro-organic mixtures. Among those reported in strictly aqueous solutions are the titrations of maleic, malic, tartaric, citric, and sorbic acids (Fig. 11.1). Their equivalence points are detected with phenolphthalein. All of them, including the polyacids just listed, exhibit only one endpoint. The endpoint color change is detected for the titration stage at which the number of added equivalents of sodium hydroxide is equal to the number of acid functions the titrand possesses. We'd like to point out the following facts about these titrations:

- the equivalence point is located in basic medium, as already described (Chap. 8);
- the pK_a values, including the highest ones of the polyacids, are sufficiently low, however, for all the acidities to be completely neutralized. In other words, all the couples RCOOH/RCOO$^-$ are located sufficiently far from the couple H_2O/OH^- on the acidity scale;
- in the polyacid cases, the acidities are too close to one another to be separately and distinctly titrated. Indeed, the difference between two successive pK_a values is always less than 4.

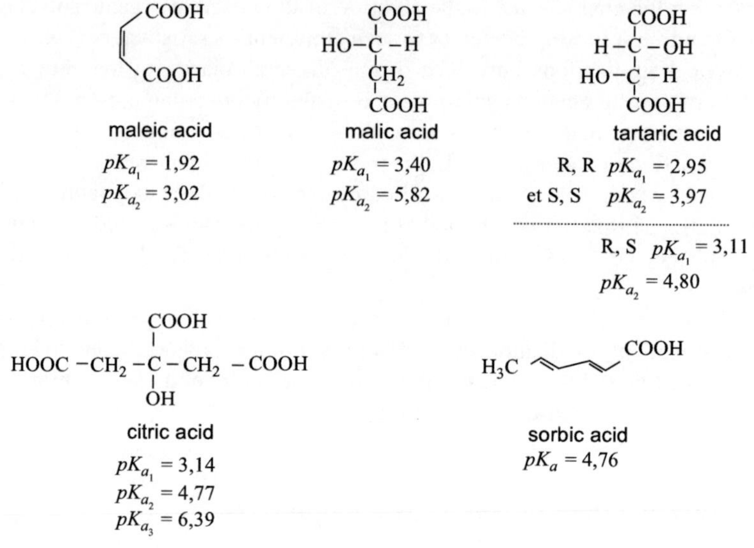

Fig. 11.1 Structures and pK_a values of some carboxylic acids

A somewhat different case is that of the α-amino acids, which contain a supplementary carboxylic group in their structure. Two examples are provided, for example, by aspartic and glutamic acids.

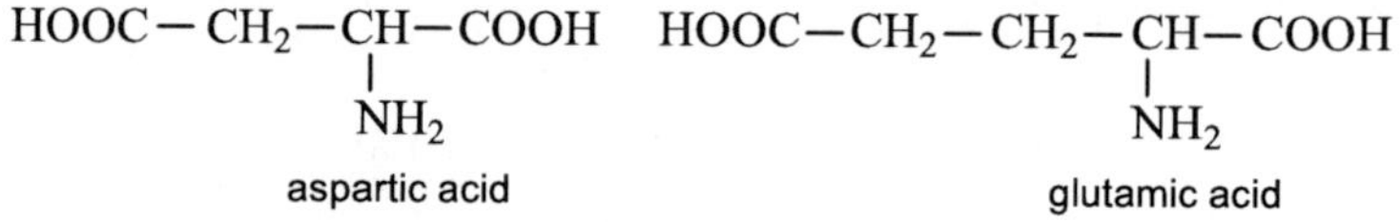

They can be titrated by sodium or potassium hydroxide solutions as monoacids instead of α-amino acids (without any supplementary function), which do not offer that possibility (see Chap. 5). The estimated values of their microscopic constants are given in Fig. 11.2.

Due to the differences in the pK_a values of each carboxylic function within each molecule, both compounds are in their neutral form. This means that at the beginning of the titration, their structure is as follows:

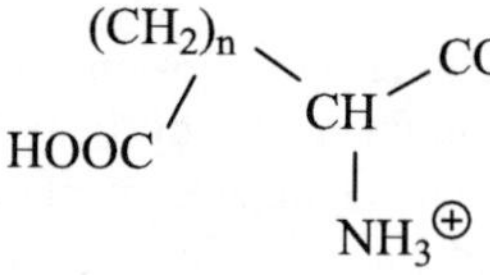

During the course of the titration, only the "supplementary" carboxylic function is neutralized. The quaternary ammonium function is too weak an acid to be neutralized by sodium hydroxide since the macroscopic constant K_{a2} is too weak (see Chap. 5).

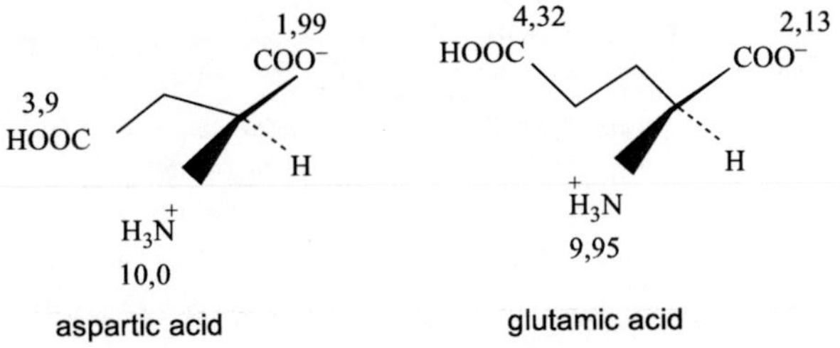

Fig. 11.2 Estimated values of the microscopic constants of aspartic and glutamic acids (triacid forms)

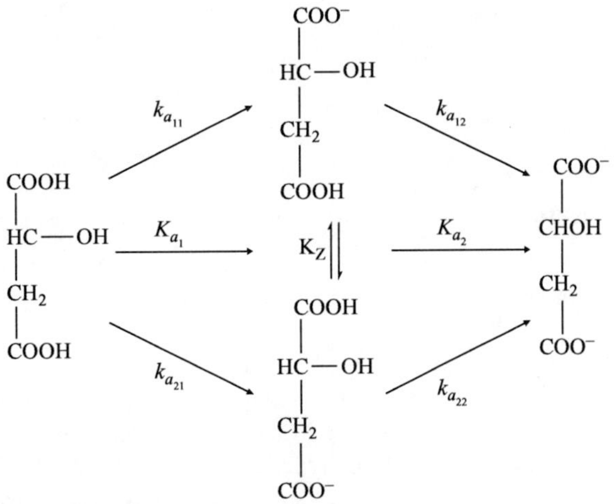

Fig. 11.3 Macroscopic and microscopic ionization constants of malic acid

Actually, its value is quasi-identical to that of the microscopic one from which it is a linear combination. The protonated amine is, indeed, the second acid function.

Recall that the behavior of polyacids depends on macroscopic and microscopic ionization constants. The former can be determined experimentally. They globally quantify the more or less simultaneous ionization of several acid functions. Their value cannot be assigned in totality to only one acidity constant if the initial acid is dissymmetric. Let us consider malic acid (Fig. 11.1). It exhibits two different microscopic ionization methods (Fig. 11.3).

The first ionization constant ($pK_{a1} = 3.40$) corresponds to the formation of the two monoacid forms whose concentrations' ratio is constant regardless of the pH value. Likewise, the two monoacid forms give the same dibase by ionization according to the second macroscopic constant ($pK_{a2} = 5.82$).

Another comment connected with the preceding example can be suggested by the fact that some compounds quoted above can exist in two enantiomeric forms. In aqueous solution, two enantiomers exhibit the same pK_a value. The corresponding racemic mixture also exhibits the same pK_a value. It is not the case for diastereoisomers. For example, the R, R and S, S tartaric acids exhibit both the values $pK_{a1} = 2.95$ and $pK_{a2} = 3.97$ since they are enantiomers. The mesotartaric acid R, S, which is

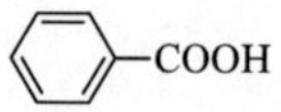

benzoic acid
$pK_a = 4{,}19$

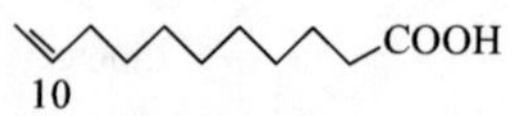

undecylenic acid
$pK_a = 4{,}50$

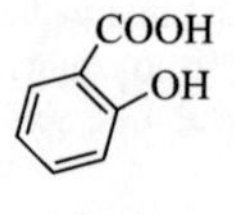

salicylic acid
$pK_{a_1} = 2{,}83$
$pK_{a_2} = 12{,}62$

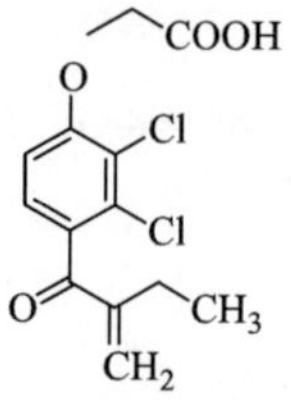

ethacrynic acid
$pK_a = 3{,}50$

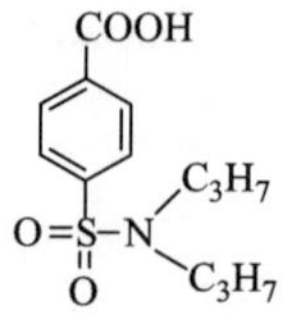

probenecid
pK_a unknown (statistically ≅ 4)

Fig. 11.4 Structures and pK_a values of some carboxylic acids titrated in hydroalcoholic mixtures

their common diastereoisomer, exhibits two very different values: $pK_{a1} = 3.11$ and $pK_{a2} = 4.80$.

Let's now study the titrations of the carboxylic acids that are too poorly soluble in water and therefore necessitate the use of mixtures of water and organic solvents. The most commonly used mixtures are the hydroalcoholic ones. Ethanol must be neutralized by a sodium hydroxide solution before being mixed with water. Benzoic, undecylenic, etacrinic, and salicylic acids as well as probenecid (Fig. 11.4) are titrated according to this methodology.

The equivalence point is detected by the color change of phenol red. In strictly aqueous medium, the color change interval is $6.8 < \text{pH} < 8.4$. Because in hydroalcoholic medium, the present species at the equivalence point is the base $RCOO^-$ as in water, it means that the color change of phenol red occurs in a medium slightly more basic in ethanol than in water. Incidentally, we follow this kind of reasoning to compare the absolute acidity levels of different solvents.

Other mixtures of water and organic solvents have also been proposed to titrate some other poorly soluble carboxylic acids. Examples are water/DMF and water/acetone. The indicators used are bromothymol blue and phenolphthalein.

An important group of pharmacologically active compounds is that containing quaternary BH^+ ammoniums resulting from the protonation of the corresponding B amines. Most of the time, they are B hydrochlorides, HCl.

The main part of this group contains derivatives from aliphatic amines: primary, secondary, and tertiary. Their pK_a values are, statistically, located in the range 8.5–10.5. Their titration with sodium hydroxide in water offers two major difficulties:

- the first is the precipitation during the titration of the liberated amine according to the reaction

$$BH^+ + OH^- \rightleftharpoons B\downarrow + H_2O;$$

The main result is that the titrations are poorly reproducible:

- the second is the lack of completeness of the titration reaction. The couples BH^+/B are located too close to the couple H_2O/OH^- on the acidity scale (see Chaps. 9 and 10). From this standpoint, the couples BH^+/B are quite comparable to orthoboric acid ($pK_a = 9.25$) and to the ammonium ion ($pK_a = 9.24$), which cannot be satisfactorily titrated with sodium hydroxide in water.

It turns out that amitriptyline and propranolol hydrochlorides can be titrated in a water/ethanol mixture with a 0.1 mol/L sodium hydroxide solution with a pH-metric equivalence point detection. Levomepromazine hydrochloride can be titrated in the same conditions but with propanol instead of ethanol as cosolvent. Propanol is preferred to ethanol because of the very high lipophilicity of the levomepromazine base.

A variation on these titrations consists of adding hydrochloric acid to the hydroalcoholic solution of the solute and then titrating them with a sodium hydroxide solution. The equivalence point is detected by pH-metry. Ephedrine, bupivacaine, desipramine, chlorprothixene, and amantadine hydrochlorides are titrated in this way. Using pH-metry is necessary in this variation since we must detect the successive equivalence points with an optimal precision. The first equivalence point corresponds to the end of the neutralization of hydrochloric acid, the second to that of the cation. The addition of hydrochloric acid facilitates the detection of the beginning of the cation neutralization reaction. It can be inferred from all these results that the addition of ethanol to water permits a satisfactory detection of the equivalence point of this kind of products, even of those that exhibit the higher pK_a values, such as the amantadine hydrochloride ($pK_a = 10.68$). This means that the reaction between the base of the titrant couple and the cation is more quantitative in these mixtures than in water. In other words, the couple BH^+/B is farther from the couple H_2O/OH^- in these mixtures than it is in water. Actually, the titrant is a mixture of the couple H_2O/OH^- and $C_2H_5OH/C_2H_5O^-$.

Some phenols can be titrated with a sodium hydroxide solution in water. This is the case of vanillin ($pK_a = 7.4$) with thymolphthalein as the indicator. It is not the case with the phenol itself ($pK_a = 10.0$). It is very likely that the difference in the strength of their acidities must be assigned to the presence of the formyl group in the vanillin molecule. This group is, indeed, a very powerful withdrawing group, by both inductor and mesomeric effects. Evidently, the methoxy group, which in ortho position on an aromatic nucleus is usually an electrodonor, does not decrease the influence of the formyl group.

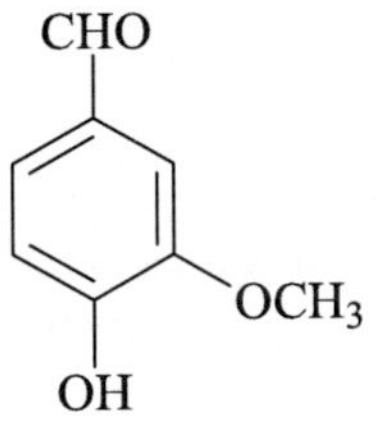

Vanillin

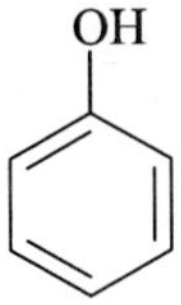

Phenol

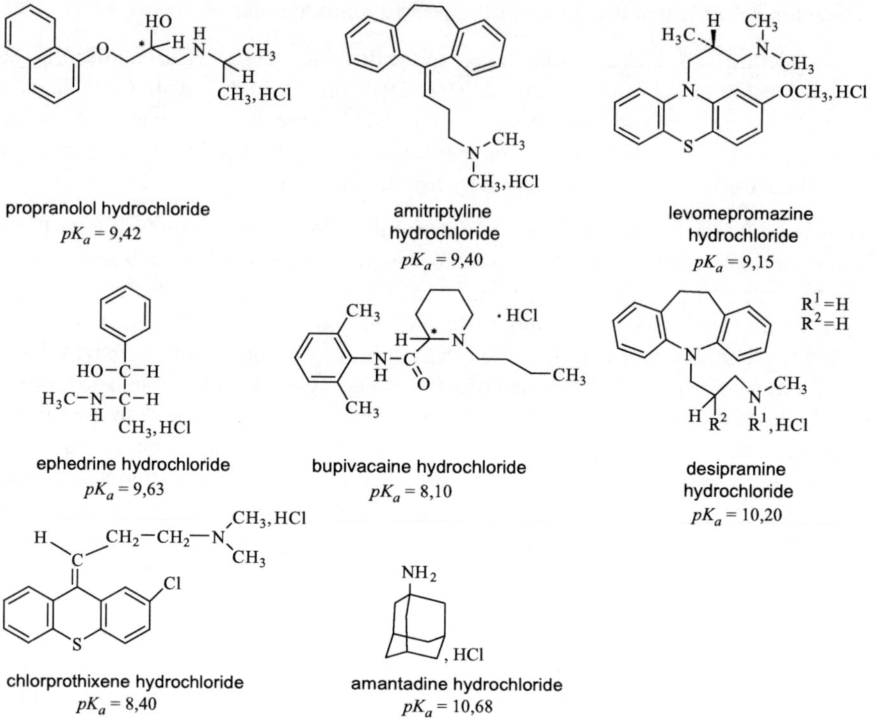

Fig. 11.5 Structures and pK_a values of some derivatives titrated in hydroalcoholic mixtures

Other kinds of organic structures exhibit a sufficiently acidic character to be titrated with a sodium hydroxide solution in water and in hydro-organic mixtures; examples include derivatives that bring a lactam function, acylureas, imides, some sulfamides, and some nitrogenous heterocycles. The pK_a of these derivatives may take any value along the entire pH range. Their acidic character would be due to the occurrence of enol or oxime forms.

11.6 Direct Titrations of Derivatives Exhibiting a Basic Character

Most of the time, derivatives exhibiting a basic character are titrated with hydrochloric or sulfuric acid solutions.

The disodium phosphate Na_2HPO_4 can be titrated with a hydrochloric solution with a mixed indicator. The mixed indicator is a mixture of bromocresol green (color-change interval: $3.8 < pH < 5.4$) and methyl red (color-change interval: $4.2 < pH < 6.3$):

$$HPO_4^{2-} + H^+ \rightarrow H_2PO_4^-.$$

The amphoteric ion dihydrogen phosphate is formed. However, the neutralization of the latter is not complete:

$$H_2PO_4{}^- + H^+ \rightleftharpoons H_3PO_4.$$

As $H_2PO_4{}^-$ is too weak a base to be titrated in water, only one equivalence point is detected with the mixed indicator. With the same indicator, borax can be titrated with hydrochloric acid. It gives orthoboric acid, H_3BO_3, according to the reaction

$$\left[B_4O_5(OH)_4\right]Na_2, 8H_2O + 2H^+ + 2Cl^- \rightarrow 4H_3BO_3 + 2Na^+ + 2Cl^- + 5H_2O.$$

Recall that borax is used as a buffer. This is one of the primary standards used for the operational definition of pH (Bates and Guggenheim's definition). The orthoboric acid formed is a very weak acid monoacid ($pK_a = 9.25$). It is evident from the standpoint of the reaction's completeness that no problem exists. If we let the acid–base couple of borax be $H_3BO_3/[B_4O_5(OH)_4]\ Na_2,8H_2O$, as the reaction above suggests, we see that it is located very far from the couple H_3O^+/H_2O on the acidity scale.

Primary, secondary, and tertiary aliphatic amines exhibit pK_a values located statistically in the range $8.5 < pH < 10.5$. Without any surprises, they can be titrated with a hydrochloric or sulfuric acid solution. The reaction is sufficiently complete. At the equivalence point, we are in the presence of a weak acid: the conjugate acid of the amine. The pH is located around 4.6. The mixed indicator is convenient for detecting it. Unfortunately, the great insolubility in water of the amines to be titrated is a very serious limit to these titrations.

The titration of aminophylline, which is the salt of theophylline (the acid) and ethylenediamine (the base), is performed in water with hydrochloric acid in the presence of bromocresol green ($3.8 < pH < 5.4$):

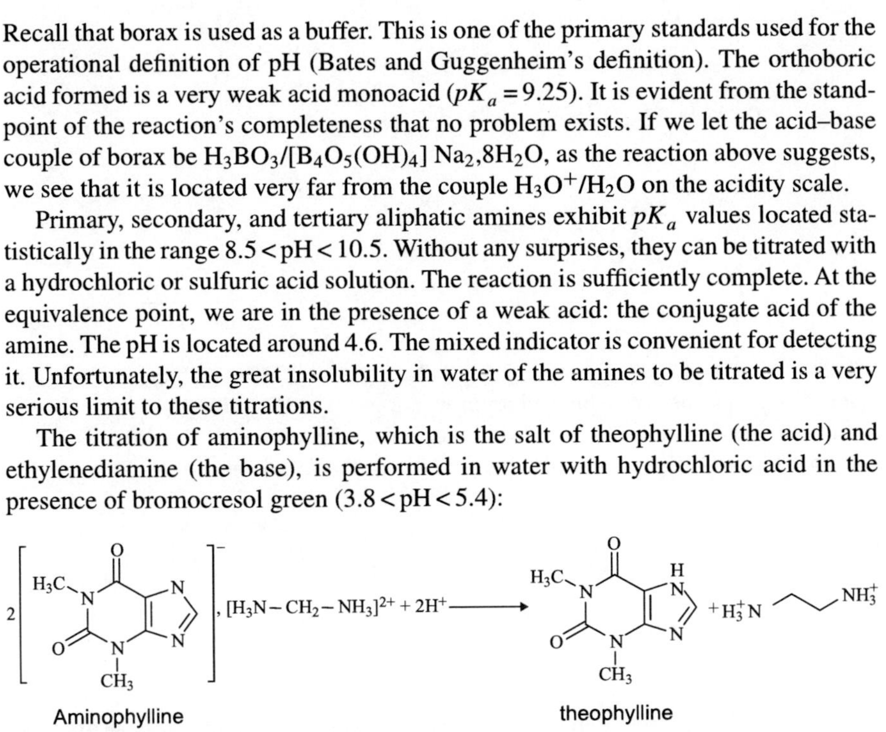

At the equivalence point, the solution in the titration vessel is a mixture of a diacid (the dication of ethylenediamine) and a monoacid, the theophylline. The pK_a values of the dication are 7.23 and 10.07, and that of theophylline 8.60. If we neglect the second acidity of the dication and consider that the pK_a value of theophylline is approximately equal to that of the first acidity of the dication, the pH at the equivalence point is approximately given by the equation $pH = 1/2\, pK_a - 1/2 \log C$. It is very likely located in the range $4.1 < pH < 4.8$ according to the initial concentration. This calculation justifies the use of bromocresol green.

For identical but symmetrical reasons as those involved for the titrations of aspartic and glutamic acids, the titrations of lysine, ornithine, and arginine, which are α-amino acids bringing a supplementary amine function, can be performed with a

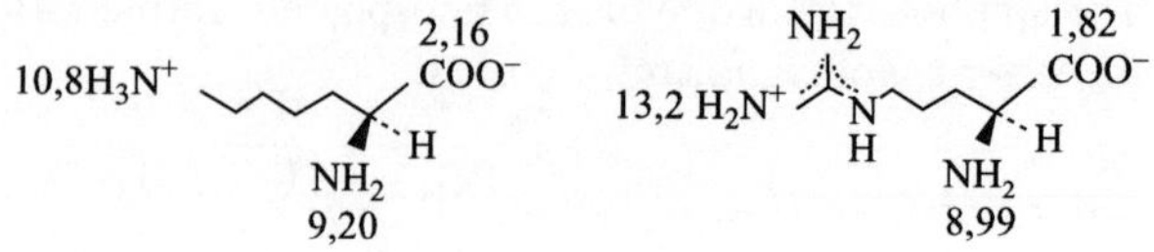

lysine arginine

Fig. 11.6 Estimated values of the microscopic ionization constants of lysine and arginine

hydrochloric acid in the presence of a neutralization indicator whose color-change interval is located in the acidic region. A mixed indicator and methyl red can be used.

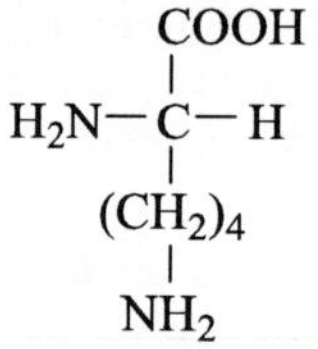

Lysine

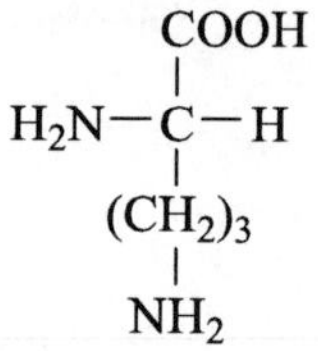

ornithine

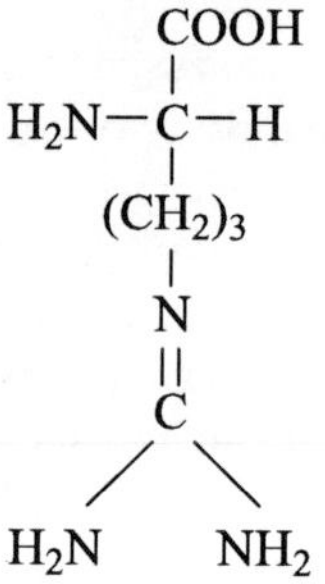

arginine

Their behavior is that of a monobase. The estimated values of the microscopic ionization constants of lysine and arginine are given in Fig. 11.6.

Due to these values, it is very likely that in the solution before the titrant is added, the structure of lysine would be either or both of the following:

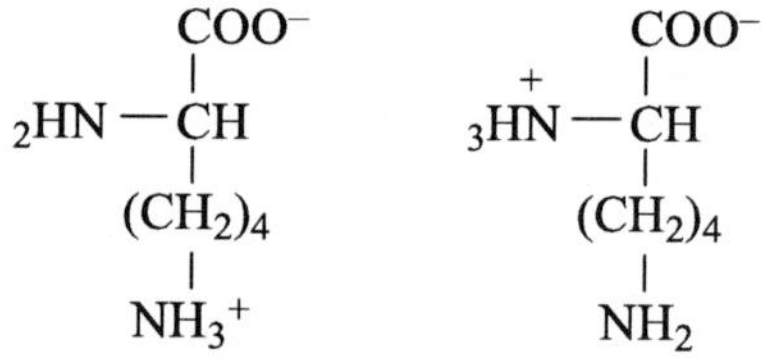

where the first one is by far in the majority because of the difference between the pK_a values of each amine function. Arginine is, without any doubt, under the following form before addition of the titrant:

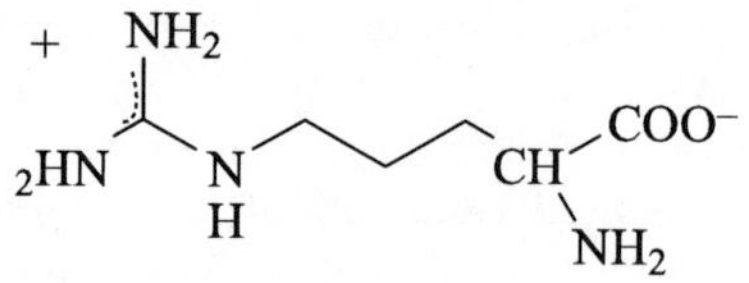

because of the very high basicity of the arginine group located at the end of the chain. Briefly, in most cases, the amine functions are always protonated during the titrations, whereas the carboxylate ones are never so. These derivatives exhibit the behavior of a monobase.

The direct titration of ammonia with an acid is often practiced to conclude the determination of nitrogen in an organic compound according to Kjeldahl's method. First, the organic matter is destroyed by mineralization by heating at ebullition with sulfuric acid, possibly in the presence of one or several adjuvants. In these conditions, nitrogen is transformed into ammonia, which is fixed by sulfuric acid in the form of ammonium salt. After treatment of the medium with a solution of strong base, ammonia is displaced from its salt, distilled, and expelled into a vessel already containing a known volume of an excess of standard sulfuric acid. Finally, the excess of the latter is titrated. This is, of course, a back titration. According to another alternative, ammonia is expelled into a boric acid solution, where, finally, it is titrated with a sulfuric acid solution. This titration methodology deserves further comment. It is the titration of a weak base (NH_3) by a strong acid. As a result, the equivalence point is located on the acid side. It occurs around pH = 5.5, and Tashiri's indicator is the most appropriate. It changes from violet to green, with a gray intermediate color. It is interesting to note that despite the great excess of boric acid, the reaction of ammonium borate formation,

$$H_3BO_3 + NH_3 + H_2O \rightleftharpoons B(OH)_4^-, NH_4^+,$$

is very far to be quantitative since its equilibrium constant value is about 1 [$pK_a(NH_4^+/NH_3) = 9.24$, $pK_a(H_3BO_3/B(OH)_4^-) = 9.25$]. At this point, the question that comes to mind is the following: Does the fixation of ammonia by boric acid distort the result of its titration with a strong acid? The answer is no. This is because the (weak) quantity of ammonia transformed into ammonium borate, which quite evidently cannot be protoned during the titration reaction with sulfuric acid, is replaced mole by mole with a borate ion, which is titrated with sulfuric acid in the same conditions.

Another titration also deserves some comment. It is that of ammonium carbonate. The prescribed methodology is close to that used to determine the concentration of carbonate in a sodium hydroxide solution. The ammonium carbonate is treated by a known quantity in excess of sodium hydroxide:

$$CO_3(NH_4)_2 + 2NaOH \rightarrow 2NH_3\uparrow + 2H_2O + Na_2CO_3.$$

Displaced ammonia is eliminated by heating. A first titration with a standard sulfuric acid solution in the presence of phenolphthalein permits us to neutralize the sodium hydroxide in excess together with the first basicity of the formed carbonate. Then the titration is continued in the presence of helianthine. The added volume between the two color changes corresponds to the neutralization of the second basicity, that is, to the carbonate concentration.

11.7 Back Titrations

Recall that back titrations are recommended if the titration reaction is slow or if it is not sufficiently quantitative. The addition of an excess of titrant increases the rate of the reaction or displaces the equilibrium toward the right (see Chap. 7).

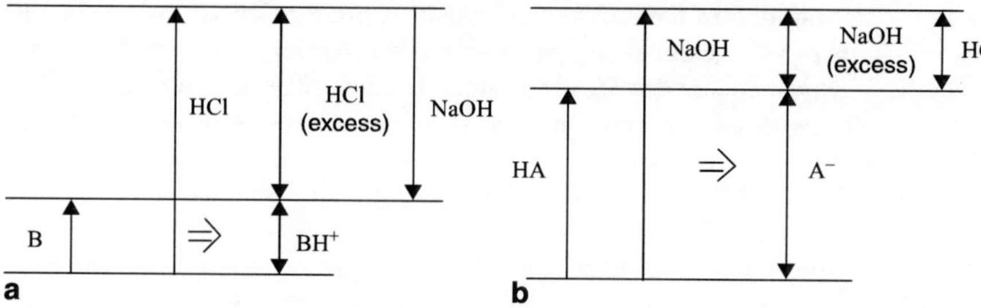

Fig. 11.7 Back titrations of an acid (**a**) and a base (**b**)

The theoretical point to be discussed here is this: In a back titration, two species exhibiting the same acidic or basic property are present in the titration vessel—the reactant excess, which is acidic or basic, and the conjugate form of the titrand. The problem is to neutralize the first species without neutralizing the second:

- in the case of an amine B, an excess of hydrochloric acid is added. The amine hydrochloride is formed according to the reaction

$$B + H^+ + Cl^- \rightarrow BH^+ + Cl^-.$$

During the back titration, a standard sodium hydroxide solution is added. Two acids are present: the hydrochloric acid in excess that we want to determine and the weak acid BH^+ not to be neutralized (Fig. 11.7a).

In the case of the titration of a weak acid HA, an excess of sodium hydroxide is added. The conjugate base A^- is formed:

$$HA + Na^+ + OH^- \rightarrow A^- + H_2O + Na^+.$$

During the back titration, we want to exclusively neutralize the sodium hydroxide excess (Fig. 11.7b).

In the first case, the conjugate acid BH^+ (Fig. 11.7a) must not be neutralized. The couple BH^+/B must hence be located sufficiently close to the couple H_2O/OH^- on the acidity scale. Aliphatic nitrogenous bases, whose conjugate acids exhibit pK_a values around 9.5–10.0, comply with the requirement of this methodology. In order to detect the equivalence point, we must use indicators whose color-change intervals are located in weakly acidic or neutral media. Using an indicator whose color change would be located in basic medium would involve beginning the neutralization of the ammonium cation, even if it is a weak acid.

In the second case, the base A^- must not react with the strong acid. The couple HA/A^- must be sufficiently close to the couple H_3O^+/H_2O on the acidity scale. Carboxylic acids, whose pK_a values are located around 4, comply with this requirement. The indicator used must exhibit its color change in neutral or weakly basic medium.

Another possibility remains. According to it, in the case of the back titration of a base B, the pK_a value of the conjugate acid BH^+ may be endowed with a median value (in the region 5.5–8.5). When this condition is complied with, the strong acid excess and the cation BH^+ may be neutralized quantitatively and successively. It is equivalent to the case of the titration with a strong base of a mixture of a strong acid and a weak acid, the strength of the latter being very far from that of the couple H_3O^+/H_2O. In such a titration, it is better to detect the equivalence points by pH-metry rather than with neutralization indicators, as the former methodology is more precise. A similar case occurs during the back titration of an acid HA possessing a median pK_a value.

Numerous back titrations exist in pharmacopeias. Several extensions of back titrations also exist (see Sect. 11.8). It is impossible to list them here. Back titrations can, of course, be applied to mineral derivatives; examples include zinc oxide and magnesium oxide. They are dissolved first in a known quantity of sulfuric acid in excess. Zn^{2+} and Mg^{2+} ions are formed according to the reactions

$$ZnO\downarrow + H_2SO_4 \rightarrow Zn^{2+} + SO_4{}^{2-} + H_2O,$$

$$MgO\downarrow + H_2SO_4 \rightarrow Mg^{2+} + SO_4{}^{2-} + H_2O.$$

The sulfuric acid in excess is titrated with a standard sodium hydroxide solution. The indicators are methyl red and helianthine.

Remark: In the case of zinc oxide, during the back titration (strictly speaking), ammonium chloride is added to avoid the precipitation of zinc hydroxide ($K_s = 10^{-17}$; see Part V). Adding ammonium chloride during the titration permits the buffer $NH_4{}^+/NH_3$ to form and prevents the pH from taking too high a value and, hence, the precipitation of the hydroxide. In another respect, the chloride ions mixed with Zn^{2+} ions to give the species $[ZnCl^+]$, $[ZnCl_2]$, $[ZnCl_3{}^-]$, and $[ZnCl_4{}^{2-}]$. This reduces the risk of precipitation.

11.8 Titrations After a Chemical Reaction (After Transformation)

This section describes titrations of products that derive from the compounds to determine after a chemical reaction. Let us remark that in a back titration there also exists a transformation by a chemical reaction of the compound to determine. This transformation is simple since it is only a proton exchange. However, in a back titration, it is not the transformed compound that is titrated, contrary to the titrations described in this section. Here, we also study the case in which the excess of reactant that has induced the transformation is titrated. This methodology is then an extension of back titrations.

It is quite evident that the different possibilities of titration under study can be used if the primitive transformation is quantitative.

There are numerous cases in which a sodium (or potassium) hydroxide solution performs the quantitative transformation. However, the reaction can be slow, contrary to that of proton exchange. When it is finished, the excess of the base is titrated with a strong acid. As in a back titration, the product formed by the transformation must not be titrated with the strong acid for the base excess to be accurately determined.

Let's mention, for example, the determination of esters in certain essential oils such as geranyl formate and, in the realm of medicinal chemistry, that of clofibrate:

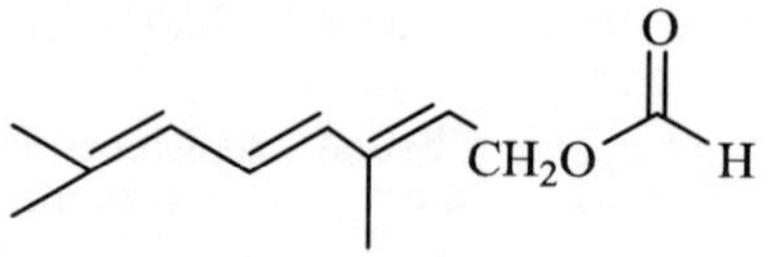

geranyl formiate

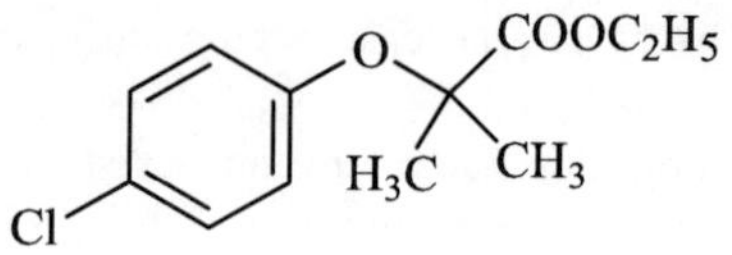

clofibrate

The transformation is called a *saponification*:

$$R-\overset{\overset{\displaystyle O}{\|}}{C}-OR' \xrightarrow{\text{NaOH}} RCOO^- + R'OH$$

The conjugate bases of carboxylic acids that are formed quantitatively are not neutralized by the strong acid due to the position of carboxylic acids on the acidity scale.

Another possibility is the neutralization of the transformation product by the acid after that of the strong base, but the titration curve must then exhibit two markedly different endpoints.

Determining the ester function is the matter of an officially recommended assay called the *saponification value* or *ester value*. Its principle is that described just above. The saponification value is the quantity of potassium hydroxide, expressed in grams, necessary to neutralize the acids liberated by the esters contained in one gram of substance. If the initial sample contained free acids, they should be taken into account in the quantity of base used.

The titrations after transformation of alkyl p-hydroxybenzoates, acetylsalicylic acid (aspirin), and lactic acid deserve particular comments.

After the saponification, p-hydroxybenzoates are in the form of the dianion

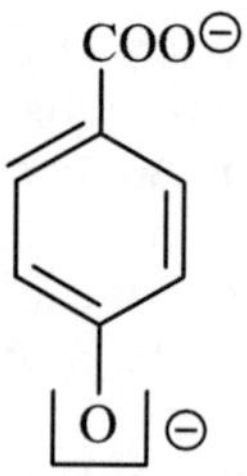

During their titrations with hydrochloric acid, the equivalence point is detected by pH-metry. It corresponds to the neutralization of the excess of hydroxide together

with that of the phenolate function (pK_a value of the corresponding phenol: 8.47). We operate with respect to a blank. That is to say, we titrate the whole quantity of sodium hydroxide added to perform the saponification. The difference between the added volumes at the equivalence points in each titration corresponds to the hydroxide concentration necessary to form the carboxylate ion. The remaining carboxylate (pK_a of the corresponding carboxylic acid: 4.53) is not protonated during this titration.

Acetylsalicylic acid is the matter of two titrations. The first is performed with sodium hydroxide at room temperature; the neutralization reaction is instantaneous. It corresponds to the free carboxylic acid. The second titration is achieved after reaction by heating with a sodium hydroxide solution for one hour. In these conditions, there is, of course, neutralization of the acid function and also, simultaneously, a saponification of the ester function.

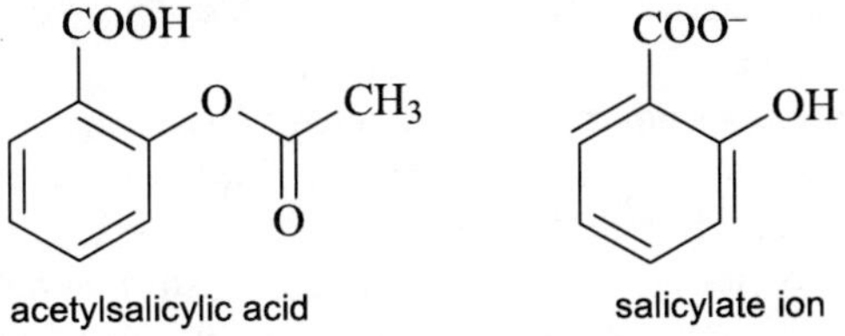

Sodium salicylate and sodium acetate are formed. Strictly speaking, the titration is the neutralization of the excess sodium hydroxide by hydrochloric acid. The comparison of the results of both titrations permits us to know the quantity of aspirin already hydrolyzed before any titration. It also permits us to know the concentration of the self-condensation products of aspirin in the sample, at least those not endowed with acidic properties. The difference in the two added volumes of hydrochloric acid at the equivalence points corresponds to the sodium hydroxide used to saponify aspirin quantitatively according to the reaction

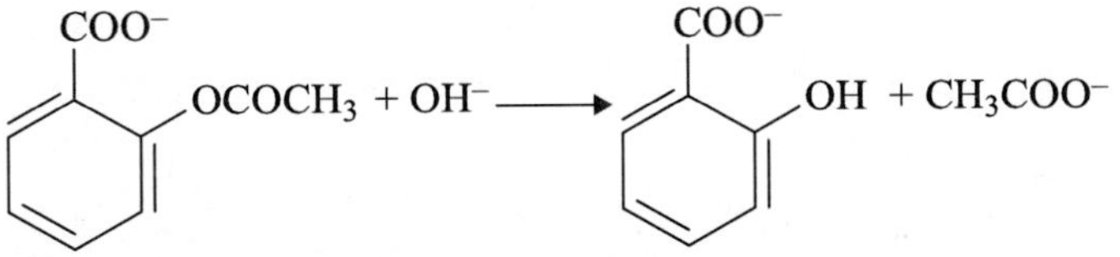

We must remark that after the saponification reaction, the sodium hydroxide does not neutralize the phenol function of the salicylate ion. This is not surprising given its pK_a value ($pK_a = 12.62$). Otherwise, neither the salicylate ion nor the acetate ion is protonated during the titration with hydrochloric acid. This is normal since these ions are carboxylate groups.

Lactic acid (2-hydroxypropionic acid) is often accompanied by condensation products called estolide, triestolide, and tetraestolide resulting from the formation of ester functions between the hydroxide of one molecule and the carboxylic group of another. Their formation induces the disappearance of some free carboxylic groups.

A cyclic lactide can also be formed. It results from the condensation of two molecules of the acid.

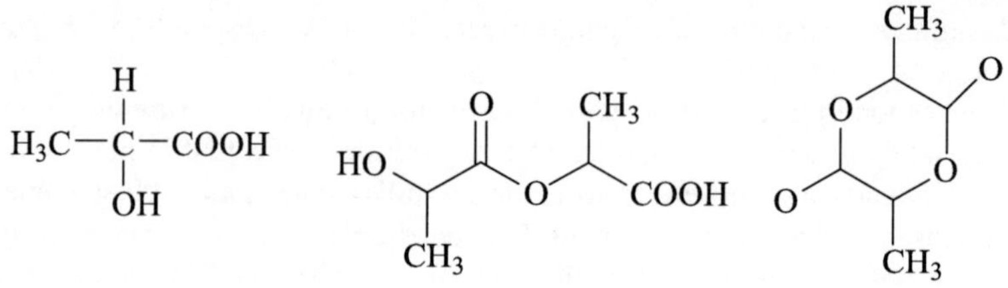

Lactic acid estolide lactide

The whole mixture is determined after reaction with sodium hydroxide in excess for 30 minutes. In these conditions, all the derivatives are quantitatively transformed into the lactate ion. It does not exhibit a sufficiently basic character to be protonated by hydrochloric acid (pK_a value of lactic acid: 3.86), the titrant that permits us to determine the sodium hydroxide excess.

Some other compounds give more complicated transformations through the action of the strong base. Examples of these are trimethadione and coumarin:

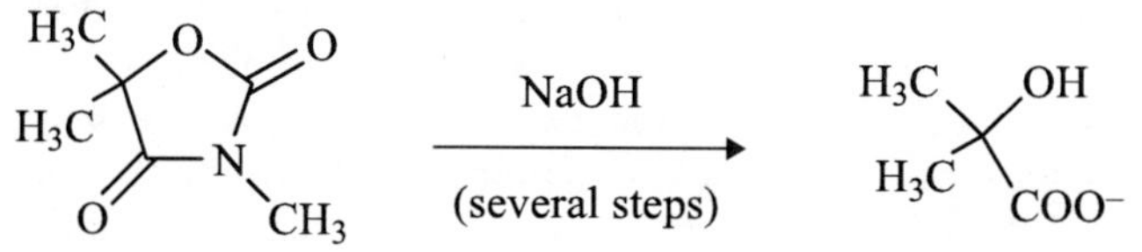

Trimethadione 2-hydroxy-2-methylpropionic acid (sodium salt)

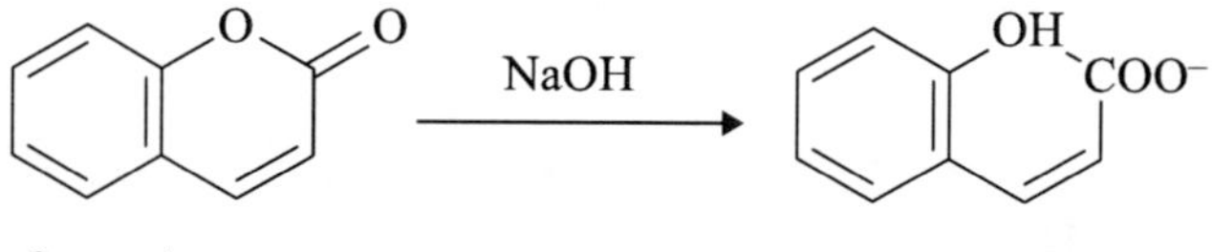

Coumarine cis-coumaric acid (sodium salt)

They are quantitatively transformed into a carboxylate anion.

Busulfan is a compound whose hydrolysis in neutral medium gives tetrahydrofuran and two molecules of methanesulfonic acid, after two steps. The latter is titrated with a sodium hydroxide solution.

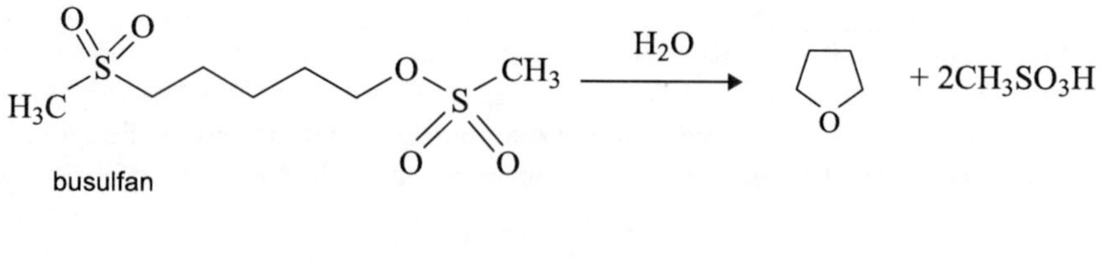

busulfan

The compounds bringing an alcohol function can be transformed quantitatively into an ester by an anhydride. The acid, which is formed mole to mole with the ester, is titrated with a sodium hydroxide solution. This gives rise to the concept of the *acetyl* or *propionyl index* or, more generally, the *esterification* or *hydroxyl index*, both terms being synonymous. The acetyl index is the number of milligrams of acetic acid necessary in order to esterify the hydroxyl groups of one gram of substance. In this assay, the alcohol is esterified by an excess of acetic anhydride in the presence of pyridine. The presence of pyridine induces the total displacement of the esterification reaction toward the right:

$$\text{R-OH} + (\text{CH}_3\text{CO})_2\text{O} \xrightarrow{\text{pyridine}} \text{ROCOCH}_3 + \text{CH}_3\text{COOH}$$

At the end of the transformation, acetic anhydride in excess is hydrolyzed in acetic acid, which is directly titrated in the reaction medium with a sodium hydroxide solution in the presence of phenolphthalein. The same operations are realized with a blank in which there is no hydroxylated derivative, all the other experimental conditions being the same as in the assay. The difference between both volumes added gives the number of moles of acetic acid used for the esterification. The "in pyridine acetylation index" is of great interest in the analysis of essential vegetable oils, fats, and numerous pharmaceutical compounds, including, for instance, the determinations of cetylic and benzylic alcohols. The propionyl index is based on the same principle, but all of the operations are performed in a nonaqueous medium.

Carbonyl derivatives can be determined quantitatively through the notion of the hydroxylamine index, whose final operation is an acid–base titration. Hydroxylamine hydrochloride reacts with the carbonyl function in judicious conditions according to the scheme:

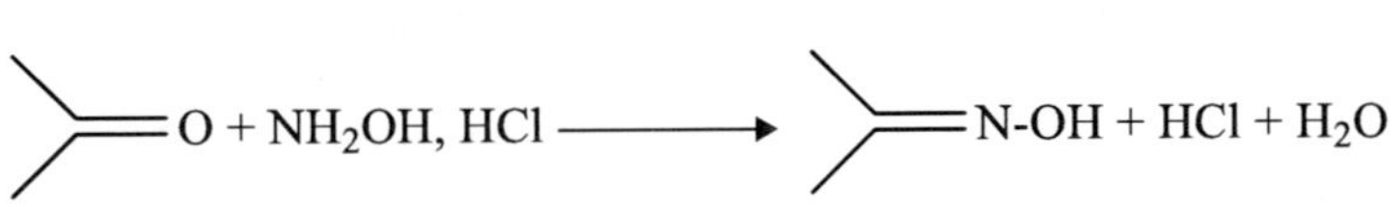

A strong acid is released during the reaction. An important point to note is that the formed oxime does not exhibit any basic character; also, the initial solution was neutral. This is easily explained by the fact that hydroxylamine is a strong base and, hence, its conjugate acid a very weak one. It is interesting also to note that the reaction medium is nonaqueous for the most part, since it only contains 10% water (in mass). Water is added essentially in order to dissolve the hydroxylamine hydrochloride. The medium also contains pyridine in order to displace the oxime formation equilibrium toward the right by formation of the pyridinium ion. The hydrochloric acid released is titrated with a methanolic sodium hydroxide solution. The indicator chosen is bromophenol blue, whose color-change interval is located on the acidic side. Actually, it is the pyridinium cation that is titrated with the methanolic sodium hydroxide solution. The pyridinium cation results from the protonation of the pyridine by the released hydrochloric acid.

Polyols are determined by periodimetry (see Chap. 18). This method is often extended by the titration of the formed formic acid. We shall see that, for example, with a polyol such as glycerol:

$$\begin{array}{l} CH_2OH \\ | \\ CHOH \\ | \\ CH_2OH \end{array} \xrightarrow{\text{periodic acid}} 2HCHO + HCOOH$$

glycerol

the primary alcohol functions are quantitatively transformed into molecules of formaldehyde and the secondary ones into molecules of formic acid under the action of periodic acid, which is simultaneously transformed into iodic acid. The reaction is performed with an excess of periodic acid. It is actually, in the medium, in the form of its conjugate base due to its pK_a value ($pK_a = 0.80$). The periodic acid excess, which would interfere with the titration of formic acid with a sodium hydroxide solution, is destroyed by the addition of ethylene glycol. (Ethylene glycol does not give formic acid in these conditions. It gives only formaldehyde and iodate ions, of course. Hence, ethylene glycol in no way interferes with the titration of formic acid.) The titration of formic acid is performed with a sodium hydroxide solution. The equivalence point is detected by pH-metry.

A particular determination is that of chloral hydrate. It also ends with the titration of the formed formic acid. Under the action of sodium hydroxide in excess, chloral hydrate is hydrolyzed to give chloroform and sodium formate:

$$CCl_3CH(OH)_2 + OH^- \rightarrow CHCl_3 + HCOO^- + H_2O.$$

The sodium hydroxide in excess is titrated with a sulfuric acid solution in the presence of phenolphthalein. The formate ion is too weak a base to be protonated during the titration by sulfuric acid. The back titration must be performed quickly to avoid the decomposition of chloroform by sodium hydroxide, giving supplementary formate ions:

$$CHCl_3 + 4OH^- \rightarrow HCOO^- + 3Cl^- + 2H_2O.$$

The occurrence of this reaction would induce too high a consumption of sodium hydroxide as a result.

Another kind of titration after transformation involves the displacement of a mobile hydrogen from some molecules (the term "mobile hydrogen" must be endowed with the meaning given in organic chemistry). The released hydrogen is titrated with a sodium hydroxide solution. Two sorts of examples are provided by "true" alkynes and also by theophylline. Some steroids of pharmacological interest are "true" alkynes. This is the case with 17-ethinyl steroids. (The term "true alkynes" means that at least one of the sp^1 carbon atoms does bring an atom hydrogen.) For their titration, they are first dissolved in tetrahydrofuran, and after the addition of an aqueous solution of silver nitrate, they are titrated with a sodium hydroxide solution. The equivalence point is detected by pH-metry, as it also can be with bromocresol green. Actually, a complex of the silver salt of the alkyne (one mole) and of silver nitrate (six moles) is formed:

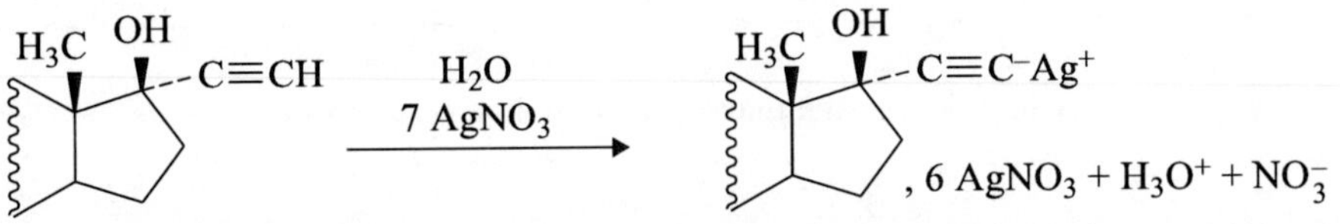

This displacement reaction is very interesting because it has the ability to transform a nonexistent acid in water [pK_a (acetylene) ≈ 25] into a completely ionized acid. In these titrations, tetrahydrofuran seems to be used only to dissolve the organic compound.

Theophylline also gives a silver salt quantitatively. It is only weakly acidic in water ($pK_a = 8.60$—see above: aminophylline). In the presence of silver nitrate, it gives a silver monosalt, which precipitates:

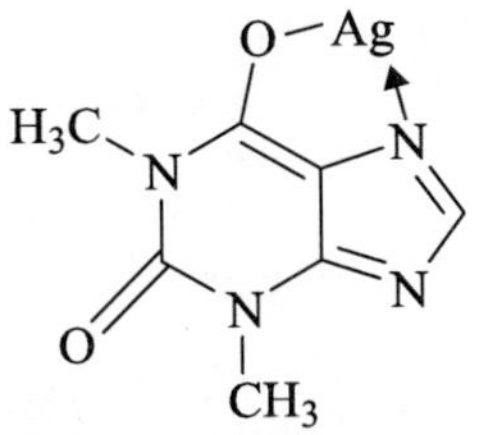

Therefore, the acid–base equilibrium is displaced toward the right. So, theophylline can be titrated by a sodium hydroxide solution with bromophenol blue as indicator.

Other organic substances also give silver salts. Their titrations can be performed in aqueous solution (as above) or in nonaqueous solvents (see other examples in Part V).

Another kind of titration after transformation is provided by the formol titration of α-amino acids (Sørensen's method). We have already seen that it is impossible to titrate them in water with sodium hydroxide, because of their macroscopic ionization constant on the order of $10^{-9.8}$. The reaction with the hydroxide ion is not sufficiently quantitative to be used as a titration reaction:

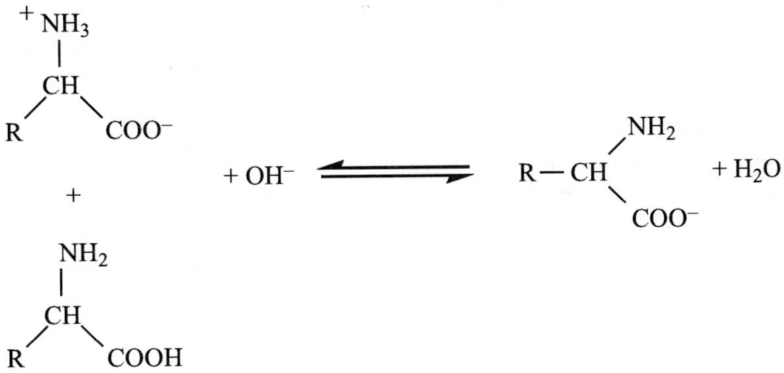

Table 11.1 Apparent pK_a values of glycocolle as a function of the percentage of added formaldehyde

CH_2O %	0	0.5	2	8	16	32
pK_a	9.75	7.8	6.8	5.7	5.4	4.8

Let's recall (see Chap. 5) that the macroscopic ionization constant K_a is defined by the following scheme:

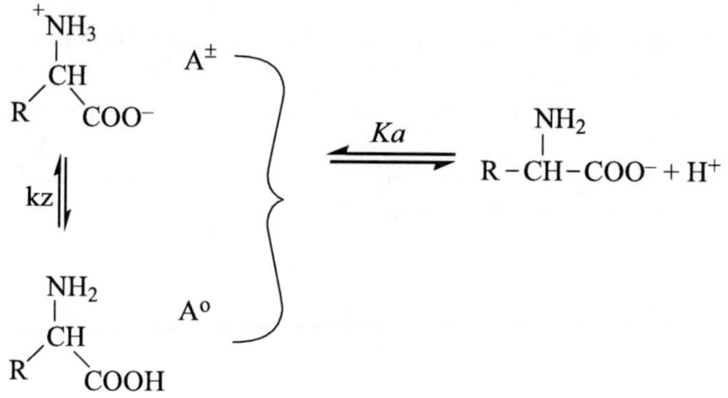

It is connected to the microscopic constants k_{12} and k_{22} by the relation

$$1/K_a = 1/k_{12} + 1/k_{22}.$$

The K_a value is about $10^{-9.8}$. By applying this relation and from the pK_a values of normal protonated aliphatic amines (8.5–10.5) and of normal carboxylic acids (3.6–5.0), it is clear that the microscopic constant k_{12} imposes its value quasi-exclusively to the macroscopic constant K_a. It is clear also that if the amine function disappears after a chemical reaction, the only acid that remains is the carboxylic function. The macroscopic constant K_a value must tend then to that of the microscopic constant k_{22} and its titration by a sodium hydroxide solution must become possible.

As a result, when the amine function is transformed after a more or less equilibrated chemical reaction, the apparent pK_a value of the amino acid must decrease. This is, in fact, the case. When a formaldehyde solution is added to that of an α-amino acid, an imine is formed according to the reaction

$$RNH_2 + CH_2O \rightleftharpoons RN{=}CH_2 + H_2O.$$

It is an experimentally proven fact that according to the percentage of added formaldehyde, the apparent pK_a of the amino acid decreases since less and less amine function remains in the solution (Table 11.1).

One analytical consequence of the given values is that glycocolle can be titrated with a sodium hydroxide solution once 0.5% formaldehyde has been added to the titration vessel. This is the principle of Sørensen's method.

The following approach permits us to estimate the part played by the different intervening parameters in the evolution of the pK_a values given in Table 11.1. It confirms that the principal factor that induces the pK_a change is the aldehyde concentration. It is an equilibrium displacement effect. In the case of a simple carboxylic acid HA, we can immediately infer the following relations:

$$[\mathrm{HA}] + [\mathrm{A}^-] = C_0$$

and

$$K_a = [\mathrm{A}^-][\mathrm{H}^+]/[\mathrm{HA}] \cdot K_a = [\mathrm{A}^-][\mathrm{H}^+]/(C_0 - [\mathrm{A}^-])$$

from the ionization constant and the mass balance expressions. Let's consider the amino acid, which dissociates according to the reaction

$$\{\mathrm{A}^{(+-)} + \mathrm{A}^0\} \rightleftharpoons \mathrm{H}^+ + \mathrm{A}^-.$$

As above, the acid dissociation constant can be written

$$K_a = [\mathrm{A}^-][\mathrm{H}^+]/[(\mathrm{A}^{(+-)} + \mathrm{A}^0)] \quad \text{or} \quad K_a = [\mathrm{A}^-][\mathrm{H}^+]/(C_0 - [\mathrm{A}^-]).$$

Let's suppose that it is the neutral form that reacts with formaldehyde (this hypothesis is probable due to the experimental conditions, in particular the pH value; however, we must be aware that there are several mechanisms of imine formation). The reaction is that mentioned above, which can be written for the sake of simplification as

$$\mathrm{RNH_2 + CH_2O \rightleftharpoons R{-}N{=}CH_2 + H_2O}$$

$$\text{or} \quad \mathrm{A^0 + F \rightleftharpoons A^0F}.$$

The mass balance in the amino acid is now

$$[(\mathrm{A}^{(+-)} + \mathrm{A}^0)] + [\mathrm{A}^0\mathrm{F}] + [\mathrm{A}^-] = C_0.$$

Let us express the concentration $[\mathrm{A}^0\mathrm{F}]$ as a function of the sum $[\mathrm{A}^{(+-)} + \mathrm{A}^0]$. Finally, this will permit us to write this sum as a function of $[C_0 - [\mathrm{A}^-]]$. If K_f is the formation constant of the imine and C_f the total concentration of formol, we can write

$$[\mathrm{CH_2O}] + [\mathrm{A}^0\mathrm{F}] = C_\mathrm{f}$$

$$[\mathrm{A}^0\mathrm{F}]/[\mathrm{CH_2O}][\mathrm{A}^0] = K_\mathrm{f}.$$

Therefore,

$$[\mathrm{A}^0\mathrm{F}]/\{(C_\mathrm{f} - [\mathrm{A}^0\mathrm{F}])[\mathrm{A}^0]\} = K_\mathrm{f},$$

$$[\mathrm{A}^0\mathrm{F}] = K_\mathrm{f} C_\mathrm{f}[\mathrm{A}^0] - K_\mathrm{f}[\mathrm{A}^0\mathrm{F}][\mathrm{A}^0].$$

By adding one clear excess of formol with respect to the amino acid, we get

$$C_f \gg [A^0F],$$

$$[A^0F] \approx K_f C_f [A^0].$$

Otherwise, we know that (see Chap. 5)

$$[A^{(+-)}]/[A^0] = k_z.$$

k_z exhibits a constant value regardless of the pH, whence we have

$$[(A^{(+-)} + A^0)] = [A^0](1 + k_z),$$

$$[A^0] = [(A^{(+-)} + A^0)]/(1 + k_z).$$

As a result,

$$[A^0F] = K_f C_f [(A^{(+-)} + A^0)]/(1 + k_z),$$

$$[(A^{(+-)} + A^0)]\{1 + (K_f C_f)/(1 + k_z)\} = C_0 - [A^-].$$

The apparent dissociation constant of the amino acid in the presence of formol, K'_a, is, by definition,

$$K'_a = [A^-][H^+]/[(A^0 + A^{(+-)})].$$

Since, after the above relation,

$$[(A^{(+-)} + A^0)] = \{C_0 - [A^-]\}/\{1 + (K_f C_f)/(1 + k_z)\},$$

we can write

$$K'_a = [A^-][H^+] \cdot \{1 + (K_f C_f)/(1 + k_z)\}/(C_0 - [A^-]).$$

When no formol is added, the expression of the ionization constant is

$$K_a = [A^-][H^+]/(C_0 - [A^-]);$$

therefore, after comparison, we have

$$K'_a = K_a\{1 + (K_f C_f)/(1 + k_z)\},$$

and since $k_z \approx 10^5$,

$$K'_a = K_a\{1 + (K_f C_f)/(k_z)\}.$$

K'_a increases with the concentration of added formol.

Part III
Redox Phenomena and Analytical Applications

Chapter 12
Generalities on Oxidation-Reduction

Balancing numerous chemical reactions often involves exchanges of electrons between different species. These reactions are called oxidation-reduction reactions, oxido-reduction reactions, or redox reactions. They can take place in a homogeneous phase, specifically within a solution. They can also take place in a heterogeneous medium, on the surface of electrodes. In the first case, we simply speak of redox reactions (it is understood in solution). In the second, we speak of electrochemical reactions.

Redox reactions exhibit some analogies with acid–base reactions. The former involve exchanges of electrons, as we shall see, and the latter involve exchanges of protons. Acid–base reactions involve the occurrence of acid–base couples, and redox reactions that of redox couples.

The concept of oxido-reduction can be applied in inorganic chemistry as well as in organic chemistry. Otherwise, the majority of cellular synthesis is issued from redox reactions, which are only possible in media endowed with well-determined redox potential values. Furthermore, biological cells exhibit a noteworthy constancy in their redox potential courtesy of the redox buffers present in them. We can affirm that all biochemical syntheses are governed by the concepts of proton and electron exchanges.

These few lines are sufficient to give us an overview of the importance of redox phenomena.

12.1 Definitions

A zinc strip dipped into a solution containing cupric ions is spontaneously (in usual conditions) covered up by metallic copper, while an equivalent quantity of zinc ions spontaneously appears within the solution. The chemical reaction that takes place is

$$Zn + {Cu_{(w)}}^{2+} \rightarrow {Zn_{(w)}}^{2+} + Cu. \qquad (12.1)$$

Clearly, it involves an electron exchange. By definition, redox reactions involve an electron exchange among several species. Hence, reaction (12.1) is a redox reaction.

J.-L. Burgot, *Ionic Equilibria in Analytical Chemistry*,
DOI 10.1007/978-1-4419-8382-4_12, © Springer Science+Business Media, LLC 2012

It can be considered to have resulted from the exchange of two electrons between the two couples Zn^{2+}/Zn and Cu^{2+}/Cu, known as *redox couples*. The couples are defined by the following reactions:

$$Zn^{2+} + 2e^- \rightleftharpoons Zn, \tag{12.2}$$

$$Cu^{2+} + 2e^- \rightleftharpoons Cu, \tag{12.3}$$

called *half-redox reactions*. [From now on, the subscript (w) of ions will be systematically omitted since it is understood that we are working in water only. Also, this subscript may be endowed with two means. First, it may symbolize a simple solvation of the ion (the case with alkaline ions, for example). Second, it may represent a true chemical bond (see Part IV, on complexes)].

Equation (12.1) is a whole redox reaction, contrary to reactions (12.2) and (12.3), which are only half-redox reactions. Reaction (12.3) results from the sum of the following two half-reactions:

$$Zn \rightleftharpoons Zn^{2+} + 2e^-,$$

$$Cu^{2+} + 2e^- \rightleftharpoons Cu.$$

In the couples Zn^{2+}/Zn and Cu^{2+}/Cu, Zn^{2+} and Cu^{2+} are the electron acceptors, whereas Zn and Cu are the donors.

During the course of reaction (12.1), Zn lost two electrons. By definition, it is said to have been *oxidized*. Zn^{2+} represents the oxidized form Ox of the couple Zn^{2+}/Zn, whereas Zn is its reduced form, Red. Likewise, Cu^{2+} captures two electrons during the course of reaction (12.1). It has been *reduced*. Cu and Cu^{2+} are, respectively, the reduced and oxidized forms of the couple Cu^{2+}/Cu. Cu^{2+} and Zn^{2+} are said to be in a state of oxidation upper than Cu and Zn. We shall see in the following section that the oxidation states are labeled by oxidation numbers (see Sect. 12.2).

These considerations are general. A redox reaction such as

$$Red_1 + Ox_2 \rightleftharpoons Ox_1 + Red_2 \tag{12.4}$$

can always be considered as resulting from an electron exchange between the two redox couples, Ox_1/Red_1 and Ox_2/Red_2. Every redox couple consists of an oxidized form and a reduced form connected by the relation

$$Ox + ze^- \rightleftharpoons Red. \tag{12.5}$$

The reagent undergoing oxidation is called the *reducing agent,* or *reductant*, and the reagent undergoing reduction is called the *oxidizing agent*, or *oxidant*. [The scheme in (12.5) is not equilibrated from the standpoint of the electrical charges for the sake of generality.]

It is important to emphasize that reaction (12.4) describes a true chemical phenomenon duly studied by experimental means, whereas the two half-reactions from which it results, that is,

$$\mathrm{Red}_1 \rightleftharpoons \mathrm{Ox}_1 + ne^-,$$

$$\mathrm{Ox}_2 + ne^- \rightleftharpoons \mathrm{Red}_2,$$

never take place independently in solution. They must be coupled. Moreover, the split of reaction (12.4) into the corresponding two half-reactions may be completely arbitrary and even can be, in some cases, considered as resulting from a simple writing game. Hence, it may not correspond to the true reaction mechanism of reaction (12.4), even if it correctly traduces the balance of the exchanged electrons.

Redox reactions are not limited to electron exchanges between two redox couples that exchange the same number of electrons in each half-reaction, as above. In most cases, they are more general than those studied so far. Therefore, with the couples

$$\mathrm{Ox}_1 + z_1\mathrm{e}^- \rightleftharpoons \mathrm{Red}_1,$$

$$\mathrm{Red}_2 \rightleftharpoons \mathrm{Ox}_2 + z_2\mathrm{e}^-,$$

where $z_1 \neq z_2$, the reaction occurring is

$$z_2\mathrm{Ox}_1 + z_1\mathrm{Red}_2 \rightarrow z_2\mathrm{Red}_1 + z_1\mathrm{Ox}_2.$$

We immediately see that it involves the exchange of $z = z_1 \, . \, z_2$ electrons, while the stoichiometries remain obeyed.

Some examples of redox couples are given in Table 12.1. Mostly, they also involve exchanges of particules different from electrons together with them. In the realm of analytic chemistry, the word "particule" means a molecule or an ion, excluding the proton. The exchange of particules between two chemical species is sometimes considered a definition of complexation reactions (see Part IV). Hence, there often exist couplings between proton exchanges and electron exchanges or, another alternative, between particule exchanges and electron exchanges. A first example is provided by the oxidation of ferrous iron into ferric iron by the potassium permanganate in acidic medium, according to the reaction

$$\mathrm{MnO_4}^- + 8\mathrm{H}^+ + 5\mathrm{Fe}^{2+} \rightarrow \mathrm{Mn}^{2+} + 5\mathrm{Fe}^{3+} + 4\mathrm{H_2O}.$$

It results from the superposition of the two half-reactions

$$\mathrm{MnO_4}^- + 8\mathrm{H}^+ + 5\mathrm{e}^- \rightleftharpoons \mathrm{Mn}^{2+} + 4\mathrm{H_2O}$$

and

$$\mathrm{Fe}^{3+} + 1\mathrm{e}^- \rightleftharpoons \mathrm{Fe}^{2+}.$$

A second example of a redox couple that can exchange one particule together with electrons is the couple hexachloroplatinate (IV)/tetrachloroplatinate (II):

$$\mathrm{PtCl_6}^{2-} + 2\mathrm{e}^- \rightleftharpoons \mathrm{PtCl_4}^{2-} + 2\mathrm{Cl}^-.$$

Table 12.1 Some examples of redox couples

$Cr_2O_7^{2-} + 14H^+ + 6e^- \rightleftharpoons 2Cr^{3+} + 7H_2O$

dichromate ion — chromic ion

$O_{2(g)} + 4H^+ + 4e^- \rightleftharpoons 2H_2O_{(l)}$

$UO_2^{2+} + 4H^+ + 2e^- \rightleftharpoons U^{4+} + 2H_2O$

uranyl cation — uranous ion

$HCHO + 2H^+ + 2e^- \rightleftharpoons CH_3OH$

formaldehyde — methanol

$CH_3COOH + 2H^+ + 2e^- \rightleftharpoons CH_3CHO + H_2O$

acetic acid — acetaldehyde

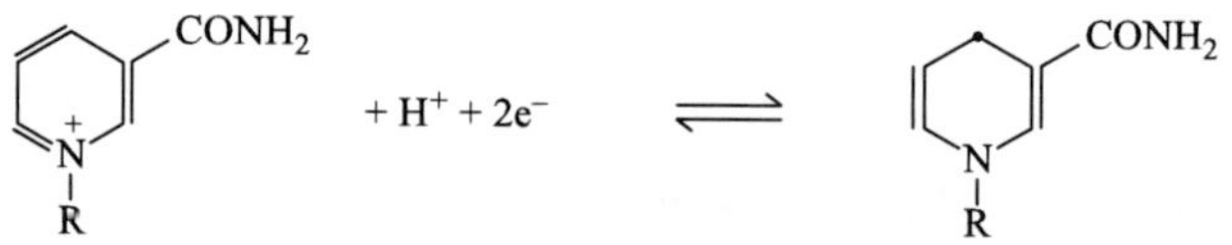

NAD^+ — NADH, H^+

nicotinamide adenine dinucleotide — reduced nicotinamide adenide dinucleotide

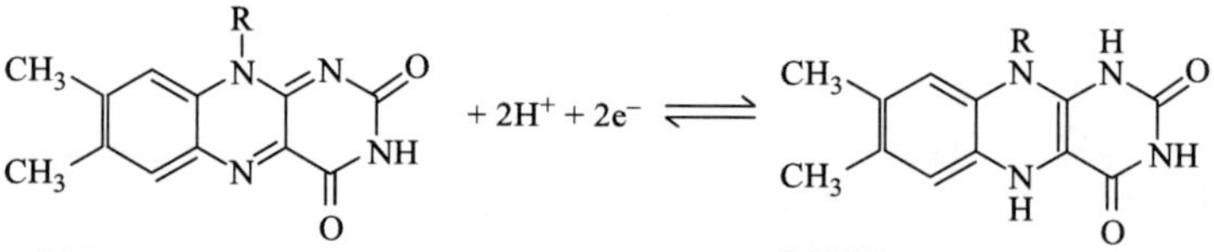

FAD — $FADH_2$

flavin adenine dinucleotide — reduced flavin adenine dinucleotide

$COO^- - CH = CH - COO^- + 2H^+ + 2e^- \rightleftharpoons COO^- - CH_2CH_2 - COO^-$

fumarate ion — succinate ion

etc.

A third example is the couple iodate (IO_3^-)/iodine chloride (ICl):

$$IO_3^- + Cl^- + 6H^+ + 4e^- \rightleftharpoons ICl + 3H_2O.$$

The two last examples also have an exchange of the chloride ion (the particule) together with that of the electrons.

Reactions redox can also take place in polyphased media. For example, the redox reaction can be accompanied by a precipitation or by a gaseous emission. Two classical examples are provided by the couple permanganate (MnO_4^-)/manganese dioxide (MnO_2) and by water:

$$MnO_4^- + 4H^+ + 3e^- \rightleftharpoons MnO_{2(s)} + 2H_2O$$
$$2H_2O \rightleftharpoons O_{2(g)} + 4H^+ + 4e^-$$

Within these limits are also located the electrochemical reactions that take place in media where at least three phases exist: both electrodes and the solution.

Finally, the preceding definitions can be extended to covalent compounds by adopting some new definitions and conventions. It is precisely in achieving this extension that the concept of oxidation number is fruitful.

12.2 Oxidation Numbers

Oxidation numbers characterize the oxidation state of one element within a chemical species. They are specified by a Roman number, positive, negative, or null, located as a superscript to the right of the symbol of the element. Alternatively, it can be located to the right and on the same line as the complete name of the species.

In most species, the element that is oxidized or reduced may not be clearly defined. For example, in the electrolytic oxidation of cyanide ion into cyanate ion,

$$CN^- + H_2O \rightleftharpoons CNO^- + 2H^+ + 2e^-,$$

who can say whether it is carbon, nitrogen, hydrogen, or oxygen that has lost the two electrons? To overcome this problem, some arbitrary conventions and definitions have been devised.

According to the IUPAC, the electrons shared by two atoms (to form the bond between them) must be assigned to the more electronegative one of the two. If the two atoms sharing them are identical, the electrons must be divided equally. The formal charge remaining on the atom, once the assignment is made, is its oxidation number. Another convention that completes and extends the preceding one is that hydrogen is positive when it is combined with a nonmetal. In the oxidation of cyanide ion, the oxidation number of carbon changes from +II to +IV, while those of nitrogen, oxygen, and hydrogen remain fixed at −III, −II, and +I, respectively.

These conventions make the oxidation numbers highly tributary of Lewis schemes, which fix the assignment of electrons of the bounds in covalent compounds. Two cases must be considered:

- simple ions containing only one atom. In this case, the oxidation number is equal to its charge. Therefore, for the ions K^+ and Th^{4+}, we can write K^{+I} and Th^{+IV}:
- molecules and complex ions. Here are some oxidation numbers:

	Oxidation numbers	
Water H_2O	O^{-II}	H^{+I}
Chromate ion $CrO_4{}^{2-}$	O^{-II}	Cr^{+VI}

The assignment of the number +VI to the chrome atom results from the comparison between the electronegativities of chrome and oxygen (respectively 1.56 and 3.50—Allred and Rochow's scale) and from the study of the Lewis structure of chromate ions. It is interesting, incidentally, to note that in the dichromate ion, the oxidation

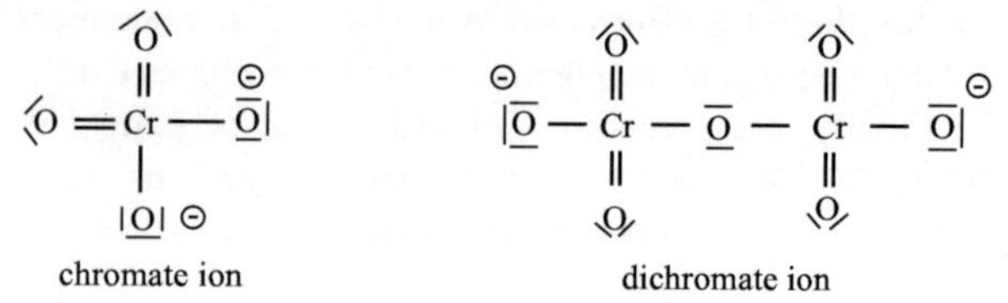

Fig. 12.1 Lewis structures of the chromate and dichromate ions

numbers of both chrome atoms are the same as that found in the chromate ion (+VI) (Fig. 12.1).

One element does not necessarily exhibit the same oxidation number when it is present several times in the same structure. For example, in the thiosulfate ion $S_2O_3^{2-}$,

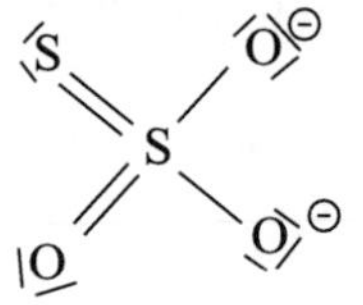

The fact that, by convention, the double bond SS does not contribute to the oxidation number of the whole ion leads to the following values: the central sulfur +IV, the other sulfur 0, and the three oxygens −II (electronegativities of sulfur 2.2 and 2.44).

A rapid means for determining an element's oxidation number in a chemical species exists, but it is not infallible. It consists of assigning the oxidation numbers +I and −II systematically to hydrogen and oxygen, respectively. This permits us immediately to deduce those of the other elements. This assertion is very often true, but it is not always the case. For example, depending on the compounds, hydrogen may exhibit the oxidation numbers $+\mathrm{I}(HCl) : 0(H_2) : -\mathrm{I}(HNa)$. These values are derived from the study of the electronegativity scale. Oxygen exhibits the following ones:

$$-\mathrm{II}(H_2O) : -\mathrm{I}(H_2O_2) : 0(O_2) : +\mathrm{II}(F_2O).$$

Considering the case of oxygen fluorine only, the value +II results from the fact that fluorine is more electronegative than oxygen.

In a molecule, the sum of the oxidation numbers is equal to zero, whereas it is equal to the electrical charge in an ion containing several elements.

From another standpoint, a positive oxidation number must not exceed the number of electrons in the external shell of the element. If it is negative, it must not exceed the number of electrons necessary to saturate the external shell. For example, for the redox couple peroxodisulfate—sulfate ($S_2O_8^{2-}/SO_4^{2-}$) assigning the value −II to oxygen leads to the oxidation number +VII for the two sulfur atoms of the ion peroxodisulfate. This result is wrong because the sulfur atom possesses six electrons only in its external shell. Actually, the peroxodisulfate ion contains a peroxo bridge in

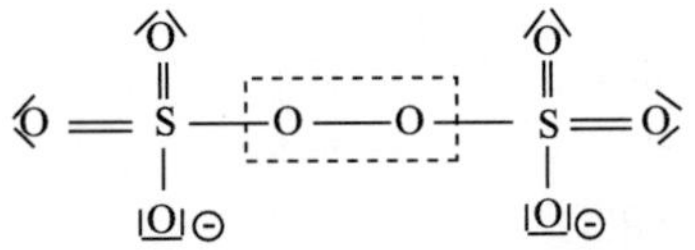

Fig. 12.2 Peroxo bridge in the peroxodisulfate ion

its structure. Its presence confers an oxidation number of −I (instead of −II usually) to two of its oxygen atoms, while others exhibit the classical value (see Fig. 12.2).

During the course of an oxidation with the peroxodisulfate ion, each oxygen atom of the peroxo bridge captures one electron while the O–O bond is breaking and, finally, two sulfate ions are formed. The half-redox reaction can be written as

$$S_2O_8{}^{2-} + 2e^- \rightleftharpoons 2SO_4{}^{2-}.$$

Finally, let's emphasize the fact that Roman numbers recall that oxidation numbers do not necessarily give the actual electrical charge of some species—species that, furthermore, may not exist. For example, manganous ions Mn^{2+} exist. This is a case in which Mn^{+II} does effectively represent the true electrical charge. On the contrary, the manganese in the permanganate ion $MnO_4{}^-$ exhibits the oxidation number +VII, but the cation Mn^{7+} does not exist.

There are advantages and drawbacks to the concept of oxidation number. Speaking of its advantages, we can say that it extends the oxido-reduction concept to covalent derivatives. This point is again discussed in the following section. It also makes the systematic description of the derivatives of an element easier by providing a logical presentation of them. This is especially clear with the presentation of transition elements. A third advantage is that it facilitates the equilibration of redox titrations. On the drawback side, let's first mention that the possible existence of several sets of oxidation numbers may occur. Moreover, the application of the oxidation number concept in the realm of organic chemistry is difficult. New conventions must be adopted. Therefore, the oxidation number concept does not bring any substantial advantage with respect to the usual reasonings and writings of the organic chemists.

12.3 Redox Titrations and Oxidation Numbers

The oxidation of an element is characterized by an increase in its oxidation number, and its reduction is represented by a decrease. An oxidant is a species whose oxidation number decreases during the course of a redox reaction, and a reductant is a species whose oxidation number increases. The coupling of two half-redox reactions for interpreting a redox reaction can be extended in terms of oxidation numbers. An example is provided by the reaction of the reduction of carbon dioxide by dihydrogen to give carbon monoxide:

$$CO_{2(g)} + H_{2(g)} \rightarrow CO_{(g)} + H_2O_{(g)}.$$

We can see that in the carbon dioxide, the carbon exhibits the oxidation number +IV, whereas in the carbon monoxide it does it in the oxidation state +II. During the course of the reaction, hydrogen goes from state 0 to state +I. The two fictitious half-redox couples involved are

$$C^{+IV} + 2e^- \rightleftharpoons C^{+II},$$

$$H_2 \rightleftharpoons 2H^+ + 2e^-,$$

the oxidation number of oxygen not having changed during the reaction.

The last sentence points out that an ambiguity can exist in the case of the oxidation or reduction of a polyatomic species without any supplementary information. Indeed, the element that undergoes oxidation or reduction must be named. Two interesting examples are provided by the couples of water. The first one corresponds to the half-redox reaction

$$O_{2(g)} + 4H^+ + 4e^- \rightleftharpoons 2H_2O.$$

Usually, we speak of water oxidation, but, actually, it is the oxidation of the oxygen element, which in water exhibits the state −II and which is in the state 0 in the dioxygen. The second couple of water is

$$2H^+ + 2e^- \rightleftharpoons H_{2(g)},$$

which can also be written according to the pH value:

$$2H_2O + 2e^- \rightleftharpoons H_{2(g)} + 2OH^-.$$

No ambiguity exists in the first half-redox equation: The redox couple under study is $H^+/H_{2(g)}$. However, according to the second half-reaction, there is an ambiguity when the reduction of water is involved. It would be more appropriate to speak of the reduction of the hydrogen element in water.

12.4 Particular Cases of Redox Reactions: Disproportionation and Retrodisproportionation Reactions

Some elements exhibit several oxidation numbers. Hence, they can successively exchange several electrons with other redox couples. By analogy with some acid–base couples, one element or a chemical species that is simultaneously the oxidant of a couple and the reductor of another is named an "ampholyte" despite the fact it exhibits the same oxidation number. Such is the case, for example, for the couple V^{3+}/V^{2+}(V^{3+}: vanadate ion; V^{2+}: hypovanadous ion):

$$V^{3+} + e^- \rightleftharpoons V^{2+}.$$

V^{3+} is an oxidant in the above reaction but is a reductant in the following one (VO^{2+}; vanadyl ion):

$$VO^{2+} + 2H^+ + 1e^- \rightleftharpoons V^{3+} + H_2O$$

A vanadate ion can react with itself through the redox reaction

$$V^{3+} + V^{3+} + H_2O \rightleftharpoons VO^{2+} + V^{2+} + 2H^+.$$

The reaction evolving from left to right is called a disproportionation reaction and that evolving from right to left a retrodisproportionation reaction or an amphoterization reaction. They are particular cases of redox reactions, and all the general considerations developed so far can be applied to them. Another example of a disproportionation reaction is provided by the reaction of dichlorine with water in the interval $3.3 < \mathrm{pH} < 7.5$. Hypochlorous acid (HClO) and chloride ions (Cl^-) are formed:

$$Cl_2 + H_2O \rightarrow HClO + H^+ + Cl^-.$$

The oxidation numbers of oxygen and hydrogen do not change in the course of this reaction, whereas that of chlorine has changed from 0 to simultaneously −I and +I.

The two involved half-redox couples are

$$^1/_2Cl_2 + 1e^- \rightleftharpoons Cl^-.$$

$$^1/_2Cl_2 + H_2O \rightleftharpoons HClO + H^+ + 1e^-.$$

One example of retrodisproportionation is provided by the element iodine in the iodate/iodide reaction:

$$IO_3{}^- + 5I^- + 6H^+ \rightarrow 3I_2 + 3H_2O.$$

The oxidation numbers of oxygen and hydrogen have not changed in the course of the reaction. However, that of iodine has changed. It went from +V (iodate ion) and −I (iodide ion) to 0 (iodine). The two redox couples are

$$IO_3{}^- + 6H^+ + 5e^- \rightleftharpoons 1/2I_2 + 3H_2O,$$

$$I^- - 1e^- \rightleftharpoons 1/2I_2.$$

The iodate/iodide reaction is very useful in analytical chemistry.

12.5 Equilibration of Redox Reactions

The equilibration of redox reactions can be done in several steps. Let's consider the oxidation reaction of iodide ions by dichromate ions. It must be written as

$$Cr_2O_7{}^{2-} + 14H^+ + 6I^- \rightarrow 2Cr^{3+} + 3I_2 + 7H_2O.$$

The first step consists of identifying the two half-reduction couples that are facing each other and affixing to them the judicious stoichiometric coefficients in such a way that the equilibrium in electrons should be satisfied. This step is called the *redox balance*. In the given example, the two half-redox reactions are

$$Cr_2O_7{}^{2-} + 6e^- + 14H^+ \rightleftharpoons 2Cr^{3+} + 7H_2O,$$

$$\text{and} \quad I^- \rightleftharpoons 1/2I_2 + 1e^-.$$

The electron balance necessarily takes into account the oxidation number values. Hence, we must multiply the first equation by the factor 1 and the second by the factor 6. We find

$$Cr_2O_7{}^{2-} + 6e^- \rightleftharpoons 2Cr^{3+},$$

$$6I^- - 6e^- \rightleftharpoons 3I_2.$$

Adding these two half-reactions finishes the redox balance process. We find

$$Cr_2O_7{}^{2-} + 6I^- \rightleftharpoons 2Cr^{3+} + 3I_2.$$

This is the end of the first step. Notice that the electrical charges are not equilibrated at this step.

The second step is devoted to the equilibration of the electrical charges on both sides of the preceding equation. This does not mean the electroneutrality of the solution (see the third step). This means only that the numbers of the positive and negative electrical charges must be the same on the two sides of the equation. We see that at the end of the first step, the left side does possess eight negative charges, while the right side does possess six positive charges. From a general standpoint, the equilibration is achieved by the addition of water molecules and protons (or hydroxide ions) according to the pH conditions of the reaction. In this example, the reaction takes place in acidic medium. In order to equalize the charges, 14 positive charges (protons) must be added to the left-hand side. This induces the appearance of seven water molecules on the right-hand side, this time in order to "balance atomically" the equation. We find

$$Cr_2O_7{}^{2-} + 6I^- + 14H^+ \rightleftharpoons 2Cr^{3+} + 3I_2 + 7H_2O.$$

Let's now emphasize the fact that the equation, written in this form, is completely satisfactory for the study of the redox phenomenon, as is the case for the study of the acid–base one that accompanies it.

A possible third step would consist of writing the equation in such a manner that the electrical neutrality of the solution should be visible. Recall that the electrical neutrality of a solution is an inescapable experimental fact. In order to do that, we must also take into account the ions that do not take place in the exchanges of the electrons and protons. They are sometimes called "spectator" ions. In this example, the dichromate and iodide ions can be dissolved as potassium salts, for example, and the acidic medium achieved by the addition of hydrochloric acid. Finally, the wholly equilibrated reaction is

$$Cr_2O_7{}^{2-} + 6I^- + 14H^+ + 8K^+ + 14Cl^- \rightarrow 2Cr^{3+} + 8K^+ + 14Cl^- + 3I_2 + 7H_2O,$$

or in a more compact form,

$$K_2Cr_2O_7 + 6KI + 14HCl \rightarrow 2CrCl_3 + 8KCl + 7H_2O + 3I_2.$$

This last writing is not realistic since some of the written species do not exist as such in water. For example, $K_2Cr_2O_7$ does not exist. The species that exist in water are the ions $Cr_2O_7{}^{2-}$ and K^+.

The described strategy can be applied to all of the redox reactions.

Exercise 1 Equilibrate the following redox reactions:

$$VO^{2+} \rightarrow VO_3{}^-,$$

$$Cr^{3+} \rightarrow Cr_2O_7{}^{2-},$$

$$Mn^{2+} \rightarrow MnO_2.$$

Answers

$$VO^{2+} + 2H_2O \rightarrow VO_3{}^- + 4H^+ + 1e^-$$

$$2Cr^{3+} + 7H_2O \rightarrow Cr_2O_7{}^{2-} + 14H^+ + 6e^-$$

$$Mn^{2+} + 4OH^- \rightarrow MnO_2 + 2H_2O + 2e^-$$

Exercise 2 The following reactions take place (in judicious experimental conditions) in the direction indicated.

$$V^{2+} + TiO^{2+} + 2H^+ \rightarrow V^{3+} + Ti^{3+} + H_2O$$

$$Bi_{(s)} + 3Fe^{3+} + H_2O \rightarrow BiO^+ + 3Fe^{2+} + 2H^+$$

$$Zn_{(s)} + 2Cr^{3+} \rightarrow Zn^{2+} + 2Cr^{2+}$$

$$2Fe^{2+} + NO_3{}^- + 3H^+ \rightarrow 2Fe^{3+} + HNO_2 + H_2O$$

$$Mg + Zn^{2+} \rightarrow Mg^{2+} + Zn$$

Decompose these global redox reactions into the corresponding half–redox reactions.

Answers

$$V^{3+} + 1e^- \rightleftharpoons V^{2+}$$

$$BiO^+ + 2H^+ + 3e^- \rightleftharpoons Bi_{(s)} + H_2O$$

$$TiO^{2+} + 2H^+ + 1e^- \rightleftharpoons Ti^{3+} + H_2O$$

$$Fe^{3+} + 1e^- \rightleftharpoons Fe^{2+}$$

$$Cr^{3+} + 1e^- \rightleftharpoons Cr^{2+}$$

$$NO_3{}^- + 3H^+ + 2e^- \rightleftharpoons HNO_2 + H_2O$$

$$Zn^{2+} + 2e^- \rightleftharpoons Zn$$

$$Mg^{2+} + 2e^- \rightleftharpoons Mg$$

Chapter 13
Redox Reactions and Electrochemical Cells

Some electrochemical cells permit a half-redox reaction to occur separately but obligatorily simultaneously with another half-redox reaction. This is not possible with redox reactions taking place within a solution. In this case, it is impossible to separate the two half-reactions into which the global reaction can be decomposed. This possibility, which comes from using electrochemical cells, permits us to quantify the oxidizing strength of redox couples through the concept of standard potentials. This is achieved with the help of Nernst's equation. Thus, predicting the direction of redox reactions and their extent is possible on the basis thermodynamics.

13.1 Electrochemical Cells and Redox Reactions: Example of Daniell's Galvanic Cell

An electrochemical cell is a device that consists of two electronic conductors (usually a metal) called *electrodes* each dipping into separate ionic solutions. They are inserted within an electrical circuit (see Chap. 2 and Figs. 2.1 and 2.2). Hence, a cell consists of two compartments, each of which consists of an electrode and a solution into which it dips. Both electrodes seldom dip into the same electrolytic solution. In the literature, the word "electrode" may refer to the whole compartment where it stands. In brief, an electrochemical cell is a system in which one of the electronic conductors acts as an electron source for particules in an ionic conductor (the solution) and the other as an electron sink that receives electrons from the ionic conductor.

From an operational standpoint, electrochemical cells in which faradaic currents are flowing are classified as either galvanic or electrolytic cells (a faradaic current occurs if charges, e.g., electrons, are transferred across the metal–solution interface—see electrochemistry). The example of Daniell's cell permits us to clarify these points.

J.-L. Burgot, *Ionic Equilibria in Analytical Chemistry*,
DOI 10.1007/978-1-4419-8382-4_13,

13.1.1 Galvanic Cell

For example, consider the redox reaction

$$Zn + Cu^{2+} \rightarrow Zn^{2+} + Cu \tag{13.1}$$

There is an electrochemical cell, called *Daniell's cell,* whose overall reaction occurring within it is (13.1). It is called the *cell reaction.* It can be decomposed into the two half-reactions

$$Zn \rightleftharpoons Zn^{2+} + 2e^- \tag{13.2}$$

$$Cu^{2+} + 2e^- \rightleftharpoons Cu \tag{13.3}$$

It is an experimental fact that in Daniell's galvanic cell (see Chap. 2), the reaction of oxidation of the zinc [reaction (13.2)] takes place at the surface of the zinc electrode dipping into the solution of zinc sulfate, while, simultaneously, the reduction of cupric ions according to (13.3) takes place at the surface of the copper electrode dipping into the solution of cupric sulfate. The zinc and copper wires are the electrodes. The two independent half-reactions (13.2) and (13.3) describe the real chemical changes onto both electrodes.

However, for the reactions to take place, an electrical circuit must exist. It is achieved as follows: On the one hand, both electrodes are connected by a metallic conductor wire; on the other hand, there is also a conductor bridge (or a liquid junction) between the two compartments. The conductor achieves the electrical contact between them. (The bridge also plays a part in preventing the solutions from mixing.)

Since they take place at the surface of the electrodes, both reactions (13.2) and (13.3) are called electrochemical reactions. They are also called electrode reactions. The cell reaction is reaction (13.1).

We already know that at the end of the redox reactions (13.2) and (13.3), the equilibrium state obtained is the same as that obtained at the end of reaction (13.1) achieved in the chemical manner. In the usual conditions ($P = 1$ bar and $T = 298$ K), reaction (13.1) takes place spontaneously, without putting any external energy into the system. An electrochemical cell whose cell reaction is a chemical reaction that occurs spontaneously is a *galvanic cell* (L. Galvani: Italian physician, 1737–1798). In such a cell, the electrons flow freely into the external circuit. The resulting current may, for example, supply energy to an electrical motor. A galvanic cell permits the direct transformation of chemical energy (stored in the zinc and copper wires for the chosen example) into work. This is the origin of its qualification as energy-producing device.

According to the laws of electrostatics, the displacement in the wire of electrons in the indicated direction (Fig. 13.1) is due to the negative charge spontaneously taken by the zinc wire, which pushes them away. The displacement is also due to the positive charge brought by the copper wire, which attracts the electrons. In brief, the electrons flow against the electrical potential.

We can easily understand that the potential difference that appears between the electrodes and induces the displacement of the electrons is due at the very beginning

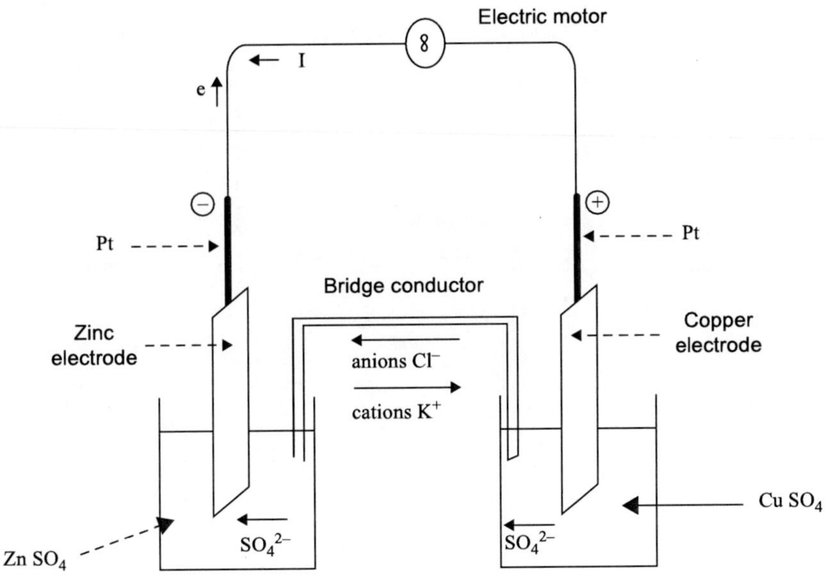

Fig. 13.1 Daniell's galvanic cell

Fig. 13.2 Spontaneous ionization of a metal in a solvent

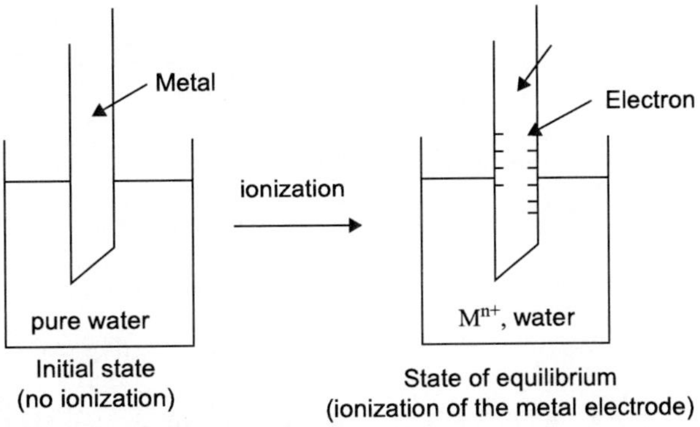

of the phenomenon to the spontaneous ionization of both metals in the electrolyte solutions, though these ionizations are not identical. In the present case, zinc is ionized more spontaneously in the zinc sulfate solution than copper is in the copper sulfate solution (Fig. 13.2).

13.1.2 Electrolytic Cell

Let's now consider an electrochemical system analogous to the galvanic Daniell cell but that differs from it by the presence of a power supply instead of the electrical

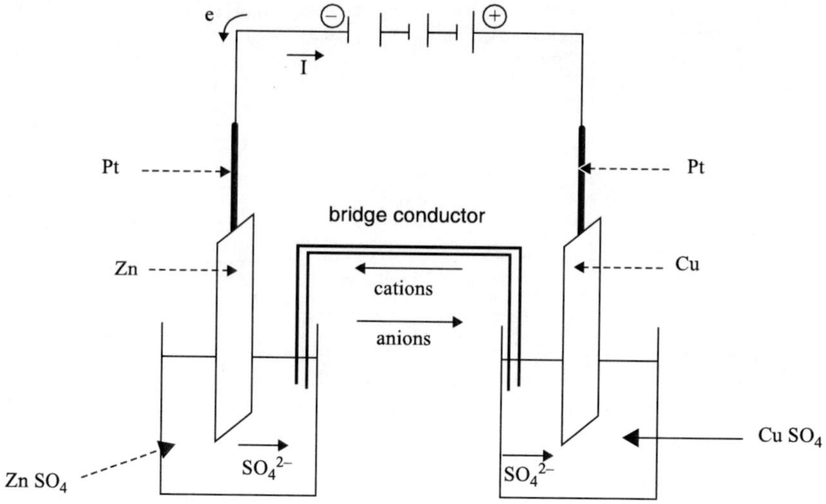

Fig. 13.3 An example of an electrolytic cell (unconventional scheme—see Sect. 13.6)

motor. It permits us to apply a potential difference sufficient to make the electrons go in a direction opposite that followed in Daniell's galvanic cell. The reactions promoted in these conditions are

$$Zn^{2+} + 2e^- \rightarrow Zn \tag{13.4}$$

at the surface of the zinc wire, and

$$Cu \rightarrow Cu^{2+} + 2e^- \tag{13.5}$$

at the surface of the copper wire. The reaction cell is

$$Zn^{2+} + Cu \rightarrow Zn + Cu^{2+} \tag{13.6}$$

It is just the opposite of that occurring in Daniell's galvanic cell. Reaction (13.6) cannot be achieved spontaneously in a pure chemical manner without receiving energy (heat, for example) from the surroundings. An electrochemical cell whose reaction cell is not spontaneous is called an *electrolytic cell* or a *substance-producing device* (Fig. 13.3).

From the electrostatic standpoint, when the cell behaves as an electrolytic cell, the zinc wire brings a negative charge and the copper wire a positive charge as in the galvanic behavior mode, but in the first case the behavior is imposed, whereas in the second, it is spontaneous.

The success of reaction (13.6) achieved in an electrochemical way is not miraculous from an energy standpoint. Both half-reactions (13.4) and (13.5) ask for external energy, the electrical energy coming from the power supply in the occurrence. Likewise, the equivalent redox reaction (13.6) achieved in a chemical manner asks for heat given from the surroundings of the system. [It is interesting to notice incidentally that the global reaction (13.6) achieved via an electrochemical means exhibits a better yield than that achieved via a chemical method.]

In brief, an electrolytic cell permits the preparation of some compounds that cannot spontaneously be obtained by a chemical means without a capture of energy, wherefrom the qualifier given above.

13.2 Nature of the Electrical Current in an Electrochemical Cell

The electrical current in the external circuit (electrodes included) is ensured by the carriage of charges by the movement of electrons, wherefrom the name "electronic (or metallic) current." In the ionic solutions of the compartments, the current is not due to the movement of electrons since their lifetime is extremely short in aqueous solutions. The current is due to the movement of ions, anions, and cations (Zn^{2+}, Cu^{2+}, $SO_4{}^{2-}$). Therefore, we speak of ionic current. The anions $SO_4{}^{2-}$, which actually do not participate in the redox phenomenon, also migrate to ensure the electrical neutrality of both compartments. The zinc compartment receives additional sulfate ions during the galvanic behavior in order to counterbalance the formation of zinc ions by ionization of the electrode. It is the opposite during the electrolytic behavior. Likewise, during the galvanic behavior, the copper compartment loses sulfate ions since the Cu^{2+} ions disappear. It is the opposite for the electrolytic mode.

Hence, two kinds of electrical current exist in an electrochemical cell: the electronic current and the ionic current. According to Kirchhoff's law, no current can accumulate at any point of the circuit. As a result, the question arises concerning the nature of the phenomenon permitting the continuity of the current. It is clear that the junction between the two kinds of current cannot be located anywhere else than at the electrode–solution interface since each member of the interface possesses its own kind of current. The answer to the question is that the current's continuity is achieved by the reactions taking place at the interfaces, that is, by the electrochemical reactions. Indeed, electron transfer between two phases is the fundamental act of electrochemistry. In the case of Daniell's galvanic cell, reactions (13.2) and (13.3) take place. During the course of reaction (13.2):

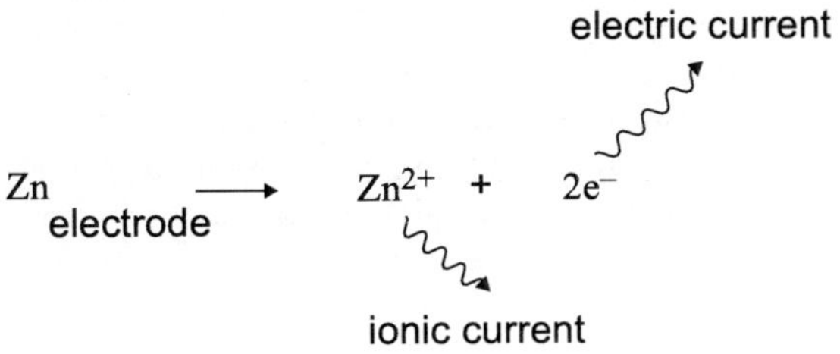

the electrons resulting from the ionization of zinc remain in the metallic circuit and will constitute the electronic (metallic) current, while zinc ions go into the solution, constituting the ionic current. Simultaneously, at the copper electrode, cupric ions are reduced according to reaction (13.3):

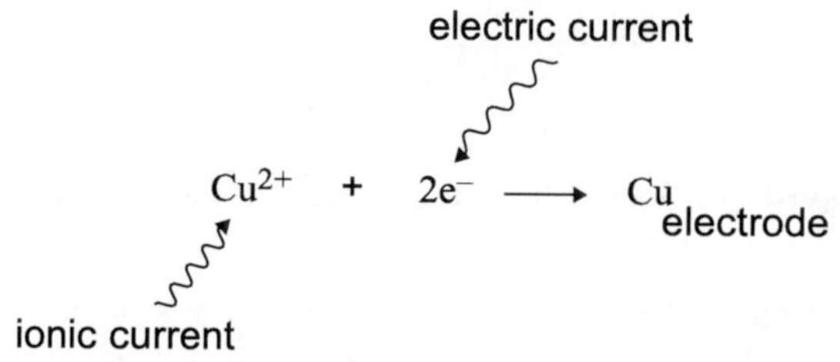

and capture the electrons coming from the zinc wire through the metallic circuit. Analogous schemes could have been presented in the case of an electrolytic cell. Then, however, the free electrons, which achieve the reduction at the surface of the zinc wire, would have come from the power supply of the external circuit, copper giving up electrons to it.

The separator enables the current stream to flow between the two compartments and avoids mixing the chemical species. The mixing would have promoted the global redox reaction solely in a chemical manner. In these conditions, no current could flow along the circuit. From a practical standpoint, for example, a separator is a conductor bridge that is a paper saturated with an aqueous solution of an inert electrolyte. It can also be a tube whose ends are constituted by sintered-glass disks. The tube is also filled with a solution of an inert electrolyte. The sintered-glass disks limit the mixing of both ionic liquids. The inert electrolyte is constituted by a solution of a salt (KNO_3–KCl) whose ions do not participate in the redox reaction. The presence of a junction between the two compartments induces supplemental potential differences in the circuit, which may or may not be negligible.

Normally, both electrodes must be connected to a metallic wire of the same nature (Pt, Cu, for example), regardless of the part played by the electrode, in order to correctly represent the electrochemical cells. Indeed, a potential difference between two identical phases is the sole potential difference to be endowed with a physical meaning. Hence, the cell potential must be the potential difference measured between two identical wires.

Remark The ionic species that undergo the electrochemical reactions move under the influence of several transport phenomena: ionic drift under an electric field (a transport phenomenon also called *conduction* or *migration*), drift under a chemical-potential gradient (*diffusion transport*), and/or a convection phenomenon. The origin of the transport of electroactive species plays a very important part in the principles of the electrochemical methods of analysis.

13.3 The Hydrated Electron

We have already noticed that the electrons in the redox phenomena do not exactly play the same part as that played by the protons in the acid–base phenomena. This is due in part to the brief lifespan of electrons in water. It is interesting, however, to know some properties of the hydrated electron.

Its lifetime is about $8 \cdot 10^{-4}$ s at pH > 5.5. The reaction that limits its lifetime is

$$e-_{(w)} + H_3O^+ \rightarrow H + H_2O.$$

(In some solvents, the solvated electron may exhibit a lifetime by far longer than in water.)

We understand why it is not judicious to systematize the writing of redox reactions by taking into account the activity of the hydrated electron ($e^-_{(w)}$) or even its concentration $[e^-_{(w)}]$, as was the case for the hydrated proton in the description of acid–base phenomena. However, some points of the chemistry of the hydrated electron are known. It is interesting to recall some of them. We note

- at very weak concentration, $e_{(w)}{}^-$ reacts with water and forms hydrogen atoms:

$$e_{(w)}{}^- + H_2O \rightarrow H + OH^-.$$

If its concentration is greater than 10^{-8} mol/L, the following reaction occurs:

$$2e_{(w)}{}^- + 2H_2O \rightarrow H_{2(w)} + 2OH^-;$$

- the hydrated electron can reduce some metallic ions. In some cases, it gives species that are unknown in classical chemistry, such as Zn^+, Pb^+, Cd^+, and Co^+. They are, of course, transitory species;
- the hydrated electron possesses a redox potential corresponding to the half-redox reaction

$$e_{(w)}{}^- + H^+ \rightleftharpoons 1/2H_2.$$

The standard potential of the couple is -2.77 V (see Sects. 13.9 and 13.10). This half-reaction differs from the following one, which defines the hydrogen electrode:

$$H^+ + e^- \rightleftharpoons 1/2H_2.$$

The difference lies in the fact that in the last reaction, the electron is a metallic one;

- the aqueous hydrogen atom and the hydrated electron form an acid–base couple according to Brønsted's theory since they are connected by the relation

$$H_{(w)} \rightleftharpoons H^+ + e_{(w)}{}^-.$$

13.4 Cathode, Anode, and Charges of Electrodes

By definition, the cathode is the electrode where the electrochemical reduction reaction occurs. The anode is that where the electrochemical oxidation occurs. The fact that the cell behaves either in a galvanic mode or in an electrolytic mode does not modify these definitions, which are quite general.

In Daniell's galvanic cell, the cathode is the copper wire and the anode is the zinc wire. In the corresponding electrolytic cell, it is the reverse: The zinc wire is the cathode and the copper wire the anode. Hence, the same electrode may be either an anode or a cathode depending on the experimental conditions.

No other definition of the anode and the cathode exists. Notice that the cathode is not necessarily the electrode toward which cations go and the anode that toward which anions go. This last point implies that the cathode is negatively charged, the anode positively, and that the cations and anions drift under the influence of the electrical fields exerted by the electrodes. This is not always true. The electrical charge brought by the electrode is not directly connected to its behavior (anodic or cathodic). The examples of Daniell's galvanic cell and of its corresponding electrolytic cell illustrate this fact. In the galvanic cell, the zinc wire is the anode by definition. Its potential V_a (a: anode) is more negative than the potential V_c (c: cathode) of the cathode since the electrons freely flow from zinc to copper (the potential of an electrode in which flows a current is symbolized by V and by E if the current is null). In this case, the Cu^{2+} cations cannot migrate toward the cathode under the influence of the electric field. In the electrolytic cell, the anode is the copper wire. However, its potential, V_a, is made positive by the power supply, whereas that of the zinc wire V_c is made negative. In this case, the above assertion is verified. The Zn^{2+} cations are effectively reduced at the cathode, the electrode being the most negative one.

One example shows that the cathodic or anodic behavior of an electrode is disconnected from its charge. Indeed, a species that undergoes reduction or oxidation is not necessarily a cation or an anion, respectively. Thus, in some conditions, the trianion hexacyanoferrate(III) (ferrocyanide ion) may be reduced on a cathode (by definition), giving the tetraanion hexacyanoferrate(II) (ferrocyanide ion) according to the scheme

$$[Fe(CN)_6]^{3-} + 1e^- \rightarrow [Fe(CN)_6]^{4-}.$$

It is quite evident that the ferrocyanide ion cannot migrate toward the cathode. Actually, in this case, it moves toward the cathode by diffusion and/or by convection.

13.5 Electrochemical Cells and Reversibility

An electrochemical cell is said to be *reversible* if the direction of the reaction cell is inverted when the electron flow is inverted. The cell is called *irreversible* if a new reaction cell appears when the electron flow is inverted.

According to this definition, Daniell's galvanic cell and the corresponding electrolytic cell constitute a reversible cell. An example of an irreversible cell is provided by a copper wire and a zinc wire dipping into a dilute sulfuric acid. Spontaneously (the behavior of a galvanic cell), the zinc is dissolved and dihydrogen evolves on the copper electrode. The two involved half-reactions are

$$Zn \rightarrow Zn^{2+} + 2e^-,$$

$$2H^+ + 2e^- \rightarrow H_2\uparrow$$

and the reaction cell is

$$Zn + 2H^+ \rightarrow Zn^{2+} + H_2\uparrow .$$

During electrolytic behavior, we experimentally find that copper is dissolved and that dihydrogen evolves on the zinc electrode. Then the cell reaction is

$$Cu + 2H^+ \rightarrow Cu^{2+} + H_2\uparrow .$$

Obviously, this reaction differs from the preceding one. The cell is irreversible.

A power supply may generate a potential difference that perfectly counterbalances that existing between the two electrodes. In these conditions, no current flows. The situation is equivalent to that encountered when the two electrodes are not connected. We say that the system is *at equilibrium*. Sometimes, the potential difference E determined at equilibrium (the zero-current cell potential) is still called the *electromotive force* of the cell. In the case of a chemically reversible cell, it is endowed with a very well-determined meaning (see Chap. 2 and later in this chapter): It permits us to obtain the standard electrode potentials. Establishing the latter is in the realm of the application of Nernst's law.

13.6 Classes of Electrodes

Several classes of electrodes or, more exactly, of classes of cell compartments exist, as follows:

- class I electrodes: a metal in equilibrium with its ions. They are constituted by a metal dipping into a solution of one of its salts, that is, into a solution of one of its ions. The ion is the oxidized form of the metal. The half-redox reaction is

  $$M^{n+} + ne^- \rightleftharpoons M_{(s)}$$

  Apart from the role it plays in the redox phenomenon, the metal also plays that of an electrical contactor. The electrodes of the first kind are written conventionally as $M|M^{n+}$. In this notation a vertical bar represents a phase boundary;
- class II electrodes: a metal in equilibrium with a saturated solution of a slightly soluble salt. They consist of a metal covered by a porous coat of one of its insoluble salts, the entire object dipping into a solution containing the anion giving the

insoluble salt of the metal. The redox couple is MX(s)/M(s), X^-. The half-redox reaction is

$$MX_{(s)} + 1e^- \rightleftharpoons M_{(s)} + X^-,$$

where X^- is the anion giving the insoluble salt of the metal. Two examples are very well known. They are the silver–silver chloride electrode, whose half-redox reaction is

$$AgCl_{(s)} + 1e^- \rightleftharpoons Ag_{(s)} + Cl^-$$

and the "calomel" electrode, whose half-redox reaction is

$$Hg_2Cl_{2(s)} + 2e^- \rightleftharpoons 2Hg_{(l)} + 2Cl^-$$

(calomel and silver chloride are very insoluble). These electrodes are written as

$$M|MX|X^-$$

- class III electrodes: two soluble species in equilibrium at an inert electrode or redox electrode. They consist of a wire of an inert (noble) metal (Pt, Au) dipping into a solution containing the two ionized members of a redox couple, for example, Fe^{2+} and Fe^{3+}. The only function of the inert electrodes is to transport the electrons to or from the ions in the solution. They are considered *potential probes*. They take the potential value of the solution into which they dip, without modifying it. The redox couple is $M^{n+}/M^{(n-p)+}$ and the half-redox reaction is

  $$M^{n+} + pe^- \rightleftharpoons M^{(n-p)+}.$$

  These electrodes are written M|Red, Ox, where M is the inert metal and the comma separates two components in one phase.

Other classes of electrodes do exist. Of fundamental importance is the hydrogen electrode. Within the electrode, gaseous dihydrogen is in equilibrium with the proton in solution in the presence of platinum. Platinum exchanges electrons with dihydrogen and with protons and takes the potential value of the solution at equilibrium. The redox couple is $H^+/H_{2(g)}$ and the half-redox reaction is

$$H_{(w)}{}^+ + 1e^- \rightleftharpoons 1/2H_2,$$

written $Pt|H_{2(g)}|H_{(w)}{}^+$ in agreement with the preceding writings. The hydrogen electrode belongs to the class of gas electrodes. We must notice the similarity of this electrode with those of the third class, where the inert metal behaves as a probe. We can also consider that the hydrogen electrode exhibits some similarity with those of the first class, dihydrogen playing the part of a metal. According to this viewpoint, the platinum adsorbs the gas on its surface, therefore facilitating the electron exchange between the two redox species and ensuring the electrical contact.

$$\mathrm{Zn} \,|\, \mathrm{Zn}^{2+} \,\vdots\, \mathrm{Cu}^{2+} \,|\, \mathrm{Cu} \qquad (13.7) \qquad\qquad \mathrm{Cu} \,|\, \mathrm{Cu}^{2+} \,\vdots\, \mathrm{Zn}^{2+} \,|\, \mathrm{Zn} \qquad (13.8)$$

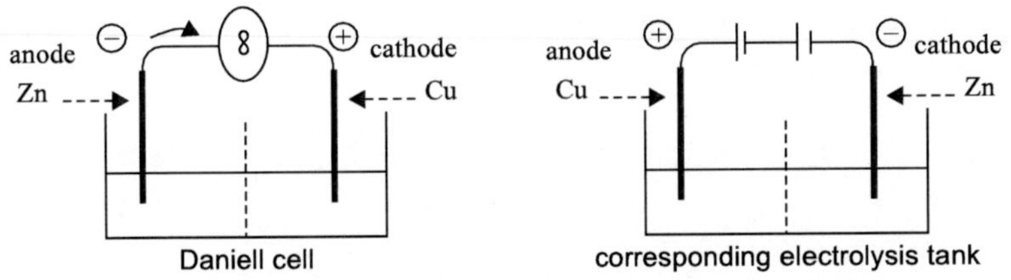

Fig. 13.4 Diagrams representing Daniell's galvanic cell and the corresponding electrolytic cell

13.7 Shorthand Notation for Electrochemical Cells

Electrochemical cells are written by juxtaposing the representations of their electrodes given above. In the adopted notation, a vertical dashed line means a junction between two miscible liquids, while two vertical lines indicates a liquid junction, whose potential difference is minimized.

By convention, the cell diagram is written such that the electrons flowing in the cell go from left to right in the metallic (external) part of the circuit, irrespective of the actual behavioral mode: galvanic or electrolytic. This means that the electrode located on the left of the diagram is necessarily the anode and that located on the right is the cathode. Daniell's galvanic cell and its corresponding electrolytic cell are represented as shown in Fig. 13.4 (diagrams 13.7 and 13.8).

Diagrams (13.7) and (13.8) are obviously incomplete. Indeed, when we are faced with them, the problem that arises is to know the cell's behavioral mode: galvanic or electrolytic. In other words, the problem is to know whether or not the cell reaction, as inferred by the diagram, is spontaneous. We must not forget that the convention implicates that the electrons flow from left to right in the external circuit. This means that the reactions implicated by the cell notation are necessarily

$$\mathrm{Zn} + \mathrm{Cu}^{2+} \rightarrow \mathrm{Zn}^{2+} + \mathrm{Cu} \quad \text{[diagram (13.7)]}$$

$$\mathrm{Cu} + \mathrm{Zn}^{2+} \rightarrow \mathrm{Cu}^{2+} + \mathrm{Zn} \quad \text{[diagram (13.8)].}$$

The answer is given by new data: the electrical charges brought by the electrodes, since the electrons cannot go anywhere other than toward positive potentials. (In Fig. 13.4, we anticipated the problem and already determined the charges.) Calculating the potentials of the different electrodes through Nernst's equation lets us know what the charges are and, finally, solves the problem.

Remark The scheme of the electrolytic cell represented in Fig. 13.3 does not obey the above conventions. It must be written in a reverse manner, as in Fig. 13.4.

13.8 Some Examples of Cells

Now, we'll give some examples of cells chosen from numerous possible combinations. Electrical charges are omitted.

- $Pb|PbSO_{4(s)}, SO_4{}^{2-}||Sn^{2+}, Sn^{4+}|Pt$.

  The two corresponding half-reactions are

$$Sn^{4+} + 2e^- \rightleftharpoons Sn^{2+},$$

$$PbSO_{4(s)} + 2e^- \rightleftharpoons Pb_{(s)} + SO_4{}^{2-}.$$

The second one corresponds to a class II electrode;

- $Pt|Fe^{2+}, Fe^{3+}||Ag^+|Ag$,

whose half-reactions are

$$Fe^{3+} + 1e^- \rightleftharpoons Fe^{2+},$$

$$Ag^+ + 1e^- \rightleftharpoons Ag_{(s)};$$

- $Pt|H_{2(g)}|H^+||Ag^+|Ag$,

  whose half-reactions are

$$2H^+ + 2e^- \rightleftharpoons H_{2(g)},$$

$$Ag^+ + 1e^- \rightleftharpoons Ag_{(s)}.$$

Not all of the chemical cells are composed of two compartments; some cells have only one compartment. A single electrolyte is common to both electrodes. This is the case of the following cell:

$$Pt|H_{2(g)}|HCl|AgCl_{(s)}|Ag.$$

It consists of a hydrogen electrode $Pt|H_2$, H^+ connected to a silver–silver chloride electrode $Cl^-|AgCl|Ag$. Both electrodes dip into the same electrolyte, a solution of hydrochloric acid that provides the two ions necessary to the cell functioning. The cell reaction is

$$2AgCl_{(s)} + H_{2(g)} \rightarrow 2Ag_{(s)} + 2H^+ + 2Cl^-.$$

The suppression of the separator is possible in the cell because this reaction, which could evolve chemically, is very slow and thus does not preclude the two electrode reactions. This cell is used in the operational definition of pH. A variant of the preceding cell is the following:

$$Pt|H_{2(g)}|HA, A^-, H^+, Cl^-|AgCl_{(s)}|Ag,$$

which is called *Harned's cell*. It is used to determine very accurate pK_a values of acid–base couples (here, the couple HA/A^-).

13.9 Electrode Potentials: Nernst's Law

The zero cell potential E is the difference between the cathode and anode potentials:

$$E = E_c - E_a$$

when no current flows within the cell.

We have seen (see Chap. 2) that electrode potential values are given by Nernst's law, which is legitimate only when no current flows through the electrodes. It calculates the cell potentials.

- The potential of a class I electrode, whose half-redox reaction is

$$M^{n+} + ne^- \rightleftharpoons M_{(s)},$$

is given by the equation

$$E = E^\circ\left(M^{n+}/M\right) - (RT/nF)\ln\left(a_M/a_{Mn+}\right),$$

where R is the perfect gas constant, T the absolute temperature of the system, F the faraday, a_M and a_{Mn+} the activities of the species at the equilibrium, and n the number of electrons exchanged in the half-reaction. $E°(M^{n+}/M)$ is the standard redox potential of the electrode (or, in other words, of the system). It is a constant at a given temperature for a given redox couple. The standard potential is the electrode potential when all the species participating in the half-redox reaction exhibit an activity value equal to unity. Per the usual conventions on the activities (see Chap. 3), $a_M = 1$.Therefore, we can write

$$E = E^\circ\left(M^{n+}/M_{(s)}\right) - (RT/nF)\ln\left(1/a_{Mn+}\right) \quad (13.9)$$

The electrode potential depends on the activity of the metallic ion;

- the potential of a class II electrode whose half-redox reaction is, for example,

$$MX_{(s)} + 1e^- \rightleftharpoons M_{(s)} + X^-,$$

which is given by the equation

$$E = E^\circ(MX_{(s)}/M_{(s)}) - (RT/F)\ln(a_{M(s)}a_{X^-}/a_{MX(s)}).$$

For the following additional example,

$$MX_{2(s)} + 2e^- \rightleftharpoons M_{(s)} + 2X^-,$$

the equation giving the potential is

$$E = E^\circ(MX_{2(s)}/M_{(s)}) - (RT/2F)\ln[a_{M(s)}(a_{X^-})^2/a_{MX2(s)}].$$

The generalization of these equations is immediate. If we take into account the conventions on the activities, the preceding equations become

$$E = E^\circ(MX_{(s)}/M_{(s)}) - (RT/F)\ln a_{X^-} \tag{13.10}$$

or

$$E = E^\circ(MX_{2(s)}/M_{(s)}) - (RT/2F)\ln(a_{X^-})^2$$

since $MX_{(s)}$, $MX_{2(s)}$, and $M_{(s)}$ each separately constitutes a pure phase. Therefore, their activities are taken to be equal to unity. We see that for this class of electrodes, the electrode potential is a function of the activity of the anion of the insoluble salt and, hence, finally, of its concentration;

- the potential of a class III electrode, whose half-redox reaction is

$$M^{n+} + pe^- \rightleftharpoons M^{(n-p)+},$$

is given by the relation

$$E = E^\circ\left(M^{n+}/M^{(n-p)+}\right) - (RT/pF)\ln\left[a_{M(n-p)+}/a_{Mn+}\right]. \tag{13.11}$$

The electrode potential is a function of the activities of the two members of the redox couple;

- the potential of the hydrogen electrode whose half-reaction is

$$H^+ + 1e^- \rightleftharpoons {}^1\!/_2\, H_{2(g)},$$

is given by the relation

$$E = E^\circ(H^+/H_{2(g)}) - (RT/F)\ln\left[(a_{H2})^{1}\!/_2\, /a_{H+}\right]. \tag{13.12}$$

We must comment on Eq. (13.12). First, the standard potential $E^0(H^+/H_2)$ is, once and for all, conventionally chosen to be equal to zero, regardless of the temperature (see also Chap. 2). Second, due to the conventions about the activities and the ideal behavior of dihydrogen in usual conditions, we can set

$$a_{H2} = P_{H2}$$

This is true in particular for $P_{H2} = 1$ bar. When this is the case, we can write

$$E = -(RT/F)\ln(1/a_{H+})$$

or

$$E = -(RT/2.03F)\text{pH} \quad (P_{H2} = 1\text{bar}).$$

In these conditions, the electrode indicates the pH of the solution.

To conclude this section, let's recall that there are other kinds of electrodes, but the generalization of the preceding is immediate. We find that the electrode potentials corresponding to the half-redox reaction

$$\text{Ox} + ne^- \rightleftharpoons \text{Red}$$

are given by expressions such as

$$E = E°(\mathrm{Ox/Red}) - (RT/nF)\ln(a_{\mathrm{Red}}/a_{\mathrm{Ox}}).$$

At $T = 298$ K, the numerical value of the term $(RT/F)\ln$ is 0.0591, that is, 0.06 once the transformation of the neperian logarithm into a decimal one has been taken into account. The preceding relation becomes

$$E = E°(\mathrm{Ox/Red}) - (0.06/n)\log(a_{\mathrm{Red}}/a_{\mathrm{Ox}}).$$

Another point must be emphasized. Some redox couples exchange protons or other chemical species together with electrons. Here are some examples:

- the couple permanganate ion/manganous cation, which also exchanges protons according to the reaction

$$MnO_4{}^- + 8H^+ + 5e^- \rightleftharpoons Mn^{2+} + 4H_2O;$$

- the couple tetrachloroaurate(III) ions/or ($AuCl_4^-/Au$), which also exchanges chloride ions, according to

$$AuCl_4{}^- + 3e^- \rightleftharpoons Au_{(s)} + 4Cl^-;$$

Nernst's equation, which gives the potentials of these electrodes, must take into account the activities of all the exchanged species. For the two preceding systems, Nernst's equations are

$$E = E°\left(MnO_4{}^-/Mn^{2+}\right) - (RT/5F)\ln\left\{a_{\mathrm{Mn2+}}/\left[a_{\mathrm{MnO4-}}(a_{\mathrm{H+}})^8\right]\right\}$$

since

$$a_{H_2O} = 1,$$

and

$$E = E°\left(AuCl_4{}^-/Au_{(s)}\right) - (RT/3F)\ln\left[(a_{\mathrm{Cl-}})^4/a_{AuCl_4{}^-}\right].$$

Therefore, the potential of one electrode depends systematically on the activities of all the species that participate in the half-redox equilibrium. Furthermore, these species are not necessarily those that exchange electrons.

Let's recall (see Chap. 2) that

- the standard potential values $E°$ of all the redox couples depend on the arbitrary choice of the value $E°(H^+/H_{2(g)}) = 0$ regardless of the temperature. This convention does not limit the use of redox potentials, at least in analytical chemistry, since the potential differences (which are experimentally accessible) do not depend on this arbitrary choice. Additionally, we have seen that these potential differences are those that quantify the affinity of the redox reaction;

Table 13.1 Some reduction electrode potentials $E°$ at 298 K

Reaction to the electrode	E°(V)	Reaction to the electrode	E°(V)
$Li^+ + e^- \rightleftharpoons Li_{(s)}$	– 3,05	$Sn^{2+} + 2e^- \rightleftharpoons Sn_{(s)}$	– 0,15
$K^+ + e^- \rightleftharpoons K_{(s)}$	– 2,93	$Pb^{2+} + 2e^- \rightleftharpoons Pb_{(s)}$	– 0,13
$Ca^{2+} + 2e^- \rightleftharpoons Ca_{(s)}$	– 2,87	$2H^+ + 2e^- \rightleftharpoons H_{2(g)}$	0,00
$Na^+ + e^- \rightleftharpoons Na_{(s)}$	– 2,71	$Cu^{2+} + 2e^- \rightleftharpoons Cu_{(s)}$	0,34
$Mg^{2+} + 2e^- \rightleftharpoons Mg_{(s)}$	– 2,37	$Hg^{2+} + 2e^- \rightleftharpoons Hg(l)$	0,79
$Mn^{2+} + 2e^- \rightleftharpoons Mn_{(s)}$	–1,18	$Ag^+ + e^- \rightleftharpoons Ag_{(s)}$	0,80
$Zn^{2+} + 2e^- \rightleftharpoons Zn_{(s)}$	– 0,76		
$Fe^{2+} + 2e^- \rightleftharpoons Fe_{(s)}$	– 0,44		

- Nernst's law permits the calculation of a redox potential when, and only when, there is equilibrium. This is the reason why the potentials calculated by the above relations are called *equilibrium potentials.* This means that the potential of by an electrode that dips into an electrochemical compartment is given by Nernst's law when no current flows through the cell ($i = 0$). With respect to this point, we must recall that the standard potentials, which differ from the equilibrium potentials due to the presence of the logarithmic terms in the latter, are obtained from zero-current cell potential values (see Chap. 2).

Even if no current flows through the electrode, it may not take the value given by Nernst's equation. This is often the case for class III electrodes. In such a case, we say that the electrochemical system is slow.

13.10 Standard Electrode Potentials and Standard Reduction Potentials of Some Redox Couples at 298 K

Tables 13.1, 13.2, and 13.3 list the standard potential values of some couples.

The preceding tables necessitate the following comments:

- These potential values are those of reduction potentials. Hence, these values are those of zero-current cell potentials in which the hydrogen electrode is located on the left and that under study on the right (see Fig. 2.5 in Chap. 2). Moreover, all the species that participate in the half-reduction equilibria are in their standard states: their activities are equal to unity. We have already seen that the hydrogen electrode necessarily plays the part of the anode and the electrode under study that of the cathode. Hence, the studied system suffers the electrodic reaction

$$\mathrm{Ox} + ne^- \rightleftharpoons \mathrm{Red},$$

from which comes the name "reduction potentials." This point is very important, especially for the calculation of the free enthalpies of the half-redox reactions.

Table 13.2 Reduction standard potential values $E°$ of some redox couples at 298 K (g: gaseous state, s: solid state, l: liquid state)

Redox semi-reaction	E°(V)
$F_{2(g)} + 2e^- \rightleftharpoons 2F^-$	2,87
$S_2O_8^{2-} + 2e^- \rightleftharpoons 2SO_4^{2-}$	2,01
$O_{3(g)} + 2H^+ + 2e^- \rightleftharpoons O_{2(g)} + H_2O$	2,075
$H_2O_2 + 2H^+ + 2e^- \rightleftharpoons 2H_2O$	1,77
$HClO + H^+ + e^- \rightleftharpoons \frac{1}{2}Cl_{2(g)} + H_2O$	1,63
$H_5IO_6 + H^+ + 2e^- \rightleftharpoons IO_3^- + 3H_2O$	1,60
$BrO_3^- + 6H^+ + 5e^- \rightleftharpoons \frac{1}{2}Br_2(l) + 3H_2O$	1,52
$BrO_3^- + 6H^+ + 6e^- \rightleftharpoons Br^- + 3H_2O$	1,44
$MnO_4^- + 8H^+ + 5e^- \rightleftharpoons Mn^{2+} + 4H_2O$	1,52
$Cl_{2(g)} + 2e^- \rightleftharpoons 2Cl^-$	1,36
$Cr_2O_7^{2-} + 14H^+ + 6e^- \rightleftharpoons 2Cr^{3+} + 7H_2O$	1,33
$O_{2(g)} + 4H^+ + 4e^- \rightleftharpoons 2H_2O$	1,23
$IO_3^- + 6H^+ + 5e^- \rightleftharpoons \frac{1}{2}I_{2(s)} + 3H_2O$	1,20
$IO_3^- + 6H^+ + 5e^- \rightleftharpoons \frac{1}{2}I_2 + 3H_2O$	1,18
$Br_2 + 2e^- \rightleftharpoons 2Br^-$	1,09
$Br_2(l) + 2e^- \rightleftharpoons 2Br^-$	1,07
$ICl_2^- + e^- \rightleftharpoons \frac{1}{2}I_{2(s)} + 2Cl^-$	1,06
$AuCl_4^- + 3e^- \rightleftharpoons Au_{(s)} + 4Cl^-$	1,00
$HNO_2 + H^+ + e^- \rightleftharpoons NO_{(g)} + H_2O$	1,00
$2Hg^{2+} + 2e^- \rightleftharpoons Hg_2^{2+}$	0,92
$Cu^{2+} + I^- + e^- \rightleftharpoons CuI_{(s)}$	0,86
$Hg_2^{2+} + 2e^- \rightleftharpoons 2Hg(l)$	0,79
$Fe^{3+} + e^- \rightleftharpoons Fe^{2+}$	0,77
$PtCl_4^{2-} + 2e^- \rightleftharpoons Pt_{(s)} + 4Cl^-$	0,73
$I_2 + 2e^- \rightleftharpoons 2I^-$	0,62
$H_3AsO_4 + 2H^+ + 2e^- \rightleftharpoons H_3AsO_3 + H_2O$	0,56
$I_3^- + 2e^- \rightleftharpoons 3I^-$	0,54
$I_{2(s)} + 2e^- \rightleftharpoons 2I^-$	0,54
$Fe(CN)_6^{3-} + 1e^- \rightleftharpoons Fe(CN)_6^{4-}$	0,36
$UO_2^{2+} + 4H^+ + 2e^- \rightleftharpoons U^{4+} + 2H_2O$	0,33
$Hg_2Cl_{2(s)} + 2e^- \rightleftharpoons 2Hg(l) + 2Cl^-$	0,27
$AgCl_{(s)} + e^- \rightleftharpoons Ag_{(s)} + Cl^-$	0,15
$Cu^{2+} + e^- \rightleftharpoons Cu^+$	0,15
$Sn^{4+} + 2e^- \rightleftharpoons Sn^{2+}$	0,15
$S_4O_6^{2-} + 2e^- \rightleftharpoons 2S_2O_3^{2-}$	0,08
$2H^+ + 2e^- \rightleftharpoons H_{2(g)}$	0,00
$V^{3+} + e^- \rightleftharpoons V^{2+}$	− 0,26
$Cr^{3+} + e^- \rightleftharpoons Cr^{2+}$	− 0,41
$U^{4+} + e^- \rightleftharpoons U^{3+}$	− 0,61

Table 13.3 Potentials $E^{\circ\prime}$ of some organic redox couples at T = 37°C and pH = 7 (biochemistry conventions: biological standard potentials)

Organic redox couples	E'°(V)
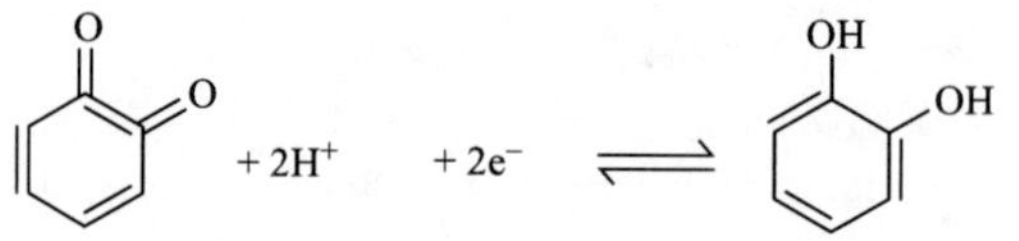 o-quinone / pyrocatechol	0,43
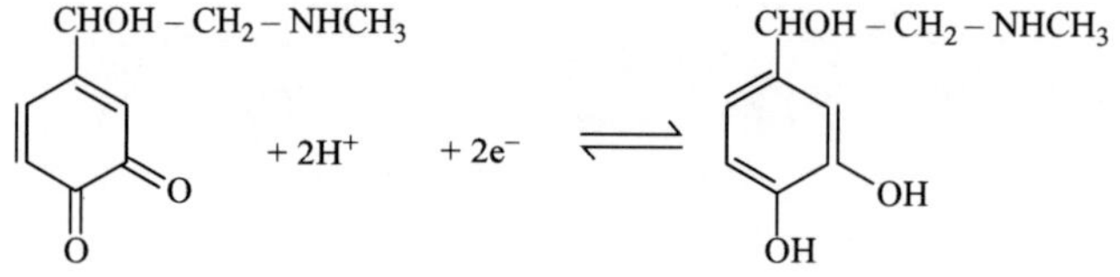 adrenochrome / adrenaline	0,38
oxidized cytochrome a+e^- ⇌ reduced cytochrome a *	0,29
oxidized cytochrome c+e^- ⇌ reduced cytochrome c	0,26
oxidized cytochrome b+e^- ⇌ reduced cytochrome b	– 0,04
met Hb (Fe^{III}) + e^- ⇌ Hb (Fe^{II}) (hemoglobine) **	0,15
fumarate + $2H^+$ + $2e^-$ ⇌ succinate	0,00
oxaloacetate + $2H^+$ + $2e^-$ ⇌ malate	– 0,17
pyruvate + $2H^+$ + $2e^-$ ⇌ lactate	– 0,18
acetaldehyde + $2H^+$ + $2e^-$ ⇌ ethanol	– 0,20
FAD + $2H^+$ + $2e^-$ ⇌ FAD H_2	– 0,22
NAD + H^+ + $2e^-$ ⇌ NADH, H	– 0,32
cysteine + $2H^+$ + $2e^-$ ⇌ Cystine	– 0,39
HCO_3^- + $2H^+$ + $2e^-$ ⇌ $HCOO^-$ + H_2O (bicarbonate / formiate)	– 0,41

* Heteroproteins in a ferroporphyrinic prosthetic group. Catalyze the oxidation of many substrates in the cell metabolism.

** Porphyrinic erythrocytic chromoproteins transporting oxygen along with one part carbon dioxide, contributing to the buffering capacity of the blood.

Let's recall that the handling of the free enthalpies of redox reactions permits us to calculate unknown standard potentials (see Sect. 2.11 in Chap. 2). This point is also important for the calculation of cell potentials, namely, for the calculation of the electrical charges brought by the electrodes (see below);

- not all redox couples appear in Tables 13.1–13.3. The standard potentials of the other couples may be calculated by using the additivity of the free enthalpies of the physical or chemical equivalent processes or with the help of Latimer–Luther's rule (see Chap. 2).

- some standard potential values given in these tables were obtained only by calculation. There are several reasons for this. Initially, one of the members of the couple does not exist at pH = 0. By convention, indeed, the standard state implicates that all the activities are equal to unity, including $a_{H+} = 1$ (i.e., pH = 0). Then it is possible that one member of the couple is too poorly soluble for its activity to be equal to unity. It precipitates before the measurement could be made. There is another result intimately connected to this poor solubility. The precipitation may change the nature of the redox couple, since one of its members has itself undergone a change of its physical state. This the case, for example, for some couples involving dibromine, such as

$$Br_{2(l)} + 2e^- \rightleftharpoons 2Br^- \quad E^\circ = 1.07V,$$

$$Br_{2(w)} + 2e^- \rightleftharpoons 2Br^- \quad E^\circ = 1.09V.$$

  It is easy to go from one standard potential to the other via a calculation involving the free enthalpy of change in the physical state;
- it is interesting to note that multiplying all the coefficients of a half-redox reaction by the same factor does not change the electrode potential value. This is due to the presence of the logarithmic term in Nernst's equation;
- Table 13.3 gives the biological standard potentials $E^{\circ\prime}$ of some organic redox couples. Those chosen are involved in the intermediary metabolism. This is why the conventions $T = 310$ K, pH = 7, and a special name for these potentials are adopted. We must know that these values only give an insight into the ability, from the thermodynamic standpoint, of performing some reactions involving them. From the electrochemistry viewpoint , we know that the electrochemical systems involving them are slow. They are called *irreversible* (see a general course in electrochemistry). Usually, an electrode of an inert metal dipping into solution containing the two members of these couples does not take the value given by Nernst's equation. This is the reason why the potentials given in the tables have been determined indirectly, either through the calculations already mentioned or through the knowledge of equilibrium constants of reactions implicating these couples;
- the metals that exhibit the most negative standard potential values are those that are the easiest to ionize. They are the most reducing metals. Those exhibiting the upper values are the most difficult to oxidize. They are called *noble metals*.

13.11 Zero-Current Electrochemical Cell Potentials—Convention

By convention, the zero-current cell potential (formerly: electromotive force) is equal, in sign and magnitude, to the potential difference between the cathode and the anode:

$$E = E_c - E_a \quad \text{(null current)}.$$

This is the potential difference between the electrode on the right and that on the left in the conventional diagrams given above since. By convention, they are written in such a manner that the electrons flow from the left to the right in the external circuit.

The convention $E = E_c - E_a$ is in agreement with the free enthalpy principle, which must decrease during the course of a spontaneous process (at constant temperature and pressure) (see Chap. 2). Indeed, in a galvanic cell, $E \geq 0$. Due to the relation

$$\Delta_r G = -nFE,$$

the free enthalpy of reaction is indeed negative, as anticipated (except at the equilibrium, for which $E_{eq} = 0$).

Applying Nernst's law immediately let us calculate the zero-current cell potential, provided the species activities are known.

Let's now consider the following cell:

$$Zn|ZnCl_2(10^{-2}\ mol/L)||CuSO_4(10^{-2}\ mol/L)|Cu.$$

As indicated by the diagram, the two half-redox reactions are

$$Zn \rightarrow Zn^{2+} + 2e^-,$$

$$Cu^{2+} + 2e^- \rightarrow Cu.$$

The potentials of both electrodes are given by Nernst's law. After mixing the activities and concentrations, we can write

$$E_{Zn} = [-0.76 - (0.059/2)\log(1/10^{-2})]V,$$

$$E_{Cu} = [-0.34 - (0.059/2)\log(1/10^{-1})]V,$$

$$E = [0.31 - (-0.82)]V,$$

$$E = 1.13V,$$

The cell potential is positive, and the cell is galvanic. The reaction cell is

$$Zn + Cu^{2+} \rightarrow Zn^{2+} + Cu.$$

The definitive diagram is

$$^{-}Zn|ZnCl_2(10^{-2}\ mol/L)||CuSO_4(10^{-1}\ mol/L)|Cu^{+}.$$

Let's now consider the cell

$$Pt|Fe^{2+}(10^{-3}\ mol/L),\ Fe^{3+}(10^{-2}\ mol/L)||Ag^{+}(3.5 \cdot 10^{-3}\ mol/L)|Ag.$$

After calculations, we find

$$E = -0.12V.$$

Due to the negative value of E, the reaction of the cell represented by the diagram is

$$Fe^{2+} + Ag^{+} \rightarrow Fe^{3+} + Ag_{(s)}.$$

It is not spontaneous. Hence, the cell written is electrolytic. The complete diagram is

$$^{+}Pt|Fe^{2+}(10^{-3}\ mol/L), Fe^{3+}(10^{-2}\ mol/L)||Ag^{+}(3.5 \cdot 10^{-3}\ mol/L)|Ag^{-}.$$

It is, of course, obvious that in the same conditions, the reverse reaction is spontaneous.

Remark In standard conditions, the cell

$$Pt|Fe^{2+}(a = 1),\ Fe^{3+}(a = 1)||Ag^{+}(a = 1)|Ag$$

is galvanic, with $E = 0.03$ V. It must be written definitively as

$$^{-}Pt|Fe^{2+}(a = 1), Fe^{3+}(a = 1)||Ag^{+}(a = 1)|Ag^{+}.$$

This is an illustration of the fact that in some cases, the direction of a redox reaction may be sharply dependent on the activities and therefore on the concentrations of the involved species. Indeed, with the preceding concentrations, the cell worked in the opposite direction.

13.12 Formal Potentials

In practice, most of the time the ionic strength of the solution in which we want to perform a redox reaction for an analytic goal is unknown. Thus, the activity coefficients are also unknown. A solution to this problem, which is a solution by default, is to mix activities and concentrations, as we did in the preceding considerations. This solution may be not very good since the medium to be analyzed can be very rich in ions and the activity values far from those of the corresponding concentrations, which are nearly always weaker. Hence, electrode potentials calculated in such a way may induce high interpretation errors.

For example, a platinum wire dipping into a solution that contains only equal and very weak concentrations of ferrous and ferric ions exhibits the potential value given by Nernst's equation; 0.77 V. In perchloric acid (1 mol/L) medium, the experimentally obtained value is 0.73 V. The difference between the two values must be assigned to the different values of the activity coefficients γ_{Fe3+} and γ_{Fe2+}, as can be inferred by applying Nernst's law to the couple Fe^{3+}/Fe^{2+}:

$$E = E°(Fe^{3+}/Fe^{2+}) - 0.059 \log \left\{(\gamma_{Fe2+}/\gamma_{Fe3+})([Fe^{2+}]/[Fe^{3+}])\right\}.$$

The activity coefficient γ_{Fe3+} is markedly smaller than γ_{Fe2+}, in agreement with Debye–Hückel's theory. In a hydrochloric solution (1 mol/L), the obtained value is

0.70 V. In this case, the formation of complexes between chloride anions and Fe^{2+} and Fe^{3+}anions plays a part. The complexes formed with Fe^{3+} are more stable than those formed with Fe^{2+}.

Another important example, by far more important than the preceding one from the standpoint of the analytical applications (see a course in analytical electrochemistry), is that of the hydrogen electrode, which can be written according to the following half-diagram:

$$\mathrm{Pt}|\mathrm{H}_{2(g)}(\mathrm{P}_{\mathrm{H2}} = 1\,\mathrm{bar})|\mathrm{H}^+(\mathrm{a}_{\mathrm{H+}} = 1)|.$$

By definition, the proton activity in this electrode is equal to unity. Hydrochloric acid has been chosen to confer this value to the solution. It turns out that the concentration in hydrochloric acid must be about 1.24 mol/L for the activity a_{H+} to be about 1. It is interesting to note that the difference between the activity and concentration values is noticeable in this case.

In order to overcome the difficulty due to the discrepancy of the values of activities and concentrations, which may be important, the concept of formal potentials $E^{\circ\prime}$ has been devised. (The symbol E° of formal potentials is, unfortunately, the same as that used for standard biological potentials.) The formal potential is that which is experimentally observed with solutions containing both Ox and Red forms of the couple at the unit concentration and that also may contain other species whose concentrations are specified. They take into account the variations in activity coefficients with the ionic strength, the acid–base equilibria involving the Ox and/or Red form(s), and their possible complexations with the other solution species. They are experimentally determined using electrochemical cells of the classes described above, after measurements of their zero-current cell potentials. The formal potentials can be used only when the experimental conditions of the redox reaction under study are the same as those under which they have been determined.

As expected, Nernst's relation stated in terms of formal potentials and concentrations

$$E = E^{\circ\prime} - (0.06/n)\log\{[\mathrm{Red}]/[\mathrm{Ox}]\}$$

is often followed. Table 13.4 gives the values of some formal potentials.

Another example of analytical interest is that provided by the couple hexacyanoferrate(III)/hexacyanoferrate(II), whose standard potential is $E^\circ = 0.36$ V. Recall that the standard potentials are determined at pH = 0, but, at this pH, both ions are protonated. However, they are protonated to different extents. The conjugate acids of the hexacyanoferrate(II) ion, namely, $H_4[Fe(CN)_6]$, $H_3[Fe\ (CN)_6]^-$, $H_2[Fe(CN)_6]^{2-}$, and so forth, are weaker than those drivating from the hexacyanoferrate(III) ion, that is, $H_3[Fe\ (CN)_6]$, $H_2\ [Fe\ (CN)_6]$, and so on. As a result, being less protonated than the hexacyanoferrate (II) ion, the hexacyanoferrate(III) ion has a greater activity than the hexacyanoferrate(II) ion. Even if both ions are dissolved initially at the same concentration in an acidic solution at pH = 0, the activity of hexacyanoferrate(III) is higher than that of hexacyanoferrate (II) because of the described phenomenon. The potential E of a class III electrode dipping into this solution is higher than the standard

Table 13.4 Some formal potential values in relationship to the medium where they have been measured

Nature of the medium	Formal potential (V)
Fe^{3+}/Fe^{2+}	
$HClO_4$ 1 mol·l^{-1}	0,73
HCl 1 mol·l^{-1}	0,70
H_2SO_4 1 mol·l^{-1}	0,68
H_3PO_4 (0,5 mol·l^{-1}) + H_2SO_4 (1 mol·l^{-1})	0,61
Recall : E° (Fe^{3+}/Fe^{2+})	0,77
Ce^{4+}/Ce^{3+}	
H_2SO_4 1 mol·l^{-1}	1,44
HNO_3 1 mol·l^{-1}	1,61
$HClO_4$ 1 mol·l^{-1}	1,70
Recall : E°(Ce^{4+}/Ce^{3+})	uncertain

potential. The system is more oxidant than anticipated. This is apparent from examining the formal potentials given by $E^{\circ\prime}$(hexacyanoferrate(III)/hexacyanoferrate (II)) = 0.71 V in 1 mol/L H_2SO_4, HCl, and $HClO_4$. We note the great difference between 0.71 V and 0.36 V.

The systematic use of formal potentials is evidently impossible. In fact, the experimental conditions in which the redox reactions may be performed for analytical purposes can vary all the way to infinity. It is impossible to reproduce them, even to imagine them!

In brief, the formal potentials are interesting from a practical standpoint, but they cannot be endowed with any thermodynamic meaning.

Chapter 14
Predicting Redox Reactions

A chemical species may be difficult to characterize, and its derivatives obtained by reduction or by oxidation may be easier to identify than the species itself. It may also turn out that a species endowed with oxidizing or reducing properties must be quantitatively determined. In such a case, it is logical to predict its redox titration. Therefore, we must necessarily know the direction and quantitative character of the predicted redox reactions.

Predicting redox reactions is not necessarily an easy task. It involves a thorough examination of the reaction's experimental conditions. In particular, we must examine the influence of such factors as the medium's acidity and the complexation and precipitation of the redox species.

The majority of this chapter is devoted to thermodynamic prediction. The last section mentions some kinetic considerations about redox reactions.

14.1 Redox Phenomena and Acidity

The medium's acidity may play a part in redox phenomena in several respects.

In the first instance, depending on its value, the pH can induce the formation of new redox couples. These new couples involve species that do not exist in other ranges of acidity. In such a respect, the pH may simply change the redox couple's nature by modifying the acido-basic status of the Ox or Red form, or both. Finally, even if the redox couple is perfectly defined notably from the acido-basic standpoint, the protonic activity of the medium may greatly change its redox power.

The example of redox equilibria involving chlorine with the oxidation numbers −I, 0, and +I illustrates the influence of the medium's acidity on their course:

- at pH < 1.2, the three species—hypochlorous acid, dichlorine, and the chloride ions Cl^-—are present. The following two equilibria occur:

$$HClO + H^+ + 1e^- \rightleftharpoons 1/2Cl_{2(w)} + H_2O, \qquad (14.1)$$

$$1/2Cl_{2(w)} + 1e^- \rightleftharpoons Cl^-. \qquad (14.2)$$

J.-L. Burgot, *Ionic Equilibria in Analytical Chemistry*,
DOI 10.1007/978-1-4419-8382-4_14, © Springer Science+Business Media, LLC 2012

In this range, chlorine can be under the oxidation states –I, 0, and +I.

- at pH > 1.2, chlorine can take only two oxidation states: +I and –I, but two pH regions must be distinguished. In the region where 1.2 < pH < 7.5, the following half-redox equilibrium must be taken into account:

$$HClO + H^+ + 2e^- \rightleftharpoons Cl^- + H_2O. \qquad (14.3)$$

In the range pH > 7.5, it is

$$ClO^- + 2H^+ + 2e^- \rightleftharpoons Cl^- + H_2O; \qquad (14.4)$$

- finally, after examination of equilibrium (14.4), it appears that the potential of the solution, in which the two species ClO^- and Cl^- coexist, is a function of pH. Indeed, Nernst's equation is

$$E = E^\circ\left(ClO^-/Cl^-\right) - 0.06pH - 0.03\log\left[(Cl^-)/(ClO^-)\right]. \qquad (14.5)$$

It must be emphasized that even if the oxidation states do not change in this case, the members of the couple do change with the pH value and, therefore, the nature itself of the couple changes. As a result, several standard potentials must be considered. For example, for the couple $HClO/ClO^-$, we find $E^\circ(HClO/ClO^-) = 1.51$ V, and for the couple ClO^-/Cl^-, we find $E^\circ(ClO^-/Cl^-) = 1.73$ V.

Calculating the value 1.73 V from the value 1.51 V is achieved according to the principles given in Chap. 2, that is, through the calculation of free enthalpies. It is sufficient to note that the reduction equilibrium of the hypochlorite ion is the sum of the reduction equilibrium of hypochlorous acid and the protonation equilibrium of the hypochlorite ion:

$$HClO + H^+ + 2e^- \rightleftharpoons Cl^- + H_2O,$$

$$ClO^- + H^+ \rightleftharpoons HClO,$$

$$ClO^- + 2H^+ + 2e^- \rightleftharpoons Cl^- + H_2O.$$

For the preceding reasons, we must often take into account the medium's acidity when predicting redox reactions. One way to do that is to introduce the concept of *normal potential* or, synonymously, that of *apparent standard potential*. Consider, for example, the oxidation of ferrous ions by permanganate ions according to the reaction

$$MnO_4{}^- + 8H^+ + 5Fe^{2+} \rightleftharpoons Mn^{2+} + 5Fe^{3+} + 4H_2O.$$

It results from the superimposition of the following two half-reduction equilibria:

$$MnO_4{}^- + 8H^+ + 5e^- \rightleftharpoons Mn^{2+} + 4H_2O, \quad E^\circ = 1.51 \text{ V},$$

$$Fe^{3+} + 1e^- \rightleftharpoons Fe^{2+}, \quad E^\circ = 0.77 \text{ V}.$$

In the first one, the protons play a part. At $T = 298$ K, Nernst's equation gives

$$E = E^\circ\left(MnO_4^-/Mn^{2+}\right) - (8 \cdot 0.06/5)\,pH - (1/5) \cdot 0.06\log\left[\left(Mn^{2+}/\left(MnO_4^-\right)\right]. \quad (14.6)$$

Let's now recall the case of a half-redox reaction with no proton exchange:

$$Ox + ne^- \rightleftharpoons Red.$$

The equilibrium potential of the solution in which the two members of the couple are present is given by the relation

$$E = E^\circ(Ox/Red) - (0.06/n)\ \log\,[(Red)/(Ox)].$$

Comparing this relation with (14.6) shows that the difference $E^\circ(MnO_4^-/Mn^{2+})$ – (8/5) · 0.06pH plays the part of a standard potential. Such differences are called apparent standard potentials, or normal potentials. They are symbolized by $E^{\circ\prime}$ (the same symbol than those of biological standard potentials and of formal potentials). In this example, we may set

$$E^{\circ\prime}\left(MnO_4^-/Mn^{2+}\right) = E^\circ\left(MnO_4^-/Mn^{2+}\right) - (8/5) \cdot 0.06pH.$$

It is the normal potential or the apparent standard potential of the couple MnO_4^-/Mn^{2+} at the pH of the solution. In the general case, for the couple

$$xOx + mH^+ + ne^- \rightleftharpoons yRed + zH_2O,$$

the apparent standard potential is

$$E^{\circ\prime}\,(Ox/Red) = E^\circ(Ox/Red) - (RT/F)(m/n)pH.$$

Normal potentials are, quite evidently, constant at a given constant pH. Naturally, they change with it. In practice, they decrease quickly when the pH increases. Thus (at $T = 298$ K), at pH = 0, $E^{\circ\prime}(MnO_4^-/Mn^{2+}) = 1.51$ V, and at pH = 5, $E^{\circ\prime}(MnO_4^-/Mn^{2+}) = 1.03$ V. This result is general: As soon as the couple exchanges protons together with electrons, the equilibrium potential of the solution decreases when the pH increases. We say that its oxidizing power decreases with the pH. This phenomenon is very interesting in chemical analysis because it permits the modulation of the oxidizing power of some couples by adjusting the medium's pH (see Sect. 14.3 and Chap. 15).

The couple quinone Q/hydroquinol QH_2, where

p- quinone hydroquinol

illustrates this phenomenon:

- in the range $0 < \text{pH} < 10$, the redox couple that must be considered is

$$Q + 2H^+ + 2e^- \rightleftharpoons QH_2,$$

$$E = E°(Q/QH_2) - 0.06\text{pH} - (0.06/2)\log[(QH_2)/(Q)] \quad \text{with}$$

$$E°(Q/QH_2) = 0.70\ \text{V},$$

  where from $E°'(Q/QH_2) = (0.70 - 0.06\text{pH})$ V;

- in the range $10 < \text{pH} < 11.5$, the redox couple that must be considered is

$$Q + H^+ + 2e^- \rightleftharpoons QH^-,$$

$$E = E°(Q/QH^-) - 0.03\text{pH} - (0.06/2)\log[(QH^-)/(Q)] \quad \text{with}$$

$$E°(Q/QH^-) = 0.40\ \text{V},$$

  where from $E°'(Q/QH^-) = (0.40 - 0.03\text{pH})$ V;

- in the range $\text{pH} > 11.5$, the couple that must be considered is

$$Q + 2e^- \rightleftharpoons Q^{2-},$$

$$E = E°(Q/Q^{2-}) - (0.06/2)\log[(Q^{2-})/(Q)] \text{ with } E°(Q/Q^{2-}) = 0.05\ \text{V}.$$

We do not need to introduce a normal potential in this range.

The case in which there are exchanges of hydroxide ions together with electrons is quite analogous to that in which there are exchanges of protons together with electrons. Some tables list potential values of redox couples that exchange hydroxide ions. These potential values are those of the normal potentials at $\text{pH} = 14$. Writing half-reduction equilibria with hydroxide ions rather than with protons is interesting because it takes into account the fact that some species may not exist at $\text{pH} = 0$, whereas they may exist at $\text{pH} = 14$. For example, the two members of the couple BrO_3^-/Br^- exist simultaneously only in the pH range 7–14. In a more acidic medium, there is a retrodisproportionation with the formation of dibrome:

$$BrO_3^- + 5Br^- + 6H^+ \rightarrow 3Br_2 + 3H_2O.$$

This reaction is of great analytical interest (see Chap. 20).

It is easy to write the expression of a half-redox equilibrium in basic medium starting from its expression in acidic medium. It is sufficient to write the latter in the direction of a reduction and to add to both members the same number of hydroxide ions as the number of protons that exist on the left-hand side. With this thought experiment, the protons are transformed into water. Thus, in order to find

Table 14.1 Examples of redox couples involving complexation or precipitation equilibria

$I_{2(s)} + 2e^-$	$\rightleftharpoons$	$2I^-$	$E° = 0{,}54\,V$
$Hg_2^{2+} + 2e^-$	$\rightleftharpoons$	$2Hg(l)$	$E° = 0{,}79\,V$
$Br_2\,(l) + 2e^-$	$\rightleftharpoons$	$2Br^-$	$E° = 1{,}07\,V$
$Cu^+ + 1e^-$	$\rightleftharpoons$	$Cu(s)$	$E° = 0{,}52\,V$

the half-redox equilibrium between bromide ions Br^- and bromate ions BrO_3^- in basic medium, starting from the writing in acidic medium, we follow this strategy:

$$BrO_3^- + 6H^+ + 6e^- \rightleftharpoons Br^- + 3H_2O,$$

$$BrO_3^- + 6H^+ + 6OH^- + 6e^- \rightleftharpoons Br^- + 6OH^- + 3H_2O,$$

$$BrO_3^- + 6H_2O + 6e^- \rightleftharpoons Br^- + 6OH^- + 3H_2O.$$

Finally, the expected chemical equation is

$$BrO_3^- + 3H_2O + 6e^- \rightleftharpoons Br^- + 6OH^-.$$

Now, the calculation of $E^{\circ\prime}$ values (pH = 14) starting from $E°$ values (pH = 0) is easily achieved by adding the free enthalpies of the processes whose resultant is equivalent to the transformation. Thus, $E^{\circ\prime}(BrO_3^-/Br^-)$ is calculated after the following two equilibria (their free enthalpy expressions are given in brackets):

$$BrO_3^- + 6H^+ + 6e^- \rightleftharpoons Br^- + 3H_2O \quad [\Delta G° = -6FE°(BrO_3^-/Br^-)],$$

$$6H_2O \rightleftharpoons 6H^+ + 6OH^- \quad [\Delta G° = -6RT\ \ln\ K_{(w)}]$$

(see Chap. 2).

14.2 Redox Phenomena, Complexation, and Precipitation

Table 13.2 of the preceding chapter also mentions redox couples that exchange chemical species other than protons (together with, of course, an exchange of electrons). The exchanged chemical species is called a *particule* in the french literature in analytical chemistry. The reactions in which an exchange of particules intervenes are more generally called *complexation reactions* (see Part IV). Table 14.1 mentions some supplementary examples.

It's important to note that some of these examples concern redox couples where one of the members is under the form of a precipitate, which constitutes a pure phase. These are couples in which the following intervene:

Table 14.2 Examples of redox equilibria accompanied by a precipitation or by a phase change

$PtCl_4^{2-} + 2e^-$ tetrachloro platinum ion^{II}	$\rightleftharpoons$	$Pt_{(s)} + 4Cl^-$	$E° = 0{,}73\,V$
$AuCl_4^- + 3e^-$ tetrachloroaurate ion^{III}	$\rightleftharpoons$	$AU_{(s)} + 4Cl^-$	$E° = 1{,}00\,V$
$I_3^- + 2e^-$ triiodide ion	$\rightleftharpoons$	$3I^-$	$E° = 0{,}54\,V$
$ICl_2^- + le^-$ ion dichloroiodate (–1)	$\rightleftharpoons$	$\frac{1}{2} I_{2(s)} + 2Cl^-$	$E° = 1{,}20\,V$
$Cu^{2+} + I^- + le^-$ cupric ion	$\rightleftharpoons$	$CuI_{(s)}$	$E° = 0{,}86\,V$
$IO_3^- + 2Cl^- + 6H^+ + 4e^-$ iodate ion	$\rightleftharpoons$	$ICl_2^- + 3H_2O$	$E° = 1{,}24\,V$
$AgCl_{(s)} + le^-$ silver chloride	$\rightleftharpoons$	$Ag_{(s)} + Cl^-$	$E° = 0{,}22\,V$
$Hg_2Cl_{2(s)} + 2e^-$ dimerized mercuric^{I} chloride	$\rightleftharpoons$	$2Hg_{(l)} + 2Cl^-$	$E° = 0{,}27\,V$
$Ag_2CrO_{4(s)} + 2e^-$ silver chromate	$\rightleftharpoons$	$2Hg_{(s)} + CrO_4^{2-}$	$E° = 0{,}45\,V$
$[Co(CN)_6]^{3-} + le^-$ ion hexacyanocobalt^{III}	$\rightleftharpoons$	$[Co(CN)_6]^{4-}$ ion hexacyanocobalt^{II}	$E° = -0{,}83\,V$

- the elements themselves, such as, for example, $I_{2(s)}/I^-$, $Br_{2(l)}/Br^-$ (Table 14.2);
- salts, as in the couple $AgCl_{(s)}/Ag_{(s)}$.

Tables of redox potentials also mention redox couples in which one of the members precipitates in the presence of protons or hydroxide ions without any complexation. It is the case, for example, for the couple $Zn(OH)_{2(s)}/Zn_{(s)}$ (see below).

The superimposition of precipitation and/or complexation equilibria to the redox equilibrium may lead to the phenomena described in Sect. 14.2.1, which are of analytical interest.

14.2.1 The Stabilization of a Redox Couple by Complexation or Precipitation of One of Its Members

In the couple

$$Au^{3+} + 3e^- \rightleftharpoons Au_{(s)},$$

the auric ion Au^{3+} is relatively unknown because it is a powerful oxidant. It oxidizes water. The standard potential of the couple $Au^{3+}/Au_{(s)}$ would be 1.40 V. The auric ion is complexed by chloride ions to give the tetrachloroaurate(III) anion $AuCl_4^-$, which is stable. In the presence of chloride ions, the couple that must be considered is

$$AuCl_4^- + 3e^- \rightleftharpoons Au_{(s)} + 4Cl^-, \quad E° = 1.00\ V.$$

Another example is provided by the cuprous ion Cu^+. It is not stable in aqueous solution no matter what its pH value is. However, it can be stabilized under the form of a precipitate. Then it can, for example, enter into a redox equilibrium with the cupric ion Cu^{2+} and with another ion. Thus, in the presence of iodide ions, we notice the formation of cuprous iodide, which precipitates. The precipitated copper (I) iodide may itself enter into a redox equilibrium. The evoked reactions are

$$Cu^{2+} + I^- + 1e^- \rightleftharpoons CuI_{(s)}, \quad E° = 0.86\ V,$$

$$CuI_{(s)} + 1e^- \rightleftharpoons Cu_{(s)} + I^-, \quad E° = -0.19\ V.$$

An analogous example is provided by the qualitative research and quantitative determination of the reducing carbohydrates with Fehling's solution. Because of the pH conditions, it is the couple Cu^{2+} (under the form of a complex with the tartrate ion)/Cu^+ (under the form of cuprous oxide) that must be considered as the antagonist couple of that of the sugar. The precipitation of cuprous oxide stabilizes the copper(I) ions.

14.2.2 *The Increase or Decrease in the Oxidizing Strength of One of Its Ox or Red Forms*

- In the couple Co^{3+}/Co^{2+}:

$$Co^{3+} + 1e^- \rightleftharpoons Co^{2+}, \quad E° = 1.81\ V.$$

The cobaltic ion Co^{3+} is very oxidizing. It oxidizes water, for example. In the couple hexacyanocobaltate(III)/hexacyanocobaltate(II), the latter ion becomes a very strong reducing species:

$$[Co(CN)_6]^{3-} + 1e^- \rightleftharpoons [Co(CN)_6]^{4-}, \quad E° = -0.83\ V.$$

Nevertheless, the oxidation numbers of cobalt in the last couple we mentioned are the same as in the preceding one. The origin of the phenomenon is the stability of both complexes. The hexacyanocobaltate(III) ion is, by far, more stable than the hexacyanocobaltate(II) ion (see Part IV).

- In the couple $Ag^+/Ag_{(s)}$, silver is a very poor reducing agent, as indicated by the standard potential of the couple ($E° = 0.80$ V). Silver is a noble metal:

$$Ag^+ + 1e^- \rightleftharpoons Ag_{(s)}, \quad E° = 0.80 \text{ V}.$$

In the presence of chloride or sulfide ions, only a very weak quantity of silver ions is not precipitated. Silver then becomes more reducing because of the following precipitation equilibria, which are superimposed into the redox equilibrium:

$$AgCl_{(s)} + 1e^- \rightleftharpoons Ag_{(s)} + Cl^-, \quad E° = 0.22 \text{ V},$$

$$Ag_2S_{(s)} + 2e^- \rightleftharpoons 2Ag_{(s)} + S^{2-}, \quad E° = -0.68 \text{ V}.$$

The silver ion activity is very weak in the presence of chloride and sulfide ions. Comparing the standard potentials of these "new" couples with that of the couple Ag^+/Ag shows the phenomenon's magnitude.

- Another example, interesting for analytical chemistry, is that of the couple hexacyanoferrate(III) (ferricyanide)/hexacyanoferrate(II) (ferrocyanide), whose standard potential is $E° = 0.36$ V. The ferrocyanide ion, in the presence of zinc ions and in some conditions, gives the zinc ferrocyanide $Zn_2[Fe(CN)_6]$. The zinc ferrocyanide is very poorly soluble in water compounds (solubility product: $K_s = 10^{-15.4}$; see Part V). Therefore, in the presence of zinc ions, ferricyanide ions, which are normally weakly oxidant, become strongly oxidant. The half-redox reaction of the couple is

$$[Fe(CN)_6]^{3-} + 2Zn^{2+} + 1e^- \rightleftharpoons Zn_2[Fe(CN)_6]_{(s)}, \quad E° = 1.28 \text{ V}.$$

The standard potentials of the new couples are calculated from those of the initial couples according to the general principles already given (see Chap. 2). The half-redox reactions in which a precipitation reaction or a complexation reaction occurs can always be decomposed into a pure redox equilibrium and into another one that can be a reaction of precipitation or of complexation. For the calculations, we must add the corresponding standard free enthalpies. Thus,

- the redox equilibrium $AuCl_4^-/Au_{(s)}$ is the sum of the two equilibria:

$$Au^{3+} + 3e^- \rightleftharpoons Au_{(s)}, \quad E° \approx 1.37 \text{ V},$$

$$AuCl_4^- \rightleftharpoons Au^{3+} + 4Cl^-, \quad 1/\beta_4 = 10^{-21},$$

$$AuCl_4^- + 3e^- \rightleftharpoons Au_{(s)} + 4Cl^-$$

(β_4 is the overall stability constant of the complex ion $AuCl_4^-$; see Part IV of this book). Here, the complexation of auric ions stabilizes them;

- the redox equilibrium $AgCl_{(s)}/Ag_{(s)}$ is the sum of the two equilibria

$$Ag^{+} + 1e^{-} \rightleftharpoons Ag_{(s)}, \quad E = 0.80\ V,$$

$$AgCl_{(s)} \rightleftharpoons Ag^{+} + Cl^{-}, \quad K_s = 10^{-9.75},$$

$$AgCl_{(s)} + 1e^{-} \rightleftharpoons Ag_{(s)} + Cl^{-},$$

where K_s is the solubility product of silver chloride. We cite this example because it is of importance in electrochemistry (electrode of second sort);

- a neighboring case is that of the precipitation of a metallic ion as a hydroxide in a definite pH range. The nature of the redox couple changes. The initial system metallic ion/metal becomes the system metallic hydroxide/metal. This is another case in which the pH value induces the appearance of a new redox couple. But now the pH plays a part indirectly by inducing a precipitation phenomenon. This was not the case with hypochlorous acid.

For example, for the couple Zn^{2+}/Zn, it is the couple $Zn(OH)_{2(s)}/Zn_{(s)}$ that must be considered in the range $6.5 < pH < 13.5$. The corresponding redox equilibrium is

$$Zn(OH)_{2(s)} + 2e^{-} \rightleftharpoons Zn_{(s)} + 2OH^{-} \quad (6.5 < pH < 13.5). \tag{14.7}$$

We can calculate the standard potential of the couple $Zn(OH)_{2(s)}/Zn_{(s)}$ by following the same principles as previously. The above equilibrium results from the superimposition of the following two equilibria (the corresponding standard free enthalpies are given in brackets):

$$Zn^{2+} + 2e^{-} \rightleftharpoons Zn_{(s)} \quad [\Delta G_1^\circ = -2FE^\circ(Zn^{2+}/Zn)]$$

$$Zn(OH)_{2(s)} \rightleftharpoons Zn^{2+} + 2OH^{-} \quad [\Delta G_2^\circ = -RT \ln K_s],$$

where $K_s = 10^{-16}$ is the solubility product of zinc hydroxide. The standard free enthalpy of the new redox couple ΔG_3° is given by the relation

$$\Delta G_3^\circ = \Delta G_1^\circ + \Delta G_2^\circ.$$

The apparent standard potential of the couple $Zn(OH)_{2(s)}/Zn_{(s)}$ at $pH = 14$ is calculated via the expression

$$\Delta G_3^\circ = -2FE^{\circ'}\ [Zn(OH)_{2(s)}/Zn_{(s)}].$$

It must be emphasized that the potential calculated in such a manner is the normal one at $pH = 14$. In order to be convinced by this assertion, it is sufficient to consider the corresponding Nernst equation:

$$E = E^{\circ'} - (0.06/2) \log \left\{ \left[(Zn_{(s)}) \left(OH^{-}\right)_2 \right] / \left[Zn(OH)_{2s} \right] \right\}.$$

We see that the potential solution takes in the presence of both solid phases equals $E^{\circ\prime}$ when the activity (OH^-) is equal to unity since the activities ($Zn(OH)_{2(s)}$) and ($Zn_{(s)}$) are also equal to unity (see Chap. 3).

The standard potential of the couple corresponds to the half-reduction equilibrium

$$Zn(OH)_{2(s)} + 2H^+ + 2e^- \rightleftharpoons Zn_{(s)} + 2H_2O. \qquad (14.8)$$

It results from the sum of the following two equilibria:

$$Zn(OH)_{2(s)} + 2e^- \rightleftharpoons Zn_{(s)} + 2OH^-, \quad \Delta G_3^\circ,$$

$$2OH^- + 2H^+ \rightleftharpoons 2H_2O, \quad \Delta G_4^\circ = -2RT\ \ln(1/K_w).$$

The standard free enthalpy of equilibrium (14.8) is given by

$$\Delta G_5^\circ = \Delta G_3^\circ + \Delta G_4^\circ.$$

The expected potential $E^\circ[Zn(OH)_{2(s)}/Zn_{(s)}]$ is given by the relation

$$\Delta G_5^\circ = -2FE^\circ\left[Zn(OH)_{2(s)}/Zn_{(s)}\right].$$

Of course, the value $E^\circ[Zn(OH)_{2(s)}/Zn_{(s)}]$ can only be obtained by calculation since zinc hydroxide does not exist at pH = 0.

Finally, when a half-redox reaction is accompanied by a complexation reaction, it is also useful to define an apparent standard potential or normal potential in order to predict redox reactions quickly. For example, in the case of the couple $AuCl_4^-/Au_{(s)}$, Nernst's equation can be written as

$$E = E^\circ\left(AuCl_4^-/Au_{(s)}\right) - (0.06 \cdot 4/3)\log\left(Cl^-\right) + 0.06/3\log\left(AuCl_4^-\right).$$

Obviously, the difference $E^\circ(AuCl_4^-/Au_{(s)}) - (0.06 \cdot 4/3)\log(Cl^-)$ plays the part of a standard potential symbolized by $E^{\circ\prime}(Au\ Cl_4^-/Au_{(s)})$. Another example of analytical interest is provided by the couple iodate IO_3^-/iodine dichloride ICl_2^-, whose half-redox equilibrium is

$$IO_3^- + 2Cl^- + 6H^+ + 4e^- \rightleftharpoons ICl_2^- + 3H_2O.$$

Nernst's equation gives the relation

$$E = E^\circ\left(IO_3^-/ICl_2^-\right) - (6 \cdot 0.06/4)\,pH + (2 \cdot 0.06/4)\log\left(Cl^-\right)$$
$$- (0.06/4)\log\left[\left(ICl_2^-\right)/\left(IO_3^-\right)\right].$$

The normal potential $E^{\circ\prime}(IO_3^-/ICl_2^-)$ is more complicated than previously since it depends on both the pH and the concentration in chloride ions. It is given by the expression

$$E^{\circ\prime}\left(IO_3^-/ICl_2^-\right) = E^\circ\left(IO_3^-/ICl_2^-\right) - (6 \cdot 0.06/4)\,pH + 0.03\log\left(Cl^-\right).$$

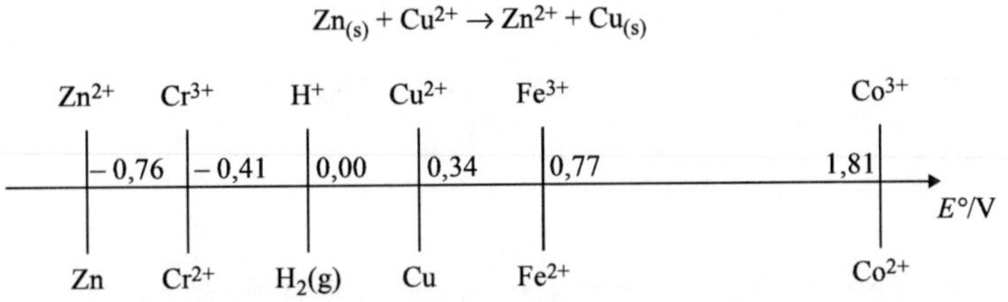

Fig. 14.1 Scale of standard potentials $E°$ with the location of some examples (water—298 K)

Remark In this case, the apparent standard potential increases with the concentration in chloride ions, whereas it always decreases when the pH increases.

14.3 Qualitative Prediction of Redox Reactions After Standard Potentials

Considering standard potentials or, if necessary, apparent standard potentials is probably the simplest way to predict the direction of a redox reaction from the qualitative standpoint. The rule is that the oxidized form of the couple that has the highest standard potential spontaneously reacts with the reduced form of the couple that has the weakest standard potential.

The redox couples are often located on a scale of increasing potentials according to their standard potential values (as acids are located on the acidity scale according to their pK_a values). The origin of the scale is fixed at the value $E = 0.00$ V, which is the standard potential of the couple $H^+/H_{2(g)}$ (Fig. 14.1). The strongest oxidants are located farthest on the right. The strongest reductants are located farthest on the left. For example, if the $Zn^{2+}/Zn_{(s)}$ and $Cu^{2+}/Cu_{(s)}$ couples are placed face to face, we can, with the rule above, predict the spontaneous reaction

$$Zn_{(s)} + Cu^{2+} \rightarrow Zn^{2+} + Cu_{(s)},$$

which evolves in the indicated direction.

When the antagonist redox couples exchange protons or some other particules (of course, in addition to that of electrons), the comparison must be accomplished by examining the apparent standard potentials, adapted, of course, to the experimental conditions. Basing ourselves on apparent standard potentials indeed also permits us to make predictions in any aqueous solution. The following example illustrates these assertions. Let's consider the reaction

$$VO_2{}^+ + I^- + 2H^+ \rightleftharpoons VO^{2+} + {}^1/_2 I_2 + H_2O,$$

which can be decomposed into the two half-equilibria

$$\underset{\text{dioxovanadium ion}}{VO_2{}^+} + 2H^+ + 1e^- \rightleftharpoons \underset{\text{vanadyl ion}}{VO^{2+}} + H_2O, \quad E^\circ = 1.00V,$$

$$\underset{\text{iodine}}{{}^1/_2I_2} + 1e^- \rightleftharpoons \underset{\text{iodide}}{I^-}.$$

The first equilibrium involves a proton exchange, while the second equilibrium does not involve such an exchange. As a result, it does not depend on the pH value until about pH < 9 (see Chap. 18 for a further explanation). The apparent standard potential of the first couple is

$$E^{\circ\prime}\left(VO_2{}^+/VO^{2+}\right) = (1.00 - 0.12\text{pH})\,V;$$

- at pH = 0, $E^{\circ\prime}(VO_2{}^+/VO^{2+}) = 1.00$ V. It is higher than $E^\circ(I_2/I^-) = 0.62$ V. The reaction goes forward in the indicated direction;
- at pH = 7 (buffered solution), $E^{\circ\prime}(VO_2{}^+/VO^{2+}) = 0.16$ V.

The normal potential is now smaller than $E^\circ(I_2/I^-)$. Then it is the inverse reaction that spontaneously occurs:

$$VO^{2+} + {}^1/_2I_2 + H_2O \rightarrow VO_2{}^+ + I^- + 2H^+.$$

This example shows the interest of the concept of apparent standard potential.

It is very important to know that the prediction rule, based solely on the consideration of standard potentials, is not infallible. It may give erroneous answers. (We shall return to this point in Sect. 14.4.) However, it turns out that the answers based on this approach are very often accurate. The origin of this state of affairs is the great difference that very often exists between the standard potential values of the antagonist couples.

14.4 Drawbacks of the Prediction Rule Based on the Sole Consideration of Standard Potentials

As we have said, the preceding rule is not absolute. We know that the reaction (see Chap. 13)

$$Ox_1 + Red_2 \rightarrow Ox_2 + Red_1, \tag{14.9}$$

which according to the preceding rule must go in the indicated direction, will actually go in this direction if, and only if, it is identical to the cell reaction of the corresponding galvanic cell whose electromotive force is given by the expression

$$\begin{aligned} E = &\{E^\circ\,(Ox_1/Red_1) - (RT/nF)\ln[(Red_1)/(Ox_1)]\} \\ &- \{E^\circ\,(Ox_2/Red_2) - (RT/nF)\ln[(Red_2)/(Ox_2)]\}. \end{aligned} \tag{14.10}$$

When reaction (14.9) involves an exchange of protons or other particules, the standard potentials used in relation (14.10) must be replaced by the apparent standard potentials, taking into account the experimental conditions. Moreover, from another standpoint, if couples 1 and 2 do not exchange the same number of electrons in their half-reduction equilibria, judicious stoichiometric coefficients must be used in Eq. (14.9).

After examining Eq. (14.10), it is obvious that the preceding rule of prediction is not perfectly accurate because it does not take into account the logarithmic terms. In other words, it is only an approximation. Certainly, it is accurate if the activities of the Ox and Red forms are equal for each couple, but this is rare. Therefore, herein lies the reason why the preceding rule, based only on the consideration of standard potentials, may fail.

For example, let's consider the reaction

$$\mathrm{Fe^{2+} + Ag^{+} \rightarrow Fe^{3+} + Ag_{(s)}}$$

for which $E°(\mathrm{Fe^{3+}/Fe^{2+}}) = 0.77$ V and $E°(\mathrm{Ag^{+}/Ag_{(s)}}) = 0.80$ V. The sole consideration of the standard potentials indicates that the reaction proceeds in the given direction. This is the case in a vast range of concentrations of species participating to the reaction. But we have already seen after examining the following cell (see the preceding chapter):

$$\mathrm{Pt_{(s)}|Fe^{2+}\left(10^{-3}\,mol/L\right), Fe^{3+}\left(10^{-2}\,mol/L\right) \mid |Ag^{+}\left(3.5 \cdot 10^{-3}\,mol/L\right) |Ag_{(s)}}$$

that the chemical reaction involving these species at the concentrations given in the diagram evolves in the following direction:

$$\mathrm{Fe^{3+} + Ag_{(s)} \rightarrow Fe^{2+} + Ag^{+}},$$

that is, in the opposite direction of that given by the rule grounded only on the standard potentials.

In summary, this rule gives a trend only. The question arises, however, of why this rule is often obeyed. The answer is simple. Nearly always, the difference between the standard potential values is sufficiently high for the logarithmic terms not to inverse the tendency given by the rule.

Exercise 1 In the mixture $\mathrm{Cu^{2+}(10^{-1}}$ mol/L), $\mathrm{Ag^{+}(2 \cdot 10^{-1}}$ mol/L) in the presence of metallic copper and silver, what reaction spontaneously proceeds? Let's build the corresponding electrochemical cell and calculate its potential.

Answer

$$E_{\mathrm{Ag+}} = 0.80 - 0.059 \log(1/0.20) = 0.76 \text{ V}$$

$$E_{\mathrm{Cu2+}} = 0.34 - (0.059/2) \log(1/0.10) = 0.31 \text{ V}$$

Let's examine the following two cells:

$$\mathrm{Cu\ |Cu^{2+}(0.10\,mol/L)||Ag^{+}(0.20\,mol/L)|Ag}, \tag{14.11}$$

$$\mathrm{Ag\ |Ag^{+}(0.20\,mol/L)||Cu^{2+}(0.10\,mol/L)|Cu}. \tag{14.12}$$

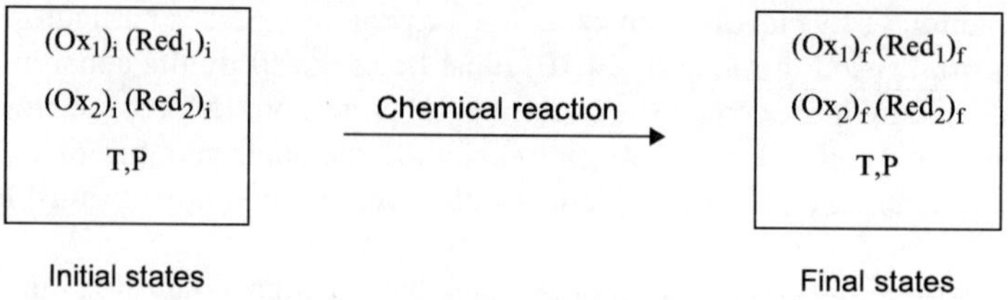

Fig. 14.2 Definition of the initial and final states of a redox reaction

zero-current potential of cell (14.1): 0.76 – 0.31 = 0.45 V,
zero-current potential of cell (14.2): 0.31 – 0.76 = –0.45 V,
Cell (14.1) is a galvanic cell. The spontaneous redox reaction is therefore

$$Cu_{(s)} + 2Ag^+ \rightarrow Cu^{2+} + 2Ag_{(s)}.$$

Exercise 2 This is the same exercise as the previous one, but with concentrations $[Ag^+] = 0.10$ mol/L and $[Cu^{2+}] = 0.20$ mol/L.

Answer It is the inverse of the preceding reaction, which spontaneously occurs.

14.5 Quantitative Character of a Redox Reaction

Let's consider the general redox reaction

$$n\mathrm{Ox}_1 + m\mathrm{Red}_2 \rightleftharpoons n\mathrm{Red}_1 + m\mathrm{Ox}_2. \quad (14.13)$$

Figure 14.2 defines the initial and final states.

The initial state of the system is defined by its temperature (which is equal to the temperature of the surroundings), its pressure (equal to that of the surroundings), and also the activities (Ox_i), (Red_i) of the different species (Fig. 14.2). These are the activities before the reaction started. We can remark that these initial conditions are quite general because of the fact that the final products Red_1 and Ox_2 already exist in the initial state. From a practical standpoint, this may mean that the reaction has been begun before and is now continuing with the addition of the species Ox_1 or Red_2. This is exactly the case with a redox titration, except at the first point. A particular case of the initial state is that in which the activities of the final species Red_1 and Ox_2 are null. As a matter of fact, it corresponds to the first point of a titration.

The final state is defined by the same temperature and pressure as the initial ones, but the activities have changed. These are the activities at equilibrium (see Chap. 2).

We know (Chaps. 2 and 13) that the cell

$$\mathrm{Pt}\ |\mathrm{Red}_2(a_{\mathrm{red2i}}), \mathrm{Ox}_2(a_{\mathrm{ox2i}})|\mathrm{Red}_1(a_{\mathrm{red1i}}), \mathrm{Ox}_1(a_{\mathrm{ox1i}})|\mathrm{Pt},$$

whose cell reaction is equivalent to (14.11), is endowed with a positive zero-current potential cell in the initial state, when the reaction proceeds toward the right. Therefore, it is a galvanic cell. It is an electrolytic cell when the reaction proceeds toward the left. The process toward the right involves the inequality:

$$E^\circ\,(\mathrm{Ox_1/Red_1}) - (RT/mF)\ln[(\mathrm{Red_{1i}})/(\mathrm{Ox_{1i}})] > E^\circ\,(\mathrm{Ox_2/Red_2})$$
$$- (RT/nF)\ln[(\mathrm{Red_{2i}})/(\mathrm{Ox_{2i}})]\,,$$

and inversely when the reaction proceeds toward the left. According to the thermodynamic considerations developed in Chap. 2, this condition is identical to that stipulating that the free enthalpy of reaction $\Delta_r G$ is negative at the initial moment; that is,

$$\Delta_r G < 0,$$

with

$$\Delta_r G = n\mu_{\mathrm{Ox1i}} + m\mu_{\mathrm{Red2i}} - n\mu_{\mathrm{Red1i}} - m\mu_{\mathrm{Ox2i}}.$$

We approached the concept of quantitative character of a chemical reaction in Part II, which is devoted to acids and bases. It can also be applied to redox reactions. Additionally, it will be thoroughly discussed later in another chapter. Now, we limit ourselves to saying that the quantitative character of a reaction may be defined by the number of moles of one of the reactants reacting when the reaction reached its state of equilibrium, that is, when it stopped. The reactant, whose modified number of moles is studied, can be chosen arbitrarily.

Determining the quantitative character necessitates evaluating the concentrations of the different species in the initial and final states. Calculating the concentrations of the different species at equilibrium is identical to that developed to determine the equilibrium potentials of solutions (see Chap. 16). It is based on handling the different mathematical relations that are obligatorily satisfied and that govern the equilibria in solution. Let's consider the redox reaction

$$\mathrm{Red_1 + Ox_2 \rightleftharpoons Ox_1 + Red_2}\ (K^\circ)$$

and let's suppose, for the sake of simplicity, that the products of the reaction—the species $\mathrm{Ox_1}$ and $\mathrm{Red_2}$—are not present at the beginning of the reaction. When the equilibrium is reached, the following relations are satisfied:

$$[(\mathrm{Ox_{1eq}})(\mathrm{Red_{2eq}})]/[(\mathrm{Red_{1eq}})(\mathrm{Ox_{2eq}})] = K^\circ,$$

$$[\mathrm{Ox_{1eq}}] + [\mathrm{Red_{1eq}}] = C_1,$$

$$[\mathrm{Ox_{2eq}}] + [\mathrm{Red_{2eq}}] = C_2,$$

$$[\mathrm{Ox_{1eq}}] = [\mathrm{Red_{2eq}}].$$

C_1 and C_2 are the mass balances on couples 1 and couples 2. The last relation expresses the equilibrium in electrons. It is called the *electron balance relation*. The equilibrium constant $K°$ is easily calculated by starting from the standard potentials of couples (see Chaps. 2 and 13). Indeed, Nernst's law permits us to write

$$RT \ \ln \ K° = nFE°;$$

that is, at 298 · 15 K with $R = 8.3143$ absolute J/deg mol and $F = 96,487$ abs coul (gram-equivalent)$^{-1}$,

$$\log K° = 16.9 \, nE°,$$

where $E°$ is the standard potential of the cell whose reaction is equivalent to that studied. The value of n that is taken is the weakest common multiple of the numbers of electrons exchanged by both redox couples.

If the stoichiometry of the studied reaction is less simple than that studied immediately above or if the products of the reaction are also present at the beginning of the reaction, some equations of the same type as those earlier can always be set up, and the resultant system of equations is always mathematically solvable (see Chap. 16).

The standard potential of the cell is equal to the difference in the standard redox potentials of both antagonistic couples. Hence, it is sufficient to read the tables of redox data.

For example, let's consider the oxidation of ferrous iron by potassium permanganate:

$$MnO_4^- + 8H^+ + 5Fe^{2+} \rightarrow Mn^{2+} + 5Fe^{3+} + 4H_2O.$$

It can be decomposed into the two half-reactions

$$MnO_4^- + 8H^+ + 5e^- \rightleftharpoons Mn^{2+} + 4H_2O, \quad E°(MnO_4^-/Mn^{2+}) = 1.51 \text{ V},$$

$$Fe^{3+} + 1e^- \rightleftharpoons Fe^{2+}, \quad E°(Fe^{3+}/Fe^{2+}) = 0.77 \text{ V}.$$

The standard potential of the cell is

$$E° = 1.51 - 0.77 = 0.75 \text{ V},$$

$$RT \ \ln K° = 5\text{F } 0.75,$$

$$K° = 3 \cdot 10^{63}.$$

When equilibrium is reached, the concentrations of the species (rigorously: their activities) are such that the following relation is satisfied:

$$\left[\left(Mn^{2+}{}_{eq}\right)\left(Fe^{3+}{}_{eq}\right)^5\right] / \left[\left(MnO_4{}^-{}_{eq}\right)\left(Fe^{3+}{}_{eq}\right)^5\left(H^+\right)^8\right] = 3 \cdot 10^{63}.$$

Due to the huge value of $K°$, the reaction may be considered to be complete. Let's recall that for a set of given experimental data, it is better to calculate the equilibrium constant by starting from formal potentials when they are known. Therefore, the calculated constant is the conditional one.

A converse exists to the calculation of equilibrium constants from the half-reduction potentials: It is the possibility to obtain the unknown redox potentials of some couples. In order to achieve it, a redox equilibrium between two couples is investigated. The equilibrium constant is determined, if the standard redox potential of one of both couples is already known. The value of the other (unknown) is immediately deduced. This strategy is, of course, of great importance in physical and analytical chemistries. It is in this way that the standard potentials of slow electrochemical systems (see electrochemistry), in particular, those of organic redox couples, have been determined.

14.6 Kinetic Considerations Concerning Redox Reactions

The preceding considerations were based only on thermodynamic bases. They only permitted us to know whether or not the reaction was possible. A negative answer is definitive. If the answer is positive, it is not definite, however, that the reaction can take place, this time for kinetic reasons. We will now somewhat develop this point.

Redox reactions may indeed be slow in water, contrary to the acid—base reactions, which can be considered immediate. If the reaction is very sluggish, no change is detectable. We are apparently in an equilibrium state that is actually a false equilibrium state. In other cases, the equilibrium may require a very long time to be reached, even if the reaction proceeds perceptibly.

Here are some examples of redox reactions that are considered to be slow:

- aqueous potassium permanganate solutions are, surprisingly, stable, whereas the permanganate ion should oxidize water according to thermodynamic predictions (see manganimetry). Some aqueous permanganate solutions have even been conserved for several years without exhibiting any change in their compositions;
- another example is provided by hydrogen peroxide solutions. According to the standard potentials of both couples of hydrogen peroxide:

$$H_2O_2 + 2H^+ + 2e^- \rightleftharpoons 2H_2O, \quad E°(H_2O/H_2O_2) = 1.77\ \text{V},$$

$$2H^+ + O_2 + 2e^- \rightleftharpoons H_2O_2, \quad E°(H_2O_2/O_2) = 0.69\ \text{V},$$

the disproportionation equilibrium

$$2H_2O_2 \rightleftharpoons 2H_2O + O_{2(g)}$$

is greatly displaced toward the right (see a graphical explanation in the next chapter). However, the disproportionation is slow. In water, the hydrogen peroxide is in a false equilibrium state, but several catalysts increase its rate of decomposition;

- aqueous dichromate solutions should also not be stable, and those of the chromous ion should not be handled for the same reason as the permanganate ion (see redox titrations);
- the peroxodisulfate ion is a very powerful oxidizing agent:

$$S_2O_8{}^{2-} + 2e^- \rightleftharpoons 2SO_4{}^{2-}, \quad E°(S_2O_8{}^{2-}/SO_4{}^{2-}) = 2.05\ V,$$

but it reacts slowly except in certain conditions.

It is an experimental fact that when the electron exchange is not accompanied by a structural change of the Ox and Red forms of the couple, the reaction is fast. For example, this is the case with the couples $Ag^+/Ag_{(s)}$, Fe^{3+}/Fe^{2+}, and $[Fe(CN)_6]^{3-}/[Fe(CN)_6]^{4-}$.

If the electron exchange is accompanied by a break of a covalent bond, the redox reaction is slow. This is the case with peroxo compounds such as hydrogen peroxide and oxidant oxanions such as $S_2O_8{}^{2-}$.

If the standard potential difference between the two antagonistic couples increases, the reaction rate increases, too. Hence, the peroxodisulfate ion is not reduced by couples whose standard potentials are higher than about 1 V. In such a case, it is in a false equilibrium. However, it is reduced by iodide ions $E°(I_2/I^-) = 0.62\ V$. The usual rules of chemical kinetics are, of course, obeyed. Thus, increases in temperature and concentrations generally increase the rate of the redox reactions.

As with every chemical reaction, the redox reaction kinetics depend on the reactional mechanisms. Several are known. Therefore, it seems that the fastest redox reactions are those in which the hydrated electron intervenes according to the reaction

$$M^{n+} + 1e^- \rightarrow M^{(n-1)+}.$$

(The simple dissolution of uranium trichloride immediately produces hydrated electrons.) The rate of these redox reactions is limited by the diffusion of the reacting species. At least two other mechanisms exist. The first, the outer-sphere mechanism, involves an electron transfer between the two reacting species without any change in their primary coordination shell. The second, the inner-sphere mechanism, involves first the displacement of a ligand between the two species and, second, an electron transfer. Both mechanisms give rise to a large range of kinetic constants' values. For example, hexacyanoferrate(III)/hexacyanoferrate(II) gives rise to the outer-sphere mechanism. The discovery of these mechanisms results from the work of Taube and Marcus, who individually won the Nobel Prize in Chemistry, in 1983 and 1992, respectively, for their work[1]. Interestingly, peroxo compounds seem to react according to several mechanisms, some of which could be radicalar ones.

---

[1] H. Taube, contemporary chemist, Stanford University; R. A. Marcus, contemporary chemist, California Institute of Technology.

Chapter 15
Predicting Redox Reactions by Graphical Means

Some graphical means permit us to predict redox reactions according to the experimental conditions. They principally allow the prediction of these reactions according to the concentrations of the different species that participate in the studied equilibrium. Quite generally, such predictions are less accurate than those based on the calculations described in the previous chapter. However, they are very useful because they immediately provide an overview of the evolution of the processes and, in particular, of the redox phenomena as a function of some experimental parameters, specifically of the pH.

15.1 Predominance Areas of a Redox Couple

Predicting redox reactions with the aid of graphical means involves the knowledge of the predominance areas of the redox couples that participate in the equilibria.

- Let's consider the couple

$$Fe^{3+} + 1e^- \rightleftharpoons Fe^{2+}.$$

Nernst's equation gives

$$E = E^\circ\left(Fe^{3+}/Fe^{2+}\right) - 0.06\ \log\left[\left(Fe^{2+}\right)/\left(Fe^{3+}\right)\right]\ (T = 298\ \mathrm{K},\ E^\circ = 0.77\ \mathrm{V})$$

We see that

$$\text{if} \quad E = E^\circ, \quad \left(Fe^{2+}\right) = \left(Fe^{3+}\right);$$

$$\text{if} \quad E > E^\circ,$$

$$\left(Fe^{3+}\right) > \left(Fe^{2+}\right): \text{ the ferric ion is predominant.}$$

$$\text{if } E < E^\circ,$$

$$\left(Fe^{2+}\right) > \left(Fe^{3+}\right): \text{ the ferrous ion is predominant.}$$

J.-L. Burgot, *Ionic Equilibria in Analytical Chemistry*,
DOI 10.1007/978-1-4419-8382-4_15,

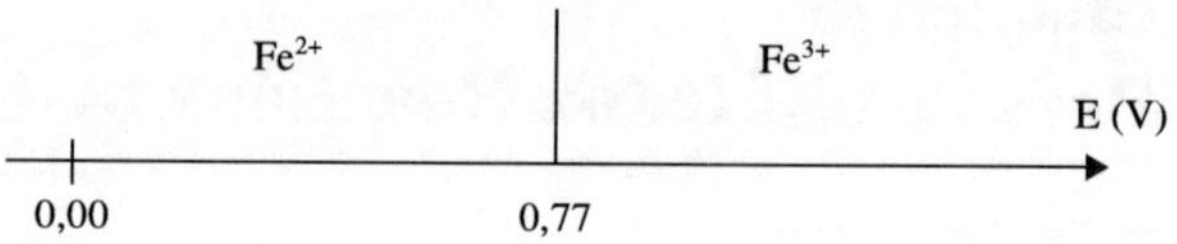

Fig. 15.1 Predominance areas for iron

The predominance-areas diagram in Fig. 15.1 comes from the above relations.

The predominance region of the ferric ion is on the right of the vertical bar $E = E^\circ = 0.77$ V, and that of the ferrous ion is situated on the left. It is important to notice that some ferrous ions remain to the right of the vertical $E = E^\circ$ (i.e., in the region of ferric ions), but this remaining quantity is weak. The farther from the bar we are located on the right, the fewer ferrous ions that remain at this point. By starting from the vertical and going into one of the two regions, we can deduce from Nernst's equation that the concentration of the other member of the couple decreases exponentially.

The example of the couple Fe^{3+}/Fe^{2+} is particularly simple for several reasons. First, both members of the couple are in solution. Second, the exponents of the activities of both members are the same in the corresponding Nernst equation. Finally, there is no supplementary species (with respect to both members of the couple) in the argument of the logarithm. All these reasons lead to the fact that, in this example, we do not have to adopt some conventions fixing one (or several) concentration(s) in order to limit the predominance areas. This is not the case with the following examples.

- Let's consider the couple $Cl_{2(w)}/Cl^-$. Nernst's equation gives

$$E = E^\circ(Cl_{2(w)}/Cl^-) - (0.06/2)\log\left[(Cl^-)^2/(Cl_{2(w)})\right],\quad T = 298\text{K},\ E^\circ = 1.40\,\text{V}.$$

The numerator and the denominator of the argument of the logarithm do not have the same exponent. In order to draw preponderance-area diagrams in such cases, we must adopt a convention concerning the concentrations of the involved species. The literature mentions two conventions that appear with the same frequency, conventions also adopted in the case of the drawing of Pourbaix diagrams (see Sect. 15.3). In the first one, we set

$$[\text{Ox}] = [\text{Red}] = C$$

on the frontier straight line, where C, expressed in mol/L, is chosen arbitrarily. (When establishing these diagrams, we assume that activities are equal to concentrations.) According to the second convention, the concentrations of the atoms of the common element of the Ox and Red forms are the same on the frontier line, whereas the total atomic concentration (of the common element) is arbitrary. In the chosen example,

- according to the first convention and with a 10^{-1} mol/L concentration, we calculate for the E value on the frontier straight line:

$$E = E^\circ\left(Cl_{2(w)}/Cl^-\right) - (0.06/2)\log\left[(10^{-1})^2/(10^{-1})\right],$$

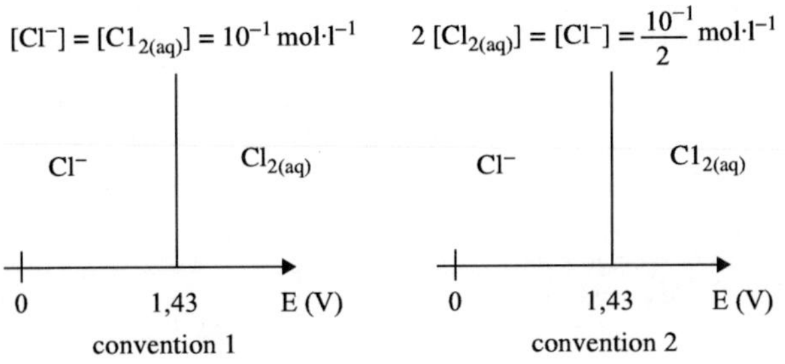

Fig. 15.2 Predominance areas of the system $Cl_{2(w)}/Cl^-$

$$E = 1.40 - 0.03 \log 10^{-1},$$

$E = 1.43$ V (frontier value: convention 1: $C = 10^{-1}$ mol/L);

- according to the second convention and with a 10^{-1} mol/L concentration, we must write

$$2[Cl_{2(S)}] + [Cl^-] = C,$$

$$2[Cl_{2(S)}] = [Cl^-] = C/2\,.$$

After handling these two equations, we find

$$E = [1.40 - 0.03 \log(10^{-1})]\,\mathrm{V},$$

$E = 1.43$ V (frontier value, convention 2, $C = 10^{-1}$ mol/L).

In this example, both conventions (with the value $C = 10^{-1}$ mol/L) give the same frontier straight line. This is not always the case.

Figure 15.2 illustrates this example.

With the help of some other examples, we could remark that generally, the frontier straight line is very close to the line $E = E°$ when they do not coincide.

- If one of the members of the couple is in a solid state, the choice of the concentration of the species in solution is arbitrary, whereas the solid species' activity is necessarily equal to unity (see Chap. 3). Therefore, with the couple $Ag^+/Ag_{(s)}$, for example, after having adopted the arbitrary concentration $[Ag^+] = 10^{-1}$ mol/L, we find $E = 0.74$ V ($E° = 0.80$ V) for the frontier straight line. From another standpoint, if one of the members of the couple is in gaseous state and the other is in solution, the pressure of the gaseous form is arbitrarily chosen to be 1 bar and the concentration of the species in solution is assigned an arbitrary value.
- Finally, if a particule different from the electron plays a part in the redox equilibrium, that is, protons or other chemical species, we must conventionally adopt a numerical value of their concentrations in order to limit the different predominance areas.

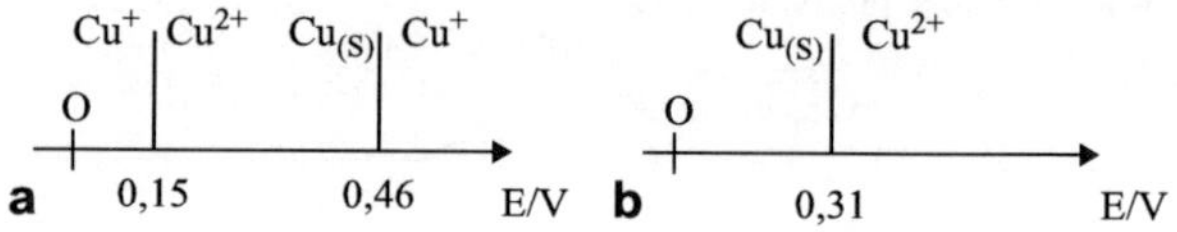

Fig. 15.3 **a** Disconnected areas of Cu^{+}. **b** True diagram

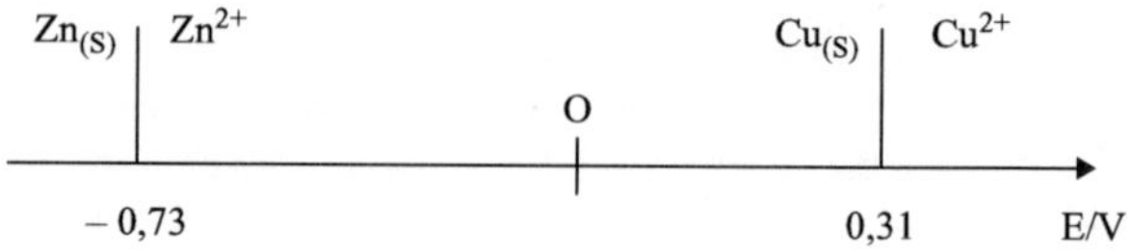

Fig. 15.4 Predominance areas of species $Zn_{(s)}$, Zn^{2+}, $Cu_{(s)}$, and Cu^{2+}

15.2 Qualitative Prediction of Redox Reactions from the Knowledge of the Predominance Areas

The superimposition of the predominance areas of the two antagonistic redox couples permits us to predict the evolution of the global redox reaction qualitatively.

- Let's study, for example, the stability of the cuprous ion Cu^{+} in aqueous solution. The redox couples in which it participates are

$$Cu^{+} + 1e^{-} \rightleftharpoons Cu_{(S)}, \quad E^{\circ} = 0.52\,V,$$

$$Cu^{2+} + 1e^{-} \rightleftharpoons Cu^{+}, \quad E^{\circ} = 0.15\,V.$$

According to convention 1, with $C = 10^{-1}$ mol/L, we find the frontier straight lines $E = 0.46\,V$ and $E = 0.15\,V$, respectively. It appears that the cuprous ion Cu^{+} belongs to two areas that are disconnected (Fig. 15.3). This is impossible. Figure 15.3a is simply the expression of the fact that the cuprous ion does not exist in water. It spontaneously disproportionates according to

$$2Cu^{+} \rightarrow Cu^{2+} + Cu_{(S)}.$$

Actually, the true diagram of the predominance area for copper in water is Fig. 15.3b, at least for pH < 3.5, for which there is no precipitation of cuprous oxide, Cu_2O.

- The diagram corresponding to the reaction

$$Zn_{(S)} + Cu^{2+} \rightarrow Zn^{2+} + Cu_{(S)}$$

is given in Fig. 15.4 ($C = 10^{-1}$ mol/L).

When we superimpose the preponderance areas of $Zn_{(s)}$ and Cu^{2+}, it appears that they are disconnected. Both species cannot remain face to face without reacting with each other. The reaction written above, that is, from left to right, takes place. Conversely, Zn^{2+} and $Cu_{(s)}$ have a portion of predominance area in common. They do not react together. The distance between the two positions of the two frontier straight lines is large. The reaction is certainly displaced strongly toward the right.

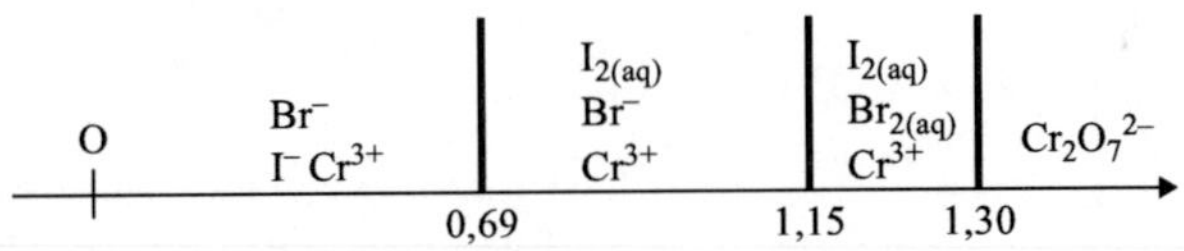

Fig. 15.5 Oxidation of iodide and bromide ions by dichromate ions

- Predominance diagrams immediately specify which reactions are possible from a thermodynamic standpoint. Let's investigate the possibility of the oxidation of a mixture of bromide and iodide ions by potassium dichromate, according to the reactions

$$Cr_2O_7{}^{2-} + 14H^+ + 6Br^- \rightarrow 2Cr^{3+} + 3Br_{2(W)} + 7H_2O, \quad (15.1)$$

$$Cr_2O_7{}^{2-} + 14H^+ + 6I^- \rightarrow 2Cr^{3+} + 3I_{2(W)} + 7H_2O. \quad (15.2)$$

According to convention 1 (with the concentrations of the different species equal to $2 \cdot 10^{-2}$ mol/L and with pH $= 0$), we find the following values for the frontier potentials:

$$E\left(Cr_2O_7{}^{2-}/Cr^{3+}\right) = 1.30\,V, \quad E\left(Br_{2(W)}/Br^-\right) = 1.15\,V,$$
$$E\left(I_{2(W)}/I^-\right) = 0.69\,V.$$

Figure 15.5 shows that the oxidation reaction of iodide ions is easier than that of the bromide ions, because the region where they predominate is farther from that of the dichromate ions than is the region where the bromide ions predominate. The diagram also indicates that when bromide ions are oxidized, iodide ions are also oxidized. The converse is not true. From thermodynamic and analytic standpoints, it is interesting to know the conditions in which the selective oxidation of iodide ions is possible (see below).

- The study of the corresponding predominance areas confirms the fact that hydrogen peroxide is instable in aqueous solution. It decomposes into water and dioxygen according to the redox reaction

$$2H_2O_2 \rightarrow 2H_2O + O_{2(g)}.$$

Hydrogen peroxide intervenes in the following two half-redox reactions:

$$H_2O_2 + 2H^+ + 2e^- \rightarrow 2H_2O, \quad E°\left(H_2O_2/H_2O\right) = 1.77\,V,$$

$$O_{2(g)} + 2H^+ + 2e^- \rightarrow H_2O_2, \quad E°\left(O_{2(g)}/H_2O_2\right) = 0.67\,V.$$

(Note, incidentally, that hydrogen peroxide may behave as a reducing agent. At first sight, this may seem paradoxical for a peroxygenated derivative.) The predominance regions of the different species involved in the above two equilibria are shown in Fig. 15.6. (This diagram was drawn with the concentration $[H_2O_2]$ fixed to the value 10^{-1} mol/L, the pH value at 4, and Po_2 at 1 bar.)

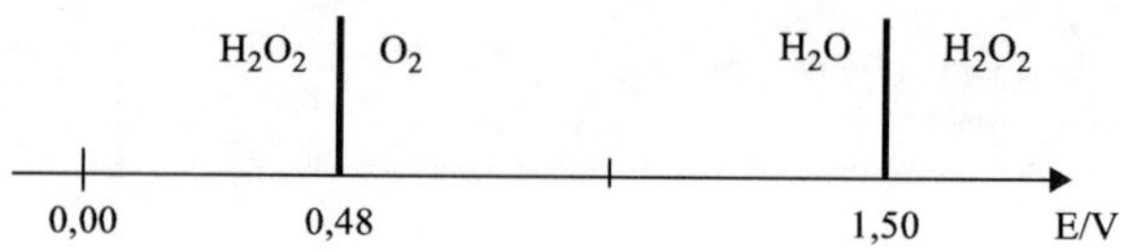

Fig. 15.6 Predominance regions of species H_2O_2, H_2O, and O_2

The predominance regions of hydrogen peroxide are disconnected. It is not surprising that hydrogen peroxide disproportionates according to the reaction already given. The reaction is actually very slow although it goes to quasi-completion toward the right (see kinetic considerations—Chap. 14).

- An interesting example of the use of the predominance areas is provided by the explanation of the attack of gold by aqua regia. The reaction may be explained in three steps:

1. first, we must recall that nitric acid alone does not attack gold. As a rule, at pH = 0, gold may intervene in the two redox couples

$$Au^+ + 1e^- \rightleftharpoons Au_{(S)}, \quad E° = 1.68\,V,$$

$$Au^{3+} + 3e^- \rightleftharpoons Au_{(S)}, \quad E° = 1.37\,V.$$

In the nitric acid solution, there are two possible oxidants:

a. the proton itself according to the equilibrium

$$2H^+ + 2e^- \rightleftharpoons H_{2(g)}, \quad E° = 0.00\,V;$$

b. the nitrate ion which may be reduced into nitrogen oxide according to

$$NO_3{}^- + 4H^+ + 3e^- \rightleftharpoons NO_{(g)} + 2H_2O, \quad E° = 0.96\,V.$$

Figure 15.7a (conventions: P_{NO} = 1 bar, $[NO_3^-]$ = 1mol/L, $[Au^{3+}]$ = 1mol/L) shows that there is no possible oxidation of gold by the nitrate ion, which is the strongest oxidant in the medium. This is true even if we consider the couple $Au^{3+}/Au_{(s)}$ which exhibits the weaker redox potential of the two couples involving gold. This is not a surprise since Au and $NO_3{}^-$ possess a large common predominance area;

2. second, we must recall that hydrochloric acid alone (at pH = 0) does not attack gold. In this medium, the only existing oxidant is the proton. However, now there is another point to consider: the appearance of two new redox couples of gold in which the element chlorine is implicated. They are

$$AuCl_2{}^- + 1e^- \rightleftharpoons Au_{(S)} + 2Cl^-, \quad E° = 1.14\,V,$$

$$AuCl_4{}^- + 3e^- \rightleftharpoons Au_{(S)} + 4Cl^-, \quad E° = 0.98\,V.$$

Figure 15.7b shows that the proton cannot oxidize gold even to give $AuCl_4{}^-$, which is, in these conditions, the more easily obtainable oxidized species of gold (convention: Cl^- = 1mol/L);

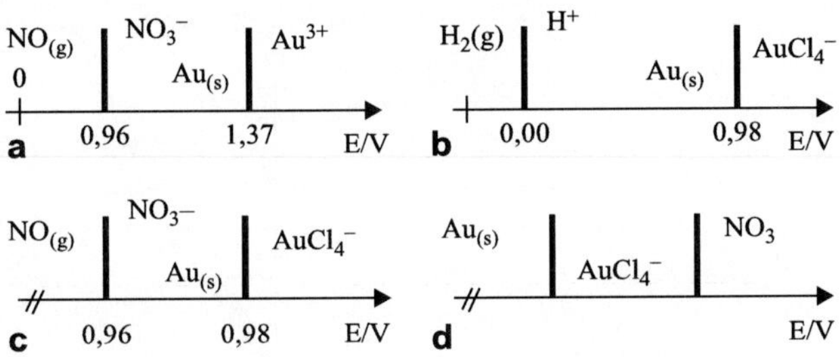

Fig. 15.7 Reaction of aqua regia with gold

3. third, it is an experimental fact that gold is attacked by a mixture of hydrochloric and nitric acids. The oxidation reaction is

$$NO_3{}^- + 4H^+ + 4Cl^- + Au_{(S)} \rightleftharpoons NO_{(g)} + AuCl_4{}^- + 2H_2O.$$

Figure 15.7c, drawn in the same conditions as before, indicates that there is a common region in the predominance areas of nitrate ion and gold. Hence, at first sight, no reaction could evolve. However, this reaction proceeds. Figure 15.7c does not predict this fact, because of the arbitrary conditions for which it was drawn, which are not realistic from an experimental standpoint. Herein lies the limit of these kinds of predictions. We now present Fig. 15.7d, which might predict a clear attack of gold by aqua regia. The regions where $Au_{(s)}$ and $NO_3{}^-$ are predominant are frankly disconnected. Briefly and to conclude, we may assert from the examination of these diagrams that the reaction of gold with aqua regia is certainly equilibrated, except if it is performed with a clear excess of it.

15.3 Frost Diagrams

The examination of another kind of diagram, the so-called Frost diagrams, provides another way to decide whether or not a redox reaction is possible, at least from a qualitative point of view. It also provides a rapid means to determine the stability of the different oxidation states of one element.

Frost diagrams figure the oxidation states of one element by points whose abscissas are equal to its oxidation numbers and whose ordinates are equal to the products of these oxidation numbers by the standard potentials of the couples. The couples consist of the element at the oxidation state under study and at the oxidation state 0.

Let's take the examples of copper and mercury. The standard potentials are

$$E°\left(Cu^+/Cu_{(S)}\right) = 0.52\,V;\ E°\left(Cu^{2+}/Cu^+\right) = 0.15\,V;$$

$$E°\left(Cu^{2+}/Cu_{(S)}\right) = 0.34\,V;$$

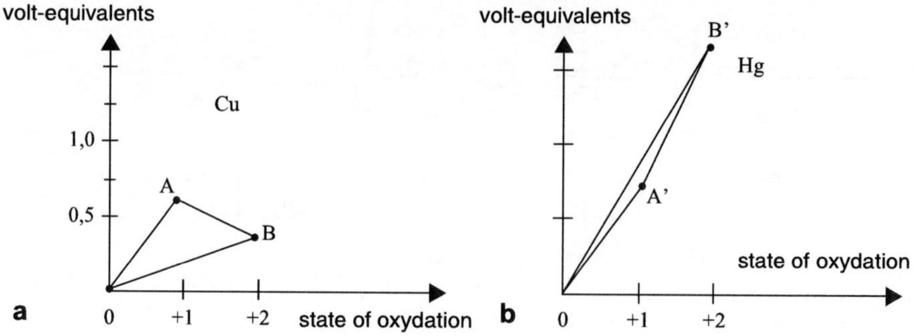

Fig. 15.8 Frost diagrams of copper and mercury (oxidation states 0, +I, +II)

$$E^\circ\left(Hg_2{}^{2+}/Hg_{(l)}\right) = 0.79\,V;\; E^\circ\left(Hg^{2+}/Hg_{(l)}\right) = 0.85\,V;$$
$$E^\circ\left(Hg^{2+}/Hg_2{}^{2+}\right) = 0.92\,V.$$

The coordinates of the state +I of copper are +1; 1 · 0.52 volts-equivalents: point A—those of copper +II are: +2; 2 ·0.15 volts-equivalents: point B (see Fig. 15.8a). For mercury, the coordinates of the oxidation states +I and +II are, respectively, (1; 1 · 0.79 volts-equivalents: point A') and (2; 2 · 0.85 volts-equivalents: point B′) (Fig. 15.8b).

Obviously, the slope values of the segments OA, OB, OA′, and OB′ are the standard potential values of the corresponding couples. The slopes of segments AB and A′B′ are, equal to the standard potentials Cu^{2+}/Cu^+ and Hg^{2+}/Hg_2^{2+}, respectively. This is a consequence of the additivity of the free enthalpies and, more precisely in this case, of Latimer–Luther's rule (see Chap. 2).

Indeed, let's consider the half-redox equilibria

$$M^{+i} + ie^- \rightleftharpoons M^\circ, \quad E_i{}^\circ = \left(M^{+i}/M^\circ\right),$$

$$M^{+j} + je^- \rightleftharpoons M^\circ, \quad E_j{}^\circ = E^\circ\left(M^{+j}/M^\circ\right),$$

with the hypothesis $j > i$. Per Latimer–Luther's rule, we find [by denoting $E^\circ_{ij} = E^\circ(M^{+j}/M^{+i})$]

$$iE^\circ_i + (j - i)E^\circ_{ij} = jE_j{}^\circ;$$

therefore,

$$E_{ij}{}^\circ = \left[jE_j{}^\circ - iE_i{}^\circ\right]/(j - i).$$

Let's now investigate the stability of Cu^{+I}. A way to solve the problem is to know its tendency to disproportionate according to the reaction

$$2Cu^+ \rightleftharpoons Cu^{2+} + Cu_{(s)}, \quad \Delta G^\circ.$$

The disproportionation results from the superimposition of the following two half-redox reactions:

$$Cu^+ - 1e^- \rightleftharpoons Cu^{2+} \quad \Delta G_1{}^\circ, \quad -E^\circ\left(Cu^{2+}/Cu^+\right),$$

$$Cu^+ + 1e^- \rightleftharpoons Cu_{(S)} \quad \Delta G_2{}^\circ, \quad E^\circ\left(Cu^+/Cu_{(S)}\right).$$

Copper +I will exhibit a tendency to disproportionate if the free enthalpy ΔG° is negative (see Chap. 2). Since the following relationships hold:

$$\Delta G^\circ = -nFE^\circ,$$

$$\Delta G_1{}^\circ = -1F\left[-E^\circ(Cu^{2+}/Cu^+)\right],$$

$$\Delta G_2{}^\circ = -1F\left[E^\circ(Cu^+/Cu_{(S)})\right],$$

and

$$\Delta G^\circ = \Delta G_1{}^\circ + \Delta G_2{}^\circ,$$

we find

$$\Delta G^\circ = [F(0.34 - 0.52)]\ \mathrm{J/mol}, \quad \Delta G^\circ < 0.$$

Copper +I is not stable. We could demonstrate that Hg +I ($Hg_2{}^{2+}$) is stable in the same manner. The demonstration we followed is equivalent to the assertion that if point A is above segment OB (Fig. 15.8a), the oxidation state it represents is not stable. When it is located under segment OB, the state it represents is stable. Mercury +I ($Hg_2{}^{2+}$) is stable. We could demonstrate that the reaction

$$Hg^{2+} + Hg_{(1)} \rightleftharpoons \text{Hg-Hg}^{2+}$$

is spontaneous from left to right. These reasons are generalizable to elements that have more than three oxidation states.

According to a rigorous line of argument, predictions based on Frost diagrams are not more accurate than those given by only considering the standard potentials of couples. In other words, the predictions given by this strategy are only accurate when the redox species involved in the reaction are in their standard states. We have already seen that these predictions may sometimes be inverted in other experimental conditions (see Chap. 14). Because standard potentials involve pH = 0, Frost diagrams are implicitly drawn for pH = 0. For other pH values, the apparent standard potentials must be used in order to build the diagrams. Actually, they are often drawn for pH = 0 and pH = 14. Whichever the case, diagrams are always drawn for a fixed pH value.

An interesting example for its analytical applications is provided by nitrous acid HNO_2 and nitrites $NO_2{}^-$ in which nitrogen is at the oxidation number +III. If, in order to simplify, we only consider the oxidation numbers +II (nitric oxide NO),

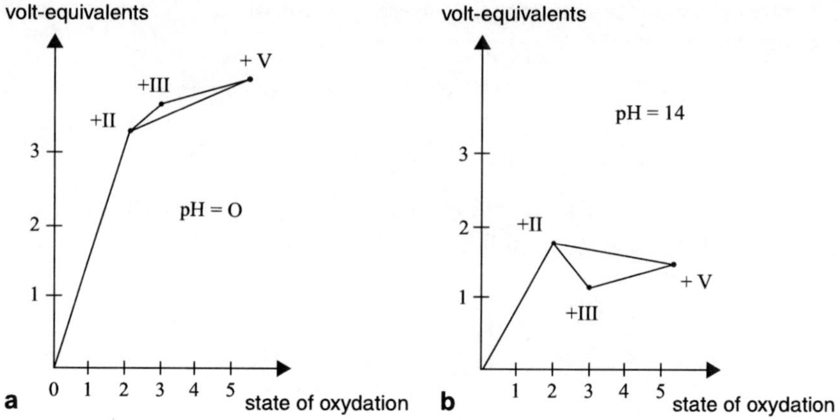

Fig. 15.9 Frost diagram of nitrogen (oxidation numbers +II, +III, and +V)

+III (nitrous acid and nitrites), and +V (nitrates) of nitrogen, the half-redox equilibria to take into account are

- at pH = 0,

$$2NO + 4e^- + 4H^+ \rightleftharpoons N_{2(g)} + 2H_2O, \quad E^\circ = 1.68\,V,$$

$$HNO_2 + 1e^- + 1H^+ \rightleftharpoons NO_{(g)} + H_2O, \quad E^\circ = 0.99\,V,$$

$$NO_3{}^- + 2e^- + 3H^+ \rightleftharpoons HNO_2 + H_2O, \quad E^\circ = 0.94\,V;$$

- at pH = 14,

$$2NO + 4e^- + 2H_2O \rightleftharpoons N_{2(g)} + 4OH^-, \quad E^{\circ\prime} = 0.84\,V,$$

$$NO_2{}^- + 1e^- + H_2O \rightleftharpoons NO_{(g)} + 2OH^-, \quad E^{\circ\prime} = -0.46\,V,$$

$$NO_3{}^- + 2e^- + H_2O \rightleftharpoons NO_2{}^- + 2OH^-, \quad E^{\circ\prime} = 0.01\,V.$$

The corresponding Frost diagram is given in Fig. 15.9.

The examination of the Frost diagram shows that nitrous acid does exhibit the tendency to disproportionate into nitrate ion and nitric oxide in acidic medium, whereas nitrite ions are stable in basic medium.

Predicting a redox reaction with the help of Frost diagrams is possible. It follows the same principle: Straight lines whose slopes are equal to the standard potentials of the antagonist couples are drawn on the same diagram. Examining the locations of the redox species gives the answer. For example, let's study the eventuality of the oxidation of stannous tin by permanganate ions to give stannic tin and manganese dioxide according to

$$3Sn^{2+} + 2MnO_4{}^- + 8H^+ \rightleftharpoons 2MnO_{2(S)} + 3Sn^{4+} + 4H_2O.$$

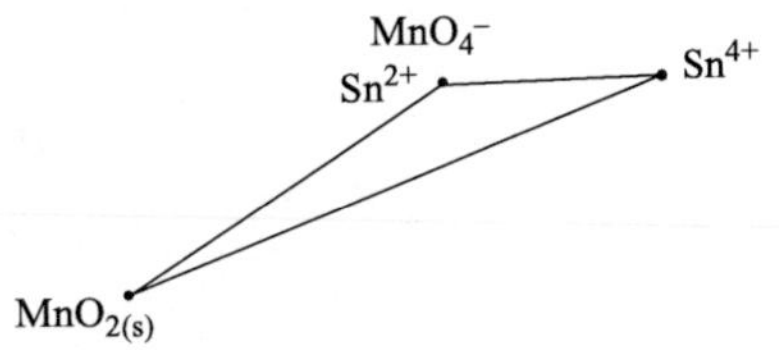

Fig. 15.10 Predicting the reaction of the permanganate ion with the stannous ion with a Frost diagram

Let's draw the segments corresponding to the two couples $MnO_4^-/MnO_{2(s)}$ and Sn^{4+}/Sn^{2+}. We find the diagram given in Fig. 15.10 after having located Sn^{2+} and MnO_4^- (initial species of the investigated reaction) at the same point.

The MnO_4^- and Sn^{2+} are located above the segment $MnO_{2(s)}/Sn^{4+}$. The reaction is possible.

In short, we can assert that the predominance areas of the antagonist redox couples permit us to predict the direction of redox reactions. They are essentially built with the standard potential values. As a result, taking into account the standard potentials or using Frost diagrams means working with the same level of prediction accuracy. Finally, both strategies lead to a strong tendency only.

15.4 *E*/pH Diagrams or Pourbaix Diagrams

E/pH diagrams, or Pourbaix diagrams,[1] are a more elaborate form of predominance diagrams than the preceding ones. They represent the predominance areas of the different redox species as a function of the potential and of the pH. Compared to the preceding diagrams in which the sole potential value was considered, they take a supplementary coordinate into account: the pH value. In the preceding example of hydrogen peroxide, we had attributed a fixed value (pH = 4) to this supplementary coordinate. Thus, it was possible to deal with this case in the preceding section. Undoubtedly, with one more dimension, Pourbaix diagrams give a supplementary element of valuation.

As for the preceding ones, building Pourbaix diagrams necessitates assuming that activities are equal to concentrations. Let's consider the couple Ox/Red in which the Ox and Red forms are related by the equation

$$x\mathrm{Ox} + m\mathrm{H}^+ + pe^- \rightleftharpoons y\mathrm{Red} + z\mathrm{H_2O}.$$

Four physico-chemical quantities, called *variables*, can be measured. They are the concentrations: [Ox], [Red], [H^+] (or the pH), and *E*, the potential of the solution. They are connected by the two independent relations

- Nernst's relation

$$E = E°(\mathrm{Ox/Red}) - (RT/pF)\ln\left[(\mathrm{Red})^y\right]/\left[(\mathrm{Ox})^x\left(\mathrm{H}^+\right)^m\right];$$

[1] M. Pourbaix (1904–1998), a Belgian (of Russian origin) biochemist, chemist, and metallurgist.

- the mass balance on the couple:

$$[\text{Red}] + [\text{Ox}] = C.$$

Therefore, we are faced with a system of two equations with four unknowns. It possesses two degrees of freedom. Hence, it is possible to choose the potential and pH values of the solution, the other variables being, *ipso facto*, fixed.

In Pourbaix diagrams, the frontier straight lines between the predominance regions are the lines *E*/pH. For example, in the general case above, the equation of the frontier line is

$$E = E°(\text{Ox/Red}) - [0.06\, m/p]\text{pH} - (0.06/p) \log [(\text{Red})^y/(\text{Ox})^x], \quad T = 298 \text{ K}.$$

As for the building of the preceding diagrams, drawing the frontier lines necessitates the adoption of conventions concerning some concentrations. They are the same as those used previously. Here we only quote some examples of Pourbaix diagrams. We shall use these diagrams later when we justify the redox titrating methods.

Example 1 Water is reduced according to the half-equilibrium

$$2\text{H}^+ + 2\text{e}^- \rightleftharpoons \text{H}_{2(g)}, \quad E°\left(\text{H}^+/\text{H}_{2(g)}\right) = 0.00\,\text{V},$$

or, equivalently, according to

$$2\text{H}_3\text{O}^+ + 2\text{e}^- \rightleftharpoons 2\text{H}_2\text{O} + \text{H}_{2(g)}.$$

(This reaction is the reduction of the hydrated proton. Actually, hydrogen exhibits the oxidation number +I before the reduction.) In basic medium, the redox half-equilibrium is

$$2\text{H}_2\text{O} + 2\text{e}^- \rightleftharpoons \text{H}_{2(g)} + 2\text{OH}^-.$$

Water is reduced since it captures two electrons. As previously, hydrogen undergoes a change in its oxidation number from +I to 0.

Water is oxidized according to the half-equilibrium

$$2\text{H}_2\text{O} \rightleftharpoons \text{O}_{2(g)} + 4\text{H}^+ + 4\text{e}^-, \quad E°\left(\text{O}_{2(g)}/\text{H}_2\text{O}\right) = 1.23\,\text{V}$$

It is oxidized since it has lost four electrons. More precisely, oxygen undergoes a change in its oxidation number from −II to 0. In basic medium, hydroxide ions are oxidized according to the half-equilibrium

$$4\text{OH}^- \rightleftharpoons \text{O}_{2(g)} + 2\text{H}_2\text{O} + 4\text{e}^-.$$

Here, also, oxygen goes from the oxidation state −II to 0. A Pourbaix diagram exhibits the following two frontier straight lines (Fig. 15.11):

$$E = (-0.06\text{pH})\,\text{V},$$

$$E = (1.23 - 0.06\text{pH})\,\text{V},$$

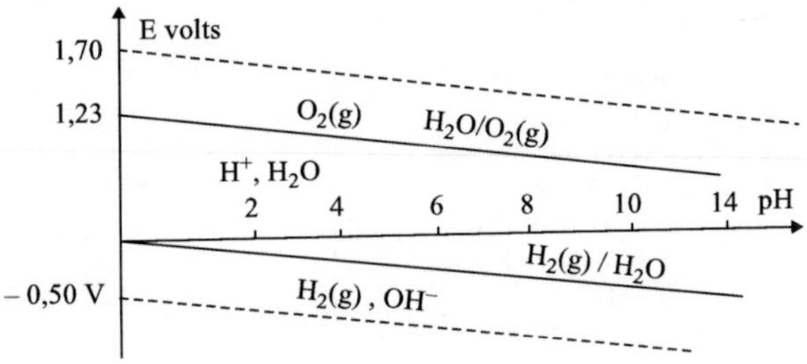

Fig. 15.11 Predominance areas of the water redox couples

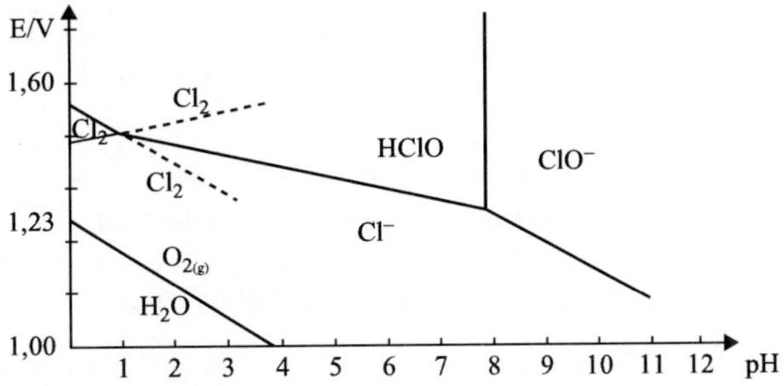

Fig. 15.12 Diagrams of the redox (Cl –I, Cl°, Cl +I) species of chlorine (convention 2 with the total concentration in atomic chlorine: 10^{-2} mol/L)

with the conventions $P_{O2} = 1$ bar and $P_{H2} = 1$ bar.

Hence, the water stability area is necessarily situated between these two parallel lines. For some pH and potential values so that their representative points should be located above the line $H_2O/O_{2(g)}$, water must be oxidized into dioxygen. For points located under the line $H_{2(g)}/H_2O$, water is reduced into dihydrogen.

Actually, water is often reduced or oxidized slowly. Thus, the practical frontiers of the stability area of water are located about 0.5 V above or under the theoretically defined lines.

Example 2 The system Cl +I/Cl°/Cl –I gives the diagram represented in Fig. 15.12. We see that chlorine exists in a very acidic mixture only, that is, for pH < 1.2–3.0 according to the adopted conventions in order to draw the diagram. At a pH value higher than this limit, chlorine disproportionates according to

$$Cl_{2(W)} + H_2O \rightarrow HClO + Cl^- + H^+.$$

The diagram shows that the couple Cl_2/Cl^- becomes more oxidant than the couple $HClO/Cl_{2(w)}$ at pH > 1.2, wherefrom the disproportionation. Chlorine of the first couple becomes the oxidant of chlorine (the reductor) of the second couple. The disproportionation is evidenced by the appearance of two disconnected areas of chlorine (dashed lines). Another interesting result may be found from the consideration of this diagram. Normally, hypochlorous acid, chlorine, and the hypochlorous ion should oxidize water with the production of dioxygen. This means that these species are not stable in water from a thermodynamic standpoint. However, here it is the contrary. They are apparently stable. This is for kinetic reasons. Likewise, at pH < 1.1,

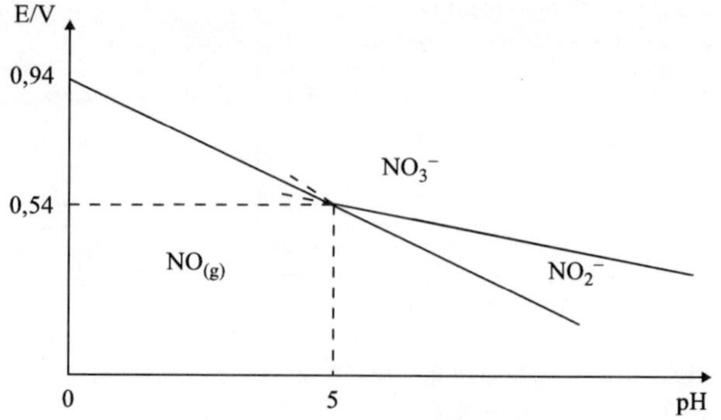

Fig. 15.13 Stability of nitrous acid [conventions: $C = 10^{-1}$ mol/L for any solute; $P_{NO} = 1$ bar; $pK_a(HNO_2/NO_2^-) = 3.3$; $E°(HNO_2/NO_{(g)}) = 1.00\,V$; $E°(NO_3^-/NO_{(g)}) = 0.96$ V; $E°(NO_3^-/HNO_2) = 0.94\,V$]

hypochlorous acid reacts with chloride ions to give chlorine. This is confirmed by examining the diagram. The latter also shows that water should be oxidized by HClO before Cl^-. Indeed, the frontier line $Cl_{2(w)}/Cl^-$ is located above that of the couple $O_{2(g)}/H_2O$. However, this is not actually the case.

Example 3 Another example is provided by the study of the stability of nitrous acid. The half-redox equilibria that must be considered are

$$\begin{array}{llll} \text{III/V} & \text{pH} < 3.3 & NO_3^- + 3H^+ + 2e^- & \rightleftharpoons HNO_2 + H_2O \\ & \text{pH} > 3.3 & NO_3^- + 2H^+ + 2e^- & \rightleftharpoons NO_2^- + H_2O \\ \text{II/III} & \text{pH} < 3.3 & HNO_2 + H^+ + 1e^- & \rightleftharpoons NO_{(g)} + H_2O \\ & \text{pH} > 3.3 & NO_2^- + 2H^+ + 1e^- & \rightleftharpoons NO_{(g)} + H_2O \\ \text{II/V} & \multicolumn{3}{l}{NO_3^- + 4H^+ + 3e^- \rightleftharpoons NO_{(g)} + 2H_2O} \end{array}$$

(our interest is focalized on the oxidation states +II, +III, and +V of nitrogen).

From an examination of Fig. 15.13, we see that in the range pH < 3.3, nitrous acid is not stable and that in the range 3.3 < pH < 5.0, nitrite ions are also not stable. They disproportionate into nitrogen oxide and nitrate ions according to the reactions

$$3HNO_2 \rightarrow 2NO_{(g)} + NO_3^- + H^+ + H_2O, \quad \text{pH} < 3.3,$$

$$3NO_2^- + 2H^+ \rightarrow 2NO_{(g)} + NO_3^- + H_2O, \quad 3.3 < \text{pH} < 5.$$

Beyond about pH = 5, nitrite ions are stable. However, we must moderate what is said about the disproportionation of nitrous acid, because of the fact that it is slow. This gives us time to carry out some analytical reactions.

Example 4 A Pourbaix diagram of copper is drawn in Fig. 15.14. Copper +I only exists at pH > 3.5 and, moreover, only in the form of a precipitate. (We note, incidentally, that from the thermodynamic point of view, cuprous oxide, Cu_2O, can

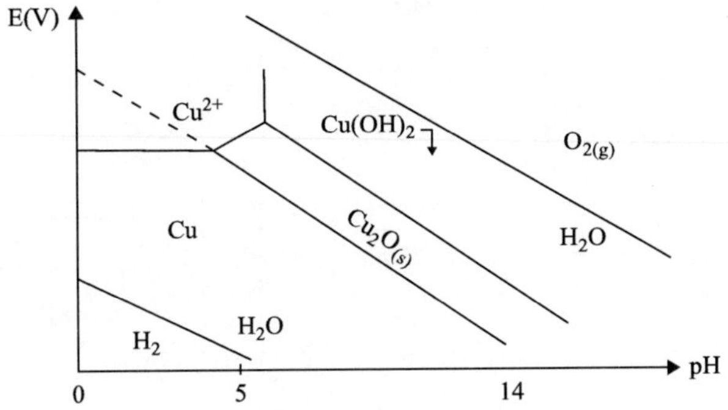

Fig. 15.14 *E*/pH diagram of copper (convention: $C = 10^{-1}$ mol/L for any species in solution)

be assimilated into cuprous hydroxide according to

$$1/2\ Cu_2O_{(S)} + 1/2\ H_2O \rightleftharpoons Cu(OH)_{(S)}$$

by virtue of the conventions on the activities.)

15.5 An Example of Application of Pourbaix Diagrams in Analytical Chemistry

The examination of a Pourbaix diagram provides a quick, although approximate, answer to the problem of the location of the pH range in which a selective oxidation or reduction of a given species must be carried out. Let's investigate the selective oxidation of iodide ions in the presence of bromide ions by potassium dichromate. We are only interested by the range $0 < \text{pH} < 4.5$. In this pH range, Cr^{+VI} is essentially like the $Cr_2O_7^{2-}$ species (for a more complete diagram of the species of chrome, see Chap. 20). Figure 15.15 shows the frontier lines of the couples under consideration (convention: any species in solution $2 \cdot 10^{-2}$ mol/L—convention 1). The equations of the frontier lines are

- for the couple:

$$Cr_2O_7^{2-} + 14H^+ + 6e^- \rightleftharpoons 2Cr^{3+} + 7H_2O,$$

$$E = (1.33 - 0.14\ \text{pH})\ \text{V},$$

with $\left[Cr_2O_7^{2-}\right] = 2 \cdot 10^{-2}$ mol/L and $\left[Cr^{3+}\right] = 2 \cdot 10^{-2}$ mol/L and $E°\left(Cr_2O_7^{2-}\right) = 1.33\ \text{V}$;

- for the couple:

$$Br_2 + 2e^- \rightleftharpoons 2Br^-$$

$$E = 1.14\ \text{V},$$

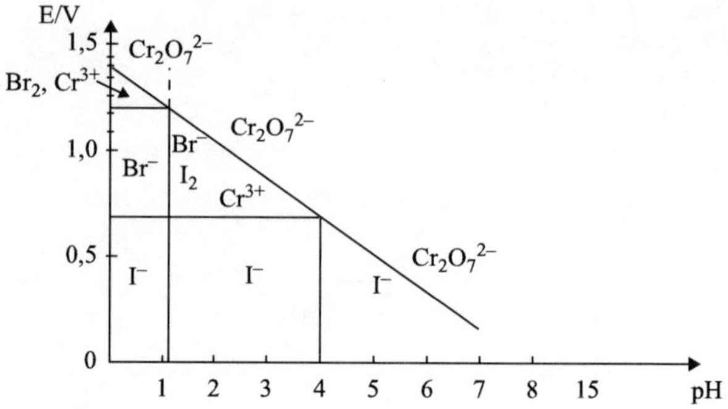

Fig. 15.15 pH range in which there is selective oxidization of iodide ions when bromide ions are present

with $[Br_2] = 2 \cdot 10^{-2}$ mol/L and $[Br^-]$

$$= 2 \cdot 10^{-2} \text{ mol/L and } E°(Br_2/Br^-) = 1.09\,V;$$

- for the couple:

$$I_2 + 2e^- \rightleftharpoons 2I^-$$

$$E = 0.69\ V,$$

with $[I_2] = 2 \cdot 10^{-2}$ mol/L and $[I^-] = 2 \cdot 10^{-2}$ mol/L and $E°(I_2/I^-) = 0.62\,V$.

The pH range in which only the iodide ions are oxidized is $1.1 < pH < 4.3$. In this range, the preponderance areas of dichromate and iodide ions are disconnected. This is not the case for the areas of dichromate and bromide ions, which are only disconnected in the range $0 < pH < 1.1$. In the latter range, the areas of dichromate and iodide ions are also disconnected. As a result, it is possible to oxidize iodide ions only by buffering the medium at about $pH = 4$ and by allowing the dichromate ions to react as long as the total oxidation of iodides lasts. It then becomes possible to oxidize bromide ions of the same sample with potassium dichromate, after acidification at $pH < 1.1$ (Fig. 15.15).

An algebraic calculation might permit us to somewhat sharpen the range of selective oxidation.

Exercise Calculate the above results algebraically. Use the numerical data given above.

The diagram of Fig. 15.15 shows that $Cr_2O_7^{2-}$ ions oxidize iodide ions when predominance areas of both species become disconnected, that is, when the frontier lines of the couples $Cr_2O_7^{2-}/Cr^{3+}$ and I_2/I^- intersect. This is the case when

$$1.33 - 0.14pH = 0.69\ V, \quad pH = 4.57.$$

This is the same state of affairs for the oxidization of bromide ions:

$$1.33 - 0.14pH = 1.14\ V, \quad pH = 1.36.$$

Iodide ions are selectively oxidized in the pH range $1.36 < pH < 4.57$.

Fig. 15.16 Diagram *E*/pL of gold (convention: $C = 10^{-2}$ mol/L for any species in solution) [data: $E°(Au^{3+}/Au) = 1.42$ V, $E°(Au^{+}/Au) = 1.68$ V, $Au^{3+} + 4Cl^- \rightleftharpoons AuCl_4^-$ $\beta_4 = 10^{22}$, $Au^{+} + 2Cl^- \rightleftharpoons AuCl_2^-$ $\beta_2 = 10^{9}$]

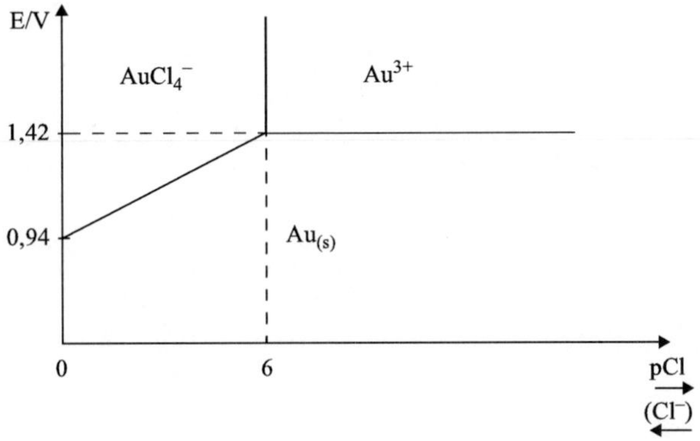

15.6 Extension of Pourbaix Diagrams

Diagrams *E*/pH can be generalized to those of the type *E*/pL, where L symbolizes a ligand or a particule (other than the proton) participating in the redox reaction. These new diagrams are interesting to consider when we study the influence of the complexation on the redox properties of a couple. Figure 15.16 shows the case of the systems of gold in the presence of chloride ions.

Starting from the thermodynamic data, it is easy to verify that the ion Au(+I) and its complex, the dichloroaurate ion [$AuCl_2^-$], cannot exist regardless of the concentration of chloride ions. They disproportionate according to the reactions

$$3Au^+ \rightarrow 2Au_{(S)} + Au^{3+},$$

$$3AuCl_2^- \rightarrow 2Au_{(S)} + AuCl_4^- + 2Cl^-.$$

Nernst's equation gives for the frontier straight line

$$E = (0.94 + 0.08\ \text{pCl})\ \text{V} \quad \text{in the range} \quad 0 < \text{pCl} < 6.$$

The normal potential, in this range, is

$$E^{\circ\prime} = (0.98 + 0.08\ \text{pCl})\ \text{V}.$$

The complexation slightly increases the reducing property of gold.

We'll close this chapter by noticing that the diagrams *E*/pH and *E*/pL are not capable of giving accurate data. However, they offer the advantage of providing a means to predict the evolution of redox phenomena as a function of the pH and pL values.

Chapter 16
Calculating Equilibrium Potentials of Solutions Containing Several Redox Couples

General Considerations Concerning Redox Titrations

Calculating the redox potentials of solutions that are mixtures of several redox couples constitutes an important matter. An interesting case of such a mixture is that in which initially only the Ox form of one couple exists together with the Red form of another. The solutions obtained during the course of redox titrations provide especially important examples of these kinds of solutions. Obviously, when they occur, such mixtures induce redox reactions. Of course, these redox reactions may (or may not) be favored from a thermodynamic or kinetic standpoint. This prompts us to investigate the conditions that must be obeyed in order to achieve satisfactory redox titrations.

16.1 Equilibrium Potentials and Electrode Potentials

To begin, we must emphasize the fact that the solution potentials calculated according to the principles mentioned in this chapter are called *equilibrium potentials.* They are given by the laws of thermodynamics, that is, by Nernst's law. Let's consider the electrodic process

$$\mathrm{Ox} + n\mathrm{e}^- \rightleftharpoons \mathrm{Red}.$$

At equilibrium, the potential the electrode takes is given by

$$E = E°(\mathrm{Ox/Red}) - (RT/nF)\ln[(\mathrm{Red})/(\mathrm{Ox})].$$

(Recall that the equilibrium potential is exhibited by the solution when the redox reaction that takes place within it has itself reached its equilibrium state. It is also the potential taken by the two compartments of a galvanic cell when it is definitively out of use—see Chap. 2.)

It often happens that the experimental electrode potential differs from that given by Nernst's law for kinetic reasons. In such a case, the electrode and the solution demand a nonnegligible time to equilibrate themselves. In this case, the electrode is said to be *slow* or *irreversible*. In the opposite case, it is said to be *fast* or *reversible,*

J.-L. Burgot, *Ionic Equilibria in Analytical Chemistry,*
DOI 10.1007/978-1-4419-8382-4_16,

or *nernstian*. The fact that we label an electrochemical process or, more simply, an electrode as being reversible or irreversible depends, of course, on the experimental conditions of the measurements used to investigate the phenomenon.

The reversibility concept is central in electrochemistry.

16.2 Potential of a Solution Containing Only One Redox Couple

Calculating the potential of a solution that contains only one redox couple is the simplest case. The potential is given by Nernst's law. For example, for a solution containing ferrous and ferric ions, the equilibrium potential is given by

$$E = E°(Fe^{3+}/Fe^{2+}) - (RT/F)\ln[(Fe^{2+})/(Fe^{3+})].$$

A particular case is that in which only one of the members of the couple Red/Ox exists alone in the solution. When Fe^{2+} is alone, a brutal application of Nernst's equation gives

$$E = E°(Fe^{3+}/Fe^{2+}) - (RT/F)\ln[(Fe^{2+})/0],$$

i.e.,

$$E \rightarrow -\infty.$$

However, a platinum electrode dipping into such a solution exhibits a finite potential value. The dissolution of a pure ferrous salt may, in a first stage, lead to such a weak potential value of the solution that water itself is reduced in a second stage. Therefore, water plays the role of an oxidant, of course. As a result, traces of ferric ions are formed. Both forms of the couple are present and the solution exhibits a finite potential value predicted by Nernst's law, which is that found experimentally. Experience indicates that the solution potential, in this case, is not lower than 0.50 V ($E° = 0.77$ V). Inversely, when the solution contains only ferric ions, the potential does not tend toward $+\infty$ as expected. In this case, water plays the role of a reductor. Traces of Fe^{2+} are formed. The potential value the solution takes does not exceed about 1.05 V.

These particular cases correspond to the first point of a redox titration. This point is, of course, not exploitable.

16.3 General Case: Equilibrium Potential of a Solution Containing Two Redox Couples

An initial mixture composed of the sole forms Red_1 and Ox_2 induces the redox reaction (16.1):

$$n_2Red_1 + n_1Ox_2 \rightleftharpoons n_2Ox_1 + n_1Red_2. \tag{16.1}$$

To have pure thermodynamic rigor, it is always equilibrated, even if it must be strongly displaced toward the right in order to achieve a satisfactory titration. The calculation of the solution equilibrium's potential is performed as a function of the analytical concentrations of species 1 and 2, C_1 and C_2, with the help of the following system of equations, which are always satisfied:

$$E = E°(\text{Ox}_1/\text{Red}_1) - (RT/n_1F)\ln[(\text{Red}_1)/(\text{Ox}_1)], \tag{16.2}$$

$$E = E°(\text{Ox}_2/\text{Red}_2) - (RT/n_2F)\ln[(\text{Red}_2)/(\text{Ox}_2)], \tag{16.3}$$

$$[\text{Red}_1] + [\text{Ox}_1] = C_1, \tag{16.4}$$

$$[\text{Red}_2] + [\text{Ox}_2] = C_2, \tag{16.5}$$

$$[\text{Red}_2]/[\text{Ox}_1] = n_1/n_2. \tag{16.6}$$

At equilibrium, the solution potential may be indifferently expressed by Eq. (16.2) or (16.3). (The case of a mixture of two weak acids is analogous. The solution's pH may be expressed as a function of the pK_a of either of the acids.) Equations (16.4) and (16.5) are the mass balances in couples (16.1) and (16.2). Note that C_1 and C_2 are the total concentrations in the forms Red_1 and Ox_2 once the mixture is achieved but before reaction (16.1) has begun. Equation (16.6) expresses the electron balance: The electrons lost by the species that is oxidized are captured by the oxidant. This relation is perhaps difficult to understand in a first approach. Some examples with n_1 values different from n_2 values will permit us to understand them better (see Chap. 17). The electron balance relation must always be written for the calculation of solution potentials. It is necessary to solve the appropriate systems of equations. Finally, it is interesting to remark that in the above system of equations, the last three are expressed in terms of concentrations, and the first two in terms of activities.

The five simultaneous Eqs. (16.2)–(16.6) constitute a system in five unknowns. It can be solved from a mathematical standpoint. When activities are assumed to be equal to the corresponding concentrations, we find Eq. (16.7) after reducing the system by substitution and elimination:

$$(n_2C_2)/(n_1C_1) = (1 + e_2)/(1 + e_1), \tag{16.7}$$

with

$$e_1 = \exp[-(n_1F/RT)(E - E°_1)],$$

$$e_2 = \exp[(n_2F/RT)(E - E°_2)],$$

$$E°_1 = E°(\text{Ox}_1/\text{Red}_1) \quad \text{and} \quad E°_2 = E°(\text{Ox}_2/\text{Red}_2).$$

Equation (16.7) is absolutely general. No simplification prevailed in setting it up, except that concerning the values of the activities [besides, when we take into account the problem of the activities, Eq. (16.7) is still at work]. Equation (16.7) has the drawback of not being explicit in the dependent variable E. However, if reaction (16.1) is strongly displaced toward the right, some approximations are possible and E can be expressed explicitly (see the following chapter).

After an examining Eq. (16.7), it appears that the solution's equilibrium potential is fixed by the ratio C_1/C_2 through the values of the constants n_1, $E°_1$, n_2, and $E°_2$, which are characteristics of the redox systems under consideration.

Remarks

1. The equilibrium constant $K°$ of reaction (16.1), defined by

$$K° = [(\text{Red}_2)^{n1}(\text{Ox}_1)^{n2}]/[(\text{Ox}_2)^{n1}(\text{Red}_1)^{n2}],$$

does not appear in the system of equations above. Actually, it is present in a hidden manner, since the equilibrium constant $K°$ is connected to the standard potentials of both redox couples through the relation (see Chap. 2)

$$E°(\text{Ox}_2/\text{Red}_2) - E°(\text{Ox}_1/\text{Red}_1) = (RT/n_1n_2)\ln K°.$$

Hence, setting up Eqs. (16.2) and (16.3) is equivalent to setting up the constant $K°$.
2. The charge balance relation (electroneutrality equation) is not necessary to calculate the solution potential. However, it is satisfied, of course.
3. There is an immediate generalization of the preceding case. It is that in which the initial mixture [that existing before reaction (16.1) has begun] contains all the redox species of both couples: Red_1, Ox_1, Red_2, Ox_2, respectively, at the analytical concentrations $[\text{Red}°_1]$, $[\text{Ox}°_1]$, $[\text{Red}°_2]$, $[\text{Ox}°_2]$. In this case, Eqs. (16.2) and (16.3) remain satisfied, whereas those expressing the mass and electron balances become

$$[\text{Red}_1] + [\text{Ox}_1] = [\text{Red}°_1] + [\text{Ox}°_1],$$

$$[\text{Red}_2] + [\text{Ox}_2] = [\text{Red}°_2] + [\text{Ox}°_2],$$

$$\{[\text{Ox}_1] - [\text{Ox}°_1]\}/\{[\text{Red}_2] - [\text{Red}°_2]\} = \frac{\mathbf{n_1}}{\mathbf{n_2}}.$$

These equations may be handled as previously. An equation, nonexplicit in E, may therefore be found in this manner.
4. The treatment of the activities problem may be achieved as it is with acids and bases (see Chap. 5). The iteration process starts with the mixing of activities and concentrations. This first operation, once achieved, begins to transform the (thermodynamic) standard potentials into potentials that are already somewhat formal ones, that is, into potentials expressed in concentration terms. The process is recommenced several times until a constant ionic strength is obtained. Therefore, the definitive potential value is a pure formal one, of course at the definitive and true constant ionic strength.

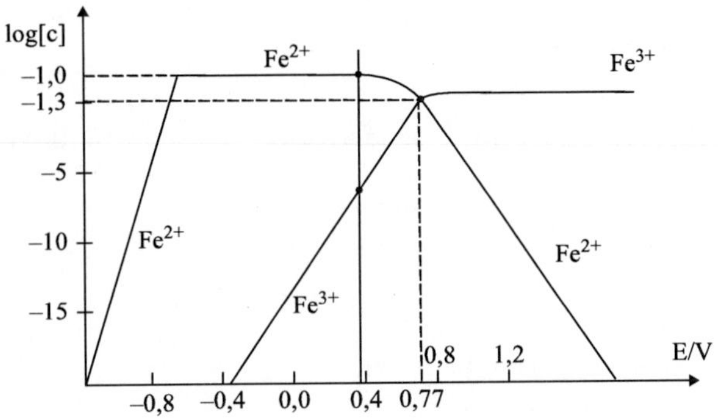

Fig. 16.1 Diagram of log (concentration)/E for the couples Fe^{3+}/Fe^{2+} and $Fe^{2+}/Fe_{(s)}$

16.4 Determining the Ox and Red Concentrations of a Couple from the Known Equilibrium Potential by Graphical Means

Calculating the equilibrium potential of a solution and calculating the redox species concentrations present in it are completely interrelated. Once the equilibrium potential is measured, it is easy to calculate the concentrations of the redox species. In order to achieve it, it is sufficient to introduce the E value to the preceding Eqs. (16.2) until (16.5).

There is a graphical means that permits a fast determination of these concentrations. However, it is an approached method. It is based on the building of the diagrams of the logarithm of redox species' concentrations (in principal activities)/solution potential (Fig. 16.1). Except in some restrained potential regions, the diagram essentially consists of straight lines whose occurrence is justified by Nernst's equation. This methodology proceeds from the same line of reasoning as that followed in the building of Hägg diagrams (see Chap. 5).

Let's consider, for example, a 0.1 mol/L solution of ferric ions. (The solution is sufficiently acidic so that the ferric ion should not be hydrolyzed. Furthermore, the pH value is fixed for the building of these diagrams. The pH- value is not a variable.) It is quite evident that when the potential of the solution decreases after, for example, the addition of a reductor, the concentration $[Fe^{3+}]$ decreases and the concentration $[Fe^{2+}]$ increases. The following two relations are satisfied:

$$E = 0.77 - 0.06\log[(Fe^{2+})/(Fe^{3+})], \quad E^\circ = 0.77\,\text{V},$$

$$[Fe^{2+}] + [Fe^{3+}] = 0.1\ \text{mol/L}.$$

The two curves $\log[Fe^{2+}]/E$ and $\log[Fe^{3+}]/E$ are deduced from them. We find

$$E = 0.77 - 0.06\log\{(Fe^{2+})/[0.10 - (Fe^{2+})]\},$$

$$E = 0.77 - 0.06\log\{[0.10 - (Fe^{3+})]/(Fe^{3+})\}$$

or

$$\log[Fe^{2+}] - \log(0.10 - [Fe^{2+}]) = -E/0.06 + 0.77/0.06,$$

$$\log[Fe^{3+}] - \log(0.10 - [Fe^{3+}]) = E/0.06 - 0.77/0.06.$$

In a very oxidizing medium such as $E \gg 0.77$ V, $[Fe^{2+}]$ is negligible compared to 0.1 mol/L. Therefore, we can write

$$\log[Fe^{2+}] = -16.7E + 11.83.$$

There is a linear relationship between log $[Fe^{2+}]$ and E. In a very reducing medium, we find the relationship

$$\log[Fe^{3+}] = 16.7E - 13.83.$$

The two straight lines intersect each other for $E = 0.77$ V and log $[Fe^{2+}]=$ log $[Fe^{3+}] = -1.3$. For values of potential close to 0.77 V, the linear relationships with E are no longer satisfied. The vertical line $E =$ system equilibrium potential intersects the two representative curves for the concentrations to be determined. Figure 16.1 summarizes these considerations. We notice that in a very reducing medium, the couple Fe^{3+}/Fe^{2+} should no longer be considered. It is the couple $Fe^{2+}/Fe_{(s)}$ ($E° = -0.44$ V) that must be considered. Nernst's equation gives the relation

$$E = -0.44 - 0.03\log[Fe^{2+}].$$

When $[Fe^{2+}] = 0.1$ V $[\log(Fe^{2+}) = -1]$, $E = -0.41$ V. The couple $Fe^{2+}/Fe_{(s)}$ is represented by the straight line located on the left of the diagram. It intersects the straight line $\log C = -1$ for $E = -0.41$ V.

16.5 A Particular Case: The Exchange of Electrons is Accompanied by an Exchange of Protons or by an Exchange of Ligands

Let's consider, for example, a mixture of potassium permanganate and ferrous ions. The reaction resulting from the mixture is

$$5Fe^{2+} + MnO_4^- + 8H^+ \rightarrow 5Fe^{3+} + Mn^{2+} + 4H_2O.$$

Nernst's equation corresponding to the couple MnO_4^-/Mn^{2+} is

$$E = E°(MnO_4^-/Mn^{2+}) - (RT/5F)\ln\{(Mn^{2+})/[(MnO_4^-)(H^+)^8]\}$$

or

$$E = E°(MnO_4^-/Mn^{2+}) + (8RT/5F)\ln(H^+) - (RT/5F)\ln[(Mn^{2+})/(MnO_4^-)].$$

If the medium is buffered, the sum of the first two terms on the right-hand side of the previous equation is constant. It is equal to a quantity called the *apparent standard (or normal) potential* $E°'$ of the couple MnO_4^-/Mn^{2+} at the given pH. Therefore, Nernst's equation may be written:

$$E = E°'(MnO_4^-/Mn^{2+}) - (RT/5F)\ln[(Mn^{2+})/(MnO_4^-)] \quad \text{(constant pH)}.$$

As a result, a system of equations absolutely analogous to those prevailing in the preceding case may be set up, provided that the standard potential is replaced with the apparent one. The potential of the solution is calculated through the equation

$$5(C_2/C_1) = (1 + e_2)/(1 + e_1),$$

with

$$e_1 = \exp[-(F/RT)(E - E°_1)],$$

$$e_2 = \exp[(5F/RT)(E - E°'_2)],$$

$$E°_1 = E°(Fe^{3+}/Fe^{2+}),$$

$$E°'_2 = E°'(MnO_4^-/Mn^{2+}).$$

This methodology can be immediately transposed to analogous cases in which there are exchanges of protons and electrons at the same time. We must emphasize the fact that the equilibrium potential of the solution depends on the pH value, since the exponential e_2 depends on $E°'_2$, which itself depends on the pH value. This is not the case when we only have an exchange of electrons.

If the redox reaction is accompanied by an exchange of ligand, the solution equilibrium potential may be calculated in the same way as earlier. An apparent standard potential is defined for the couple whose members exchange a ligand together with the electrons. It is constant if the concentration in the ligand is held constant. The equilibrium potential depends on this concentration.

16.6 Case in Which One of the Species Redox is Polynuclear

Actually, reaction (16.1) is not absolutely general. It is not rare among the redox couples used in chemical analysis for some of them to have a Red or Ox member that possesses several atoms of the element undergoing a change in its oxidation state. This is the case, for example, for the following half-equilibria:

$$Br_2 + 2e^- \rightleftharpoons 2Br^-,$$

$$I_2 + 2e^- \rightleftharpoons 2I^-,$$

$$Cr_2O_7^{2-} + 14H^+ + 6e^- \rightleftharpoons 2Cr^{3+} + 7H_2O.$$

In this case, the calculation of the equilibrium potential as a function of the concentrations C_1 and C_2 is far more difficult than previously. Let's consider, for example, the oxidation of ferrous ion by bromine (bromine liquid or in solution; it doesn't matter) according to

$$Br_2 + 2Fe^{2+} \rightleftharpoons 2Br^- + 2Fe^{3+}.$$

The system of equations to solve is

$$E = E°(Br_2/Br^-) - (RT/2F)\ln\{[Br^-]^2/[Br_2]\},$$
$$E = E°(Fe^{3+}/Fe^{2+}) - (RT/F)\ln\{[Fe^{2+}]/[Fe^{3+}]\},$$
$$[Br_2] + 1/2[Br^-] = C_2,$$
$$[Fe^{2+}] + [Fe^{3+}] = C_1,$$
$$[Fe^{3+}] = [Br^-],$$

where C_2 is the mass balance in bromine expressed in bromine. From a mathematical standpoint, the system is soluble, but because in one of Nernst's equations the numerator of the argument of one of the logarithms exhibits an exponent different from that of the denominator, it is more complicated to solve. The general solution issuing from the system and permitting the calculation of the solution's equilibrium potential is

$$4C_1{}^2/(1 + e_1)^2 + 2C_1/(1 + e_1)\, e_2 - 4C_2/e_2 = 0,$$

where $e_1 = \exp[(-F/RT)(E - E°_1)]$ and $e_2 = \exp[(2F/RT)(E - E°_2)]$,

with $E°_1 = E°(Fe^{3+}/Fe^{2+})$ and $E°_2 = (Br_2/Br^-)$.

Neither E nor the ratio C_2/C_1 can be expressed explicitly. However, the E value can be calculated from the above equation by numerical calculus. Fortunately, the above equation may be considerably simplified when the reaction is strongly displaced toward the right (see Chap. 17).

Another example, which is well known for its analytical applications, is provided by the so-called fundamental reaction of iodometry (see Chap. 18). It is the reaction of tri-iodide anions with thiosulfate ions:

$$I_3{}^- + 2S_2O_3{}^{2-} \rightarrow 3I^- + S_4O_6{}^{2-}.$$

Whatever the example may be, the general equation becomes considerably simpler when the reaction is strongly displaced toward the right.

16.7 Equilibrium Potential of a Solution When it Contains an Ampholyte

An example of an equilibrium potential of a solution when it contains an ampholyte is provided by the reaction of the hypovanadous ion V^{2+} with the vanadyl ion VO^{2+} to give the vanadate ion V^{3+} according to

$$V^{2+} + VO^{2+} + 2H^+ \rightleftharpoons H_2O + 2V^{3+}$$

From a general standpoint, let's consider the following two couples:

$$\mathrm{Red_1} - n_1\mathrm{e^-} \rightleftharpoons \mathrm{A} \quad E°(\mathrm{A/Red_1}),$$

$$\mathrm{Ox_2} + n_2\mathrm{e^-} \rightleftharpoons \mathrm{A} \quad E°(\mathrm{Ox_2/A}).$$

Let's also consider the mixture of both forms Red_1 and Ox_2 in any proportion. The redox reaction that occurs between them is

$$n_2\mathrm{Red_1} + n_1\mathrm{Ox_2} \rightleftharpoons (n_1 + n_2)\mathrm{A}.$$

The calculation of the solution equilibrium potential is achieved after resolving the system of the following relations:

$$E = E°(\mathrm{A/Red_1}) - (RT/n_1F)\ln[(\mathrm{red_1})/(\mathrm{A})],$$

$$E = E°(\mathrm{Ox_2/A}) - (RT/n_2F)\ln[(\mathrm{A})/(\mathrm{Ox_2})],$$

$$[\mathrm{Red_1}] + [\mathrm{A_1}] = C_1,$$

$$[\mathrm{Ox_2}] + [\mathrm{A_2}] = C_2,$$

$$[\mathrm{A_1}] + [\mathrm{A_2}] = [\mathrm{A}],$$

$$[\mathrm{A_1}] = (n_2/n_1)[\mathrm{A_2}].$$

$[A_1]$ and $[A_2]$ are the concentrations of A, which, from a formal standpoint, come from the oxidation of Red_1 and from the reduction of Ox_2, respectively. A_1 and A_2 are, of course, the same species. They cannot be distinguished from one another. The previous set of equations leads to the equation

$$C_2/C_1 = \{e_1[n_1(1 + e_2) + n_2e_2]/[n_1 + n_2(1 + e_1)]\},$$

with $e_1 = \exp[-(n_1F/RT)(E - E°_1)]$, $e_2 = \exp[(n_2F/RT)(E - E°_2)]$,

where $E°_1 = E°(\mathrm{A/Red_1})$ and $E°_2 = E°(\mathrm{Ox_2/A})$.

Again, it is impossible to explicitly express the equilibrium potential as a function of the ratio C_2/C_1, whereas the inverse is true. The above relationship differs markedly from Eq. (16.7).

In the example of the two couples of vanadium, the equilibrium potential is given by the relation

$$C_2/C_1 = [e_1(1 + 2e_2)/(2 + e_1)],$$

where e_1 and e_2 correspond to the couples V^{3+}/V^{2+} and VO^{2+}/V^{3+}. Concerning the latter, the potential to be taken into account in order to calculate the potential e_2 is the apparent standard potential $E°'$, since the half-reduction potential

$$\mathrm{VO^{2+}} + 2\mathrm{H^+} + 1\mathrm{e^-} \rightleftharpoons \mathrm{V^{3+}} + \mathrm{H_2O}$$

involves an exchange of protons. It is easy to verify, with $C_1 = C_2$, that

$$E = [E°(\mathrm{A/Red_1}) + E°'(\mathrm{Ox_2/A})]/2.$$

16.8 Potential of a Solution Containing a Mixture of the Reduced Polyfunctional Member of a Couple and of the Oxidized Member of Another Couple

We'll call Red_1 the most reduced form of a polyfunctional compound that can be successively oxidized according to the three reactions

$$Red_1 - n_{11}e^- \rightleftharpoons Z_1, \quad E^\circ{}_{11},$$
$$Z_1 - n_{12}e^- \rightleftharpoons Z_2, \quad E^\circ{}_{12},$$
$$Z_2 - n_{13}e^- \rightleftharpoons Ox_1, \quad E^\circ{}_{13}.$$

Z_1 and Z_2 are the species endowed with the intermediary oxidation states (ampholytes). The problem is to calculate the potential of the solution containing a mixture of the oxidized form Ox_2 and the form Red_1. Ox_2 belongs to the couple whose half-reaction is

$$Ox_2 + ne^- \rightleftharpoons Red_2.$$

Due to the existence of the mixture, the following redox reactions will occur:

$$n_2Red_1 + n_{11}Ox_2 \rightleftharpoons n_2Z_1 + n_{11}Red_2, \tag{16.8}$$

$$n_2Z_1 + n_{12}Ox_2 \rightleftharpoons n_2Z_2 + n_{12}Red_2, \tag{16.9}$$

$$n_2Z_2 + n_{13}Ox_2 \rightleftharpoons n_2Ox_1 + n_{13}Red_2. \tag{16.10}$$

Once the equilibria are reached, the equations that must be satisfied are

$$E = E^\circ{}_{11} - (RT/n_{11}F)\ln[(Red_1)/(Z_1)],$$
$$E = E^\circ{}_{12} - (RT/n_{12}F)\ln[(Z_1)/(Z_2)],$$
$$E = E^\circ{}_{13} - (RT/n_{13}F)\ln[(Z_2)/(Ox_1)],$$
$$E = E^\circ{}_2 - (RT/n_2F)\ln[(Red_2)/(Ox_2)],$$
$$[Red_1] + [Z_1] + [Z_2] + [Ox_1] = C_1,$$
$$[Red_2] + [Ox_2] = C_2,$$
$$[Red_2] = (n_{11}/n_2)\{[Z_1] + [Z_2] + [Ox_1]\}$$
$$+ (n_{12}/n_2)\{[Z_2] + [Ox_1]\} + (n_{13}/n_2)[Ox_1],$$

where C_1 and C_2 are the analytical concentrations. The last relation expresses the electron balance. It can be justified by the following reasoning: The species Red_2 is formed simultaneously with the species Z_1, Z_2, and Ox_1 because of the three reactions (16.8), (16.9), and (16.10). Obtaining the ultimate species Ox_1 necessitates the oxidation of Red_1 into Z_1, then that of Z_1 into Z_2, and finally that of Z_2 into Ox_1.

The concentration $[Red_2]$, which has formed during the oxidation of Red_1 into Ox_1, is

$$[Red_2] = [n_{11}/n_2 + n_{12}/n_2 + n_{13}/n_2][Ox_1] \quad (Ox_1 \text{ formation, only})$$

(this transformation has indeed demanded three steps). The concentration $[Z_2]$ demands the formation of the following concentration $[Red_2]$ only (the first two steps only):

$$[Red_2] = [n_{11}/n_2 + n_{12}/n_2][Z_2] \quad (Z_2 \text{ formation only})$$

etc.

The system of equations leads to the general equation

$$(n_2C_2/C_1) = [n_{11}e_{11} + (n_{11} + n_{12})e_{11}e_{12} + (n_{11} + n_{12} + n_{13})e_{11}e_{12}e_{13}][1 + e_2] \\ /[1 + e_{11} + e_{11}e_{12} + e_{11}e_{12}e_{13}] \quad (16.11)$$

where $e_{11} = \exp[(n_{11}F/RT)(E - E°_{11})]$, $e_{12} = \exp[(n_{12}F/RT)(E - E°_{12})]$,

$e_{13} = \exp[(n_{13}F/RT)(E - E°_{13})]$, $e_2 = \exp[(n_2F/RT)(E - E°_2)]$.

For an easy interpretation of Eq. (16.11), note that

$$[Red_1] = C_1/D; \quad [Z_1] = C_1e_{11}/D; \quad [Z_2] = C_1e_{11}e_{12}/D;$$
$$[Ox_1] = C_1e_{11}e_{12}e_{13}/D,$$
$$\text{with} \quad D = 1 + e_{11} + e_{11}e_{12} + e_{11}e_{12}e_{13}.$$

Equation (16.11) is complicated. However, it permits us to solve the set up problem, that is, to calculate the equilibrium potential of the solution (hidden in the exponentials) as a function of the ratio C_2/C_1. We notice, again, that the potential E does not appear explicitly. This is the reason why it is considerably easier to calculate the ratio C_2/C_1 as a function of E.

When the initial derivative (to oxidize) is only monofunctional, Eq. (16.11) becomes very simplified. It is indeed reduced to

$$(n_2C_2/n_{11}C_1) = [e_{11}(1 + e_2)]/(1 + e_{11}),$$

which is strictly identical to Eq. (16.7) since in the present case e_{11} is the inverse of e_1 from the preceding case.

In the following chapter, we shall see an example of the use of Eq. (16.11) when we study the redox titration curve of hypovanadous ions V^{2+} by permanganate ions MnO_4^-.

Of course, by following the same kind of reasoning, a relationship analogous to (16.11) would be obtained for a redox reaction occurring in an inverse direction.

16.9 Potential of a Solution Containing a Mixture of an Oxidized Form of a First Couple and of Two Reduced Forms Belonging to Two Other Different Redox Couples

Let's consider a solution containing two reduced forms belonging to two different redox couples Red_{11} and Red_{12} and an oxidized form Ox_2 of a third couple. The half-redox reactions are

$$Ox_{11} + n_{11}e^- \rightleftharpoons Red_{11}, \quad E°_{11},$$
$$Ox_{12} + n_{12}e^- \rightleftharpoons Red_{12}, \quad E°_{12},$$
$$Ox_2 + n_2e^- \rightleftharpoons Red_2, \quad E°_2.$$

The chemical reactions induced by the mixing of the two reduced forms and of the oxidized form, which occur simultaneously and more and less completely, are

$$n_2Red_{11} + n_{11}Ox_2 \rightleftharpoons n_2Ox_{11} + n_{11}Red_2,$$
$$n_2Red_{12} + n_{12}Ox_2 \rightleftharpoons n_2Ox_{12} + n_{12}Red_2.$$

The equilibrium potential is calculated from the system of the following equations, which must be obeyed:

$$E = E°_{11} - (RT/n_{11}F)\ln[(Red_{11})/(Ox_{11})],$$
$$E = E°_{12} - (RT/n_{12}F)\ln[(Red_{12})/(Ox_{12})],$$
$$E = E°_2 - (RT/n_2F)\ln[(Red_2)/(Ox_2)],$$
$$[Red_{11}] + [Ox_{11}] = C_{11},$$
$$[Red_{12}] + [Ox_{12}] = C_{12},$$
$$[Red_2] + [Ox_2] = C_2,$$
$$[Red_2] = (n_{11}/n_2)[Ox_{11}] + (n_{12}/n_2)[Ox_{12}].$$

C_{11}, C_{12}, and C_2 are the analytical concentrations. Handling the equations gives the general relationship (16.12):

$$[C_2/(1 + e_2)] = (n_{11}/n_2)[C_{11}/(1 + e_{11})] + (n_{12}/n_2)[C_{12}/(1 + e_{12})], \qquad (16.12)$$

where

$$e_{11} = \exp[-(n_{11}F/RT)(E - E°_{11})], \quad e_{12} = [-(n_{12}F/RT)(E - E°_{12})],$$
$$e_2 = \exp[(n_2F/RT)(E - E°_2)].$$

Some supplementary considerations concerning this case are given in Chap. 17 during the study of the sequential titration of Sn^{2+} and Fe^{2+} by the ceric ion, Ce^{IV}.

16.10 General Considerations Concerning Redox Titrations

A redox titration permits us to determine the concentration of a form Ox_1 or Red_1 by adding antagonistic solution of Red_2 or Ox_2. The redox reaction between the antagonistic species, which is the titration reaction, induces a change in the equilibrium potential of the titrand solution for each added volume of titrant solution. A redox titration curve is the diagram of the solution's potential/added volume of titrant solution. Studying the redox titration curves and, especially, calculating the equilibrium redox potential at each point of the curve, with the help of the preceding relations, enable us to know its value at the equivalence point, among other interests. Knowing the equivalence point potential permits the judicious choice of a color indicator. However, first, we must specify the necessary conditions to obtain a satisfactory titration. There are two kinds: thermodynamic and kinetic.

16.11 Thermodynamic Condition for a Redox Titration Reaction

The thermodynamic condition is an absolute necessity: A titration reaction must be as complete as possible, regardless of the kind of titration: acid–base, redox, etc. This condition must be satisfied in order to weaken the titration error (see Chap. 7).

If the titration reaction is

$$n_2\mathrm{Red}_1 + n_1\mathrm{Ox}_2 \rightleftharpoons n_2\mathrm{Ox}_1 + n_1\mathrm{Red}_2, \qquad (16.13)$$

its quantitative character depends on the value of its equilibrium constant $K°$:

$$K° = [(\mathrm{Ox}_1)^{n2}(\mathrm{Red}_2)^{n1}]/[(\mathrm{Red}_1)^{n2}(\mathrm{Ox}_2)^{n1}].$$

$K°$ itself is connected to the potential difference $\Delta E°$ of both couples 1 and 2 by the relationship

$$\log K° = 16.9\, n_1 n_2 \Delta E° \quad \text{(at 25°C)},$$

where $\Delta E° = E°(\mathrm{Ox}_2/\mathrm{Red}_2) - E°(\mathrm{Ox}_1/\mathrm{Red}_1)$ and n_1 and n_2 are the numbers of electrons that each of the redox couples exchanges (see previous chapters). The extent of a reaction also depends on the concentration of the titrand and its stoichiometry.

As we've already noted, it is rarely realistic to use standard redox potentials to calculate equilibrium constants since the standard conditions often do not prevail. It is more judicious to use apparent standard potentials and formal potentials when they are known. In this case, the calculated equilibrium constants are no longer the thermodynamic ones, but the apparent ones. They are given by the relation

$$\log K = 16.9\, n_1 n_2 \Delta E°' \quad \text{(at 25°C)},$$
$$(\Delta E°' = E°_2 - E°'_1).$$

In the simple case in which $n_1 = n_2$, the condition most often quoted in the literature in order for a redox titration to be satisfactory is

$$\Delta E° \geq 240\,\text{mV}.$$

This limiting value corresponds to the equilibrium constant value $K = 10^4$. One may find with this value that only 1% of the species remains to be titrated at the equivalence point (see immediately below). With a difference in standard potentials of 360 mV, which imposes an equilibrium constant value $K = 10^6$, only 0.1% of the species remains to be titrated at the equivalence point.

The percentages of not yet titrated substance at the equivalence point as a function of the equilibrium constants are easily calculated. For example, let's consider the reaction

$$\text{Red}_1 + \text{Ox}_2 \rightleftharpoons \text{Red}_2 + \text{Ox}_1,$$

with C_O the initial concentration in Red_1 (the species to be titrated) and $\varepsilon\ C_O$ its concentration not having reacted at the equivalence point, due to the fact that the reaction is slightly equilibrated. ε (which is the parameter to calculate now) is the fraction of titrand that has not reacted at the equivalence point. It is a dimensionless number, $0 \leq \varepsilon \leq 1$. By neglecting the dilution of the titrand solution with that of the titrant, we immediately find the relation,

$$K° = [(1-\varepsilon)^2 C_O{}^2]/[C_O{}^2\varepsilon^2],$$

$$K° = [(1-\varepsilon)^2]/\varepsilon^2$$

after using the mass law. The value $K° = 10^4$ immediately gives $\varepsilon = 0.01$. An identical calculation would have been achieved in the case of acid–bases.

We must emphasize the fact that the numerical values given above ($\varepsilon = 1\%$ for $K° = 10^4$, etc.) are only correct for titration reactions of the type (16.1) in which $n_1 = n_2$. For the other classes of reactions, they are no longer so. However, in a first approximation, the rule $\Delta E° \geq 240\,\text{mV}$ may be adopted as a safe criterion, regardless of the type of titration reaction.

16.12 Kinetic Conditions in Order to Achieve a Satisfactory Redox Titration Reaction

From a practical standpoint, a titration reaction must be achieved as fast as possible. It is not always the case for redox titrations. The rate of redox reactions is more difficult to predict than their quantitative character, essentially because of the fact that very often it depends on the mixture's composition, which changes with the extent of the reaction and, finally, depends on its rate. It also depends on the reactional mechanism. Moreover, in chemical analysis, the composition of the mixture is unknown.

Recall that the rate of a redox reaction may be considerably increased with the use of catalysts. Additionally, some titrations imply them.

16.13 Detection of the Equivalence Point of a Redox Titration

At the equivalence point of a satisfactory redox titration, there is a sharp change in the equilibrium potential of the titrated solution. Several means exist for its detection. They are based on the current use

- of internal redox indicators;
- of special redox species. The latter are purely and simply one (or several) of the reactants of the redox titration under consideration. They also play the role of indicators. For example, it is the case with iodine–iodide and of potassium permanganate solutions (see Chaps. 18 and 20);
- of specific indicators that exhibit a particular color in the presence of a given reactant or product. It is the case for a starch solution in the presence of iodine and iodide (see Chap. 18);
- of several physical methods of analysis, the most important of which is probably zero-current potentiometry. This is an electrochemical method that, as a rule, measures the potential of a solution.

The only means of detection studied in this chapter is the use of the internal redox indicators.

16.14 General Considerations on Internal Redox Indicators

The internal redox indicators are dissolved in the titrand solution. They are organic dyes whose oxidized and reduced forms are different colors. They are derivatives for which their oxidation-reduction is achieved rapidly. Their half-redox equilibrium may be symbolized by

$$\mathrm{In_{ox}} + n\mathrm{e}^- \rightleftharpoons \mathrm{In_{red}}$$

They are added in a very weak concentration in order not to change the redox potential of the solution, since they are redox couples themselves.

Nernst's equation concerning them gives the following relation:

$$E = E^\circ(\mathrm{In_{ox}}/\mathrm{In_{red}}) - (RT/nF)\ln[(\mathrm{In_{red}})/(\mathrm{In_{ox}})].$$

In order to conveniently locate the potential interval change in the conditions of the titration, it is more judicious to use the formal potentials of the internal redox indicators than their standard potentials. Therefore, Nernst's equation is written:

$$E = E^{\circ\prime}(\mathrm{In_{ox}}/\mathrm{In_{red}}) - (RT/nF)\ln\{[\mathrm{In_{red}}]/[\mathrm{In_{ox}}]\}$$

in which the Ox and Red forms intervene through their concentrations.

The potential interval through which the color change occurs is given by the following rule, which is derived from experimental measurements. It stipulates that

Fig. 16.2 1,10-Phenanthroline and its ferrous complex

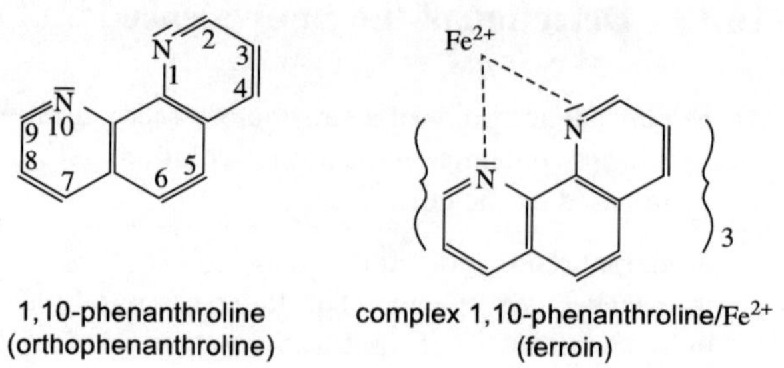

when both members of the couple exhibit colorations of the same intensity, the potential change interval corresponds to a change in the concentrations' ratio of the indicator such as

$$1/10 < [\mathrm{In_{ox}}]/[\mathrm{In_{red}}] < 10 \quad \text{when we consider increasing potentials.}$$

In potential values, this corresponds to the interval

$$E^{\circ\prime} - 0.059/n < \text{interval change} < E^{\circ\prime} + 0.059/n \quad (\text{at } 25^{\circ}\mathrm{C})$$

in which n is the number of electrons exchanged by the couple. Hence, the potential change interval is approximately $(0.12/n)$V. For an indicator that exchanges two electrons, the potential change interval is about 0.06 V. This is the potential interval necessary for the color change to be complete. If the intensities of the colors of both members of the indicator couple are not about the same, the intermediary color is located for a potential value that differs from its formal or standard potentials. Finally, according to some authors, the standard or formal potentials of the indicator must differ from those of the couples participating in the titration by at least 0.15 V. This condition leads to the important analytical conclusion that if the standard potentials of the two redox couples participating in the titration do not differ by more than the limiting value of 0.24 V (see above), it is better to use a physical method to detect the equivalence point.

16.15 Some Internal Redox Indicators

The most commonly used internal redox indicators are derivatives of 1,10-phenanthroline, diphenylamine, phenothiazine, and diphenylpyrazine.

16.15.1 1,10-Phenanthroline

1,10-Phenanthroline is also called orthophenanthroline. It forms a very stable complex with Fe^{2+}, intensively colored dark red, called ferroin (Fig. 16.2).

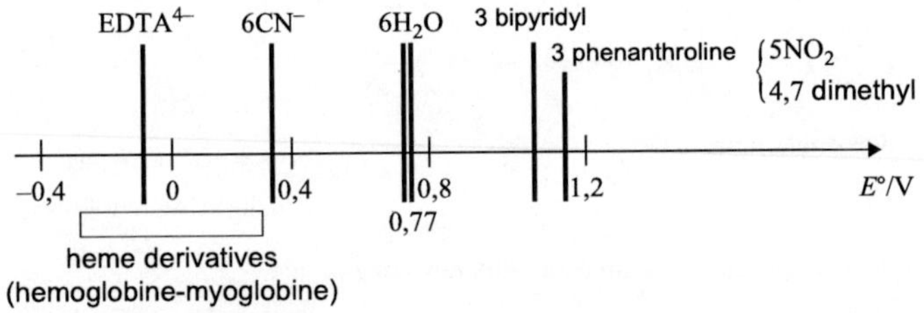

Fig. 16.3 Standard potentials of varied couples of Fe^{+III}/Fe^{+II}

In a very oxidizing medium, the orthophenanthroline/Fe^{3+} complex of a pale blue color is formed from ferroin. The corresponding half-redox equilibrium is

$$\underset{\text{pale blue}}{[Fe(C_{12}H_8N_2)_3{}^{3+}]} + 1e^- \rightleftharpoons \underset{\text{dark red}}{[Fe(C_{12}H_8N_2)_3{}^{2+}]}.$$

The standard potential of the couple is $E° = 1.14$ V. In 1 mol/L hydrochloric acid, its formal potential is $E°' = 1.06$ V. However, the color change is detected for $E = 1.12$ V since its reduced form is far more colored than the oxidized one.

Ferroin is a quasi-ideal indicator. Indeed, it reacts quickly and reversibly. ("Reversibly" means that for each potential range, it always exhibits the same chemical structure. From a structural standpoint, it does not change. In other words, its sole chemical behavior consists of releasing or capturing one electron without any change in its structure.) The color change is strong. Its solutions are stable. They are easily prepared by a stoichiometric mixing of 1,10-phenanthroline and ferrous sulfate. Its high standard potential value explains why its color change occurs in very oxidizing media.

The standard potential (and, of course, the formal potential) exhibited by ferroin may be appreciably modified by the introduction of varied substituents on the 1,10-phenanthroline nucleus. The most important indicators of this series are the derivatives of the 5-nitro ($E° = 1.25$ V), 5-methyl ($E° = 1.02$ V), and 4,7-dimethyl ($E° = 0.88$ V) orthophenanthroline.

Incidentally, it is interesting to note that the complexations of Fe^{2+} and Fe^{3+} ions give redox couples whose standard potentials may strongly differ from the value 0.77 V of the reference couple, in this case, the couple Fe^{3+}/Fe^{2+}. This is the case for ferroin. Actually, in the couple Fe^{3+}/Fe^{2+}, both ions are themselves complexed by six water molecules attached to the metallic ions by true chemical bonds. The name for this kind of complex is *aqua* (or aquo) *complexes* (see Part IV). Figure 16.3 gives some values of standard potentials of varied complexes in which ferric and ferrous ions are present. It shows the strong influence of the substituent on the standard potential value.

Figure 16.3 shows that negatively charged ligands tend to stabilize more Fe^{3+} than Fe^{2+} complexes, from which stems the decrease in their standard potential values related to that of the couple $[Fe(OH_2)_6{}^{3+}]/[Fe(OH_2)_6{}^{2+}]$. Inversely, ligands such as phenanthroline or bipyridine tend to stabilize more Fe^{2+} than Fe^{3+} complexes.

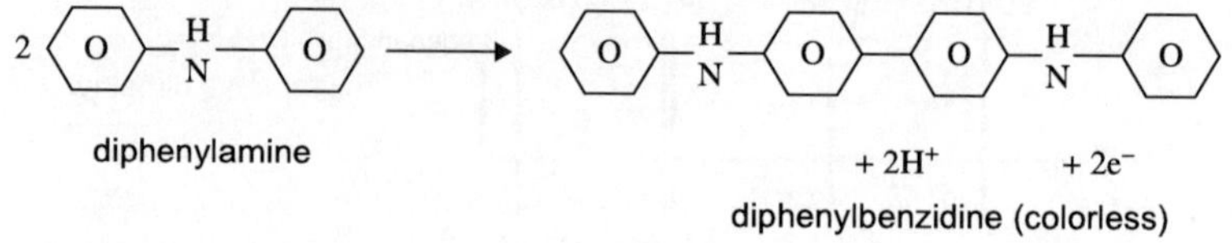

Fig. 16.4 Structures of diphenylamine and diphenylbenzidine

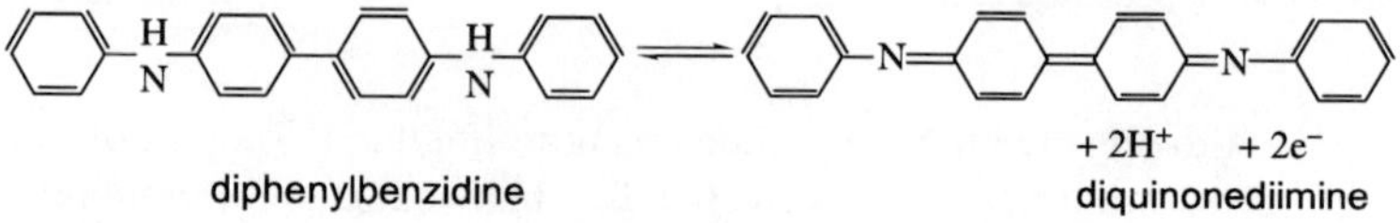

Fig. 16.5 Oxidation of diphenylbenzidine into the corresponding diquinonediimine

Hemic derivatives such as hemoglobin and myoglobin, which are biological electron exchangers, may be located in this class of derivatives.

16.15.2 Diphenylamine

Some derivatives of diphenylamine are internal redox indicators that have been used very often. Diphenylamine dissolved in diluted acidic medium exhibits the formal potential $E^{\circ\prime} = 0.76\,\text{V}$. In the presence of a strong oxidizing agent, it first undergoes an irreversible chemical oxidation to give the colorless diphenylbenzidine (Fig. 16.4).

Diphenylbenzidine can be reversibly oxidized according to a bielectronic process to give a colored diquinonediimine (Fig. 16.5).

Finally, the formed diquinonediimine can be oxidized once more, but this time irreversibly (from a chemical standpoint—see the concept of reversibility in electrochemistry) when it stays too long in the presence of the oxidizing solution. Diphenylamine can be used in the case of the titration of ferrous ions by potassium dichromate. It is poorly soluble in water. This is the reason why the barium and sodium salts of diphenylaminesulfonic acid are used. Another interesting indicator deriving from diphenylamine is variamine blue. The structures of its reduced (colorless) and oxidized (blue-violet) forms are

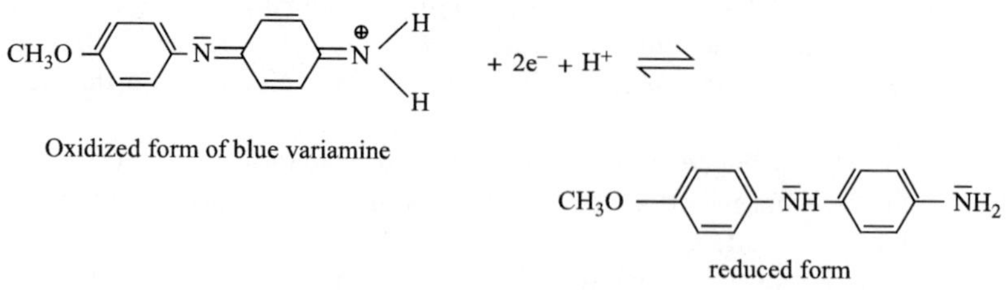

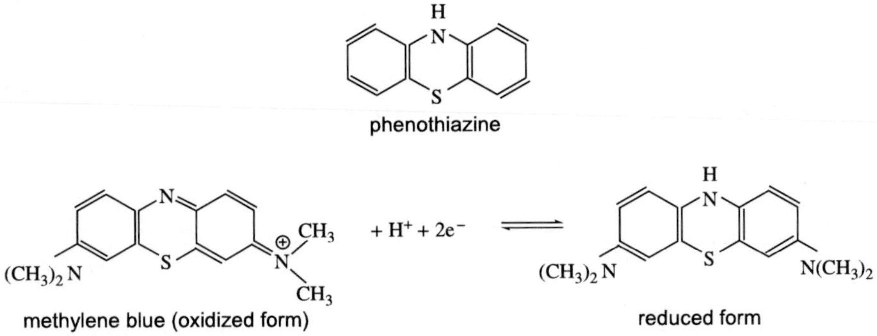

Fig. 16.6 Structure of phenothiazine and redox equilibrium of methylene blue

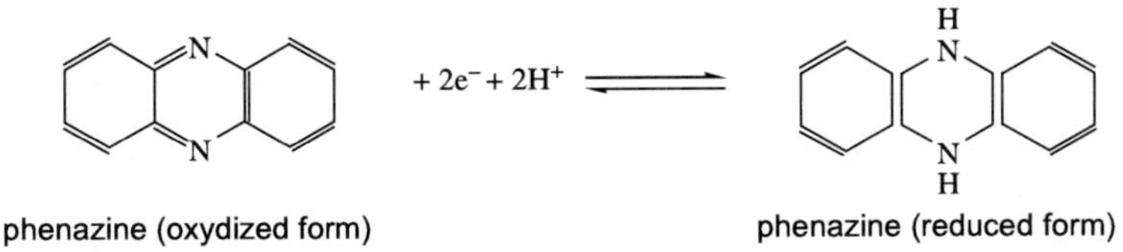

Fig. 16.7 Redox couple of phenazine

The mechanism of the color change does not involve the formation of a benzidine. Its standard potential is 0.74 V.

16.15.3 Methylene Blue

Methylene blue derives from phenothiazine (Fig. 16.6).

(The reduced form can also undergo two further protonations on the dimethylamino rests depending on the medium's pH.) The standard potential of methylene blue is $E° = 0.53$ V.

16.15.4 Diphenylpyrazine

The class of diphenylpyrazine (phenazine) also provides internal redox indicators. These derivatives, as well as phenazine itself, exchange two electrons and one or two protons depending on the pH value (Fig. 16.7).

The standard potentials of phenazines are located in the range 0.25–0.40 V.

Chapter 17
A Study of Some Redox Titration Curves

The study of redox titration curves necessitates knowing their general equations and the simplified mathematical relations derived from them. We can already notice that the general equations are obtained from all the relations necessarily satisfied in aqueous solutions without any simplification. They describe all the stages of the titration. These equations enable us to calculate the equilibrium potential of the solution in the titration vessel. The study of these curves allows us to find the parameters contributing to a satisfactory detection of the equivalence point. It also explains the origin of the relations, frequently quoted in the literature, obtained by simplifying the general equations.

This chapter investigates several classes of titrations: symmetrical and asymmetrical titrations, and those during which a proton exchange occurs together with that of electrons. Moreover, we shall also study titrations involving polynuclear species, plurifunctional species, and mixtures.

The study given here is uncommon since it is systematically based on the general equations governing the redox titration curves and, more precisely, by starting with them. This is not the case in the literature, in which the general equations are rarely mentioned.

17.1 Titration of the Ferrous Ion by the Ceric Ion, One of the Simplest Examples of a Redox Titration

The titration reaction for a ferrous ion by a ceric ion is as follows:

$$Fe^{2+} + Ce^{IV} \rightarrow Fe^{3+} + Ce^{III}. \tag{17.1}$$

This is the simplest form of the quasi-general redox titration reaction

$$n_2\mathrm{Red}_1 + n_1\mathrm{Ox}_2 \rightarrow n_2\mathrm{Ox}_1 + n_1\mathrm{Red}_2 \tag{17.2}$$

The titrations for which $n_2 = n_1$ are sometimes called *symmetrical titrations*. Moreover, in the above example, $n_2 = n_1 = 1$.

J.-L. Burgot, *Ionic Equilibria in Analytical Chemistry*,
DOI 10.1007/978-1-4419-8382-4_17, © Springer Science+Business Media, LLC 2012

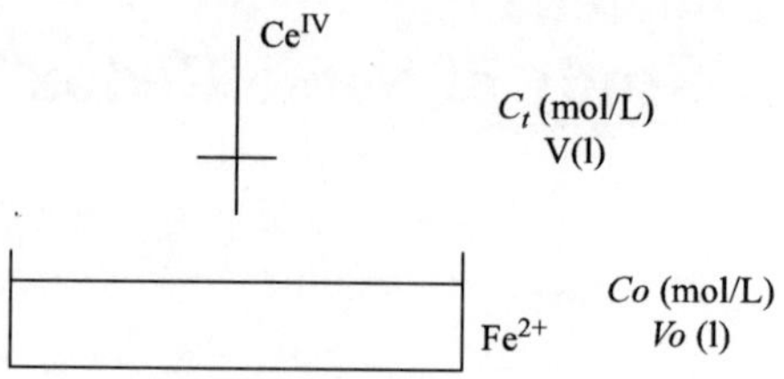

Fig. 17.1 Titration of Fe^{2+} with Ce^{IV}: symbolism used (symbolism used for all the studied titrations)

The reaction is achieved in strong acidic medium (most of the time in 1 mol/L sulfuric acid medium) in order to avoid the hydrolysis and polymerization of the ceric ion Ce^{4+}. In sulfuric acid medium, the ions Ce^{4+} and Ce^{3+} give a set of complexes of the types $[CeSO_4{}^{2+}]$, $[Ce(SO_4)_2]$, $[Ce(SO_4)_3{}^{2-}]$, and $[CeSO_4{}^{+}]$. The symbolism already used results from the ability to complex Ce^{4+} and Ce^{3+}. Ce^{IV} and Ce^{III} respectively symbolize all the species of cerium with the oxidation state + IV and all those with the oxidation number + III. The formal potential value $E^{\circ\prime}(Ce^{4+}/Ce^{3+}) = 1.44$ V takes these complexations into account. From another standpoint, the formal potential value $E^{\circ\prime}(Fe^{3+}/Fe^{2}) = 0.68$ V differs significantly from the standard potential value $E^{\circ}(Fe^{3+}/Fe^{2+}) = 0.77$ V. This is due to an ionic strength effect. The activity coefficient of Fe^{3+} is far weaker than that of Fe^{2+} due to the fact that the more charged an ion is, the more sensitive it becomes to ionic strengths.

The symbolism used to study the titration is given in Fig. 17.1.

The general and rigorous equation describing the titration results from the following relations, which are necessarily satisfied all along its course (see Chap. 16):

$$E = E^{\circ\prime}(Fe^{3+}/Fe^{2+}) - (RT/F)\ln\{[Fe^{2+}]/[Fe^{3+}]\} \quad (17.3)$$

$$E = E^{\circ\prime}(Ce^{4+}/Ce^{3+}) - (RT/F)\ln\{[Ce^{III}]/[Ce^{IV}]\} \quad (17.4)$$

$$[Fe^{2+}] + [Fe^{3+}] = C_o V_o/(V_o + V) = X, \quad (17.5)$$

$$[Ce^{III}] + [Ce^{IV}] = C_t V/(V_o + V) = Y, \quad (17.6)$$

$$[Fe^{3+}] = [Ce^{III}]. \quad (17.7)$$

Relations (17.5) and (17.6) are the mass balance equations. Through the denominators of fractions X and Y, relations (17.5) and (17.6) take into account the dilution of the initial volume V_o by the added volume V of the titrant solution. (The symbols X and Y are only used for the sake of simplicity when writing the relations.)

17.1.1 Common Simplified Theoretical Study

Commonly, the study of a redox titration curve is achieved as follows. The first step is to recognize the quantitative character of the titration reaction, which must be quasi-complete. For the chosen example, $\Delta E^{\circ\prime} = 0.76\,\text{V}$. This value is far above 0.24 V, the limiting value under which the reaction can no longer be considered quasi-complete (see Sect. 16.11). Applying the relationship between the difference in standard potentials and the equilibrium constant gives $K = 10^{12.8}$ (at 25 °C) for the chosen example. Reaction (17.1) can be considered complete. Due to this fact, some approximations may rightfully be realized in order to simply describe the titration curve before and after the equivalence point.

17.1.1.1 Before the Equivalence Point

Before the equivalence point, $Y < X$. The present species are the remaining ions Fe^{2+} and the Fe^{3+} and Ce^{III} formed. Applying relations (17.5)–(17.7) and assuming activities equal to the corresponding concentrations give

$$Ce^{III} = Y,$$
$$Fe^{3+} = Y,$$
$$Fe^{2+} = X - Y,$$

and after using (17.3), we have

$$E = E^{\circ\prime}\left(Fe^{3+}/Fe^{2+}\right) - (RT/F)\ln[(X - Y)/Y] \quad \text{(before the equivalence point)} \tag{17.8}$$

or

$$E = E^{\circ\prime}\left(Fe^{3+}/Fe^{2+}\right) - (RT/F)\ln[(1 - \varphi)/\varphi], \quad \varphi < 1. \tag{17.9}$$

As a result, if we accept the hypothesis that the titration reaction is complete, we see that the equilibrium potential value and the shape of the curve depend only on the couple Fe^{3+}/Fe^{2+} through its standard potential value $E^{\circ\prime}(Fe^{3+}/Fe^{2+})$. Some authors say that the potential value is imposed by the couple Fe^{3+}/Fe^{2+}. The first point ($\varphi = 0$) is a particular one. The only species existing in the titration vessel is Fe^{2+}. The potential value is ill defined and is of no practical consequence.

17.1.1.2 After the Equivalence Point

After the equivalence point ($Y > X$; $\varphi > 1$), the present species are Fe^{3+}, Ce^{III}, and Ce^{IV} (which is in excess). Fe^{2+} no longer exists since the titration reaction may be considered complete. After handling Eqs. (17.4)–(17.7), we find

$$E = E^{\circ\prime}\left(Ce^{IV}/Ce^{III}\right) - (RT/F)\ln[X/(Y - X)] \quad \text{(after the equivalence point)} \tag{17.10}$$

or

$$E = E^{\circ\prime}\left(Ce^{IV}/Ce^{III}\right) - (RT/F)\ln\left[1/(\varphi - 1)\right], \quad \varphi < 1. \tag{17.11}$$

The potential of the solution is imposed by the couple Ce^{IV}/Ce^{III}.

17.1.1.3 At the Equivalence Point

At the equivalence point, we can think, a priori, that the potential E_{ep} must depend on both couples, that is, on the values $E^{\circ\prime}(Fe^{3+}/Fe^{2+})$ and $E^{\circ\prime}(Ce^{IV}/Ce^{III})$, since just before it, the potential value is imposed only by one couple and just after it by the other one. The equivalence point is defined exactly by the relations X = Y and $\varphi = 1$. Due to the quantitative character of the titration reaction, it is clear that only a very weak concentration $[\varepsilon Fe^{2+}]$ of ferrous ions has not reacted. Thus, according to Eqs. (17.5)–(17.7), there exists an equal concentration $[\varepsilon Ce^{IV}]$ of Ce^{IV} that has not reacted at the equivalence point:

$$\left[\varepsilon Ce^{IV}\right] = \left[\varepsilon Fe^{2+}\right].$$

Both Nernst's relations applied at the equivalence point are

$$E_{ep} = E^{\circ\prime}\left(Fe^{3+}/Fe^{2+}\right) - (RT/F)\ln\left\{\left[\varepsilon Fe^{2+}\right]/\left[Fe^{3+}\right]\right\},$$
$$E_{ep} = E^{\circ\prime}\left(Ce^{IV}/Ce^{III}\right) - (RT/F)\ln\left\{\left[Ce^{III}\right]/\left[\varepsilon Ce^{IV}\right]\right\}.$$

After adding these two relations, we find

$$2E_{ep} = E^{\circ\prime}\left(Fe^{3+}/Fe^{2+}\right) + E^{\circ\prime}\left(Ce^{IV}/Ce^{III}\right)$$
$$- (RT/F)\ln\left\{\left([\varepsilon Fe^{2+}][Ce^{III}]\right)/\left([Fe^{3+}][\varepsilon Ce^{IV}]\right)\right\};$$

and as a result:

$$E_{ep} = \left[E^{\circ\prime}\left(Fe^{3+}/Fe^{2+}\right) + E^{\circ\prime}\left(Ce^{IV}\right)/\left(Ce^{III}\right)\right]/2, \quad \varphi = 1 \qquad (17.12)$$

since the logarithm argument is equal to unity.

The equivalence point potential is the average of the formal potentials of both couples. The fact that the equivalence point potential is equal to their half-sum is due to the equality $n_1 = n_2$. This result cannot be generalized to asymmetrical titrations. Notice also for this example that the equivalence point potential is independent of the titrand concentration (see later).

The titration curve is shown in Fig. 17.2. We can make the following comments:

- before the equivalence point, it exhibits the S shape characteristic of the mathematical function $y = \log[x/(1-x)]$;
- at the half-neutralization, $E = E^{\circ\prime}(Fe^{3+}/Fe^{2+})$. Near this value, the potential change is weak. It is the buffered potential zone or the potentiostatic zone;
- the potential change at the equivalence point is abrupt. Relation (17.12), which gives its value, explains this result. The detailed reason for this state of things is the great difference between the formal potential values of the couples $E^{\circ\prime}(Fe^{3+}/Fe^{2+})$ and $E^{\circ\prime}(Ce^{IV}/Ce^{III})$ (see below);
- at the double equivalence ($\varphi = 2$), $E = E^{\circ\prime}(Ce^{IV}/Ce^{III})$, as shown by relation (17.11).

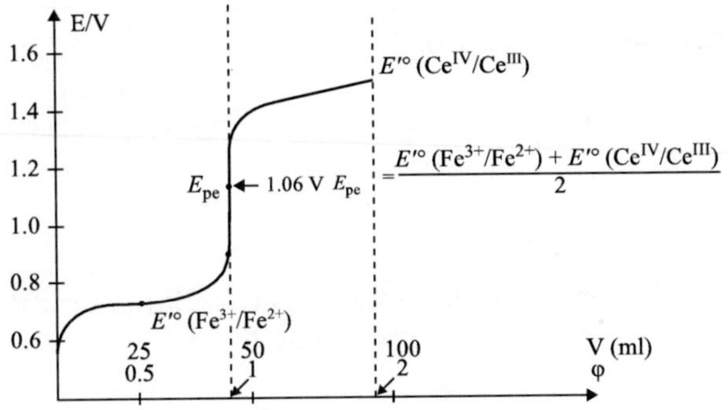

Fig. 17.2 Titration curve of Fe^{2+} by Ce^{IV} ($V_o = 50$ ml, $C_o = 0.1$ mol/L, $Ct = 0.1$ mol/L, $T = 25$ °C)

From a practical standpoint, we must choose an internal redox indicator whose color change is located in the vertical zone of the curve around $E = E_{pe}$, where φ does not differ significantly from unity. Then the titration error is negligible (see Sect. 17.2). Ferroin and its derivatives are suitable.

17.1.2 Rigorous Study

Describing the titration curve by the three different Eqs. (17.9) [or (17.8)], (17.12), and (17.10) [or (17.11)] is not fully satisfactory. Indeed, they implicate that both antagonistic couples do not play a part simultaneously in the fixation of the equilibrium potential values (except at the equivalence point). This is a kind of negation of the very existence of the titration. Moreover, such a methodology also implicates that the titration curve is discontinuous, since three different equations are used. Finally, with this description, it is not possible to study the slope of the curve $dE/d\varphi$ about the equivalent point. This is a drawback because it is this slope that governs the success of a titration, as we saw when we considered acid–base titrations. This is the basis for the sharpness index concept.

A general equation exists that applies itself to any stage of the titration. It can be easily deduced from Eqs. (17.3)–(17.7). It is the following:

$$\varphi = \left\{1 + \exp[(F/RT)(E - E^{\circ\prime}{}_2)]\right\} / \left\{1 + \exp[-(F/RT)(E - E^{\circ\prime}{}_1)]\right\} \quad (17.13)$$

or

$$\varphi = (1 + e_2)/(1 + e_1),$$

with $E^{\circ\prime}{}_2 = E^{\circ\prime}(Ce^{IV}/Ce^{III})$, $E^{\circ\prime}{}_1 = E^{\circ\prime}(Fe^{3+}/Fe^{2+})$, $e_1 = \exp[-(F/RT)(E - E^{\circ\prime}{}_1)]$, $e_2 = \exp[(F/RT)(E - E^{\circ\prime}{}_2)]$ (in the following, the subscript 2 is systematically connected to the titrant).

Equation (17.13) is explicit in φ, but not in E. It is very easy to check that relation (17.9) describing the first part of the curve is equivalent to Eq. (17.13) in which the exponential e_2 would be negligible compared to unity. Likewise, Eq. (17.11) describing the curve after the equivalence point is equivalent to Eq. (17.13) in which the exponential e_1 is negligible with respect to unity. Finally, it is sufficient to set $\varphi = 1$ in Eq. (17.13) to find relation (17.12) again. This means that both exponentials possess the same numerical value at the equivalence point.

17.2 Further Considerations Concerning Symmetrical Titrations: Titration Error

17.2.1 Consideration 1

The preceding results are valid for all symmetric titrations ($n_1 = n_2$, with, in particular, $n_1 = n_2 > 1$). The general (and rigorous) equation of the curve describing the titration corresponding to reaction (17.2) is

$$\varphi = \left\{1 + \exp[(n_2F/RT)(E - E^{\circ\prime}{}_2)]\right\} / \left\{1 + \exp[(-n_1F/RT)(E - E^{\circ\prime}{}_1)]\right\} \tag{17.14}$$

which can be written in abbreviated form as above:

$$\varphi = (1 + e_2)/(1 + e_1).$$

17.2.2 Consideration 2

When the difference $\Delta E^{\circ\prime}$ is sufficient ($\Delta E^{\circ\prime} > 0.24$ V), the curve portions before and after the equivalence point can be, as already seen, described by the relations

$$E = E^{\circ\prime}{}_1 - (RT/n_1F)\ln[(1 - \varphi)/\varphi] \tag{17.15}$$

$$E = E^{\circ\prime}{}_2 - (RT/n_2F)\ln[1/(\varphi - 1)] \tag{17.16}$$

which are analogous to relations (17.9) and (17.11). At the equivalence point, the equilibrium potential is, as before, given by relation (17.12):

$$E_{\mathrm{ep}} = (E^{\circ\prime}{}_1 + E^{\circ\prime}{}_2)/2, \tag{17.17}$$

for all values $n_1 = n_2$. It is independent of any concentration.

Now, it is convenient to make clear the conditions in which the general Eqs. (17.13) and (17.14) can be reduced to the simplified ones (17.15) and (17.16), that is, to make clear the conditions in which the exponentials e_1 and e_2 are successively negligible with respect to unity in Eqs. (17.13) and (17.14). It is obvious that the exponential numerical values depend on the difference of the formal potentials

$\Delta E^{\circ\prime} = E^{\circ\prime}{}_2 - E^{\circ\prime}{}_1$, on the values $E^{\circ\prime}{}_2$ and $E^{\circ\prime}{}_1$ themselves, and also on the fraction titrated φ, which plays a part in the E value through Eqs. (17.13) and (17.14). A calculation shows that when $\Delta E^{\circ\prime} > 0.24$ V, the exponential e_2 (relative to the titrant) is perfectly negligible with respect to unity, as long as the fraction titrated is very close to unity and under it; that is,

$$e_2 \ll 1 \quad \text{for} \quad \varphi \leq 1 - \varepsilon \quad (\Delta E^{\circ\prime} \geq 0.24 \text{ V}).$$

Likewise, the exponential e_1 relative to the titrand is perfectly negligible with respect to unity as soon as the fraction titrated becomes slightly greater than unity; that is,

$$e_1 \ll 1 \quad \text{for} \quad \varphi > 1 + \varepsilon \quad (\Delta E^{\circ\prime} \geq 0.24 \text{ V}).$$

For example, for the difference $\Delta E^{\circ\prime} = 0.24$ V and for the arbitrary values $E^{\circ\prime}{}_1 =$ 0.94 V and $E^{\circ\prime}{}_2 = 1.18$ V, the exact and approached potential values for $\varphi = 0.993$ are

$E = 1.050$ V [exact value calculated through relation (17.14)],
$E = 1.067$ V [approached value calculated through (17.15)].

The difference between them is very weak. A priori, before this calculation, one could think that the exponential e_2 should be higher than it is actually for this φ value and, thus, that it was playing a greater part in the potential value. The same phenomenon exists for the exponential e_1 (relative to the titrand) after the equivalence point. It becomes negligible as soon as the fraction titrated becomes slightly higher than unity. The greater the difference $\Delta E^{\circ\prime}$ is, the more negligible the exponentials e_1 and e_2 will be near the equivalence point.

17.2.3 Consideration 3

It is interesting to notice that assuming that the titration reaction is complete corresponds to neglecting 1 with respect to the constant K°; that is,

$$K^\circ - 1 \approx K^\circ.$$

For the example of the titration of Fe^{2+} by Ce^{IV}, handling Eqs. (17.5)–(17.7) and introducing the equilibrium constant K° give

$$K^\circ = \left[\left(Ce^{III}\right)\left(Fe^{3+}\right)\right] / \left[\left(Ce^{IV}\right)\left(Fe^{2+}\right)\right].$$

Then, assuming the activities equal to the concentrations gives the equation in $[Fe^{3+}]$:

$$\left[Fe^{3+}\right]^2 - \{K^\circ/(K^\circ - 1)\}(X + Y)\left[Fe^{3+}\right] + \{K^\circ/(K^\circ - 1)\}XY = 0.$$

Indeed, this equation becomes independent of K° if 1 becomes negligible compared to it. When this is the case, the resolution of the above second-order in $[Fe^{3+}]$ equation gives $[Fe^{3+}] = Y$ for $Y < X$ (before the equivalence point) and $[Fe^{3+}] = X$ for $Y > X$ (after the equivalence point). These equalities imply that the titration reaction is complete, as we have seen.

17.2.4 Consideration 4

The calculation of the derivative $d\varphi/dE$ gives the relation

$$d\varphi/dE = (F/RT)\left\{[e_2(1+e_1)+e_1(1+e_2)]/(1+e_1)^2\right\} \quad (n_1 = n_2) \quad (17.18)$$

The derivative is always positive and, as a result, so is its inverse $dE/d\varphi$. The solution potential is always increasing with the fraction titrated. (During an inverse redox titration, the potential is always decreasing.)

17.2.5 Consideration 5

The calculation of the second-order derivative of $d^2E/d\varphi^2$ shows that, rigorously speaking, the curve E/φ does not exhibit an inflection point for $e_2 = e_1$ exactly, that is, at the equivalence ($\varphi = 1$). However, when the difference between the formal potentials is higher than 240 mV, the exponential values e_1 and e_2 are negligible versus unity (around the equivalence point, their values are about 10^{-3}). As a result, this condition is satisfied, and after simplifying the expression of the second derivative, we find

$$d^2E/d\varphi^2 = 0 \quad \text{for } e_1 = e_2, \quad \text{i.e., for } \varphi = 1, \text{ with } e_2 \ll 1 \text{ and } e_1 \ll 1.$$

The inflection point of the curve E/φ is then located at the equivalence. In this case, it is easy to demonstrate that the slope of the curve is maximum at this point. The sharpness index of the titration curve E/φ is maximum at the equivalence point. We can deduce from these results that the equivalence point may be detected by locating the inflexion point and that by following this strategy, the perpetrated titration error is negligible (see below). Indeed, this strategy is very often followed when potentiometry is used to detect the equivalence point (see electrochemistry).

17.2.6 Consideration 6

The preceding results underline the importance of the formal (or standard) potential values for the success of a redox titration. The following reasoning points out the part their difference $\Delta E^{\circ\prime}$ plays. Expression (17.17) of the first derivative gives for the equivalence point ($\varphi = 1$ and $e_1 = e_2$)

$$(dE/d\varphi)_{\mathrm{ep}} = (RT/F)\left\{[1+\exp(-F\Delta E^{\circ\prime}/2RT)]/[2\exp(-F\Delta E^{\circ\prime}/2RT)]\right\},$$

with $\Delta E^{\circ\prime} = E^{\circ\prime}{}_2 - E^{\circ\prime}{}_1$.

It is easy to find that

$$d\left\{[dE/d\varphi]_{\mathrm{ep}}\right\}/d\Delta E^{\circ\prime} = (F/RT)(1/\exp[-F\Delta E^{\circ\prime}/2RT]).$$

The slope of the titration curve at the equivalence point increases with the difference between the standard (or formal) potentials $\Delta E^{\circ\prime}$ of both couples.

17.2.7 Consideration 7

The calculation of the titration error is directly connected to the preceding considerations. It is calculated as follows. Let E_{fp} be the potential at the final point (when the indicator color changes). The E_{fp} value is slightly different from the theoretical one E_{ep} (that at the equivalence point). Let φ_{pf} be the fraction titrated at the final point. According to the definition of the titration error, already given when we considered acids and bases, the absolute titration error is $\varphi_{fp} - 1$ since 1 is the fraction titrated at the equivalence point. The general expression of $\varphi - 1$, which is valid for any point of the titration curve, is (with $n_1 = n_2$)

$$\varphi - 1 = \left\{[1 + \exp((F/RT)/(E - E^{\circ\prime}{}_2))]/[1 + \exp(-(F/RT)(E - E^{\circ\prime}{}_1))]\right\} - 1.$$

The final point φ_{fp}, E_{fp} is a point on this curve. Thus, we can write

$$\varphi_{fp} - 1 = \left\{[1 + \exp((F/RT)/(E_{fp} - E^{\circ\prime}{}_2))]/[1 + \exp(-(F/RT)(E_{fp} - E^{\circ\prime}{}_1))]\right\} - 1$$

or, after rearranging,

$$\begin{aligned}\varphi_{fp} - 1 = \{&[\exp(F/RT)(E_{fp} - E^{\circ\prime}{}_2)] \\ &- [\exp(-(F/RT)(E_{fp} - E^{\circ\prime}{}_1)]\}/\{1 + \exp[-(F/RT)(E_{fp} - E^{\circ\prime}{}_1)]\}.\end{aligned}$$

This relationship can be expressed as a function of $\Delta E^{\circ\prime}$:

$$\begin{aligned}\varphi_{fp} - 1 = \exp[-F\Delta E^{\circ\prime}/RT]\{&\exp[F(E_{fp} - E^{\circ\prime}{}_1)/RT] \\ &- \exp[-F(E_{fp} - E^{\circ\prime}{}_2/RT)]\}/\{1 + \exp[-F\Delta E^{\circ\prime}/RT] \\ &\times \exp[-F(E_{fp} - E^{\circ\prime}{}_2)/RT]\}.\end{aligned}$$

The absolute titration error (by definition due to the false detection of the equivalence point) depends on the difference in formal potentials $\Delta E^{\circ\prime}$. The last expression clearly shows that when $\Delta E^{\circ\prime}$ is very high, the error tends toward zero since the numerator itself tends toward zero and the denominator simultaneously tends toward unity. For example, when $\Delta E^{\circ\prime} = 0.24$ V and when the color redox indicator changes for the potential value $E_{fp} = (E_{ep} - 0.01)$V, the relative error expressed in percentages is about -0.67%. When $\Delta E^{\circ\prime} = 0.30$ V, with the same difference $E_{fp} - E_{ep}$, the error decreases to -0.21%. In the case of the titration of Fe^{2+} by Ce^{IV}, the titration error remains about -0.01 V, even if the indicator color changes for a potential value weaker than that at the equivalence point by a difference as great as 0.20 V. This remarkable result must be attributed to the great difference $\Delta E^{\circ\prime}$ ($\Delta E^{\circ\prime} = 0.76$ V).

17.3 Study of the Titration Curve of Stannic Ions by Chromous Ions—Generalization to All Asymmetrical Titrations

The titration reaction is

$$2Cr^{2+} + Sn^{4+} \rightleftharpoons 2Cr^{3+} + Sn^{2+} \quad K. \tag{17.19}$$

The titrant is a solution of chromous ions. The titration reaction (17.19) is a reduction reaction. It is an example of an asymmetrical titration since n_2 ($n_2 = 2$) differs from n_1 ($n_1 = 1$). From an experimental standpoint, it is achieved in a 1 mol/L hydrochloric acid solution.

First, it is important to notice that the exchange of the two electrons of the couple Sn^{4+}/Sn^{2+} takes place simultaneously and not successively. This possibility is a characteristic of some redox processes that markedly distinguishes them from acid—base (and complexation; see Part IV) phenomena. In acid–base phenomena, indeed, there is never a fully simultaneous exchange of several protons at the same pH.

The equations necessarily satisfied during the course of the titration are

$$E = E^{\circ\prime}\left(Cr^{3+}/Cr^{2+}\right) - (RT/F)\ln\left[\left(Cr^{2+}\right)/\left(Cr^{3+}\right)\right] \quad E^{\circ\prime}\left(Cr^{3+}/Cr^{2+}\right) = -0.38\,V,$$

$$E = E^{\circ\prime}\left(Sn^{4+}/Sn^{2+}\right) - (RT/2F)\ln\left[\left(Sn^{2+}\right)/\left(Sn^{4+}\right)\right], \quad E^{\circ\prime}\left(Sn^{4+}/Sn^{2+}\right) = 0.14\,V,$$

$$[Cr^{2+}] + [Cr^{3+}] = C_t V/(V_o + V) = Y,$$

$$\left[Sn^{4+}\right] + \left[Sn^{2+}\right] = C_o V_o/(V_o + V) = X,$$

$$\left[Cr^{3+}\right] = 2\left[Sn^{2+}\right],$$

in which, as earlier, the symbols Y and X concern the titrant and the titrand, respectively. The rigorous general equation describing the titration is obtained by reducing the above system of equations. We find

$$Y/2X = [1 + \exp(-F/RT)(E - E^{\circ\prime}{}_2)]/[1 + \exp(2F/RT)(E - E^{\circ\prime}{}_1)],$$

in which $E^{\circ\prime}{}_1 = E^{\circ}(Sn^{4+}/Sn^{2+})$ and $E^{\circ\prime}{}_2 = E^{\circ\prime}(Cr^{3+}/Cr^{2+})$. The ratio Y/2X is the fraction titrated:

$$\varphi = Y/2X.$$

Hence, the general equation of the titration curve is

$$\varphi = [1 + \exp(-F/RT)(E - E^{\circ\prime}{}_2)]/[1 + \exp(2F/RT)(E - E^{\circ\prime}{}_1)] \tag{17.20}$$

or $\varphi = (1 + e_2)/(1 + e_1)$, a relation identical to (17.13) from a formal standpoint. However, it is important to notice that the signs of the exponentials of (17.20) are inverted with respect to those of (17.13). This is due to the fact that the titrand is an oxidized species. The factor 2 in the fraction titrated is issued from the titration reaction stoichiometry. At the equivalence point, at which, by definition, all the Sn^{4+} ions are supposed to have disappeared (as is also the case with Cr^{2+} ions), the titrated fraction should be equal to unity. Hence, we must add two moles of Cr^{2+} ions for one mole of Sn^{4+} ions in order to attain the equivalence point; that is,

$$\mathrm{Y} = 2\mathrm{X} \quad \text{(equivalence point).}$$

The following reasons are identical to those already described.

The general Eq. (17.20) may be considerably simplified in the present case since the titration reaction is very strongly displaced toward the right. Indeed, the equilibrium constant K calculated through the expression

$$\log K = 16.9 \cdot 2[0.14 - (-0.38)]$$

exhibits a value close to $10^{17.6}$. Reaction (17.19) can be considered going to completion. With this condition,

- before the equivalence point, the exponential $e_2 = \exp(-F/RT)\ (E - E^{\circ\prime}{}_2)$ is negligible, and the simplified relation fully equivalent to the general one is

  $$E = E^{\circ\prime}\left(\mathrm{Sn}^{4+}/\mathrm{Sn}^{2+}\right) - (RT/2F)\ln[\varphi/(1-\varphi)], \quad \varphi < 1.$$

  The solution potential is imposed by the couple of the titrand. A particular point is that corresponding to the half-neutralization, namely, at which $\varphi = \frac{1}{2}$. Then we find

  $$\mathrm{Y} = \mathrm{X} \quad (\varphi = 1/2) \quad \text{and}$$

  $$E = E^{\circ\prime}\left(\mathrm{Sn}^{4+}/\mathrm{Sn}^{2+}\right), \quad (\varphi = 1/2);$$

- at the equivalence point, the general Eq. (17.20) gives, with $\varphi = 1$,

  $$e_1 = e_2.$$

  As a result,

  $$E_{\mathrm{ep}} = (E^{\circ\prime}{}_2 + 2E^{\circ\prime}{}_1)/3 \quad \text{(equivalence).}$$

  This relation comes from neither simplification nor assumption. It comes directly from relation (17.20). In the present case, $E_{\mathrm{ep}} = -0.03$ V. Notice that E_{ep} does not depend on any concentration (see later);
- after the equivalence point, the fact that reaction (17.19) may be considered to be complete permits us to neglect the exponential e_1 and to write

  $$\varphi = 1 + \exp(-F/RT)(E - E^{\circ\prime}{}_2) \quad \text{or equivalently:}$$

  $$E = E^{\circ\prime}\left(\mathrm{Cr}^{3+}/\mathrm{Cr}^{2+}\right) - (RT/F)\ln[(\varphi - 1)/1], \quad \varphi > 1.$$

The solution potential may be considered as being imposed by the couple Cr^{3+}/Cr^{2+}. The point curve for which $\varphi = 2$ (i.e., Y = 4X) is a particular point since then $E = E^{\circ\prime}(Cr^{3+}/Cr^{2+})$.

The titration curve is shown in Fig. 17.3.

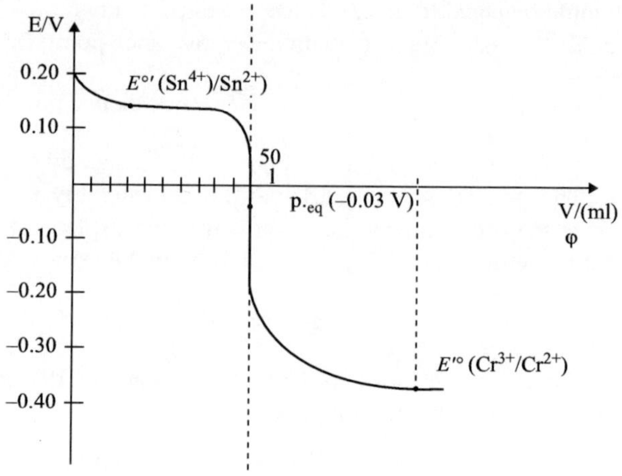

Fig. 17.3 Titration curve of Sn^{4+} ($C_o = 0.1$ mol/L), ($V_o = 50$ ml) by a solution of Cr^{2+} ($C_t = 0.1$ mol/L)

From a practical standpoint, because no satisfactory indicator in such a reducing medium exists like the one at the equivalence point ($E_{ep} \approx -0.03$ V), the latter must be detected by potentiometry.

These considerations may be generalized to any asymmetrical titration, whose reaction is

$$n_2\text{Red}_1 + n_1\text{Ox}_2 \rightarrow n_2\text{Ox}_1 + n_1\text{Red}_2.$$

What remains to do now is to generalize the preceding formalism.

Considering the titration reaction as again being an oxidation reaction [with the result that the signs of exponentials e_2 and e_1 are the same as in relations (17.13) and (17.14)], the general equation is

$$\varphi = \left\{1 + \exp[(n_2F/RT)(E - E^{\circ\prime}{}_2)]\right\} / \left\{1 + \exp[(-n_1F/RT)(E - E^{\circ\prime}{}_1)]\right\}$$

or, in the same manner, $\varphi = (1 + e_2)/(1 + e_1)$, an expression analogous to (17.13). The fraction titrated is defined by

$$\varphi = n_2\text{Y}/n_1\text{X}.$$

At the equivalence,

$$n_2\text{Y} = n_1\text{X};$$

- before the equivalence point, the exponential $e_2 = \exp[(n_2F/RT)\ (E - E^{\circ\prime}{}_2)]$ is negligible and the curve may be described by the equation

$$E = E^{\circ\prime}(\text{Ox}_1/\text{Red}_1) - (RT/n_1F)\ln[(\text{Red}_1)/(\text{Ox}_1)], \quad \varphi < 1,$$

or

$$E = E^{\circ\prime}(\text{Ox}_1/\text{Red}_1) - (RT/n_1F)\ln[(1 - \varphi)/\varphi], \quad \varphi < 1;$$

- after the equivalence point, the exponential $e_1 = \exp[(-n_1F/RT)(E - E^{\circ\prime}_1)]$ is negligible and the curve may be described by the equation

$$E = E^{\circ\prime}(Ox_2/Red_2) - (RT/n_2F)\ln[(Red_2)/(Ox_2)], \quad \varphi > 1,$$

$$E = E^{\circ\prime}(Ox_2/Red_2) - (RT/n_2F)\ln[1/(\varphi - 1)], \quad \varphi > 1,$$

- the curve E/φ is always increasing;
- the equivalence point potential is given by the relation

$$E_{ep} = (n_1E^{\circ\prime}_1 + n_2E^{\circ\prime}_2)/(n_1 + n_2);$$

- the equivalence point potential value is displaced toward that of the formal (or standard) potential of the couple exchanging the greater electron number in its half-redox reaction. It does not depend on any concentration;
- at half-neutralization, the solution potential is equal to the standard (or formal) potential of the titrand couple, whereas at twice the equivalence, the equilibrium potential is equal to the standard (or formal) potential of the titrant couple;
- it can be demonstrated that the inflection point is not located at exactly the equivalence point. The difference in the solution potentials at these points is about ± 0.01 V. It depends on n_1 and n_2 values;
- it can also be demonstrated that the slopes of the curves E/φ are slightly weaker if one of the antagonistic couples exchanges several electrons rather than only one.

17.4 Redox Titrations in Which a Simultaneous Exchange of Electrons and Protons or Other Particules Exists

An example of a redox titration in which a simultaneous exchange of electrons and protons occurs is provided by the titration of the ferrous ion by the permanganate ion according to the reaction

$$5Fe^{2+} + MnO_4^- + 8H^+ \rightarrow 5Fe^{3+} + Mn^{2+} + 4H_2O.$$

In the preceding chapter, we introduced the concept of the apparent normal potential for the redox couples exchanging electrons and protons simultaneously. In the present case, the apparent normal potential $E^{\circ\prime}(MnO_4^-/Mn^{2+})$ is given by the relation

$$E^{\circ\prime}\left(MnO_4^-/Mn^{2+}\right) = E^{\circ}\left(MnO_4^-/Mn^{2+}\right) + \left(8RT/5F\right)\ln(H^+).$$

The investigated titration may be studied like other asymmetrical titrations after the standard (or formal) potentials are replaced with the apparent one at the working pH, provided the latter is a constant. The general equation of the titration curve is

$$\varphi = \left\{1 + \exp[(5F/RT)(E - E^{\circ\prime}_2)]\right\} / \left\{1 + \exp[(-F/RT)(E - E^{\circ\prime}_1)]\right\}$$

or $\varphi = (1 + e_2)/(1 + e_1)$, analogously to (17.13). $E^{\circ\prime}_1 = E^{\circ\prime}(Fe^{3+}/Fe^{2+})$ is the formal potential of the couple Fe^{3+}/Fe^{2+}, and $E^{\circ\prime}_2 = E^{\circ\prime}(MnO_4^-/Mn^{2+})$ is the apparent or normal potential of the couple MnO_4^-/Mn^{2+}, $\varphi = 5Y/X$, and X and Y are the total concentrations in iron and manganese, respectively. Based on the arguments already given and from the difference between the standard potentials of both couples ($E°_1 = 0.77$ V and $E°_2 = 1.52$ V), the titration reaction may be considered complete provided the pH value remains low (pH < 5). As a result,

- before the equivalence point, the above-mentioned general equation can be legitimately simplified and becomes

$$E = E°(Fe^{3+}/Fe^{2+}) - (RT/F)\ln[(1 - \varphi)/\varphi].$$

The equilibrium solution potential is said to be imposed by the couple Fe^{3+}/Fe^{2+};

- after the equivalence point, the general equation can be reduced to the relation

$$E = E°(MnO_4^-/Mn^{2+}) - (RT/5F)\ln[1/(1 - \varphi)].$$

The solution potential is imposed by the couple MnO_4^-/Mn^{2+};

- at the equivalence point, the equilibrium potential is given by the relation

$$E_{ep} = (E°'_1 + 5E°'_2)/6.$$

The striking difference with the preceding cases lies in the fact that the solution equilibrium potential value at the equivalence point depends on the pH value, since $E°'_2$, which is an apparent standard potential, itself depends on the pH. Thus,

- at pH = 0, $E_{ep} = 1.36$ V;
- at pH = 1, $E_{ep} = 1.31$ V;
- at pH = 3, $E_{ep} = 1.13$ V.

The change is sharp. The results obtained with this example can be immediately generalized to all the titrations in which there is simultaneously an exchange of electrons together with an exchange of protons or any other particule (see Part IV). In the latter case, the apparent standard (or normal) potential takes the ligand concentration into account. It is constant when the latter is constant. An example is provided by the couple Ce^{IV}/Ce^{III} in sulfuric acid medium, the ligand being the sulfate ion.

17.5 Cases in Which the Equivalence Potential Values Depend on the Concentration of One of the Reactants

Let's consider, for example, the following titration:

$$2Fe^{2+} + Br_{2(w)} \rightarrow 2Fe^{3+} + 2Br^-.$$

The reaction differs from reaction (17.2) because one of the reacting redox species, in the bromine occurrence, appears to be a kind of dimer of the element that suffers a change in its oxidation number during the reaction. By setting the mass balances in iron and in bromine, we obtain

$$[Fe^{2+}] + [Fe^{3+}] = C_o V_o/(V_o + V) = X.$$

$$[Br_{2(w)}] + 1/2[Br^-] = C_t V/(V_o + V) = Y.$$

We then find the general equation

$$2X/(1 + e_1)^2 + 1/(1 + e_1)e_2 - \varphi/e_2 = 0 \quad (17.21)$$

in which $\varphi = 2Y/X$ is the fraction titrated, and

$$e_1 = 1 + \exp[-(F/RT)(E - E°_1)] \quad \text{with } E°_1 = E°'(Fe^{3+}/Fe^{2+}),$$

$$e_2 = 1 + \exp[(2F/RT)(E - E°_2)] \quad \text{with } E°_2 = E°'(Br_{2(w)}/Br^-).$$

Equation (17.21) describes the titration curve. It is not explicit in φ or in E. Moreover, it still contains the parameter X. This was not the case for the preceding titrations. Hence, it is difficult to handle. The fact that φ cannot be explicitly expressed as a function of e_1 and e_2, as it could be previously, results from the presence of the squared term $(Br^-)^2$ in Nernst's equation. However, Eq. (17.21) may be solved numerically. Moreover, it can easily be solved via a simplified approach. After the preceding considerations, the titration reaction may indeed be considered complete [$E°(Br_{2(w)}/Br^-) = 1.07$ V and $E°(Fe^{3+}/Fe^{2+}) = 0.77$ V]. Therefore,

- before the equivalence point, the exponential e_2 (relative to the titrant) is very low, as in the previous cases. Equation (17.21) can be reduced to

$$\varphi \approx 1/(1 + e_1), \quad \varphi < 1$$

or to the equivalent relation

$$E = E°'(Fe^{3+}/Fe^{2+}) - (RT/F)\ln[(1 - \varphi)/\varphi], \quad \varphi < 1.$$

The solution potential is imposed by the couple Fe^{3+}/Fe^{2+};

- after the equivalence point, the exponential e_1 is very weak. The general Eq. (17.21) can be reduced to the expression

$$\varphi - 1 = 2Xe_2, \quad \varphi > 1,$$

or, in the same manner,

$$E = E°'(Br_{2(w)}/Br^-) + (RT/2F)\ln(\varphi - 1) - (RT/2F)\ln X.$$

The equilibrium solution potential depends on the initial concentration of Fe^{2+}, through the X value;

- at the equivalence point, the calculation of the equilibrium potential may be approximately achieved as follows. The general Eq. (17.21), with $\varphi = 1$, becomes

$$2X_{ep} + (1 + e_1)/e_2 = \left(1 + 2e_1 + {e_1}^2\right)/e_2.$$

Bearing in mind that the reaction is quasi-complete, we can assert that the exponential e_1 is endowed with a very weak numerical value, as is also the case with the exponential e_2. As a result, the square ${e_1}^2$ is negligible and we immediately find the new relation that is approached:

$$E_{ep} = (E^{\circ\prime}{}_1 + 2E^{\circ\prime}{}_2)/3 - (RT/3F)\ln(2X_{ep}), \quad \varphi = 1.$$

Actually, the term in X_{ep} is negligible with respect to the first one on the right-hand side of the relation. For example, in the titration of a 10^{-1} mol/L$^-$ solution of Fe^{2+} ions, the difference between the E_{ep} values is about 0.01 V whether or not this term has been taken into account.

Another interesting case is provided by the titrations with potassium dichromate, $K_2Cr_2O_7$. For example, consider the titration of ferrous ions with dichromate ions ${Cr_2O_7}^{2-}$ according to the reaction

$$6Fe^{2+} + {Cr_2O_7}^{2-} + 14H^+ \rightarrow 6Fe^{3+} + 2Cr^{3+} + 7H_2O.$$

The relations that are necessarily satisfied are

$$E = E^{\circ\prime}\left(Fe^{3+}/Fe^{2+}\right) - (RT/F)\ln\left[\left(Fe^{2+}\right)/\left(Fe^{3+}\right)\right],$$
$$E = E^{\circ\prime}{}_2 - (RT/6F)\ln\left[\left(Cr^{3+}\right)^2/\left({Cr_2O_7}^{2-}\right)\right],$$

with $E^{\circ\prime}{}_2 = E^\circ({Cr_2O_7}^{2-}/Cr^{3+}) + (RT/6F)\ln(H^+)^{14}$,

$$\left[Fe^{2+}\right] + \left[Fe^{3+}\right] = X,$$
$$\left[{Cr_2O_7}^{2-}\right] + 1/2\left[Cr^{3+}\right] = Y,$$
$$\left[Cr^{3+}\right] = 1/3\left[Fe^{3+}\right].$$

The general equation coming from this system of relations is not explicit in φ or in E. The study of the standard potential values of both couples shows that the titration reaction equilibrium constant K° is about $6 \cdot 10^{56}$ in sulfuric acid medium. The titration reaction may certainly be considered complete. In this condition and after having followed the arguments already developed, we find the simplified relations

- before the equivalence point:

$$E = E^\circ\left(Fe^{3+}/Fe^{2+}\right) - (RT/F)\ln[(1-\varphi)/\varphi], \quad \varphi < 1$$

or $\varphi = 1/(1 + e_1)$, with $e_1 = \exp[-(F/RT)(E - E^\circ{}_1)]$, $E^\circ{}_1 = E^{\circ\prime}(Fe^{3+}/Fe^{2+})$, and $\varphi = 6Y/X$.

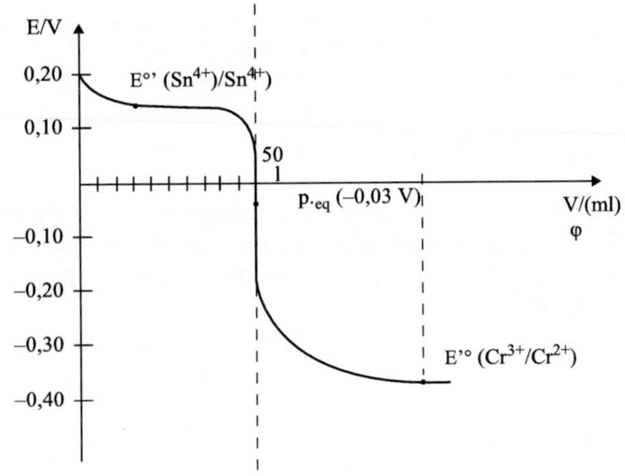

Fig. 17.4 Titration curve of a 10^{-1} mol/L ferrous ion solution with a 1/6 mol/L potassium dichromate solution in sulfuric acid solution (1 mol/L): ($V_o = 50$ ml, $E^{\circ\prime}{}_1 = 0.68$ V, $E^{\circ\prime}{}_2 = 1.33$ V)

The solution potential is imposed by the couple Fe^{3+}/Fe^{2+};

- after the equivalence point:

$$E = E^{\circ\prime}{}_2 + (RT/6F)\ln(\varphi - 1) - (RT/6F)\ln(2X/3) \quad (\varphi > 1)$$

or $\varphi - 1 = 2Xe_2/3$, with $e_2 = \exp[(6F/RT)(E - E^{\circ\prime}{}_2)]$;

- at the equivalence point:

$$E_{ep} = (6E^{\circ\prime}{}_2 + E^{\circ\prime}{}_1)/7 - (RT/7F)\ln(2X_{ep}/3) \quad (\varphi = 1).$$

Figure 17.4 represents the titration curve.

We can remark that the potential at the equivalence point E_{ep} is very close to the apparent standard potential of the couple $Cr_2O_7^{2-}/Cr^{3+}$. The E_{ep} change with X is actually weak. However, it depends markedly on the pH value for two reasons. First, $E^{\circ\prime}{}_2$ changes with the pH value since it is an apparent standard potential. The change amounts to 0.12 V by unity pH. Second, when the pH value increases, the following supplementary equilibria occur:

$$Cr_2O_7^{2-} + H_2O \rightleftharpoons 2HCrO_4^-,$$

$$HCrO_4^- \rightleftharpoons CrO_4^{2-} + H^+,$$

$$\text{and } Cr^{3+} + H_2O \rightleftharpoons CrOH^{2+} + H^+.$$

This induces changes in the formal and apparent standard values of the couple $Cr_2O_7^{2-}/Cr^{3+}$.

Another example is provided by the reaction known as the basic reaction of iodometry (see Chap. 18). It is the reaction of periodide ions I_3^- with thiosulfate ions $S_2O_3^{2-}$ according to

$$I_3^- + 2S_2O_3^{2-} \rightarrow S_4O_6^{2-} + 3I^-.$$

The potential at the equivalence point E_{ep} depends on the concentrations (activities) of the reactants and products existing and subsisting in the titration vessel at this point. It is given by the relation

$$E_{ep} = (E°_1 + E°_2)/2 + (RT/4F)\ln\left\{[S_4O_6{}^{2-}{}_{ep}][I_3{}^-{}_{ep}]\right\}\left\{[S_2O_3{}^{2-}{}_{ep}][I^-{}_{ep}]^3\right\},$$

where $E°_1$ and $E°_2$ are the standard potentials of the couples $S_4O_6{}^{2-}/S_2O_3{}^{2-}$ and $I_3{}^-/I^-$. A calculation shows that the logarithmic term induces a negligible difference between the values calculated whether or not we take it into account. The difference is about 0.08 V.

Briefly, each titration of this kind does indeed need a special theoretical study.

17.6 Titration of the Hypovanadous Ion by the Permanganate Ion

Hypovanadous ions are oxidized by permanganate ions in a multistep process according to the following reactions:

$$5V^{2+} + MnO_4{}^- + 8H^+ \rightarrow 5V^{3+} + Mn^{2+} + 4H_2O,$$

$$5V^{3+} + MnO_4{}^- + H_2O \rightarrow 5VO^{2+} + Mn^{2+} + 2H^+,$$

$$5VO^{2+} + MnO_4{}^- + H_2O \rightarrow 5VO_2{}^+ + Mn^{2+} + 2H^+,$$

The vanadium compounds, which are successively formed, are the vanadate, vanadyl, and dioxovanadium ions. This is an example of the titration of a polyfunctional reduced species with an oxidant. The redox couples that must be taken into account are

$$V^{3+} + 1e^- \rightleftharpoons V^{2+}, \quad E°_{11} = -0.255\text{ V},$$

$$VO^{2+} + 2H^+ + 1e^- \rightleftharpoons V^{3+} + H_2O, \quad E°_{12} = +0.337\text{ V}$$

$$VO^{2+} + 2H^+ + 1e^- \rightleftharpoons VO^{2+} + 2H_2O, \quad E°_{13} = 1.000\text{ V}.$$

The successive standard potential values are markedly different. By transposing relation (16.11) of the preceding chapter to this titration and noting that $n_{11} = n_{12} = n_{13} = 1$, we find the relation

$$\varphi = [(e_{11} + 2e_{11}e_{12} + 3e_{11}e_{12}e_{13})(1 + e_2)] / (1 + e_{11} + e_{11}e_{12} + e_{11}e_{12}e_{13}) \quad (17.22)$$

with $\varphi = 5Y/X$,

$$X = [V^{2+}] + [V^{3+}] + [VO^{2+}] + [VO_2{}^+],$$

$$Y = [MnO_4{}^-] + [Mn^{2+}],$$

and $e_{11} = \exp[(F/RT)(E - E°_{11})]$, $e_{12} = \exp[(F/RT)(E - E°_{12})]$, and so forth.

Equation (17.22) is general and describes the whole curve. It enables the calculation of the solution equilibrium potential (which is hidden in the different exponentials) at any φ value. The important task is to calculate the potentials at the successive equivalence points and to deduce from these calculations the ability to distinguish each of the latter from the others. In order for us to solve this problem, at the first equivalence point, the concentration $[V^{3+}]$ must be at its maximum and at the second, $[VO^{2+}]$ must become its maximum. This hypothesis is in agreement with chemical intuition. In order to achieve a satisfactory titration, vanadate ions V^{3+} must be quasi-alone at the first equivalence point. They are formed from hypovanadous ions V^{2+} before being oxidized into vanadyl ions VO^{2+}. This assumption will probably be satisfied in this case because of the standard potentials $E°_{11}$, $E°_{12}$, and $E°_{13}$, which are markedly different.

17.6.1 First Equivalence Point

The concentration $[V^{3+}]$ of the first equivalence point is maximum when the following relationship is satisfied:

$$d\left[V^{3+}\right]/d\varphi = 0,$$

a condition identical to the following one:

$$d\left\{e_{11}/(1 + e_{11} + e_{11}e_{12} + e_{11}e_{12}e_{13}\right\}/d\varphi = 0$$

(see Chap. 16). After the chain rule of derivation, the preceding condition can be written as

$$\{d[e_{11}/(1 + e_{11} + e_{11}e_{12} + e_{11}e_{12}e_{13})]/dE\}\{dE/d\varphi\} = 0.$$

Calculating the derivative $d\varphi/dE$ from the general equation shows that it never vanishes. As a result and according to the mentioned assumption, the first equivalence point potential is such that

$$\{d[e_{11}/(1 + e_{11} + e_{11}e_{12} + e_{11}e_{12}e_{13})]/dE\} = 0.$$

Calculating the derivative gives

$$e_{11} - e_{11}{}^2 e_{12} - 2e_{11}{}^2 e_{12}e_{13} = 0. \quad (17.23)$$

As the different standard potentials are markedly different, we can assume that the first equivalence point potential is well below $E°_{13}$ and, thus, that the exponential e_{13} is quasi-null:

$$e_{13} \approx 0.$$

With this assumption, relation (17.23) gives

$$e_{11}e_{12} = 1 \text{ (first equivalence point).} \quad (17.24)$$

After the development of the exponentials e_{11} and e_{12}, we find

$$E_{1\mathrm{ep}} = (E^\circ{}_{11} + E^\circ{}_{12})/2 \quad (e_{13} \text{ negligible}).$$

Hence, with the preceding hypothesis, the potential value at the first equivalence point is given by the relation

$$E_{1\mathrm{ep}} = \left[E^\circ\left(V^{3+}/V^{2+}\right) + E^\circ\left(VO^{2+}/V^{3+}\right)\right]/2.$$

We can notice that the fact that relation (17.24) is satisfied implicates the following equality:

$$[V^{2+}] = [VO^{2+}].$$

To verify the accuracy of this equality, it is sufficient to write the equilibrium solution potential with two Nernst relations, one involving the couple V^{3+}/V^{2+} and the other the couple VO^{2+}/V^{3+}.

In terms of the titration curve before the first equivalence point, its general Eq. (17.22) may be simplified by setting $e_{13} = 0$ and $e_2 = 0$ (no titrant in the titration vessel). We find

$$\varphi = [e_{11}(1 + 2e_{12})]/[1 + e_{11}(1 + 2e_{12})] \quad \text{(equation curve before the first e. p.)} \tag{17.25}$$

This last relation, which already results from several simplifications, may be further simplified. Indeed, until the first equivalence point (not included), the exponential e_{12} may be considered as being negligible with respect to e_{11} because of the great difference between the standard potentials $E^\circ{}_{12}$ and $E^\circ{}_{11}$. Therefore, we find from (17.25) that

$$\varphi = e_{11}/(1 + e_{11}), \quad \varphi < 1. \tag{17.26}$$

Equation (17.26) is identical to Eq. (17.14), where the exponential $\exp[(n_2F/RT)(E - E^\circ{}_2)]$ is negligible with respect to unity.

From a chemical standpoint, the preceding considerations lead to the following considerations:

- before the first equivalence point, simply legitimately neglecting the exponentials e_{12}, e_{13}, and e_2 (once more because of the standard potential values) is equivalent to ascertaining that the only evolving reaction is

$$5V^{2+} + MnO_4{}^- + 8H^+ \rightarrow 5V^{3+} + Mn^{2+} + 4H_2O, \quad \varphi < 1.$$

The solution potential is imposed by the couple V^{3+}/V^{2+}. It is given by the relation

$$E = E^\circ\left(V^{3+}/V^{2+}\right) - (RT/F)\ln\left[\left(V^{2+}\right)/\left(V^{3+}\right)\right], \quad \varphi < 1,$$

or

$$E = E^\circ\left(V^{3+}/V^{2+}\right) - (RT/F)\ln\left[(1-\varphi)/\varphi\right], \quad \varphi < 1,$$

a relation identical to (17.26). At half of the first equivalence,

$$E = E^\circ(V^{3+}/V^{2+}) \quad \text{(half first equivalence);}$$

- at the first equivalence point,

$$[V^{2+}] = [VO^{2+}],$$

as it is involved in relation (17.24). Actually, it is more accurate to write

$$\left[\varepsilon V^{2+}\right] = \left[\varepsilon VO^{2+}\right].$$

These concentrations are indeed very weak since the titration reaction may be considered complete due to the standard potential values. At the first equivalence point, the solution in the titration vessel may be considered to be the solution of the ampholyte V^{3+}, which weakly disproportionates into V^{2+} and VO^{2+} according to

$$2V^{3+} + H_2O \rightleftharpoons VO^{2+} + V^{2+} + 2H^+.$$

It is interesting to notice that at the equivalence point, the potential value given by the relation

$$E_{1\text{ep}} = (E^\circ{}_{11} + E^\circ{}_{12})/2$$

is the same as that of the corresponding ampholyte solution with $n_{11} = n_{12}$ (see Chap. 16).

We must recall that all these results are derived from the initial hypothesis stipulating that the species V^{3+} exhibits its maximal concentration at the first equivalence point. We must now specify the condition in which the exponential e_{13} is negligible. Rearranging relation (17.23) (which results solely from the hypothesis that, at this point, the derivative vanishes) gives

$$e_{11} = e_{11}{}^2 e_{12}(1 + 2e_{13}).$$

As a result, the exponential e_{13} must be such that

$$e_{13} \ll \frac{1}{2}.$$

Equivalently, we must have

$$E^\circ{}_{13} - E_{1\text{ep}} \geq 0.19\ \text{V}.$$

This condition is widely satisfied in the chosen example. Actually, the values of the exponentials e_{11}, e_{12}, and e_{13}, calculated with the value $E_{1\text{ep}} = 0.041$ V calculated itself in the framework of the adopted hypothesis, are, respectively, $e_{11} = 100855$; $e_{12} = 9.915 \cdot 10^{-6}$; and $e_{13} = 6.147 \cdot 10^{-7}$. Hence, these calculations are autocoherent. (Let's also notice, incidentally, that according to these values, $e_{11}e_{12} = 1$.)

17.6.2 Second Equivalence Point

Lines of reasoning analogous to the preceding ones may be followed. A derivative calculation demonstrates that the concentration $[VO^{2+}]$ is maximum for $e_{12}e_{13} = 1$ provided that $e_{11} \gg e_{12}$. Introducing this value into the general Eq. (17.22) gives $\varphi = 2$. Equation (17.22) together with $\varphi = 2$ gives the solution equilibrium potential value at the second equivalence point:

$$E_{2ep} = (E^\circ_{12} + E^\circ_{13})/2.$$

This is the potential value of a solution of the ampholyte VO^{2+}. Between the first and second equivalence points, the general equation of the curve (17.22) can be reduced to

$$\varphi = 1 + 1/(1 + e_{12}), \quad 1 < \varphi < 2.$$

In this range the exponential e_{11} indeed exhibits a very high numerical value and the e_{13} and e_2 values remain low. (For $E_{2ep} = 0.67$ V, $e_{11} = 4.09 \cdot 10^{15}$; $e_{12} = 401608$; $e_{13} = 2.49 \cdot 10^{-6}$; and $e_{12}e_{13} \approx 1$.)

From the chemical standpoint, these results mean that the sole reaction evolving is

$$5V^{3+} + MnO_4^- + H_2O \rightarrow 5VO^{2+} + Mn^{2+} + 2H^+, \quad 1 < \varphi < 2.$$

For $\varphi = 1.5$, we find $E = E^\circ_{12}$.

17.6.3 Third Equivalence Point

In this case, the exponentials e_{11} and e_{12} exhibit very high numerical values, and before the equivalence point, e_2 remains negligible. Therefore, the general equation becomes

$$\varphi = 2 + e_{13}/(1 + e_{13}), \quad 2 < \varphi < 3.$$

This relation indicates that the sole reaction to evolve is

$$5VO^{2+} + MnO_4^- + H_2O \rightarrow 5VO_2^+ + Mn^{2+} + 2H^+.$$

In particular, for $\varphi = 2.5$, $E = E^\circ_{13}$. At the third equivalence point ($\varphi = 3$), the general equation becomes

$$3 = (2 + 3e_{13})(1 + e_2)/(1 + e_{13}),$$

which results from simplifications legitimated by the numerical values of the exponentials. After calculations, we find

$$E_{3ep} = 1.42 \text{ V} \quad (\varphi = 3).$$

Fig. 17.5 Titration curve of a 0.1 mol/L solution of V^{2+} (50 ml) by MnO_4^- (0.1 mol/L) (pH = 0)

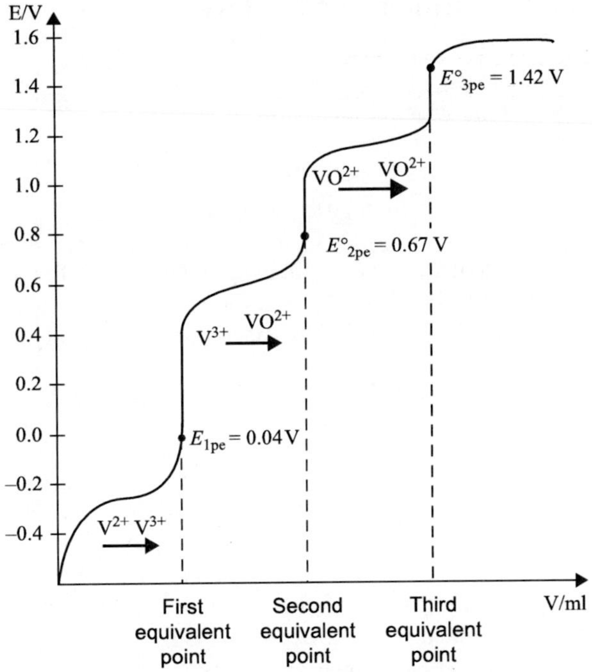

After the equivalence point, the general equation may be further simplified since the exponential e_{13} exhibits a very high numerical value. We find

$$\varphi = 3(1 + e_2) \quad (\varphi > 3).$$

The solution potential value is imposed by the couple MnO_4^-/Mn^{2+}. This last equation is equivalent to the relation

$$E = E^{\circ\prime}(MnO_4^-/Mn^{2+}) + (RT/5F)\ln[(\varphi - 3)/3].$$

Figure 17.5 shows the solution potential change during the titration of the hypovanadous ion with a potassium permanganate solution.

Taking all of these considerations together, we can deduce that the possibility of the satisfactory sequential titration of the different functions of a polyfunctional compound is, above all, based on the values of the standard potential differences: $E^\circ_{12} - E^\circ_{11}$, $E^\circ_{13} - E^\circ_{12}$. These differences permit us to neglect some exponentials according to the stages of the titration. In the case of the titration of multifunctional redox derivatives, it experimentally turns out that a difference amounting to 0.24 V is sufficient to detect the successive functions. The difference $\Delta E^\circ = 0.24$ V has already been encountered. It was during the study of the titration of a monofunctional redox compound, but the difference concerned the standard potentials of the titrand and the titrant, and not only that of the titrand as here.

Of course, the standard potential value of the titrant E°_2 must be sufficiently high. It governs the e_2 value.

17.7 Titration of a Mixture

Let's consider the titration of the two species Red_{11} and Red_{12} by the compound Ox_2. The two titration reactions are

$$n_2Red_{11} + n_{11}Ox_2 \rightarrow n_2Ox_{11} + n_{11}Red_2,$$

$$n_2Red_{12} + n_{12}Ox_2 \rightarrow n_2Ox_{12} + n_{12}Red_2.$$

By adopting the same symbolism as that used in the preceding chapter (Sect. 16.8) but by replacing C_{11}, C_{12}, and C_2 with X_1, X_2, and Y in order to take the dilution into account, we find for the general equation of the titration curve,

$$Y = [n_{11}X_1/n_2(1 + e_{11}) + n_{12}X_2/n_2(1 + e_{12})](1 + e_2), \tag{17.27}$$

where $X_1 = C_{11}V_o/(V_o + V)$, $X_2 = C_{12}V_o/(V_o + V)$, $Y = C_2V/(V_o + V)$,

$$e_{11} = \exp[-(n_{11}F/RT)(E - E°_{11})]; \quad e_{12} = \exp[-(n_{12}F/RT)(E - E°_{12})],$$

$$e_2 = \exp[(n_2F/RT)(E - E°_2)].$$

Equation (17.27) enables the calculation of the solution equilibrium potential as a function of Y through the exponentials e_{11}, e_{12}, and e_2. It is difficult to handle, but it can be simplified in some conditions.

The principal goal of this study is to find the conditions that must prevail for the sequential titrations of the species Red_{11} and Red_{12} to be satisfactory. As we shall see, it will be possible if, in some parts of the titration curve, some exponentials become negligible.

For the first species Red_{11} to be solely titrated, the following inequality:

$$n_{11}X_1/n_2(1 + e_{11}) >> n_{12}X_2/n_2(1 + e_{12}) \tag{17.28}$$

must be imperatively satisfied in Eq. (17.27). When this is the case, Eq. (27) becomes

$$Y = (n_{11}/n_2)[X_1/(1 + e_{11})](1 + e_2) \quad \text{(before the first equivalence point).}$$

Moreover, there is, of course, no excess of titrant, and the solution potential is, by far, lower than $E°_2$. (Recall that $E°_2$, the standard potential of the titrant couple, must be markedly higher than the standard potentials $E°_{11}$ and $E°_{12}$ in order to achieve a satisfactory titration.) The exponential e_2 is negligible with respect to unity; Eq. (17.27) may be further simplified and becomes

$$Y = (n_{11}/n_2)[X_1/(1 + e_{11})] \quad \text{(before the first equivalence point).} \tag{17.29}$$

This equation is purely and simply that describing the titration of only one reduced form before the equivalence point with $\varphi_1 = n_2Y/n_{11}X_1$ (see Sect. 17.1; φ_1 is the fraction titrated of species 1). In the case in which (17.28) is satisfied, the solution equilibrium potential is imposed by the couple Ox_{11}/Red_{11}. It is given by relations (17.30) and (17.31):

$$E = E°_{11} - (RT/n_{11}F)\ln[(Red_{11})/(Ox_{11})] \quad \text{(before the first equivalence point)} \tag{17.30}$$

or by introducing the fraction titrated φ_1 of the species Red_{11}:

$$\varphi_1 = n_2Y/(n_{11}X_1),$$

$$E = E°_{11} - (RT/n_{11}F)\ln[(1-\varphi_1)/\varphi_1], \quad 0 < \varphi_1 < 1. \tag{17.31}$$

Continuing with the assumption that relation (17.28) is prevailing, we can intuitively assert that at the first equivalence point, the term X_2 can no longer be negligible (as it was before), whereas the exponential e_2 remains negligible for the same reason as before. Therefore, we can write relation (17.27) as

$$(n_2Y/n_{11}X_1)(1+e_{11}) = 1 + (n_{12}X_2/n_{11}X_1)[(1+e_{11})/(1+e_{12})].$$

But at the first equivalence point, according to our hypothesis concerning the relative magnitudes of the terms X_1 and X_2:

$$n_2Y/n_{11}X_1 \approx 1,$$

we find

$$e_{11}(1+e_{12})/(1+e_{11}) = n_{12}X_2/n_{11}X_1, \quad \varphi_1 = 1.$$

If $E°_{12}$ is markedly higher than $E°_{11}$ (this is the essential condition to achieve a satisfactory sequential titration—see below), the following two inequalities are simultaneously satisfied:

$$e_{11} \ll 1 \quad \text{and} \quad e_{12} \gg 1 \quad (\varphi_1 = 1).$$

Hence, at the first equivalence point, after rearranging, we find

$$E_{1ep} = (n_{11}E°_{11} + n_{12}E°_{12})/(n_{11}+n_{12}) - [RT/F(n_1+n_2)]\ln(n_2X_2/n_{11}X_1). \tag{17.32}$$

The part of the curve describing the titration of the second species Red_{12} before the second equivalence point (e_2 remaining negligible) is, according to (17.27) (since e_{11} is negligible),

$$Y - n_{11}X_1/n_2 = n_{12}X_2/[n_2(1+e_{12})]$$

$$\text{(titration of the second species before the equivalence point)} \tag{17.33}$$

or $\varphi_2 = n_{12}/n_2(1+e_{12})$,

with, by definition,

$$\varphi_2 = [Y - n_{11}X_1/n_2]/X_2.$$

φ_2 is the fraction titrated of the second species Red_{12} defined from the first equivalence point. Equation (17.33) describes the titration of the species Red_{12} by the species Ox_2 after we hypothesized that Red_{12} is alone before the second equivalence point. The solution equilibrium potential is imposed by the couple Ox_{12}/Red_{12}. The quasi-sole reaction evolving is

$$n_2Red_{12} + n_{11}Ox_2 \rightarrow n_2Ox_{12} + n_{11}Red_2 \quad (0 < \varphi_2 < 1)$$

and $E = E°_{12} - (RT/n_{12}F)\ln[(1 - \varphi_2)/\varphi_2] \qquad E°_2 > E°_{12} \gg E°_{11}$.

The second equivalence point potential is given by the relation

$$E_{2ep} = (n_{12}E°_{12} + n_2E°_2)/(n_{12} + n_2), \quad \varphi_2 = 1,$$

since, by hypothesis, the first species does not interfere and since, from the second equivalence point, the exponential e_2 related to the titrant can no longer be considered negligible. After the second equivalence point, the titration curve equation is

$$\varphi_2 = n_{12}(1 + e_2)/n_2.$$

The solution potential is imposed by the couple Ox_2/Red_2.

All the preceding conclusions are dependent on the condition

$$n_{11}X_1/(1 + e_{11}) >> n_{12}X_2/(1 + e_{12})$$

This equation largely depends on the difference of standard (or formal) potentials $\Delta E° = E°_{12} - E°_{11}$, as we shall see, and, to a lesser extent, on the stoichiometric coefficients n_{11}, n_{12} together with the concentrations X_1 and X_2. In order to facilitate the study of the influence of the difference in standard potentials, let's set $X_1 = X_2$ and $n_{11} = n_{12}$. Quite evidently, these assumptions are plausible.

After studying Eq. (17.27), we can see that in order to get a satisfactory sequential titration, the exponential e_{11} must be negligible with respect to unity in the term containing the concentration X_1, whereas, simultaneously, e_{12} must be much higher than 1. Therefore, the term in X_2 is negligible. With these two conditions, $n_2Y \approx n_{11}X_1$ at the first equivalence point ($\varphi_1 = 1$). Let's set the following conditions, arbitrarily:

$$e_{11} \leq 10^{-2} \quad \text{and} \quad e_{12} \geq 10^2.$$

We find

$$\exp[-(n_{11}F/RT)(E - E°_{11})] << 10^{-2},$$

$$\exp[-(n_{12}F/RT)(E - E°_{11} - \Delta E°)] >> 10^2$$

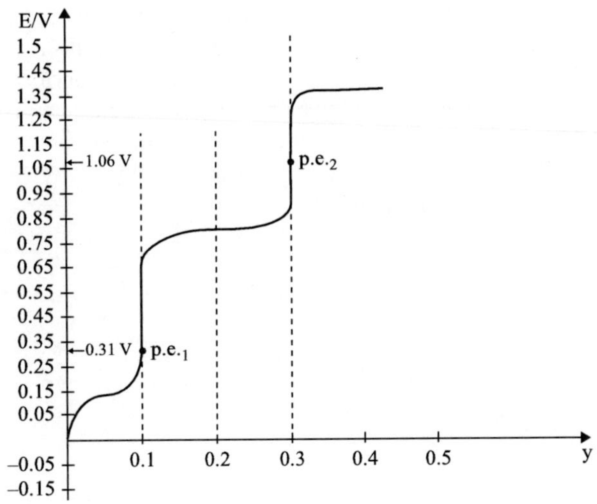

Fig. 17.6 Curve obtained during the titration of a mixture of stannous ion ($C_{11} = 5 \cdot 10^{-2}$ mol/L) and of ferrous ion ($C_{12} = 2 \cdot 10^{-1}$ mol/L) ($V_o = 50$ ml) with a solution of ceric ion (0.5 mol/L)

since, by hypothesis, $n_{12} = n_{11}$:

$$\exp[-(n_{11}F/RT)(E - E^\circ{}_{11})] \exp[(n_{11}F/RT)\Delta E^\circ] \gg 10^2,$$

$$\exp[(n_{11}F/RT)\Delta E^\circ] \geq 10^4,$$

$$\Delta E^\circ \geq 0.24 \text{ V} \quad (\text{with } n_{11} = n_{12} = 1).$$

This difference has already been encountered several times.

Figure 17.6 shows the curve obtained during the titration of a mixture of stannous and ferrous ions by ceric ions. The successive titration reactions are

$$Sn^{2+} + 2Ce^{IV} \rightarrow Sn^{4+} + Ce^{III},$$

$$Fe^{2+} + Ce^{IV} \rightarrow Fe^{3+} + Ce^{III},$$

From the standard potential values $E^\circ{}_{11}(Sn^{4+}/Sn^{2+}) = 0.14$ V, $E^\circ{}_{12}(Fe^{3+}/Fe^{2+}) = 0.68$ V, and $E^\circ{}_2(Ce^{IV})/(Ce^{III}) = 1.44$ V, the sequential titration is satisfactory.

In order to conclude this chapter, the same remark as that already given for acid–base titrations (see Chap. 7) can be made. Obtaining the titer of a solution of a reductor (or of an oxidant) with the help of a titration curve is based on determining the equivalence point. Thus, it is based on only one piece of information, that given by the location of the equivalence point. Obtaining this information imposes some constraints concerning the values of the standard or formal potentials in order to obtain satisfactory titrations. We have examined them. Microinformatics permits us to simultaneously obtain both analytical data (unknown concentrations) and physico-chemical data (redox potentials) from several points on one titration curve. The strategy to be followed consists of researching the best fit between the experimental curve and the calculated curve. The latter curve is calculated by giving judiciously chosen numerical values to the parameters that must be determined. According to this method, the definitively retained numerical values of the parameters under study are those that lead to the best fit. The interest in this methodology lies in the fact that the results are obtained from several pieces of experimental information, since all the titration curve points may be processed together. Its major drawback lies in the fact that it is inevitably based on the use of computer programs that must be specifically dedicated to the process under study. They are not necessarily too difficult to conceive or, subsequently, to write.

Chapter 18
Oxidoreductimetry: Direct and Indirect Iodometries

In this chapter, we begin our description of the principal titrimetric methods involving a redox reaction. Such methods are sometimes studied in the part of analytical chemistry named *oxidoreductimetry*. The first titrimetric methods we shall study are direct and indirect *iodometries*, two methods that are very closely related. They constitute an important part of the methods involving iodine and iodide ions as redox reagents.

18.1 Oxidoreductimetry

The word "oxidoreductimetry" concerns titrations based on the use of reducing or oxidizing reactants that are able to react quantitatively with mineral and organic substances. The word "oxidoreductimetry" is essentially encountered in the french literature.

Let us briefly recall that, as with any titration reaction, the redox reactions involved in oxidoreductimetry must be sufficiently quantitative (see Chaps. 16 and 17). They must also be sufficiently fast, and the equivalence points must be detected easily. The latter two conditions are required for practical reasons.

Frequently, it turns out that the species to determine must be in a suitable oxidation state in order to get a satisfactory titration. Therefore, most of the time, the titration follows a first step consisting of the transformation of the compound to determine the ad hoc state that may make the subsequent redox reaction satisfactory.

As with other kinds of titrations, the notion of a normal solution is still used in everyday practice, although it is not recommended by the IUPAC. Normal oxidizing or reducing solutions may capture or give an electron-gram per liter of solution. The redox equivalent is the molar mass (formerly called the gram molecular weight) of the species divided by the number of electrons exchanged during the titration reaction. Recall that the advantage of the normalities system lies in the fact that the titrant and titrand solutions exhibit similar volumes at the equivalence point for the same normality. In other words, with the system of normalities, the titration reaction of the analyte is complete when identical volumes of sample and titrant solutions

J.-L. Burgot, *Ionic Equilibria in Analytical Chemistry*,
DOI 10.1007/978-1-4419-8382-4_18, © Springer Science+Business Media, LLC 2012

Table 18.1 Some names of redox titration methods involving the iodine element

Iodometry (U.S.)	: *direct iodometric titrations*	*Iodimetry* (IUPAC)
Iodimetry (U.S.)	: *indirect iodometric titrations*	*iodometrie* (Germany)
Iodometry in an alkaline medium		
Iodatometry		
Periodimetry		

have been brought together. Its drawback lies in the fact that it is dependent on the number of electrons exchanged during the titration reaction. Hence, it depends on the reaction itself. The inverse of the number of electrons exchanged is called the equivalence factor f_{eq}. For example, for the oxidation of iodide ions with iodate ions in acidic medium:

$$IO_3^- + 5I^- + 6H^+ \rightarrow 3I_2 + 3H_2O,$$

the potassium iodate equivalent is $M(KIO_3)/5$, that is, $f_{eq} = 1/5$. For the oxidation reaction of hydrazine with potassium iodate in a strongly hydrochloric acid medium (see Chap. 19) according to the reaction:

$$IO_3^- + NH_2NH_2 + 2H^+ + 2Cl^- \rightarrow ICl_2^- + 3H_2O + N_2\uparrow$$

the potassium iodate equivalent this time is $M(KIO_3)/4$; that is, $f_{eq} = 1/4$. Therefore, the same chemical derivative may exhibit several different equivalents according to the titration reactions in which it participates.

18.2 Nomenclature of the Titration Methods Involving the Use of Iodine or the Formation of Iodine

According to the IUPAC, the term *iodimetry* refers to titrations with a standard solution of iodine or of the tri-iodide ion I_3^- (see below) and also deals with the titration of iodine liberated in chemical reactions.

However, some ambiguities exist concerning this nomenclature. In Germany, the term "iodimetry" endowed with the IUPAC meaning is called *iodometry*. In France, both terms are endowed with a more restrained meaning. *Iodimetry* solely involves the titration of the formed iodine after reaction of the compound to determinate with iodide ions, whereas the term "iodometry" solely concerns the titrations with solutions of iodine or tri-iodide ions. Moreover, in Great Britain and the United States, the meanings of "iodometry" and "iodimetry" are exactly the inverse of their meanings in the french literature. Furthermore, some authors now use the terms "direct iodometric titrations" and "indirect iodometric titrations" in lieu of the french "iodometry" and "iodimetry."

Table 18.1 summarizes the titrimetric methods involving the iodine element studied in this book. In addition to direct and indirect iodometries, other interesting methods exist from the standpoint of analytical chemistry. These include iodometry in alkaline medium, iodatometry, and periodimetry.

This classification is far from being perfect because of the fact that the methods listed do not exclude the others. For example, a periodimetric titration ends with an indirect iodometry. It is the same case for iodometry in alkaline medium.

18.3 Some Physicochemical Properties of Iodine

Iodine is weakly soluble in water (1.18×10^{-3} mol/L at 25°C). Its solubility is increased by the formation of the tri-iodide ion I_3^-, which may be considered a complex (sometimes also called the periodide ion) according to the equilibrium

$$I_2 + I^- \rightleftharpoons I_3^-, \quad (K^\circ = 710 \text{ at } 25°\text{C}).$$

Aqueous solutions of I_3^- have an intense yellow to brown color. The vapor pressure of iodine is important even when it is in solution. Over time, aqueous solutions of iodine exhibit a spontaneous decrease in their concentrations by vaporization. Iodine is freely soluble in organic solvents. As a practical consequence of these properties, tri-iodide ions are used in direct iodometric titrations rather than aqueous solutions of I_2. Due to the iodine/tri-iodide equilibrium, whose formation constant is weak, I_3^- solutions contain more and less free iodine I_2, which is more and less liberated as the titration reaction evolves further. In a first approximation, tri-iodide ion solutions behave like iodine solutions. The formation of this ion complex has two interests. On the one hand, it (apparently) increases the solubility of iodine, permitting the attainment of I_2 concentrations that allow practical titrations. On the other hand, the fact that iodine is in large part under the form of the tri-iodide complex rather than under the free form somewhat precludes its vaporization. However, the latter fact remains nonnegligible.

18.4 Predominance Areas of Some Species of Iodine

Figure 18.1 shows the diagram *E*/pH of some species containing the iodine element. It only describes the species that are in true equilibrium. This is the reason why iodine at oxidation numbers +I and +VII is not mentioned in this diagram (see iodometry in alkaline medium and periodimetry).

We can make the following comments about Fig. 18.1:

- at oxidation state –I, iodide ions could be thought of as being possibly protonated in iodhydric acid. Actually, iodhydric acid plays no part in the redox equilibria since it is a strong acid. The only acid—base couple included in the diagram is that of iodic acid HIO_3, which may be considered as being completely dissociated in the whole pH range, due to its *pKa* value: $pKa(HIO_3/IO_3^-) = 0.80$;

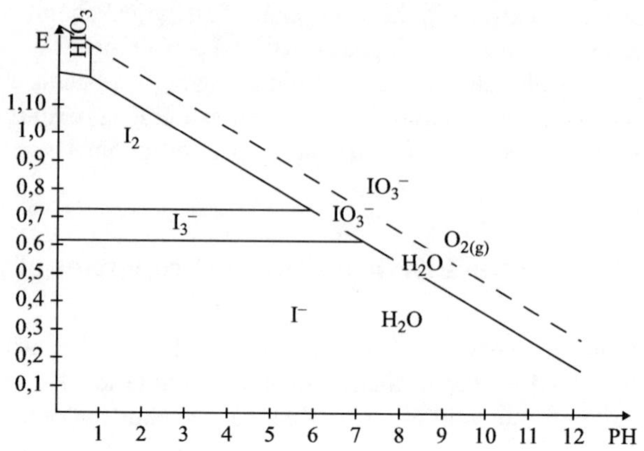

Fig. 18.1 Diagram *E*/pH of some species derived from iodine (species in true equilibrium) (second convention: equipartition of iodine atoms on frontiers, total concentration in iodine atoms: 10^{-2} mol/L)

- because iodide ions exhibit no basic character, the pH value is devoid of any influence on the oxidizing property of a solution containing the couple I_2/I^-. The half-redox equilibrium

$$I_2 + 2e^- \rightleftharpoons 2I^-$$

cannot be perturbed by the reaction

$$I^- + H^+ \rightarrow HI,$$

which does not exist in water. Indeed, Nernst's relation applied to the redox equilibrium does not contain the proton activity (H^+). However, the diagram shows that at pH > about 6.5,[1] the equilibrium between tri-iodide ions and iodine tends to disappear, and iodide and iodate ions tend to exist solely. This means that beyond this pH value, iodine disproportionates into iodide and iodate (−I and +V oxidation states) ions according to the reaction

$$3I_2 + 6OH^- \rightleftharpoons IO_3^- + 5I^- + 3H_2O \quad pH > 6.5.$$

Actually, when we alkalinize an iodine solution by adding a strong base, iodide and iodate ions are not formed the first time as we would expect. The species initially formed is, for kinetic reasons, iodine at the oxidation number +I (hypoiodous acid HIO and hypoiodites IO^-), which is formed by disproportionation (see Chap. 19);
- Figure 18.1 also shows that, regardless of the pH value, the predominance areas of dioxygen and iodide ions are disconnected. Therefore, iodide solutions are sensitive to air. The following reaction occurs:

$$4I^- + O_2 + 4H^+ \rightarrow 2I_2 + 2H_2O.$$

Iodide ions are transformed into iodine.

[1] We must not pay too much attention to this pH value since the diagrams are built on some conventional values (see the legend for Fig. 18.1).

18.5 Interesting Features Exhibited by the Couple I_2/I^- for Its Use in Titrimetry

18.5.1 Stability

Standardized solutions of I_3^- ions are stable, but they lose iodine by vaporization. Hence, they cannot be used as primary standard solutions. In the presence of 1.2×10^{-1} mole of iodide ions, 5×10^{-2} mole of iodine can be dissolved in one liter of water.

18.5.2 Coloration

Aqueous solutions of tri-iodide ion I_3^- exhibit a very sensitive yellow to brown color that is detectable by the human eye down to a concentration as low as 5×10^{-5} mol/L. The solution of I_3^- may be its own indicator. During a direct iodometric titration, the equivalence point may be detected by the appearance of a persistent yellow color. Another way to detect the equivalence point is to use starch as an indicator. Starch reacts with iodine in the presence of iodide ions to form a blue-colored complex. The coloration is intense. It is still perceptible for a concentration as low as 10^{-5} mol/L. The color sensitivity decreases with increasing temperature.

The intensely blue color is due to a charge-transfer complex formed by the tri-iodide ion and starch. It results from the inclusion of tri-iodide ions into the straight-chain fractions of α-amylose of starch, which form a helix. The formed clathrate exhibits a narrow band of charge transfer near 620 nm (Fig. 18.2).

18.5.3 Solubilities

Iodine dissolves freely into organic solvents that are immiscible or poorly miscible with water, such as chloroform and carbon tetrachloride, with the formation of a violet color. This property is sometimes used in chemical analysis.

18.5.4 Standard Potential Values of I_2/I^- Couples

The standard potentials of couples intervening in iodometric titrations are

$$I_{2(w)} + 2e^- \rightleftharpoons 2I^-, \quad E^\circ(I_2/I^-) = 0.615\,V,$$
$$I_{2(s)} + 2e^- \rightleftharpoons 2I^-, \quad E^\circ(I_{2(s)}/I^-) = 0.530\,V,$$
$$I_3^- + 2e^- \rightleftharpoons 3I^-, \quad E^\circ(I_3^-/I^-) = 0.536\,V.$$

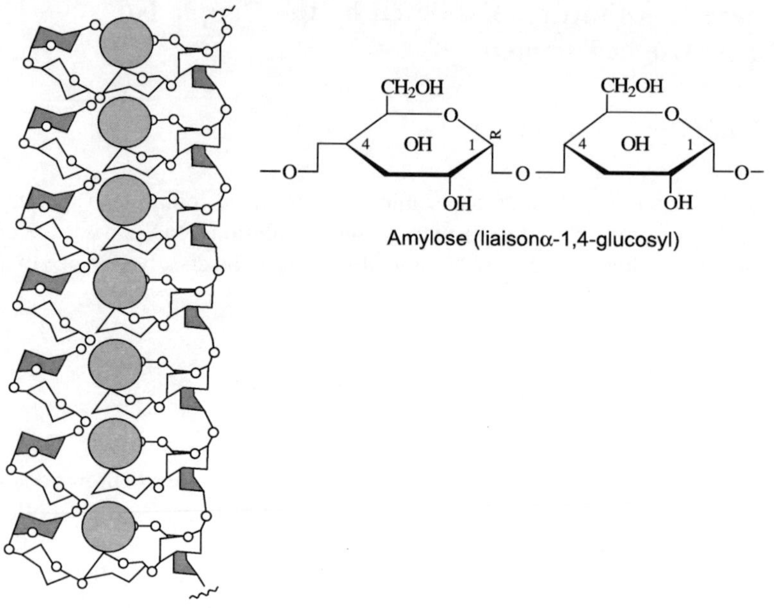

Fig. 18.2 Clathrate I_3^-—starch

The tri-iodide ion formation slightly decreases the apparent oxidizing power of iodine. From a practical standpoint, the value $E° = 0.536$ V must be principally retained since, most of the time, the used couple is I_3^-/I^- for the sake of solubility. Going from the standard potential of one of these three couples to the two others is achieved according to the principle given in Chap. 2, that is, by handling the corresponding standard free enthalpies.

The values of these potentials are located in the middle of the standard potentials scale. As a result,

- iodine and tri-iodide ions are weakly oxidizing agents. For example, they are less oxidizing than other halogens, permanganate and dichromate ions, the latter being considered the weakest of the strong oxidants. However, they oxidize stannous tin, arsenic, and antimony (at oxidation state +III), sulfurous acid and sulfite ions, sulfides and hydrogen sulfides, thiosulfate ions, etc.;
- conversely, iodide ions are rather strong reductants. For example, they reduce dioxygen, other halogens, iodate and bromate ions, hydrogen peroxide, and permanganate and dichromate ions, with the formation of iodine and tri-iodide ions when some iodide ions are present in the medium. Iodide ions are stable in aqueous solutions provided they do not contain oxidants and that they are exempt of dioxygen.

18.5.5 The Influence of pH

As iodhydric acid is a strong acid, the pH value is of no influence on the couple I_2/I^-.

Most redox systems exhibit an apparent standard potential that decreases when the pH increases (for example, this is the case of the couples MnO_4^-/Mn^{2+} and $Cr_2O_7^{2-}/Cr^3$; see Chap. 20). By comparison with these couples, iodine becomes a strong oxidant in slightly acidic and weakly alkaline media. We must recall that in alkaline media (about pH = 9), iodine disproportionates into hypoiodous acid (or into hypoiodites) and into iodide ions, these species being in false equilibrium. Moreover, the second time, iodine disproportionates into iodate and iodide ions that are in true equilibrium. Hence, the potential of the couple I_2/I^- is independent of pH in the approximate range $0 \le pH \le 9$.

18.5.6 Existence of the Fundamental Reaction of Direct and Indirect Iodometries

The existence of the fundamental reaction of direct and indirect iodometries permits us to determine an excess of iodine or iodine liberated from iodide ions. This reaction is very useful (see Sect. 18.6).

18.5.7 Detection of the Equivalence Point

The usefulness of the fundamental reaction of direct and indirect iodometries has its roots, at least in part, in the fact that it is accompanied by the disappearance or appearance of the yellow-brown color in the solution due to tri-iodide ions. Tri-iodide ions are their proper indicator. Hence, its equivalence point detection is particularly easy. In some difficult cases, starch may be used. The partitioning of iodine into an organic phase at the equivalence point may also be used. Finally, some internal indicators of intermediary standard potential values such as variamine blue ($E° \approx 0.60\,V$) may also be used. The difficult cases are those in which the tri-iodide coloration is masked by that of the solution.

18.6 The Fundamental Reaction of Iodometries

The fundamental reaction of iodometries consists of the action of iodine or tri-iodide ions on thiosulfate ions $S_2O_3^{2-}$ giving tetrathionate ions $S_4O_6^{2-}$ according to the scheme

$$I_2 + 2S_2O_3^{2-} \rightleftharpoons 2I^- + S_4O_6^{2-}.$$

Thiosulfate and tetrathionate ions are colorless. The half-equilibrium $S_4O_6^{2-}/S_2O_3^{2-}$ obeys the equation

$$S_4O_6^{2-} + 2e^- \rightleftharpoons 2S_2O_3^{2-}, \quad E°(S_4O_6^{2-}/S^2O_3^{2-}) = 0.17 \text{ V}.$$

Thiosulfate ions are a good reducing species. A normal solution of thiosulfate contains one mole per liter since the half-redox reaction involving it exchanges one electron for one $S_2O_3^{2-}$ molecule. Let's recall that in thiosulfate ions, one of the sulfur atoms exhibits the +IV oxidation state, while the other sulfur exhibits the null value. From another standpoint, we can consider that each sulfur atom exhibits the value +II. Likewise, in tetrathionate ions, we can consider that, on average, the oxidation state of each sulfur atom is 2.5. If we consider the structure of the tetrathionate ion:

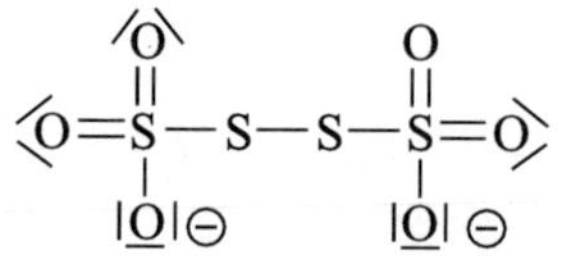

we may find that it is more judicious to consider the two sulfur atoms of the disulfide bridge at oxidation state –I and the sulfur of the sulfonate groups at oxidation state +VI. For the couple iodine/iodide, the redox equivalent of iodine is half its molar weight.

The iodine/thiosulfate reaction can be used only in a slightly acidic medium, that is, in the approximate range $2 < \text{pH} < 5$. In a strongly acidic medium ($\text{pH} < 2$), thiosulfuric acid, which forms in these conditions, decomposes into sulfurous acid and sulfur according to the reactions

$$S_2O_3^{2-} + 2H^+ \rightarrow H_2S_2O_3$$

and

$$H_2S_2O_3 \rightarrow H_2SO_3 + S\downarrow$$

We must recall that one mole of sulfurous acid reacts with one mole of iodine according to the equation

$$H_2SO_3 + I_2 + H_2O \rightarrow SO_4^{2-} + 4H^+ + 2I^-$$

to give one mole of sulfate ions, whereas half a mole of iodine reacts with one mole of thiosulfate. The stoichiometry of the reaction to perform and that of the parasitic one are not the same. (However, a strongly acidic iodine solution can be titrated normally if the thiosulfate solution is added slowly into it under vigorous stirring. The inverse titration must be achieved imperatively in the pH conditions given above.) From another standpoint, for $\text{pH} > 8$, hypoiodous acid or hypoiodite ions more or less oxidize thiosulfate ions into sulfate ions:

$$4IO^- + S_2O_3^{2-} + 2OH^- \rightarrow 4I^- + 2SO_4^{2-} + H_2O.$$

In these conditions, the half-redox reaction involving thiosulfate ions is

$$2SO_4^{2-} + 8e^- + 5H_2O \rightleftharpoons S_2O_3^{2-} + 10OH^-, \quad E° = 1.09 \text{ V}$$

Quite evidently, we are no longer in the field of the fundamental reaction of iodometries.

The tri-iodide/thiosulfate reaction is complex. In a first reaction, the intermediary colored species $S_2O_3I^-$ is formed according to the fast and reversible reaction

$$S_2O_3^{2-} + I_2 \rightleftharpoons S_2O_3I^- + I^-.$$

In a second reaction, a second molecule of thiosulfate reacts with the intermediary species

$$S_2O_3^{2-} + S_2O_3I^- \rightarrow S_4O_6^{2-} + I^-.$$

The latter one may also react with iodide ions to give tetrathionate and tri-iodide ions:

$$2S_2O_3I^- + I^- \rightarrow S_4O_6^{2-} + I_3^-.$$

This may be the explanation of the fact that tri-iodide ions appear again about the equivalence point of the iodine/thiosulfate reaction.

18.7 Iodine Solutions

Iodine is not a standard because of its vapor pressure, which is nonnegligible even in solution. Standardized solutions of iodine (containing tri-iodide ions also) are prepared approximately at the desired concentration after weighing iodine and potassium iodide. They are standardized after. The most common method of standardization is to use arsenious oxide, which is a good primary standard. In water, it gives arsenious acid:

$$As_2O_3 + 3H_2O \rightleftharpoons 2H_3AsO_3.$$ [2]

Arsenious acid is a weak monoacid:

$$H_3AsO_3 \rightleftharpoons H^+ + H_2AsO_3^-, \quad pK_a\left(H_3AsO_3/H_2AsO_3^-\right) = 9.22.$$

In arsenious oxide and acid, arsenic is at oxidation number +III. Arsenious acid and arsenic acid H_3AsO_4 (As +V) form a half-redox couple that involve an exchange of two electrons. Arsenic acid is a triacid according to the three equilibria

$$H_3AsO_4 \rightleftharpoons H^+ + H_2AsO_4^-, \quad pK_a\left(H_3AsO_4/H_2AsO_4^-\right) = 2.2,$$

$$H_2AsO_4^- \rightleftharpoons H^+ + HAsO_4^{2-}, \quad pK_a\left(H_2AsO_4^-/HAsO_4^{2-}\right) = 6.9,$$

$$HAsO_4^{2-} \rightleftharpoons H^+ + AsO_4^{3-}, \quad pK_a\left(HAsO_4^{2-}/AsO_4^{3-}\right) = 11.5.$$

[2] It is probable that the structure of the acid is $As(OH)_3$. It does not seem to have been isolated in a pure state.

Fig. 18.3 Standard potential of the couple I_3^-/I^- and apparent standard potentials of the couple As + V/ As+III in relation to the pH value

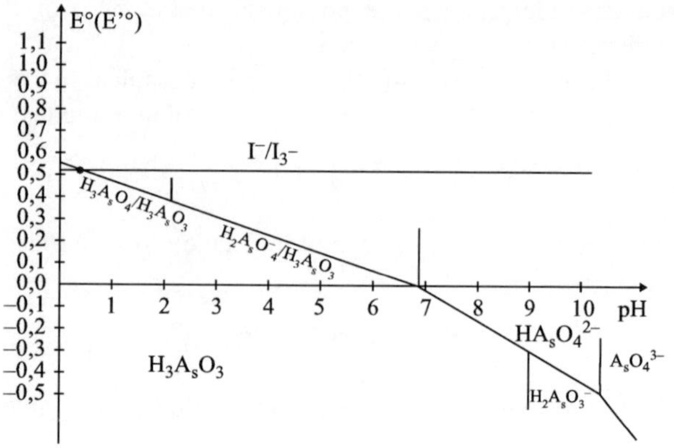

According to the solution's pH, the redox couples As +V/As +III are

$$H_3AsO_4 + 2H^+ + 2e^- \rightleftharpoons H_3AsO_3 + H_2O, \quad E^\circ = 0.56\ \text{V}, \quad 0 < \text{pH} < 2.2,$$

$$H_2AsO_4^- + 3H^+ + 2e^- \rightleftharpoons H_3AsO_3 + H_2O, \quad E^\circ = 0.63\ \text{V}, \quad 2.2 < \text{pH} < 6.9,$$

$$HAsO_4^{2-} + 4H^+ + 2e^- \rightleftharpoons H_3AsO_3 + H_2O, \quad E^\circ = 0.83\ \text{V}, \quad 6.9 < \text{pH} < 9.2,$$

$$HAsO_4^{2-} + 3H^+ + 2e^- \rightleftharpoons H_2AsO_3^- + H_2O, \quad E^\circ = 0.56\ \text{V}, \quad 9.2 < \text{pH} < 11.5,$$

$$AsO_4^{3-} + 4H^+ + 2e^- \rightleftharpoons H_2AsO_3^- + H_2O, \quad E^\circ = 0.90\ \text{V}, \quad 11.5 < \text{pH}.$$

At pH < 9, the limiting value beyond which the system I_2/I^- no longer remains the same (see before), the standardization reactions depend on the pH value:

$$I_2 + H_3AsO_3 + H_2O \rightleftharpoons 2I^- + H_3AsO_4 + 2H^+, \quad \text{pH} < 2.2,$$

$$I_2 + H_3AsO_3 + H_2O \rightleftharpoons 2I^- + H_2AsO_4^- + 3H^+, \quad 2.2 < \text{pH} < 6.9,$$

$$I_2 + H_3AsO_3 + 4OH^- \rightleftharpoons 2I^- + HAsO_4^{2-} + 3H_2O, \quad 6.9 < \text{pH} < 9.0.$$

We must emphasize the fact that these three reactions are equilibrated. Actually, they can evolve either forward or backward depending on the pH value. The simple superimposition of the standard and normal redox couples I_2/I^- and As +V/As +III immediately shows this point for the first one (Fig. 18.3).

As a result, the pH must be sufficiently high in order for the standardization reaction to be sufficiently displaced toward the right. Moreover, it appears that the higher the pH value is, the more displaced toward the right the redox reaction gets. Additionally, when the pH increases, the difference in standard and normal potentials $[E^\circ(I_3^-/I^-) - E^{\circ\prime}(\text{As} + \text{V/As} + \text{III})]$ also increases. From a practical standpoint, the chosen pH range is 4.0 < pH < 9.0. In these conditions, the standardization reactions are the last two mentioned above. According to some authors, the best value seems to be pH = 6.5. The reaction is achieved in a buffered solution. The most commonly used buffers are sodium dicarbonate, sodium tetraborate with boric acid, or the couple dihydrogenophosphate/monohydrogenophosphate.

Remark It is always possible to standardize an iodine solution with a thiosulfate solution provided the latter has recently been standardized (see Sect. 18.8).

18.8 Thiosulfate Solutions

Thiosulfate solutions must be standardized before their use. Sodium thiosulfate $Na_2S_2O_3,\ 5H_2O$ is readily obtainable in a state of high purity. It is efflorescent and its water content is always uncertain. Therefore, it is unsuitable as a primary standard. Its aqueous solutions, prepared with ordinary distillated water, are not stable. Under the influence of carbon dioxide, hydrogen sulfite ions and sulfur form according to the reaction

$$S_2O_3^{2-} + H^+ \rightarrow HSO_3^- + S\downarrow$$

Finally, sodium thiosulfate may be decomposed by microbiological action under the influence of *thiobacillus thioparus*.

As iodine and tri-iodide ions solutions are not primary standard solutions, the standardization of thiosulfate solutions may be achieved with the help of iodine solutions extemporaneously prepared from primary standards. Iodine is usually prepared by the iodate/iodide reaction or by the oxidation of iodides with potassium dichromate, but other possibilities exist.

Potassium iodate can be obtained in a state of at least 99.9% purity. It reacts with potassium iodide in acidic solution to liberate iodine:

$$IO_3^- + 5I^- + 6H^+ \rightarrow 3I_2 + 3H_2O.$$

According to the diagram in Fig. 18.1, this is an amphoterization reaction. It decomposes into the following two half-redox reactions:

$$IO_3^- + 6H^+ + 5e^- \rightleftharpoons \tfrac{1}{2}I_2 + 3H_2O,$$

$$I^- \rightleftharpoons \tfrac{1}{2}I_2 + 1e^-.$$

During this reaction, the oxidation state of iodate ions goes from +V to 0. It is a process sometimes called a *five-electron iodatometry* (see the next chapter). In order to standardize thiosulfate solutions and since potassium iodate is a primary standard, its concentration must be the limiting factor of the reaction. This means that there must be an excess of iodide ions and of protons in comparison to iodate ions, when the reaction stoichiometry is taken into account. In other words, for exactly one mole of potassium iodate weighed, we must add more than five moles of iodide ions and more than six moles of protons. It is not necessary to know their exact numbers provided they obey the above conditions. In these conditions, exactly three moles of iodine are prepared from one mole of iodate. Figure 18.4 summarizes these considerations.

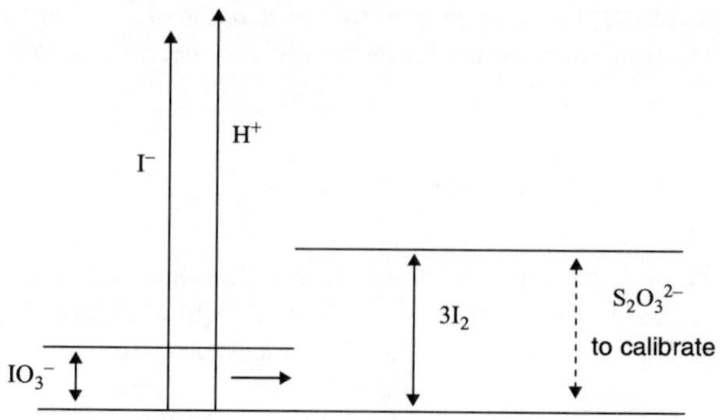

Fig. 18.4 Standardization of a thiosulfate solution by a solution of iodine prepared by the iodate/iodide reaction. (Notice the symbolism: A vertical segment limited by two horizontal lines represents an exactly weighed quantity.)

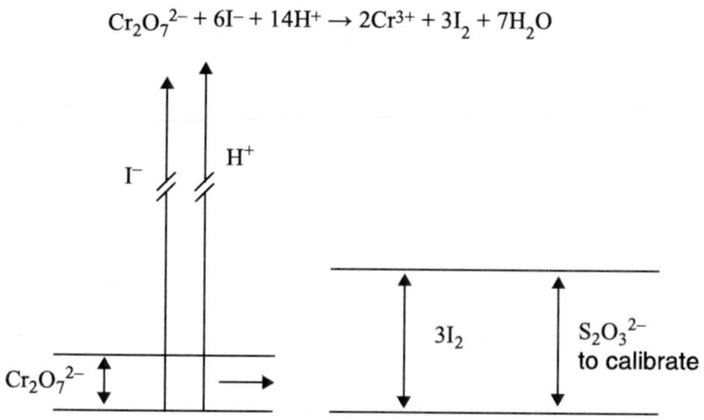

Fig. 18.5 Standardization of a thiosulfate solution by an iodine solution prepared according to the $Cr_2O_7^{2-}/I^-$ reaction (limiting factor: dichromate concentration)

Remark The limiting factor may also be either the medium's acidity or the iodide ions' concentration. In the first case, it is sufficient to add an excess of potassium iodate and of potassium iodide. This process is a means of standardizing an acidic solution by titrating the iodine liberated under its influence. Iodine is titrated by a solution of As +III or of thiosulfate.

Iodine permitting the standardization of thiosulfate solutions can also be extemporaneously prepared by the oxidation of iodides with potassium dichromate in acidic medium:

$$Cr_2O_7^{2-} + 6I^- + 14H^+ \rightarrow 2Cr^{3+} + 3I_2 + 7H_2O.$$

Potassium dichromate is a primary standard. Of course, its concentration must be the limiting factor. Moreover, we must add more than 6 moles of iodides and 14 moles of protons for 1 mole, weighed exactly, of potassium bichromate. Figure 18.5 summarizes these considerations. However, this reaction exhibits some drawbacks: The iodine formation is not instantaneous and iodide ions are easily oxidized with air dioxygen in the presence of chromic ions.

Other extemporaneous preparations of iodine solutions have been described in the literature. We briefly list the following:

- the oxidation of iodides with ceric sulfate in well-determined conditions:

$$2Ce^{4+} + 2I^- \rightarrow 2Ce^{3+} + I_2;$$

- the oxidation of iodides with hexacyanoferrate(III) according to the reaction

$$2Fe(CN)_6{}^{3-} + 2I^- \rightleftharpoons 2Fe(CN)_6{}^{4-} + I_2.$$

  This reaction, which was already cited in this book, can be equilibrated. In strongly acidic medium, it evolves from left to right. It is the inverse in neutral or basic medium. It can be displaced toward the right in neutral medium by the addition of a zinc salt. There is a formation of a poorly soluble double salt, the zinc and potassium ferrocyanide, which precipitates

$$2\left[Fe(CN)_6\right]^{4-} + 8K^+ + 3Zn^{2+} \rightarrow K_2Zn_3\left[Fe(CN)_6\right]_2\downarrow + 6K^+;$$

- the oxidation of iodide ions with permanganate ions (see manganimetry Chap. 20):

$$MnO_4{}^- + 5I^- + 8H^+ \rightarrow 5/2I_2 + Mn^{2+} + 4H_2O.$$

  Although this method has been recommended by several authors, its drawback is that it involves the use of a permanganate solution, which is not a primary standard itself;
- the oxidation of iodides with potassium peroxodisulfate $K_2S_2O_8$ available in a good state of purity and that is recommended as a primary standard. The couple peroxodisulfate/sulfate is endowed with a high standard potential value $E°(S_2O_8{}^{2-}/SO_4{}^{2-}) = 2.01$ V. The half-redox reaction is

$$S_2O_8{}^{2-} + 2e^- \rightleftharpoons 2SO_4{}^{2-}.$$

  Recall (see Chap. 12) that each oxygen atom of the peroxo bridge gains one electron. The reaction of iodine formation is

$$S_2O_8{}^{2-} + 2I^- \rightarrow 2SO_4{}^{2-} + I_2.$$

It is achieved in neutral medium.

18.9 Examples of Titration by Direct Iodometry

18.9.1 Determination of Sulfurous Acid, Hydrogen Sulfites, and Sulfites

The standard potential of the couple

$$SO_4{}^{2-} + 2e^- + 4H^+ \rightleftharpoons H_2SO_3 + H_2O, \quad E°\left(SO_4{}^{2-}/H_2SO_3\right) = 0.17\text{ V}$$

is markedly lower than that of the I_2/I^- couple. The titration reactions are as follows:

$$SO_3^{2-} + I_2 + H_2O \rightarrow SO_4^{2-} + 2H^+ + 2I^-,$$

$$HSO_3^- + I_2 + H_2O \rightarrow SO_4^{2-} + 3H^+ + 2I^-.$$

The solutions must be very diluted in order to obtain accurate results. Moreover, the sulfite solution must be added to the iodine solution (and not the inverse) and must have as little contact with air as possible. They are back titrations. A known volume of the sulfite solution is added to a solution of iodine in excess, slightly acidified ($pH \approx 3$). The excess of iodine is titrated with a thiosulfate solution. This is the principle of determining sulfites in wine.

18.9.2 Determination of Hydrogen Sulfide, Hydrogen Sulfides, and Sulfides

The standard potential of the couple S/H_2S corresponds to the half-equilibrium

$$S\downarrow + 2e^- + 2H^+ \rightleftharpoons H_2S_{(g)},$$

$E°(S_{(s)}/H_2S_{(g)}) = 0.14$ V. The determination is achieved in acidic and diluted medium. The titration reaction is

$$H_2S + I_2 \rightarrow 2H^+ + 2I^- + S\downarrow$$

In basic medium, the following reaction also evolves:

$$4I_2 + 4H_2O + SH^- \rightarrow SO_4^{2-} + 8I^- + 9H^+,$$

which is the oxidation of hydrogen sulfide into sulfate ions. In order to obtain satisfactory results, we must add the diluted sulfide solution to the acidified one of iodine in excess. This avoids a loss of hydrogen sulfide. Another process involves the addition of an excess of sodium arsenite to that of S–II to be determined and an acidification. Arsenic +III sulfide precipitates. Arsenious acid in excess is determined by direct iodometry:

$$2H_3AsO_3 + 3H_2S \rightarrow As_2S_3 + 6H_2O.$$

18.9.3 Determination of Alkaline Cyanides: Fordos and Gelis's Method

In neutral or weakly alkaline medium, iodine reacts with cyanide ions with the formation of cyanogen iodide, according to the equations

$$I_2 + CN^- \rightarrow ICN + I^-,$$

$$\text{or } I_2 + HCN \rightarrow ICN + H^+ + I^-.$$

In too alkaline a medium, hypoiodite ions are formed and the reaction exhibits a different course. Cyanate ions are formed:

$$CN^- + IO^- \rightarrow CNO^- + I^-.$$

The titration is usually achieved in hydrogen carbonate—diluted medium. As cyanogen iodide is colorless, the equivalence point is detected by the appearance and persistence of a yellow-brown color since it is a direct titration.

The examination of the titration reactions shows, without any ambiguity, that iodine plays the part of an oxidizing species. Cyanide ions, at the global oxidation number −I, become oxidized into the species CN +I, as stoichiometry shows. The global redox reaction may be divided into the following two half-redox reactions:[3]

$$I_2 + 2e^- \rightleftharpoons 2I^-,$$

$$CN^- - 2e^- \rightleftharpoons CN(+\,I).$$

The reaction equilibrium constant is known:

$$K° = (I_2)(HCN)/[(ICN)(H^+)(I^-)],$$

$$K° = 1.37 \quad (25\ °C).$$

In order for the reaction to be sufficiently displaced toward the right, it must be carried out in slightly alkaline medium.

18.9.4 Determination of Hydrazine and Its Derivatives

Hydrazine belongs to the couple $N_{2(g)}/NH_2\text{-}NH_3{}^+$:

$$N_{2(g)} + 5H^+ + 4e^- \rightleftharpoons NH_2\text{-}NH_3{}^+, \quad E°\left(N_{2(g)}/N_2H_5{}^+\right) = 0.2\ V.$$

The titration reaction is

$$NH_2\text{-}NH_2 + 2I_2 \rightarrow N_{2(g)} + 4I^- + 4H^+.$$

The reaction is achieved in hydrogen carbonate buffer. It is a back titration. The excess iodine is titrated with thiosulfate. Phenylhydrazine, semicarbazide, and thiosemicarbazide are determined according to the same principle:

$$\underset{\text{phenylhydrazine}}{C_6H_5\text{-}NH\text{-}NH_2} + 2I_2 \rightarrow N_{2(g)} + \underset{\text{iodobenzene}}{C_6H_5I} + 3H^+ + 3I^-,$$

$$\underset{\text{semicarbazide}}{NH_2\text{-}NH\text{-}CO\text{-}NH_2} + 2I_2 + H_2O \rightarrow N_{2(g)} + CO_2 + NH_4{}^+ + 3H^+ + 4I^-,$$

$$\underset{\text{thiosemicarbazide}}{NH_2\text{-}NH\text{-}CS\text{-}NH_2} + 6I_2 + 5H_2O$$

$$\rightarrow N_{2(g)} + SO_4{}^{2-} + 12I^- + 14H^+ + \underset{\text{cyanic acid}}{HCNO}.$$

[3] We have not found the standard potential of the couple CN(+I)/CN⁻ in the literature.

18.9.5 Determination of Arsenicals

Some organic arsenicals, where arsenic exhibits the +III oxidation state, can be titrated directly with an iodine solution. They are methylarsine CH_3AsH_2, methyldiiodoarsine CH_3AsI_2, and methylarsine oxide CH_3AsO:

$$CH_3AsH_2 + 3I_2 + 3H_2O \rightarrow CH_3AsO(OH)_2 + 6H^+ + 6I^-,$$

$$CH_3AsI_2 + I_2 + 3H_2O \rightarrow CH_3AsO(OH)_2 + 4H^+ + 4I^-,$$

$$CH_3AsO + I_2 + 2H_2O \rightarrow CH_3AsO(OH)_2 + 2H^+ + 2I^-.$$

Methylarsinic acid is formed. In methylarsine and methyldiiodoarsine, arsenic is at oxidation state +III, and both hydrogen atoms and both iodine atoms bound to arsenic atom must be considered at oxidation state −I. Some other organic arsenicals must be mineralized before their arsenic content can be determined (see Sect. 18.10.7).

18.9.6 Determination of Derivatives of Antimony +III

Antimony +III in the antimonyl cation SbO^+ belongs to the redox couple Sb_2O_5/SbO^+ and obeys the half-redox equilibrium

$$Sb_2O_{5(w)} + 6H^+ + 4e^- \rightleftharpoons 2SbO^+ + 3H_2O, \quad E^\circ\left(Sb_2O_{5(w)}/SbO^+\right) = 0.58\,V.$$

The reaction of the oxidation of Sb +III with iodine in acidic medium is

$$2SbO^+ + 3H_2O + 2I_2 \rightarrow Sb_2O_5 + 4I^- + 6H^+,$$

which is quite analogous to that of the oxidation of As +III with iodine. It is equilibrated as expected after inspection of the standard potential values. It can evolve forward and backward according to the pH value. Going forward (oxidization) implicates working in slightly alkaline medium (hydrogen carbonate buffer) in order to displace the equilibrium. Some authors have recommended ammonium and antimonyl tartrate (tartar emetic) as standard for iodine solutions.

COOSbO
|
$(CHOH)_2, \frac{1}{2}H_2O$
|
COOK

tartar emetic

18.9.7 Determination of Stannous Tin

The standard potential of the couple Sn^{4+}/Sn^{2+} is $E^\circ(Sn^{4+}/Sn^{2+}) = 0.14$ V. The titration reaction with iodine is

$$Sn^{2+} + I_2 \rightarrow Sn^{4+} + 2I^-.$$

The oxidation of tin + II by the air dioxygen may lead to an inaccurate value. The reaction of titration of tin + II with iodine provides an example of determining a metallic salt "at its minimum" (of oxidation).

18.9.8 Determination of Mercurous Salts: Extension to the Determination of Mercuric Salts and to That of Reducing Organic Substances

Mercury(I) salts (mercurous salts) such as the dichlorodimercury(II) Hg_2Cl_2 (improperly called calomel or mercurous chloride) and other metallic salts "at their minimum" can be titrated with a iodine solution in the presence of iodide ions according to the reaction

$$Hg_2Cl_{2(s)} + I_2 + 6I^- \rightarrow 2[HgI_4]^{2-} + 2Cl^-.$$

The complex tetraiodomercurate(II) ion $[HgI_4{}^{2-}]$ is formed. It is a back titration. The excess iodine is titrated with a thiosulfate solution. We notice that mercury(I) is oxidized into mercury(II), an oxidation state that it exhibits in the ion complex tetraiodomercurate. Simultaneously, one molecule of iodine is reduced to two iodide ions. We notice also that the iodide ions formed are not free after the reaction, as in a classical direct iodometry, but are complexed. Now, we must wonder why this reaction is possible from a thermodynamic standpoint. There is indeed, at first sight, a sort of contradiction between the occurrence of this reaction and the values of the standard potentials of systems under study:

$$2Hg^{2+} + 2e^- \rightleftharpoons Hg_2{}^{2+}, \quad E°(Hg^{2+}/Hg_2{}^{2+}) = 0.92 \text{ V},$$

$$I_2 + 2e^- \rightleftharpoons 2I^-, \quad E°(I_2/I^-) = 0.62 \text{ V},$$

Hg(II) should not be formed. But the studied reaction is more complicated than the simple redox reaction between the couples $Hg^{2+}/Hg_2{}^{2+}$ and I_2/I^- that can be predicted at first sight. The above redox reactions are coupled to the reactions

$$Hg_2Cl_{2(s)} \rightleftharpoons Hg_2{}^{2+} + 2Cl^-, \quad K_s = (Hg_2Cl_2) = 10^{-17.9},$$

$$Hg^{2+} + 4I^- \rightleftharpoons [HgI_4]^{2-}, \quad \beta_4\,(HgI_4{}^{2-}) = 10^{30}.$$

Calculations achieved according to the principles mentioned in Chap. 2 give the value $K = 10^{31}$ for the global constant of the studied reaction. The reaction toward the right can be considered complete. The value $K_s = 10^{-17.9}$ of the solubility product of the "mercurous chloride," which would command the mercury to stay in the solution as mercury(I), is completely surmounted by the value of the global constant of complexation, $\beta_4 = 10^{30}$.

The calculation of the standard potential of the couple $HgI_4{}^{2-}/Hg_{(l)}$, carried out according to the principles mentioned in Chap. 2, gives the value

$E°(HgI_4^{2-}/Hg_{(l)}) = -0.03$ V (complex formation constant used $\beta_4 = 10^{30}$; see Part IV). The complexation of mercury(II) as tetraiodomercurate transforms Hg^{2+} of the couple $Hg^{2+}/Hg_{(l)}$, which is a good oxidizing species [$E°(Hg^{2+}/Hg_{(l)}) = 0.88$ V], into a markedly weaker oxidizing species [HgI_4^{2-}], oxidant in the couple $HgI_4^{2-}/Hg_{(l)}$.

This reaction has given rise to a double extension involving both mercuric salts. In the first one, the determination of mercuric salts is proposed after the reduction of mercury(II) into mercury (I or 0). The mercury is then again oxidized with an excess of a titrated solution of iodine in the presence of iodide ions. The excess iodine is titrated with a thiosulfate solution. In the second extension, an identical scheme of processes is followed, but, finally, the quantity of the reductor is determined. An example is provided by the determination of formaldehyde (the most reducing aldehyde) and some other aldehydes. In these extensions, the determination is achieved with respect to a blank. The "reoxidization" of metallic mercury can be written

$$Hg + I_2 + 2I^- \rightarrow \left[HgI_4^{2-}\right].$$

18.9.9 Determination of Thiocyanates

Thiocyanates can be titrated by iodine:

$$SCN^- + 4I_2 + 4H_2O \rightarrow SO_4^{2-} + 7I^- + ICN + 8H^+.$$

Cyanogen iodide and sulfate ions are formed. It is a back- titration. The excess iodine is titrated with a thiosulfate solution in hydrogen carbonate medium. When the medium is acidified before the titration of the excess iodine, the reaction follows a different course. It becomes

$$SCN^- + 3I_2 + 4H_2O \rightarrow SO_4^{2-} + HCN + 7H^+ + 6I^-.$$

18.9.10 Determination of Thiols

Thiols exhibit reducing properties. They are oxidized into disulfides by iodine. The couple disulfide/thiol obeys the half-redox equilibrium

$$\text{R-S-S-R} + 2e^- + 2H^+ \rightleftharpoons 2RSH.$$

The standard potential somewhat depends on the structure R. The reaction can be observed in water and in ethanol, and the titrations can be direct or indirect.

Electrochemical studies show that, from a mechanistic standpoint, the reaction is complex. From a theoretical point of view, the value of the oxidation state of the sulfur atoms in the thiol and in the disulfide is questionable, due to the similar electronegativity values of sulfur and carbon atoms. In the thiol, the sulfur may be considered at oxidation state −I or −II. In the disulfide, it may be in oxidation state 0 or −I.

We now give some examples of such determinations:

- the determinations of cysteine (2-amino-3-mercaptopropanoic acid) and of glutathione (γ-L-glutamyl-L-cysteinylglycine), which are physiological derivatives, respectively an amino acid and a tripeptide:

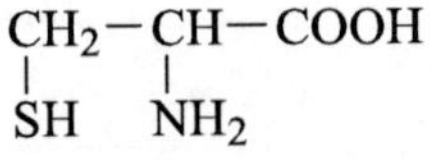

Cystein

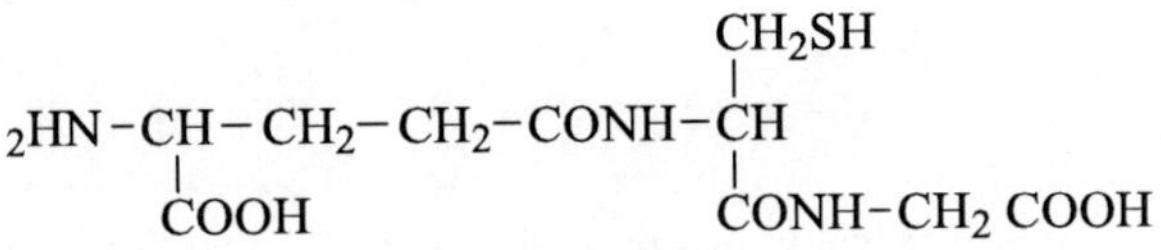

reduced glutathione

Their titrations are achieved in hydroalcoholic media;

- the determination of 1,2-dimercaptopropanol:

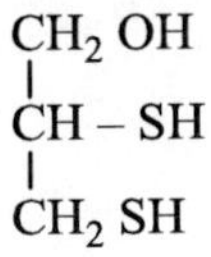

This is a historical example (see Chap. 31).

18.9.11 Determination of Xanthogenates and Derivatives: Determination of Hydrazoic Acid and of Azides

Xanthogenates, as salts, are oxidized by iodine with the formation of a disulfide bridge.

xanthogenate

The reactions of iodine with hydrazoic acid HN_3 and azides N_3^-, when they are performed in the presence of carbon disulfide, must be considered as being very close to the preceding one. Actually, iodine reacts slowly with hydrazoic acid alone. Some catalysts can be used. With carbon disulfide, azidothiocarbamic acid is formed according to the reaction

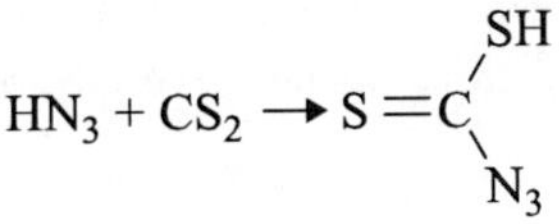

azidodithiocarbamic acid

Azidodithiocarbamic acid is oxidized into the corresponding disulfide in the presence of an excess of iodine.

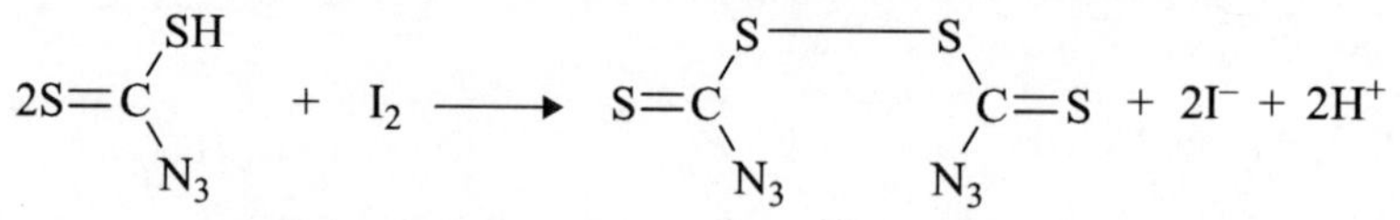

The disulfide spontaneously decomposes to give nitrogen and simultaneously regenerates carbon disulfide.

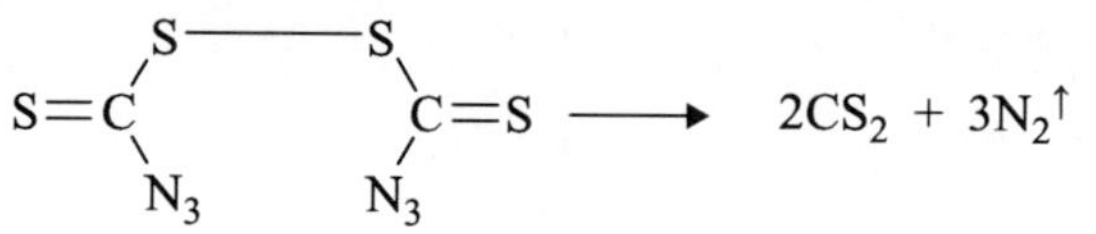

The global reaction is

$$2HN_3 + I_2 \rightarrow 3N_2\uparrow + 2H^+ + 2I^-.$$

18.9.12 *Determination of Hydroquinol*

Iodine oxidizes hydroquinol into p-quinone in hydrogen carbonate medium. Actually, the reaction is equilibrated:

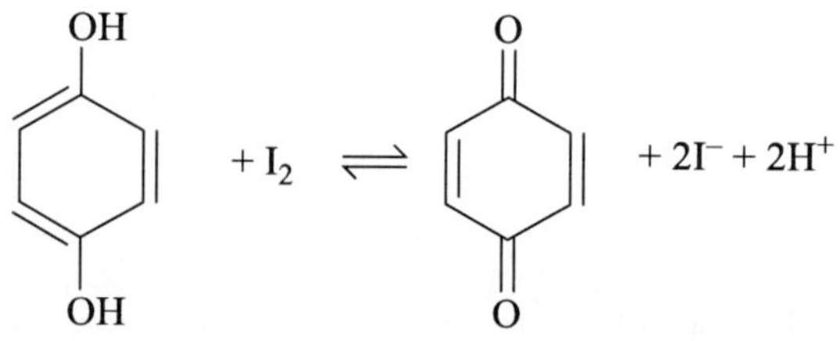

The reaction is quite similar to that of iodine with arsenious acid. The hydrogen carbonate buffer is used to displace the reaction toward the right. The standard potential of the couple p-quinone/hydroquinol is $E° = 0.70$ V. Resorcinol (meta-diphenol) and pyrocatechol (ortho-diphenol) do not give any reaction of analytical interest with iodine.

18.9.13 Determination of Vitamin C

Vitamin C, or ascorbic acid,[4] exhibits the following structure:

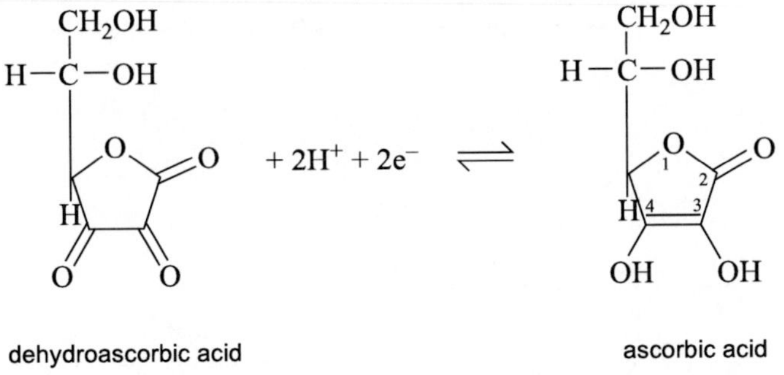

It uniquely possesses an *enediol group* (also called a *reductone group*) in positions 3, 4. Through oxidization, vitamin C gives dehydroascorbic acid after the exchange of two electrons and two protons. The exchange of two electrons is achieved in two steps. The intermediary species is the following radical:

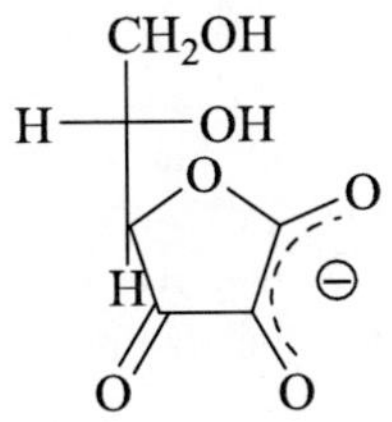

It is presumed to play a very important part in some vital processes. Each redox step involves exchanges of one electron and one proton. The standard potential of the couple dehydroascorbic acid/ascorbic acid is $E° = 0.185$ V. The titration is achieved in acidic medium with an iodine solution as oxidant in the presence of starch. Ascorbic acid has been proposed as a reductant in some titrations. It is easily oxidized by air.

We find it interesting to highlight the fact that the metabolism of numerous species is achieved *in vivo* through redox processes. For example, riboflavin (vitamin B_2) undergoes a two-electron reduction with a two-proton gain to offer the corresponding reduced form:

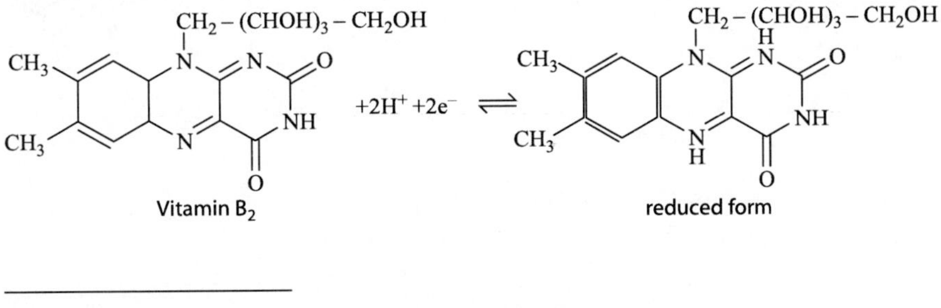

[4] (5R)-5-[(S)-1,2-dihydroxyethyl]-3,4-dihydroxy-2(5 H)-furanone.

The reduction may be achieved *in vitro* by zinc in acidic medium. The reduction is achieved *in vivo* in several steps, during which semi-quinonic radicals appear.

Another interesting example is provided by derivatives of the phenothiazine nucleus. It gives rise to a stable radical after oxidation. For example, chlorpromazine reversibly loses an electron through the action of light and dioxygen:

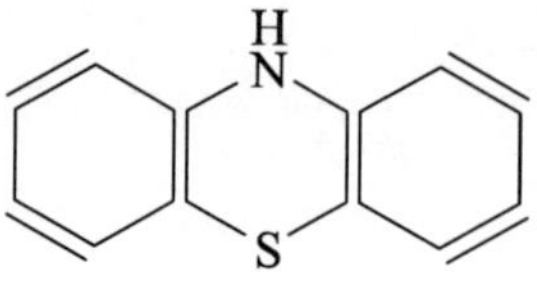

phenothiazine

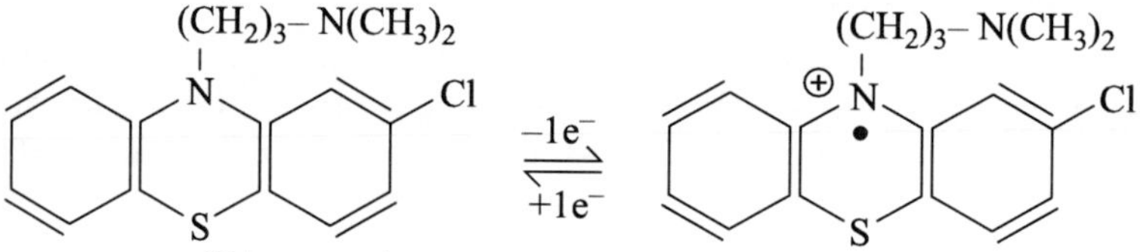

Chlorpromazine

Of course, several limit formulas contribute to its actual structure and, hence, to its stability. The radical formed may evolve by disproportionation, simultaneously giving chlorpromazine and the two-electron product of oxidization. Finally, the latter may evolve in several ways. For example, it may be transformed into the corresponding sulfoxide:

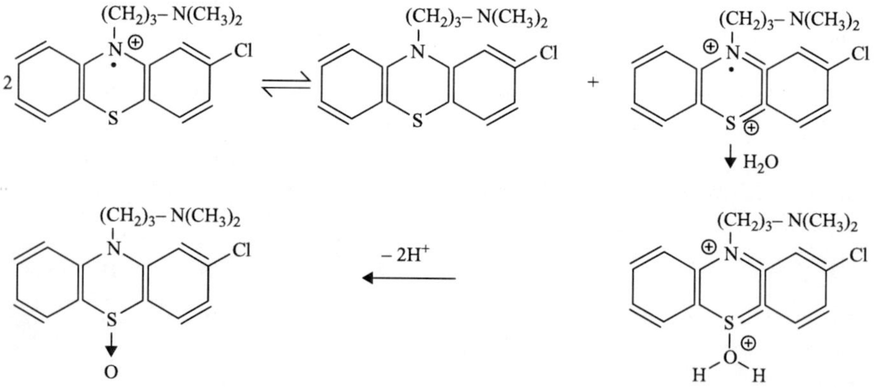

In the realm of medicine, we find the direct iodometric titration of the sodium or potassium salt of benzylpenicillin (called penicillin G).

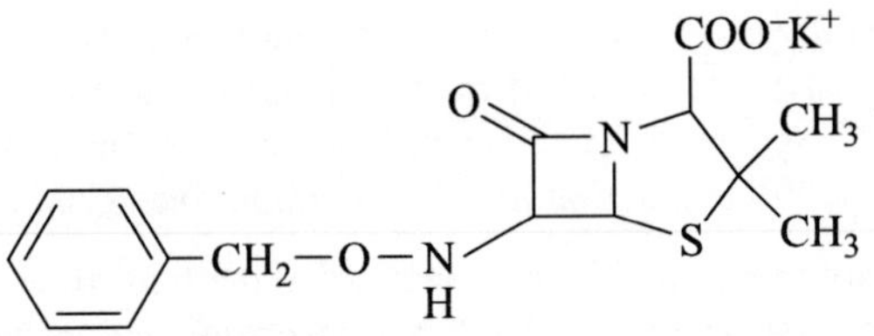

Recall that the β-lactam cycle is not very stable and that it is opened in alkaline medium. The titration is achieved in alkaline medium with respect to a blank. From the practical standpoint, penicillin and 1 M sodium hydroxide are brought together for 15 min. Then penicilloic acid is formed.

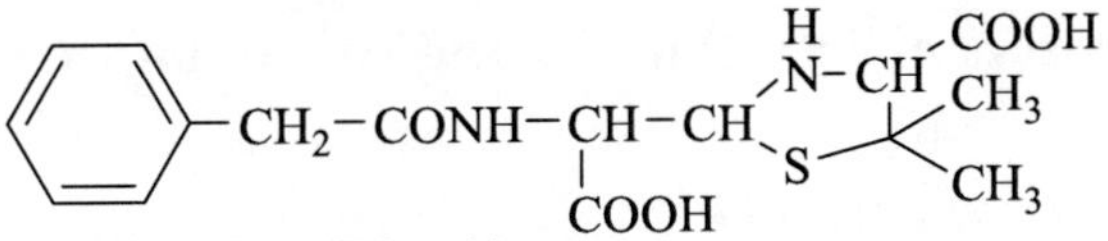

Penicilloic acid

After neutralization with a normal solution of sulfuric acid in the presence of phenolphthalein, a decinormal iodine solution is added and the reaction is allowed to evolve over 15 min. The excess iodine is titrated with a 10^{-2} mol/L thiosulfate solution. According to some authors, opening the β-lactam cycle does not stop at the stage of penicilloic acid. It leads to penaldic acid and to penicillamine in equimolar quantities.

pendaldic acid

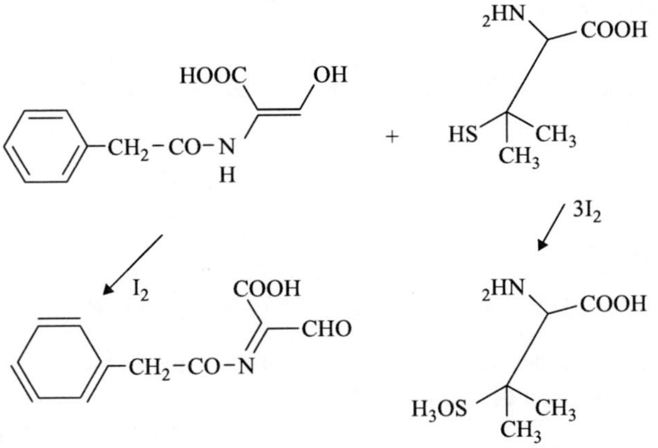

penicillamine

On the one hand, penaldic acid consumes one molecule of iodine, to give the corresponding imine. On the other hand, penicillamine consumes three molecules of iodine, to give the corresponding sulfonic acid. The direct iodometric method furnishes results that are in agreement with the microbiological ones.

Before closing this section devoted to direct iodometry, we'll mention that iodine can be introduced to the solution to be titrated not only by an iodine/iodide solution but also by coulometry (see electrochemistry). Moreover, iodine can also be generated *in situ* by the iodate/iodide reaction in acidic medium (see Chap. 19).

Remark We do not mention here the quantitative methods based on the fixation of iodine on the molecules to be determined. A good example is provided by the fixation of iodine on the double bonds, as is the case in the determination of an iodine index (see food analysis). It is difficult, indeed, to classify these reactions as being redox reactions.

18.10 Examples of Titrations by Indirect Iodometry

18.10.1 Recall of Information

At first, we must again mention some usual species that quantitatively oxidize iodide ions to give iodine. They permit the extemporaneous standardization of thiosulfate solutions (see Sect. 18.8). It is the case of dichromate, ceric, hexacyanoferrate(III), permanganate, and peroxodisulfate ions. Quite evidently, they can be titrated by indirect iodometry. This is also the case with iodate ions, which is the matter of Chap. 19.

18.10.2 Titration of Nitrous Acid and Nitrites

Nitrous acid and nitrites oxidize iodide ions, as can be predicted by examining the corresponding standard potentials:

$$HNO_2 + 1e^- + H^+ \rightleftharpoons NO_{(g)} + H_2O, \quad E^\circ\left(HNO_2/NO_{(g)}\right) = 0.98\ V.$$

The reaction is

$$2HNO_2 + 2I^- + 2H^+ \rightarrow 2NO_{(g)} + 2H_2O + I_2.$$

It must be achieved under very precise experimental conditions in order to avoid the formation of nitrogen dioxide NO_2 from nitric oxide NO by reaction with the air dioxygen. Nitric acid itself is formed by the oxidation reaction of iodide ions. Nitrogen oxides, present in the medium, catalyze the oxidization of iodide ions by dioxygen and result in too great a release of iodine.

In terms of determining nitrites by indirect iodometry, the oxidization of iodide ions by nitrous acid may be used to indicate the equivalence point of a titration reaction involving nitrous acid. This is the case with one of the general methods of analysis in the french pharmacopeia named *nitritometrie*. It concerns the determination of primary aromatic amines with nitrous acid. The reaction is achieved at $pH = 0$

in the presence of iodide ions. The titration reaction is the formation of the corresponding diazonium salt. Once the equivalence point is reached, the excess nitrous acid (added as sodium nitrite) can no longer react with the amine group. Therefore, it oxidizes iodide ions into iodine. Most of the time, a sharp potential change is detected with an electrochemical method of analysis.

18.10.3 Determination of Halogens

Chlorine and bromine may be titrated by indirect iodometry due to their standard potential values:

$$Cl_{2(g)} + 2e^- \rightleftharpoons 2Cl^-, \quad E°(Cl_2/Cl^-) = 1.36\text{ V},$$

$$Br_{2(w)} + 2e^- \rightleftharpoons 2Br^-, \quad E°(Br_2/Br^-) = 1.09\text{ V}.$$

The global reactions are

$$Cl_2 + 2I^- \rightarrow 2Cl^- + I_2,$$

$$Br_2 + 2I^- \rightarrow 2Br^- + I_2.$$

The reactions are achieved in the presence of an excess of potassium iodide. The iodine released is titrated with thiosulfate. It is interesting to notice that chlorine and bromine cannot be directly titrated by thiosulfate, as might be, a priori, predicted from the sole standard potential values. In these conditions, thiosulfate is transformed partly into tetrathionate and into sulfate. The two reactions are of different stoichiometries (see the beginning of the chapter).

The oxidization of iodide ions with bromine is of great interest in practice for the analysis of organic substances. Some organic structures may indeed fix bromine atoms after reacting with bromine added in excess. The excess is determined by indirect iodometry.

18.10.4 Determination of Hypochlorites

Most of the time, hypochlorites are determined by indirect iodometry. These determinations are of great practical interest since, for example, bleaching chlorides (calcium hypochlorite) and bleaching solutions (sodium hypochlorite) are determined in such a manner. Chlorine is used industrially for its oxidizing, bleaching, and antiseptic properties, among others. Its use is difficult and dangerous. Hypochlorites are sufficiently unstable to easily release hypochlorous acid, which, during its decomposition, exhibits the behavior of chlorine. For this goal, one uses calcium hypochlorite, sodium hypochlorite, and some organic compounds that also generate hypochlorous acid.

Bleaching solutions are aqueous sodium hypochlorite. One example is "Javel water"[5] in France. It can be prepared by reacting chlorine with a solution of sodium hydroxide at room temperature according to

$$Cl_{2(g)} + 2NaOH \rightarrow NaClO + NaCl + H_2O.$$

Hence, Javel water may be considered a mixture of equal quantities of sodium hypochlorite and sodium chloride. Concentrated and more diluted solutions are commercially available. The latter are called "ordinary Javel waters." For pharmaceutical use, there is Dakin solution, which is a diluted Javel water to which has been added some sodium hydrogen carbonate to weaken the alkaline character of the solution, which is irritating for tissues.

Calcium hypochlorite is prepared by reacting chlorine with calcium hydroxide according to

$$2Ca(OH)_2 + 2Cl_{2(g)} \rightarrow Ca(ClO)_2 + CaCl_2 + 2H_2O.$$

This is the reason why it is improperly called "slaked lime chloride." Actually, the commercial calcium hypochlorite is essentially a mixture of calcium hypochlorite (*sensu stricto*) and basic calcium chloride, that is, in fact, $Ca(ClO)_2$, $CaCl_2$, and $Ca(OH)_2$, H_2O. It is used as a white powder called "bleaching powder."

Recall (Chap. 15, Fig. 15.12) that the species HClO, ClO^-, and Cl_2 are not stable in water. They oxidize it. However, the reactions are slow. This is the reason why neutral and alkaline hypochlorite solutions exhibit some undeniable stability despite their strong oxidizing properties. Oxidization of water[6] by hypochlorous acid and by hypochlorites is written as follows according to the pH value:

$$\begin{aligned} &4HClO \rightarrow O_2 + 2Cl_2 + 2H_2O, && pH < 3.5,\\ &2HClO \rightarrow O_2 + 2Cl^- + 2H^+, && 3.5 < pH < 7.5,\\ &2ClO^- \rightarrow O_2 + 2Cl^-, && pH > 7.5, \end{aligned}$$

Let's also recall (see the same diagram) that hypochlorite ions decompose in acidic medium and liberate chlorine in the presence of chlorides. This fact leads to the concepts of active chlorine and chlorometric degrees. When we consider the retrodismutation reaction of hypochlorous acid in acidic medium and in the presence of chloride ions:

$$HClO + Cl^- + H^+ \rightarrow H_2O + Cl_2,$$

we see that one mole of hypochlorous acid gives one mole of chlorine. It is said that one mole of hypochlorous acid has the same oxidizing power as one mole of

---

[5] In the past, Javel water was a solution of potassium hypochlorite.

[6] These equations do not make it evident that water is oxidized. It becomes so when both half-redox reactions are written. After the addition of both half-reactions, water appears on the right- and left-hand sides of the equations, and a writing simplifica is possible by substraction.

chlorine. This equivalence is the origin of the concept of *active chlorine*. We notice that active chlorine truly evolves only when the decomposition of the hypochlorite takes place in hydrochloric acid medium.

The *chlorometric degree* is the number of liters of gaseous chlorine evolved in normal conditions that may be released by one liter of hypochlorite solution or by one kilogram of product in the presence of hydrogen chloride. For example, concentrated Javel water exhibits 47–50 chlorometric degrees. An ordinary Javel water exhibits 15 chlorometric degrees. The Dakin solute is at 1.6 chlorometric degrees. The commercial calcium hypochlorite has a titer of about 110 chlorometric degrees. More concentrated products may also be prepared by industry. They can exhibit as many as 200 chlorometric degrees.

The active chlorine of a sample or available chlorine is defined as being the weight of chlorine that liberates the same quantity of iodine (from iodide ions) as that liberated by the sample under the same conditions.

The determination of hypochlorites is based on their oxidizing power. Two reductors are used:

- iodide ions. The solution (or the suspension) of hypochlorite is treated with an excess of a potassium iodide solution and is added to a solution of acetic acid. Iodine is released and is titrated with thiosulfate:

$$ClO^- + 2I^- + 2H^+ \rightarrow Cl^- + I_2 + H_2O.$$

This is Bunsen's method. When the solution is too strongly acidified, the small quantity of chlorates contained by the hypochlorite solution oxidizes iodide ions according to the reaction

$$ClO_3^- + 6I^- + 6H^+ \rightarrow Cl^- + 3I_2 + 3H_2O.$$

Chlorate contained by the sample of hypochlorite comes from the disproportionation of the latter, according to:

$$3ClO^- \rightarrow ClO_3^- + 2Cl^-.$$

Organic compounds generating hypochlorous acid often possess a mono or gem dichloro amine or amide group. They are more oxidizing than hypochlorite. For example, dichloramine T (N-dichloroparatoluenesulfonamide):

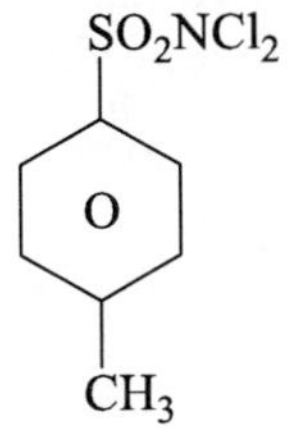

dichloramine T

liberates bromine from a neutral solution of potassium bromide [$E°(Br_2/Br^-) = 1.09$ V]. They can be determined by indirect iodometry. For example, chloramine T liberates iodine from iodide ions according to the reaction

$$CH_3\text{-}C_6H_4SO_2\text{-}\overset{\overset{\displaystyle Na}{|}}{N}Cl + 2I^- + 2H^+$$

$$\rightarrow CH_3\text{-}C_6H_4SO_2NH_2 + I_2 + Na^+ + Cl^-;$$

- the other often used reductor is arsenious acid at a pH value near neutrality.

The reaction is

$$ClO^- + H_3AsO_3 \rightarrow H_2AsO_4{}^- + H^+ + Cl^-.$$

It is a back titration. The excess arsenite is titrated by an iodine solution. This is Gay-Lussac's method. Quite evidently, it is no longer an indirect iodometry.

Bunsen's and Gay-Lussac's titrations can easily be justified by examining Pourbaix diagrams of the different couples.

18.10.5 Determination of Halogens at Oxidation Numbers +III, +V, +VII

The determination of halogens at oxidation states +III, +V, +VII may be achieved by indirect iodometry. We just mentioned that for the chlorate ions' case in which chlorine is at oxidation state +V. We have also already seen that iodate ions quantitatively oxidize iodide ions in acidic medium. That was for the standardization of thiosulfate solutions. Likewise, periodate ions and periodic acids quantitatively oxidize iodide ions into iodine (see the next chapter).

An interesting example of selective analysis operations is provided by the determination of hypochlorite ClO^-, chlorite $ClO_2{}^-$ (Cl + III), and chlorate ions in admixture. This is a mixture encountered in a bleaching powder due to the disproportionation of hypochlorite, transformation +I → +V that is achieved through oxidation state + III. The analysis involves three indirect iodometric titrations:

- one in strongly acidic medium in which the three ions are oxidants;
- one in acetic acid medium in which chlorite and hypochlorite ions are oxidants;
- one in neutral or slightly alkaline medium in which hypochlorite is the sole oxidant.

18.10.6 Determination of Metallic Salts "at Their Maximum"

Metallic salts "at their maximum" may be determined by indirect iodometry.

18.10.6.1 Cupric Ions

Cupric ions oxidize iodide ions according to the equation

$$2Cu^{2+} + 4I^- \rightarrow 2CuI_{(s)} + I_2.$$

The liberated iodine is titrated with thiosulfate.

As the examination of this equation shows, this is not a simple redox reaction. There is also a precipitation reaction together with the redox one. Half of the iodide ions added are indeed oxidized into iodine, while the other half remains at state-I. The part that the latter plays is to precipitate Cu + I as cuprous iodide. Cu + I is formed by the redox reaction between Cu^{2+} and I^-:

$$2Cu^{2+} + 2I^- \rightarrow 2Cu^+ + I_2,$$
$$Cu^+ + I^- \rightarrow CuI_{(s)}.$$

The precipitation of cuprous as cuprous iodide permits its stabilization (see Fig. 15.14 of Chap. 15). The cuprous ions that are not precipitated (or complexed) do not exist in aqueous solution. They disproportionate according to

$$2Cu^+ \rightleftharpoons Cu^{2+} + Cu_{(s)}, \quad K^\circ = 5.4 \cdot 10^5 \text{ at } 25\ ^\circ\text{C}.$$

Moreover, the precipitation makes the redox reaction possible from a thermodynamic standpoint. Indeed, the simple consideration of the standard potentials of both couples:

$$Cu^{2+} + 1e^- \rightleftharpoons Cu^+, \quad E^\circ(Cu^{2+}/Cu^+) = 0.15 \text{ V},$$
$$I_2 + 2e^- \rightleftharpoons 2I^-, \quad E^\circ(I_2/I^-) = 0.62 \text{ V}$$

seems to show that the redox reaction that would evolve is the opposite of that studied:

$$I_2 + 2Cu^+ \rightarrow 2Cu^{2+} + 2I^-.$$

This is not the case. Because of the precipitation of cuprous iodide, the redox reaction that must be taken into account is

$$Cu^{2+} + 1e^- + I^- \rightleftharpoons CuI_{(s)},$$

whose standard potential is $E^\circ(Cu^{2+}/CuI_{2(s)}) = 0.86$ V. This value is calculated from $E^\circ(Cu^{2+}/Cu^+)$ and from the solubility product $K_{s(CuI)} = 1.1 \times 10^{-12}$ according to the principle already seen in Chap. 2.

From a practical standpoint, it must be noticed that the precipitate of cuprous iodide may adsorb the liberated iodine; as a result, an error in the cupric determination occurs. To avoid this drawback, a soluble thiocyanate is added to the solution just before the equivalence point. Cuprous thiocyanate is still less soluble than cuprous iodide, and it adsorbs less iodine than the latter.

The Cu^{2+}/I^- reaction is the basis of several methods of the determination of reducing sugars. They are based on the reducing properties of these sugars, due to the presence of an aldehyde or an enediol function in their structure. Their reducing power may be estimated by cuprimetry. Sugars may reduce cupric ions to give cuprous oxide. (This reaction is not stoichiometric and must be achieved under strictly specified experimental conditions.) The cuprimetric determination of sugars consists of a back titration. One excess of a cupric salt is added initially and excess cupric ions (after reaction) are determined. Therefore, excess iodide ions are added and the released iodine titrated with thiosulfate. This methodology is called Lehman and Grimbert's method. We'd like to mention, incidentally, that there exists another iodometric method of determination of reducing sugars. It is that described by Hagedorn and Jensen. The sugar is oxidized by an excess of potassium ferricyanide (this reaction is not stoichiometric, as is that given by sugars and cupric ions). Ferricyanide ions in excess are determined by indirect iodometry (see Sect. 18.8 and Chap. 13).

18.10.6.2 Ferric Ions

Ferric ions react with iodide ions to give ferrous ions and iodine:

$$2Fe^{3+} + 2I^- \rightleftharpoons 2Fe^{2+} + I_2.$$

The examination of standard potentials [$E°(Fe^{3+}/Fe^{2+}) = 0.77$ V and $E°(I_2/I^-) = 0.62$ V] shows that the reaction tends to react spontaneously toward the right. Some conditions must be respected. The solution must be strongly acidic to preclude ferric ion hydrolysis, which displaces the reaction toward the left. Iodide ions must be in large excess to favor the reaction toward the right. The reaction is slow and, hence, we must wait the necessary time for it to be complete.

Reciprocally, since the above reaction is equilibrated, ferrous ions may be titrated by iodine. The reaction (toward the left) is made quantitative by complexation of the formed ferric ions with fluoride or pyrophosphate ions. Then $FeF_5{}^{2-}$ and $Fe(P_2O_7)_2{}^{5-}$ ions are formed, among others.

18.10.7 Determinations of Arsenic and Antimony at Oxidation State +V

The reaction to determine arsenic and antimony at oxidation state +V is the inverse of that already seen concerning the standardization of iodine solution by a solution of arsenious anhydride. It is

$$H_3AsO_4 + 2H^+ + 2I^- \rightleftharpoons H_3AsO_3 + I_2 + H_2O.$$

Some experimental conditions must be respected in order to obtain satisfactory results. The hydrochloric acid concentration must be high, about 4 mol/L. The oxidization of iodide ions by the air dioxygen must be avoided and we must wait five minutes, at least, before titrating the formed iodine. An analogous process permits us to determine antimony at oxidation state +V.

In the French pharmacopeia, two curious determinations exist. They demonstrate the necessity to involve the reactions $As(III) + I_2$ and $As(V) + I^-$ successively, that is, the same reaction in both directions. These determinations are those of two organic arsenicals: neoarsphenamine and acetarsol:

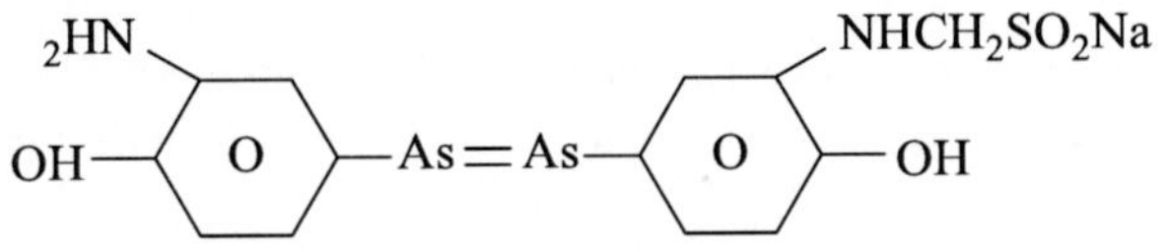

neoarsphenamine

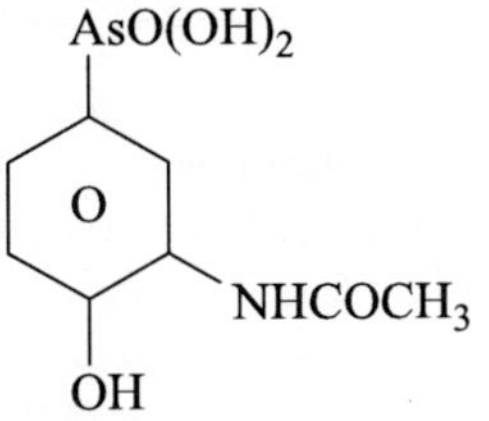

acetarsol

The determination of arsenic in these derivatives is imperatively required by the pharmacopeia. To achieve it, the molecule is first mineralized by the sulfonitric mixture, which has strong oxidizing properties. The organic arsenic of the initial sample becomes mineralized and transformed into arsenic acid H_3AsO_4. The latter is brought in the presence of a solution of iodide ions in concentrated hydrochloric acid. In these conditions, As(+V) quantitatively oxidizes iodide ions to give iodine. The reaction evolves in the direction $As(+V) \rightarrow As(+III)$. The liberated iodine is titrated later with an arsenious acid solution in hydrogen carbonate medium. Therefore, at this time, the reaction $As(+III) \rightarrow As(+V)$ is used.

18.10.8 Determination of Hydrogen Peroxide and of Peroxy Salts

Hydrogen peroxide, peroxy salts, and suroxygenated salts may be titrated by indirect iodometry.

Hydrogen peroxide reacts with iodine ions in acidic solution according to the reaction

$$H_2O_2 + 2H^+ + 2I^- \rightarrow I_2 + 2H_2O.$$

The reaction is slow, but the rate increases with the concentration of the acid. Adding ammonium molybdate makes the reaction quasi-instantaneous.

Zinc and magnesium peroxides ZnO_2 and MgO_2, monoperoxycarbonates MOC(O)COOM, peroxycarbonates $M_2C_2O_6$ (MOC(O)OOC(O)OM) with $M^+ = Na^+$ or another alkaline ion, sodium perborate $Na_2[B_2(O_2)_2(OH)_4]\ 6H_2O$, are titrated as the same way hydrogen peroxide is. Concerning this subject, let's recall that peroxodisulfate ions oxidize iodide ions very slowly. Minium Pb_3O_4 and plumbic oxide PbO_2 also quantitatively liberate iodine from potassium iodide in acidic medium. Therefore, in acetic acid medium,

$$PbO_2 + 4H^+ + 2I^- \rightarrow Pb^{2+} + 2H_2O + I_2.$$

18.10.9 Determination of Aqueous Dioxygen by Winkler's Method

The determination of aqueous dioxygen can be achieved by indirect iodometry. This is Winkler's method. It is based on the oxidation of iodide ions by MnO_2 (manganese dioxide). The first time, there is a reaction between dioxygen and a suspension of manganese(II) hydroxide in strongly alkaline medium according to the reaction

$$2Mn^{2+} + 4OH^- + O_{2(w)} \rightarrow 2MnO_2 + 2H_2O.$$

(The suspension of manganous hydroxide is prepared by dissolution of manganous sulfate $MnSO_4$ in sodium hydroxide.) After acidification and in the presence of iodide ions, MnO_2 is reduced by the latter with the liberation of iodine:

$$MnO_2 + 4H^+ + 2I^- \rightarrow I_2 + Mn^{2+} + 2H_2O.$$

This determination is achieved comparatively as a blank.

18.10.10 Determination of Peroxides and Hydroperoxides

In organic solvents, iodide ions reduce organic peroxides and hydroperoxides with the liberation of iodine. The reactions may be schematized as follows:

$$R_1\text{-O-O-}R_2 + 2I^- + 2H^+ \rightarrow I_2 + R_1OH + R_2OH,$$

$$R\text{-OOH} + 2I^- + 2H^+ \rightarrow I_2 + ROH + H_2O.$$

The liberated iodine is titrated by thiosulfate. The accuracy of the results strongly depends on the structure of the peroxide and on experimental conditions.

18.10.11 Determination of Diverse Organic Compounds

Diverse organic compounds may be determined by indirect iodometry. We give only one example here, that of ethacridine lactate, whose determination is described in the european pharmacopeia.

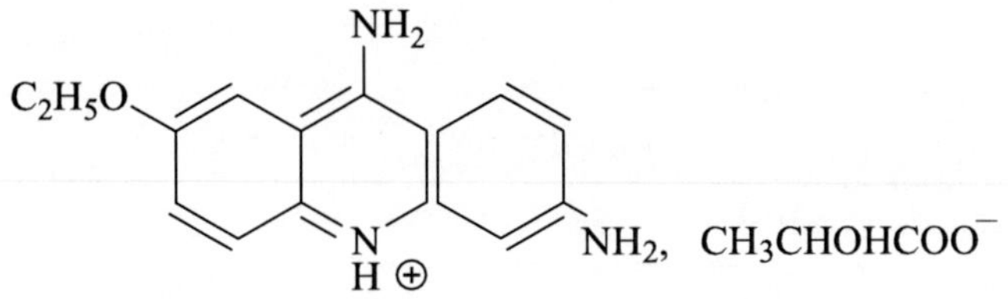

The compound is precipitated by an excess of a standardized solution of potassium bichromate. The excess bichromate is titrated by adding sodium iodide. The liberated iodine is titrated by thiosulfate. Thus, the quantitative reaction exhibited by ethacridine lactate is not a redox reaction. It is a precipitation reaction.

Chapter 19
Iodometry in Alkaline Medium, Iodatometry, Periodimetry, and Bromometry

In this chapter, we continue to study titrations with halogens at their most common oxidation states.

19.1 Iodometry in Alkaline Medium

19.1.1 General Considerations

Iodometry in alkaline medium is sometimes called *hypoiodometry*. It is based on the use of iodine at oxidation state + I, that is, on the use of hypoiodous acid HIO and hypoiodite ions IO^-.

Hypoiodous acid may be considered a result of the hydrolysis of iodine according to the reaction

$$I_2 + H_2O \rightleftharpoons HIO + I^- + H^+.$$

The formation equilibrium of hypoiodous acid is displaced toward the right in alkaline medium. At pH = 0, the apparent equilibrium constant is 2.0×10^{-13} mol^2/L^2. It is, by far, in favor of iodine. At pH = 14, the constant of the equilibrium

$$I_2 + 2HO^- \rightleftharpoons IO^- + H_2O + I^-$$

is 30 mol/L, in favor of hypoiodite. Hypoiodous acid's pK_a is

$$pK_a(HIO/I^-) = 11.0.$$

The above formation reactions are disproportionation reactions of iodine resulting from the superimposition of the following half-reduction equilibria:

$$\tfrac{1}{2} I_2 + 1e^- \rightleftharpoons I^-,$$

$$\tfrac{1}{2} I_2 + H_2O \rightleftharpoons HIO + 1e^- + H^+$$

J.-L. Burgot, *Ionic Equilibria in Analytical Chemistry*,
DOI 10.1007/978-1-4419-8382-4_19, © Springer Science+Business Media, LLC 2012

Fig. 19.1 Pourbaix diagram of species I_2, I_3^-, HIO, I^-, and IO^- in false equilibrium

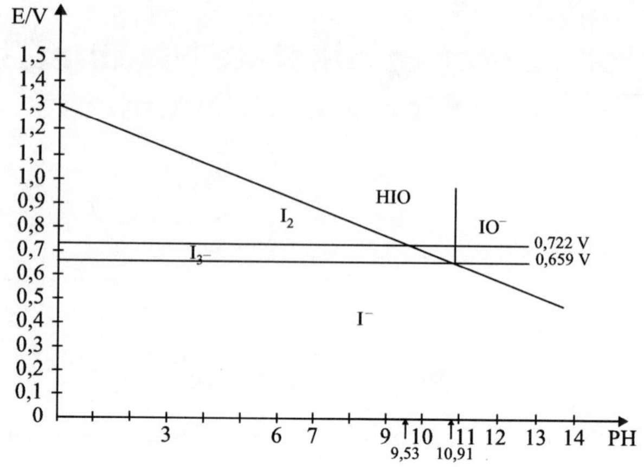

or

$$\tfrac{1}{2}\, I_2 + 1e^- \rightleftharpoons I^-,$$

$$\tfrac{1}{2}\, I_2 + 2OH^- \rightleftharpoons IO^- + e^- + H_2O.$$

Hence, we can safely say that hypoiodous acid and hypoiodites are formed by the disproportionation of iodine in alkaline medium. An examination of a Pourbaix diagram (Fig. 19.1) built on the following half-redox equilibria:

$$HIO + 2H^+ + 2e^- \rightleftharpoons I_2 + 2H_2O, \qquad E°(HIO/I_2) = 1.35\,V,$$

$$3HIO + 3H^+ + 4e^- \rightleftharpoons I_3^- + 3H_2O, \qquad E°(HIO/I_3^-) = 1.21\,V,$$

$$IO^- + 2H^+ + 2e^- \rightleftharpoons I^- + H_2O, \qquad E°(IO^-/I^-) = 1.31\,V,$$

$$\tfrac{3}{2}\, I_2 + 1e^- \rightleftharpoons I_3^-, \qquad E°(I_2/I_3^-) = 0.79\,V,$$

$$I_3^- + 2e^- \rightleftharpoons 3I^-, \qquad E°(I_3^-/I^-) = 0.536\,V$$

shows that the disproportionation occurs in the range $9 < pK < 11$, according to the conventions adopted to draw the diagram. However, the phenomena are more complicated than described above. Indeed, hypoiodous acid and hypoiodite ions are not stable from a thermodynamic standpoint. Their ephemeral lifespan (in minutes) must be credited to kinetic reasons. Frost diagrams (Fig. 19.2) clearly show that hypoiodite ions disproportionate into iodate and iodide ions according to the reaction

$$3IO^- \rightleftharpoons 2I^- + IO_3^-.$$

The equilibrium constant of this reaction is about 10^{20}. This reaction results from the superimposition of the following half-redox equilibria:

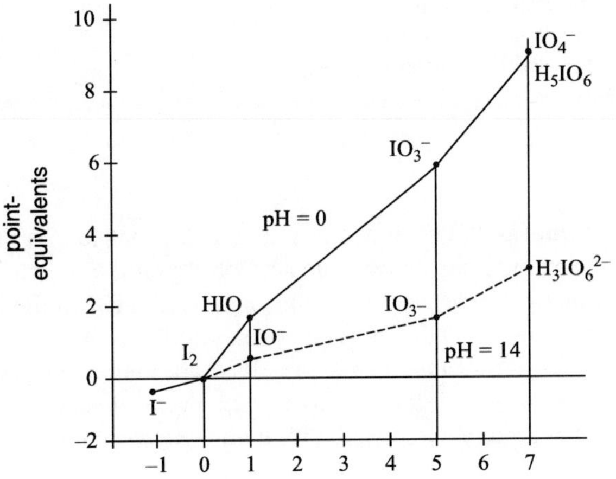

Fig. 19.2 Frost diagram of iodine

$$IO^- + 2H^+ + 2e^- \rightleftharpoons I^- + H_2O,$$

$$IO^- + 2H_2O \rightleftharpoons IO_3{}^- + 4H^+ + 4e^-.$$

In the pH range $0.8 < pH < 11$, the disproportionation reactions are

$$3IOH + 3OH^- \rightarrow 2I^- + IO_3{}^- + 3H_2O$$

and

$$3IOH \rightarrow 2I^- + IO_3{}^- + 3H^+$$

For a pH value smaller than 0.8, hypoiodous acid disproportionates into iodine and iodic acid according to the reaction

$$5HIO \rightarrow HIO_3 + 2I_2 + 2H_2O,$$

$$pK_a(HIO_3/IO_3{}^-) = 0.8.$$

The instability of hypoiodite ions is clearly apparent in the complete Pourbaix diagram when we consider the frontier straight lines $IO_3{}^-/IO^-$ and $IO_3{}^-/HIO$. If we limit ourselves to the zone for which $pH > 11$ by drawing a vertical line on the diagram, say at $pH = 12$, we obtain the diagram in Fig. 19.3.

Preponderance areas of hypoiodite ions IO^- are unconnected, which is not possible. Hence, hypoiodite ions disproportionate into iodate and iodide ions.

Finally, it is interesting to note that hypoiodous acid and hypoiodite ions are more oxidizing than iodate, tri-iodide ions, and iodine in the range where they can (temporarily) coexist.

Fig. 19.3 Virtual preponderance diagram of species I^-, IO^-, and IO_3^- at pH = 12

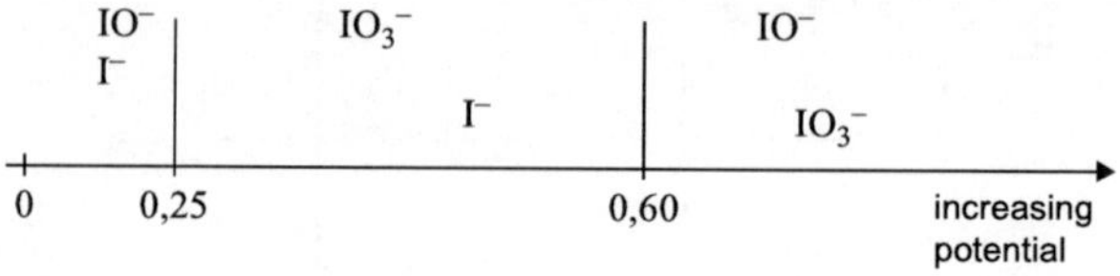

Remark We shall see in iodatometry that iodine (+I) may be stabilized in other forms than as hypoiodous acid or hypoiodite ions. Indeed, the species ICl (iodine monochloride), ICl_2^- (iodine dichloride), IBr (iodine monobromide), IBr_2^- (iodine dibromide), and ICN[1] (cyanogen iodide) exist.

In the oxidation reaction by hypoiodite ions, we consider that the reacting species is hypoiodite ion. For example, during the oxidation of some aldehydes into carboxylate ions, the following two reactions occur:

$$I_2 + 2OH^- \rightarrow IO^- + I^- + H_2O,$$

$$OH^- + RCHO + I^- \rightarrow RCOO^- + I^- + H_2O,$$

that is, globally,

$$RCHO + I_2 + 3OH^- \rightarrow RCOO^- + 2I^- + 2H_2O.$$

From a practical standpoint, the titration is achieved comparatively as a blank:

- in the titration vessel containing the titrand, a iodine (tri-iodide ions) solution in excess and sodium hydroxide are added. The yellow-brown solution (I_3^-) fades due to the formation of hypoiodite ions. We make the reaction of IO^- ions with the titrand run for a few minutes. Hypoiodite ions in excess disproportionate into iodide and iodate ions. An important point to note is that iodate ions cannot oxidize iodide ions at the pH at which we are working (see Chap. 18 Fig. 18.1). After some minutes, the medium is acidified. Iodate ions become oxidizing and oxidize iodide ions into iodine and tri-iodide ions, which, finally, are titrated with thiosulfate. The sequences of reactions occurring in the reaction vessel are
  - before acidification:

$$3I_2 + 6OH^- \rightarrow 3I^- + 3IO^- + 3H_2O \quad \text{(hypoiodite ions' formation)},$$

$$3IO^- \rightarrow IO^- + 2IO^-.$$

One mole of IO^- reacts with the species to titrate. Two moles of IO^- disproportionate according to

$$2IO^- \rightarrow 4/3\ I^- + 2/3\ IO_3^-;$$

[1] This denomination is not very judicious. It can indeed lead us to thinkthat iodine is at oxidation state –I.

– after acidification:

$$2/3IO_3^- + 10/3I^- + 4H^+ \rightarrow 2I_2 + 3H_2O \quad \text{(amphoterization)}$$

Finally, two moles of iodine remain for three moles initially used;

- for the blank, the same experimental conditions are retained. The only difference consists of the replacement of the volume of the titrand solution by the same volume of water. In these conditions, the three formed hypoiodite moles disproportionate according to

$$3IO^- \rightarrow 2I^- + IO_3^-,$$

and by acidification there are the amphoterization (retrodismutation) and recuperation of the three initial moles of iodine:

$$IO_3^- + 5I^- + 6H^+ \rightarrow 3I_2 + 3H_2O.$$

The difference between the added volumes of thiosulfate solution in the two experiments permits us to determine the number of moles that reacted.

Hypoiodometry is usually used in quantitative organic analysis. We now give some examples of its use.

19.1.2 Applications

19.1.2.1 Aldehydes

We have already encountered the determination of aldehydes with the formation of the corresponding carboxylate ions. A good example is provided by formaldehyde. The oxidation reaction of formol into formiate ion can be written as

$$HCHO + I_2 + 3NaOH \rightarrow HCOONa + 2NaI + 2H_2O.$$

Another example is provided by aldoses, for which it is generally admitted that, in alkaline medium, their structure is opened. In other words, in such a condition, they do possess a free aldehyde function. Hence, with glucose,

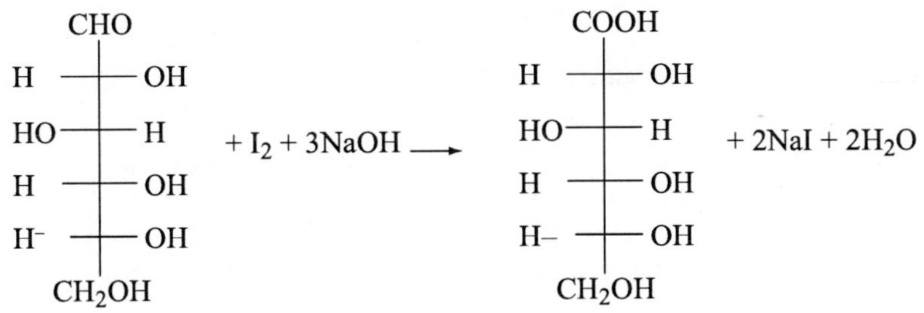

There is the formation of the sodium salt of the corresponding gluconic acid. Galactose, arabinose, and maltose react exactly as glucose. We must emphasize the fact that this reaction is one of the rare stoichiometric ones given by sugars. Chloral may also be determined by iodometry in alkaline medium according to the reaction

$$CCl_3CH(OH)_2 + I_2 + 3OH^- \rightarrow CCl_3COO^- + 2I^- + 3H_2O.$$

However, acetaldehyde exhibits a different behavior from that of chloral. It gives the iodoform reaction (see below). Unfortunately, the hypoiodite ions/aldehydes reaction is not selective. Several organic compounds react with hypoiodite ions; as a result, their determination may interfere with this reaction with aldehydes.

19.1.2.2 Dimethyl Ketone and the Iodoform Reaction

Dimethyl ketone (acetone) may be titrated by hypoiodite ions, with the formation of iodoform:

$$CH_3COCH_3 + 3IO^- \rightarrow CHI_3 + CH_3COO^- + 2OH^-.$$

It is the same reaction with acetaldehyde:

$$CH_3CHO + 3IO^- \rightarrow CHI_3 + HCOO^- + 2OH^-.$$

Several derivatives give a positive iodoform reaction. They are methyl ketones, monoiodinated methyl ketones, and diiodinated methyl ketones, and they exhibit the structural chain

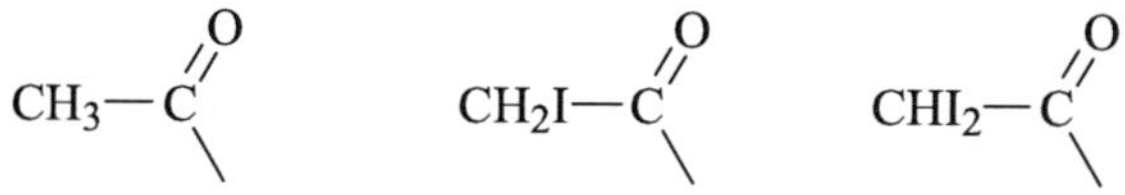

All the compounds that give a methyl ketone or acetaldehyde by hydrolysis or by oxidization also give a positive iodoform reaction. It is the case for ethanol, isopropanol, lactic acid, and β-dicarbonyl derivatives after acid hydrolysis.

19.2 Iodatometry

19.2.1 General Considerations

The term "iodatometry" groups titration methods involving potassium iodate as the oxidant. Potassium iodate is available in at least a 99.9% state of purity. Before use in iodatometry, it must be dried at 120 °C. The anhydrous salt does not tend to absorb the ambient moisture. It is a good primary standard. However, because of the high number of electrons it can exchange during the course of redox reactions (see below), it exhibits a small equivalent and the error in weighing it may be appreciable. This is its sole drawback.

Iodic acid may be considered a strong acid: pK_a (HIO_3/IO_3^-) = 0.80. Iodates are stable at every pH. They do not disproportionate into periodate ions (or periodic acid) and iodine or iodide ions (see Fig. 19.2). Recall that for about pH > 7, iodine and tri-iodide ions disproportionate into iodate and iodide ions. For these pH values, iodate ions are weakly oxidizing. For example, they do not oxidize iodide ions into iodine and tri-iodide ions (see Fig. 18.1 of Chap. 18). Conversely, in acidic medium, there is retrodismutation of iodate and iodide ions with the formation of iodine:

$$IO_3^- + 5I^- + 6H^+ \rightarrow 3I_2 + 3H_2O.$$

This reaction, as we already mentioned, is of very great analytical interest. A Pourbaix diagram (Fig. 18.1 of Chap. 18) clearly shows that iodate ions (and iodic acid) become powerful oxidizing agents. The course of the reaction is closely governed by the experimental conditions and the nature of the reducing agent. More precisely, it is the number of exchanged electrons by iodate that may change according to the reaction under consideration.

19.2.1.1 Iodatometry at Five Electrons

Without any particular condition (see below) and in moderately acidic medium (10^{-1}–2 mol/L hydrochloric acid), the redox couples to which iodate ions belong are the following:

$$IO_3^- + 5e^- + 6H^+ \rightleftharpoons I_2 + 3H_2O, \quad E°(IO_3^-/I_2) = 1.18\,V$$

$$\text{or} \quad HIO_3 + 5e^- + 5H^+ \rightleftharpoons I_2 + 3H_2O.$$

There is an exchange of five electrons. Therefore, we speak of "iodatometry at five electrons." The best-known example is probably that provided by the generation of iodine from iodide ions in acidic medium. We must note that in this case, in the three moles of liberated iodine, five iodine atoms come from iodide ions.

19.2.1.2 Iodatometry at Six Electrons

With very strong reducing agents such as titanous salts Ti^{3+}, an iodatometry at six electrons may occur. In the present case, titanic salts Ti^{4+} and iodide ions may form:

$$IO_3^- + 6Ti^{3+} + 6H^+ \rightarrow I^- + 6Ti^{4+} + 3H_2O,$$

The corresponding half-redox equilibria are

$$Ti^{4+} + 1e^- \rightleftharpoons Ti^{3+}, \quad E°Ti(+IV)/Ti(+III) = 0.03\,V,$$

$$IO_3^- + 6e^- + 6H^+ \rightleftharpoons I^- + 3H_2O, \quad E°(IO_3^-/I^-) = 1.09\,V.$$

19.2.1.3 Iodatometry at Four Electrons

In very concentrated hydrochloric acid (3–6 mol/L), an iodatometry at four electrons occurs. This is the base of Andrews' method. Iodine monochloride is formed. The corresponding half-redox equilibrium is

$$IO_3^- + 6H^+ + Cl^- + 4e^- \rightleftharpoons ICl + 3H_2O.$$

Iodine monochloride forms a stable ion complex with chloride ions, iodine dichloride:

$$ICl + Cl^- \rightleftharpoons ICl_2^-, \quad K = 1.7 \cdot 10^2.$$

The global half-redox reaction is

$$IO_3^- + 6H^+ + 2Cl^- + 4e^- \rightleftharpoons ICl_2^- + 3H_2O, \quad E°(IO_3^-/ICl_2^-) = 1.24\,V.$$

In these conditions, potassium iodate becomes a powerful oxidizing agent. As we said previously, iodine at oxidation state +I can be stabilized by forming miscellaneous derivatives. Iodine monochloride and iodine dichloride provide examples of this property.

From a mechanistic standpoint, it is interesting to notice that in hydrochloric acid, the oxidization with four exchanged electrons by IO_3^- first occurs through an iodatometry with six exchanged electrons. This means that the first half-redox equilibrium to be operative is

$$IO_3^- + 6H^+ + 6e^- \rightleftharpoons I^- + 3H_2O.$$

Iodide ions are then oxidized by iodate ions themselves:

$$IO_3^- + 5I^- + 6H^+ \rightleftharpoons 3I_2 + 3H_2O.$$

Finally, iodine is oxidized by iodate ions in hydrochloric acid. This reaction gives iodine monochloride:

$$IO_3^- + 6H^+ + 4e^- + Cl^- \rightleftharpoons ICl + 3H_2O.$$

An argument in favor of the first step involving six electrons is the fact that the iodide ions formed have been trapped by forming a chemical species duly characterized before their oxidization by iodate ions into iodine and iodine chlorides. Thus, it is an experimental fact that in the intermediary chemical species, iodine exhibits oxidization state −I. Hence, the determination of arsenious acid by potassium iodate in hydrochloric acid may usually be achieved by an iodatometry at four electrons according to the reaction

$$IO_3^- + 2H_3AsO_3 + 2H^+ + Cl^- \rightarrow ICl + 2H_3AsO_4 + H_2O.$$

Arsenic acid is formed in agreement with the standard potential values. But in the presence of a mercuric salt, the reaction exhibits another course. It becomes

$$4IO_3^- + 12H_3AsO_3 + Hg^{2+} \rightarrow HgI_4^{2-} + 12H_3AsO_4.$$

In this case, iodine has obviously gone from oxidation state +V to − I.

The detection of the equivalence point is often achieved by adding some drops of carbon tetrachloride or chloroform to the aqueous solution in the titration vessel. The equivalence point is detected by the disappearance of the purple coloration of the organic phase, which was due to the presence of iodine. (This is further proof of the sequence of reactions given above.) The color of an iodine monochloride is pale yellow.

19.2.2 Applications

The iodatometry at four electrons is the most commonly used iodatometry.

19.2.2.1 Determination of Iodide Ions

In 7 mol/L hydrochloric acid, the reaction is

$$2I^- + IO_3^- + 6H^+ + 6Cl^- \rightarrow 3ICl_2^- + 3H_2O.$$

19.2.2.2 Determination of Iodine

In the same conditions, iodine is oxidized by iodate ions according to

$$2I_2 + IO_3^- + 6H^+ + 10Cl^- \rightarrow 5ICl_2^- + 3H_2O.$$

Another methodology consists of operating in the presence of cyanide ions. Iodine is quantitatively oxidized into iodine cyanide (which may also be called cyanogen iodide). The half-redox reaction is

$$2ICN + 2H^+ + 2e^- \rightleftharpoons I_2 + 2HCN, \quad E°(ICN/I_2) = 0.63\,V.$$

The global reaction is

$$IO_3^- + 5HCN + 2I_2 + H^+ \rightarrow 3H_2O + 5ICN.$$

The advantage of this methodology is to avoid a medium too rich in hydrochloric acid; a 1 mol/L concentration is sufficient. Therefore, it is possible to use starch as the indicator. This was not the case in the preceding conditions.

These reactions do not seem to present much interest, because of the several existing titration methods of the species mentioned above. However, they at least present one point of interest: the case in which both species must be titrated in a mixture. In this case, the analysis is achieved in two steps. In the first one, an iodatometry in very concentrated hydrochloric acid is achieved. Therefore, iodide ions and iodine are oxidized. If the solution contains x moles of iodide and y moles of iodine per liter, and if m moles of iodate ions are added at the final point, we can write

$$2m = x + y.$$

In the second step, a titration with thiosulfate permits us to directly obtain y.

19.2.2.3 Determination of Hydrazines and Derivatives

Hydrazine N_2H_4 belongs to the half-redox couple

$$N_{2(g)} + 4e^- + 5H^+ \rightleftharpoons NH_2{}^-NH^{3+}, \quad E°(N_2/N_2H_5{}^+) = -0.23\,V.$$

Nitrogen goes from oxidization state 0 in dinitrogen to state – II in hydrazine. The value of the standard potential shows that hydrazine is a very reducing agent. It is oxidized in Andrews' conditions by potassium iodate:

$$IO_3{}^- + N_2H_5{}^+ + H^+ + Cl^- \rightarrow ICl + N_{2(g)} + 3H_2O.$$

The titration of isoniazid or isonicotinic acid hydrazid, a pharmaceutical of great interest, is based on this titration reaction:

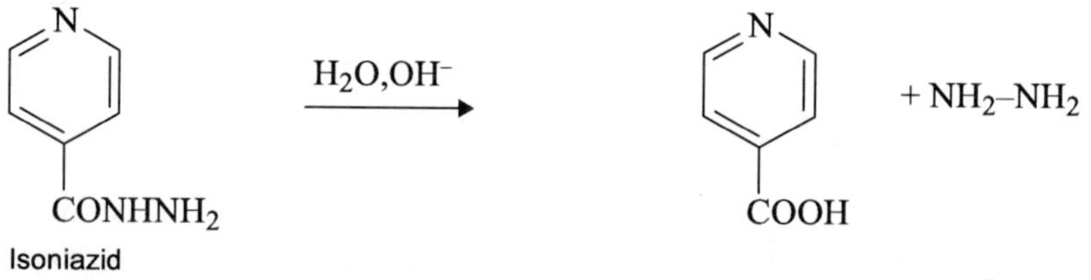

In a first step, isoniazid is quantitatively hydrolyzed to give isonicotinic acid and hydrazine, which is titrated by the iodatometry at four electrons.

19.2.2.4 Determination of Some Quaternary Ammoniums

The determination of quaternary ammoniums is based on the formation of an ion pair between the quaternary ammonium cation $^+NR_1R_2R_3R_4$ and the iodide ion. Hence, the ion pair is $^+NR_1R_2R_3R_4,I^-$. The initial quaternary salt (that to be determined) $^+NR_1R_2R_3R_4,X^-$ (with $X^- \neq I^-$) in aqueous solution is set up in the presence of an exactly known concentration of iodide ions in excess, chloroform, and sodium hydroxide. The ion pair $^+NR_1R_2R_3R_4,I^-$ is formed quantitatively and extracted into chloroform. This phase is thrown out. The iodide ions remaining in the aqueous solution are titrated with potassium iodate in the presence of concentrated hydrochloric acid. The European pharmacopeia urges the determination of cetrimide and benzalkonium chloride[2] according to this principle:

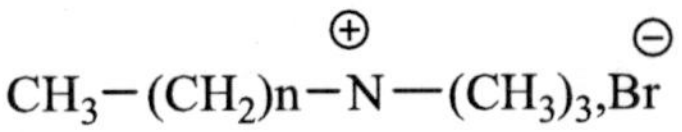

Cetrimide

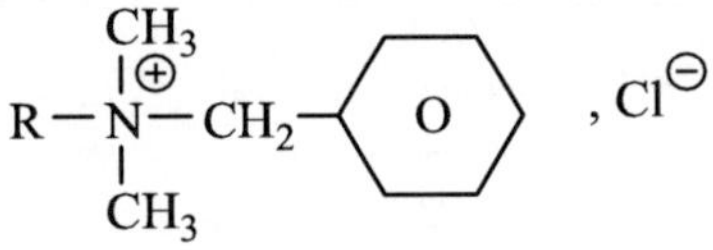

benzalkonium chloride[2]

The interest in the ion-pair extraction in the presence of sodium hydroxide lies in the fact that quaternary ammoniums of the $^+HNR_1R_2R_3$ type cannot exist in these conditions. If that were the case, they would also be extracted as ion pairs such as $^+HNR_1R_2R_3,I^-$, which would consume more iodide ions than those used to extract the product $^+NR_1R_2R_3R_4$ to be determined. Indeed, due to their pK_a values ($8.8 < pK_a < 10.2$), $^+HNR_1R_2R_3$ cations cannot exist in strongly basic medium. We must also notice that tertiary amines $NR_1R_2R_3$, which are often impurities of quaternary ammoniums, are extracted by the organic phase.

Unfortunately, all the quaternary ammoniums cannot be extracted as ion pairs into an organic solvent immiscible with water.

19.2.2.5 Determination of Antimony(III)

The reaction that determines antimony(III) is the following one:

$$IO_3^- + 2[SbCl_4^-] + 6H^+ + 5Cl^- \rightarrow ICl + 2[SbCl_6^-] + 3H_2O.$$

Antimony is oxidized into Sb(+V). The titration must be achieved in concentrated hydrochloric acid, 2.5–3.5 mol/L.

19.2.2.6 Determination of Mercury(+I) and Mercury(+II) by Extension

Mercury is precipitated as dimercury(+I) dichloride (calomel), which is oxidized with potassium iodate in hydrochloric acid medium:

$$IO_3^- + 2Hg_2Cl_2 + 6H^+ + 13Cl^- \rightarrow ICl + 4[HgCl_4]^{2-} + 3H_2O.$$

The formation of the complex tetrachloromercurate(II) facilitates the reaction evolving toward the right. An examination of standard potentials of the couples IO_3^-/ICl

---

[2] n and R are variable substituents. Actually, this product is a mixture.

Table 19.1 Nomenclature of periodic acids

Name		Dehydration vs. paraperiodic acid
H_5IO_6	Ortho or paraperiodic acid	
HIO_4	Periodic or metaperiodic acid	H_5IO_6 $(-2H_2O)$
H_3IO_5 ou $(H_6I_2O_{10})$	Mesoperiodic acid or diperiodic acid	$2H_5IO_6$ $(-2H_2O)$
$H_7I_3O_{14}$	Triperiodic acid	$3H_5IO_6$ $(-4H_2O)$

and Hg^{2+}/Hg_2^{2+} (see the preceding chapter) shows that the oxidization of mercurous mercury with iodate ions is favored, regardless of the conditions. (The global formation constant of the tetrachloromercurate complex is $\beta_4 = 10^{16}$. It is less stable than the complex tetraiodomercurate(II); see Chap. 18.)

To conclude our study of iodatometries, let's recall that, obviously, the equivalents of potassium iodate are not the same according to the occurring iodatometry. They may be the potassium iodate molar mass divided by six, five, or four.

19.3 Periodimetry

The name "periodimetry" groups all the titrimetric methods based on the oxidizing power of periodic acids.[3]

In periodic acids, iodine is at oxidation state +VII. Periodic acids correspond to the hypothetical peroxidized anhydride I_2O_7 depending on the hydration degrees of I_2O_7. Table 19.1 gives the nomenclature of periodic acids.

The most used are paraperiodic acid H_5IO_6, prepared *in situ* by dissolving trisodium paraperiodate $Na_3H_2IO_6$ into a $5 \cdot 10^{-2}$ mol/L solution of sulfuric acid and metaperiodic acid HIO_4 prepared *in situ* by dissolving the sodium salt into sulfuric acid. Their concentration is generally about 10^{-1} mol/L. Actually, several equilibria exist among the different species: They are dissociation acid–base, dehydration, and dimerization equilibria. We now give some pK_a values at 25 °C:

$$H_5IO_6 \rightleftharpoons H_4IO_6{}^- + H^+, \quad pK_a = 3.29,$$

$$H_4IO_6{}^- \rightleftharpoons H_3IO_6{}^{2-} + H^+, \quad pK_a = 8.31,$$

$$H_3IO_6{}^{2-} \rightleftharpoons H_2IO_6{}^{3-} + H^+, \quad pK_a = 11.6.$$

In another respect, paraperiodate ion $H_4IO_6{}^-$ dehydrates according to the equilibrium

$$H_4IO_6{}^- \rightleftharpoons IO_4{}^- + 2H_2O, \quad K = 29(25°C),$$

and $H_3IO_6{}^{2-}$ dimerizes according to

$$2H_3IO_6{}^{2-} \rightleftharpoons H_2I_2O_{10}{}^{4-} + 2H_2O, \quad K = 820(25°C),$$

[3] The names of L. Malaprade, who discovered it in 1928, and P. Fleury and J. Lange, who thoroughly studied it, must be associated with this method.

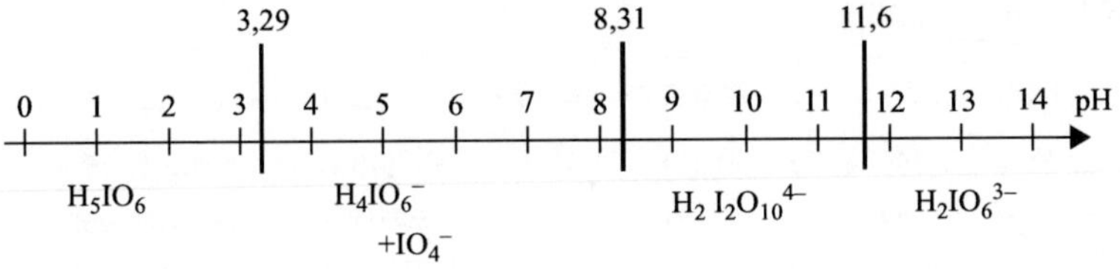

Fig. 19.4 Predominance area of periodic acids and of their conjugate bases

Fig. 19.5 A Pourbaix diagram of paraperiodic acid (second convention: $C = 10^{-2}$ mol/L)

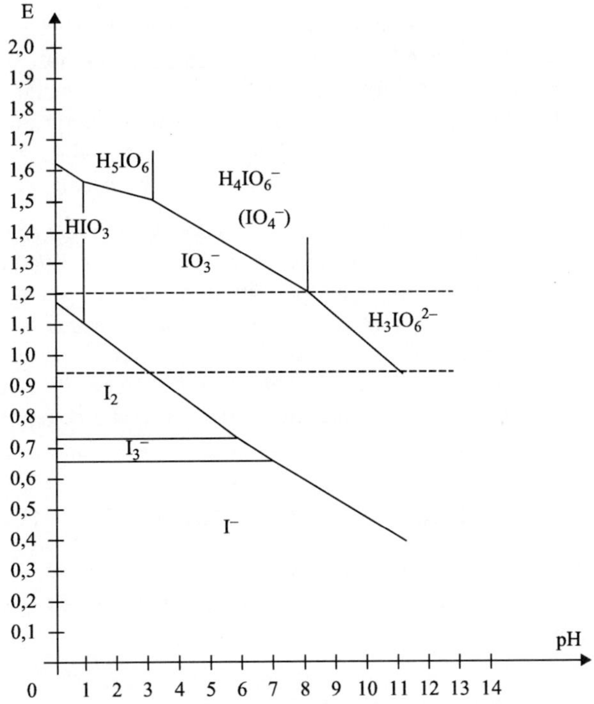

As a result, in the range 3.29 < pH < 8.31, the two species $H_4IO_6^-$ and IO_4^- predominate, whereas for pH > 8.31, it is the predominance area of $H_2I_2O_{10}^{4-}$ (Fig. 19.4).

The examination of the Pourbaix diagram of paraperiodic acid (Fig. 19.5) permits us to make the following conclusions:

- it is a very strong oxidizing agent. Paraperiodate ions remain very oxidant in basic media;
- during the course of the redox reaction, periodic acid and paraperiodate ions are usually reduced into iodic acid or iodate ions according to the pH value at which the titration is achieved. Therefore, most of the time, there is a two-electron exchange. Schematically, iodine goes from oxidation state + VII to state + V.

The half-redox equilibria are

$$H_5IO_6 + H^+ + 2e^- \rightleftharpoons IO_3^- + 3H_2O, \quad E°(H_5IO_6/IO_3^-) = 1.60\,V,$$

$$HIO_4 + H^+ + 2e^- \rightleftharpoons IO_3^- + H_2O, \quad E°(HIO_4/IO_3^-) = 1.65\,V,$$

$$H_4IO_6^- + 2H^+ + 2e^- \rightleftharpoons IO_3^- + 3H_2O, \quad E°(H_4IO_6^-/IO_3^-) = 1.70\,V,$$

$$H_3IO_6^{2-} + 3H^+ + 2e^- \rightleftharpoons IO_3^- + 3H_2O, \quad E°(H_3IO_6^{2-}/IO_3^-) = 1.95\,V,$$

$$H_2IO_6^{3-} + 4H^+ + 2e^- \rightleftharpoons IO_3^- + 3H_2O, \quad E°(H_2IO_6^{3-}/IO_3^-) = 2.30\,V.$$

The high standard potential values confirm the strong oxidizing power of periodic acids and paraperiodate ions. A good illustration of this assertion is the fact that periodic acids oxidize manganous ions into permanganate ions in acidic medium. If the pH value is lower than 5 and if the reductor is sufficiently strong, periodic acid may be reduced into iodine or even into iodide ions. We must also notice that oxidization reactions with periodic acids are usually slow.

- from a strict thermodynamic standpoint, periodic acids and periodate ions are not stable in aqueous solution regardless of the pH value. However, the oxidization of water is slow. This permits us to prepare standardized solution of periodic acids that are relatively stable. They are aqueous solutions of periodic acid or of sodium or potassium metaperiodates extemporaneously dissolved in sulfuric acid medium. They can be standardized with arsenious acid in neutral medium or in a slightly alkaline one (hydrogen carbonate or borax buffer). The global redox reaction is

$$H_4IO_6^- + H_3AsO_3 + 2OH^- \rightarrow IO_3^- + HAsO_4^{2-} + 4H_2O,$$

which is a back titration. Arsenious acid in excess is titrated by iodometry. The solutions of periodic acids can also be standardized by Fleury and Lange's method (see below).

Concerning this point, the only species derivating from the iodine element to be stable (from a thermodynamic standpoint) in acidic medium are iodine, tri-iodide and iodide ions, iodic acid, and iodate ions. In basic medium, only iodate and iodide ions are stable.

From a practical standpoint, titrations are only back titration achieved in strictly specified experimental conditions (time, temperature, pH, and light). They vary with the substrates. After the reaction has evolved, the excess of periodic acid is determined. Usually, there is an exchange of two electrons, and the equivalence factor $f_{eq} = ½$.

In order to appreciate the excess of periodic acid, two methodologies are used:

- that of Fleury and Lange: After the reaction in sulfuric acid medium has evolved, the solution's pH is increased and is fixed in the range $7 < pH < 8$ by the addition of potassium hydrogen carbonate. Then potassium iodide is added. In these conditions, iodate ions do not oxidize iodide ions (connected areas—Fig. 19.5), but paraperiodate ions oxidize them according to the reactions

$$H_4IO_6^- + 2H^+ + 2I^- \rightarrow IO_3^- + I_2 + 3H_2O,$$

$$H_3IO_6^{2-} + 2I^- \rightarrow IO_3^- + I_2 + 3OH^-.$$

One iodine molecule is formed for each periodate molecule in excess. The liberated iodine is titrated with an arsenite solution, due to the pH value;

- that of Malaprade: An excess of iodide ions is added to the strongly acidified medium in which the oxidization reaction by periodic acid has evolved. In these conditions, periodic acid as well as iodate ions and iodic acid oxidize iodide ions according to the reactions (see Fig. 19.5)

$$H_5IO_6 + 2I^- + H^+ \rightarrow IO_3^- + I_2 + 3H_2O,$$

$$IO_3^- + 5I^- + 6H^+ \rightarrow 3I_2 + 3H_2O$$

$$(\text{or} \quad HIO_3 + 5I^- + 5H^+ \rightarrow 3I_2 + 3H_2O).$$

The liberated iodine is titrated with a thiosulfate solution. Parallel to this process, a blank titration is achieved exactly in the same experimental conditions, except that the volume of the solution to study is replaced by the same volume of water. Iodine liberated by the blank solution is also appreciated with thiosulfate. The sought-after concentration is determined after comparing the added volumes of thiosulfate solution to the sample and to the blank. The basis of Malaprade's methodology lies in the fact that the reactions of oxidization of iodide ions with periodic acid and with iodic acid are endowed with different stoichiometries.

If n moles of paraperiodic acid are used, n moles of iodine are liberated by the blank since the acid has not been transformed and since it liberates iodine mole to mole. In the sample, x moles of the substance to determine give x moles of iodic acid (if we admit that the product is oxidized with a loss of two electrons—if this is not the case, the reaction stoichiometry must be taken into account), which, in turn, liberates $3x$ moles of iodine. Hence, $n-x$ moles of paraperiodic acid remain. As a result, the sample liberates on the whole

$$(n - x) + 3x \quad \text{moles of iodine.}$$

n and $n+2x$ are appreciated with the help of the thiosulfate volumes. x is immediately deduced.

We remark that with both methodologies, the periodimetry ends with an indirect iodometry.

Another methodology exists to follow a periodimetric titration. It is based on quite a different principle. We shall see (see immediately below) that in some well-defined organic structures, a secondary alcohol group gives one mole of formic acid, quantitatively. This transformation can be schematically written as

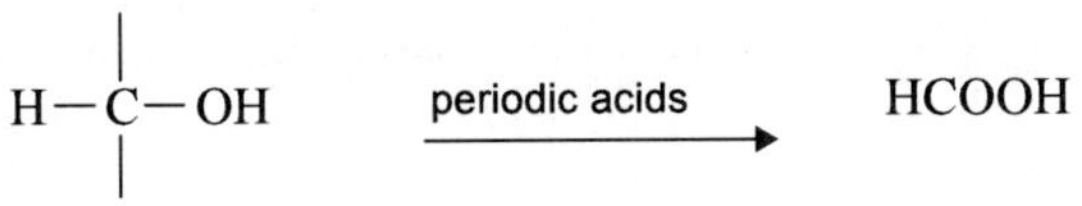

The other methodology involves the protometric titration of the formic acid that was formed. It necessitates first adding ethylene glycol to the solution once the oxidization reaction has finished. Ethylene glycol destroys the excess of periodic acid, which is transformed into iodate ions. Formic acid titration is achieved by pH-metry, the equivalent point being taken for pH = 7. Beyond pH = 7, the equilibria between the different periodate ions induce the formation of hydroxide ions, which distort the determination.

The realm of applications of periodimetry lies in the quantitative organic analysis. It is based on *Malaprade's reaction.* Periodic acids and periodate ions selectively oxidize α-glycols of the same sort as that represented just below to give the two corresponding carbonylated derivatives. The reaction may be written as

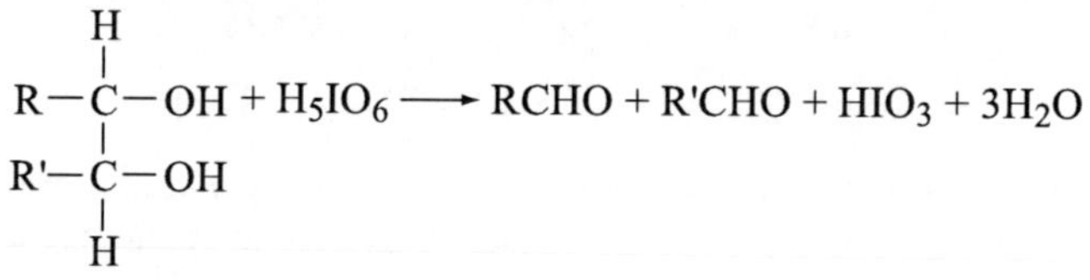

α-glycol

Of course, according to the pH value, periodic and iodic acids may be ionized. In this equation, R and R' may be hydrogen atoms or organic groups. We may admit that in these redox reactions the antagonistic redox couple of the couple periodic acid (periodate)/iodate is the couple (RCHO + R'CHO)/RCHOH–CHOHR', for which we may write the following half-redox reaction:

$$\begin{matrix} \mathrm{R\,CHO} \\ + \\ \mathrm{R'\,CHO} \end{matrix} + 2H^+ + 2e^- \rightleftharpoons \begin{matrix} \mathrm{RCHOH} \\ | \\ \mathrm{R'\text{-}CHOH} \end{matrix}$$

In other words, if in the initial diol the carbon atoms bringing the alcohol functions exhibited oxidation numbers x and y, after oxidization they would exhibit the numbers $x + 1$ and $y + 1$; schematically, this looks like

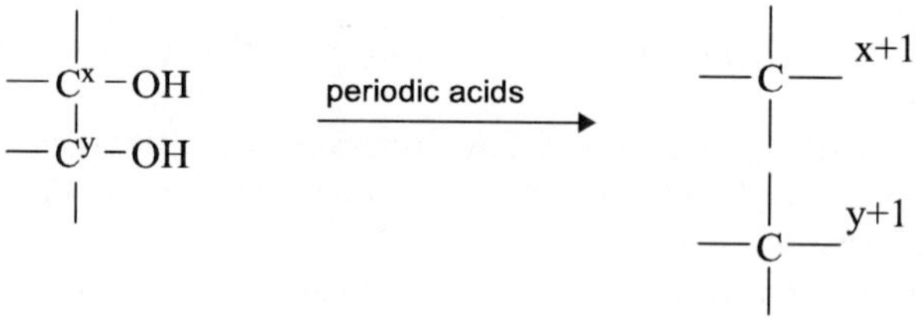

The reaction evolves in a few minutes at room temperature in pH ranges located between very acidic media and mild basic media. Monoalcohols are not appreciably oxidized in these conditions.

An easy way to predict the structures (on paper) of the obtained products after the reaction of periodic acids (or periodate ions) with diols consists of replacing the bond that linked the two carbons, bringing the alcoholic function by two new C–OH bonds, one for each carbon. With this approximate rule, we usually find aldehyde or ketone hydrates, which lose one molecule of water from each gem-dihydroxy derivative virtually formed. Therefore, for example, one molecule of 2,3-butanediol gives two molecules of acetaldehyde, which is the final product.

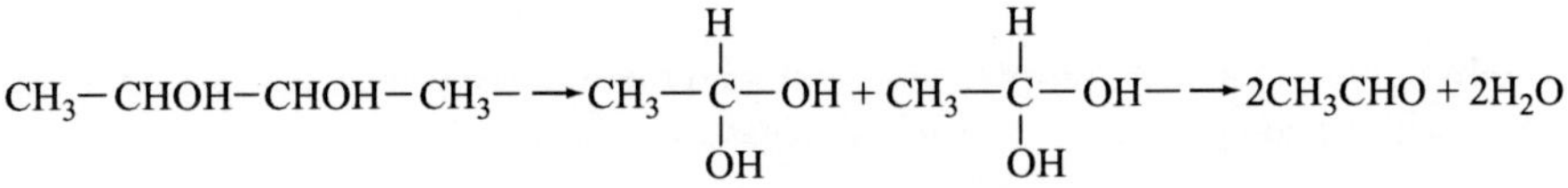

The periodic oxidization of glycerol globally gives two molecules of formaldehyde and one molecule of formic acid:

$$CH_2OH{-}CHOH{-}CH_2OH \xrightarrow{H_5IO_6} 2HCHO + HCOOH$$

glycerol

The above rule easily predicts this result. We must take note that the secondary alcohol group is transformed into formic acid, which can be determined by protometry (as we have seen). Likewise, with the same rule, we immediately find that 2,3-dimethylbutane-2,3-diol gives two molecules of acetone:

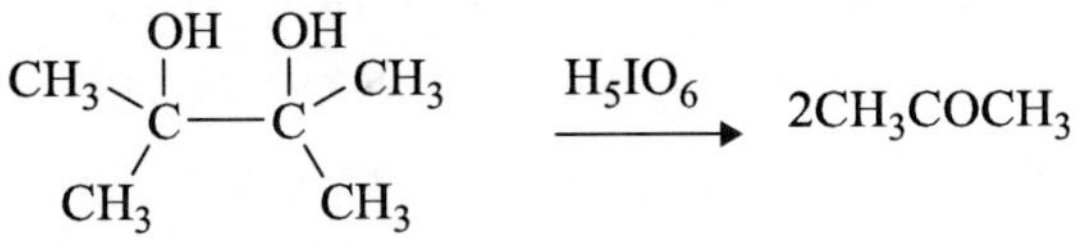

We emphasize that it is only a trick of writing and is in no case a reactional mechanism.

Malaprade's reaction is interesting for the determination of sugars and polyols. Thus, aldoses give the following reaction:

$$CH_2OH{-}(CHOH)_4{-}CHO + 5H_5IO_6 \rightarrow HCHO + 5HCOOH + 5HIO_3 + 10H_2O.$$

(As could be predicted, the primary alcohol group gives one formol molecule and the four secondary alcohol groups four molecules of formic acid.) The aldehyde function also gives one molecule of formic acid. Therefore, the reaction follows the above explicative scheme. Ketoses can also be determined by periodimetry. Thus, fructose gives the following reaction:

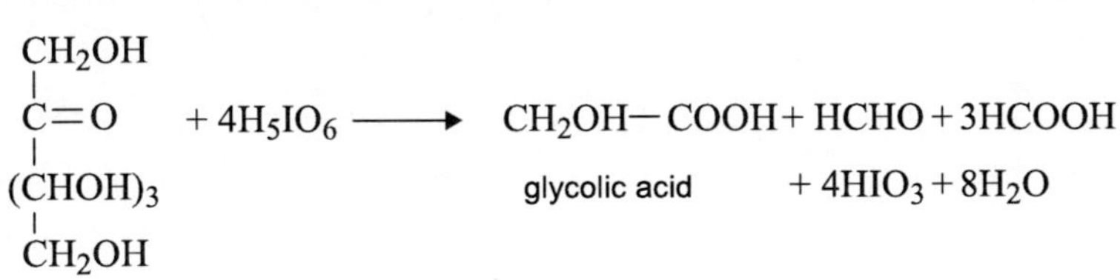

Fructose

In this example, we notice that glycolic acid coming from the cetol group of fructose does not transform further. Polyols derivating from sugars, such as sorbitol and mannitol, are oxidized as predicted. Therefore, for these two polyols, which are isomers, the reaction is

$$CH_2OH–(CHOH)_4–CH_2OH + 5H_5IO_6 \rightarrow 2HCHO + 4HCOOH + 5HIO_3 + 12H_2O.$$

Methyl glycols, such as propylene glycol, and methyl pentoses, such as rhamnose, react with periodic acids to give acetaldehyde according to the reactions

$$CH_3–CHOH–CH_2OH + H_5IO_6 \rightarrow CH_3CHO + HCHO + HIO_3 + 3H_2O,$$
$$CH_3–(CHOH)_4–CHO + 4H_5IO_6 \rightarrow CH_3CHO + 4HCOOH + 4HIO_3 + 8H_2O.$$

A means exists here to appreciate propylene glycol in the presence of glycerol and methyl pentoses in mixture with hexoses.

Some other structures also suffer a quantitative Malaprade reaction. They are α-cetols of the general structure RCOCHOHR', α-diketones, α-ketoaldehydes, α-hydroxyaldehydes, and glyoxal. The reactions are, respectively,

$$CH_3COCHOHCH_3 + H_5IO_6 \rightarrow CH_3COOH + CH_3CHO + HIO_3 + 2H_2O,$$
$$CH_3COCOCH_3 + H_5IO_6 \rightarrow 2CH_3COOH + HIO_3 + H_2O,$$
$$CH_3COCHO + H_5IO_6 \rightarrow CH_3COOH + HCOOH + HIO_3 + H_2O,$$
$$CH_2OH–CHO + H_5IO_6 \rightarrow HCHO + HCOOH + HIO_3 + 2H_2O,$$
$$CHO–CHO + H_5IO_6 \rightarrow 2HCOOH + HIO_3 + H_2O.$$

The rule given above enables us to find the structures of the formed products without any difficulty. Periodic acids also react quantitatively and rapidly with α-amino alcohols in which the amine function is primary or secondary. Globally, the reaction is

$$\begin{array}{l} \quad\;\; H \\ \quad\;\; | \\ R-C-OH \\ \quad\;\; | \\ R'-C-NH_2 \\ \quad\;\; | \\ \quad\;\; H \end{array} + H_5IO_6 + H^+ \rightarrow RCHO + R'CHO + NH_4^+ + HIO_3 + 2H_2O$$

Two typical examples are provided by ethanolamine and serine. Serine is an amino acid:

$$CH_2OH–CH_2NH_2 + H_5IO_6 + H^+ \rightarrow 2HCHO + NH_4^+ + HIO_3 + 2H_2O,$$
$$CH_2OH–CH(NH_2)COOH + H_5IO_6 + H^+ \rightarrow HCHO + CHO–COOH + NH_4^+ + HIO_3 + 2H_2O.$$

The European pharmacopeia recommends the titrations of glycerol, sorbitol, mannitol, calcium gluconate, adenosine triphosphate, and riboflavin (vitamin B2) by periodimetry:

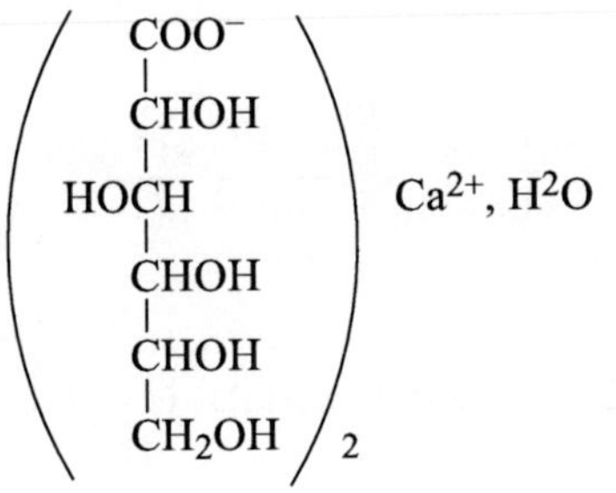

Calcium gluconate

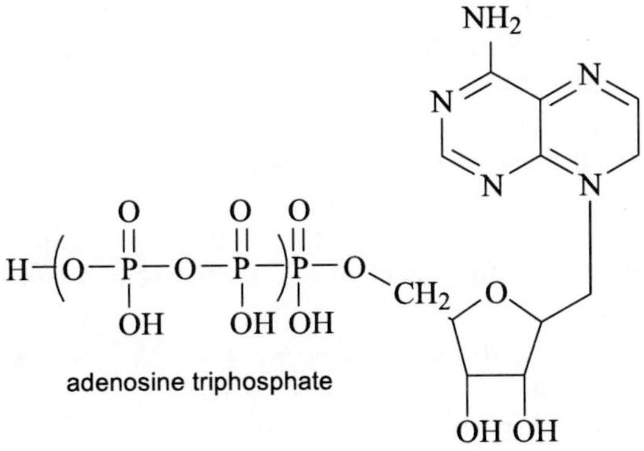

adenosine triphosphate

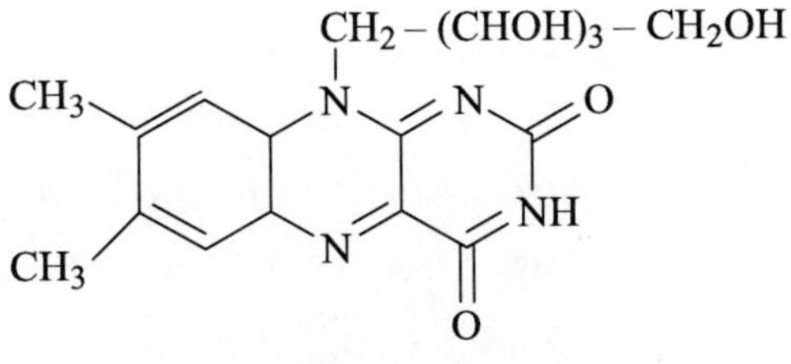

Riboflavine (vitamine B_2)

It is easily conceivable that, due to its specificity toward some substrates, the periodimetry has presented an interest in qualitative organic analysis for the determination of the structures. Knowledge of the fragment structures permits that of the initial structure. Today it has been completely supplanted by some spectroscopic methods, in particular, by NMR.

The high selectivity of the mechanism of the oxidization reaction comes from the fact that periodic acids form cyclic esters with α-diols according to the scheme

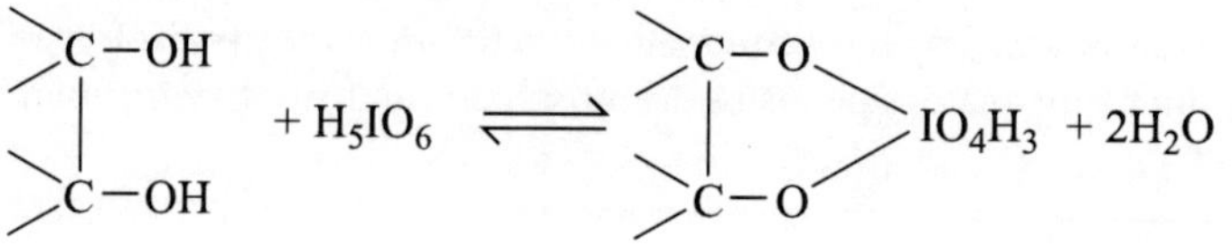

However, this hypothesis does not explain all the experimental facts. The formation of oxygenated radicals is also envisaged, at least in some cases. This hypothesis is in agreement with the result that often some light conditions must be respected.

19.4 Bromatometry, Hypobromometry, and Bromometry

19.4.1 General Considerations

In this section, we study the titration methods involving the bromine element.

Bromine Br_2 is a red-brown liquid at ordinary temperature. It is very volatile. It gives very dense red-brown vapors that are highly toxic to mucous membranes. In the liquid state, it is particularly corrosive. It results in very serious burns on the skin.

Its chemical analysis exhibits great analogies with iodine.

Dibromine aqueous solutions are very unstable, because of bromine's volatility. They cannot be more concentrated than 0.015 mol/L for solubility reasons. Bromine is more soluble in the presence of chloride ions by formation of the complex ion $[Br_2Cl^-]$ and in the presence of bromide ions by formation of complex ions $[Br_3{}^-]$ and $[Br_5{}^-]$. Its solubility is 0.4 mol/L in a 0.4 mol/L NaCl solution and 1.4 mol/L in a 1.0 mol/L KBr solution.

The standard potentials of the couples involving the bromine element and that are interesting for the chemical analysis are

$$Br_{2(l)} + 2e^- \rightleftharpoons 2Br^-, \quad E°(Br_{2(l)}/Br^-) = 1.065\,V,$$

$$Br_{2(w)} + 2e^- \rightleftharpoons 2Br^-, \quad E°(Br_{2(w)}/Br^-) = 1.087\,V,$$

$$BrO_3{}^- + 6H^+ + 5e^- \rightleftharpoons \tfrac{1}{2}Br_{2(l)} + 3H_2O, \quad E°(BrO_3{}^-/Br_{2(l)}) = 1.520\,V,$$

$$BrO_3{}^- + 6H^+ + 6e^- \rightleftharpoons Br^- + 3H_2O, \quad E°(BrO_3{}^-/Br^-) = 1.440\,V,$$

$$HBrO + H^+ + 1e^- \rightleftharpoons \tfrac{1}{2}Br_2 + H_2O, \quad E°(HBrO/Br_2) = 1.600\,V,$$

$$HBrO + H^+ + 2e^- \rightleftharpoons Br^- + H_2O, \quad E°(HBrO/Br^-) = 1.341\,V.$$

Bromine is markedly more oxidizing than iodine.

A Pourbaix diagram of bromine is quite analogous to that of iodine except for the fact that the frontier straight lines are displaced. For pH values lower than about 6.2 and considering decreasing potential values, the stable species in aqueous solutions are bromic acid or bromates $[pK_a(HBrO_3/BrO_3{}^-) = 0.70]$, bromine, and bromide ions. For higher pH values, bromine slowly disproportionates into bromate and bromide ions according to the reactions

$$3Br_2 + 3H_2O \rightleftharpoons BrO_3^- + 5Br^- + 6H^+,$$

$$3Br_2 + 6OH^- \rightleftharpoons BrO_3^- + 3H_2O + 5Br^-.$$

These reactions are true equilibria more and less displaced toward the right depending on the pH value. In acidic medium, bromate ions oxidize iodide ions to give bromine according to the reaction

$$BrO_3^- + 5Br^- + 6H^+ \rightarrow 3Br_2 + 3H_2O.$$

This is a very important reaction. It permits us to generate bromine *in situ* from potassium bromate, which is a primary standard.

As iodine, bromine disproportionates into hypobromite BrO^- and bromide ions for pH values higher than about 8.5:

$$Br_2 + 2OH^- \rightleftharpoons BrO^- + H_2O + Br^-, \quad (K = 2.10^8 \text{ mol/L at pH} = 14).$$

This is a false equilibrium and hypobromites and hypobromous acid HBrO ($pK_a = 8.6$) exist only for kinetic reasons. Hypobromite ions disproportionate themselves into bromide and bromate ions, which constitute the true thermodynamic reaction limit involving these species at basic pH values for which hypobromite ions are transitorily formed:

$$3BrO^- \rightleftharpoons 2Br^- + BrO_3^- \quad (K = 10^{15}),$$

The constant K is very high in favor of the bromate ions' formation. However, the reaction is rather slow at room temperature (in the family of halogens, hypochlorous acid and hypochlorites are the hypohalogenites, which disproportionate more slowly at room temperature).

Potassium perbromate $KBrO_4$ and perbromic acid $HBrO_4$ are known. Although the redox couple

$$BrO_4^- + 2H^+ + 2e^- \rightleftharpoons BrO_3^- + H_2O$$

has the very high standard potential value $E°(BrO_4^-/BrO_3^-) = 1.85$ V, perbromate ions are known for their sluggish reactivity, probably for kinetic reasons. They do not seem to have received analytical applications.

Bromate ions are stable and hence do not disproportionate into perbromate ions and bromine or bromide ions in acidic medium (see Fig. 19.6).

This is the case for the analogous species of iodine (see Fig. 19.2). For the sake of comparison, let's say that this is not the case for chlorate ions ClO_3^-, which disproportionate (very slowly) into chloride and perchlorate ions.

From a purely analytical standpoint, we now mention some quantitative reactions. They are classified under the following headings:

- oxidization reactions with bromate and hypobromite ions and with bromine;
- fixation reactions of bromine on organic molecules either by substitution or by addition. This latter class of reaction can also be classified as being redox reactions, but this concept is difficult to handle in this case.

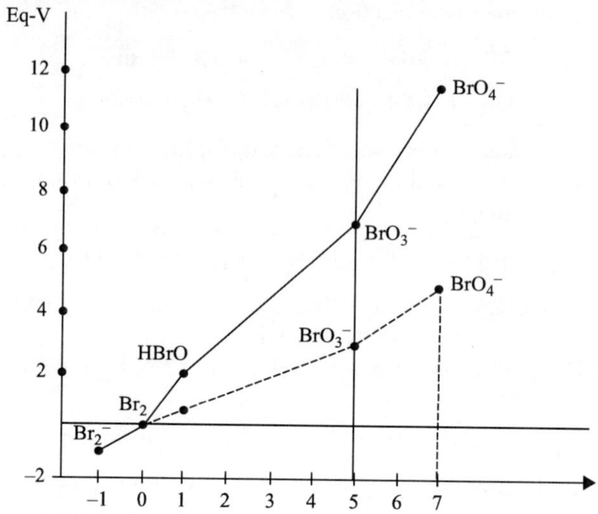

Fig. 19.6 Frost diagram of species derivating from bromine

19.4.2 Oxidization Reactions

- Bromate ions are strong oxidizing agents. During the course of the reaction, they are reduced into bromide ions according to the half-redox equilibrium

$$BrO_3^- + 6H^+ + 6e^- \rightleftharpoons Br^- + 3H_2O, \quad E°(BrO_3^-/Br^-) = 1.44\,V.$$

The potassium bromate equivalent is its molar weight divided by 6. For example, in hydrochloric acid, potassium bromate oxidizes arsenious acid, hydrazine, hexacyanoferrate(II) (ferrocyanide ion), and hydroxylamine, according to the following respective reactions:

$$BrO_3^- + 3H_3AsO_3 \rightarrow Br^- + 3H_3AsO_4,$$

$$2BrO_3^- + 3N_2H_4 \rightarrow 2Br^- + 3N_{2(g)} + 6H_2O,$$

$$BrO_3^- + 6[Fe(CN)_6]^{4-} + 6H^+ \rightarrow Br^- + 6[Fe(CN)_6]^{3-} + 3H_2O,$$

$$BrO_3^- + NH_2OH \rightarrow Br^- + NO_3^- + H^+ + H_2O.$$

We can check that there is an exchange of six electrons by the couple bromate/bromide ions by considering the oxidization states of the members of the antagonistic couples. (In hydroxylamine, the nitrogen atom must be considered to be at oxidation state – I. Because of its two bonds with hydrogen atoms, indeed, at first sight, it exhibits oxidation state – II, but simultaneously, it is also bound to an oxygen atom. Since oxygen is more electronegative than nitrogen, the bound N–O must be considered as fictitiously ionized into N^+ and O^-. Hence, the definitive oxidation state – I of nitrogen is established.) From another standpoint, the direction of the above reactions is in agreement with the tendency given by the examination of standard potentials. [The standard potential of the couple NO_3^-/NH_3OH^+ is $E°(NO_3^-/NH_3OH^+) = 0.73$ V. The half-redox equilibrium is

$$NO_3^- + 6e^- + 8H^+ + 1/2H_2O \rightarrow 5/2\ H_2O + NH_3OH^+.$$

The reaction with arsenious acid allows the verification of the titer of bromate solutions. Likewise, the reaction with iodide ions,

$$BrO_3^- + 6I^- + 6H^+ \rightarrow 3I_2 + Br^- + 3H_2O,$$

allows the determination of an excess of bromate ions by indirect iodometry, iodine being titrated with thiosulfate ions.

When the oxidization reaction has reached its equivalent point, there is, of course, the appearance of an excess of bromate ions, which react with bromide ions in solution and therefore give free bromine. The solution becomes yellow. The reaction is

$$BrO_3^- + 5Br^- + 6H^+ \rightarrow 3Br_2 + 3H_2O.$$

The equivalence point may also be detected by using color indicators (methyl orange, methyl red, naphthalene black, xylidine ponceau, and fuchsine), which exhibit their usual color before the equivalence point and after become discolored by an irreversible chemical transformation. The transformation may be an oxidization of the indicator or a fixation of bromine on it, since it is generated once the equivalence point is reached. In hydrochloric acid, which is the most commonly used medium, we may consider that after the equivalence point, bromate ions oxidize chloride ions [$E°(Cl_2/Cl^-) \approx 1.38$ V] according to the reaction

$$BrO_3^- + 5Cl^- + 6H^+ \rightarrow 5/2\ Cl_2 + \tfrac{1}{2}Br_2 + 3H_2O.$$

Then the formed chlorine could bleach the indicator. We must note that the last two reactions are a bromatometry at five electrons and not at six electrons as earlier [the standard potential of the couple BrO_3^-/Br_2 ($E° = 1.52$ V) is slightly higher than that of the couple BrO_3^-/Br^-]. We should also note that the indicators used are not redox indicators.

It is possible to see in a course devoted to analytical electrochemistry that it is very easy to detect an excess of bromine regardless of its origin. It may come from the titrant itself or it may be generated just after the equivalence point, which is the case for the present example.

- now concerning the oxidization by hypobromite ions, let's recall that they are prepared by adding a bromine solution to a solution of sodium hydroxide. The solution, once prepared, must also contain bromide and bromate ions since hypobromite ions disproportionate into both preceding ones. Their solutions are unstable and must be prepared extemporaneously. To experimentally proceed, we add the unknown solution to a known quantity in excess of hypobromite solution.

After about 30 minutes of reaction, solutions of hydrochloric acid and potassium iodide are added. Hypobromite ions in excess oxidize iodide ions into iodine, which is then titrated with a thiosulfate solution:

$$BrO_3^- + 2I^- + 2H^+ \rightarrow H_2O + I_2 + Br^-.$$

Another pathway to explain the oxidization is possible. In a first reaction, a disproportionation of hypobromite ions into bromate and bromide ions would occur according to

$$3BrO^- \rightarrow 2Br^- + BrO_3^-.$$

Iodide ions are then oxidized by bromate ions since the solution has been acidified:

$$BrO_3^- + 6I^- + 6H^+ \rightarrow 3I_2 + 3H_2O + Br^-.$$

In fact, this scheme is equivalent to the preceding one.

Hypobromite is a very good oxidizing agent in spite of the fact that it must be used in basic medium. We notice the following potential value:

$$BrO^- + 2e^- + H_2O \rightleftharpoons Br^- + 2OH^-, \quad E = 0.76\,V, \quad (pH = 14).$$

For the sake of comparison, for the same pH value, we notice that $E = 0.89$ V for the couple ClO^-/Cl^- and that $E = 0.49$ V for the couple IO^-/I^-. Hypobromite ions are of relatively little use in titrimetric analysis. However, it is used in the titrations of:

– urea:

$$3BrO^- + O{=}C\begin{matrix} \diagup NH_2 \\ \diagdown NH_2 \end{matrix} \longrightarrow CO_2 + N_{2(g)} + 2H_2O + 3Br^-$$

From the standpoint of the oxidation states, we can admit that in the urea molecule, each of the two nitrogen atoms is at state – III. This reaction has been frequently used in analytical biochemistry. It finishes with a measurement of the evolved dinitrogen in the gaseous state in well-determined conditions, of course;

– formic acid:

$$HCOOH + BrO^- \rightarrow Br^- + CO_2 + H_2O.$$

In this reaction, carbon passes from oxidization state + II to state+ IV. In the presence of hypobromite ions, formic acid plays the part of a reductant. The reaction is interesting since it is the basis for the determination of methanol, formaldehyde, and formic acid in mixtures. The analysis, in fact, is achieved through three different titrations:

– the first one consists of titrating the sole formol by a hypoiodite solution (see Sect. 19.1);

- the second consists of determining the sum formol + formic acid by hypobromite. The BrO^-/formic acid reaction is described just above. Hypobromite ions react with formol as do hypoiodite ions:

$$OH^- + HCHO + BrO^- \rightarrow HCOO^- + Br^- + H_2O.$$

Next, formiate ions are oxidized into carbon dioxide;

- the third consists of determining the sum of the three components by reaction with dichromate in acidic medium. In a first reaction, chromic acid oxidizes methanol into formaldehyde, and, finally, the three components are oxidized until the carbon dioxide stage. The oxidization reaction of methanol into formol is

$$3CH_3OH + Cr_2O_7^{2-} + 8H^+ \rightarrow 2Cr^{3+} + 3HCHO + 7H_2O;$$

- bromine water is not used for the reasons already given. In its place, bromine, extemporaneously prepared in the solution containing the species to determinate by the reaction bromate/bromide in acidic medium, is used:

$$BrO_3^- + 5Br^- + 6H^+ \rightarrow 3Br_2 + 3H_2O.$$

Bromine may also be generated by coulometry. Titrations by oxidization with bromine are back titrations. When the oxidization reaction is achieved, bromine in excess is appreciated by the addition of potassium iodide (in excess itself compared to the bromine excess) and after a possible acidification according to the pH of the solution. Iodine liberated by the reaction

$$Br_2 + 2I^- \rightarrow I_2 + 2Br^-$$

is titrated with a thiosulfate solution. Some examples include

- the determination of glycerol. Dihydroxyacetone is formed:

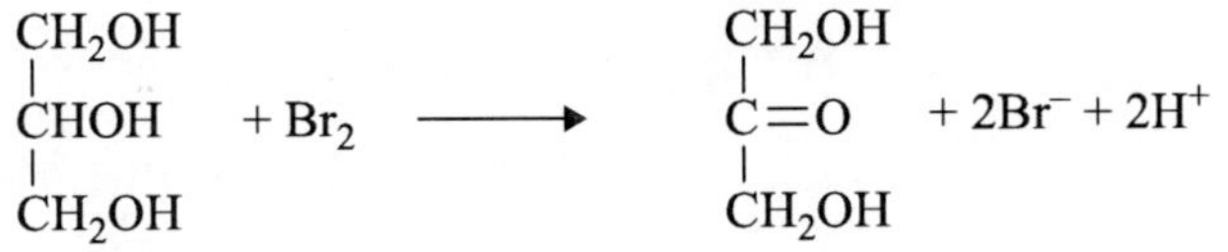

This example and the preceding one (determination of methanol) show that secondary and primary alcohols belong to the redox couples, whose oxidized members are the corresponding aldehyde and ketones. These couples may be classified as rather reducing ones since they exhibit negative values of standard potentials (see Chap. 13). They constitute slow electrochemical systems.

- in the pharmaceutical field, let us mention the titrations of
  - disulfiram [disulfandiylbis(diethylcarbothioamide)]

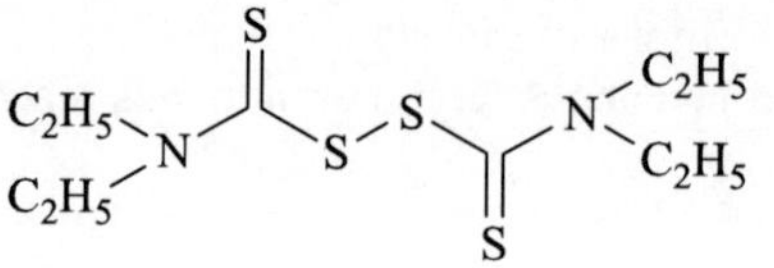

$$+\ 13Br_2 + 20H_2O \longrightarrow 30\,H^+ + 26Br^- + 4HSO_4^- + 2HCOOH + 2HN(C_2H_5)_2$$

In this reaction, the four sulfur atoms and the carbon atoms of the thiocarbonyl groups are oxidized. The sulfur atoms of the disulfide bridge may be considered as passing from oxidation state +I to state +VI, the two others passing from state +II to state+ VI, the oxidized carbon atoms having passed from state – II to state +II.

– isoniazid

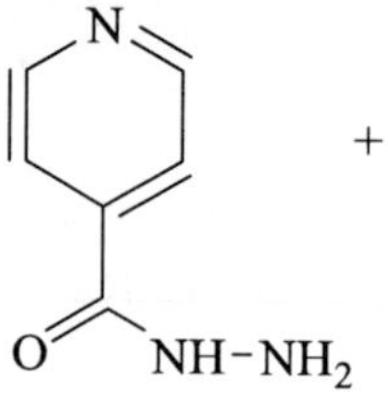

$$+\ 2Br_2 + H_2O \longrightarrow$$

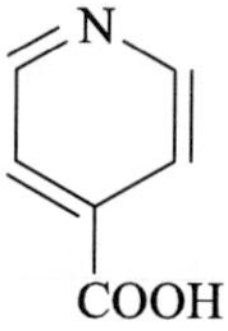

$$+\ N_{2(g)} + 4H^+ + 4Br^-$$

We may admit that there is a hydrolysis in a first step with the formation of hydrazine. Then its nitrogen atoms are oxidized.

– methionine

$$CH_3{-}\overline{\underline{S}}{-}(CH_2)_2{-}\underset{\displaystyle NH_2}{\underset{|}{CH}}{-}COOH + H_2O + Br_2 \longrightarrow CH_3{-}\overset{\oplus}{\overline{S}}(\underset{\ominus\,|\underline{O}|}{|}){-}(CH_2)_2{-}\underset{\displaystyle NH_2}{\underset{|}{CH}}{-}COOH + 2Br^- + 2H^+$$

There is formation of the corresponding sulfoxyde. In some other conditions, the reaction limit may also be the corresponding sulfone. We may admit that methionine is the reduced form of the couple

$$\underset{\displaystyle O}{\underset{\downarrow}{CH_3{-}S}}{-}(CH_2)_2{-}\underset{\displaystyle NH_2}{\underset{|}{CH}}{-}COOH + 2e^- + 2H^+ \longrightarrow CH_3{-}S{-}(CH_2)_2{-}\underset{\displaystyle NH_2}{\underset{|}{CH}}{-}COOH + H_2O$$

Incidentally, we note that when there is formation of the sulfone, there is an exchange of four protons and four electrons. The oxidization of a disulfide into a sulfoxyde or into a sulfone by bromometry or bromatometry is a classical means of quantitatively determining these compounds.

19.4.3 Determinations by Fixing bromine into an Organic Substrate Either by Substitution or by Addition

Bromine also exhibits the property to fix itself on some organic structures. This can be achieved either by substitution or by addition. The compound to determinate is faced with a solution of bromine in excess, bromine being generated previously by the reaction bromate/bromide or by coulometry. Bromine in excess is determined by indirect iodometry. Of all the titrations involving the bromine element, they are the most numerous.

19.4.3.1 Bromine Fixation Reaction by Substitution

These reactions are actually electrophilic substitutions. They concern aromatic derivatives in which the aromatic nucleus is activated by electron-donating substituents. Among them are the amino and phenol groups. The determination of phenols by fixation of bromine is named Koppeschaar's reaction. Phenol itself settles three bromine atoms.

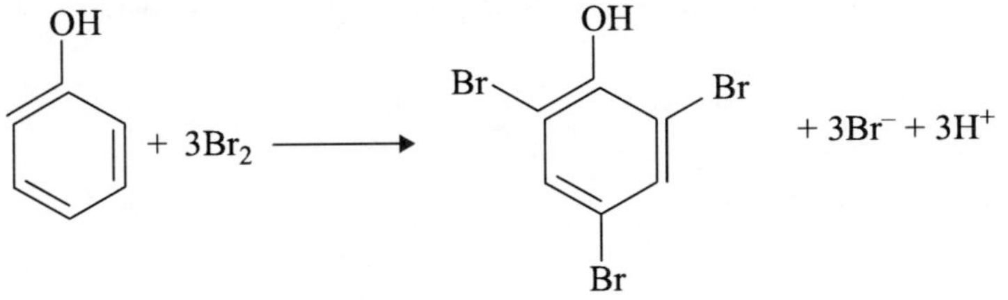

Thymol fixes two bromine atoms.

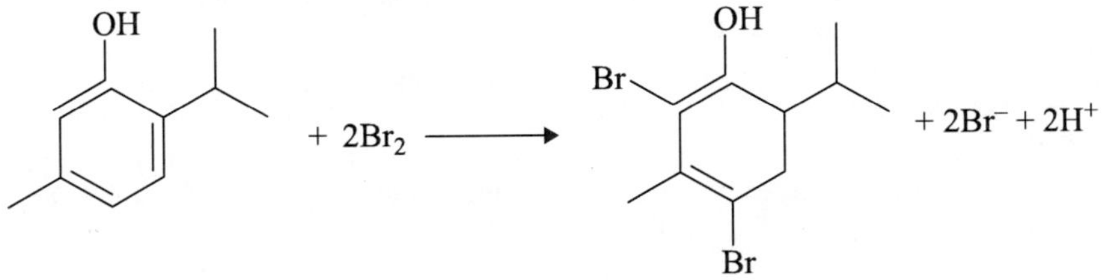

Let's also mention the determination of chlorocresol, resorcinol, phenolsulfonephthalein, and some lesser-known pharmaceutical derivatives such as hydroxyethyl salicylate and etilefrin.

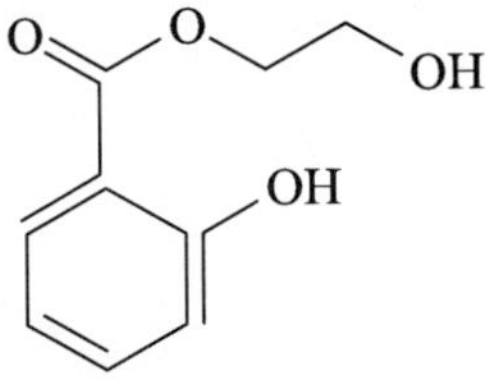

Hydroxyethylsalicylate

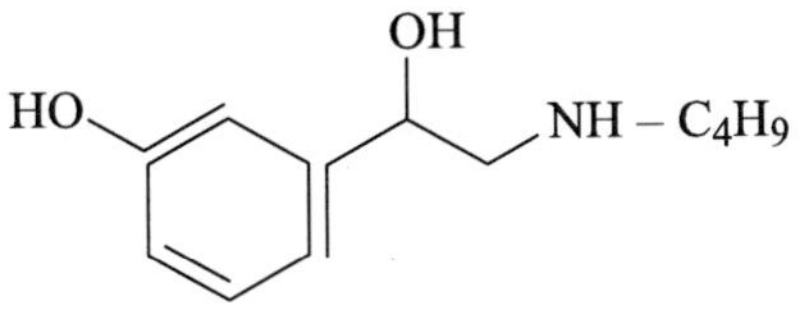

etilefrin

Of course, the number and location of bromine atoms fixed on the organic rest depend on the nature, location, and number of the initial substituents of the compound under study.

Aniline, as phenol, fixes three bromine atoms:

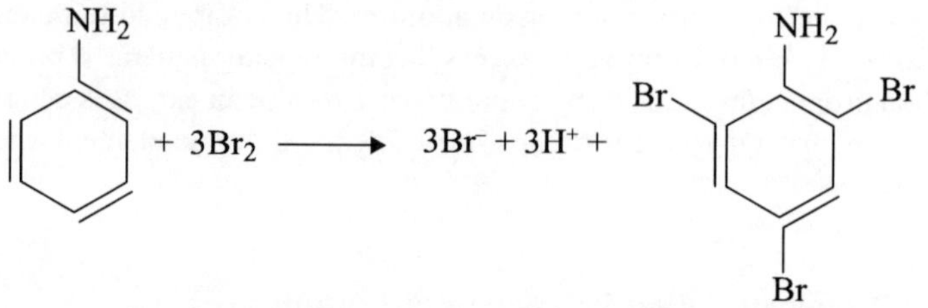

An example, found in the field of medicinal chemistry, is provided by the antibacterial sulfonamides (sulfamides). These derivatives are aromatic compounds generally possessing a primary amino group in para of the sulfonamide rest. Although the last one is a deactivating group toward an electrophilic substitution, the primary amino rest is sufficiently activating to counterbalance the effect of the preceding one and even to activate the aromatic nucleus in these compounds toward this sort of reaction. Hence, the fixation of bromine is easy. It is achieved at room temperature according to the reaction

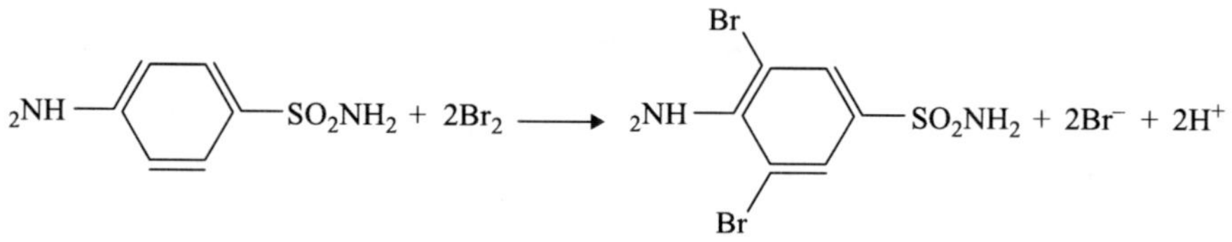

It is quasi-instantaneous. Notice that in this case (as in several others), bromine can be generated by coulometry (see a general course in analytical electrochemistry).

Some other heterocycles easily fix bromine by substitution.

Another important analytical application of bromine fixation is the determination of several metallic ions through the formation of metallic "oxinates". Oxine or 8-hydroxyquinoleine reacts with two bromine molecules to give 5,7-dibromo-8-hydroxyquinoleine:

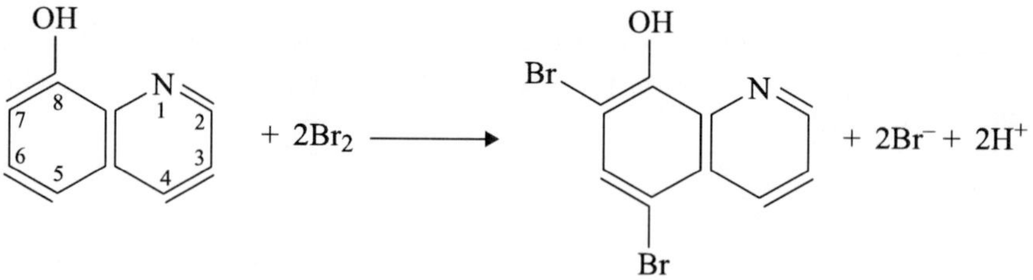

From another standpoint, oxine forms insoluble precipitates with several metallic ions called "oxinates" (see Part V). The precipitation reaction may be quantitative if the pH conditions are judiciously chosen. They depend on the nature of the metallic ion. The formula of the precipitates are $M(C_9H_6ON)_n$ where *n* is the valency of the metallic ion. Exceptions to this rule are rare. From these considerations, it results a methodology of determination of metallic ions. The metallic ion is quantitatively precipitated by oxine at the judicious pH value. After filtration of the precipitate, this one is treated by a strong acid that displaces oxine from it. The liberated oxine is titrated by an excess of a bromate/bromide mixture in acidic medium. The bromine in excess is titrated by indirect iodometry.

Actually, complexometric titrations are simpler to achieve (see Part IV).

19.4.3.2 Determination After Bromine Fixation by Addition

In this section, we consider the determinations based on the addition of bromine on double bonds. Most of the time, these addition reactions take place in nonaqueous media. We limit ourselves to recall the notions of iodine value and fat indices. Some drugs are determined by the addition of bromine on a double bond that their structures contain. We'll first consider cyclobarbital and hexobarbital, which are derived from barbituric acid:

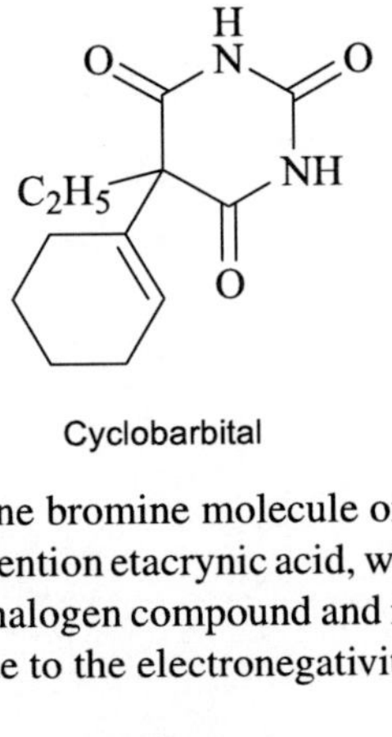

Cyclobarbital

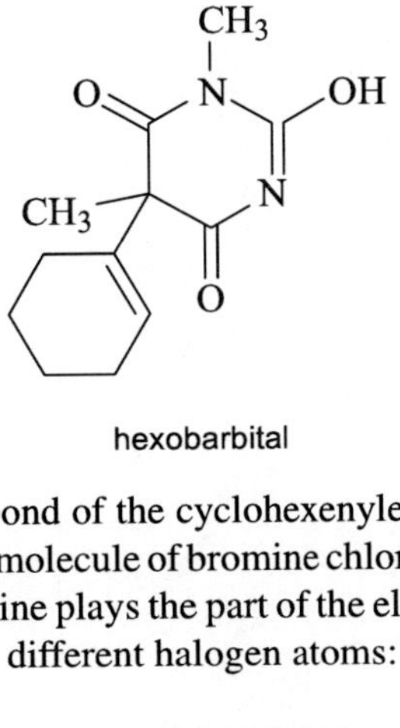

hexobarbital

Both fix one bromine molecule on the double bond of the cyclohexenyle cycle. We can also mention etacrynic acid, which fixes one molecule of bromine chloride, which is an interhalogen compound and in which bromine plays the part of the electrophilic species due to the electronegativities of the two different halogen atoms:

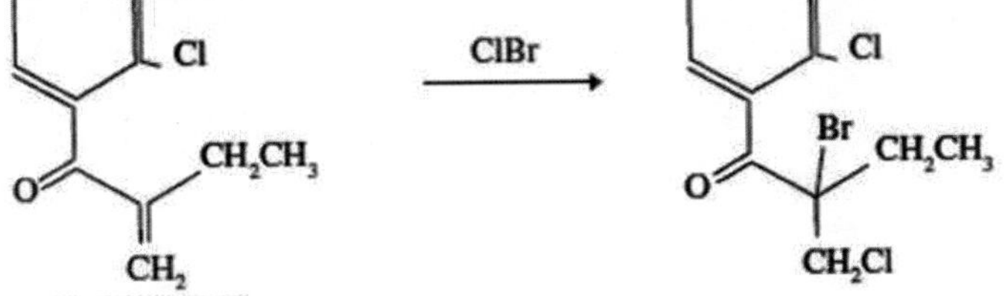

etacrynic acid

Bromine chloride is prepared by the reaction of potassium bromide with potassium bromate in concentrated hydrochloric acid according to the reaction

$$2Br^- + BrO_3{}^- + 6Cl^- + 6H^+ \rightarrow 3BrCl + 3Cl^- + 3H_2O.$$

It is an oxidization reaction of bromide ions into bromine at oxidation state +I, stabilized as a chloride derivative. Bromine chloride is a powerful oxidizing agent [$E°(BrCl/Br^-) = 1.20$ V]. Therefore, we can predict that it displaces iodine from iodide ions. This fact permits us to determinate it by indirect iodometry.

Chapter 20
Oxidizations with Permanganate, Dichromate, and Ceric Ions Some Titration Methods Involving a Reduction Reaction

We begin this chapter with the study of three important titration methods involving quantitative oxidization reactions. They are based on oxidizations with permanganate, dichromate, and ceric ions. These methods are sometimes (notably in the french literature) called manganimetry, chromimetry, and cerimetry, respectively. Next, we very briefly mention three titration methods involving reduction reactions.

20.1 Oxidization with Permanganate Ions

20.1.1 General Considerations

20.1.1.1 About the Oxidation States of Manganese

The term "manganimetry" groups all the titrimetric determinations based on the oxidizing power of permanganate ions MnO_4^- toward several reductors. Manganimetry is usually achieved in acidic medium, but some determinations are also performed in neutral and alkaline media.

Manganese exhibits the following oxidation states:

- $+$VII in permanganate ion MnO_4^- (permanganic acid is a strong acid);
- $+$VI in manganate ion MnO_4^{2-};
- $+$V in hypomanganous ion MnO_4^{3-} and in $MnOCl_3$;
- $+$IV in manganese dioxide MnO_2;
- $+$III in manganic oxide Mn_2O_3;
- $+$II in the manganous cation Mn^{2+} and in the manganous hydroxide $Mn(OH)_2$ (the oxide MnO is not stable from a thermodynamic standpoint);
- 0 in the metal.

(The $+$V state is of no interest in chemical analysis. it is the same thing for state $+$II. They will no longer be evoked here except for Mn^{2+} ions.) The half-redox equilibria used in chemical analysis are

$$MnO_4^- + 8H^+ + 5e^- \rightleftharpoons Mn^{2+} + 4H_2O, \qquad E°(MnO_4^-/Mn^{2+}) = 1.507\,V,$$

J.-L. Burgot, *Ionic Equilibria in Analytical Chemistry*,
DOI 10.1007/978-1-4419-8382-4_20, © Springer Science+Business Media, LLC 2012

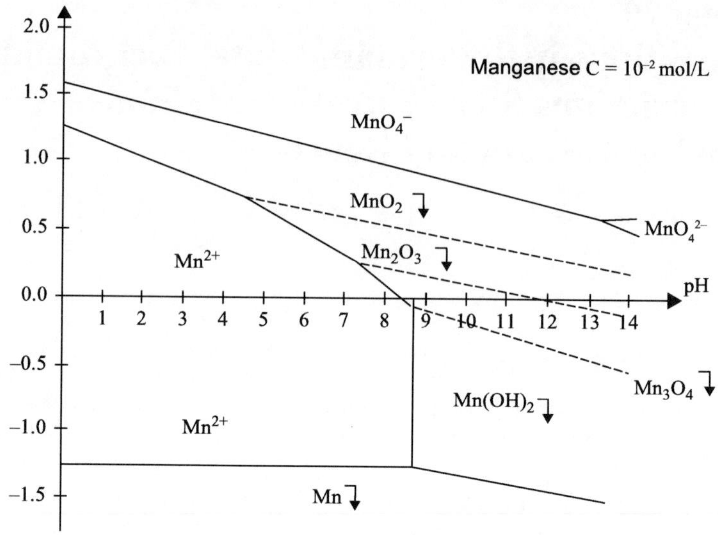

Fig. 20.1 Pourbaix diagram of manganese (first convention: $C = 10^{-2}$ mol/L)

$$MnO_4{}^- + 4H^+ + 3e^- \rightleftharpoons MnO_{2(s)} + 2H_2O, \qquad E°(MnO_4{}^-/MnO_{2(s)}) = 1.70\,V,$$

or

$$MnO_4{}^- + 2H_2O + 3e^- \rightleftharpoons MnO_{2(s)} + 4OH^-$$

and

$$MnO_4{}^- + e^- \rightleftharpoons MnO_4{}^{2-}, \qquad E°(MnO_4{}^-/MnO_4{}^{2-}) = 0.56\,V.$$

The first equilibrium is used in acidic medium, the second and third ones in weakly acidic, neutral, or weakly alkaline media, and the last one in strongly alkaline medium. A Pourbaix diagram (Fig. 20.1) gives evidence of the following points:

- in strongly acidic medium, permanganate ions are reduced to manganous ions Mn^{2+}, due to the fact that the predominance area of the dioxide MnO_2 is narrow for high potential values. The couple $MnO_4{}^-/MnO_2$ is not effective in these conditions. The value $E°(MnO_4{}^-/Mn^{2+}) = 1.51$ V allows us to assert that permanganate ion $MnO_4{}^-$ is a strong oxidizing agent. Its oxidizing power decreases quickly when the pH value increases. Its normal apparent potential $E°'$ is given by the relation

$$E°'(MnO_4{}^-/Mn^{2+}) = E°(MnO_4{}^-/Mn^{2+}) - 0.06 \cdot 8/5pH = (1.51 - 0.06 \cdot 8/5pH)\,V.$$

However, in the pH range in which the couple is used, the apparent potentials of several other redox couples also decrease when the pH value increases and, relatively, permanganate ion remains a strong oxidizing agent;

- A Pourbaix diagram also shows that permanganate ions must react with manganous ions, with the formation of manganese dioxide. Their preponderance areas are indeed disconnected. The reaction, which is a retrodismutation (or amphoterization) reaction, is

$$2MnO_4^- + 3Mn^{2+} + 2H_2O \rightarrow 5MnO_2 + 4H^+.$$

However, in some conditions, it is very slow and can safely be considered to be negligible (see later);

- in weakly acidic, neutral, or weakly alkaline medium, the whole of the predominance areas of oxides $MnO_{2(s)}$, $Mn_2O_{3(s)}$, and $Mn_3O_{4(s)}$ (mangano-manganic oxides) is important. Permanganate ions, while oxidizing, are reduced to the oxidation state + IV. In these conditions, permanganate ion remains a strong oxidizing agent. Its apparent standard potential is $E^{\circ\prime}(MnO_4^-/MnO_{2(s)}) = 0.97$ V at pH = 9;
- in strongly alkaline medium (pH ≈ 14), manganate ions MnO_4^{2-} do exist. In a lesser alkaline medium, they disproportionate according to the reaction

$$3MnO_4^{2-} + 2H_2O \rightarrow 2MnO_4^- + MnO_2 + 4OH^-;$$

- for pH values higher than about 8,6, manganous ions give rise to the formation of manganous hydroxide $Mn(OH)_2$. Let's recall (see iodometry) that in Winkler's method of determination of dioxygen, manganous hydroxide, first transformed into manganese dioxide, oxidizes iodide ions. This is one of the rare examples of the use of the $MnO_{2(s)}/Mn(OH)_2$ couple in oxido-reductimetry:

$$2Mn(OH)_2 + O_{2(w)} \rightarrow 2MnO_{2(s)} + 2H_2O.$$

During the oxidization of iodide ions into iodine in acidic medium, there is, again, formation of manganese at oxidation state + II:

$$MnO_{2(s)} + 4H^+ + 2I^- \rightarrow I_2 + Mn^{2+} + 2H_2O.$$

The respective locations of the effective redox couples on the potentials scale justify these reactions [$E^{\circ\prime}(MnO_2/Mn(OH)_2) \approx -0.10$ V at pH = 14];

- permanganate ions are not stable in aqueous solutions. Indeed, the predominance areas of MnO^{4-} and H_2O are disconnected regardless of the solution's pH. As a result, the following reaction should occur:

$$4MnO_4^- + 4H^+ + 6H_2O \rightarrow 8H_2O + 4MnO_2 + 3O_{2(g)}.$$

The stability area of MnO_4^- is markedly located above the frontier line corresponding to the oxidization of water into dioxygen under atmospheric pressure. From the same standpoint, $O_{2(g)}$ and $MnO_{2(s)}$ exhibit a common area. Despite all these considerations, pure aqueous solutions of the permanganate ion are surprisingly stable. Some of them have been stored for several years! However, when manganous ions

Mn^{2+} are in the solution, the decomposition rate of the permanganate ion considerably increases. The catalyst is manganese dioxide itself, which is initially formed by the oxidization of traces of Mn^{2+} by MnO_4^-;

- finally, the stability area of manganese metal is markedly located under that of water. The metal is very unstable when it is brought together with water. It reacts with water to give dihydrogen and manganous ions:

$$2H^+ + Mn \rightarrow H_{2(g)} + Mn^{2+}.$$

However, this reaction is slow due to the strong overpotential of dihydrogen evolving onto the metal (this is a problem of kinetics—see a general course in electrochemistry). As a result, manganese is easily dissolved in acidic or neutral medium. Mn^{2+} gives the oxides Mn_3O_4, Mn_2O_3, and MnO_2 in a slightly oxidizing medium.

Solutions of permanganate ions are purple. This is one of the advantages of manganimetry. The reactant is its own indicator. During the titration of a colorless or slightly colored reductor with a permanganate solution, a drop in excess imparts a pale-pink color to the titrand solution. The error made by detecting the equivalent point in this manner is negligible, with 10^{-1} mol/L solutions. A 2×10^{-6} mol/L solution in permanganate exhibits a pink color that is still detectable. If the solution to be titrated is too colored, we can use an internal redox indicator endowed with a high standard (or formal) potential, such as, for example, ferrous ortho-phenanthroline.

Sulfuric acid is the most convenient acid to achieve oxidizations in acidic medium. It does not react with permanganate ions in diluted solutions. In more concentrated solutions, water is oxidized according to the reaction

$$2MnO_4^- + 5H_2O + 6H^+ \rightarrow 2Mn^{2+} + 8H_2O + 5/2O_{2(g)}.$$

In hydrochloric acid, the following reaction occurs:

$$2MnO_4^- + 10Cl^- + 16H^+ \rightarrow 2Mn^{2+} + 5Cl_2 + 8H_2O.$$

This is not surprising since $E°(Cl_{2(g)}/Cl^-) = 1.36\,V$. As a result, permanganate ions are consumed in excess. However, in some concentration conditions, the latter reaction may be minimized and may even be negligible. Hence, there exist titrations such as those of As^{+III}, Sb^{+III}, and H_2O_2 that can be achieved in a hydrochloric medium. In principle, nitric acid is convenient, but it may contain nitrites and nitrous derivatives, which are reducing agents. For this reason, its use must be avoided.

20.1.1.2 Stability and Standardization of Permanganate Solutions

Although it is available in the pure state, potassium permanganate cannot be a standard for two reasons:

- the already evocated stability of its aqueous solutions;
- the possible presence in water and of reducing substances in the bottle of the solution, especially organic ones. As a result, a lowering of the awaited permanganate concentration occurs, but after some time it becomes stabilized.

Hence, particular precautions must be taken when we prepare permanganate solutions. The most often used solution contains 1/50 mol/L (decinormal solution). They must be standardized before use. For their standardization, we can use some standard reducing inorganic substances such as ferrous salts, arsenious acid, and sodium thiosulfate. We can also use reducing organic substances such as oxalic acid and its derivatives.

20.1.1.2.1 Standardization with Ferrous Iron (Mohr's Salt)

Mohr's salt is the iron + II and ammonium double sulfate, $FeSO_4$, $(NH_4)_2SO_4$, $6H_2O$, commercially available in a state of great purity. The standardization reaction is

$$MnO_4{}^- + 5Fe^{2+} + 8H^+ \rightarrow Mn^{2+} + 5Fe^{3+} + 4H_2O,$$

which is in agreement with the value $E°(Fe^{3+}/Fe^{2+}) = 0.77$ V. The reaction is achieved in diluted sulfuric acid. Some authors recommend adding phosphate ions $PO_4{}^{3-}$ in order to complex the formed ferric ions.

20.1.1.2.2 Standardization with Arsenious Acid

The reaction to standardize arsenious acid is slow. It occurs in a concentrated acid medium. It is catalyzed by chloride ions in sufficiently high concentrations. The reaction is

$$2MnO_4{}^- + 5H_3AsO_3 + 6H^+ \rightarrow 2Mn^{2+} + 5H_3AsO_4 + 3H_2O.$$

We notice that it is not necessary to achieve the reaction in neutral or weakly alkaline medium as with iodine for arsenious acid to be sufficiently reductant. We also notice that arsenious acid is oxidized before chloride ions. This is normal because of the standard potential values of the different couples brought together.

20.1.1.2.3 Standardization with a Thiosulfate Solution

Sodium thiosulfate cannot be directly titrated with potassium permanganate because a mixture of tetrathionate and sulfate ions is formed. The titration is indirect. Potassium iodide is oxidized by permanganate ions in sulfuric acid medium. The liberated iodine is titrated with thiosulfate. Let's recall that the used thiosulfate solution must have been quite recently standardized.

20.1.1.2.4 Standardization by Oxalic Acid

Oxalic acid crystallized with five water molecules is used in the standardization of oxalic acid. It is commercially available in a state of guaranteed purity. Sodium, potassium, and ammonium salts of oxalic acid are also used. The corresponding half-redox reaction is

$$2CO_{2(g)} + 2H^+ + 2e^- \rightleftharpoons (COOH)_2.$$

The equivalent is the molar mass divided by 2, as in protometry. This is purely coincidental. The standard potential of the couple is $E°[CO_{2(g)}/(COOH)_2] = 0.49$ V. The standardization reaction is

$$2MnO_4{}^- + 5(COOH)_2 + 6H^+ \rightarrow 2Mn^{2+} + 10CO_{2(g)} + 8H_2O.$$

At the beginning of the titration, the reaction is slow. Hence, we must warm up the titration vessel. The formed Mn^{2+} ions then catalyze the reaction. Manganous sulfate may also be added to play this role.

20.1.2 Applications of Manganimetry in Acidic Medium

20.1.2.1 Determination of Inorganic Reductors

20.1.2.1.1 Determination of Nitrites

Nitrites are oxidized by permanganate ions in acidic medium with the formation of nitrate ions. The determination consists of a back titration. The nitrite solution is progressively added into the solution of permanganate ions in excess. The medium consists of a diluted sulfuric acid solution heated at 50°C since the reaction is slow. It is

$$2MnO_4{}^- + 5NO_2{}^- + 6H^+ \rightarrow 2Mn^{2+} + 5NO_3{}^- + 3H_2O.$$

It is in good accordance with the standard potential value of the couple $NO_3{}^-/HNO_2$:

$$NO_3{}^- + 3H^+ + 2e^- \rightleftharpoons HNO_2 + H_2O, \qquad E°(NO_3{}^-/HNO_2) = 0.94\,V.$$

Permanganate ions in excess are titrated with a Mohr's salt solution.

20.1.2.1.2 Determination of Ferrous Salts and, by Extension, of Ferric Salts

Ferrous salts can be determined with potassium permanganate in sulfuric acid medium (see standardization, Sect. 20.1.1.2). The reaction may also be achieved in hydrochloric acid medium after the addition of a manganous sulfate (dissolved in a sulfuric and phosphoric solution—the Zimmermann–Reinhardt solution) into the medium. Manganous sulfate would have the property to decrease the apparent standard potential of the couple MnO_4^-/Mn^{2+}. Hence, it would be less oxidizing versus chloride ions. From another standpoint, phosphoric acid combines with ferric ions, which are formed during the titration reaction. As a result, the equivalence point is more easily detected.

An extension of this determination is the titration of ferric ions after their reduction into ferrous ions [$E°(Fe^{3+}/Fe^{2+}) = 0.77$ V]. After elimination of the excess reducing agent, ferrous ions are determined by manganimetry. The reduction is achieved with sulfur dioxide or with a hydrogen sulfite solution in weakly acidic medium:

$$2Fe^{3+} + HSO_3^- + H_2O \rightarrow 2Fe^{2+} + SO_4^{2-} + 3H^+$$

[the standard potential of the couple SO_4^{2-}/H_2SO_3 is $E°(SO_4^{2-}/H_2SO_3) = 0.17$ V and the half-redox reaction is $SO_4^{2-} + 4H^+ + 2e^- \rightleftharpoons H_2SO_3 + H_2O$].
After the reduction, the sulfur dioxide in excess is expelled by a stream of carbon dioxide. The reduction of ferric ions may also be achieved with stannous chloride in excess [$E°(Sn^{4+}/Sn^{2+}) = 0.15$ V]:

$$2Fe^{3+} + Sn^{2+} \rightarrow 2Fe^{2+} + Sn^{4+}.$$

The Sn^{2+} ions in excess are eliminated by the addition of a mercuric chloride solution. The formed "mercurous chloride" precipitates and does not have any influence on the subsequent oxidization of ferrous ions by manganimetry:

$$Sn^{2+} + 2HgCl_2 \rightarrow Hg_2Cl_2 \downarrow + Sn^{4+} + 2Cl^-.$$

We remark that this reaction is also a redox one. The precipitation of Hg_2Cl_2 induces the reduction of Hg^{+II} to the stage Hg^{+I}. It is interesting to notice that the redox reaction between Sn^{2+} and Fe^{3+} given above allows the determination of Sn^{2+} through that of Fe^{2+}. This is an example of a titration by substitution. We also notice that the reduction of Fe^{3+} may also be achieved with solid cuprous oxide Cu_2O (it is very poorly soluble). The occurring redox reaction is

$$Cu_2O_{(s)} + 2Fe^{3+} + 2H^+ \rightarrow 2Fe^{2+} + 2Cu^{2+} + H_2O.$$

Chloride ions that might disrupt the reaction are eliminated by the formation of cuprous chloride from the excess of cuprous oxide. This reaction is the basis of Bertrand's method of determination of reducing sugars (see ahead).

20.1.2.1.3 Determination of Hydroxylamine

This determination is an indirect one. It is achieved through the formation of a ferrous salt that is titrated with potassium permanganate. Hydroxylamine reduces ferric ions in boiling sulfuric acid according to the reaction

$$2NH_2OH + 4Fe^{3+} \rightarrow N_2O_{(g)} + 4Fe^{2+} + H_2O + 4H^+.$$

It results from the superimposition of the following two half-redox reactions:

$$Fe^{3+} + 1e^- \rightleftharpoons Fe^{2+}.$$

$$N_2O_{(g)} + 4e^- + 4H^+ + H_2O \rightleftharpoons 2NH_2OH, \quad E°(N_2O_{(g)}/NH_2OH) = -0.05\,V.$$

20.1.2.1.4 Determination of Persulfate Ions

Persulfate ion is a powerful oxidizing agent [$E°(S_2O_8{}^{2-}/SO_4{}^{2-}) = 2.01$ V]. It easily oxidizes ferrous ions into ferric ions:

$$S_2O_8{}^{2-} + 2Fe^{2+} + 2H^+ \rightarrow 2Fe^{3+} + 2HSO_4{}^-.$$

The determination is a back titration. Ferrous ions in excess are determined by manganimetry. An analogous titration can be achieved by replacing ferrous ions by oxalic acid.

20.1.2.1.5 Determination of Hydrogen Peroxide

Hydrogen peroxide is a reducing agent with respect to permanganate ions. The half-redox couple is

$$O_{2(g)} + 2H^+ + 2e^- \rightleftharpoons H_2O_2, \qquad E°(O_{2(g)}/H_2O_2) = 0.68\,V.$$

The titration reaction is the following one:

$$2MnO_4{}^- + 5H_2O_2 + 6H^+ \rightarrow 5O_{2(g)} + 2Mn^{2+} + 8H_2O.$$

The titer of a hydrogen peroxide solution is often expressed in volume. This is the number of liters of liberated oxygen per liter of solution at 0°C and under the pressure of 760 mmHg (101,325 kPa) when it decomposes according to the reaction

$$H_2O_2 \rightarrow H_2O + 1/2O_2.$$

The above titration reaction is stoichiometric when the titer is less than six volumes. Thirty-four grams of H_2O_2 liberate 11.2 L of dioxygen; that is, 3 g liberate 1 L. A 30% solution (in masses) of H_2O_2 contains 300 g for 1000 g. It is a "100 volume" hydrogen peroxide solution.

From the mechanistic point of view, it is interesting to know that the reaction between permanganate ions and hydrogen peroxide is a bimolecular one and that all the liberated dioxygen comes from the latter, as is ascertained by the reaction achieved with marked oxygen. The chronological sequence of the formed species is

$$H_2O_2 \rightarrow HO_2{}^- \rightarrow HO_2 \rightarrow O_2{}^- \rightarrow O_2.$$

There are successively and alternatively losses of protons and electrons.

20.1.2.1.6 Determination of Peroxysalts

Peroxysalts can be back-titrated with permanganate ions. This is the case of MnO_2, PbO_2, and Pb_3O_4. The peroxysalt is reduced in acidic medium. The reducing agent used is ferrous iron or oxalic acid, the excess of which is titrated with permanganate ions. Both redox reactions are (for example, with MnO_2)

$$MnO_2 + 4H^+ + 2Fe^{2+} \rightarrow Mn^{2+} + 2Fe^{3+} + 2H_2O,$$

$$MnO_2 + 2H^+ + (COOH)_2 \rightarrow 2CO_{2(g)} + Mn^{2+} + 2H_2O.$$

Another methodology consists of taking into account the fact that metallic peroxides such as alkaline and alkaline-earth dioxides Na_2O_2, BaO_2, and K_2O_2 and also zinc dioxide ZnO_2 behave like salts of hydrogen peroxide considered an acid. The salts liberate hydrogen peroxide in sulfuric acid solution, which is titrated with permanganate ions. In order to avoid the dioxygen loss occurring during the dissociation of the salt into water, they are, in a first reaction, dissolved in a boric acid solution. Therefore, perboric acid $H_2B_2(O_2)_2(OH)_4$ is formed. With water, the latter liberates boric acid and hydrogen peroxide:

$$H_2B_2(O_2)_2(OH)_4 + 2H_2O \rightarrow 2H_3BO_3 + 2H_2O_2.$$

20.1.2.2 Determination of Organic Reductants

20.1.2.2.1 Determination of Oxalic Acid and Calcium Ion

(With regard to the determination of oxalic acid, see the standardization of potassium permanganate, above.) An interesting extension of the oxidization reaction of oxalic acid is the determination of calcium ions. The latter are precipitated as calcium oxalate in hot acetic acid (solubility product $K_s = 10^{-8.6}$). The precipitate is separated out from the solution, washed, and dissolved into a 2 mol/L sulfuric acid solution. After heating, the liberated oxalic acid is titrated with permanganate. In fact, oxalic acid is simply liberated by displacement of the precipitation equilibrium:

$$C_2O_4Ca_{(s)} \rightleftharpoons C_2O_4{}^{2-} + Ca^{2+},$$

$$C_2O_4{}^{2-} + 2H^+ \rightarrow C_2O_4H_2.$$

20.1.2.2.2 Determination of Reducing Sugars According to Bertrand's Method

This is a cuprimetric determination. The cuprous oxide obtained by the action of a reducing carbohydrate on Fehling's solution in strictly experimental conditions is separated, after precipitation, out from the solution with the help of a sintered glass. (Fehling's solution is a solution of a cupric salt in alkaline medium. Cupric salt is kept in solution by complexation.) The precipitate is washed and then brought together with a sulfuric solution of ferric sulfate. The ferrous ions formed according to the following redox reaction:

$$Cu_2O_{(s)} + 2Fe^{3+} + 2H^+ \rightarrow 2Cu^{2+} + 2Fe^{2+} + H_2O$$

are determined with permanganate ions. The reaction between the reducing sugar and the copper(II) of Fehling's solution is not a stoichiometric one. However, there are correspondence tables for the quantity of reducing sugar added vs. the volume of permanganate solution. They are established in very well-specified experimental conditions, which, of course, must be obeyed.

20.1.3 Manganimetry in Neutral and Weakly Alkaline Media

Recall that the redox couple of manganese playing a part in these conditions is

$$MnO_4^- + 4H^+ + 3e^- \rightleftharpoons MnO_{2(s)} + 2H_2O,$$

or

$$MnO_4^- + 2H_2O + 3e^- \rightleftharpoons MnO_{2(s)} + 4OH^-.$$

Let's also recall that this couple is a strong oxidant, despite the weakly alkaline medium [at pH = 9, $E^{\circ\prime}(MnO_4^-/MnO_{2(s)}) = 0.97$ V].

20.1.3.1 Determination of Manganous Ions

Manganous ions are formed according to

$$3Mn^{2+} + 2MnO_4^- + 2H_2O \rightarrow 5MnO_{2(s)} + 4H^+.$$

Obviously, this is an amphoterization (or retrodismutation) reaction. This methodology is called Guyard and Volhard's method. The reaction occurs in neutral medium. The manganese dioxide, which is forming during this reaction, catalyzes it. It is quasi-instantaneous in neutral medium. The precipitate of manganese dioxide that forms during the course of the reaction tends to adsorb manganous ions, which therefore partly escape from the oxidization. Hence, we must add an indifferent ion such as Zn^{2+} that is adsorbed in place of Mn^{2+}.

20.1.3.2 Determination of Iodide Ions

Whereas in acidic medium, permanganate ions oxidize iodide ions to give iodine, in neutral and weakly alkaline medium, iodate ions are formed. This is logical since iodine, which could possibly be thought to be formed, disproportionates for pH values higher than 7. Of course, iodide ions formed due to the disproportionation are again oxidized with permanganate ions, and so forth. The oxidization reaction is

$$I^- + 2MnO_4^- + H_2O \rightarrow IO_3^- + 2OH^- + 2MnO_{2(s)}.$$

20.1.3.3 Organic Analysis

In organic analysis, the oxidizing power of permanganate ions in neutral or weakly alkaline medium permits the determination of some alcohols, oxalic and formic acids, and their salts with the production of carbon dioxide. In these pH conditions, permanganate ions react faster than in acidic medium.

It is not possible to draw general rules indicating the organic structures that undergo quantitative oxidization with potassium permanganate. The stability of the compounds to be determined and the experimental conditions considerably influence the course of the oxidization reaction. We mention some examples here taken from the field of pharmaceutical analysis:

- the determination of undecylenic acid. This one is quantitatively oxidized into the corresponding α-diol. The α-diol function is located in place of the initial double bond. The formed α-diol quantitatively liberates formaldehyde after reaction with sodium periodate. Formaldehyde is determined in different manners:

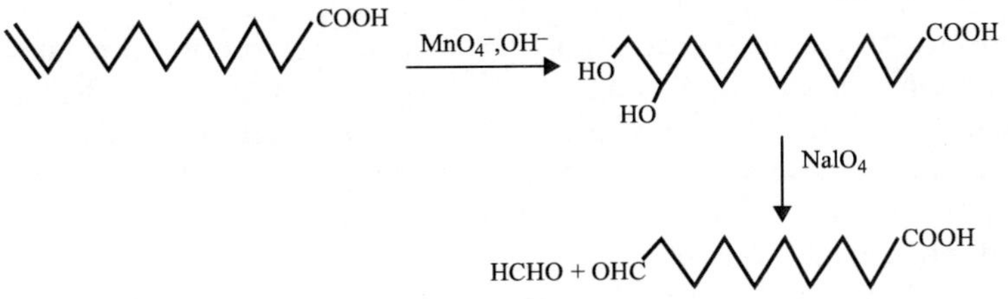

Undecylenic acid

- the determination of dextromoramide. It gives diphenylketone after reaction with potassium permanganate in slightly alkaline medium:

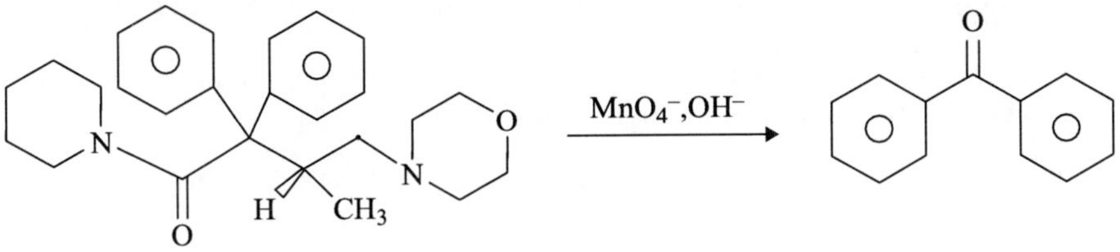

Dextromoramide

The formed benzophenone is determined by absorptiometry.

20.1.4 Manganimetry in Strongly Alkaline Medium

In strongly alkaline medium, the reduction of permanganate ions into manganese dioxide takes place in two consecutive steps:

1. the reduction into manganate ions. This step is fast:

$$MnO_4^- + 1e^- \rightleftharpoons MnO_4^{2-};$$

2. the reduction of manganate ions into manganese dioxide. This step is slow:

$$MnO_4^{2-} + 2e^- + 2H_2O \rightleftharpoons MnO_2 + 4OH^-.$$

(The succession of these two steps is equivalent to the already seen reduction at three electrons of permanganate into manganese dioxide.) The standard potentials of the two new redox couples are $E°(MnO_4^-/MnO_4^{2-}) = 0.56$ V and $E°(MnO_4^{2-}/MnO_2) = 0.60$ V.

Stamm's method is based on rate differences existing between both reduction reactions in alkaline medium. In the method, only the first reaction plays a part. The slower transformation $MnO_4^{2-} \rightarrow MnO_2$ is, moreover, precluded by the addition of barium nitrate or chloride to the titrand solution. Adding the barium salt has two effects: At first, its speeds up the first step; and second, it precipitates poorly soluble barium manganate ($K_s = 2.5 \times 10^{-10}$). Therefore, once precipitated, it cannot undergo the second reduction step since it is out of the reaction medium. [At this point, it is interesting to note that barium permanganate $Ba(MnO_4)_2$ is soluble in water.] The precipitation of barium manganate displaces the first equilibrium (first step) toward the right. The couple MnO_4^-/MnO_4^{2-} does not lose its oxidizing power when the pH value is increasing. For the sake of comparison, this is not the case with the couple IO_4^-/IO_3^-, whose standard potential (pH = 0) is, by far, higher than that of the couple MnO_4^-/MnO_4^{2-}.

From the experimental point of view, we operate with an excess of potassium permanganate. After reaction, it is back-titrated with a sodium formiate solution:

$$2MnO_4^- + 3OH^- + HCOO^- + 2Ba^{2+} \rightarrow 2BaMnO_{4(s)} + CO_3^{2-} + 2H_2O.$$

The equivalence point is indicated by the complete disappearance of the color. The potassium permanganate equivalent is its molar mass;

- in inorganic analysis, we can mention the following determinations:
  - determination of iodides:

$$I^- + 8Ba^{2+} + 8OH^- + 8MnO_4^- \rightarrow 8BaMnO_{4(s)} + IO_4^- + 4H_2O.$$

The noteworthy point is the oxidization of iodide ions up to the periodate stage and not only to the iodine or iodate stages, as is the case in acidic and weakly alkaline media:

– determination of iodate ions

They are oxidized into periodate ions according to the reaction

$$2MnO_4^- + IO_3^- + 2OH^- + 2Ba^{2+} \rightarrow IO_4^- + 2MnO_4Ba_{(s)} + H_2O;$$

- in organic analysis, we mention the determination of
  - oxalic and malonic acids;
  - hydroxylated, ketonic, and unsaturated acids such as lactic, pyruvic, and fumaric acids;
  - alcohols, aldehydes, and ketones such as methanol, glycerol, paraldehyde, dimethyl ketone, etc.;
  - aromatic acids such as benzoic and cinnamic acids;
  - phenols.

The result obtained with the oxidization of salicylic acid is surprising: The aromatic nucleus is indeed broken down. This means that the stabilization energy due to the aromaticity of the nucleus has been surmounted. The reaction is as follows:

$$28MnO_4^- + 42OH^- + 28Ba^{2+} + C_6H_4(OH)COOH \rightarrow 7CO_3^{2-}$$
$$+28MnO_4Ba_{(s)} + 24H_2O.$$

[From the viewpoint of the equilibration of this redox reaction, we first notice that there are 28 exchanged electrons. We can admit that in salicylic acid there are four carbon atoms in oxidation state − I (those substituted by hydrogen atoms), one in state 0 (that bringing the carboxylic group), and two in oxidation state + I (those bringing the phenol group and that of the carboxylic group). After oxidization, they are in oxidation state + IV in the carbonate ion].

20.1.5 Determination of Organic Matters in Water

Among the proposed methods to determine organic matters in water, probably the most practical is to perform a permanganic oxidization of the sample under study. According to some authors, the oxidization might be directed toward organic matters from vegetable or animal origin according to the sample's pH. In acidic medium, we would principally determine the first ones and in alkaline media, the second. This is the reason why the oxidization in alkaline medium is prescribed for drinking water. In hydrogen carbonate medium and by warming for a well-determined time, the organic matter is oxidized with an excess of permanganate ions. The permanganate in excess is brought together with an excess of ferrous salt after acidification with sulfuric acid. The ferrous salt in excess is determined with a permanganate solution. The results are satisfactory only if a great excess of permanganate ions is used.

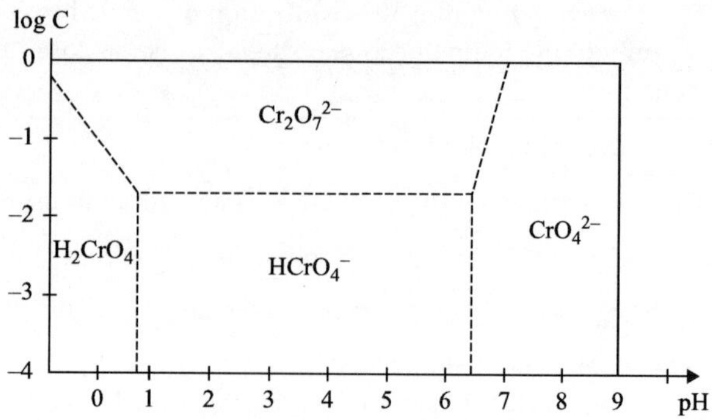

Fig. 20.2 Predominance area of the different ions of Cr^{+VI} (C: chromium concentration in atom-grams—see text)

20.2 Titrations with Dichromate Ions: Chromimetry

20.2.1 Definition

The term "chromimetry" refers to all the titrimetric methods using standardized solutions of potassium dichromate $K_2Cr_2O_7$. Titrations with chromous ions Cr^{2+} are not taken into account in the term "chromimetry."

20.2.2 General Considerations

The most interesting oxidation states of chromium in chemical analysis are states +VI, +III, and +II, but states +V and +IV are known.

Cr^{+VI} shows a marked tendency to polymerization, as indicated by the occurrence of the equilibrium dichromate $Cr_2O_7^{2-}$/acid chromate $HCrO_4^-$:

$$Cr_2O_7^{2-} + H_2O \rightleftharpoons 2HCrO_4^-, \quad K = 10^{-2.2}$$

on which the following two acid–base equilibria are superimposed:

$$H_2CrO_4 \rightleftharpoons HCrO_4^- + H^+, \quad pK_{a1} = 4.10,$$
$$HCrO_4^- \rightleftharpoons CrO_4^{2-} + H^+, \quad pK_{a2} = 6.45.$$

For a total concentration in chromium (expressed in atom-grams) higher than $10^{-1.68}$ mol/L on the one hand and with a pH value higher than 0.75 on the other hand, chromium at oxidation state +VI is either as dichromate ions $Cr_2O_7^{2-}$ or as chromate ions CrO_4^{2-} depending on the pH value (see Fig. 20.2). Both species are linked by the acid–base equilibrium:

$$1/2Cr_2O_7^{2-} + 1/2H_2O \rightleftharpoons CrO_4^{2-} + H^+, \qquad (pK_a = 7.2).$$

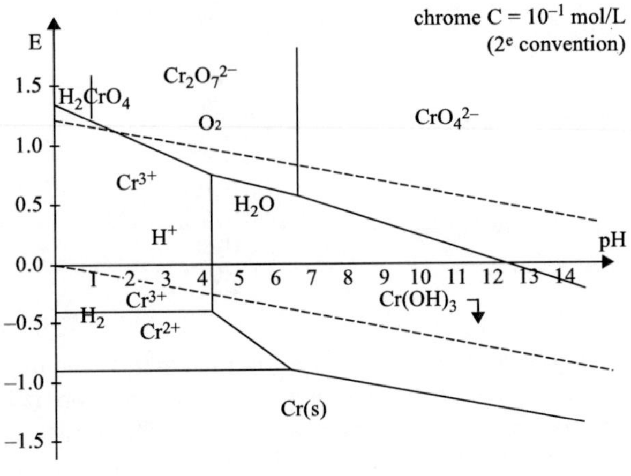

Fig. 20.3 Predominance-area diagram of different species of chromium ($C = 10^{-1}$ mol/L, second convention)

For a total concentration less than $10^{-1.68}$ mol/L, Cr^{+VI} is, depending on the pH value, either as H_2CrO_4 or $HCrO_4^-$ or CrO_4^{2-}. (We shall put ourselves in the case in which Cr^{+VI} is either as dichromate ions $Cr_2O_7^{2-}$ or as chromate ions CrO_4^{2-}.) Lewis's formulas of chromate and dichromate ions are given below. In both cases, chromium is at oxidation state +VI.

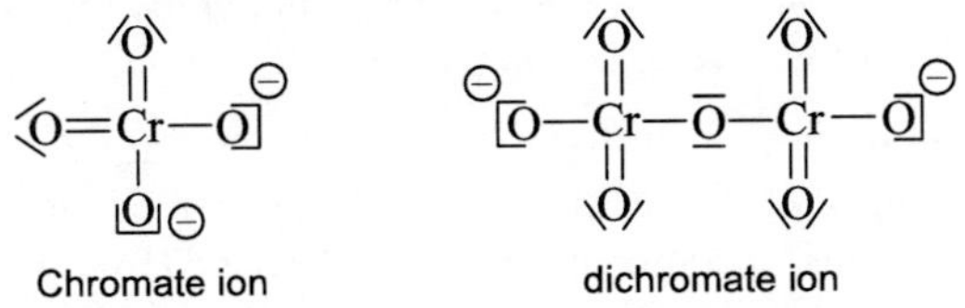

Chromium at oxidation state +III is under the form of chromic ions Cr^{3+} Chromic hydroxide $Cr(OH)_3$ already precipitates at weakly acidic pH [$K_s(Cr(OH)_3) = 10^{-30}$]. The predominance diagram E/pH of the different species of chromium is given in Fig. 20.3. The couple used in chromimetry is $Cr_2O_7^{2-}/Cr^{3+}$. Hence, it is practiced only for pH < 4. The half-redox reaction is

$$Cr_2O_7^{2-} + 14H^+ + 6e^- \rightleftharpoons 2Cr^{3+} + 7H_2O.$$

Its standard potential is $E°(Cr_2O_7^{2-}/Cr^{3+}) = 1.333$ V. It is quasi-identical to that of the couple H_2CrO_4/Cr^{3+}:

$$H_2CrO_4 + 6H^+ + 3e^- \rightleftharpoons Cr^{3+} + 4H_2O, \quad E°(H_2CrO_4/Cr^{3+}) = 1.335\,V.$$

Diacid H_2CrO_4 exists somewhat at pH = 0. Dichromate is a strong oxidant in acidic medium. Some authors consider it the weaker of the strong oxidizing agents. Its oxidizing power strongly decreases when the pH increases. Its apparent normal potential $E°'$ is given by the relation

$$E°' = (E° - 0.14\text{pH})\,V.$$

Table 20.1 Formal potentials of the couples $Cr_2O_7^{2-}/Cr^{3+}$

Acid (mol/L)	$E^{\circ\prime}(Cr_2O_7^{2-}/Cr^{3+})/V$
0.1 HCl	0.93
1.0 HCl	1.00
2.0 HCl	1.05
4.0 HCl	1.10
1.0 H_2SO_4	1.03
2.0 H_2SO_4	1.11
4.0 H_2SO_4	1.15
8.0 H_2SO_4	1.35
1.0 $HClO_4$	1.03

The system $Cr_2O_7^{2-}/Cr^{3+}$ is a slow system from chemical and electrochemical standpoints. From a chemical standpoint, dichromate ions react with a noticeable rate only for markedly acidic pH values. Once $pH > 2$, the oxidization reaction becomes slow. From an electrochemical standpoint, the system is slow (or irreversible). An inert metal electrode dipping into a solution containing the two members of the couple ($Cr_2O_7^{2-}$ and Cr^{3+}) does not immediately take the potential given by Nernst's relation.

There also exist important changes in the formal potentials of the couple in relation to the nature and concentration of the acid in the medium (see Table 20.1).

Another characteristic point of the oxidization by $Cr_2O_7^{2-}$ is the complexity of the oxidization mechanisms. They involve chromium at the intermediary oxidation states +IV and +V during several coupled reactions.

- a Pourbaix diagram shows that $Cr_2O_7^{2-}$, CrO_4^{2-}, Cr^{3+} are stable ions in aqueous solution. In strongly acidic medium, H_2CrO_4 is formed. It reacts with water, since for $pH < 0.8$, the predominance areas of H_2O and H_2CrO_4 are disconnected. Likewise, Cr^{2+} and Cr are never stable (when they are brought together with water);
- the equivalent of potassium dichromate is its molar mass divided by 6;
- chromimetry presents two advantages:
  - potassium dichromate solutions are very stable. A 10^{-1} mol/L solution, stored in a well-corked bottle, may be considered indefinitely stable;
  - chloride ions do not interfere provided their concentration is lower than 1 mol/L. They are not oxidized with dichromate ions. This is not surprising when we consider the standard and formal potentials of the couples:

$$E^\circ(Cl_2/Cl^-) = 1.36\,V \quad \text{and} \quad E^{\circ\prime}(Cr_2O_7^{2-}/Cr^{3+})$$
$$= 1.33\,V \quad \text{(in 1 mol/L HCl medium).}$$

This result is interesting from a practical standpoint: Oxidizations with dichromate ions can be achieved in hydrochloric acid solutions;

- the drawback of chromimetry lies in the fact that it is difficult to detect the equivalence point by perceiving the color change by the eye. The use of internal redox indicators is often necessary. The most commonly used ones are

N-phenylanthralinic and diphenylaminesulfonic acids as sodium salts. Their standard potentials are, respectively, about 0.85 V and 0.75 V;

- the standardized solutions are prepared by weighing potassium dichromate and dissolution. Potassium dichromate can be considered a primary standard. If necessary, the standardization of a solution may be achieved:
  - either with Mohr's salt. The reaction is

$$Cr_2O_7^{2-} + 6Fe^{2+} + 14H^+ \rightarrow 2Cr^{3+} + 6Fe^{3+} + 7H_2O.$$

When phosphoric acid, which complexes ferric iron, is added, there is a decrease in the equivalence point potential value and the color change of the sulfonated diphenylamine at the equivalence point becomes very sharp;

  - either by indirect iodometry. The reaction takes place in sulfuric acid solutions. We add an excess of potassium iodide to a known volume of the solution under study. The formed iodine (three moles for one mole of dichromate) is titrated with a thiosulfate solution:

$$Cr_2O_7^{2-} + 6I^- + 14H^+ \rightarrow 2Cr^{3+} + 3I_2 + 7H_2O.$$

20.2.3 Applications

Chromimetry is used in inorganic and organic analysis. We mention the following examples:

- determination of ferrous iron (see above);
- determination of ferric iron. This is an extension of the preceding one. Fe^{3+} ions are quantitatively reduced into Fe^{2+} ions, which are then titrated by dichromate ions. Fe^{3+} ions are reduced as in manganimetry, that is, by an excess of stannous chloride. The stannous ions in excess are oxidized into stannic ions with mercury(II) chloride;
- determination of uranyl salts. Uranyl salts can be determined by chromimetry but only after reduction into uranous ions U^{4+}. The involved redox couple UO_2^{2+}/U^{4+} obeys the equilibrium

$$UO_2^{2+} + 4H^+ + 2e^- \rightleftharpoons U^{4+} + 2H_2O, \qquad E°(UO_2^{2+}/U^{4+}) = 0.33\,V$$

The titration reaction is

$$Cr_2O_7^{2-} + 2H^+ + 3U^{4+} \rightarrow 2Cr^{3+} + 3UO_2^{2+} + H_2O.$$

The reaction is slow. The equivalence point is detected with diphenylamine sulfonic acid. Another methodology is to treat the uranous cation solution with an excess of iron + III chloride, which, in so doing, is reduced into iron + II ions:

$$U^{4+} + 2Fe^{3+} + 2H_2O \rightarrow UO_2^{2+} + 4H^+ + 2Fe^{2+}.$$

Iron +II is titrated with potassium dichromate. The initial reduction of the uranyl salt is achieved with a Jones reductor (amalgamated zinc), which is a very powerful reductor. Uranium +VI goes, in a first stage, from oxidation state +VI into state +III (hypouranous cation U^{3+}), which is spontaneously oxidized into U^{4+} by air dioxygen.

In organic analysis, most of the time the determinations involving dichromate ions are back titrations. The dichromate in excess is determined by indirect iodometry. Several organic compounds can be determined by chromimetry, including ethylenics, alcohols, carboxylic acids, and aldehydes. The reaction products are usually water and carbon dioxide. Ethanol gives acetic acid. This reaction is used to determine the alcohol level in blood.

In a first step, ethanol is separated from blood by distillation with some adjuvants such as picric acid, which precipitates proteides. The distillate is titrated with an excess of potassium dichromate in nitric acid solution. This is the principle of Cordebard's method. The alcohol oxidization reaction is

$$3CH_3CH_2OH + 2Cr_2O_7{}^{2-} + 16H^+ \rightarrow 3CH_3COOH + 11H_2O + 4Cr^{3+}.$$

The half-redox couple CH_3COOH/CH_3CH_2OH obeys the equilibrium

$$CH_3COOH + 4H^+ + 4e^- \rightleftharpoons CH_3CH_2OH + H_2O,$$
$$E°(CH_3COOH/CH_3CH_2OH) = 0.05\,V$$

The dichromate in excess is determined by indirect iodometry.

In the field of medicine, we notice the one titration of isoniazid again:

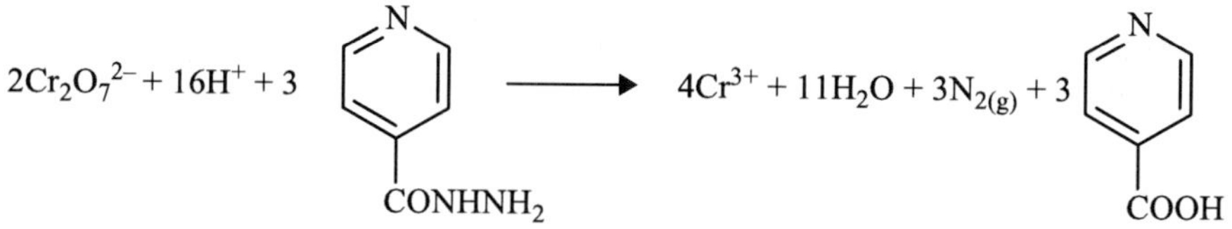

Both nitrogen atoms of the hydrazide group, which were initially in oxidation state −II are oxidized in state 0.

Let's also recall the determination of ethacridine (see Chap. 18). The precipitated dichromate ions are determined by indirect iodometry. Quinine and quinidine salts can be determined in the same manner.

20.3 Titrations with Ceric Ions

Titrations with ceric salts, symbolized here by Ce^{IV}, are sometimes grouped under the generic term "cerimetry." Most of the time, ceric sulfate is used.

Table 20.2 Influence of the nature of the acid on the formal potential values

	$E^{\circ\prime}(Ce^{IV})/(Ce^{III})$
Perchloric acid	1.70 V
Nitric acid	1.61 V
Sulfuric acid	1.44 V
Hydrochloric acid	1.28 V

20.3.1 Some Properties of Cerous and Ceric Salts

In acid solutions, ceric salts are powerful oxidizing agents. Simultaneous to the oxidization, they are reduced into cerous salts according to the half-redox equilibrium

$$Ce^{IV} + 1e^- \rightleftharpoons Ce^{III}.$$

The redox potential of the couple Ce^{IV}/Ce^{III} strongly depends on the nature and concentration of the present acid, as the different values of formal potentials indicate (see Table 20.2).

We notice that changes are important. We notice also, and overall, that Ce^{IV} is a very strong oxidizing agent.

One of the characteristics of the chemistry of ceric salts is the formation of complexes by ceric ions Ce^{4+}. For example, the complexes $[Ce(SO_4)_4{}^{4-}]$, $[Ce(SO_4)_3{}^{2-}]$, etc., and $[Ce(NO_3)_6{}^{2-}]$, $[Ce(NO_3)_5{}^{-}]$ are formed in sulfuric and nitric acid, respectively. Even with perchlorate ions, which are not known to give complexes easily, Ce^{IV} gives the complex $[Ce(ClO_4)_6{}^{2-}]$ in perchloric acid medium.

Likewise, Ce^{3+} also gives complexes. For example, with sulfate ions, there is formation of the complex $[Ce(SO_4)^+]$.

Another characteristic of cerous and ceric salts is their hydrolysis even at acid pH values. Ceric ions are already hydrolyzed at pH = 0. The following equilibrium is effective:

$$Ce^{4+} + H_2O \rightleftharpoons Ce(OH)^{3+} + H^+.$$

Hydroxyceric ions are yellow. At higher pH values, a second equilibrium is established:

$$Ce(OH)^{3+} + H_2O \rightleftharpoons Ce(OH)_2{}^{2+} + H^+.$$

The hydroxide $Ce(OH)_4$ precipitates at pH values about 1 in sulfuric acid solution. The cerous ion Ce^{3+} precipitates as hydroxide $Ce(OH)_3$ at about pH = 7.5.

The practical conclusion that can be drawn from these properties is that for ceric hydroxide formation to be avoided, we must work in strongly acidic medium in order to displace equilibria toward the left. However, in so doing, we favor the complexation phenomena with the anions of the strong acids, already evocated. There is possibly another complication. Some authors also consider the possibility of the dimerization of Ce^{IV} ions in some conditions to give the species $(Ce–O–Ce)^{6+}$, especially in perchloric medium.

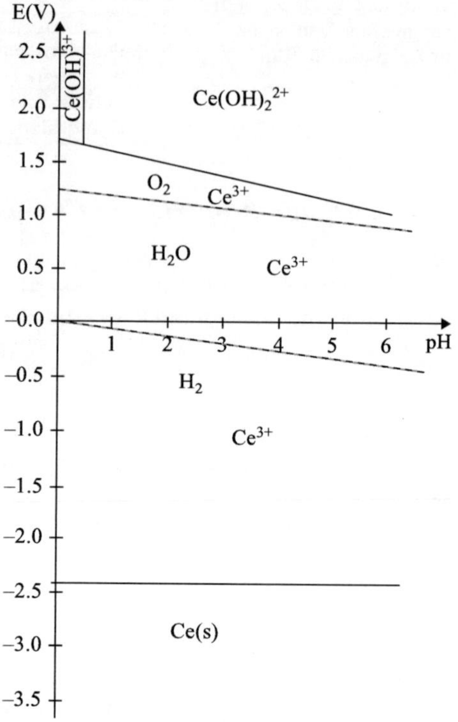

Fig. 20.4 *E*/pH diagram of cerium (simplified diagram)

All these phenomena explain the important changes in the formal potential value of the couple and justify the symbolism Ce^{IV} and Ce^{III}. The simplified Pourbaix diagram of cerium is represented in Fig. 20.4. In the pH range of the diagram (0–6), ceric salts are not stable from a thermodynamic standpoint. They must attack water with liberation of dioxygen, but the reaction is slow.

20.3.2 *Advantages of Cerimetry*

Solutions of ceric sulfate are remarkably stable. This assertion is in contradiction with the *E*/pH diagram. The explanation lies in the occurrence of a kinetic phenomenon. The kinetics of the reaction of water oxidization by ceric ion are complex. The formation of the species $Ce(OH)^{3+}$ seems to prevent any fast kinetic. From another standpoint, another advantage is that ceric sulfate solutions do not need to be protected against light.

Ceric salts may be used in hydrochloric media without any noticeable oxidization of chloride ions. The explanation of this result is simple. In hydrochloric acid, the formal potential of the Ce^{IV}/Ce^{III} couple is 1.28 V (see Table 20.2), whereas the standard potential of the couple Cl_2/Cl^- is 1.36 V.

Ceric salts are yellow or yellow-orange according to the pH. Hence, using internal indicators is not necessary. Such titrations give a clear visual change even in diluted solutions. If necessary, we can use ferrous ortho-phenantroline or the N-phenylanthranilic acid, whose standard potential is about 1.08 V.

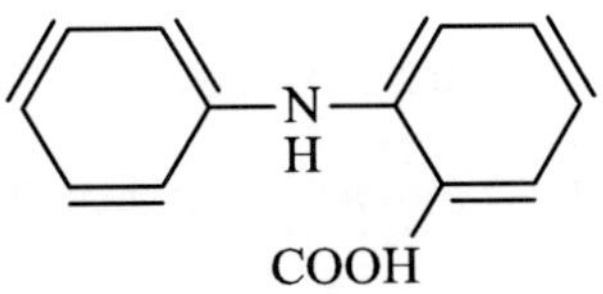

N-phenylanthranilic acid

20.3.3 *Standardized Solutions*

The equivalent is the molar mass of the species containing one atom of cerium in its structure. They are prepared

- either by starting from ceric sulfate, which is not a standard,
- or by starting from ammonium and cerium double sulfate, which may be purchased in a state of warranted purity,
- or from hexanitratocerate IV, also called ceriammonic nitrate $[Ce(NO_3)_6](NH_4)_2]$, which is a nitratocomplex of cerium IV. It exhibits the interesting property to be soluble in water. It may be considered a primary standard. It is dissolved into a 1 mol/L sulfuric acid in order to prepare solutions.

The standardization may be achieved

- with a solution of arsenious acid. The reaction is

$$2Ce^{IV} + H_3AsO_3 + H_2O \rightarrow 2Ce^{III} + H_3AsO_4 + 2H^+.$$

The reaction is slow, but it is accelerated when osmium tetraoxide is present in the reaction vessel. It is a catalyst for the reaction. Oxide As_2O_3 is dissolved in a sodium hydroxide solution, and the solution is acidified by the addition of sulfuric acid. Then some drops of an osmium tetraoxide solution are added and the solution under study is titrated with a ceric sulfate solution;

- with a Mohr's salt solution. We can also use a ferric salt that has previously been reduced with stannous chloride, as we already saw. The titration of Fe^{2+} with Ce^{IV} can be achieved in either sulfuric or hydrochloric acid;
- with an iodide solution. Iodide ions are oxidized into iodine, which is titrated with a thiosulfate solution:

$$2Ce^{IV} + 2I^- \rightarrow I_2 + 2Ce^{III}.$$

The titration is achieved in hydrochloric acid with an excess of potassium iodide;

- with a sodium oxalate solution, which gives oxalic acid in acidic medium. Oxalic acid is a reducing agent (see manganimetry, Sect. 20.1.1.2.4). The ceric salt oxidizes oxalic acid into carbon dioxide:

$$2Ce^{IV} + (COOH)_2 \rightarrow 2Ce^{III} + 2CO_{2(g)} + 2H^+.$$

The titration is achieved at 70°C with ferroin as the indicator.

20.3.4 *Applications of Cerimetry*

20.3.4.1 In Inorganic Analysis

We will list the determinations of

- ferrous salts (see standardization);
- hydrazine. The reaction is $N_2H_4 + 4Ce^{IV} \rightarrow N_{2(g)} + 4H^+ + 4Ce^{III}$.

Hydrazine is oxidized with an excess of boiling ceric sulfate in excess. The excess is titrated with a ferrous salt $[E°(N_2H_5{}^+/N_2) = -0.23\ V]$;

- hydrazoic acid and hydrazoates;

This is a back titration. The oxidization reaction of hydrazoic acid is

$$2N_3H + 2Ce^{IV} \rightarrow 3N_2 + 2Ce^{III} + 2H^+.$$

The excess of ceric salt is titrated with ferrous sulfate in sulfuric acid with ferroin as the indicator;

- nitrites;

The nitrite solution is added to a known quantity in excess of ceric sulfate. The excess is titrated with a ferrous salt solution or with an oxalate solution with ferroin or N-phenylanthranilic acid as the indicator. The oxidization reaction is

$$2Ce^{IV} + HNO_2 + H_2O \rightarrow 2Ce^{III} + NO_3{}^- + 3H^+;$$

- hydroxylamine;

This is a back titration. The ceric salt oxidizes hydroxylamine according to the reaction

$$2NH_2OH + 4Ce^{IV} \rightarrow N_2O + 4Ce^{III} + 4H^+;$$

Nitrous oxide N_2O forms $[E°(N_2O/NH_3OH^+) = -0.05\ V]$. The excess of ceric salt is titrated with an arsenious acid solution with ferroin as the indicator;

- hexacyanoferrate II (ferrocyanide);

This is a direct titration that is achieved in 1 mol/L sulfuric acid:

$$Fe(CN)_6{}^{4-} + Ce^{IV} \rightarrow Ce^{III} + Fe(CN)_6{}^{3-}.$$

Ferroin is used as the indicator: [$E^{\circ\prime}(Fe(CN)_6{}^{3-}/Fe(CN)_6{}^{4-} = 0.71$ V];

- hydrogen peroxide;

This is a direct titration with ceric sulfate with ferroin or N-phenylanthranilic acid as the indicator:

$$2Ce^{IV} + H_2O_2 \rightarrow 2Ce^{III} + O_{2(g)} + 2H^+.$$

The determination can be achieved in nitric, sulfuric, and chlorhydric acid media [$E°(O_{2(g)}/H_2O_2) = 0.69$ V];

- peroxodisulfate (or persulfate);

This is an indirect titration. Peroxodisulfate ions cannot be directly oxidized with ceric ions [$E°(S_2O_8{}^{2-}/SO_4{}^{2-}) = 2.01$ V]. From another standpoint, they cannot be directly titrated with ferrous ions since the following reaction is too slow:

$$S_2O_8{}^{2-} + 2Fe^{2+} \rightarrow 2SO_4{}^{2-} + 2Fe^{3+}.$$

As a result, the following methodology is used: A known quantity in excess of ferrous ions is brought together with peroxodisulfate ions. After the necessary time for the reaction to quantitatively proceed, the ferrous ions in excess are determined with ceric sulfate;

- antimony III;

Antimony III can be directly titrated with ceric sulfate with ferroin as the indicator. The reaction must be achieved at 50°C. The solution in the titration vessel must be sufficiently acidic to prevent the hydrolysis of the antimonyl cation SbO^+ existing in these conditions. The oxidization reaction is

$$4Ce^{IV} + 2SbO^+ + 3H_2O \rightarrow 4Ce^{III} + Sb_2O_{5(s)} + 6H^+.$$

Let's recall that the half-redox equilibrium corresponding to the couple Sb^{+V}/Sb^{+III} is

$$Sb_2O_{5(s)} + 6H^+ + 4e^- \rightarrow 2SbO^+ + 3H_2O, \quad E°(Sb^V/Sb^{III}) = 0.58\,V;$$

- uranium +VI;

The principle of the titration is the same as in chromimetry.

20.3.4.2 In Organic Analysis

- Determination of organic acids;

The ceric oxidization of organic acids has been the matter of several studies. Except for oxalic acid, which can be directly titrated, other organic acids are back-titrated. They are oxidized with an excess of ceric sulfate in a sulfuric acid medium heated with a heat bath for about one hour. After cooling at room temperature, the ceric sulfate in excess is titrated with a solution of Mohr's salt with N-phenylanthranilic acid as the indicator.

- determination of hydroxylated derivatives;

They are also back titrations. The excess of ceric salt is titrated with a standardized solution of oxalate with nitroferroin as indicator. The precise experimental conditions depend on the nature of the substrate. It is not possible to issue a general rule predicting the structure of the obtained derivatives in relation to the structure of the compound to determine.

- determination of hydroquinol

Hydroquinol is quantitatively and fastly oxidized into p-quinone with ceric sulfate [$E°(Q/QH_2) = 0.70$ V].

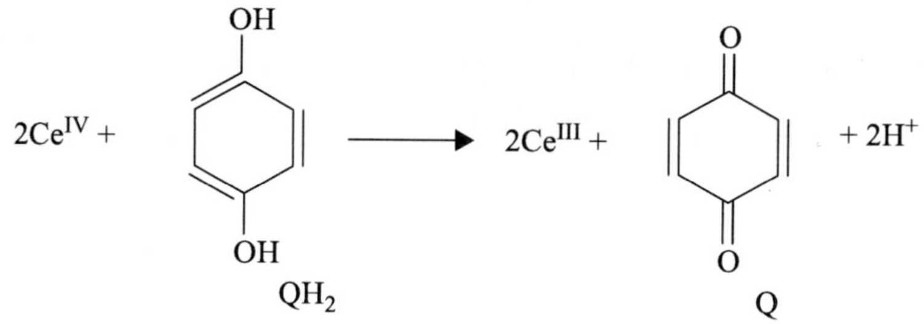

Several organic compounds interfere.

In the field of the determination of drugs, we can cite

- determination of menadione (vitamin K_3);

In a first step, it involves the reduction of menadione into the corresponding diphenol with zinc in acidic medium:

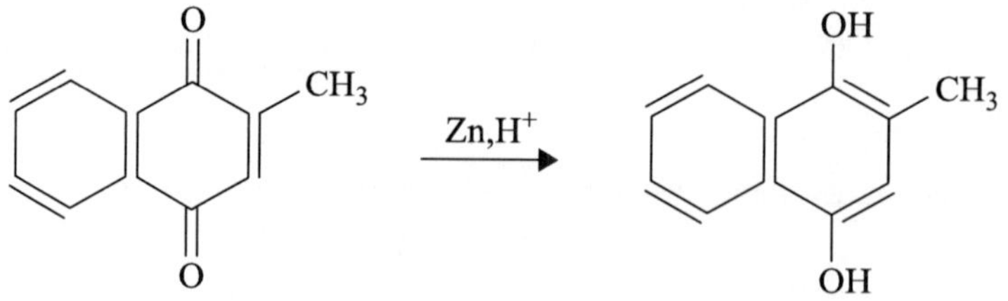

The formed diphenol is oxidized by cerimetry. The titration is a direct one with p-ethoxychrysoidine as the indicator:

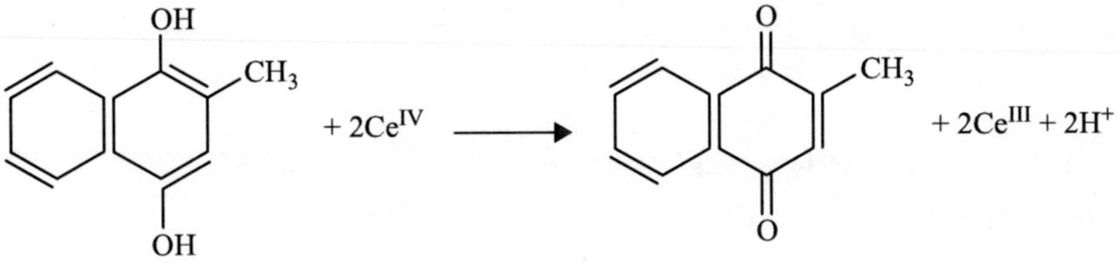

A possible titration of tocopherols is schematized as follows:

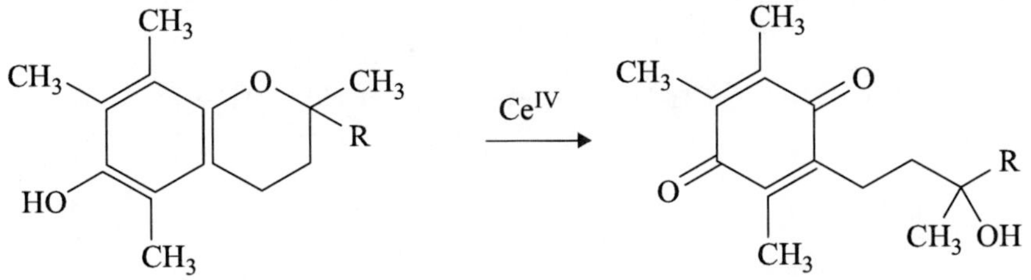

It is achieved by ceric oxidization. There is the quantitative formation of the corresponding p-quinone;

- determination of vitamin C;

There is formation of dehydroascorbic acid (see direct iodometry, Chap. 18) by direct titration with ceric sulfate:

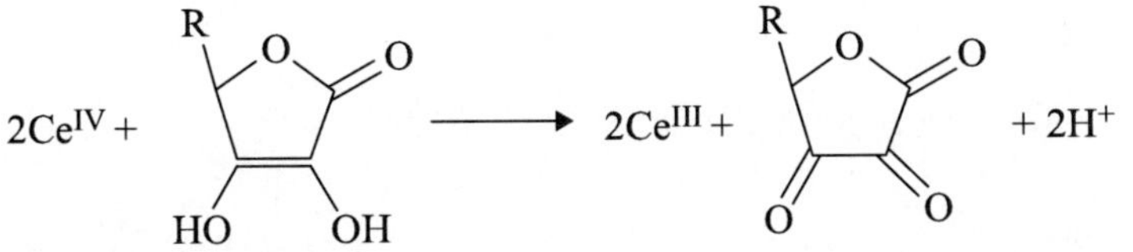

The titration is achieved with ferroin as the indicator;

- determination of paracetamol;

At first, paracetamol is quantitatively hydrolyzed into 4-aminophenol in sulfuric acid medium. Then 4-aminophenol is directly titrated with the ceric salt to give 1,4-benzoquinonimine:

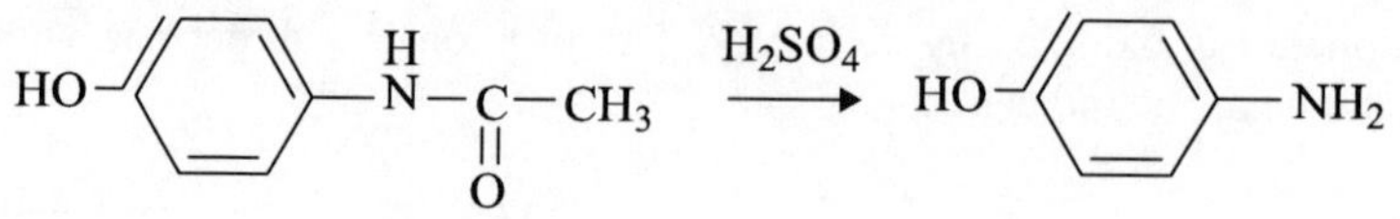

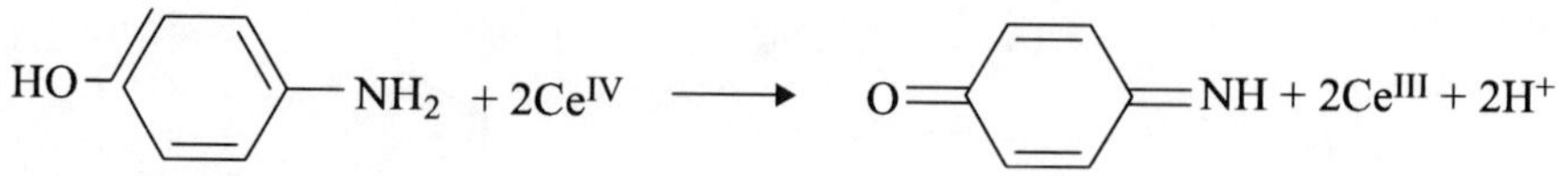

- determination of isoniazid;

As in other redox titrations of isoniazid, nitrogen of the hydrazide group passes from oxidation state – II to 0.

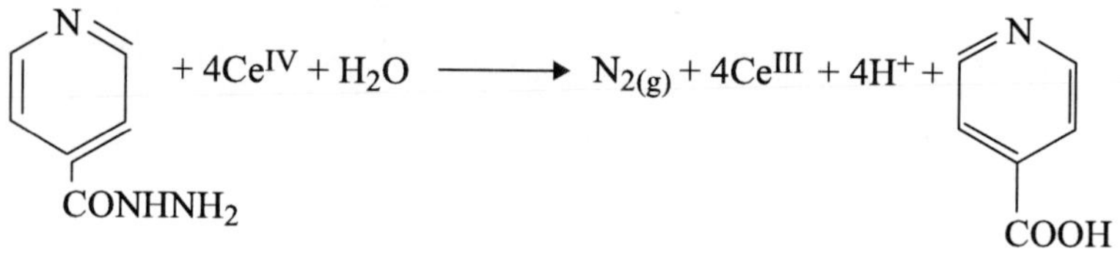

20.4 Some Other Oxydoreductimetric Titration Methods

Some other titrating reagents generating quantitative redox reactions are mentioned from time to time in the literature. Here we only list titrations involving titanous and chromous salts as well as ascorbic acid. They are three reducing agents.

20.4.1 Titrations with Titanium III Salts

The standard potential of the couple TiO^{2+}/Ti^{3+} is $E°(TiO^{2+}/Ti^{3+}) = 0.1$ V. It corresponds to the half-redox couple

$$TiO^{2+} + 2H^+ + 1e^- \rightleftharpoons Ti^{3+} + H_2O.$$

Ti^{3+} is a fairly strong reducing agent. The tendency of TiO^{2+} to form complexes transforms TI^{3+} into a strong reductor by displacement of the equilibrium toward the left. Titane III chloride and sulfate are prone to aerial oxidization. Generally, determinations are back titrations. The titane III salts in excess are titrated with a ferric salt with potassium thiocyanate used as the indicator.

20.4.2 Titrations with Chromium II Salts

Chromous ions Cr^{2+} enter into the half-redox equilibrium

$$Cr^{3+} + 1e^- \rightleftharpoons Cr^{2+}, \quad E°(Cr^{3+}/Cr^{2+}) = -0.41\,V.$$

The standard potential value indicates that chromium II ions are a very strong reducing agent. A Pourbaix diagram shows that they are not stable in water. Protons oxidize them according to the reaction (see Fig. 20.3)

$$2Cr^{2+} + 2H^+ \rightarrow 2Cr^{3+} + H_{2(g)}.$$

Likewise, they are oxidized by the air dioxygen. They are one of the most reducing titrants. They are easily prepared by the reduction of chromic ions with zinc. They reduce several inorganic and organic species.

20.4.3 Titrations with Ascorbic Acid

Ascorbic acid is a rather strong reducing agent: $E° = 0.185$ V (see iodometry, Chap. 18). It reacts with dioxygen. When it is in solutions with edta and formic acid, its solutions are stabilized and working in an inert atmosphere no longer is necessary. Traces of heavy metals catalyze its oxidization. Hence, the role played by edta is explained.

Chapter 21
Some Applications of Redox Reactions in Qualitative Analysis

Numerous applications of redox reactions exist in qualitative analysis. Some fall in the realm of organic chemistry, and others in that of inorganic chemistry.

21.1 Organic Analysis

21.1.1 Colorimetric Analysis

The formation of colored derivatives often results from one or several redox reactions.

- Phenols and amines endowed with a free ortho or para position can be characterized with 3-methylbenzothiazoline-2-one hydrazone (mbth);

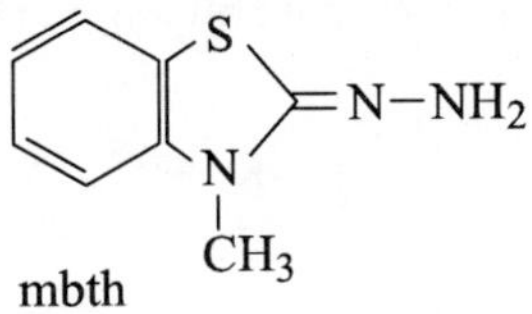

mbth

mbth gives a coupling reaction with phenol in an oxidizing medium. A colored azo derivative forms. The coupling reaction may be achieved in acidic or alkaline medium in the presence, for example, of cerium^{IV} and ammonium double sulfate:

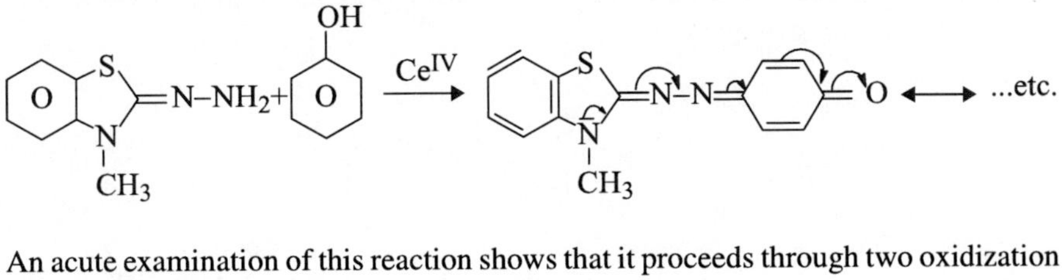

An acute examination of this reaction shows that it proceeds through two oxidization steps. The first one involves the oxidization of mbth, which gives a strongly electrophilic diazonium salt:

J.-L. Burgot, *Ionic Equilibria in Analytical Chemistry*,
DOI 10.1007/978-1-4419-8382-4_21, © Springer Science+Business Media, LLC 2012

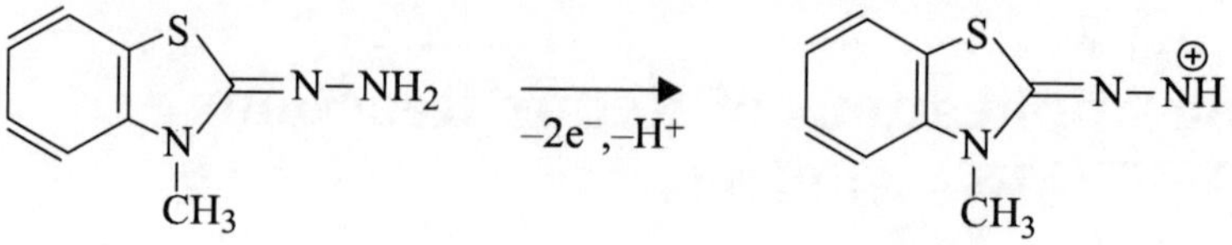

diazonium salt

The actual coupling reaction (that occurring between the diazonium salt and phenol) is the second step. It is characterized by a loss of two electrons and by a loss of one or two protons according to the acido-basic form phenol was under before the reaction. At least formally, one may imagine a pathway through the intermediary product

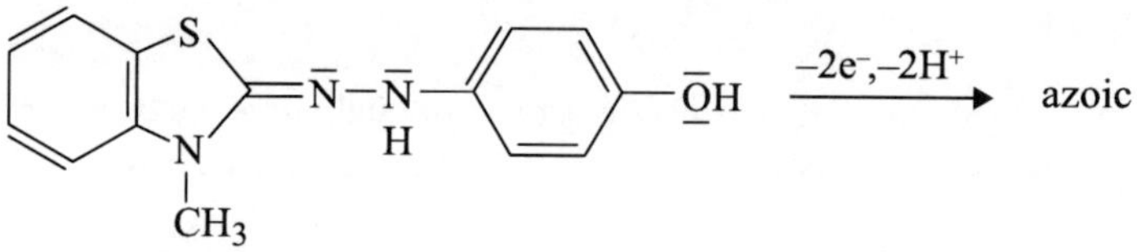

hypothetic scheme

which loses two electrons. We notice, however, that the two successive reactions have not been seen so far. Aromatic amines give colored derivatives of the same kinds as those obtained with phenols through an analogous reaction. Their formation provides evidence of their presence. They may be formed under the influence of the oxidizing power of Fe^{III}. The analogy of behavior exhibited by amines and phenols is not surprising since the formation of colored derivatives in both cases involves an electrophilic substitution reaction, which is easy because of the very comparable electron-donating effects of hydroxy and amino groups.

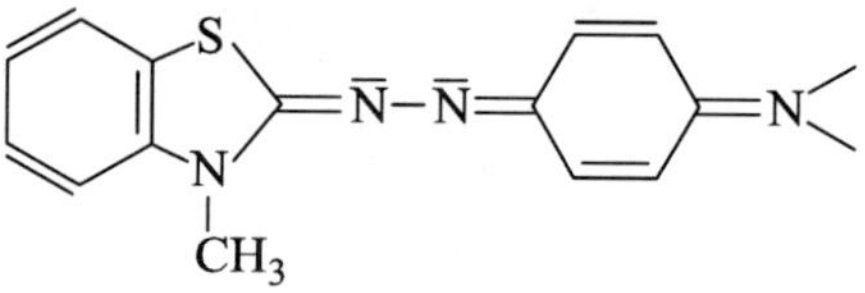

- The detection of aldehydes with mbth also involves several redox reactions. The formed colored derivatives are cations. Taken as a whole, they result from the condensation of two mbth molecules with one of aldehyde. One of their possible limit structures is the following one:

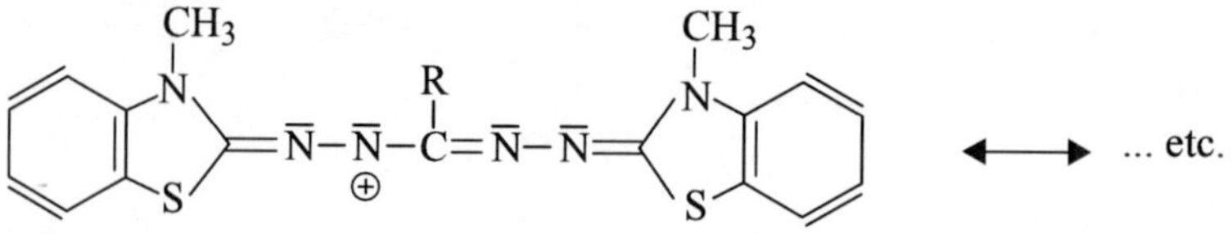

It is rational to think that in a first step, a first mbth molecule is condensing with the aldehyde molecule to give the corresponding hydrazone.

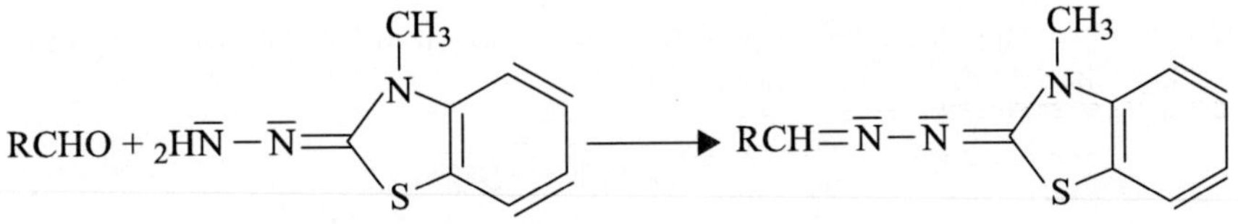

Furthermore, in the oxidizing medium (presence of Fe^{III}), the hydrazone is oxidized, giving one molecule of the diazonium salt already described:

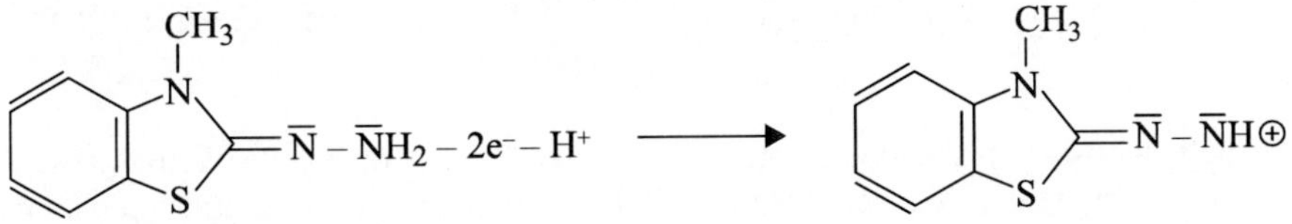

It is at this point that the first oxidization occurs. Then the diazonium salt attacks the carbon atom of the imino group, which is electron-rich, as indicated by the following limit structures:

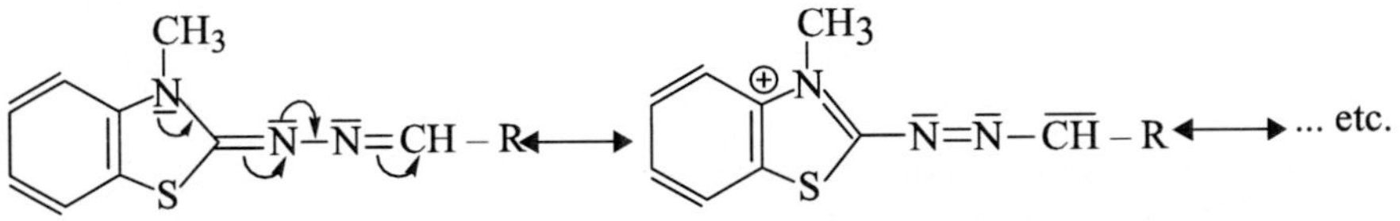

The condensation of the diazonium salt is accompanied by a loss of two electrons and two protons. This is the second oxidization step. This reaction is of great analytical importance because not only does it allow the characterization of aldehydes, but it also permits that of derivatives, which can be transformed in aldehydes. This is the case for

- primary alcohols,
- alkenes of the type $RCH=CR'R''$,
- glycerides, which in a first step are saponified in glycerol, which is then submitted to the periodic acid action, thereby giving formol,
- sugars and polyols, which, likewise, lead to formaldehyde, glycolic acid, etc.. after reaction with a periodate salt,
- methylpentoses.

We notice that all the preceding reactions are redox and particularly oxidization ones.

- The reduction of colorless 1,2,3,4-tetrazolium salts to give colored formazans allows the determination of reductive organic compounds such as aldehydes, thiols, and ketols ($RCOCH_2OH$):

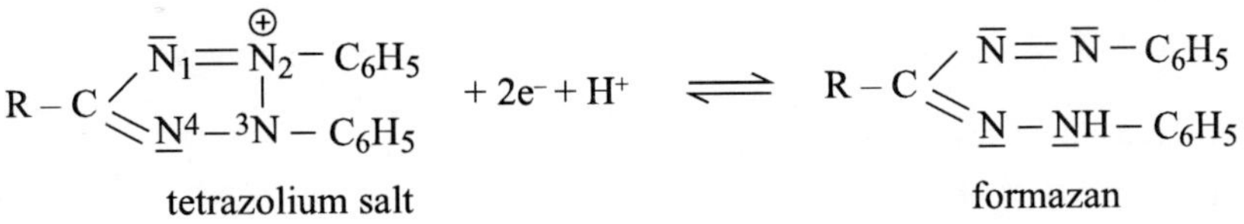

Hence, for example, the reaction of triphenyltetrazolium chloride with a ketol in a weakly alkaline medium may be written as

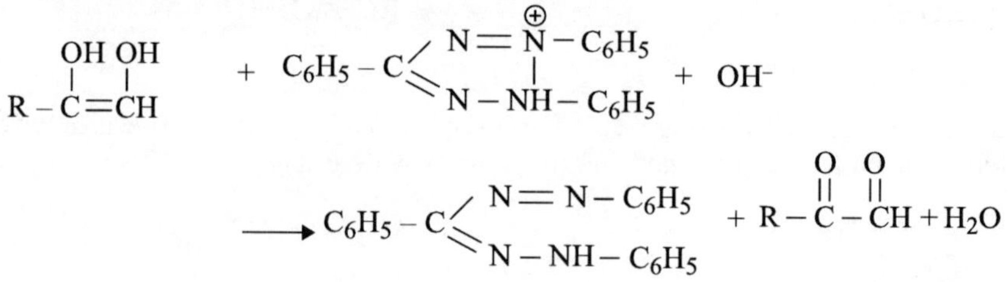

Simultaneously with the oxidization of the ketol into the corresponding ketoaldehyde, the tetrazolium salt is reduced in triphenylformazan. Of course, some other reductants react with tetrazolium salts according to an analogous scheme as that given above. It is interesting to notice incidentally that tetrazolium salts can also be reduced by some free radicals. For example, alcoxy radicals produced by the reaction of organic peroxides with Fe^{2+} according to

$$Fe^{2+} + \underset{\text{hydroperoxide}}{R - OOH} \rightarrow Fe^{3+} + OH^- + \underset{\text{alcoxy radical}}{R\dot{O}}$$

oxidize ethanol, giving another radical:

$$R\dot{O} + CH_3CH_2OH \rightleftharpoons CH_3\underset{\bullet}{C}HOH + ROH.$$

The corresponding redox half-equilibria are

$$R\dot{O} + H^+ + e^- \rightleftharpoons ROH,$$

$$CH_3\underset{\bullet}{C}HOH + H^+ + e^- \rightleftharpoons CH_3CH_2OH.$$

The free radical reduces tetrazolium blue:

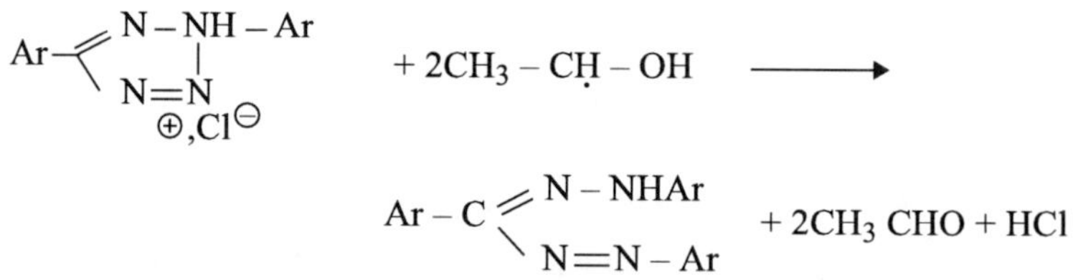

The redox half-equilibrium that is then operative is

$$CH_3CHO + H^+ + e^- \rightleftharpoons CH_3\underset{\bullet}{C}HOH.$$

The ketol function is present in the structure of several derivatives of utmost importance from a pharmacological standpoint, such as, for example, cortisone.

Another interesting analytical application of tetrazolium salts is the detection in different media of antioxidants of the hydroquinol type. The detection is performed, in particular, with the following tetrazolium salt:

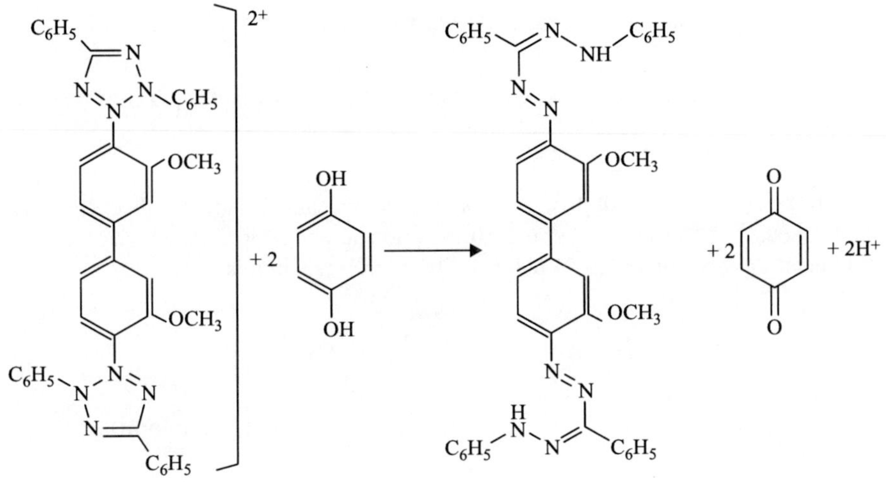

p-quinone is formed.

Remark A formazan can be obtained from the reaction between a phenyldiazonium salt with a phenylhydrazone:

$$RCH{=}N{-}NH{-}C_6H_5 + N{\equiv}\overset{+}{N}{-}C_6H_5 \longrightarrow R{-}C\begin{matrix} {\diagup} N{=}N{-}C_6H_5 \\ {\diagdown\!\!\diagdown} N{-}NH{-}C_6H_5 \end{matrix} + H^+$$

The phenyldiazonium salt itself results from the oxidization of the phenylhydrazine, used to prepare the hydrazone, by hexacyanoferrate-(III).

- Thiols react with the colored 2,6-dichloroindophenol. They are oxidized in the corresponding disulfide, while the indophenol is reduced in the corresponding colorless diphenolamine:

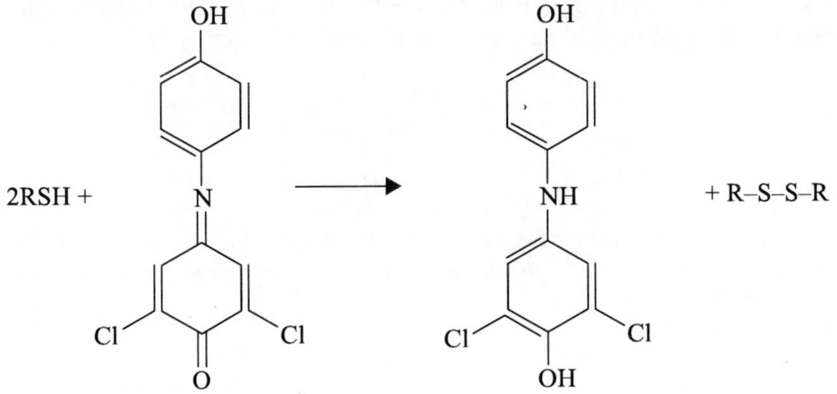

21.1.2 Detection in Chromatography

Numerous redox reactions are used for detection in thin-layer and paper chromatographies.

• Thiols can be characterized with Ellman's reagent [5,5'-dithio-bis-(2-nitrobenzoic)diacid (dtnb)]. It converts them into a mixed disulfide. Simultaneously, it gives the colored 2-nitro-5-mercaptobenzoic acid, which colors the thin layer or paper in yellow at pH = 8. The overall reaction is

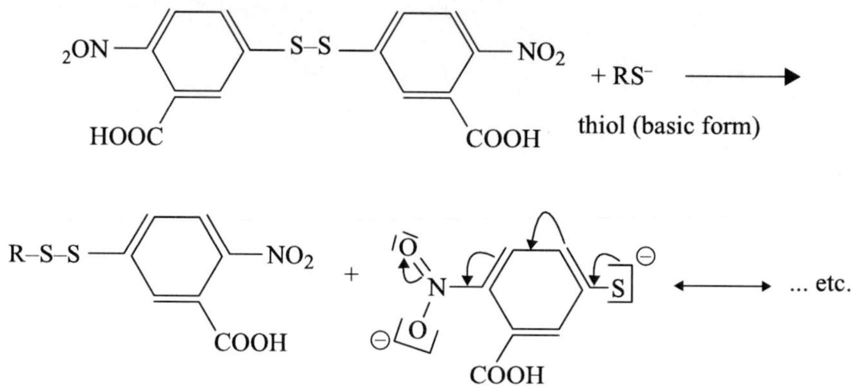

mixed disulfide colored 2-nitro-5-mercaptobenzoic acid

(The existence of several resonant limiting forms describing the structure of the formed nitromercaptobenzoic acid explains the color.)

• Compounds whose structure possesses a weakly acidic NH group, such as primary amines, amides, imides, and lactam can be detected by performing a redox reaction through the formation of intermediary N-chloro derivatives. Potassium permanganate reacting with hydrochloric acid generates chlorine (probably at oxidation state +I), which transforms the invoked derivatives in N-chloroamines. The latter derivatives oxidize iodide ions into iodine detected with a starch solution:

$$Cl_2,$$
$$RNH_2 \rightarrow RNHCl,$$
$$RNHCl + 2I^- + H_2O \rightarrow RNH_2 + I_2 + Cl^- + OH^-.$$

This reaction is also used for the characterization of peptides. Treating the thin layer (after migration) by sodium hypochlorite gives N-chloroamide groups in place of the initial peptide bonds:

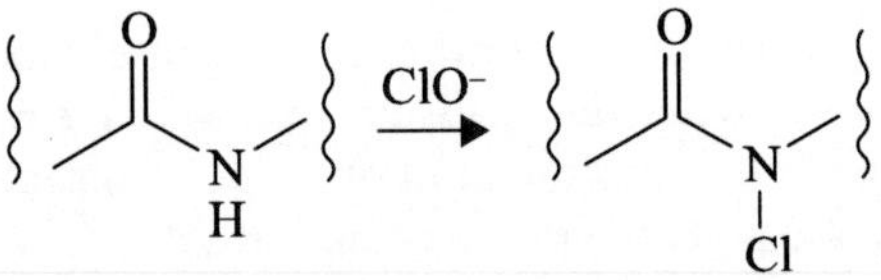

Spraying an iodide solution on the layer in the presence of a starch solution gives the blue color due to iodine formed by oxidization of iodide ions by the N-chloroamide groups.

- Primary aliphatic amines can be detected through a spot test by using a colorless paper impregnated with silver nitrate and manganous sulfate. The paper turns gray in the presence of the amine vapors. The gray color is due to a deposit of metallic silver. The manganous cation is oxidized by the silver cation in the presence of aliphatic amines with the formation of manganese dioxide, metallic silver, and the conjugate acid of the amine:

$$Mn^{2+} + 2Ag^{+} + 4RNH_2 + 2H_2O \rightarrow MnO_{2(s)} + 2Ag_{(s)} + 4RNH_3{}^{+}.$$

(It is interesting to note that the reaction does not take place with aromatic amines, whose basicity is not strong enough. In another respect, Mn^{2+} can be replaced by Fe^{2+} in order to carry out the reaction.) Evidently, the amine shifts the equilibrium of the redox reaction toward the right. Concerning this point, we'll also mention the fact that the reaction only occurs at neutral or weakly alkaline pH values.

21.1.3 Titration Reactions for Which the Equivalence Point Is Detected Through the Occurrence of a Redox Reaction

Under this heading, titration reactions themselves are not redox ones or, in any case if they are, are not implicated in the detection of the equivalence point. One example is given by the method sometimes called "nitritometry" evoked in connection with the "indirect iodometry." Nitrous acid in excess appearing once the equivalence point is reached during the titration of aromatic primary amines is detected by the formation of bromine. Bromide ions, added in the medium before the beginning of the titration, are indeed oxidized by nitrous acid in excess. The titration reaction itself is

$$ArNH_2 + HNO_2 + HCl \rightarrow Ar - \overset{+}{N}\equiv N + Cl^{-} + 2H_2O.$$

At the equivalence point, the following redox reaction occurs:

$$2HNO_2 + 2H^{+} + 2Br^{-} \rightarrow 2NO_{(g)} + Br_2 + 2H_2O.$$

Bromine formation is detected in several manners, especially by an electrochemical method. Notice that the corresponding standard potentials are not in favor of the formation of bromine according to the above reaction:

$$HNO_2 + H^{+} + e^{-} \rightleftharpoons NO_{(g)} + H_2O, \quad E°(HNO_2/NO_{(g)}) = 1.00\,V,$$

$$Br_2 + 2e^{-} \rightleftharpoons 2Br^{-}, \quad E°(Br_2/Br^{-}) = 1.07\,V.$$

At first sight, the reaction would evolve in the reverse sense of the given one. However, at the equivalence point, the reaction is displaced toward the right since the bromine concentration is quasi-null and the equilibrium redox potential of the solution is weaker than the standard potential of the couple Br_2/Br^-.

21.1.4 Functional Analysis

The presence of some functions in the structures of organic derivatives may be detected by some redox reactions. Many of them have been evoked in the preceding chapters.

Primary alcohols are oxidized in well-defined conditions by chromic anhydride CrO_3, giving the corresponding carboxylic acids, while chromium reduces from state + VI to + III. In the same conditions, secondary alcohols give the corresponding ketones and tertiary alcohols give orange-colored chromic esters. This color contrasts with the green color of chromic salts obtained with both previous alcohols.

Aldehydes are reductants. They are oxidized, for example, by the complex ion tetraiodomercurate-(II) (Nessler's reagent). The mercury of the complex is reduced to the metallic state. Aldehydes also reduce the ammoniacal silver nitrate (Tollens' reagent) with the formation of metallic silver. Concerning this point, let's us recall that aldoses are also reductants. For example, they reduce cupric copper to cuprous copper (Fehling's reagent). Ketoses also exhibit this property. In this case, this is due to the presence of the group enediol in their structure.

21.2 Inorganic Analysis

In this section, we give only some examples concerning the dichotomic method of ion analysis. In this way, depending on the reactions into which they may or may not participate, metallic ions are classified into groups. The discriminating reactions are those invoking

- some particular reactant,
- organic reactants,
- redox phenomena.

Otherwise, anions are the matter of reactions performed at the end of the sample analysis when the species to be analyzed have already been classified into a group or subgroup as just described.

- Hg^{II} is displaced from its solutions by metallic copper. As a result, a copper blade becomes covered with a brilliant layer when it is put in the presence of Hg^{II}.The redox reaction is

$$Hg^{2+} + Cu_{(s)} \rightarrow Hg_{(s)} + Cu^{2+}.$$

It is in perfect agreement with the standard potential values: $E°(Hg^{2+}/Hg) = 0.85$ V and $E°(Cu^{2+}/Cu) = 0.34$ V;

- arsenic can be detected by a reduction reaction. As^{III} and As^{V} present in the initial solution as arsenous and as arsenic acids are reduced into arsine by zinc in sulfuric acid. The involved arsenic couples are

$$H_3AsO_4 + 2H^+ + 2e^- \rightleftharpoons H_3AsO_3 + H_2O, \quad E^\circ(As^V/As^{III}) = 0.56\,V,$$

$$H_3AsO_3 + 3H^+ + 3e^- \rightleftharpoons As_{(s)} + 3H_2O, \quad E^\circ(As^{III}/As_{(s)}) = 0.25\,V,$$

$$As_{(s)} + 3H^+ + 3e^- \rightleftharpoons AsH_{3(g)}, \quad E^\circ(As_{(s)}/As^{III}) = -0.60\,V.$$

It is interesting to note that the standard potential of the couple Zn^{2+}/Zn is $E^\circ(Zn^{2+}/Zn) = -0.76$ V. At first glance, it is surprising that the reduction reaction of arsenic goes until arsine according to the reaction

$$2As + 3Zn + 6H^+ \rightarrow 2AsH_{3(g)} + 3Zn^{2+}$$

due to the standard potential values. This is the inverse reaction that would occur. The explanation is accounted for by the fact that simultaneous to the arsenic reduction also occurs that of water with the formation of dihydrogen:

$$Zn_{(s)} + 2H^+ \rightarrow Zn^{2+} + H_{2(g)},$$

whose vapors drive hydrogen arsenide and displace the formation equilibrium toward the right. Once formed, hydrogen arsenide may be characterized in several ways:

- antimony-III and V are reduced by tin-II [$E^\circ(Sn^{2+}/Sn) = -0.14$ V]. In this way, they give a deposit of metallic arsenic in a slightly acidic medium. Involved antimony couples are
  - antimonic oxide Sb_2O_5/antimonyl cation SbO^+:

$$Sb_2O_5 + 6H^+ + 4e^- \rightleftharpoons 2SbO^+ + 3H_2O, \quad E^\circ(Sb_2O_5/SbO^+) = 0.58\,V;$$

  - antimonyl cation/antimony:

$$SbO^+ + 2H^+ + 3e^- \rightleftharpoons Sb_{(s)} + H_2O, \quad E^\circ(SbO^+/Sb) = 0.21\,V;$$

- stannic tin is reduced by iron powder according to

$$Sn^{4+} + Fe_{(s)} \rightarrow Sn^{2+} + Fe^{2+}.$$

The corresponding couples are

$$Sn^{4+} + 2e^- \rightleftharpoons Sn^{2+}, \quad E^\circ(Sn^{4+}/Sn^{2+}) = 0.15\,V,$$

$$Fe^{2+} + 2e^- \rightleftharpoons Fe_{(s)}, \quad E^\circ(Fe^{2+}/Fe_{(s)}) = -0.44\,V.$$

Sn^{2+} formed in this way is detected by the bleaching of an iodine/iodide/starch solution since iodine is reduced into iodide ions [$E^\circ(I_2/I^-) = 0.62$ V)]:

$$I_2 + Sn^{2+} \rightarrow 2I^- + Sn^{4+}.$$

Sn^{2+} can also be characterized by its property to reduce methylene blue into hydrochloric acid, giving the corresponding leuco base:

$$Sn^{2+} + \text{methylene blue} + H^+ \rightarrow Sn^{4+} + \text{methylene blue (leuco base)}.$$

The methylene blue/leuco base couple's standard potential is [$E°$(methylene blue/leuco base) = 0.53 V]:

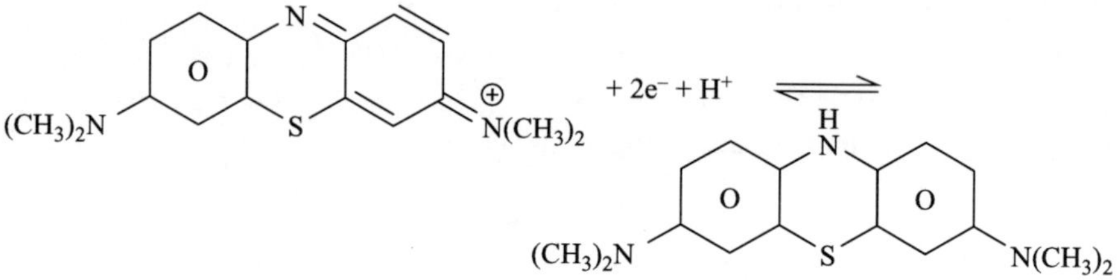

(We notice incidentally that methylene blue is a phenothiazine.)

- Bi^{III} is detected in alkaline medium by stannite anions. Therefore, bismuth stands in the solution as bismuth hydroxide $Bi(OH)_3$. At pH = 14, the involved bismuth couple is $Bi(OH)_3/Bi_{(s)}$ according to the half-redox reaction

$$Bi(OH)_3 + 3e^- \rightleftharpoons Bi_{(s)} + 3OH^-, \quad E°(Bi(OH)_3/Bi_{(s)}) = -0.46\,V).$$

At the same pH value, the half-redox Sn^{IV}/Sn^{II} equilibrium that must be considered is

$$SnO_3{}^{2-} + 3H^+ + 2e^- \rightleftharpoons HSnO_2{}^- + H_2O, \quad E°(SnO_3{}^{2-}/HSnO_2{}^-) = -0.93\,V.$$

The overall redox reaction at pH = 14 is

$$2Bi(OH)_3 + 3HSnO_2{}^- + 3OH^- \rightarrow 2Bi_{(s)} + 3SnO_3{}^{2-} + 6H_2O.$$

Remark Bi^V, as bismuthic acid $HBiO_3$, is a very strong oxidant in acidic medium. It enters into the half-redox equilibrium through which the cation bismuthyle BiO^+ is formed:

$$HBiO_3 + 3H^+ + 2e^- \rightleftharpoons BiO^+ + 2H_2O, \quad E°(HBiO_3/BiO^+) = 1.70\,V.$$

For example, bismuthic acid (metabismuthate in acidic medium) oxidizes Ce^{III} into Ce^{IV} and manganous ions into permanganate ions;

- chromic ion Cr^{3+} can be oxidized by peroxodisulfate ions in acidic medium, as can be predicted by examining the corresponding standard potentials:

$$S_2O_8{}^{2-} + 2e^- \rightleftharpoons 2SO_4{}^{2-}, \quad E°(S_2O_8{}^{2-}/SO_4{}^{2-}) = 2.01\,V,$$

$$Cr_2O_7{}^{2-} + 14H^+ + 6e^- \rightleftharpoons 2Cr^{3+} + 7H_2O, \quad E°(Cr_2O_7{}^{2-}/Cr^{3+}) = 1.23\,V.$$

The overall reaction is

$$2Cr^{3+} + 3S_2O_8{}^{2-} + 7H_2O \rightarrow Cr_2O_7{}^{2-} + 6SO_4{}^{2-} + 14H^+.$$

However, it is slow. The reactional vessel must be warmed and a catalyst such as the silver ion must be added;

- manganous ions are detected by their oxidization into permanganate ions $MnO_4{}^-$ in the presence of Ag^+ acting as a catalyst. The oxidizing agent is, as before, the peroxodisulfate ion. The reaction occurs in a slightly sulfuric or nitric acid medium:

$$2Mn^{2+} + 5S_2O_8{}^{2-} + 8H_2O \rightarrow 2MnO_4{}^- + 10SO_4{}^{2-} + 16H^+.$$

The detection of Mn^{2+} may interfere with that of Cr^{3+};

- chloride ions are detected by their oxidization with permanganate ions in sulfuric acid medium:

$$MnO_4{}^- + 5Cl^- + 8H^+ \rightarrow Mn^{2+} + 5/2Cl_2 + 4H_2O.$$

Manganese bioxide MnO_2 and Pb^{IV} oxide give chlorine only in more acidic media than that used with permanganate (see Fig. 20.1 of Chap. 20). The corresponding redox reactions are

$$MnO_{2(s)} + 4H^+ + 2Cl^- \rightarrow Mn^{2+} + Cl_{2(g)} + 2H_2O,$$

$$PbO_{2(s)} + 4H^+ + 2Cl^- \rightarrow Pb^{2+} + Cl_{2(g)} + 2H_2O.$$

The corresponding standard potential values are $E°(Cl_2/Cl^-) = 1.36$ V, $E°(MnO_4{}^-/Mn^{2+}) = 1.51$ V, $E°(MnO_{2(s)}/Mn^{2+}) = 1.23$ V, $E°(PbO_2/Pb^{2+}) = 1.46$ V;

- hypochlorite and hypobromite anions are reduced by arsenous acid or arsenites in hydrogen carbonate medium (see Sect. 18.10):

$$ClO^- + H_3AsO_3 \rightarrow Cl^- + HAsO_4{}^{2-} + 2H^+;$$

- chlorate ions are also detected with the help of a redox reaction. We should consider that they are often accompanied by hypochlorite and chloride ions. As a result, to carry out this determination, hypochlorite ions are eliminated by the reaction with arsenites and chloride ions are eliminated as silver chloride. Nitrites are then added to the remaining solution. Chlorate ions oxidize them with the formation of chlorides, which are detected by the addition of silver nitrate. The oxidization reaction of nitrite ions by chlorate ions is

$$ClO_3{}^- + 3HNO_2 \rightarrow Cl^- + 3NO_3{}^- + 3H^+.$$

The involved redox couples are

$$NO_3^- + 3H^+ + 2e^- \rightleftharpoons HNO_2 + H_2O, \quad E^\circ(NO_3^-/HNO_2) = 0.94\,V,$$
$$ClO_3^- + 6H^+ + 6e^- \rightleftharpoons 3H_2O + Cl^-, \quad E^\circ(ClO_3^-/Cl^-) = 1.45\,V;$$

- Bromide ions are detected by the formation of bromine after their oxidation by permanganate in acetic acid medium:

$$2MnO_4^- + 10Br^- + 16H^+ \rightarrow 5Br_2 + 2Mn^{2+} + 8H_2O.$$

The standard potential value $E^\circ(Br_2/Br^-)$ is 1.09 V. It is independent of the pH value until about pH = 8, beyond which hypobromite ions and hypobromous acid are generated. Iodide ions are awkward in this characterization since they react like bromide ions. In order to circumvent the difficulty, iodide ions are first oxidized by iodate ions at pH = 4.8. Iodine is formed. An examination of Fig. 18.1 of Chap. 18 shows that iodate ions at this pH value are unable to oxidize bromides. The iodine formed is expelled by volatilization. The search for bromide ions may then begin;

- hypobromite ions may be characterized by their property to oxidize hydrogen peroxide at high pH values. This pH condition is indeed one of the necessary ones for which hypobromite ions can exist. Redox couples and the corresponding standard reduction potentials are

$$BrO^- + H_2O + 2e^- \rightleftharpoons Br^- + 2OH^-, \quad E^{\circ\prime} = 0.78\,V\ (pH = 14),$$
$$O_{2(g)} + 2H^+ + 2e^- \rightleftharpoons H_2O_2, \quad E^{\circ\prime} = -0.07\,V\ (pH = 14).$$

The redox reaction is

$$BrO^- + H_2O_2 \rightarrow Br^- + H_2O + O_{2(g)};$$

- bromate ions oxidize helianthine, which is discolored during the reaction. Numerous oxidants exhibit the same reaction. Otherwise, some reductants reduce bromate ions once they are in acidic medium. Therefore, one oxidization reaction by bromate ions is not necessarily conclusive. (The acidic medium is necessary because as iodate ions, bromate ions are sufficiently oxidizing in acidic pH, only. At pH = 7, the apparent standard potential of the couple $BrO_3^-/Br_{2(l)}$ is about 1.02 V [$E^\circ(BrO_3^-/Br_{2(l)}) = 1.52$ V]; see Chap. 19.) Awkward reductants are arsenite, thiocyanate, thiosulfate, ferrocyanide, sulfide, bromide, and chloride ions. They are eliminated by oxidization with permanganate in alkaline medium, which remains a strong oxidizing reagent in these conditions (see Fig. 20.1 of Chap. 20). Permanganate in excess and manganous dioxide, the latter also being a possible oxidant, are eliminated after reaction with hydrogen peroxide in alkaline medium. Indeed, they should give oxidization reactions analogous to those given by bromate ions, which are searched for. It is interesting to notice that because of the standard potential values of the couples $BrO_3^-/Br_{2(l)}$ and $O_{2(g)}/H_2O_2$ [$E°(O_{2(g)}/H_2O_2) = 0.69$ V], bromate ions would oxidize hydrogen peroxide even at high pH values and therefore would not be characterized. However, the reaction

$$2BrO_3^- + 5H_2O_2 \rightarrow Br_{2(l)} + 5O_{2(g)} + 4H_2O + 2OH^-$$

does not occur for kinetic reasons. The involved redox couple of hydrogen peroxide is

$$O_{2(g)} + 2H^+ + 2e^- \rightleftharpoons H_2O_2;$$

- iodide ions are characterized by iodine formation after reaction with nitrous acid. Iodine is extracted with carbon tetrachloride, which hence becomes violet-colored. The redox reaction that occurs is

$$2HNO_2 + 2H^+ + 2I^- \rightarrow 2NO_{(g)} + 2H_2O + I_2.$$

Its direction is in agreement with the standard potential values of both couples. The above experiment is specific to iodide ions. Iodine can also be liberated from iodide ions by chlorate, bromate, iodate, periodate, and persulfate ions provided that some pH conditions, necessary for the oxidization reaction of iodide ions to be successful, are fulfilled;

- iodate ions can be characterized by their oxidizing power:

$$IO_3^- + 6H^+ + 5e^- \rightleftharpoons 1/2I_2 + 3H_2O, \quad E°(IO_3^-/I_2) = 1.19\,V.$$

Potassium iodate oxidizes iodide ions for pH values weaker than about 7 (see Fig. 18.1 in Chap. 18). The liberated iodine is detected by its partitioning into carbon tetrachloride or chloroform;

- periodate ions IO_4^- are characterized by their property to oxidize manganous ions into permanganate ions in nitric acid (1 mol/L) according to the reaction

$$5H_5IO_6 + 2Mn^{2+} \rightarrow 5IO_3^- + 2MnO_4^- + 11H^+ + 7H_2O.$$

It is the result of the following half-redox reactions:

$$H_5IO_6 + H^+ + 2e^- \rightleftharpoons IO_3^- + 3H_2O, \quad E°(H_5IO_6/IO_3^-) = 1.60\,V,$$

$$MnO_4^- + 8H^+ + 5e^- \rightleftharpoons Mn^{2+} + 4H_2O, \quad E°(MnO_4^-/Mn^{2+}) = 1.51\,V;$$

- sulfide ions (or hydrogen sulfide ions according to the pH value) can be characterized after oxidization into sulfur, which precipitates. The oxidization can be carried out with

– iodine in acidic medium:

$$H_2S + I_2 \rightarrow S_{(s)} + 2I^- + 2H^+;$$

– by hexacyanoferrate-(III) at pH = 9:

$$2[Fe(CN)_6]^{3-} + SH^- \rightarrow S_{(s)} + H^+ + 2[Fe(CN)_6]^{4-};$$

– by ferric iron in acidic medium:

$$2Fe^{3+} + H_2S \rightarrow 2Fe^{2+} + 2H^+ + S_{(s)};$$

• thiosulfate or hyposulfite anions $S_2O_3^{2-}$ are characterized by oxidization with iodine (see Sect. 18.6);

• Sulfites and sulfurous anhydride are also characterized by their reducing power. Let's recall that sulfurous anhydride is a gas in usual conditions. In aqueous solutions it behaves like a diacid ($pK_{a1} = 1.8$ and $pK_{a2} = 7.1$). Sulfite ions are involved in the following two half-redox equilibria:

$$SO_4{}^{2-} + 4H^+ + 2e^- \rightleftharpoons H_2SO_3, \quad E°(SO_4{}^{2-}/H_2SO_3) = 0.15\,V,$$

$$S_2O_6{}^{2-} + 4H^+ + 2e^- \rightleftharpoons 4H_2SO_3, \quad E°(S_2O_6{}^{2-}/H_2SO_3) = 0.49\,V.$$

Sulfite and hydrogen sulfite ions reduce iodine into iodides (see iodometry). They also reduce permanganate ions. They are simultaneously converted into sulfate ions (Bunsen's method) or into dithionite ions $S_2O_6{}^{2-}$ depending on the experimental conditions. The reaction must be carried out in a very acidic medium in order for the + II oxidation state of manganese to be reached and also to avoid the precipitation of manganese dioxide. Reduction reactions of permanganate are

$$2MnO_4{}^- + 5H_2SO_3 \rightarrow 2Mn^{2+} + 5SO_4{}^{2-} + 3H_2O + 4H^+,$$

$$2MnO_4{}^- + 10H_2SO_3 \rightarrow 2Mn^{2+} + 5S_2O_6{}^{2-} + 8H_2O + 4H^+.$$

Dichromate ions also oxidize sulfite ions into sulfate ions in acidic medium:

$$Cr_2O_7{}^{2-} + 3H_2SO_3 + 2H^+ \rightarrow 3SO_4{}^{2-} + 2Cr^{3+} + 4H_2O;$$

• peroxodisulfate ions $S_2O_8{}^{2-}$ are characterized by their strong oxidizing power. We have already seen that they oxidize manganous and chromic ions until the ultimate permanganate and dichromate states, respectively. [Recall that the peroxodisulfate ion is among the strongest oxidizing species together with fluorine [$E°(F_2/F^-) = 2.85$ V], ozone [$E°(O_3/O_2) = 2.07$ V], and perxenate anions [$E°(HXeO_6{}^{3-}/XeO_3) = 2.10$ V], whose half-redox reaction is

$$HXeO_6{}^{3-} + 5H^+ + 2e^- \rightleftharpoons XeO_3 + 3H_2O$$

in which xenon passes from the + VIII to + VI state.] Another reaction of peroxodisulfates is their ability to oxidize diphenylamine in neutral or acetic medium. There is formation of the blue-violet oxidized diphenylbenzidine:

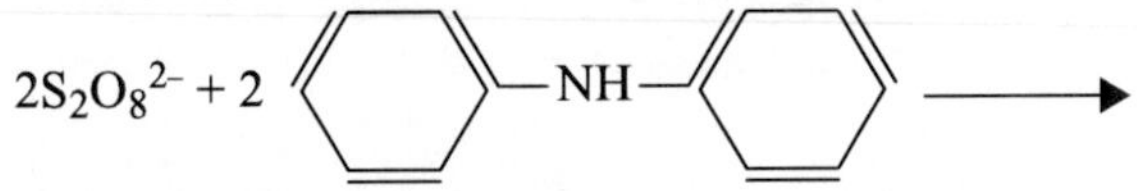

diphenylamine

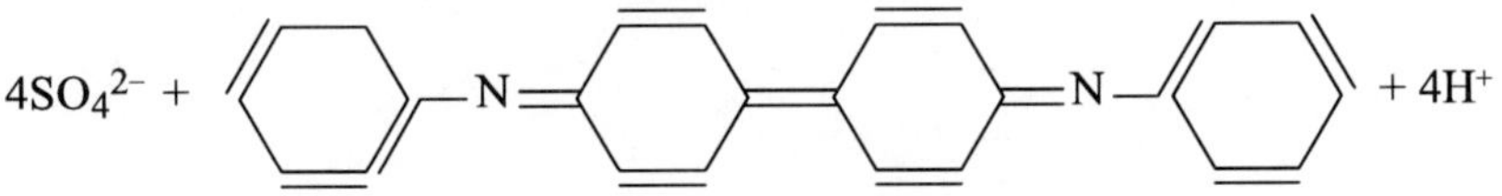

oxidized diphenylbenzidine

The standard potential value of the diphenylbenzidine/diphenylamine couple is 0.75 V. One admits that the indicator is actually the reduced diphenylbenzidine, which results from the rearrangement of two molecules of diphenylamine occurring with an oxidization of them according to the half-redox equilibrium

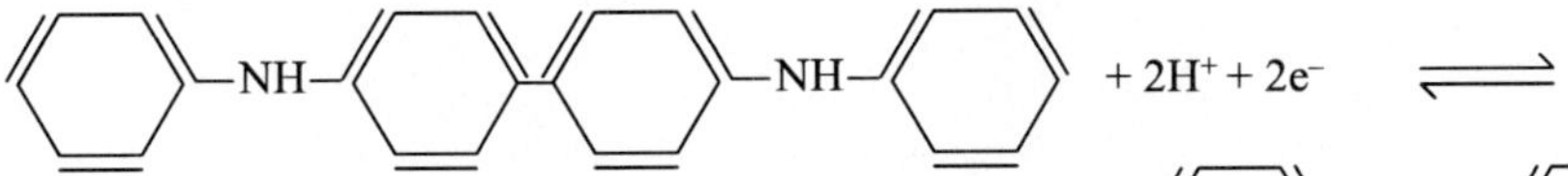

reduced diphenylbenzidine

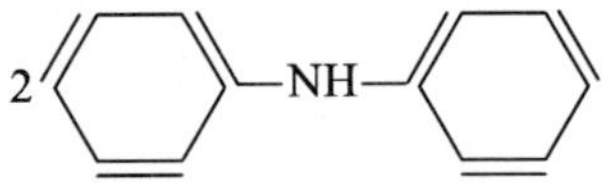

The reduced form of the oxidized diphenylbenzidine is diphenylbenzidine according to the half-redox equilibrium

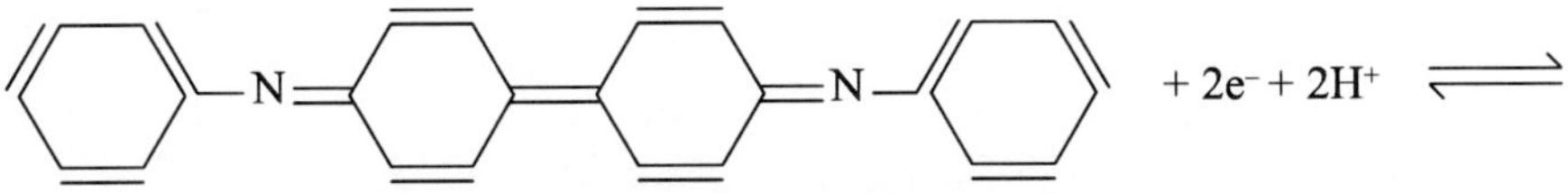

violet oxidized diphenylbenzidine

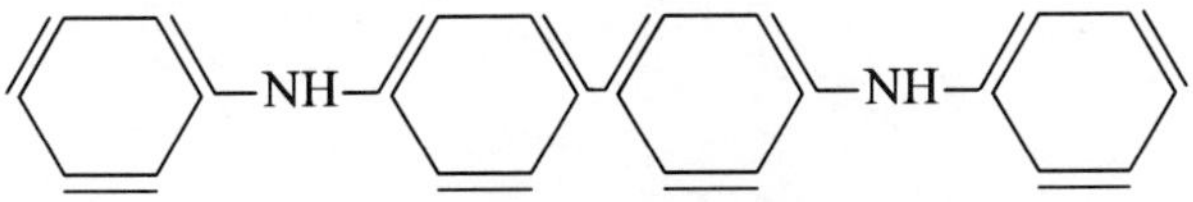

reduced diphenylbenzidine

- we should also notice that nitrate ions are able to oxidize diphenylamine, hence giving oxidized benzidine. Unfortunately, this reaction is given by numerous oxidants.

Part IV
Complexation Reactions—Analytical Applications

Chapter 22
General Definitions Concerning Complexes Rules of Nomenclature and Writing

The chemistry of complexes is particularly interesting. It can indeed be considered from different standpoints, all of whose properties confer great interest in them. Let's begin by recalling that a very large number of complexes now exists. Hence, it is a very rich chemistry from the standpoint of the number of compounds. This characteristic results from the fact that they implicate the majority of the elements and very numerous organic or inorganic ligands. The discovery of new complexes is routine today and seems to be incessant. Their structures are sometimes unexpected and then pose real new problems concerning the nature of the chemical bonds that condition their stabilities. This point makes this chemistry particularly attractive in the theoretical field. Another characteristic of the chemistry of complexes is its location at the intersection of the ways followed by inorganic and organic chemistries. Finally, let's also recall (without being too emphatic!) that life, at least as we know it, would not be possible without the existence *in vivo* of some definite complexes, in particular some enzymes. Therefore, it is not surprising to find the study of some complexes also in bioinorganic chemistry.

The formation of some complexes constitutes the basis of important methods of the chemical analysis. The important turning point in this field was the synthesis of particular aminocarboxylic acids near the end of the 1930s and the subsequent discovery of their analytical possibilities.

Several definitions of complexes exist. In this chapter, we consider the general definitions only. We shall see the definitions concerning their analytical possibilities and their applications in Chap. 24.

22.1 General Definition of Complexes

According to the IUPAC, a *coordination entity*, or *complex*, is composed of a central atom, usually a metal atom, to which is attached a surrounding array of other atoms or groups of atoms, each of which is called a *ligand*. The central atom is also called the *nucleus*.

J.-L. Burgot, *Ionic Equilibria in Analytical Chemistry*,
DOI 10.1007/978-1-4419-8382-4_22, © Springer Science+Business Media, LLC 2012

Hence, for example, in the complex ion hexamminecobalt(III), $[Co^{III}(NH_3)_6]^{3+}$ (Werner's luteocobaltic complex), the central nucleus is the cobalt atom and the ammoniac molecules are the ligands. We can already notice that in the formula of a complex, the complex is written between brackets (for details about the writing and nomenclature of complexes, see Sect. 22.4).

The IUPAC definition of a complex necessitates the following comments:

- quite evidently, a complex is of a structural nature;
- as mentioned before, the definition only concerns complexes possessing one unique central atom. They are named *mononuclear* complexes for this reason. However, complexes possessing several central atoms also exist. They are called *binuclear*, *trinuclear*, or *polynuclear* complexes. A simple example of a binuclear complex is provided by dichlorodimercury(I) Hg_2Cl_2 (formerly called calomel or incorrectly called mercury (I) chloride—incorrectly because it is a complex and not a salt);
- according to the definition, the central atom seems to be obligatorily a metallic one. A less general definition of complexes results from this assertion, which is that a complex may be considered to be a metallic salt that is not completely dissociated. An example of this conception is provided by mercury(II) chloride (incorrect name), which is not a salt, as the name suggests, but a complex. The correct name is dichloromercury(II): $[Hg^{II}Cl_2]$;
- the limits of the complexes are difficult to specify precisely. After considering the structure of some compounds, it is difficult indeed to decide whether or not to classify them as complexes. This point is discussed in the following two sections.

22.2 Complexes as Compounds Resulting from the Interaction of Electron-Donating and Electron-Accepting Species

Complexes are also considered entities resulting from the reaction of electron-donating species with electron-accepting ones. Central atoms are electron-acceptor and ligands electron-donor species. This viewpoint is not incompatible with the preceding definition. On the contrary, it clarifies it. In order to study this definition more deeply, we'll start by acknowledging that in a complex, the ligands are bound to the central atom through an electron pair that they bring. For example, in the group of the iodinated complexes of cadmium [iodocadmium(II) $[Cd^{II}I]^+$, diiodocadmium(II) $[Cd^{II}I_2]$, triiodocadmiate(II) $[Cd^{II}I_3]^-$, tetraiodocadmiate(II) $[Cd^{II}I_4]^{2-}$], iodide ions (the ligands) give an electron pair to the cadmium ion Cd^{2+}. In the complex hexamminecobalt(III) $[Co^{III}(NH_3)_6]^{3+}$, each ammoniac molecule binds the Co^{3+} ion through the nitrogen atom's electron pair. The result of this viewpoint of complexes is that the central atom of a complex is a Lewis acid and the ligands are Lewis bases. In other words, to use the language of organic chemists, we can say that central atoms are *electrophilic* and ligands *nucleophilic* species.

This definition is also structural in nature. Like the preceding one, it is not perfect for several reasons. First, the formation of bonds in a complex cannot be reduced to a simple sharing of an electron pair (see the end of Chap. 23). Second, it poses the problem of the limits of the set (see Sect. 23.3). Furthermore, it is not perfect because it cannot easily be applied to polynuclear complexes. Finally, we can also say that it is not perfect because, in some complexes, it is now well accepted that the central metal ion behaves essentially as an electron donor rather as a simple acceptor. In the latter case, we speak of classical complexes. It seems that the latter are the sole type to be used in chemical analysis.

The example of iodinated complexes of cadmium also leads to an interesting remark from a structural standpoint. The central atom and the ligands are not necessarily of opposite electrical charges, although the definition of complexes in terms of electron-donor and -acceptor species implicates that they are, at least at first sight. Iodinated complexes of cadmium form according to the successive reactions

$$Cd^{2+} + I^- \rightleftharpoons [CdI]^+,$$

$$[CdI]^+ + I^- \rightleftharpoons [CdI_2],$$

$$[CdI_2] + I^- \rightleftharpoons [CdI_3]^-,$$

$$[CdI_3]^- + I^- \rightleftharpoons [CdI_4]^{2-}.$$

We see that iodine ions bind positive, neutral, and negative species, successively.

22.3 Limits of the Set of Complexes

The object of this section is to clarify the preceding definitions, at least to decide whether or not to classify a given compound in the group of complexes.

We can consider that in the general IUPAC definition, the meaning of "complex" usually goes beyond only the presence of an electron-acceptor metallic central atom and allows us to consider unusual cases in which the coordinator ion is not metallic. An example is provided by orthoboric acid H_3BO_3. It reacts with some polyols, such as glycerol and mannitol, according to the general scheme

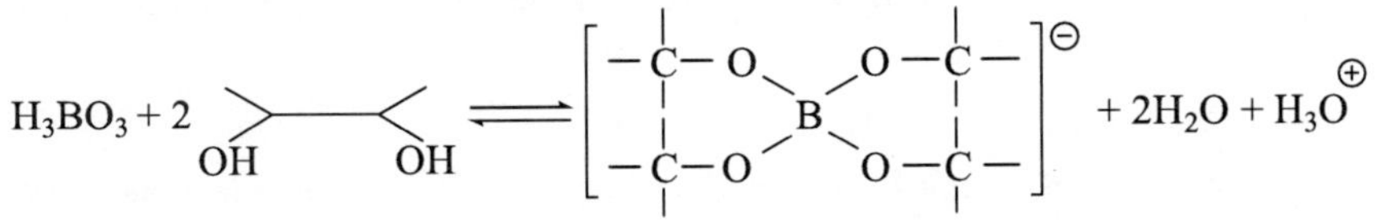

We remark that orthoboric acid behaves like an electrophilic species and that the oxygen atoms of the polyol constitute nucleophilic centers because of the existence of their lone electronic pairs.

This example is of analytical interest. The formation of these complexes allows us to ascertain the configuration of polyols, in particular of sugars. From another standpoint, it confers acidic properties to the solution where it has been formed. This provides an indirect means to study it through the simple measurements of the solution's pH. Finally, it is used in some separation methods, such as certain chromatographic techniques.

Another interesting example is that of triiodide ions I_3^- (also called periodure ions). Formally, they may be considered as resulting from the interaction of iodine at the oxidation state +I, playing the role of the central metal atom (involved in the preceding definitions), with iodide ions as ligands. Of course, here again, the central atom is not a metallic one!

In this book, we consider these compounds to be complexes.

Another case is that of ion pairs. They can exist in water under certain conditions of concentrations. This is the case of zinc and calcium sulfates. Ion pairs are species where the positively charged metallic ion is linked to an anion through an electrostatic bond. The problem is to know if the electrostatic bond can be considered as resulting from the sharing of electrons. Regardless, in this book we consider them to be complexes.

Another problem is that of classifying as complexes the species resulting either from the formation of hydrogen bonds or from the occurrence of charge transfers. These compounds may be considered complexes since, as the second definition proscribes, in both cases, some constitutive atoms or groups of atoms are electron donors. This is the reason why we call them "complexes." They are typically encountered in organic chemistry. Likewise, we also name complexes *Meisenheimer's compounds*[1] when their lifespan is sufficiently long. By doing so, we follow the denomination of organic chemists. Usually, their lifespan is very brief, in which case they are considered reactional intermediaries, a status that is closer to that of an activated complex than to a true complex. However, some of them are sufficiently stable to be considered here, and, furthermore, their formation is of interest in organic analysis (see Chap. 32). They are often responsible for colored reactions, which are the basis of quantitative absorptiometric measurements.

The classification in terms of electron-donor or -acceptor species actually allows us to consider many pluriatomic compounds as being complexes, as we just saw. Some pluriatomic inorganic ions may be in this group, including fluoboric BF_4^-, fluosilicic SiF_6^{2-}, nitrate NO_3^-, sulfate SO_4^{2-}, and phosphate PO_4^{3-}. Sulfate ions, for example, may be considered as resulting from the reaction of oxide ion O^{2-}, which is an electron- donor with sulfur trioxide, which is the acceptor, according to the reaction

$$SO_3 + O^{2-} \rightarrow SO_4^{2-}.$$

Usually, these species do not belong to the set of complexes, although some atoms or groups of atoms from which they are formed are considered ligands. This is the case, for example, with O^{2-}.

[1] For details, refer to a text on organic chemistry.

22.4 Writing and Systematic Nomenclature of Complexes

We'll now give some basic rules of the nomenclature and writing of complexes. We will discuss only the classical coordination compounds here.

The metal must be written first in the formula of a complex. It must be followed by the symbols of the ligands ranked according to their electrical charges, which must decrease in negativity from left to right. The entire complex must be enclosed in square brackets. The global electrical charge of the complex must be indicated outside the brackets as a right superscript, with the number before the sign. For example,

$$[Pt^{IV}(NH_3)_6]^{4+} \quad \text{hexammineplatinum(IV) cation,} \tag{22.1}$$

$$[Pt^{II}(Cl_2)(NH_3)(H_2O)] \quad \text{ammineaquadichloroplatinum(II) complex,} \tag{22.2}$$

$$[Al(H_2O)_6]^{3+} \quad \text{hexaaquaaluminium(III) cation,} \tag{22.3}$$

$$[Fe^{III}(CN)_5(NO)]^{2-} \quad \text{pentacyanonitrosylferrate(III) anion,} \tag{22.4}$$

$$[Cu^{II}(en)_3]^{2+} \quad \text{tris-(ethylenediamine)copper(II) cation,} \tag{22.5}$$

$$[Cr^{III}(NCS)_4(NH_3)_2]^{-} \quad \text{diamminetetraisothiocynatochromate(III).} \tag{22.6}$$

Roman numerals indicate the oxidation state of the metal in the complex. To simplify the writing, some ligands may be represented by standard symbols in the formula. This is the case of complex (22.5), in which the symbol "en" represents the ligand ethylenediamine:

$$_2HN - CH_2 - CH_2 - NH_2.$$

The rule concerning the names of the complexes is as follows: We must begin by listing the ligands in alphabetic order before writing the name of the central atom. The prefixes *mono*, *bi*, *tri*, *tetra*, and so on are used in expressions of complexes. The central nucleus is designated next. When the complex is an anion, the name of the central element must be terminated with the ending "ate," whereas no distinguishing termination is used for cationic or neutral coordination entities. The oxidation number of the central atom is indicated by appending a Roman numeral to the central atom's name. This number is enclosed in parentheses next to the name of the central atom. The net electrical charge must be written in Arabic figures outside the brackets. The neutral ligands must be designated by their usual names, and the anionic ones must have their usual names with the suffix "o." Therefore, the ligands H_2O, NH_3, CO, NO, and OH^- are respectively named *aqua*, *ammine*, *carbonyl*, *nitroxyl*, and *hydroxo*.

The symbol η (hapto^n) gives a supplementary topological description of the complex by indicating the connectivity of the ligand and of the central atom. It is written before the ligand to which it is related and the *n*, written as a superscript, indicates the number of ligand atoms bound to the central atom. Hence, in the complex amminedichloro-(η^2-ethene)platinum(II) $[Pt^{II}Cl_2(C_2H_4)(NH_3)]$:

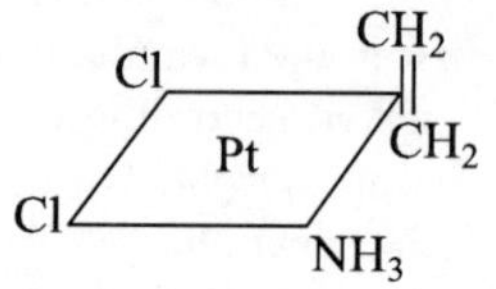

the ligand ethene (ethylene) is bound to platinum through both carbon atoms. This explains the prefix η^2.

Remark The representation of a complex between brackets may be confused with its concentration, which is also systematically written between brackets, and also, in some cases, with its activity. A complex's concentration is not represented by several sets of brackets. In this case, only the brackets of the concentration exist, but the complex's charge must be included within the brackets.

22.5 Electrical Charge of an Ion Complex

The electrical charge of an ion complex is equal to the algebraic sum of the charges of the central ion and the ligands. For example, in the case of the hexacyanoferrate(III) complex $[Fe(CN)_6]^{3-}$, the iron atom is supposed to be at oxidation state +III and the cyanide at the state −I, as in potassium cyanide. The rule is somewhat arbitrary because the oxidation state of the central metallic ion is not always easy to determine.

Chapter 23
Some Elements Concerning the Chemistry of Complexes

In this chapter, we recall some points concerning the chemistry of complexes. First, we briefly consider their isolation. Then we give the structure of some typical ligands. We also mention some ligands that have quite an extraordinary structure. Next, we give some elements concerning the stereochemistry of complexes. In particular, we recall some points of Werner's theory. Werner was indeed one of the chemists who began the modern study of coordination compounds, also named complexes. Finally, we will recall some experimental facts concerning the states of the ions in aqueous solution, from which the complexes are most often formed, at least in chemical analysis.

23.1 Attaining Complexes

Now we must deal with an important practical aspect of the chemistry of complexes, namely, the difficulty to isolate them in pure state. Of course, ionic complexes can only be isolated as electrically neutral salts. Thus, the hexaammineplatinum(IV) cation, the hexaaquaaluminium(III) cation, and the tris(ethylenediamine)-copper(II) cation (refer to the preceding chapter) have been respectively isolated as chloride, chloride, and sulfate and the anions pentacyanonitroxyl ferrate(III) and diamminetetraisothiocyanatochromate(III) as sodium and ammonium salts.

In order to isolate a complex, the choice of the counterion is often crucial. Furthermore, it is possible that the counterion is a complex itself. Such an example occurs in the precipitation of the tetraammineplatinum(II) cation as tetrachloroplatinate(II):

$$\left[Pt^{II}(NH_3)_4\right]\left[Pt^{II}Cl_4\right]$$

One of the causes of this difficulty stems from the fact that admixtures of several complexes often form simultaneously, most commonly if the same ligand can be bound to the same central ion with different stoichiometries (see the example of the cadmium iodocomplexes in Chap. 22). Some ionic complexes have never been isolated in pure state. Their existence has only been inferred through a critical study of some experimental results. For example, the formation of the diiodocadmium(II)

J.-L. Burgot, *Ionic Equilibria in Analytical Chemistry*,
DOI 10.1007/978-1-4419-8382-4_23, © Springer Science+Business Media, LLC 2012

$[Cd^{II}I_2]$ complex that occurs when Cd^{2+} and I^- ions are put together in solution must be realized through the inferior complex $[CdI]^+$,$^+$ which consequently must exist in solution. Indeed, the complexes that possess several identical ligands cannot be formed without involving successive steps. The calculations carried out on some analytical data concerning solutions allow the determination of the formation equilibrium constants of these intermediary complexes, even if they have not been isolated (see Chap. 24)!

23.2 Some Ligands Found in Classical Complexes

Here we consider only the ligands that are electron-donating species. We give just a few examples because they are so numerous. We classify them in different groups according to the number of coordination points with the central ion (or, in other words, according to the number of electron-donating groups) they exhibit per molecule. If there is only one coordination point, the ligand is said to be *monodentate;* otherwise, it is called *bidentate* ("two-toothed"), etc. More generally, it is described as *polydentate*. At this point, the concept of chelate must be introduced. A *chelate* is a complex formed by the coordination of several groups belonging to the same ligand to the same metallic central ion with the formation of a ring. This concept is very important in analytical chemistry (see Chap. 28). We'll just mention now that chelates are considerably more stable than the corresponding complexes formed from separate ligands of the same basicity as those forming the chelate and from the same central atom, as indicated by the values of their stability constants (see Chap. 24).

23.2.1 *Some Monodentate Ligands*

We can distinguish among the following monodentate ligands:

- mineral anions, such as cyanide CN^-, thiocyanate SCN^-, fluoride F^-, chloride Cl^-, bromide Br^-, iodide I^-, azide $N_3{}^-$, isocyanate NCO^-, hydroxide OH^-, nitrite $NO_2{}^-$, nitrate $NO_3{}^-$, isothiocyanate NCS^-, sulfide S^{2-}, carbonate $CO_3{}^{2-}$, sulfate $SO_4{}^{2-}$, and thiosulfate $S_2O_3{}^{2-}$ ions;
- organic anions, such as amidure $NR_2{}^-$, alcoolate RO^-, carboxylate $RCOO^-$, and thiolate RS^- ions;
- mineral molecules, such as ammoniac NH_3 and water H_2O;
- organic molecules, such as pyridine, aniline, phosphines PR_3, phosphites $P(OR)_3$, and arsines AsR_3.

23.2.2 *Some Polydentate Ligands*

In terms of polydentate ligands, we can mention

- dipyridine and terpyridine, which coordinate to the central atom through the electron pairs of the nitrogen atoms:

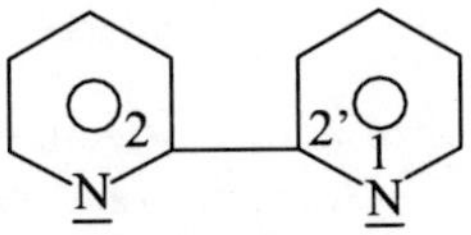

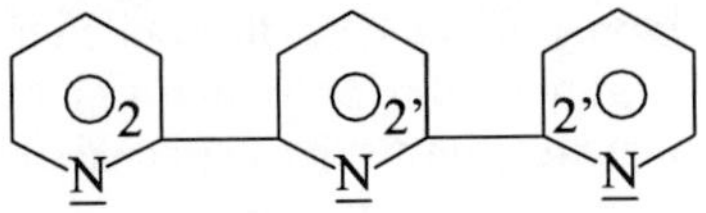

- ethylenediamine (en), which is bidentate:

$$_2HN - CH_2 - CH_2 - NH_2;$$

- ethylenediaminetetraacetic acid (H4EDTA) and its derivatives. Its tetraanion is

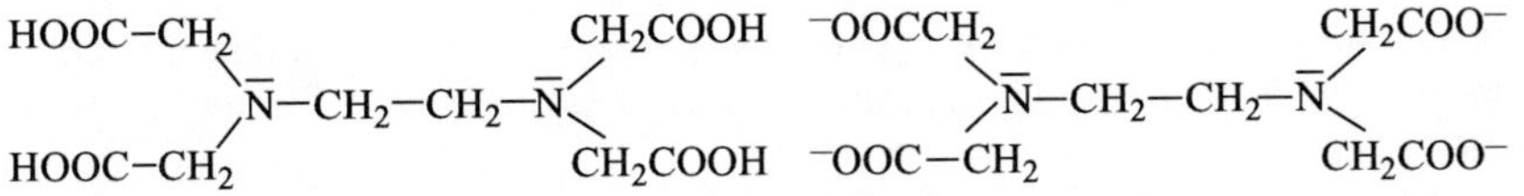

  hexadentate (four pairs brought by carboxylate groups and two brought by nitrogen atoms). This ligand is of the utmost importance in chemical analysis (see Chap. 28);
- α-amino acids, which are bidentate through the lone electron pair of the nitrogen atom and through the carboxylate rest;

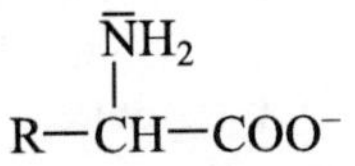

- acids-alcohols, such as tartaric acids COOH–CHOH–CHOH–COOH, citric acid COOH–CH_2–C(OH) (COOH)–CH_2–COOH, and malic acid COOH–CH_2–CH(OH)–COOH;
- organic polyacids, such as oxalic acid and succinic acid.

Incidentally, the formation of complexes with derivatives from the last four groups of derivatives listed here is accompanied by the expulsion of one or several protons. This point is developed in Chap. 25, which is devoted to the superimposition of acid–base and complexation phenomena.

Crown ether ligands deserve particular mention, although so far they have not been used much in analysis. They are quasi-planar macrocyclic compounds with

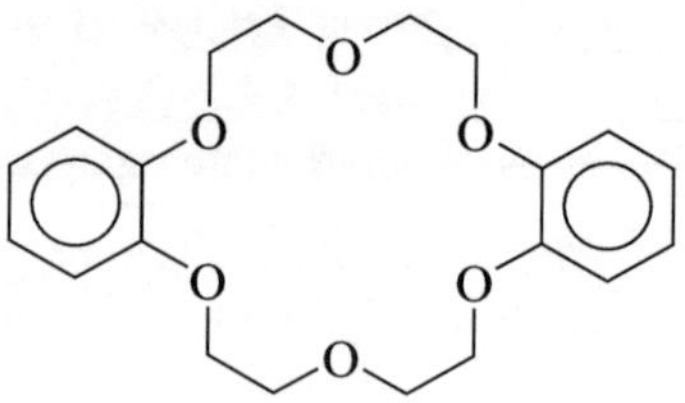

Fig. 23.1 Stucture of dibenzo [18]-crown-6

several nitrogen atoms engaged in their structure. Figure 23.1 shows the structure of dibenzo[18]-crown-6. The systematic nomenclature of these derivatives is more cumbersome than that used above. As a result, the latter nomenclature is used. In the simplified name given here, [18] indicates the total number of atoms constituting the macrocycle and 6 is the number of oxygen atoms included in it.

Numerous other crown ether derivatives are described. They differ from one another in the size of the macrocycle and the number of oxygen atoms. These derivatives sequestrate metallic ions more or less selectively in proportion to the size of the cavity existing inside the cycle. Electrostatic interactions between the sequestrated metallic ion and the oxygen atoms of the cycle also play a part in the formation of this kind of complexes. This is why crown ether ligands, although recently discovered,[1] can be considered classical ligands.

Some natural products are crown ether derivatives. They can sequestrate alkali and alkaline earth ions. These polypeptides include valinomycin, monactin, and monensin. They play an important biological role: They bind K^+ in preference to Na^+ ions, except for monensin, where the tendency is reversed. They also contribute to the maintenance of the different concentrations of these ions inside and outside the cells and, consequently, to that of the potential differences across cell membranes. Beyond these biological aspects, valinomycin is the basis for an ion-selective electrode for K^+.

Some sequestrating agents are even more powerful ligands than crown ether compounds. They are also electron-donating species. Called *cryptands*, *spherands*, *sepulchrates*, and *sarcophagines*, some of them exhibit exceptional selectivity. These derivatives are in the realm of supramolecular chemistry, in which the molecular systems are bound together through intermolecular forces and not through covalency bonds. Yet these derivatives seem to have no analytical applications.

Let's also recall the existence of ligands deriving from tetrapyrrole that are very interesting from a biological standpoint. More precisely, they are derivatives of porphin that exist in chlorophylls and in hemic proteins such as hemoglobin, myoglobin, and cytochromes. In chlorophylls, a magnesium atom is located at the center of the macrocycle deriving from porphin, whereas in heme, it is iron that is surrounded with another derivative of porphin. In vitamin B12, the cobaltic ion is also inserted within a tetrapyrrole derivative called corrin.

---

[1] C. J. Pedersen, J. M. Lehn, and D. J. Craw won the 1987 Nobel Prize in Chemistry for the development and use of crown ether and cryptand derivatives of high selectivity.

Finally, we'll also mention that some extraordinary ligands exist, the word "extraordinary" meaning that these ligands were totally unexpected. Yet while they are of no analytical interest, they are sufficiently interesting from a theoretical standpoint to deserve some further comments. Their very existence is a perfect illustration of the richness of the chemistry of complexes. Dihydrogen H_2 behaves in some circumstances like a ligand. Such is the case in the following complex of tungsten: $[W(CO)_3(\eta^2 - H_2)(PPr^i{}_3)_2]$. A hydrogen atom itself intervenes as a ligand in some complexes, in which it exhibits coordination numbers that vary from 1 to 6! Dioxygen O_2 is also engaged in some complexes as a ligand, such as in the following complex of iridium: $[Ir(CO)Cl(O_2)(PPh_3)_2]$. This complex can reversibly lose dioxygen, and one can easily imagine the considerable interest that a dioxygen carrier compound may generate. Because of this property, it may be compared to metalloproteins such as hemoglobin and myoglobin. Today we know that the perchlorate ion ClO_4^-, which was long considered incapable of giving complexes, can behave as a monodentate or bidentate ligand. Numerous organic molecules also give complexes with transition metal ions, usually through a multiple bond existing within them. The complex ammine dichloro(ethane)platine (see Chap. 22) provides an example of such a complex.

23.3 Some Aspects of the Chemistry of Complexes

Werner[2] was the first chemist to systematize many experimental facts concerning the chemistry of complexes in the form of some postulates that remain true today even if they have been clarified and explained from a theoretical standpoint in light of recent knowledge concerning the chemical bond. According to Werner,

1. Complexes are built around a central element called the *coordinator element* (we saw this point in the preceding chapter).
2. Coordinator elements exhibit two kinds of valences: the principal or usual valence, which is ionic, and a secondary valence that is nonionic. This concept is undoubtedly outdated today and has been totally revisited in light of modern theories concerning the chemical bond. The notion of principal ionic valence has been replaced by that of oxidation number. The oxidation number of the central element is defined as being the electrical charge it should bring if the electrons it shared with the ligands (within the complex) had been assigned to the more electronegative atom constituting a bond with the central element.

The fact that the notion of oxidation number may be ambiguous (see Chap. 12) has already been noticed. In the case of a complex, the ambiguity may have two origins. The first ambiguity lies in the fact that the electronegativities of the ligand and of the central atom are the same. In this case, the legitimacy of the separation of the

[2] Alfred Werner, Swiss chemist born in Mulhouse, Alsace (then part of France), in 1866, died in Zurich in 1919; won the Nobel Prize in Chemistry in 1913.

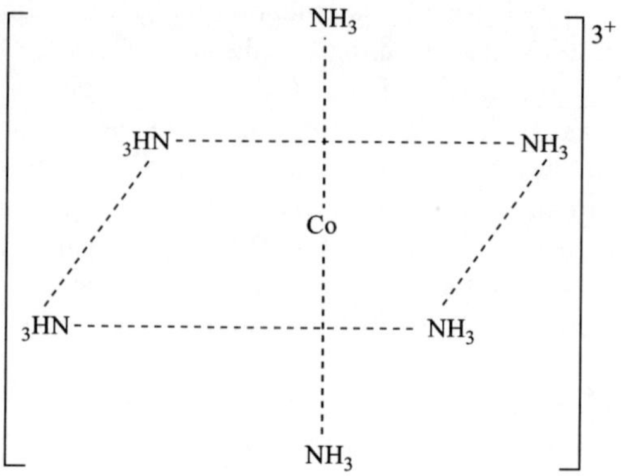

Fig. 23.2 Octahedral structure of the hexamminecobalt(III) complex

charges in favor of the ligands is questionable. The second lies in the fact that the ligand may, in some cases, be coordinated according to different structures. As a result, in some cases, a ligand may share different charges with the central ion. This is the case of the complex [tris(1,2-diphenyl-1,2-dithiolate)rhenium(VI)] (form a) in which rhenium is at oxidation state +VI, but in form b, it appears with oxidation state 0.

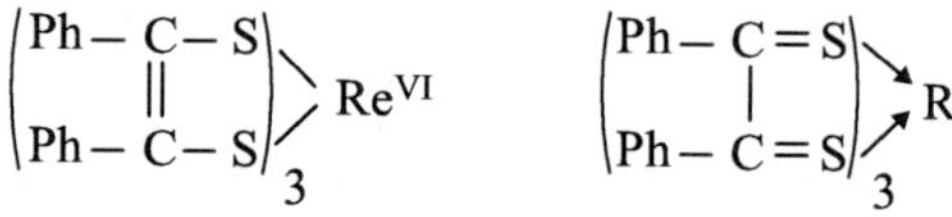

3. In a complex, the sum of the numbers of the principal and secondary valences is constant. This number is the coordination number, or the *coordinence*, of the metallic ion. It is the maximum number of monodentate ligands that can be bound to it. Most of the time, the coordinence is equal to 6, but it can take the values 2, 4, 5, 8, and more rarely the values 3, 7, 9, and even sometimes higher ones. For example, the cobaltic ion Co^{3+} exhibits the coordinence 6 in the hexamminecobalt(III) complex $[Co^{III}(NH_3)_6]^{3+}$. In this entity, however, it keeps its principal valence, which is ionic: 3^+. In a complex, the principal ionic valence may be neutralized partially or totally by some negatively charged ligands. An example, taken from numerous others, is provided by the pentaamminemonochlorocobalt (III) cation $[Co^{III}(NH_3)_5Cl]^{2+}$ (Werner's purpureocobaltic complex). It is easy to check that its global electrical charge is equal to the algebraic sum of the charges of cobalt and of the ligands.
4. According to Werner, ligands are spatially distributed around the metallic ion. If, for example, we are considering the complexes of coordinence 6, they are distributed at the summits of a regular octahedra whose center is occupied by the coordinator metal. This is the case with the hexamminecobalt(III) cation $[Co(NH_3)_6]^{3+}$ (Werner's luteocobaltic complex) (Fig. 23.2). In general, a coordinence number is associated with one or several particular geometries of the complex. For example, the complexes of Cu(I), Ag(I), Au(I), and Hg(II) are

Fig. 23.3 Geometrical isomerism of a square complex

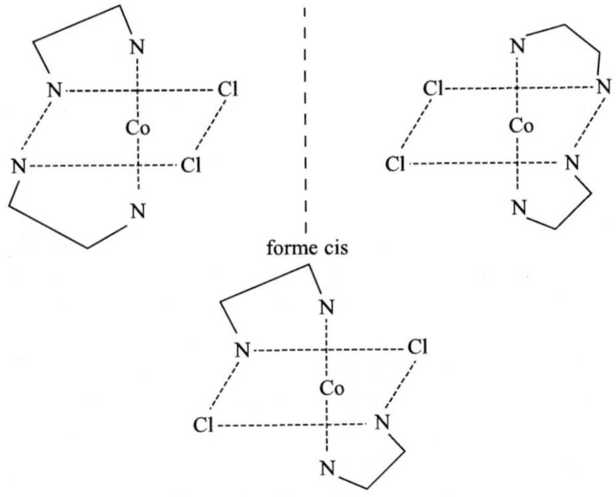

Fig. 23.4 Geometrical and optical isomerisms of dichlorobis (ethylenediamine)cobalt(III)

linear. A coordinence 4 is in agreement either with a square geometry such as that encountered in the tetrachloroplatinate(II) $[Pt^{II}Cl_4]^{2-}$ complex or with a tetrahedral geometry exhibited, for example, by the tetracarbonylnickel (0) complex: $[Ni^0(CO)_4]$.

5. According to these considerations, it is not surprising that some complexes may present several sorts of isomerisms. We shall restrict ourselves to the cases of geometrical and optical isomerisms. They occur in the square and octahedral complexes. One can distinguish *cis* and *trans* derivatives with the square complexes of structure Ma_2b_2 (Fig. 23.3), in which a and b are monodentate ligands.

It is important to underline the fact that a tetrahedral complex, which would have the same global formula as the preceding one, cannot exhibit this isomerism. Hence, for example, the isolation of two isomers of the diamminedichloroplatinum(II) complex $[Pt^{II}(Cl)_2(NH_3)_2]$ has been a very strong argument in favor of its planar structure. The octahedral hexacoordinated complex dichlorobis(ethylenediamine)cobalt(III) $[Co^{III}(Cl_2)(en)_2]^+$ exists in the *cis* and *trans* forms (Fig. 23.4).

This example also shows that optical isomerisms may exist in complexes. According to Fig. 23.4, it appears that the *cis* derivative exists as two enantiomers. It is chiral. Hence, two *cis* forms exist. This is not the case with the *trans* derivative, which possesses a symmetry center. It is achiral. Another example is provided by the complex formed by EDTA and cobalt(III): $[Co(EDTA)]^-$, which can be resolved in two enantiomers (Fig. 23.5).

This aspect of the chemistry of complexes is seldom mentioned in the analytical literature.

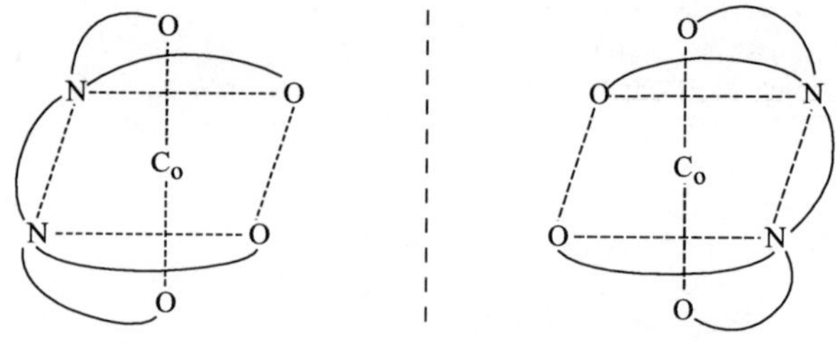

Fig. 23.5 Optical isomerism of the ethylenediaminotetraacetatocobalt(III) complex anion

23.4 State of the Ions in Aqueous Solution and Consequences

In water, ions are not naked. They are more or less energetically solvated by water molecules. Here the adjective "solvated" must be taken to have its usual meaning. The solvation of ions implicates phenomena that may be a simple ion–dipole interaction between the ion and water as well as a true chemical bond between both, as is the case in aqua complexes. Whatever the case, the reactivity of ions in aqueous solutions is strongly influenced by these solvation phenomena. As a result, it is also the case of the formation of complexes starting from the central metallic ion.

The ions of alkali metals, which are the most electropositive elements, generally exhibit the weakest tendency to form complexes. As an example, we can mention the fact that they give poorly stable complexes with EDTA, which is a powerful complexating agent of numerous metallic ions (see Chap. 28). (However, let us recall that these ions are complexed by crown ether compounds and by cryptands, but according to a particular mechanism.) In water, alkali ions must be considered as being only solvated by water in the restrictive meaning, that is, by ion–water dipole interactions for the first sheath of solvation and by water–dipoles interactions for the second.

Cations of alkaline earth metals and those of lanthanides give few complexes but more than the preceding ones, although they are only a little less electropositive than them. In some conditions, they give complexes with EDTA whose stability is sufficient to receive analytical applications (see Chap. 28).

Ions issuing from transition metals and from the first actinides yield a huge number of complexes, particularly with water, which in this case takes the status of a ligand. Octahedral complexes form for the bi- and trivalent ions of the first series of transition metals of the general formula $[M(H_2O)_6]^{2+}$ and $[M(H_2O)_6]^{3+}$. Some of them are in the form of regular octahedra, whereas others are in the form of an irregular octahedron (Fig. 23.6). For example, Ti^{3+} gives a regular tetrahedron and Cu^{2+} an irregular one.

Of course, other kinds of aqua complexes are encountered. For example, Co^{2+} yields both octahedral and tetrahedral aqua complexes.

All these aqua complexes are nothing more than a particular case of coordination compounds, which we have already studied. From a structural standpoint, indeed, nothing fundamental distinguishes the complex $[Cr^{III}(H_2O)_6]^{3+}$ from the complex hexaamminechrome(III) $[Cr^{III}(NH_3)_6]^{3+}$ or from the complex hexathiocyanatochromate(III) $[Cr^{III}(NCS)_6]^{3-}$, for example.

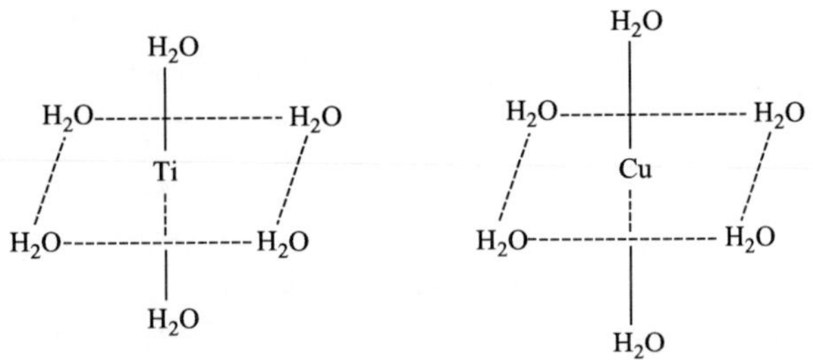

Fig. 23.6 Some examples of aqua complexes (note that in the cupric complex, vertical bonds are longer than horizontal ones)

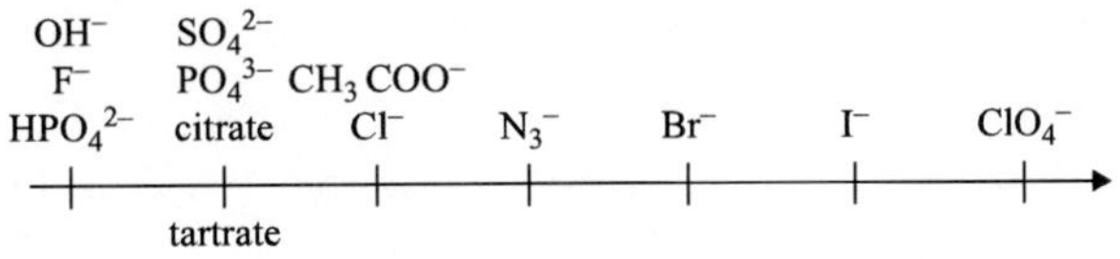

Fig. 23.7 Hofmeister's series: increasing hydrophobic character of anions

The formation of aqua complexes induces an important consequence for the formation of coordination compounds. When a transition metal ion is mixed with its future ligands in aqueous solution, the formation reaction is actually a nucleophilic substitution or consists of several successive substitution reactions in which the leaving group is the ligand water according to

$$M(H_2O)_n + L \rightleftharpoons M(H_2O)_{n-1}L + H_2O$$

$$M(H_2O)_{n-1}L + L \rightleftharpoons M(H_2O)_{n-2}L_2 + H_2O$$

$$M(H_2O)L_{n-1} + L \rightleftharpoons ML_n + H_2O$$

(For the sakes of generality and simplicity, the electrical charges of the metal and of the ligand are omitted.) In this reactional scheme, n is the coordination number of the metallic ion. The complex ML_n is called the ultimate complex.

Incidentally, it is interesting to note that a coordination compound can also be prepared from another one, which is different from an aqua complex.

The state of anions in water is poorly known. What must be underlined first is that, finally, few monoatomic anions do exist. This is the case with halides. They are hydrophilic and nucleophilic species. However, a great number of oxoanions exist, such as nitrate, perchlorate, permanganate, sulfate, and so forth. Due to the diversity of their stoichiometries and of their stereochemistries, it is difficult to give general rules concerning their solvation and their reactivity in water. However, one can affirm that they are less nucleophilic than the preceding ones. From the standpoint of nucleophilicity, the pseudo-halides CN^- and CNS^- are located in an intermediary position between the two preceding classes. Generally speaking, the less solvated the anion is, the more nucleophilic it is. The hydration extent of anions is given by Hofmeister's scale, which classifies them by increasing hydrophobic character, that is, according to weaker and weaker hydration (see Fig. 23.7).

We'll finish this chapter by saying that until now, it hasn't been possible to totally rationalize the strength of the ion–water interactions in aqua complexes.

Chapter 24
Stability of Complexes: Some Elements Concerning the Kinetics of Their Formation

The stability of complexes is described by their stability constants, which can take several forms. They permit the calculation of the complex's concentrations and, hence, those of other species after one or several complexation reaction(s) have occurred, provided the equilibrium of the system has been reached. They also permit us to predict ligand exchange reactions between ligands that are free and those that are already bound in complexes. These possibilities are, of course, of considerable importance in chemical analysis. Finally, we can also say that their knowledge permits us, on the strength of numerical values, to identify the structural factors that are at the origin of the formation of complexes. Of course, complex formations also depend on kinetic factors.

24.1 Definition of Complexes in the Context of Analysis

The general definition of complexes, which we've already seen, can be somewhat refined in the context of the chemical analysis. For this purpose, we can say that

> Complexes are derivatives resulting from the association of one or several metallic ions of one given element with molecules or organic and inorganic ions. These associations are such that in solutions, some of the chemical reactions that are characteristic of the complex's constituents do not occur further. The constituents are said to be *masked* or *dissimulated* to their usual reagents.

Some examples of such behavior are given by

- mercuric ions. They give a precipitate of yellow mercuric oxide HgO after the addition of sodium or potassium hydroxide, according to the reaction

$$Hg^{2+} + 2OH^- \rightarrow HgO\downarrow + H_2O.$$

This is one of the characterization reactions of mercuric ions. Before the addition of sodium hydroxide, if a sufficient quantity of iodide ions has been added to the solution of mercuric ions, the mercuric oxide does not precipitate as above. In

J.-L. Burgot, *Ionic Equilibria in Analytical Chemistry*,
DOI 10.1007/978-1-4419-8382-4_24,

this case, the tetraiodomercurate ion complex (formerly called mercuritetraiodide ion) is formed:

$$Hg^{2+} + 4I^- \rightarrow \underset{\text{tetraiodomercurate(II)anion}}{[HgI_4]^{2-}} .$$

It masks mercuric ions to hydroxide ions;

- copper ions in the presence of ammonia. The tetrammine copper(II) (*cupritetrammine*) cation complex $[Cu(NH_3)_4]^{2+}$ forms when a sufficient quantity of ammonia is added to a solution of copper(II). Adding sodium hydroxide to this solution does not produce the blue gelatinous precipitate of copper hydroxide $Cu(OH)_2$, whose formation is characteristic of Cu^{2+}. In the $[Cu(NH_3)_4]^{2+}$complex, Cu^{2+} is masked to the hydroxide ions.

24.2 Stability of Complexes: Perfect and Imperfect Complexes

The constituents of some complexes are dissimulated to the quasi-totality of their usual reagents. They are called *perfect complexes*, other complexes being called *imperfect complexes*. Of course, between both groups exists the full possible gradation of complexes.

In the group of those that can be considered perfect, we can mention

- the hexacyanoferrate(III) ions $[Fe(CN)_6]^{3-}$ (ferricyanide ions). Most of the characteristic reactions of free Fe^{3+} ions in aqueous solutions are negative in the presence of this complex. This is the case, for example, of the ferric hydroxide $Fe(OH)_3$ formation reaction;
- the hexamminecobalt(III) ions $[Co(NH_3)_6]^{3+}$. Their internal dissociation according to the reaction:

$$[Co(NH_3)_6]^{3+} \rightleftharpoons Co^{3+} + 6NH_3$$

is very weak. The internal dissociation of a complex must not be confused with its external or total dissociation. In the present case, the complex is isolated as the trichloride salt $[Co(NH_3)_6]Cl_3$. Its external dissociation corresponds to the following process:

$$\underset{\text{solid}}{[Co(NH_3)_6]Cl_3} \rightarrow \underset{\text{solution}}{[Co(NH_3)_6]^{3+}} + 3Cl^-;$$

- the pentaamminechlorocobalt(III) ions $[CoCl(NH_3)_5]^{2+}$ (Werner's purpureocobaltic complex). It is isolated as a dichloride. It is also a quasi-perfect complex. The addition of silver nitrate to its dichloride solution only precipitates two thirds of the chlorine contained in the solution. This means that one atom of chlorine is firmly engaged in the complex.

Among imperfect complexes, we can again deal with the cases of

- the tetraiodomercurate(II) complex. The masking is imperfect since the addition of formaldehyde to its solution induces the reduction of Hg^{II} engaged in the complex in metallic mercury, absolutely as if Hg^{II} were free in solution as Hg^{2+} ions (we have already seen in Chap. 18 that this reaction is one determination method of formaldehyde);
- the tetraamminecopper(II) complex $[Cu(NH_3)_4]^{2+}$. Bubbling dihydrogen sulfide into its solution leads to a precipitate of copper sulfide CuS absolutely as if free Cu^{2+} ions were present in the solution. However, if a sufficient quantity of cyanide ions had been added (as potassium cyanide) to the tetraamminecopper(II) solution before bubbling H_2S, no copper(II) sulfide would have precipitated. In this case, another complex, this time a perfect one, would have been formed, that is, $[Cu(CN)_4]^{3-}$. Another argument indicates that $[Cu(CN)_4]^{3-}$ is a perfect complex: The CuS solubility product value is very weak ($10^{-35.2}$), and this would favor its formation from any complex in which it is already engaged, by destruction of the latter. This is not the case with $[Cu(CN)_4]^{3-}$ (as we have seen). From another (but very near the preceding one) standpoint, the formation of the latter complex from cupritetrammine constitutes an example of what has already been said (Chap. 23): Complexes may be prepared by displacement of ligands already engaged in an initial complex, by other ligands.

24.3 Formation or Stability Constants of Complexes

Let's recall that metallic ions often stand as aqua complexes in aqueous solutions. Complexes' formation reactions are, therefore, exchange reactions between the water ligand and the incoming ligand. We symbolize these reactions as follows:

$$\left[M(H_2O)_{m-i}L_i\right] + L \rightleftharpoons \left[M(H_2O)_{m-i-1}L_{i+1}\right] + H_2O.$$

As an example, taken among others, we can mention the hexamminechromium(III): $[Cr(NH_3)_6]^{3+}$. The successive complexes that form until the preceding one are $[Cr(NH_3)(OH_2)_5]^{3+}$, $[Cr(NH_3)_2(OH_2)_4]^{3+}$, and so forth.

Two sets of formation or stability constants are used: the stepwise stability constants (or successive stability constants) and the overall stability constants. The first type is defined by the following expressions:

$$K_1 = (ML)/(M)(L); \quad K_2 = (ML_2)/(ML)(L); \quad K_n = (ML_n)/(ML_{n-1})(L)$$

For the sake of simplicity, the electrical charges and the water molecules engaged as ligands are omitted in these expressions. The overall stability constants are defined by the following expressions:

$$\beta_1 = (ML)/(M)(L); \quad \beta_2 = (ML_2)/(M)(L)^2; \quad \beta_n = (ML_n)/(M)(L)^n.$$

Table 24.1 Stepwise formation constants of some complexes (not an exhaustive list—the mentioned values were not necessarily obtained under the same conditions of ionic strength)

Ligand	Central ion	Logarithm of the equilibrium constant					
		K_1	K_2	K_3	K_4	K_5	K_6
Cl^-	Ag^+	2.85	1.87	0.32	0.86		
	Cd^{2+}	1.32	0.90	0.09	−0.45		
	Hg^{2+}	6.74	6.48	0.85	1.00		
	Pb^{2+}	0.88	0.61	−0.40	−0.15		
	In^{3+}	1.42	0.81	1.00	−0.20		
	Fe^{3+}	0.62	0.11	−1.40	−1.92		
	Fe^{2+}	0.36	0.04	–	–		
Br^-	Ag^+	4.15	2.96	0.84	0.94		
	Cd^{2+}	1.56	0.46	0.23	0.41		
	Hg^{2+}	9.05	8.28	2.41	1.26		
I^-	Cd^{2+}	2.08	0.87	2.09	1.59		
	Hg^{2+}	12.87	10.95	3.78	2.23		
CN^-	Cd^{2+}	5.48	5.14	4.56	3.58		
	Hg^{2+}	18.00	16.70	3.83	2.98		
SCN^-	Ag^+	4.59	3.70	1.77	1.20		
	Ni^{2+}	1.18	0.46	0.17	–		
	Cr^{3+}	3.1	1.70	1.0	0.3	−0.7	−1.6
	Fe^{3+}	1.96	2.02	−0.41	−0.14	−1.57	−1.51
$S_2O_3{}^{2-}$	Ag^+	8.82	4.64	0.69	–		

We notice that

$$\beta_1 = K_1$$

$$\text{and} \quad \beta_n = K_1 K_2 \ldots K_n.$$

Knowledge of stepwise or overall stability constants permits us to calculate the concentrations of all the species in solution provided the complexation equilibria are reached. We give numerical values of stepwise constants of some complexes in Table 24.1.

We shall mention some other values later. (A third group of stability constants exists, but they can be used only when one of the reagents or one of the formed complexes is precipitated; see Chap. 36.)

24.4 General Methodology of Determining Stability Constants

Before investigating the calculations that permit us to obtain the concentrations of the different species from the values of stability constants, it is interesting to briefly recall some general principles of their determination in solutions. Of course, the principles followed to determine the stability constants of complexes are the same as those followed for any other sort of equilibrium constants.

In order to achieve these determinations, the concentration (the activity) of at least one species participating in the equilibrium (a reagent, the complex) must be, in a first stage, experimentally determined, when the system equilibrium has been reached, of course. In a second stage, the equations governing the concentrations (activities) in solution (mass and charge balances, mass law, especially that involving the complex's formation) must be handled in order to find a mathematical relation permitting the calculation of the value experimentally determined above as a function of the experimental parameters of the system (analytical concentrations, for example). This relation also contains the constant to be determined. In a third stage, arbitrary numerical values are given to the equilibrium constant to be determined, and comparing the calculated values with the experimental ones permits us to choose its best value. This methodology is called *curve-fitting*. It is probably the best method, but it can be very complicated and laborious and often requires the use of computers. A good means of comparing the experimental and calculated values is to use a least-squares process. One difficulty that often occurs with this sort of approach is to have an exact view of all the chemical phenomena taking place simultaneously. In other words, the chemical model must be exact. If it is not exact, the comparison between the calculated and the experimental values is fully meaningless since the calculated and experimental values do not correspond to the same phenomenon. Fortunately, the algorithms used in the comparison are sufficiently powerful to detect such a discrepancy. In 1928, less powerful but shrewd methodologies, such as Job's method or that based on the study of the formation curves, were used.

From a purely analytical standpoint, the experimental determination of a (or several) concentration(s) (first stage) is achieved with instrumental methods of analysis such as UV-visible spectrophotometry, varied electrochemical methods, calorimetric methods, and so forth. For example, one electrode may be sensitive to the metal that is complexed. Hence, potentiometry is recommended.

A distinction between direct and indirect methods has been made. In the first group, the concentration of one of the species participating in the equilibrium is determined. In the second group, the concentration of a species that does not participate in this equilibrium is determined. In this case, of course, the complex's formation reaction must induce a change in the followed species concentration. This may be, for example, a change in the pH value. In this case, using pH-metry is recommended.

The introduction of activities induces serious difficulties in the previous methodologies. They are overcome by working at constant ionic strength, about 1–4 mol/L, since, for these values, the formal constants do not differ greatly from the thermodynamic ones. Another possibility is to use Debye–Hückel's equations provided the ionic strengths are not too high.

The determination of kinetic constants proceeds from the same methodology. However, in some cases, the kinetics may be very fast and the instrumental methods of analysis evoked above would be no longer suitable. Then some special methods such as flux and relaxation techniques would be required.

24.5 Some Examples of Calculations Carried Out with Stability Constants

In this section, we study only the case of mononuclear complexes.

If only one ligand is bound to the central metallic ion, the calculation of the complex's concentration is easy. For example, let's take the case of the thiocyanatocobalt(II) complex, $[Co(SCN)]^+$($K_1 = \beta_1 = 10$). Let C_L and C_M be the analytical concentrations in the ligand and metallic ion, respectively. By assimilating activities to the corresponding concentrations, the following three relations are satisfied:

$$[Co(SCN)^+] = \beta_1[Co^{2+}][SCN^-],$$
$$[Co(SCN)^+] + [Co^{2+}] = C_M,$$
$$[Co(SCN)^+] + [SCN^-] = C_L.$$

This is an easily solvable system of three equations in three unknowns.

The calculations are far more laborious when several complexation equilibria occur simultaneously. Let's consider the formation of the following complexes: $[CdCl]^+$, $[CdCl_2]$, $[CdCl_3]^-$, and $[CdCl_4]^{2-}$, which occur when hydrochloric acid and cadmium nitrate (which is completely ionized) are brought together. The relations that are satisfied are as follows:

$$[CdCl^+] = K_1[Cd^{2+}][Cl^-] \quad [CdCl_2] = K_2[CdCl^+][Cl^-],$$
$$[CdCl_3{}^-] = K_3[CdCl_2][Cl^-] \quad [CdCl_4{}^{2-}] = K_4[CdCl_3{}^-][Cl^-].$$

The mass balance equations are

$$[Cl^-] + [CdCl^+] + 2[CdCl_2] + 3[CdCl_3{}^-] + 4[CdCl_4{}^{2-}] = C_{Cl},$$
$$[Cd^{2+}] + [CdCl^+] + [CdCl_2] + [CdCl_3{}^-] + [CdCl_4{}^{2-}] = C_{Cd}.$$

C_{Cl} and C_{Cd} are the analytical concentrations of chlorine and cadmium. From the standpoint of mathematics, the system is determined since there is a system of six equations in six unknowns. It can be reduced to the following system of two equations in two unknowns:

$$[Cd^{2+}]\left\{1 + \beta_1[Cl^-] + \beta_2[Cl^-]^2 + \beta_3[Cl^-]^3 + \beta_4[Cl^-]^4\right\} = C_{Cd},$$
$$[Cl^-]\{1 + \beta_1[Cd^{2+}] + 2\beta_2[Cd^{2+}][Cl^-] + 3\beta_3[Cd^{2+}][Cl^-]^2$$
$$+ 4\beta_4[Cd^{2+}][Cl^-]^3 = C_{Cl}.$$

It is possible to "extract" $[Cd^{2+}]$ from the first of the last two equations and to "inject" it into the latter. Then an eighth-order equation in $[Cl^-]$ appears that can be solved only by numerical calculus. A means to avoid this difficulty (which is one of the most frequently used tricks in analytical chemistry to overcome this recurrent difficulty) consists of adding a large excess of one of the reactants, usually the ligand.

In this experimental condition, one may consider that the concentration of the ligand remaining free (uncomplexed) after the varied equilibria were reached is equal to its analytical concentration. The larger the excess of ligand is, the more legitimate this approximation is. Hence, when a great excess of chloride ions is added, the following assumption is made:

$$[Cl^-] \approx C_{Cl}.$$

The penultimate equation permits us, then, to obtain $[Cd^{2+}]$ quickly and after the complexes' concentrations through the expressions of the constants K_1 through K_4.

Exercise 1 A total of 10^{-2} mol of cadmium nitrate is dissolved in 1 L of hydrochloric acid of concentration 1 mol/L. Calculate the concentrations of the different species. The stability constants' values are given in Table 24.1.

According to the data,

$$C_{Cd} = 10^{-2}\ \text{mol/L} \quad \text{and} \quad C_{Cl} = 1\ \text{mol/L}.$$

The approximation is to set $[Cl^-] = 1$ mol/L after complexation. The values obtained are, therefore,

$$[Cd^{2+}] = 2.16 \times 10^{-5}\ \text{mol/L}; \quad [CdCl^+] = 4.55 \times 10^{-4}\ \text{mol/L};$$
$$[CdCl_2] = 3.58 \times 10^{-3}\ \text{mol/L}; \quad [CdCl_3{}^-] = 4.40 \times 10^{-3}\ \text{mol/L};$$
$$[CdCl_4{}^{2-}] = 1.54 \times 10^{-3}\ \text{mol/L}.$$

A rigorous calculation gives the value $[Cl^-] = 0.97$ mol/L. The above results are exact within a 3% error.

One of the numerical processes used to solve the preceding system exactly, a process that is applicable to other systems, consists of using concentrations found through the preceding approximate calculation to obtain a new and better estimate of the initial concentration $[Cl^-]$ and to reiterate the process with this new initial value until the concentrations no longer differ from those found at the end of the preceding iteration. Often, calculations converge quickly and four or five loops are sufficient to stabilize the concentration values. The difficulties that are sometimes encountered in the convergence process often result from a bad choice of the initial approximation.

Conversely, of course, knowledge of stability constants also permits us to appreciate the dissociation of a complex in a solution. The same equations as above are operative.

Exercise 2 A total of 10^{-2} mol of the $[CdCl_2]$ complex is dissolved in 1 L of water. What are the different species' concentrations? (See the stability constants in Table 24.1.)

The problem is solved by using the same equations as in the preceding exercise. The mass balances are

$$[Cl^-] + [CdCl^+] + 2\,[CdCl_2] + 3\,[CdCl_3{}^-] + 4\,[CdCl_4{}^{2-}] = 0.02\ \text{mol/L},$$
$$[Cd^{2+}] + [CdCl^+] + [CdCl_2] + [CdCl_3{}^-] + [CdCl_4{}^{2-}] = 0.01\ \text{mol/L}.$$

By virtue of Ostwald's law (the more diluted a dissociable solute is, the more dissociated it is), it is reasonable to set the hypothesis that cadmium chloride, once dissolved in solution, is more and less dissociated under the dilution effect. As a result, a reasonable approximation is to consider as predominant species Cd^{2+} and $CdCl^{+}$. Therefore, the system that must be solved becomes

$$[Cl^-] + [CdCl^+] = 0.02 \text{ mol/L}^{-1},$$
$$[Cd^{2+}] + [CdCl^+] = 0.01 \text{ mol/L}^{-1},$$
$$[CdCl^+] = K_1[Cl^-][Cd^{2+}].$$

The values obtained are $[Cd^{2+}] = 7.35 \times 10^{-3}$ mol/L, $[CdCl^+] = 2.67 \times 10^{-3}$ mol/L, and $[Cl^-] = 1.73 \times 10^{-2}$ mol/L. With these values and by using the constants K_2, K_3, and K_4, we find for the other complexes the concentrations $[CdCl_2] = 3.65 \times 10^{-4}$ mol/L, $[CdCl_3{}^-] = 7.76 \times 10^{-6}$ mol/L, and $[CdCl_4{}^{2-}] = 4.70 \times 10^{-8}$ mol/L. Effectively, $CdCl_2$ is strongly dissociated. This is due not only to the weak initial concentration but also to the weak values of the constants K_1, K_2, K_3, and K_4. An iterative calculation starting from these values does not significantly ameliorate the previous results. They are accurate within a 5% error.

Remark In these exercises, the charge balance (electroneutrality) relation has not been taken into account. Of course, it is still satisfied, but its use is not required when no acid–base phenomenon is operating.

It is very important to emphasize the fact that the preceding calculations are correct because no other chemical reaction or other physical phenomenon but that of complexation exists in solution. The system of equations used accurately described the phenomena: The chemical model was exact. In the above-studied case, the hydrolysis of Cd^{2+}, for example, could have been considered to be possible (see Chap. 26). This is not the case.

24.6 Distribution Diagrams

The stability constants permit us to draw the distribution diagrams of the different complexes. They are valuable diagrams since they permit, at first glance, the identification of the predominant species in a given solution.

A first sort of diagram consists of drawing the fraction α of the metal present (the ratio of the concentration of the considered species and the sum of concentrations of all the species containing the metallic ion) as a function of the free ligand concentration or of its decimal co-logarithm. Hence, in the case of cadmium complexes (investigated above), the different fractions α are

$$\alpha_0 = [Cd^{2+}]/C, \quad \alpha_1 = [CdCl^+]/C, \quad \alpha_2 = [CdCl_2]/C,$$
$$\alpha_3 = [CdCl_3{}^-]/C, \quad \alpha_4 = [CdCl_4{}^{2-}]/C.$$

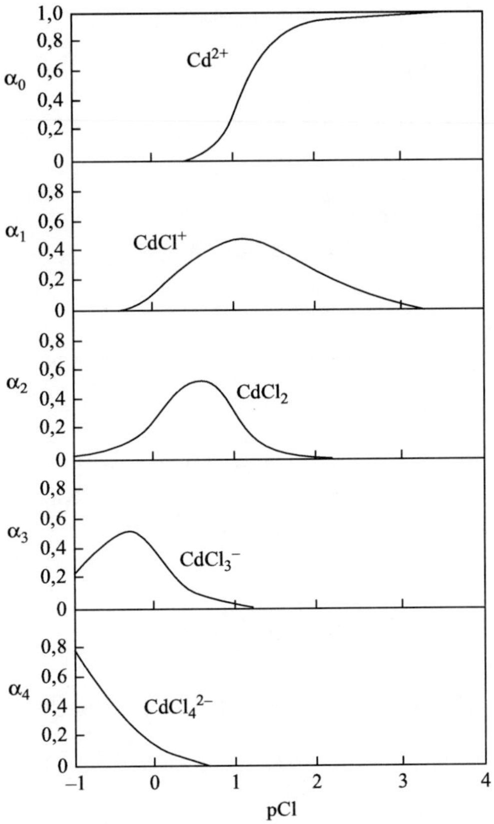

Fig. 24.1 Distribution diagram of the different chlorinated complexes of cadmium

We already encountered the following expression:

$$C_{\mathrm{Cd}} = [\mathrm{Cd}^{2+}]\left\{1 + \beta_1[\mathrm{Cl}^-] + \beta_2[\mathrm{Cl}^-]^2 + \beta_3[\mathrm{Cl}^-]^3 + \beta_4[\mathrm{Cl}^-]^4\right\}.$$

Therefore,

$$\alpha_0 = 1/\left\{1 + \beta_1[\mathrm{Cl}^-] + \beta_2[\mathrm{Cl}^-]^2 + \beta_3[\mathrm{Cl}^-]^3 + \beta_4[\mathrm{Cl}^-]^4\right\},$$

$$\alpha_1 = \beta_1[\mathrm{Cl}^-]\alpha_0, \quad \alpha_2 = \beta_2[\mathrm{Cl}^-]^2\alpha_0,$$
$$\alpha_3 = \beta_3[\mathrm{Cl}^-]^3\alpha_0, \quad \alpha_4 = \beta_4[\mathrm{Cl}^-]^4\alpha_0.$$

It is clear that the cadmium fraction of each species is independent of the total cadmium concentration. Distribution diagrams are given in Fig. 24.1 (the subscripts 0 through 4 are the numbers of ligands bound to the metallic ion).

Another sort of distribution diagram is presented in Fig. 24.2. It is the diagram showing the cumulated fractions as a function of the free ligand concentration or of its decimal antilogarithm. In the case of the preceding example, it exhibits the

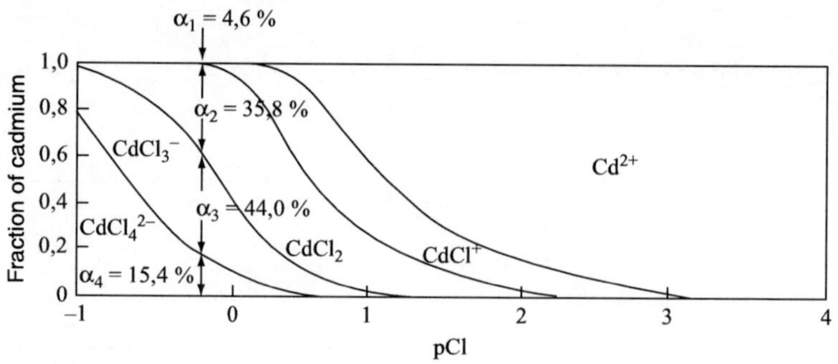

Fig. 24.2 Cumulated fractions of the different complexes of cadmium as a function of *p*Cl

cumulated fractions $\alpha_4, \alpha_4 + \alpha_3, \alpha_4 + \alpha_3 + \alpha_2, \alpha_4 + \alpha_3 + \alpha_2 + \alpha_1$ as a function of *p*Cl. On these diagrams, a vertical line (for the same abscissa *p*Cl) is divided into segments limited by their intersections with the different curves. The lengths of these segments are proportional to the fraction of each species.

24.7 Formation Curve

The extent of the *formation* of a complex, denoted by the symbol *n*, is defined as being the average number of ligands bound to the metallic ion. It is equal to the ratio of the difference between the total ligand concentration and that remaining free, on the one hand, and the analytical concentration of the metallic ion on the other. In the case of the cadmium complexes, it is written as

$$n = (C_{\mathrm{Cl}} - [\mathrm{Cl}^-])/C_{\mathrm{Cd}}.$$

It is easy to check, according to the preceding considerations, that

$$n = \{[\mathrm{CdCl}^+] + 2[\mathrm{CdCl_2}] + 3[\mathrm{CdCl_3}^-] + 4[\mathrm{CdCl_4}^{2-}]\}/C_{\mathrm{Cd}}$$

and that

$$n = \alpha_1 + 2\alpha_2 + 3\alpha_3 + 4\alpha_4.$$

The extent *n* can be obtained by experimental means.

The diagram *n* as a function of the decimal antilogarithm of the ligand activity (concentration) of the free ligand after complexation (here *p*Cl) is known as the *formation curve*. The mathematical study of the formation curve permits us to determine the equilibrium constants, K_1, K_2, K_3, K_4, etc. (see Sect. 24.3). The formation curve concept is due to J. Bjerrum.[1] It can be extended to other phenomena in solution.

[1] Jannik Bjerrum (1879–1958), professor of chemistry at the University of Copenhagen; the majority of his work was devoted to the physical chemistry of aqueous solutions.

24.8 The Complexes as Particle Donors

Another definition of complexes exists that is of some utility in analytical chemistry. It must be ascribed to G. Charlot[2] et al.: "A complexe is an entity which is able to donate a particle, ionized or not." (Of course, the word "particle" must not be endowed with the nuclear meaning but, simply, with the meaning of "ligand.")

Hence, a complex is a particle donor and the latter is the ligand. To the complex (donor) corresponds a conjugate species called the *particle acceptor*. The donor and the acceptor are linked by the relation

$$\underset{\text{(donor)}}{\text{complex}} \rightleftharpoons \text{acceptor} + \underset{\text{(ligand)}}{\text{particle } p}$$

For example, the tetraamminecopper(II) complex is a donor of the particle ammoniac, whereas the copper ion is the acceptor. The proton and the electron whose exchanges define the acid–base and redox phenomena are excluded from the set of particles implied in Charlot's definition, as are the nuclear ones. With this definition, the analogy of the complexation phenomena with the acid–base and the redox ones becomes evident.

Let's consider the following two compounds, the dichloromercury(II) $[HgCl_2]$ and the monoamminesilver(I) $[Ag(NH_3)^+]$ complexes. According to the following two relations:

$$[HgCl_2] \rightleftharpoons [HgCl]^+ + Cl^-,$$

$$[AgNH_3]^+ \rightleftharpoons Ag^+ + NH_3,$$

they are complexes since they donate the particles Cl^- and NH_3. In order to pursue the analogies with the acid–base and redox phenomena, one may consider that the above two reactions are half-complexation reactions analogous to the reactions

$$HA \rightleftharpoons H^+ + A^-,$$

$$Ox + e^- \rightleftharpoons Red.$$

A global complexation reaction becomes, with these considerations, a competition for the particle p between two donor–acceptor couples.

A particular case is that of the particle solvation. For consistency in the definition, the following reaction is indeed warranted:

$$Cl^- + mH_2O \rightleftharpoons [Cl(H_2O)_m]^-.$$

It is also a half-complexation reaction in which water plays the part of a particle acceptor. Therefore, the dissociation in water of the dichloromercury(II) complex according to the reaction

$$[HgCl_2] + mH_2O \rightleftharpoons [HgCl^+] + [Cl(H_2O)_m]^-$$

[2] Gaston Charlot (1904–1994), French analyst, professor of analytical chemistry at the Faculty of Sciences of Paris and at the School of Industrial Physics and Chemistry of Paris.

may be considered as resulting from the superimposition of the following half-equilibria:

$$[\mathrm{HgCl}]^+ + \mathrm{Cl}^- \rightleftharpoons [\mathrm{HgCl_2}],$$
$$[\mathrm{Cl(H_2O)}_m]^- \rightleftharpoons \mathrm{Cl}^- + m\mathrm{H_2O}.$$

The great advantage of the theory is to permit the immediate transposition of the acid–base and redox methods of reasoning and formulas to the complexation phenomena. Therefore,

- One may define a constant χ analogous to the acid-base constants K_a. Let's consider, for example, the reaction

$$[\mathrm{HgCl_2}] + m\mathrm{H_2O} \rightleftharpoons [\mathrm{HgCl}]^+ + [\mathrm{Cl(H_2O)}_m]^-$$
$$\chi = [\mathrm{HgCl^+}][\mathrm{Cl(H_2O)}_m{}^-]/[\mathrm{HgCl_2}], \quad \chi = 10^{-18}.$$

  pχ is analogous to pK_a. It is interesting to note that $\chi = 1/\beta_1$, where β_1 is the stability constant;
- one may also write

$$\mathrm{pX} = \mathrm{p}\chi - \log\{(\mathrm{donor})/(\mathrm{acceptor})\},$$

  where X is the particle, and so pX is analogous to pH. In particular, if one dissolves both members of a couple so that the concentrations are the following ones:

$$[\mathrm{donor}] = C,$$
$$[\mathrm{acceptor}] = C',$$

  and if C and C' are not too weak, one finds

$$\mathrm{pX} = \mathrm{p}\chi - \log(C/C').$$

  This relation is analogous to Henderson's relation applicable to buffer solutions;
- one may calculate the concentrations of the different species (donor, acceptor, and particle), as we've already seen, that is, by starting with the system of the following equations that must be satisfied:

$$[\mathrm{M}][\mathrm{X}]/[\mathrm{MX}] = \chi,$$
$$[\mathrm{M}] + [\mathrm{MX}] = C,$$
$$[\mathrm{X}] + [\mathrm{MX}] = C',$$

where MX is the complex and M the acceptor (the central ion). If one dissolves one complex (donor) in a solution at concentration C and if only one sort of complex is in the solution, one finds

$$[\mathrm{X}]^2/(C - [\mathrm{X}]) = \chi,$$

and if the complex is sufficiently stable,

$$C - [\mathrm{X}] \approx C;$$

$$\text{hence, } [\mathrm{X}]^2 = \chi$$

$$\text{and } \mathrm{pX} = \frac{1}{2}\mathrm{p}\chi - \frac{1}{2}\log C.$$

This relation is analogous to the one that permits the pH calculation of a weak acid solution in some conditions of concentration;

- one may find a relation analogous to that found with ampholytes. We'll consider the case in which an acceptor 2 (a central ion) is added to a solution of donor 1 (complex 1). The following equilibrium occurs:

$$\underset{\text{donor2}}{\mathrm{M_2X}} + \underset{\text{acceptor1}}{\mathrm{M_1}} \rightleftharpoons \underset{\text{acceptor2}}{\mathrm{M_2}} + \underset{\text{donor1}}{\mathrm{M_1X}}.$$

In the solution at equilibrium, particle X's concentration may be expressed as a function of the couple 1 as well as a function of couple 2:

$$\mathrm{pX} = \mathrm{p}\chi_2 + \log\left([\mathrm{M_2}]/[\mathrm{M_2X}]\right),$$
$$\mathrm{pX} = \mathrm{p}\chi_1 + \log\left([\mathrm{M_1}]/[\mathrm{M_1X}]\right),$$

from which we have

$$2\mathrm{pX} = \mathrm{p}\chi_1 + \mathrm{p}\chi_2 + \log\left([\mathrm{M_2}]\,[\mathrm{M_1}]/[\mathrm{M_2X}]\,[\mathrm{M_1X}]\right).$$

If we mixed donor 2 with acceptor 1 in equivalent quantities, the following relation would be satisfied:

$$[\mathrm{M_2X}] + [\mathrm{M_2}] = [\mathrm{M_1X}] + [\mathrm{M_1}].$$

As a result, we would find

$$\mathrm{p}\chi = \frac{1}{2}\left(\mathrm{p}\chi_1 + \mathrm{p}\chi_2\right).$$

This is the ampholytes formula.

- one may build a pX scale on which each couple (for example, $[HgCl_2]/[HgCl^+]$) is located at the value pX = pχ for which [donor] = [acceptor];

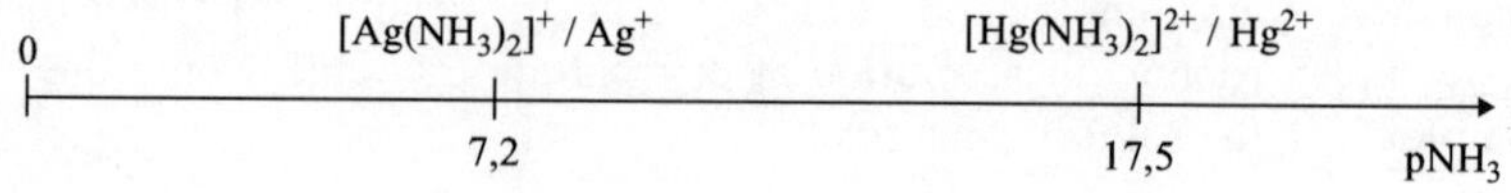

Fig. 24.3 Prediction of the ligand NH_3 exchange reaction between the couples $[Ag(NH_3)_2]^+/Ag^+$ and $[Hg(NH_3)_2]^{2+}/Hg^{2+}$

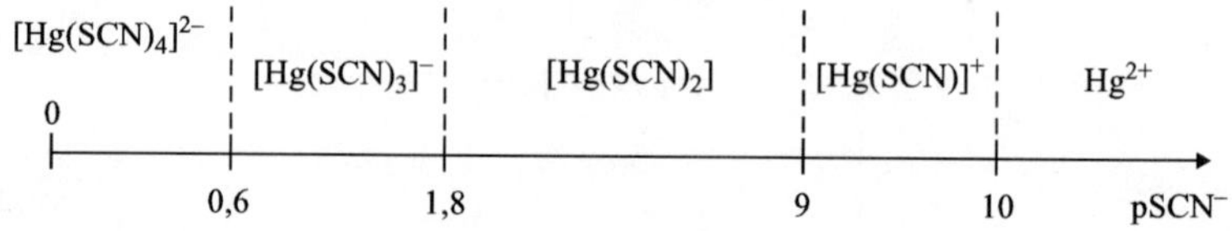

Fig. 24.4 Predominance area of the different complexes formed by the mercuric and thiocyanate ions

- one may see that the dissociation constants χ permit us to predict reactions between complexes. Consider, for example, the case of the following two ammine complexes: diamminesilver(I) $[Ag(NH_3)_2]^+$ and diamminemercury(II) $[Hg(NH_3)_2]^{2+}$, which dissociate according to the equilibria

$$\left[Ag(NH_3)_2\right]^+ \rightleftharpoons Ag^+ + 2NH_3, \quad p\chi_1 = 7.2,$$
$$\left[Hg(NH_3)_2\right]^{2+} \rightleftharpoons Hg^{2+} + 2NH_3, \quad p\chi_2 = 17.5.$$

The exchange reaction,

$$\left[Ag(NH_3]^+ + Hg^{2+} \rightleftharpoons \left[Hg(NH_3)_2\right]^{2+} + Ag^+,$$

is thermodynamically favored. Its equilibrium constant value is

$$10^{(p\chi 2 - p\chi 1)} = 10^{10.3}.$$

As for the other phenomena, of course, for the complex to be formed, no kinetic problem must occur (see Sect. 24.10);
- one also may predict the direction of this exchange reaction by locating both couples on the scale pNH_3 for the values $pNH_3 = p\chi_1 = 7.2$ and $pNH_3 = p\chi_2\chi = \chi 17.5$ (Fig. 24.3).
As the silver couple is located to the left of the mercury couple, the $[Ag(NH_3)_2]^+$ complex is the donor of the NH_3 particle to mercury. The distance of both couples on the scale measures the quantitative character of the reaction;
- finally, as in the case of polyacids, one can draw the predominant area of the different species on the pX scale when there are several complexes that may successively form with the same particle. Let's consider, for example, the different thiocyanatomercurate(II). The predominant areas of the different species are located as indicated in Fig. 24.4, which is the *p*SCN^- scale.

From a more general standpoint and from the viewpoint of the solution chemistry, this theory appears to be an attempt to reunify the reactions that may occur in aqueous solutions:

$$\text{donor} \rightleftharpoons \text{acceptor} + \begin{array}{l} e^- \text{ electrons} \\ H^+ \text{ protons} \\ X \text{ particle} \end{array}$$

Finally, it brings nothing new, but it may perhaps permit us to reason easily at the time of the study of complexes' formations by making use of the reasoning already followed to study acid–base and redox phenomena. However, this theory appears to be artificial, since there are great differences in the different particles' behaviors:

- The proton (under the form H_3O^+ which is solvated itself) does exist in water. It is endowed with a particular stability;
- the proton does not flow through an electrical circuit;
- the electron has no stability in aqueous solution;
- the particle solvates, usually without gaining any exceptional stability. This difficulty appears markedly when we consider the following half-complexation equilibrium (already evoked):

$$Cl^- + mH_2O \rightleftharpoons \left[Cl(H_2O)_m\right]^-.$$

  A well-defined species of the formula $[Cl(H_2O)_m]^-$, in which true strong bonds between the chloride ion and the water molecules would exist, is unknown. This half-equilibrium corresponds more to a solvation process than to a true chemical reaction.
- Finally, there is no universal particle X whose recurring appearance is a major feature of the chemistry of solutions. On the contrary, it is the case of the proton or electron. This is not the role of NH_3, Br^-, and any other ligand. The result is that several pX scales, of limited interest, exist.

24.9 Factors Influencing the Stability of Complexes

Complexes may be more or less stable, more or less perfect. Now, it is interesting to find the factors that govern the ligands' reactivities toward the central metallic ions and those that govern the stability of complexes. The ability to complex metallic ions and also ligands is now considered. (The external factors playing a part in the complexes' formation, such as the superimposition of varied equilibria, are investigated only in the next chapter.)

In terms of the ability of metallic ions to form complexes, Schwarzenbach[3] distinguished two categories: classes A and B.

[3] Gerold Schwarzenbach (1904–1978), Swiss chemist, professor of chemistry in Zurich. He initiated works concerning polyaminopolycarboxylic acids in the 1940s.

According to Schwarzenbach, class A metals are Lewis's acids that exhibit an affinity for halide ions according to the following sequence:

$$F^- \gg Cl^- > Br^- > I^-.$$

Generally, in aqueous solutions, they form the most stable complexes with the first members of each group of donor atoms of the periodic table (N, O, F).

Class B metals form complexes far more easily with iodide ions than with fluoride ions and, in aqueous solutions, give the most stable complexes with the heaviest donor atoms of a group of the periodic table (P, S, Cl).

According to this classification, Schwarzenbach distinguishes three sorts of acceptor metallic ions:

- the cations exhibiting the electronic structure of the preceding noble gas. They are the alkaline cations, the alkaline earth cations, and also the Al^{3+} cation. They are all of class A. The bonds within these complexes are essentially of an electrostatic origin. Interactions between ions of a small size and of high electrical charges are particularly strong. For example, the aluminum cation forms very stable complexes with fluoride anions and very weak ones with iodide ions. With these cations, water gives stabler complexes than ammonia does;
- the cations that have their d-shell totally filled. Typical cations are Cu^{I}, Ag^{I}, and Au^{I}. They are of class B. The bonds they form with ligands exhibit a noticeable covalent character. The more noble (the less oxidizable by air) the metal is and the less electronegative the donor atom of the ligand is, the more stable the complex is. For example, Cd^{2+} and Hg^{2+} form very stable complexes with iodide and cyanide ions but very weak ones with fluoride ions;
- transition cations with an incomplete d-shell. They belong either to class A or to class B. They are on the borderline of the preceding two classes.

It is impossible to give general rules governing the ability of ligands to form complexes. However, among the characteristics that must be retained in this context, we must consider their basicity, but, here again, predictions are difficult to make. For example, it is true that the perchlorate ion ClO_4^-, which is a null base in water, exhibits a very weak trend to form complexes. On the contrary, the chloride ion, which is a null base in water, gives very stable complexes.

The concept of hard and soft acids and bases (*HSAB concept*) is of some utility to characterize the behavior of metallic ions and of ligands. Hard acids are located in class A and soft acids in class B. Ligands are also classed in two categories, A and B. Those located in class A are hard bases, while those of class B are soft bases. Ligands, whose donor atoms are oxygen, nitrogen, or fluorine, are hard bases. Those whose donor atoms are sulfur, phosphorus, or halogens, to the exclusion of fluorine, are soft bases. The basic principle of the HSAB concept is that the homogeneous pairs hard acid/hard base, on the one hand, and soft acid/soft base, on the other, give the most stable complexes. The heterogeneous pairs hard acid/soft base and vice versa give the least stable ones. Table 24.2 mentions some hard and soft acids.

Actually, the constants K_n depend in a complicated way on the nature of the metallic ion and the ligand. The principal component of the metal–ligand bond is

Table 24.2 Hardness and softness of some metallic ions

	log K_1			
Class A	> 5 4 to 5 1 to 4	Be^{2+}, Sc^{3+}, Zr^{4+}, Th^{4+}, U^{4+} Sn^{2+}, Cr^{3+}, Fe^{3+}, Y^{3+} Mg^{2+}, La^{3+}, Ce^{3+}, Ac^{3+}	hard acids	↓ decreasing hardness
Class B	0 to 1 0 0 to −1	Pb^{2+}, Fe^{2+}, Co^{2+}, Ni^{2+}, Cu^{2+} Tl^{+} Cd^{2+}	limit of hard and soft acids	
		Ag^{+}, Hg^{2+} Pd^{2+}, Pt^{2+}	soft acids	↓ increasing softness

the σ bond between them, which itself depends on the donor and acceptor character of the ligand and metallic ion, respectively. Sometimes there is a supplementary complication: the existence of a component π in the metal ligand bond. It adds itself to the σ component. Finally, we must keep in mind the fact that the stability constants K_n in some way reflect the difference in affinity between water and the ligand for the same metallic ion. The reason is that the constants K_n correspond to the replacement of the water ligand present in the initial aqua complex by the ligand engaged in the definitive complex. We must know that it is difficult to rationalize water–metallic ion interactions.

From another standpoint, we must also recall that stability constants are nothing more than an expression of the free standard enthalpy of the complex's formation by virtue of the relation (see Chap. 2)

$$\Delta G^\circ = -RT\ln K.$$

A more elaborate study necessitates the knowledge of the formation enthalpy and the entropy of the complex since a free enthalpy is, by definition, a linear combination of these two functions. In some cases, complex formation is a process that is quasi-exclusively enthalpy-driven. As an example of such a case, we mention the tetraamminecopper(II) complex $[Cu(NH_3)_4]^{2+}$. In some other cases, the formation is quasi-exclusively entropy-driven. This is the case with hexafluoroaluminate(III) $[AlF_6]^{3-}$. In most cases, the phenomenon is driven simultaneously by both functions.

Another point to stress is that, generally, the numerical values of the successive stability constants K_n decrease regularly as n increases. Table 24.3 gives some values that illustrate this assertion.

Exceptions exist to this "rule," but they are rare. An irregular decrease in K_n with n may be the mark of a change in the coordination number (for example, from an octahedral complex to a tetrahedral or linear one). It may also be the mark of a change in the magnetic moment reflecting a crystal field effect. The latter may indeed change at a given stage of the successive replacements of the initial ligands by the definitive ones. Incidentally, crystal field effects may explain the fact that complexes

Table 24.3 Decrease in the successive stability constants with the index n: examples of octahedral complexes

Successive stability constants	Al^{3+}/F^-	Co^{2+}/NH_3	Cr^{3+}/NCS^-
$\log_{10} K_1$	6.1	2.1	3.0
$\log_{10} K_2$	5.0	1.6	1.7
$\log_{10} K_3$	3.9	1.1	1.0
$\log_{10} K_4$	2.7	0.7	0.3
$\log_{10} K_5$	1.6	0.2	−0.7
$\log_{10} K_6$	0.5	−0.6	−1.3

of metallic ions at oxidation state + III are more stable than those of the same metals at oxidation state + II. The stability of complexes of bivalent metals of the first series of transition metals, for a same ligand, changes according to the order

$$Mn^{2+} < Fe^{2+} < Co^{2+} < Ni^{2+} < Cu^{2+}.$$

This is the Irving–Williams's series. The order tends to indicate that in these complexes, the metal–ligand bond is largely of an electrostatic nature.

24.10 Stability of Chelates: Chelate and Macrocyclic Effects

Recall that chelates are particular complexes in which a multidentate coordinate consisting of an ion or organic molecule contracts all its bonds with the same central ion. The latter is therefore tied in a cyclic structure as if it were in a claw (*chelatos* = claw). Chelates are sometimes called *internal complexes.*

The striking characteristic of these particular complexes is their very high stability. They are incomparably more stable than complexes built with monodentate ligands of basicity (and probably of nucleophilicity) close to that exhibited by the teeth of the chelate and built with the same central ion. To be convinced, it is sufficient to compare the global stability constants β_6 of the hexamminenickel(II) $[Ni(NH_3)_6]^{2+}$ complex and β_3 of the triethylenediamminenickel(II) $[Ni(en)_3]^{2+}$, which is a chelate (Fig. 24.5). In both compounds, it is the same central ion, and one may easily admit that the basicities and the nucleophilicities of both sorts of ligands are about the same. The difference between the stability constants, $\beta_3 = 10^{18}$ and $\beta_6 = 3.1 \times 10^8$, is tremendous.

The difference in stabilities exhibited by the complex that could, a priori, be considered analogous to the chelate and by the chelate itself is called the *chelate effect.* Concerning the stability of the chelates, Schwarzenbach underlined the following stability factors:

1. the number of pentagonal cycles in which the metallic ion is engaged. In the case of the chelate $[Ni(en)_3]^{2+}$, the number is three. The greater their number is, the higher the chelate stability is. The series given in Fig. 24.6 illustrates this observation. $ML_4{}^{n+}$ is less stable than $M(L{-}L)_2{}^{n+}$, itself less stable than $M(LLLL)^{n+}$, where L is a monodentate ligand, and L−L bidentate and L−L−L−L tetradentate ones;

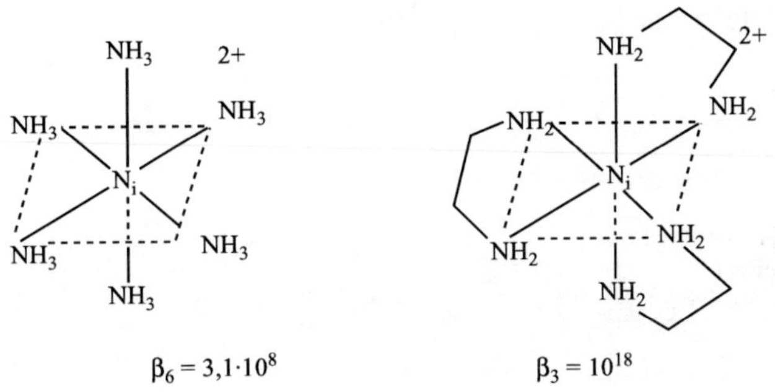

Fig. 24.5 Overall stability constants of complexes $[Ni(NH_3)_6]^{2+}$ and $[Ni(en)_3]^{2+}$

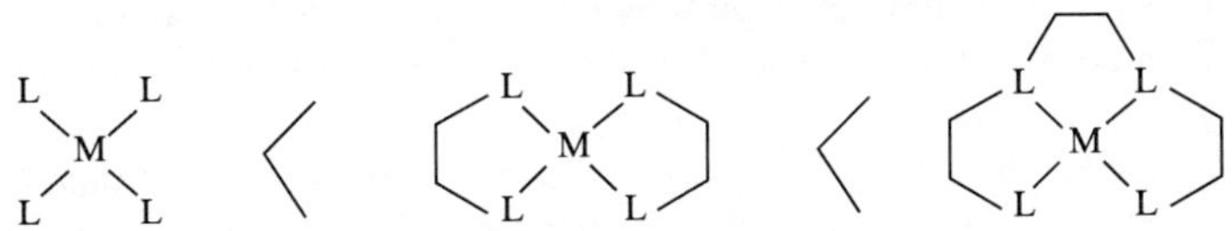

Fig. 24.6 Stability of complexes and chelates as a function of the number of pentagonal cycles

Table 24.4 Overall stability constants as a function of the number of pentagonal cycles

	$4NH_3$ $\log_{10} \beta_4$	2en $\log_{10} \beta_2$	1trien $\log_{10} \beta_1$
Cd^{2+}	7.5	10.1	10.8
Co^{2+}	4.7	10.7	11.2
Cu^{2+}	13.1	19.9	20.5

Table 24.4 mentions the overall stability constant values in the cases of chelates of Cd^{2+}, Co^{2+}, and Cu^{2+} with ammonia (in these cases, they are simple complexes), ethylenediammine (en), and triethylenediammine (trien).

Within the chelate effect, we can distinguish the macrocyclic effect. The comparison of the stabilities of the following chelates provides an argument in favor of what may be called the *macrocyclic effect*. The stability of the chelate possessing the maximal number of pentagonal cycles is higher than that of the structurally analogous chelate, notably endowed with the same denticity but with one fewer pentagonal cycle (see Figs. 24.7 and 24.8).

Concerning the role of the pentagonal cycles in the stability of chelates, we can note that cycles with four summits are less stable than those with five summits. It is the same state of affairs with cycles possessing seven summits compared to cycles with six summits and a fortiori for cycles with more than seven summits;

2. the nature of the atoms that donate electronic pairs. Oxygen and nitrogen atoms give the most stable chelates;
3. the ratio of the number of chelator molecules/number of coordinating ions existing in the chelate once formed. The optimal value is 1. In the chelate $[Ni(en)_3]^{2+}$, the value is 3.

Fig. 24.7 Triethylendiammine (trien)

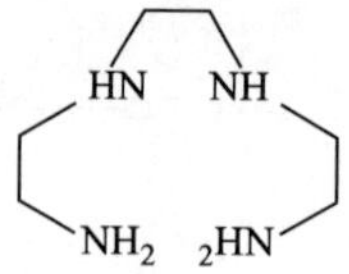

Fig. 24.8 An example of the macrocyclic effect

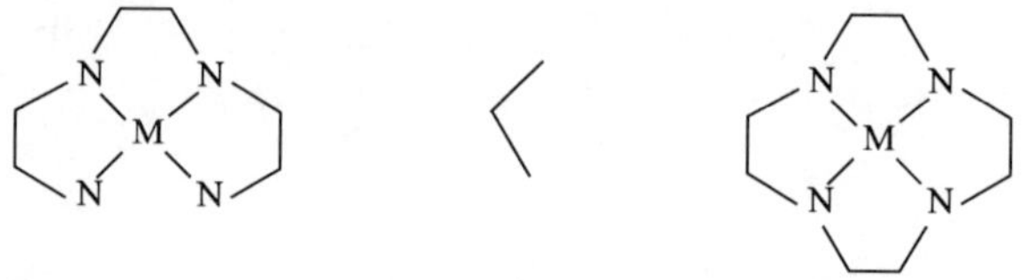

Beyond these observations of experimental origin,the chelate effect has been the matter of numerous works. Several authors ascribe it to an entropic effect. The reasoning is as follows:

At the time of the chelate formation, the replacement of monodentate ligands (water, for example) by one or several chelator molecule(s) (whose number must be inferior to that of ligands) induces the appearance of a number of free molecules greater than in the initial state. Hence, the system entropy increases because of these liberated molecules. For example, as a result of the relation

$$\left[\mathrm{Cd(NH_3)_4}\right]^{2+} + \mathrm{en} \rightarrow \left[\mathrm{Cd(NH_3)_2\,(en)}\right]^{2+} + 2\mathrm{NH_3},$$

we notice the formation of two free ammoniac molecules for one ethylenediamine molecule engaged in the complex. The changes in the thermodynamic functions accompanying this reaction are $\Delta G^{\mathrm{o}} = -5.0$ kJ/mol, $\Delta H^{\mathrm{o}} = 0.4$ kJ/mol, and $T\Delta S^{\mathrm{o}} = 5.4$ kJ/mol. The process is effectively entropy-driven.

Some supplementary comments concerning the chelate effect are interesting. Some authors dispute the preceding explanation. Some others even call into question the very existence of the chelate effect. It is also another argument of a thermodynamic nature that is against it. According to them, the β_4 value of say $[Cd(NH_3)_4]^{2+}$ is by no means comparable to β_2 in the chelate $[Cd(en)_2]^{2+}$ if one reasons in that the concentrations β_4 and β_2 are not expressed in the same dimension. If, in order to obviate this problem, other standard states are chosen, the stability increase accompanying the crossing from an ordinary complex to a chelate of analogous structure, namely, the chelate effect, no longer exists. The structures exhibit the same stability! However, it remains true that the displacement of ammoniac molecules from the above complex by ethylenediamine well and truly exists. According to Schwarzenbach, the chelate effect's origin lies in the fact that, because of its structure, the second tooth of the chelator (for example, the second amine group of ethylenediamine) stands near the central ion. Its concentration appears to be far greater than that in the case of a chemical reaction occurring between a ligand and a more or less already complexed metallic ion when a complex (and not a chelate) is forming. During the formation of a chelate, the collision probability between the tooth and the central ion is much higher than during the formation of an ordinary complex.

24.11 Kinetics of Complexes' Formation: Labile and Inert Complexes

Of course, for complexes to be formed, not only must they be of sufficient stability, but their rate of formation must also be sufficiently fast. This leads to the concept of labile and inert complexes. The rate with which a given complex exchanges its ligands with other complexes present in the solution is a mark of the *lability* property. Those for which these reactions are fast are said to be *labile*. Those for which they are slow are said to be *inert*. The lability concept only concerns reaction rates. Hence, it only concerns the kinetic aspects of complexation reactions. It has nothing to do with their thermodynamic aspect, that is, with the stability of complexes. A good example of the distinction between these two aspects is provided by the hexamminecobalt (III) complex $[Co(NH_3)_6]^{3+}$. It can survive for several days in acidic aqueous solution although it is not stable. (The fact that a compound is not stable is purely and simply in the realm of thermodynamics.) This is because it is inert. Its instability is proved by the equilibrium constant value of the following reaction:

$$\left[Co(NH_3)_6\right]^{3+} + 6H^+ + 6H_2O \rightleftharpoons \left[Co(H_2O)_6\right]^{3+} + 6NH_4{}^+, \quad K = 10^{25}.$$

Another example, but, a contrario, is provided by the tetracyanonickelate(II) complex $[Ni(CN)_4]^{2-}$. It is very stable:

$$Ni^{2+} + 4CN^- \rightleftharpoons \left[Ni(CN)_4\right]^{2-}, \quad \beta_4 = 10^{22}.$$

The exchange rate of the ligand CN^- with its counterpart radiolabeled on the carbon ($^*CN^-$) is extremely fast. Actually, it cannot be measured.

From a more general standpoint, what is striking in the kinetics of the complexes' formation is the extent of the range in which the rates of ligand exchanges are located. It is very large. Hence, for first-order reactions involving the exchange of water, the values are spread from 10^{-6}/s for Rh^{3+} until 10^{10}/s for Cu^{2+}.

Qualitatively speaking, a labile complex is a complex that forms in the time interval necessary to carry out the mixture of reagents, that is, immediately. For example, the hexamminecopper(II) complex $[Cu(NH_3)_6]^{2+}$ is immediately formed by mixing ammonia with a copper(II) solution. An inert complex is a complex whose half-life is around several hours and even, sometimes, several days, at usual temperatures.

The inertness or lability of a complex is under the control of several factors. At first glance, we can say that

- the elements of the main groups form labile complexes;
- the elements of the first transition row form labile complexes, except Cr^{3+} and Co^{3+};
- those of the second and third rows tend to form inert complexes.

Note that the reactions evoked in this section are substitution reactions, which, for octahedral complexes, may be symbolized quite generally as follows:

$$\left[ML_5L'\right]^{n+} + L'' \rightarrow \left[ML_5L''\right]^{n+} + L'.$$

Chapter 25
Superimposition of Varied Equilibria to Complexation Equilibria

Most of the time, various equilibria are superimposed onto those of complexation. Their occurrence may strongly modify the course of the expected complexation reaction(s). They may be of several origins: formation of other complexes, acid–base equilibria, formation of polynuclear complexes, and, finally, precipitation and redox reactions.

These parasitic reactions cause the yield from the expected complexes to be weaker than that predicted on the basis of the sole consideration of the studied complexation equilibrium.

25.1 Superimposition of Several Complexation Equilibria

Let's consider the very simple following example. Frequently, the metallic ion M (electrical charge omitted for simplicity) we want to engage into a complex with the ligand L (electrical charge omitted) is initially in the presence of another potential ligand L′. For example, L′ may be in the initial solution in order to maintain the metallic ion in the solution. As another example, the presence of a buffer is often necessary for some complexes to be sufficiently stable. In this case, L′ may be one of the members of the buffer. In these conditions, the two superimposing equilibria are

$$\mathrm{ML'} \rightleftharpoons \mathrm{M} + \mathrm{L'}, \quad 1/\beta',$$
$$\mathrm{M} + \mathrm{L} \rightleftharpoons \mathrm{ML}, \quad \beta,$$

and as a result,

$$\mathrm{ML'} + \mathrm{L} \rightleftharpoons \mathrm{ML} + \mathrm{L'}, \quad \beta/\beta'.$$

β and β' are the stability constants of complexes ML and ML′. The equilibrium constant K of the global reaction, namely, of the reaction of the formation of the complex ML from the complex ML′, is such that

$$K = \beta/\beta'.$$

J.-L. Burgot, *Ionic Equilibria in Analytical Chemistry*,
DOI 10.1007/978-1-4419-8382-4_25,

It is evident that the constant K must be inferior to the constant β since, by hypothesis, the parasitic complex ML′ exists; that is, $\beta' > 1$. We also notice that, as a limiting case, the constant K may be inferior to the unity ($\beta' > \beta$, where the initial complex ML′ is more stable than ML). Then the formation of the complex ML searched for is inhibited.

Therefore, from a general standpoint, ML is obtained with a weaker yield than would be obtained if there were no interference with another complexation equilibrium. An example, among numerous others, is provided by the complexation of Ni^{2+} by the complexone(IV), Y^{4-}. The expected (and occurring) reaction is

$$Ni^{2+} + Y^{4-} \rightleftharpoons [NiY]^{2-}.$$

To obtain satisfactory yields, the reaction must be carried out at pH $\geq$ 9 (see Sect. 25.4). Hence, the ammonia buffer is used. In these conditions and due to the fact that a buffer must be used in a rather important concentration, the following ammine complexes form: $[Ni(NH_3)]^{2+}$, $[Ni(NH_3)_2]^{2+}$, ..., $[Ni(NH_3)_6]^{2+}$, whose overall formation constants are $\beta_1, \beta_2, \ldots, \beta_6$. As a result, the formation reactions of the complex of interest must be written as

$$[Ni(NH_3)]^{2+} + Y^{4-} \rightleftharpoons [NiY]^{2-} + NH_3,$$

$$[Ni(NH_3)_2]^{2+} + Y^{4-} \rightleftharpoons [NiY]^{2-} + 2NH_3,$$

and so on.

We shall see that the yield in the complex $[NiY]^{2-}$ may be considerably lower than expected after having considered only the following simple reaction:

$$Ni^{2+} + Y^{4-} \rightleftharpoons [NiY]^{2-}.$$

25.2 An Important Particular Case of Parasitic Reactions: Formation of the Complexes Hydroxo, Oxo, and so Forth from the Hydrolysis of Metallic Ions

The complexes hydroxo, oxo, and so forth result from the hydrolysis of metallic ions, which is ineluctable in aqueous solutions.

25.2.1 Hydrolysis of Metallic Ions

The hydrolysis of metallic ions may be defined from two viewpoints, which are equivalent for our purpose.

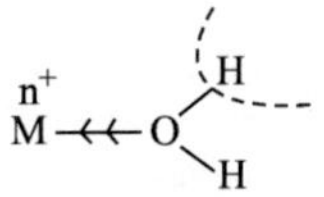

Fig. 25.1 Loss of a proton by a withdrawing inductor effect

According to the first one, the hydrolysis may be considered as resulting from the reaction of the metallic ion with hydroxide anions OH^- to give the corresponding hydroxo complex according to

$$M^{n+} + OH^- \rightleftharpoons [M(OH)]^{(n-1)+} \; K_1.$$

Actually, it is the aqua complex that undergoes this reaction:

$$[M(H_2O)_6]^{n+} + OH^- \rightleftharpoons [M(H_2O)_5(OH)]^{(n-1)+} + H_2O \; K. \quad (25.1)$$

Several successive substitutions by hydroxide ions may occur. This definition of hydrolysis permits us to consider it as a true complexation phenomenon. Reaction (25.1) is a true reaction of a complex formation since it involves the reaction of the hydroxo ligand. Indeed, it consists of the replacement of a water molecule from an initial aqua complex by the hydroxo ligand.

According to the second viewpoint, the hydrolysis of a metallic cation results from the acidity of its aqua complex. For example, the formation of the above hydroxo complex may also result from the prototropic equilibrium

$$[M(H_2O)_6]^{n+} + H_2O \rightleftharpoons [M(H_2O)_5(OH)]^{(n-1)+} + H_3O^+ \; K_a. \quad (25.2)$$

After comparing Eqs. (25.1) and (25.2), we immediately see that

$$K_a = K K_w.$$

The literature mentions the acidity of numerous metallic ions, some of whose acid dissociation constants are given now: $pK_a(Li^+) = 13.9$; $pK_a(Na^+) = 14.7$; $pK_a(Ag^+) = 11.9$; $pK_a(Mg^{2+}) = 11.4$; $pK_a(Fe^{3+}) = 2.0$; $pK_a(Pd^{2+}) = 1.4$; $pK_a(U^{4+}) = 1.0$; and $pK_a(Ce^{IV}) = 0$.

The pK_a values of metallic ions are located in a very large pH range. There is a very noticeable relation between the pK_a value and the ion's charge. The more charged the ion is, the more acidic it gets. A simple reasoning explains the phenomenon: The positive charge brought by the hydrated metallic ion favors the loss of a proton from the H_2O ligand by the inductor effect $-I$. The more positive the ion is, the stronger the effect is (Fig. 25.1).

However, this theory does not explain all the values found.

Regardless of the definition adopted for the hydrolysis, it appears that this phenomenon is accompanied by an acidification of the solution [see Eqs (25.1) and (25.2)]. Table 25.1 mentions some values of successive stability constants of some hydroxo complexes. For example, cadmium gives the complexes hydroxocadmium(II) $[Cd(OH)]^+$, dihydroxocadmium(II) $[Cd(OH)_2]$, trihydroxocadmiate(II)

Table 25.1 Stability constants of some mononuclear hydroxo complexes

Central ion	Logarithm of some stability constants			
	K_1	K_2	K_3	K_4
Hg^{2+}	10.3	11.4		
Ag^{+}	2.3	1.7		
Ni^{2+}	3.4	6.8	2.8	−0.4
Fe^{2+}	5.7	3.4	0.9	
Co^{2+}	1.8	7.4	1.3	
Mn^{2+}	3.4	3.4	1.0	1.3
Zn^{2+}	4.15	6.0	4.1	−0.3
Cd^{2+}	4.2	4.2	0.7	
Mg^{2+}	2.6			

$[Cd(OH)_3]^-$, and tetrahydroxocadmiate(II) $[Cd(OH)_4]^{2-}$. Likewise, mercury(II) gives the complexes hydroxomercury(-II) $[Hg(OH)]^+$ and dihydroxomercury(-II) $[Hg(OH)_2]$.

The hydrolysis phenomenon may play a considerable role during the formation of complexes for several reasons. The first one could have been studied in the first section of this chapter. Indeed, the hydrolysis phenomenon induces a competition between the ligand L and the hydroxo ions L′ for the metallic ion. The second reason results from the fact that hydroxo complexes may precipitate as insoluble hydroxides or oxides. Hence, in this case, a phenomenon of precipitation also occurs, and we can speak of a complexation–precipitation interaction. Of course, all these processes depend on the pH value since the hydroxo ligand and the hydroxo complexes intervene in the complexation phenomenon.

25.2.2 Competition Between the Hydroxo Complexes and Other Ligands

In order to facilitate the discussion, it is interesting first to know the concentrations of the hydroxo complexes resulting from the hydrolysis phenomenon, no other ligand being present at that time. With the mercuric ion Hg^{2+}, we calculate (see Exercise 1 ahead) that in an aqueous solution containing 10^{-4} mol/L of mercuric nitrate and whose acidity is such that pH = 3.00, the concentrations are $[Hg^{2+}] = 5.9 \times 10^{-5}$ mol/L, $[Hg(OH)^+] = 1.2 \times 10^{-5}$ mol/L, and $[Hg(OH)_2] = 2.9 \times 10^{-5}$ mol/L. This is an interesting result. It shows that in these conditions, only half of the mercuric ions remain as such, although $Hg(NO_3)_2$ is totally ionized.

Remarks

- The total concentration $[Hg^{II}]$ is too weak and the pH value too low for HgO to precipitate (see ahead);
- the nitrate ion does not give a complex with Hg^{II}. (We'll point out, however, incidentally, that it is exceptional that an anion does not give a complex with any metallic ion.)

Now, in order to study the competition between the ligand L and the hydroxo ions for the metallic ion, let's investigate the formation of chlorinated complexes of Hg(II) in aqueous solution. The following complexes form with chloride ions:

$$Hg^{2+} + Cl^- \rightleftharpoons [HgCl]^+,$$

$$[HgCl]^+ + Cl^- \rightleftharpoons [HgCl_2],$$

$$[HgCl_2] + Cl^- \rightleftharpoons [HgCl_3]^-,$$

$$[HgCl_3]^- + Cl^- \rightleftharpoons [HgCl_4]^{2-}.$$

But since it is an aqueous solution, the hydroxo complexes $[HgOH]^+$ and $[Hg(OH)_2]$ also exist. Hence, it is legitimate to think that a fraction of the quantity of the first chlorinated complex that forms results from the equilibrated reaction

$$[Hg(OH)]^+ + Cl^- \rightleftharpoons [HgCl]^+.$$

There is more or less a competition between the OH^- and Cl^- ligands. We calculate the following concentrations (see Exercise 2) in a solution containing chloride ions at the concentration 2×10^{-1} mol/L and mercuric ions at the concentration 10^{-1} mol/L:

$$[Cl^-] = 3.40 \times 10^{-2} \text{ mol/L}, \qquad [Hg(OH)^+] = 4.96 \times 10^{-10} \text{ mol/L},$$

$$[HgCl^+] = 7.36 \times 10^{-7} \text{ mol/L}, \qquad [Hg(OH)_2] = 7.86 \times 10^{-7} \text{ mol/L},$$

$$[HgCl_2] = 7.56 \times 10^{-2} \text{ mol/L}, \qquad [H^+] = 10^{-5.80} \text{ mol/L},$$

$$[HgCl_3{}^-] = 1.82 \times 10^{-2} \text{ mol/L}, \qquad [OH^-] = 10^{-8.20} \text{ mol/L},$$

$$[HgCl_4{}^{2-}] = 6.19 \times 10^{-3} \text{ mol/L}, \qquad pH = 5.80,$$

$$[Hg^{2+}] = 3.94 \times 10^{-12} \text{ mol/L}.$$

In this example, the fraction of mercury(II) escaping from complexation to chloride ions due to the hydrolysis phenomenon is negligible. This is due to the great value of the overall stability constant of the $[HgCl_2]$ complex on the one hand and, on the other, to the fact that the concentration $[OH^-]$ intervening in expressions (25.5) and (25.6) (see Exercise 2) is weak. In this example, the hydrolysis of the metallic ion (Hg^{2+}) does not preclude the formation of the expected complexes.

Exercise 1 Calculate the concentrations of all the species of a solution containing 1.0×10^{-4} mol/L of mercuric nitrate whose pH = 3.00 (mercuric nitrate is totally ionized and the oxide HgO does not precipitate in these conditions).

The relations that are satisfied at equilibrium are

$$\frac{[Hg(OH)^+]}{[Hg^{2+}][OH^-]} = 2.0 \times 10^{10}; \quad \frac{[Hg(OH)_2]}{[Hg(OH)^+][OH^-]} = 2.5 \times 10^{11};$$

$$[Hg^{2+}] + [Hg(OH)^+] + [Hg(OH)_2] = 1.0 \times 10^{-4} \quad \text{(mass balance on mercury).}$$

With $[H^+] = 10^{-3}$ (activities and concentrations mixed), we find $[OH^-] = 1.0 \times 10^{-11}$, $[Hg(OH)^+] = 0.20\ [Hg^{2+}]$, and $[Hg(OH)_2] = 0.50[Hg^{2+}]$. From the relation $[Hg^{2+}](1 + 0.20 + 0.50) = 1.0 \times 10^{-4}$ mol/L, we find

$$[Hg^{2+}] = 5.9 \times 10^{-5} \text{ mol/L}, [Hg(OH)^+] = 1.2 \times 10^{-5} \text{ mol/L},$$

$$[Hg(OH)_2] = 2.9 \times 10^{-5} \text{ mol/L}.$$

Exercise 2 Calculate the concentrations of the different species when 10^{-1} mol of mercuric nitrate and 2×10^{-1} mol of chloride ions are dissolved in water to give 1 L of solution. (In these conditions, HgO and $[HgCl_2]$ do not precipitate, and mercuric nitrate is totally ionized.)

The following equilibrium expressions are satisfied:

$$\frac{[HgCl^+]}{[Hg^{2+}][Cl^-]} = 10^{6.74}, \tag{25.3}$$

$$\frac{[HgCl_2]}{[HgCl^+][Cl^-]} = 10^{6.48}, \tag{25.4}$$

$$\frac{[HgCl_3{}^-]}{[HgCl_2][Cl^-]} = 10^{0.85}, \tag{25.5}$$

$$\frac{[HgCl_4{}^{2-}]}{[HgCl_3{}^-][Cl^-]} = 10^{1.00}, \tag{25.6}$$

$$\frac{[Hg(OH)^+]}{[Hg^{2+}][OH^-]} = 10^{0.30}, \tag{25.7}$$

$$\frac{[Hg(OH)_2]}{[Hg(OH)^+][OH^-]} = 10^{11,40}. \tag{25.8}$$

The mass balances on mercury and on chloride ions are, respectively,

$$[Hg^{2+}] + [HgCl^+] + [HgCl_2] + [HgCl_3{}^-] + [HgCl_4{}^{2-}] + [Hg(OH)^+] + [Hg(OH)_2] = 10^{-1} \text{ mol/L}, \tag{25.9}$$

$$[Cl^-] + [HgCl^+] + 2[HgCl_2] + 3[HgCl_3{}^-] + 4[HgCl_4{}^{2-}] = 2 \times 10^{-1} \text{ mol/L}. \tag{25.10}$$

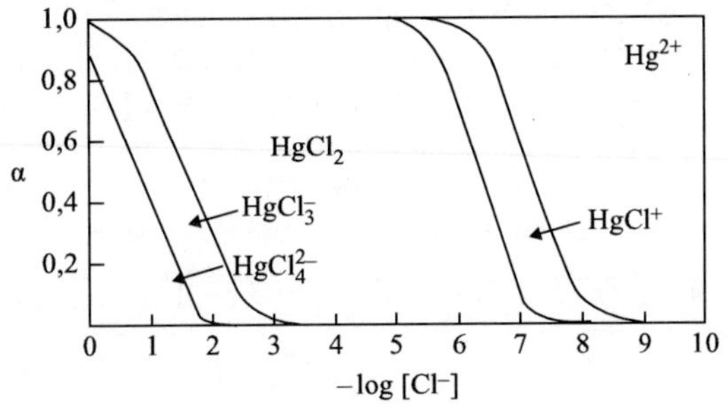

Fig. 25.2 Distribution diagram of the chlorinated complexes of mercury

The system to solve is a system in 10 unknowns, which are $[Hg^{2+}]$, $[HgCl^+]$, $[HgCl_2]$, $[HgCl_3{}^-]$, $[HgCl_4{}^{2-}]$, $[Hg(OH)^+]$, $[Hg(OH)_2]$, $[Cl^-]$, $[H^+]$, and $[OH^-]$. Two supplementary relations (independent of the two preceding ones) are necessary for the system to be determined. They are

- the ion product of water:

$$[H^+][OH^-] = 10^{-14};$$

- the charge balance relation:

$$[H^+] + [HgCl^+] + [Hg(OH)^+] + 2[Hg^{2+}]$$
$$= [OH^-] + [Cl^-] + [HgCl_3{}^-] + 2[HgCl_4{}^{2-}].$$

The last two reactions are also required because acid–base equilibria occur. The combination of the charge balance with both mass balance conditions gives the simpler relation (called the *proton condition relation*)

$$[H^+] = [OH^-] + [Hg(OH)^+] + 2[Hg(OH)_2].$$

The system seems to be difficult to solve. However, some judicious simplifications make its resolution considerably easier. Hence, the hydrolysis phenomenon, which induces an acidification of the medium, incites neglecting $[OH^-]$ in the proton condition relation. From another standpoint, the stability constant K_2 of $[Hg(OH)_2]$ also incites neglecting $[Hg(OH)^+]$ in the same expression, which therefore becomes

$$[H^+] \approx 2[Hg(OH)_2]. \qquad (25.11)$$

Moreover, an examination of the distribution diagram of chlorinated complexes of mercury (see Fig. 25.2) shows that it is reasonable to make it sure that $[HgCl_2]$ is the predominant complex.

Calculating the concentrations of the different species is carried out iteratively by going backward and forward to Eq. (25.10) from Eq. (25.9). To begin with, we consider that $[HgCl_2] \approx 10^{-1}$ mol/L and $[Cl^-] \approx 10^{-1}$ mol/L. This leads to the value $[Hg^{2+}] = 10^{-12.22}$ after the use of Eqs. (25.3) and (25.4). Then, through

Eqs. (25.3)–(25.6) and (25.10), a new value of $[Cl^-]$ is calculated. (Incidentally, by calculating $[H^+]$ through Eqs. (25.11), (25.7), and (25.8) with $[Hg^{2+}] = 10^{-12.22}$, we notice that $[Hg(OH)_2]$ and a fortiori $[Hg(OH)^+]$ are negligible.)

The process is then reiterated with the new value $[Cl^-]$, which, introduced into (25.9), leads to a new value $[Hg^{2+}]$. After convergence of the successive iterations, the results are those given in the text above.

Remark In this exercise, the overall constants of displacement reactions such as

$$[Hg(OH)^+] + Cl^- \rightleftharpoons [HgCl^+]{+}OH^-$$

are not used. Actually, they would be redundant since they result from ratios or products of stability constants already used.

25.2.3 Complexation—Precipitation Interaction: Formation of Insoluble Oxides and Hydroxides

The complex's formation may be inhibited by a precipitation phenomenon, or, in other words, a complex may be destroyed by a precipitation phenomenon. For example, the diamminesilver(I) complex is destroyed in certain conditions of concentrations by the addition of chloride ions. The precipitate of silver chloride is formed:

$$[Ag(NH_3)_2]^+ + Cl^- \rightleftharpoons AgCl\downarrow + 2NH_3.$$

Inversely, a precipitate of silver chloride may be dissolved by adding ammonia.

In this section, we are interested in the particular case of the interaction of the precipitation of insoluble metal oxides and hydroxides whose metallic ions also form complexes. Within this case, a frequent phenomenon exists: the dissolution of an insoluble hydroxide or oxide after adding an excess of precipitating ions. This case is also investigated here. (We shall investigate the treatment of the complexation–precipitation interaction more deeply in Part V of this book.)

It is an experimental fact that metallic hydroxides in usual experimental conditions spontaneously give insoluble oxides. Hence, mercuric ions give rise to the following equilibria:

$$Hg^{2+} + 2OH^- \rightleftharpoons [Hg(OH)_2],$$

$$[Hg(OH)_2]_{saturation} \rightleftharpoons HgO\downarrow + H_2O.$$

When $Hg(OH)_2$ is at saturation, the solid phase in equilibrium with the liquid phase consists of HgO. It is the same state of affairs with silver ions Ag^+:

$$Ag^+ + OH^- \rightleftharpoons AgOH,$$

$$2AgOH_{saturation} \rightleftharpoons AgO\downarrow + H_2O.$$

(From a thermodynamic standpoint, it is quite equivalent to handle MO(solid) and $M(OH)_2$(solid) in the calculations, because their activities are equal to unity since they constitute a pure phase and because water, which participates in their equilibria, possesses an activity equal to unity since the saturated solutions are very diluted.)

Before studying the precipitation–complexation interaction, it is essential first to deal briefly with precipitation phenomena from a quantitative standpoint (see also Part V).

For example, let's consider the case of solid silver chloride, which dissociates in water to give the cation Ag^+ and the anion Cl^-. The equilibrium is

$$AgCl_{(solid)} \rightleftharpoons Ag^+ + Cl^-.$$

It is governed by the constant K_s, called the *solubility product* and defined by the expression

$$K_{(S)} = (Ag^+)(Cl^-),$$

where (Ag^+) and (Cl^-) are activities. In the most general case of the equilibrium,

$$M_zX_y \rightleftharpoons zM^{+y} + yX^{-z},$$

the solubility product is defined by

$$(M^{+Y})^z(X^{-z})^y = K_s.$$

(The solubility-product concept has already been discussed in Chaps. 2 and 14.) Actually, solubility products would be written as

$$K_S = \frac{(Ag^+)(Cl^-)}{(AgCl)_{solid}}, \quad K_s = \frac{(M^{+y})^z(X^{-z})^y}{(M_zX_y)_{solid}},$$

but since (AgCl)solid and $(M_zX_y)_{solid}$ are activities of substances constituting one phase only, they are equal to unity (see Chap. 2) and hence are omitted.

The solubility of a salt is the maximum number of moles that can be dissolved in 1 L of solution. It is closely related to the solubility-product value. Strictly speaking, the solubility S of, for example, silver chloride in water is

$$S = [AgCl\ ]_{saturation} + [Ag^+],$$

$$S = [AgCl\ ]_{saturation} + [Cl^-],$$

where $[AgCl]_{saturation}$ is the concentration at saturation of undissociated silver chloride. The concentrations $[Ag^+]$ and $[Cl^-]$ are equal since, by dissolution and by dissociation, silver chloride gives one mole of Ag^+ for one mole of Cl^-. It is an experimental fact that the undissociated silver chloride in solution at saturation

$[AgCl]_{saturation}$ is negligible. In the present case, $[AgCl]_{saturation} = 10^{-6.7}$ mol/L, whereas

$$[Ag^+] = [Cl^-] = 10^{-4.9} \text{ mol/L}$$

since

$$[Ag^+] = [Cl^-] = K_s{}^{\frac{1}{2}}.$$

In the case of a poorly soluble metallic hydroxide MOH, the precipitation equilibrium is

$$MOH\downarrow \rightleftharpoons M^+ + OH^-.$$

It is governed by the solubility product

$$K_s(MOH) = (M^+)(OH^-).$$

Now, we can study the superimposition of the precipitation and complexation phenomena.

Frequently, a metallic ion forms a series of complexes with the same ligand, one of which is poorly soluble. This is the case with some hydroxo and chloro complexes. We shall see that a new sort of equilibrium constant permits us to investigate the superimposition of complexation and precipitation phenomena.

The silver ion, for example, may give the successive complexes [AgOH], $[Ag(OH)_2]^-$ according to the equilibria

$$Ag^+ + OH^- \rightleftharpoons [AgOH\,] \quad \beta_1,\ K_1,$$

$$[AgOH\,] + OH^- \rightleftharpoons [Ag(OH)_2]^- \ K_2,$$

where [AgOH], which is a nonionic species, is poorly soluble.

Mercury ions give the complexes $[HgOH^+]$ and $[Hg(OH)_2]$ according to the equilibria

$$Hg^{2+} + OH^- \rightleftharpoons [HgOH^+],$$

$$[HgOH^+] + OH^- \rightleftharpoons [Hg(OH)_2].$$

Again, it is the nonionic species $[Hg(OH)_2]$ that is poorly soluble.

Other ligands may also give rise to the same phenomenon. For example, recall that chloride ions give the successive complexes [AgCl], $[AgCl_2]^-$, $[AgCl_3]^{2-}$, and $[AgCl_4]^{3-}$. (Other examples are given in Chap. 35.)

The poorly soluble molecular species, here $[Hg(OH)_2]$ and [AgOH], may be in equilibrium with a solid phase constituted by the corresponding oxide (rather than by the hydroxide), but this is not always the case (see above and Table 25.2). The

Table 25.2 Formation constants of mononuclear hydroxo complexes of some metallic ions from the corresponding insoluble oxides or hydroxides

Central ion	Logarithm of stability constants					
	K_{s0}	K_{s1}	K_{s2}	K_{s3}	K_{s4}	Solid phase
Hg^{2+}	−25.32	−15.02	−3.62			HgO
Ag^+	−7.73	−5.43	−3.71			Ag_2O
Ni^{2+}	−17.2	−13.8	−7.0	−4.2		$Ni(OH)_2$
Zn^{2+}	−17.15	−13.0	−7.0	−2.92	−1.66	$Zn(OH)_2$
Cd^{2+}	−13.66	−9.50	−5.37	−4.68	−5.00	$Cd(OH)_2$
Mg^{2+}	−10.74	−8.16				$Mg(OH)_2$
Ca^{2+}	−5.26	−3.75				$Ca(OH)_2$

complexes [AgOH], $[Ag(OH)_2]^-$, and so forth are, of course, governed by the usual formation constants β and K, that is,

$$\frac{[AgOH]}{[Ag^+][OH^-]} = K_1 = \beta_1,$$

$$\frac{[Ag(OH)_2{}^-]}{[AgOH][OH^-]} = K_2 \quad \text{and} \quad \frac{[Ag(OH)_2{}^-]}{[Ag^+][OH^-]^2} = \beta_2.$$

It must be understood that all the species whose concentrations are handled in the above expressions, K_1, K_2, β_1, β_2, are only in solution. Another set of constants is used when the complexes are formed from an insoluble salt and from the anion entering the insoluble salt. For example, silver hydroxide, which is very poorly soluble, may react with the hydroxo ligand to give the dihydroxosilverate (I) complex $[Ag(OH)_2]^-$:

$$AgOH\downarrow \rightleftharpoons Ag^+ + OH^-, \quad K_{so} = [Ag^+][OH^-] \quad \text{(saturated solution)},$$

$$AgOH\downarrow \rightleftharpoons AgOH, \quad K_{s1} = [AgOH] \quad \text{(saturated solution)},$$

$$K_{s1} = \mathrm{S} \quad \text{(see below)},$$

$$AgOH\downarrow + OH^- \rightleftharpoons [Ag(OH)_2]^-, \quad K_{s2} \quad \text{(see below)}.$$

The nomenclature followed here is that of J. N. Butler.[1] The indices 0, 1, and 2 of the K_s constants give the number of ligands linked to the metallic ion, once the complex is formed from the insoluble salt. These constants can be used only when the solution is saturated (in AgOH or Ag_2O for the chosen example). The relations

[1] See the bibliography at the end of the book.

between the constants K_n or β_n and K_{sn} are found by handling K_n or β_n and the solubility products. Now, we'll make these relations explicit:

$$K_{so} = [Ag^+][OH^-], \quad K_1 = \beta_1 = \frac{[AgOH]}{[Ag^+][OH^-]}.$$

$$\text{At saturation:} \quad [AgOH] = S = K_{s1} \quad \text{and} \quad [Ag^+][OH^-] = K_{so};$$

$$\text{as a result} \quad K_1 = \frac{K_{s1}}{K_{s0}}.$$

When no precipitation has already occurred, the constants K_2 and β_2 are defined as usual:

$$K_2 = \frac{[Ag(OH)_2{}^-]}{[AgOH][OH^-]} \quad \text{and} \quad \beta_2 = \frac{[Ag(OH)_2{}^-]}{[Ag^+][OH^-]^2}.$$

At saturation, the following equilibrium holds:

$$AgOH\downarrow + OH^- \rightleftharpoons [Ag(OH)_2]^-.$$

By definition,

$$K_{s2} = \frac{[Ag(OH)_2{}^-]}{[OH^-]};$$

that is,

$$K_{s2} = K_2 K_{s1} \quad \text{since} \quad K_2 = \frac{[Ag(OH)_2{}^-]}{[AgOH][OH^-]},$$

and as a result,

$$K_{s2} = K_2 K_1 K_{s0} \quad K_{s2} = \beta_2 K_{s0}.$$

Another example is provided by the chlorinated complexes of the silver cation. The constants K_{s0}, K_{s1}, K_{s2}, K_{s3}, and K_{s4} are defined as immediately below. They govern the following equilibria:

$$AgCl\downarrow \rightleftharpoons Ag^+ + Cl^-, \qquad [Ag^+][Cl^-] = K_{s0}$$

$$AgCl\downarrow \rightleftharpoons AgCl_{solution}, \qquad [AgCl]_{solution} = K_{s1},$$

$$AgCl\downarrow + Cl^- \rightleftharpoons [AgCl_2]^-, \qquad [AgCl_2{}^-] = K_{s2}[Cl^-],$$

$$AgCl\downarrow + 2Cl^- \rightleftharpoons [AgCl_3]^{2-}, \qquad [AgCl_3{}^{2-}] = K_{s3}[Cl^-]^2,$$

$$AgCl\downarrow + 3Cl^- \rightleftharpoons [AgCl_4]^{3-}, \qquad [AgCl_4{}^{3-}] = K_{s4}[Cl^-]^3.$$

Quite evidently, this nomenclature is generalized to other complexes (see Chap. 35). Table 25.2 mentions formation constant values of mononuclear hydroxo complexes, the complexes being formed from the corresponding insoluble oxides or hydroxides. These constants permit us to calculate the concentrations of the formed complexes when there is a precipitation. They also permit us to calculate the number of moles that are precipitated.

Exercise 3 Mercuric ions are hydrolyzed in water with the result that their solutions are acidic. Calculate the pH of the solution prepared by dissolution of 1.0×10^{-2} mol of mercuric nitrate $Hg(NO_3)_2$ in 1 L of water. At saturation, $[Hg(OH)_2] = 2.4 \times 10^{-4}$ mol/L.

There is an uncertainty in the data of this problem. It depends on whether or not the precipitation of the oxide HgO resulting from the equilibrium

$$[Hg(OH)_2] \rightleftharpoons H_2O + HgO\downarrow$$

(resulting from the hydrolysis of the mercuric ion) takes place.

Let's check the hypothesis that the precipitate is forming. The mass balance on mercury is

$$[Hg^{2+}] + [HgOH^+] + [Hg(OH)_2] + P = 1.0 \times 10^{-2} \text{ mol/L},$$

where P is the "concentration" (mol/L) of mercury precipitated as mercuric oxide. The other relations that are satisfied are

$[Hg(OH)_2] = 2.40 \times 10^{-4}$ mol/L since there is saturation by hypothesis,

$[Hg^{2+}] \times [OH^-]^2 = 4.8 \times 10^{-26}$ relations which are legitimate since there is

$[HgOH^+] \times [OH^-] = 9.16 \times 10^{-16}$ precipitation of HgO (see Table 25.2),

$2\,[Hg^{2+}] + [H^+] + [HgOH^+] = [OH^-] + [NO_3{}^-]$ charge balance,

$[NO_3{}^-] = 2 \times 10^{-2}$ mass balance on nitrate ions,

$[H^+][OH^-] = 10^{-14}$.

Since, because of the hydrolysis, the solution becomes acidic, it appears reasonable to neglect the hydroxide concentration in the charge balance relation. Using the preceding two relations, the latter becomes

$$\frac{2 \times 4.8 \times 10^{-26}}{[OH^-]^2} + \frac{9.6 \times 10^{-16}}{[OH^-]} + \frac{10^{-14}}{[OH^-]} = 2 \times 10^{-2}.$$

$$\text{Thus,} \quad [OH^-] = 2.47 \times 10^{-12},$$

$$pH = 2.39.$$

By using all of the above relations, we find

$$[HgOH^+] = 3.9 \times 10^{-4}\ \text{mol/L},\ [Hg^{2+}] = 7.87 \times 10^{-3}\ \text{mol/L},$$

$$P = 1.50 \times 10^{-3}\ \text{mol/L and of course } [Hg(OH)_2] = 2.4 \times 10^{-4}\ \text{mol/L}.$$

These results are by no means absurd. They are in agreement with the data in the problem. In particular, the "concentration" P is positive and is less than the initial concentration of mercuric nitrate. The hypothesis about the precipitation of HgO turns out to be exact.

Exercise 4 Mercuric oxide HgO may precipitate when mercuric nitrate is dissolved in water, even in a slightly acidic solution. What is the maximum pH value that the solution must exhibit to avoid the precipitation of a 10^{-2} mol/L solution of mercuric nitrate?

Just at the saturation, the mass balance on mercury is

$$[Hg^{2+}] + [HgOH^+] + [Hg(OH)_2] = 10^{-2}\ \text{mol/L}.$$

(We notice that no term denoting a precipitated concentration is mentioned in the relation. This is normal since, by hypothesis, we are just at saturation.) In order to solve the problem, we can use two sets of constants. The first is that of constants K_1, K_2 (or β_1, β_2) since no precipitate exists yet. The second set is that using constants K_{sn} since the saturation is reached and, therefore, the solubility product is satisfied.

With the first set, we can write

$$\frac{[Hg(OH)^+]}{[Hg^{2+}][OH^-]} = 10^{10.3}, \quad \frac{[Hg(OH)_2]}{[HgOH^+][OH^-]} = 10^{11.4},$$

$$[Hg(OH)_2] = 2.4 \times 10^{-4}\ \text{mol/L (saturation)}$$

These relations lead to the quadratic equation

$$\frac{4.79 \times 10^{-26}}{[OH^-]^2} + \frac{9.55 \times 10^{-16}}{[OH^-]} = 9.76 \times 10^{-3}.$$

The second set of constants (Table 25.2) takes into account the constants $K_{s0} = 10^{-25.32}$, $K_{s1} = 10^{-15.02}$, and $K_{s2} = 2.4 \times 10^{-4}$ mol/L since we can consider that we are just at the very beginning of the precipitation. In these conditions, we find

$$[Hg^{2+}] = \frac{10^{-25.32}}{[OH^-]^2}, \quad [HgOH^+] = \frac{10^{-15.02}}{[OH^-]}.$$

By taking into account the mass balance on mercury, we find a relation strictly identical to the above quadratic one.

This exercise is particularly interesting because it permits us to specify the conditions that must be respected in order to carry out the titration of chloride, bromide, thiocyanate, and cyanide ions with a mercuric salt according to Votocek–Dubsky's method (see Chap. 27). The titrant solution is a mercuric nitrate (or sulfate) solution in which the precipitation of HgO resulting from the hydrolysis of Hg^{2+} is precluded by addition of nitric acid.

Exercise 5 In Votocek–Dubsky's titration method, the titrant solution may be prepared by dissolution of mercury oxide in a nitric acid solution. What concentration of nitric acid must be used in order to dissolve 3.5×10^{-2} mol of HgO in 1 L of solution? What are the concentrations of the other species of the solution?

This problem differs from the preceding one. Here we must calculate the nitric acid concentration, which must be added so that no mercuric oxide exists. This calculation must take into account the fact that the dissolution of mercuric oxide is accompanied by acid–base equilibria. The latter are

$$HgO\downarrow + H_2O \rightleftharpoons [HgOH^+] + [OH^-],$$

$$[HgOH^+] \rightleftharpoons Hg^{2+} + OH^-,$$

$$OH^- + H^+ \rightleftharpoons H_2O.$$

In the preceding problem, it was sufficient to calculate the required pH to have no precipitation without taking any account of the protons' origin (hydrolysis and/or acid addition). Here the addition of protons displaces the above equilibria in the direction of supplementary dissociation and dissolution. To solve the problem, we'll write all the relations that are satisfied when the last trace of oxide is dissolved:

$$[Hg^{2+}] + [HgOH^+] + [Hg(OH)_2] = 3.5 \times 10^{-2} \quad \text{mass balance on mercury,}$$

$$\frac{[Hg(OH)^+]}{[Hg^{2+}][OH^-]} = K_1, \quad K_1 = 10^{10.3} \quad \text{(Table 25.1),}$$

$$\frac{[Hg(OH)_2]}{[HgOH^+][OH^-]} \times K_2 = 10^{11.4} \quad \text{(Table 25.1),}$$

$$[Hg(OH)_2] = 2.4 \times 10^{-4} \quad \text{(saturation),}$$

$$[H^+] + 2[Hg^{2+}] + [HgOH^+] = [OH^-] + [X^-] \quad \text{charge balance.}$$

$[X^-]$ is the concentration of added nitrate ions, that is, the concentration we are seeking. The first four relations give the optimal pH value for which there is no precipitation (until now, this exercise has been quite equivalent to the preceding one). We find

$$[H^+] = 8.42 \times 10^{-3} \text{ mol/L} \quad \text{or} \quad pH = 2.07.$$

We immediately deduce

$$[HgOH^+] = 8.04 \times 10^{-4}\ \text{mol/L},$$

$$[Hg(OH)_2] = 2.40 \times 10^{-4}\ \text{mol/L} \quad \text{by hypothesis},$$

$$[Hg^{2+}] = 3.39 \times 10^{-2}\ \text{mol/L}.$$

The charge balance gives

$$[X^-] = 7.71 \times 10^{-2}\ \text{mol/L}.$$

This is the concentration of nitric acid that must be added. We must add more nitric acid than the value pH = 2.07 seems to indicate at first glance.

25.3 Formation of Polynuclear Complexes; Polymerization

The hydrolysis of metallic ions often leads to hydroxo polynuclear complexes. In water, for example, Fe(III) gives rise to the following equilibria (simplified writing):

$$Fe^{3+} + H_2O \rightleftharpoons [Fe(OH)]^{2+} + H^+, \quad K_{11} = 10^{-3.05},$$

$$[Fe(OH)]^{2+} + H_2O \rightleftharpoons [Fe(OH)_2]^+ + H^+, \quad K_{12} = 10^{-3.26},$$

$$2Fe^{3+} + 2H_2O \rightleftharpoons [Fe_2(OH)_2]^{4+} + 2H^+, \quad \beta_{22} = 10^{-2.91}.$$

Notice the symbolism of equilibrium constants K_{ij} and β_{ij}, where i is the number of metallic ions and j that of bound ligands. The structure of the di-µ-hydroxooctaaquadiiron(III) complex involves the occurrence of bridges through hydroxyl groups linking the two ferric ions:

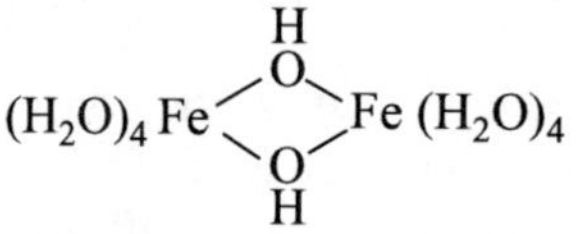

soit $[Fe_2(OH)_2(H_2O)_8]^{4+}$

It may also be considered as resulting from the dimerization of the di(hydroxo)-iron(II) complex $[Fe(OH)_2]^+$ according to the reaction

$$2[Fe(OH)_2]^+ + 2H^+ \rightleftharpoons [Fe_2(OH)_2]^{4+} + 2H_2O.$$

Several metallic ions give bridged ions: beryllium(II), tin(II), bismuth(III), and copper(II), for example.

These phenomena find expression in the presence of supplementary equilibria that actually govern them and that interfere with those corresponding to the complex under interest. Constants K_{ij} (or β_{ij}), together with those already handled with mononuclear complexes, permit us to calculate the concentrations of the different species, including those of the polynuclear hydroxo complexes.

Exercise 6 Find the concentrations of the different species in a 1 L solution containing 10^{-1} mol of ferric nitrate and 10^{-2} mol of nitric acid (the nitrate ion does not complex ferric ion and no precipitation occurs in these conditions).

The equilibria that must be taken into account are

$$Fe^{3+} + H_2O \rightleftharpoons [Fe(OH)]^{2+} + H^+, \quad K_1 = 10^{-3.05},$$

$$[Fe(OH)]^{2+} + H_2O \rightleftharpoons [Fe(OH)_2]^+ + H^+, \quad K_2 = 10^{-3.26},$$

$$2Fe^{3+} + 2H_2O \rightleftharpoons [Fe_2(OH)_2]^{4+} + 2H^+, \quad \beta_{22} = 10^{-2.91},$$

$$Fe(OH)_3 + 3H^+ \rightleftharpoons Fe^{3+} + 3H_2O, \quad K_{10} = 10^{-3.96}.$$

The relations that must be satisfied are

$$\frac{[Fe(OH)^{2+}][H^+]}{[Fe^{3+}]} = \beta_{11} = 10^{-3.05}, \quad \frac{[Fe(OH)_2{}^+][H^+]^2}{[Fe^{3+}]} = \beta_{12} = 10^{-6.31},$$

$$\frac{[Fe_2(OH)_2{}^{4+}][H^+]^2}{[Fe^{3+}]^2} = \beta_{22} = 10^{-2.91},$$

$$[Fe^{3+}] + [Fe(OH)^{2+}] + [Fe(OH)_2{}^+] + 2[Fe_2(OH)_2{}^{4+}] = 0.10 \text{ mol/L}.$$

The proton condition that results from the charge balance and the mass balance on Fe(III) is

$$[H^+] = [OH^-] + 10^{-2} + [Fe(OH)^{2+}] + 2[Fe(OH)_2{}^+] + 2[Fe_2(OH)_2{}^{4+}].$$

Since the dimer formation occurs with the expulsion of two protons, as is the case for the formation of $[Fe(OH)_2^+]$, the solution is certainly acidic and $[OH^-]$ can be neglected in the proton condition relation. A reasonable assumption is to consider that very few ferric ions are hydrolyzed because of the presence of nitric acid in the solution, which displaces the equilibria toward unhydrolyzed Fe^{3+}. As a result, we can consider that

$$[Fe^{3+}] \approx 0.1 \text{ mol/L}.$$

Taking into account these assumptions and using the first three equilibria constants and the proton condition, we obtain a third-degree equation in $[H^+]$. The acceptable solution is

$$[H^+] = 3.43 \times 10^{-2}\ \text{mol/L} \quad \text{(temporary result)}.$$

With this value, the following concentrations are easily calculated: $[Fe(OH)^{2+}] = 2.60 \times 10^{-3}$ mol/L, $[Fe(OH)_2{}^+] = 4.16 \times 10^{-5}$ mol/L, and $[Fe_2(OH)_2{}^{4+}] = 1.05 \times 10^{-2}$ mol/L (temporary results). It is possible to start a new calculation again after having used the value

$$[Fe^{3+}] = 7.64 \times 10^{-2}\ \text{mol/L}$$

as the initial value, which markedly differs from the previous one. This new initial value comes from the mass balance on iron after introducing it to the concentration values of the hydroxo complexes found previously. The mass balance on iron is

$$[Fe^{3+}] + [Fe(OH)^{2+}] + [Fe(OH)_2{}^+] + 2[Fe_2(OH)_2{}^{4+}] = 10^{-1}\ \text{mol/L}.$$

The definitive results are

$$[H^+] = 3.10^{-2}\ \text{mol/L},\ [Fe^{3+}] = 8.0 \times 10^{-2}\ \text{mol/L},$$

$$[Fe(OH)^{2+}] = 2.4 \times 10^{-3}\ \text{mol/L},\ [Fe(OH)^{2+}] = 4.4 \times 10^{-5}\ \text{mol/L},$$

$$[Fe_2(OH)_2{}^{4+}] = 8.85 \times 10^{-3}\ \text{mol/L}.$$

The fraction of dinuclear complex compared to that of free ferric ions is about 10%. It is the most important one after that of Fe^{3+}. (Note that the ferric hydroxide solubility product $K_s = 2.96 \times 10^{-39}$ is not reached. Actually, no precipitation occurs.)

The concept of a formation curve presents great interest when there is formation of polynuclear complexes. The study of its evolution as a function of the total concentration of the metallic ion permits us to detect the presence of polynuclear complexes. We have already seen (Chap. 24, Sects. 24.6 and 24.7) that when no polynuclear complex is formed, the curve is independent of the total concentration in the metallic ion. Inversely, this is no longer the case when there are polynuclear complexes. For the example of the hydroxo complexes of ferric ion, the curve formation is the curve **n** as a function of $[OH^-]$ or n as a function of pH. By definition,

$$\mathbf{n} = \frac{\{[Fe(OH)^{2+}] + 2[Fe(OH)_2{}^+] + [Fe_2(OH)_2{}^{4+}]\}}{C}$$

where C is the analytical concentration of Fe^{III}. The number of hydroxo ligands divided by the number of iron atoms in the dimer is 1. This explains why the dimer

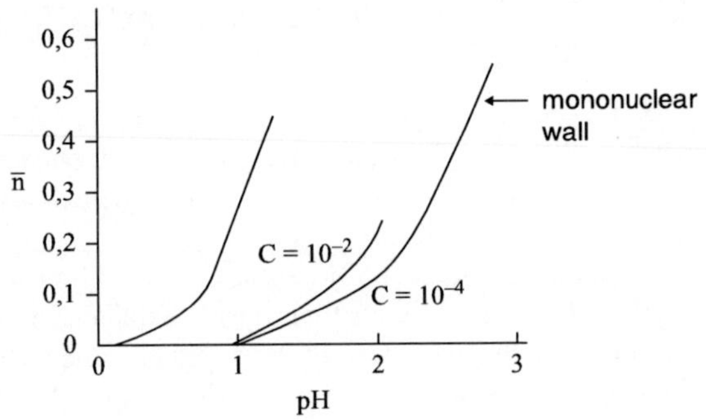

Fig. 25.3 Formation curves of ferric hydroxides as a function of analytical concentrations in Fe^{III}

is multiplied by the coefficient 1 in the above formula. By introducing the overall constants β_{11}, β_{12}, and β_{13} into the expression of **n**, we find

$$\mathbf{n} = \frac{\{\beta_{11}[H^+] + 2\beta_{12} + \beta_{22}[Fe^{3+}]\}}{\{[H^+]^2 + \beta_{11}[H^+] + \beta_{12} + 2\beta_{22}[Fe^{3+}]\}}.$$

Quite evidently, **n** depends on the concentration $[Fe^{3+}]$ and therefore on its analytical concentration *C*. The formation curve **n**/pH loses its shape with *C* (see Fig. 25.3).

However, there is an interesting particular case: that of a weak analytical concentration in metallic ion. Then the terms being composed of the concentration $[Fe^{3+}]$ in the expression of **n** become negligible. In this case, the expression of **n** reduces to

$$\mathbf{n} \approx \frac{\{\beta_{11}[H^+] + 2\beta_{12}\}}{\{[H^+]^2 + \beta_{11}[H^+] + \beta_{12}\}}.$$

Binuclear species are no longer present. It is said that a "mononuclear wall" has been reached. From a practical standpoint, the change of the formation curve with *C* constitutes a convenient means to detect the presence of polynuclear complexes. Fortunately, from a strictly analytical point of view, numerous analytic experiments are carried out in diluted solutions and, very often, the polynuclear complexes' formation may be neglected.

A way to avoid polymerization and even precipitation phenomena (see Sect. 25.4) consists of masking the disturbing metallic ion with another ligand. From a theoretical standpoint, this methodology is still tractable (in an analogous way to that followed here) by introducing conditional constants (see Chap. 26).

Numerous hydroxo aqua cations exhibit a strong tendency to polymerize. The chemistry of polynuclear complexes is very varied and largely remains poorly understood.

25.4 Ability of Ligands to Complex Metallic Ions and Acidity of Solution

The acidity of a solution may have an influence on the ability of the ligands to complex. Herein lies another kind of influence developed by the pH value on the complexation equilibria.

Ligands are nucleophilic entities. Very often, but not necessarily, their nucleophilic character parallels their basic character. In this case, they may be protonated according to the pH of the solution, and they hence lose their ability to complex. The following rule summarizes this consideration: "A decrease in the pH value decreases the stability of complexes, the ligand of which is the conjugate base of a weak acid. The more protonated the ligand is (i.e., the lower the pH value), the more the decrease in stability is." (Obviously, in this rule, we must understand the word "stability" as the apparent stability since the complex's stability is quantified once and for all with its stability constant value.) The rule does not apply to the conjugate bases of strong acids. Then the pH is deprived of any influence on the (apparent) stability of the complexes they form. This is the case with the ligands Cl^-, Br^-, I^-, SCN^-, and so on. Qualitatively speaking, the rule corresponds to a phenomenon of equilibrium displacement. An example is provided by the hydroxide ligand. Let's consider the formation of hydroxo complexes of mercury(-II) according to the reactions

$$Hg^{2+} + OH^- \rightleftharpoons [HgOH]^+,$$

$$[HgOH]^+ + OH^- \rightleftharpoons [Hg(OH)_2],$$

and also consider the superimposition of the equilibrium of the dissociation of water onto the preceding ones:

$$H_2O \rightleftharpoons H^+ + OH^-.$$

After having carried out their sum, we find the following overall equilibria:

$$Hg^{2+} + H_2O \rightleftharpoons [HgOH]^+ + H^+$$

$$\text{and} \quad [HgOH]^+ + H_2O \rightleftharpoons [Hg(OH)_2] + H^+.$$

When examining them, we see that decreasing the pH displaces the equilibrium toward the left: The complex is dissociating. If we consider Exercise 1 again, with the same experimental conditions but with pH = 1 instead of pH = 3, we find $[Hg^{2+}] = 9.98 \times 10^{-5}$ mol/L. The Hg^{2+} ions are totally free. At pH = 3, they were free at the level of 59%, the remaining mercury being hydroxo complexes. Actually, this example is particularly striking because of the fact that the hydroxide ion is the strongest base existing in water.

Quantitatively speaking, the influence of the pH on the complexation equilibria through the protonation of the ligand is difficult to grasp because several phenomena may superimpose onto it. The formation of hydroxo complexes, and even possibly that of polynuclear complexes, may indeed be superimposed onto the complexation equilibrium between the metallic ion and the basic ligand. Furthermore, precipitations are also possible. The complexation of gadolinium ions by fluoride ions provides a rather simple example since it only gives one fluorinated complex and one hydroxo complex. The effective equilibria are

$$Ga^{3+} + F^- \rightleftharpoons [GaF]^{2+} \frac{[GaF]^{2+}}{[Ga^{3+}][F^-]} = 10^{5.1},$$

$$Ga^{3+} + OH^- \rightleftharpoons [Ga(OH)]^{2+} \frac{[Ga(OH)]^{2+}}{[Ga^{3+}][OH^-]} = 10^{11.3},$$

$$HF \rightleftharpoons H^+ + F^-, \quad K_a = 6.76 \times 10^{-4},$$

The studied phenomenon is represented by the overall equation

$$Ga^{3+} + HF^- \rightleftharpoons [GaF]^{2+} + H^+.$$

Exercise 7 Gadolinium nitrate and sodium fluoride are both dissolved at the analytical concentration 10^{-2} mol/L. Calculate the percentage of the fluorogadolinium(-III) complex as a function of pH.

In addition to the expressions of equilibrium constants given above, the mass balances are

$$[Ga^{3+}] + [GaF^{2+}] + [GaOH^{2+}] = 10^{-2},$$

$$[F^-] + [GaF^{2+}] + [HF] = 10^{-2}.$$

As a result, we can write

$$10^{-2} = [Ga^{3+}] + 10^{5.1}[Ga^{3+}][F^-] + [Ga^{3+}]\frac{K_w}{[H^+]}10^{11.1},$$

$$10^{-2} = [Ga^{3+}]\left[1 + \frac{10^{-2.9}}{[H^+]}\right] + 10^{5.3}[Ga^{3+}][F^-],$$

$$10^{-2} = [F^-]\left[1 + \frac{[H^+]}{6.76 \times 10^{-4}}\right] + 10^{5.1}[Ga^{3+}][F^-].$$

The elimination of the product $10^{5.1}[Ga^{3+}][F^-]$ from these two relations gives a quadratic equation in $[Ga^{3+}]$ or $[F^-]$ as a function of $[H^+]$, which provides the solution to the problem. Table 25.3 and Fig. 25.4 summarize the results.

Table 25.3 Percentages of the different gadolinium species formed with fluoride ions as a function of pH

pH	$[GaF^{2+}]$	$[Ga^{3+}]$	$[Ga(OH)^{2+}]$	[F]	[X]
3	93.5	2.86	3.6	2.6	3.00×10^{-2}
2	88.8	9.95	1.25	0.7	3.80×10^{-2}
1	70.9	28.8	0.36	0.2	1.24×10^{-1}
0	35.4	64.5	0	0	1.01

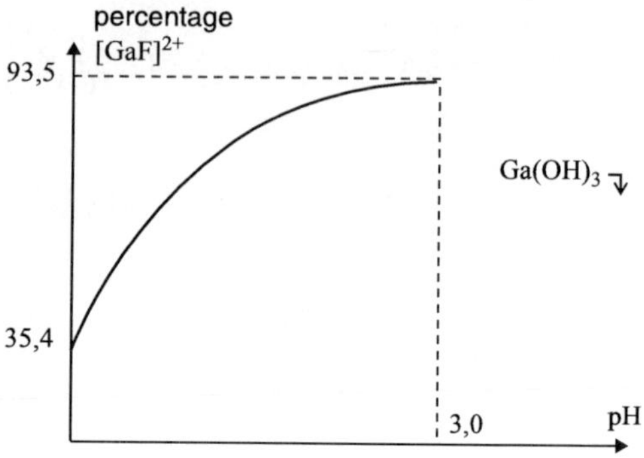

Fig. 25.4 Percentages of the different gadolinium species complexed with fluoride ions as a function of pH

The basic reason is the protonation of the F^- and OH^- ligands. Note that beyond pH = 3, the solubility product is reached. Also, note that in order to carry out these calculations, it is not necessary to use the charge balance relation. Indeed, one liberty degree remained: the concentration $[H^+]$. However, the use of this relation would be necessary to calculate the concentration of a strong acid—whose anion would not complex gadolinium—necessary to reach a given pH value. This relation is

$$3[Ga^{3+}] + 2[GaF^{2+}] + 2[GaOH^{2+}] + [H^+] + [Na^+]$$

$$= [OH^-] + [F^-] + [NO_3{}^-] + [X^-],$$

where $[X^-]$ is the analytical concentration of the strong acid that might be added. The values are given in Table 25.3. The value $[X^-]$ is always greater than 10^{-pH}. This was a predictable result. The reason lies in the fact that the fluoride ion is a base according to the reaction

$$F^- + H_2O \rightleftharpoons HF + OH^-.$$

It must be neutralized before reaching the desired pH value.

From an analytical standpoint, a very important example of the superposition of complexation—acido-basic phenomena is provided by the formation of complexes with H_4EDTA and its anions (see Chap. 28).

According to the pH value of the solution, the reactions of EDTA with metallic ions may be written as

$$M^{n+} + H_4EDTA \rightleftharpoons [MEDTA]^{(n-4)} + 4H^+ \quad \mathbf{K_1} \qquad (25.12)^2$$

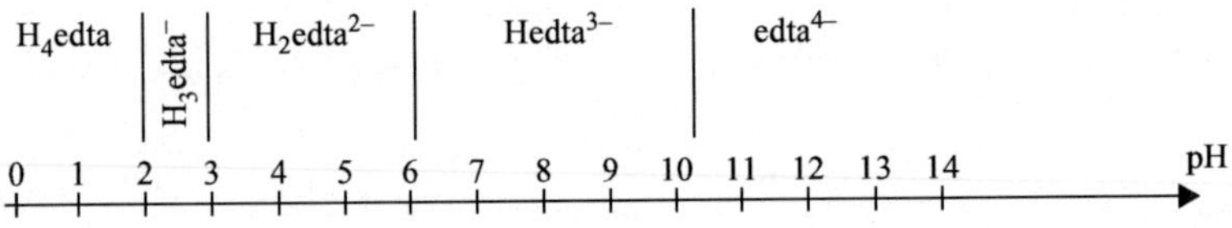

Fig. 25.5 Predominance area of the species deriving from EDTA as a function of pH

$$M^{n+} + H_3EDTA^- \rightleftharpoons [MEDTA]^{(n-4)+} + 3H^+ \quad \mathbf{K_2} \tag{25.13}$$

$$M^{n+} + H_2EDTA^{2-} \rightleftharpoons [MEDTA]^{(n-4)+} + 2H^+ \quad \mathbf{K_3} \tag{25.14}$$

$$M^{n+} + HEDTA^{3-} \rightleftharpoons [MEDTA]^{(n-4)+} + H^+ \quad \mathbf{K_4} \tag{25.15}$$

$$M^{n+} + EDTA^{4-} \rightleftharpoons [MEDTA]^{(n-4)+} \quad K_5 \tag{25.16}$$

Even several of these reactions may occur simultaneously. Indeed, depending on the pH value, it is H_4EDTA or one of its anions or, even, a mixture of some of them which is (are) predominant. The examination of the predominance area of the species deriving from EDTA (Fig. 25.5) permits us to specify the reactions that are occurring.

Therefore, at pH about 2.40, it is the species H_3EDTA^- that essentially reacts with the metallic ion according to reaction (25.13), but the species H_4EDTA and H_2EDTA^{2-} also somewhat react according to reactions (25.12) and (25.14). We'll specify that the equilibrium constants (25.12)–(25.15) are overall equilibrium constants. By definition,

$$K_5 = \frac{[MEDTA^{(n-4)+}]}{[M^{n+}][EDTA^{4-}]};$$

Notice that K_5 is not highlighted (the protons do not intervene in the equilibrium it represents);

$$\mathbf{K_4} = \frac{[MEDTA^{(n-4)+}][H^+]}{[HEDTA^{3-}][M^{n+}]};$$

$$\mathbf{K_4} = \frac{K_5[EDTA^{4-}][H^+]}{[HEDTA^{3-}]};$$

$$\mathbf{K_4} = K_5 K_{a4}$$

As we've already noticed, K_5 and the other (bold) constants $\mathbf{K_1}$–$\mathbf{K_4}$ are not endowed with the same significance. K_5 is the "true" stability formation constant

[2] Some of these constants are highlighted (see just below).

of the $[MEDTA^{(n-4)+}]$ complex, whereas the constants $\mathbf{K_1}$–$\mathbf{K_4}$ must be considered constants of the displacement equilibria (25.12)– (25.15) in which the protons intervene.

In any case, the important point lies in the fact that the numerical values of all these constants differ greatly. For example, with Ni^{2+}, we find $K_5 = 10^{18.56}$, $\mathbf{K_4} = 10^{8.3}$, $\mathbf{K_3} = 10^{2.14}$, $\mathbf{K_2} = 0.30$, and $\mathbf{K_1} = 3.0 \times 10^{-3}$ (values calculated for pH $= 0$). From these considerations, we deduce two points:

- First, if we want to form the $[NiEDTA]^{2-}$ complex quantitatively, the medium must be as basic as possible so that the reaction occurring may be reaction (25.16) or for lack of this one, reaction (25.15);
- only the metallic ions that give the most stable complexes with $EDTA^{4-}$, can give the latter in more acidic media. Most of the time, the reactions are carried out in the range $8 < pH < 11$. This is the reason why they are carried out in the ammonia buffer. The simple examination of the equilibria (25.12)–(25.15) evidently shows that the more acidic the medium is, the more displaced toward the dissociation the equilibrium is.

The complexes formed with EDTA may themselves exhibit an acidic or a basic character. For example, ferric ions with EDTA give the $[Fe^{III}(EDTA)]^-$ complex, which is endowed with a basic character, as the following reaction indicates:

$$[Fe(EDTA)]^- + H^+ \rightarrow [Fe(H)EDTA].$$

If we consider how easily the $[Fe^{III}(EDTA)]^-$ complex forms as a function of the experimental conditions, in particular and overall as a function of pH, we realize that an acidic medium favors its formation in accordance with the last reaction. This is not necessarily true, because, simultaneously, the acidic medium induces the protonation of EDTA, as relations (25.12)–(25.15) indicate. Therefore, it is difficult to predict which of the two effects is more important. Only the stability constant knowledge of all the superimposing equilibria allows this prediction, at least on thermodynamic grounds. To reinforce these considerations, we'll mention the fact that the $[Fe(EDTA)]^-$ complex is also a diacid since it may accept two hydroxide ions (it is a Werner diacid):

$$[Fe(EDTA)]^- + OH^- \rightleftharpoons [Fe(OH)(EDTA)]^{2-},$$

$$[Fe(EDTA)]^- + 2OH^- \rightleftharpoons [Fe(OH)_2(EDTA)]^{3-}.$$

Of course, this property makes the a priori prediction of the best experimental conditions even more difficult. Using conditional constants permits us to solve this problem elegantly (see Chap. 26).

Chapter 26
Conditional Stability Constants

The introduction of conditional stability constants permits us, from a theoretical standpoint, to treat a complexation equilibrium to which are superimposed other equilibria of any nature as if it were alone, at least in some conditions. Conditional stability constants were introduced by G. Schwarzenbach 1957 to take into account acido-basic phenomena occurring simultaneously with complexation phenomena. The first case for which conditional stability constants were introduced concerned the titration of some metallic ions with EDTA tetraanion. The scope of their application scope was extended to other parasitic phenomena by A. Ringbom 1959.

26.1 Species Existing in Solution When a Metallic Ion Is Titrated with EDTA

In a typical titration of a metal cation with EDTA tetraanion, the solution in the reaction vessel, among other species, contains the formed complex $[MEDTA]^{(n-4)+}$, free remaining metallic ions M^{n+}, hydroxo complexes resulting from their hydrolysis, and some complexes formed with a possibly present buffer. Usually, the latter are those formed with ammonia and with acids-alcohols (see Chap. 28).

We have become aware of the complexity of the titrand solution's composition and of the difficulty in calculating the different species' concentrations after all the reactions in the reaction vessel have taken place. It indeed appears to be a true challenge. Fortunately, the formation of polynuclear complexes, which would complicate the calculations even further, may be neglected since the concentrations of these derivatives are usually very weak close to the equivalence point. This is due to the strong complexing power of EDTA. But the titrand solution also contains other species, which this time come from the addition of the titrant. They are the protonated forms of EDTA: $HEDTA^{3-}$, H_2EDTA^{2-}, H_3EDTA^{-}, and H_4EDTA. The complex under study, $[MEDTA]^{(n-4)+}$, may possibly capture a proton or a hydroxide ion, therein exhibiting a basic or acidic character. Finally, we shall see that a secondary complexing agent may be added (see Chap. 28). This once more complicates our calculations.

J.-L. Burgot, *Ionic Equilibria in Analytical Chemistry*,
DOI 10.1007/978-1-4419-8382-4_26, © Springer Science+Business Media, LLC 2012

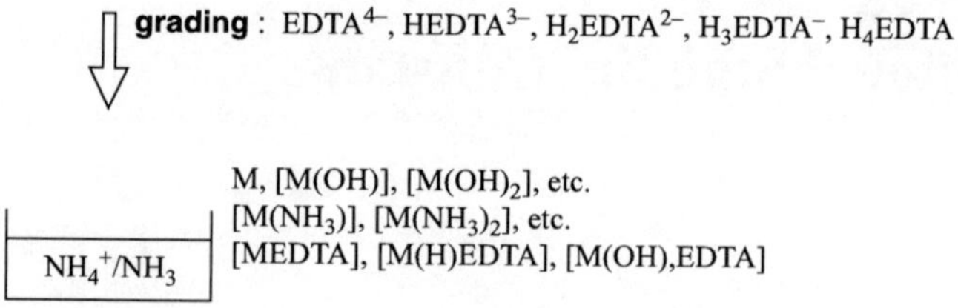

Fig. 26.1 Various species that may be present in the titration of a metallic ion M with EDTA or with its anions (the possible presence of a secondary complexing agent is not mentioned)

26.2 Conditional Constants, Parasitic Reaction Coefficients, and Apparent Concentrations

Using conditional constants permits us easily and rapidly to quantitatively predict the course of complexes' formation reactions that would occur in the experimental conditions depicted in the previous section. However, the prediction can be performed only at the cost of some approximations, which can be legitimized.

The underlying principle of the conditional constant concept is that in analysis, the concentration of utmost importance is that of the anticipated complex, and not necessarily that of any other species in the reaction vessel.

Let's consider the classical titration of a metallic ion[1] M with H_4EDTA or with one of its anions in a medium buffered with the couple NH_4^+/NH_3. Suppose that the metallic ion also gives hydroxo and ammine complexes. Both behaviors are normal. We'll also suppose that the complex under study, [MEDTA], exhibits an acidic character by capturing a hydroxide anion to give [MEDTA(OH)] or a basic character by capturing a proton to give [MEDTAH], or even exhibits both characters. (The latter two behaviors are less typical than the previous one.) These possibilities are summarized in Fig. 26.1.

Let [M]′ be the total concentration of the metallic ion for whatever species it is engaged in except for the exact complex in which we want to engage it and also the hydroxylated or protonated forms of this complex; that is,

$$[\mathrm{M}]' = [\mathrm{M}] + [\mathrm{M(NH_3)}] + [\mathrm{M(NH_3)_2}] + \cdots + [\mathrm{M(OH)}] + [\mathrm{M(OH)_2}] + \cdots .$$

Let's call [EDTA]' the total H_4EDTA concentration (tetraacid and its anions) except for the concentration of the complex under study:

$$[\mathrm{EDTA}]' = [\mathrm{EDTA}]^{4-} + [\mathrm{HEDTA}]^{3-} + [\mathrm{H_2EDTA}]^{2-} + [\mathrm{H_3EDTA}]^{-} + [\mathrm{H_4EDTA}]$$

and let [MEDTA]′ be the total concentration of the mononuclear complex under study:

$$[\mathrm{MEDTA}]' = [\mathrm{MEDTA}] + [\mathrm{MHEDTA}] + [\mathrm{M(OH)EDTA}].$$

[1] Henceforth, for the sakes of generality and simplicity, we will omit electrical charges except in some particular cases.

(Polynuclear complexes are of little interest in analysis. However, they can also be taken into account in analysis through conditional constants. This will not be the case here.) [M]′, [EDTA]′, and [MEDTA]′ are called the *apparent concentrations* of the metallic ion, of EDTA, and of the complex, respectively.

We shall demonstrate that apparent concentrations are given by the following expressions:

$$[\mathrm{M}]' = \alpha_{\mathrm{M}}[\mathrm{M}],$$
$$[\mathrm{EDTA}]' = \alpha_{\mathrm{EDTA}}[\mathrm{EDTA}^{4-}],$$
$$[\mathrm{MEDTA}]' = \alpha_{\mathrm{MEDTA}}[\mathrm{MEDTA}],$$

where α_{M}, α_{EDTA}, and α_{MEDTA} are coefficients that depend on the nature of the species and on the experimental conditions. According to the IUPAC, they must be called *parasitic reaction constants*. Taking the stability constants of the different complexes into account, it is easy to check, after the expressions that define [M]′, [EDTA]′, and [MEDTA]′, that, for example,

$$\alpha_{\mathrm{M}} = 1 + \beta_1{}^{\mathrm{NH3}}[\mathrm{NH_3}] + \beta_2{}^{\mathrm{NH3}}[\mathrm{NH_3}]^2 + \cdots + \beta_1{}^{\mathrm{OH}}[\mathrm{OH^-}] + \beta_2{}^{\mathrm{OH}}[\mathrm{OH^-}]^2 + \cdots,$$

where $\beta_1{}^{\mathrm{NH3}} = [\mathrm{M(NH_3)}]/[\mathrm{M}][\mathrm{NH_3}]$, $\beta_2{}^{\mathrm{NH3}} = [\mathrm{M(NH_3)_2}]/[\mathrm{M}][\mathrm{NH_3}]^2$,

$$\beta_1{}^{\mathrm{OH}} = [\mathrm{M(OH)}]/[\mathrm{M}][\mathrm{OH^-}], \text{ etc.,}$$

and $\alpha_{\mathrm{Y}} = 1 + \beta_1{}^{\mathrm{Y}}[\mathrm{H^+}] + \beta_2{}^{\mathrm{Y}}[\mathrm{H^+}]^2 + \cdots\cdots$. (with Y ≡ EDTA for the sake of brevity), $\alpha_{\mathrm{MY}} = 1 + K_{\mathrm{MHY}}[\mathrm{H^+}]$, where $\beta_1{}^{\mathrm{NH3}}$, $\beta_2{}^{\mathrm{NH3}}$, $\beta_1{}^{\mathrm{OH}}$, etc. are the overall stability constants of the ammine and hydroxo complexes. $\beta_1{}^{\mathrm{Y}}$ is the inverse of the ionization constant K_{a4} of EDTA:

$$\beta_1{}^{\mathrm{Y}} = 1/K_{\mathrm{a4}},$$
$$\beta_1{}^{\mathrm{Y}} = [\mathrm{HEDTA^{3-}}]/[\mathrm{H^+}][\mathrm{EDTA^{4-}}].$$

$\beta_2{}^{\mathrm{Y}}$ and so on are defined in the same manner.

The interest in conditional constants lies in the fact that they can be easily calculated. Indeed, for α_{Y} and α_{MY} it is sufficient to know the solution's pH, the constants $\beta_1{}^{\mathrm{Y}}, \ldots,$ and K_{MHY} being determined once and for all. The calculation of α_{M} appears to be more complicated since the unknown concentration of free ammonia $[\mathrm{NH_3}]$ intervenes in its expression. We only know its analytical concentration, which is given by the expression

$$C_{\mathrm{NH3}} = [\mathrm{NH_3}] + [\mathrm{M(NH_3)}] + 2[\mathrm{M(NH_3)_2}] \ldots.$$

However, if a great excess of ammonia compared to the metallic ion concentration is added, the ammine complexes' concentrations remain negligible with respect to the free ammonia concentration. Therefore, we can write

$$[\mathrm{NH_3}] \approx C_{\mathrm{NH3}}.$$

In these conditions, the calculation of α_M is immediate since the stability constants intervening in its expression are known. (Let's recall incidentally that the strategy consisting of adding a great excess of reagent to simplify calculations is a method frequently used in analytical chemistry. The closer we are located to the equivalence point, the better the condition $[NH_3] \approx C_{NH3}$ is since, by the definition of a satisfactory titration, the cation and ammine complexes' concentrations must be very weak at this point.)

The conditional constant K' prevailing in the above experimental conditions is calculated from the thermodynamic constant K as follows. The expression of the thermodynamic constant K is

$$K = [MY]/[M][Y].$$

It does not take into account phenomena other than that of the complexation under study. In the framework of the introduction of conditional constants, K' governing the equilibrium under study is defined by the relation

$$K' = [MY]'/[M]'[Y]'.$$

This is the constant of prime interest to the analyst. It indeed governs the stability of the complex as a function of the ligand and metallic ion concentrations regardless of their form in the experimental conditions. By comparing the expressions K and K', we find

$$K' = K\alpha_{MY}/\alpha_M\alpha_Y.$$

Hence, K' is easily calculable since K is known definitively, and the α coefficients are calculated as explained above.

The first point to underline is the fact that the conditional constant is really a constant in the given experimental conditions. The second point to stress is that the coefficients α must be greater than 1 due to their polynomial structure containing unity. If the complex under study exhibits no acidic or basic character, $\alpha_{MY} = 1$ and

$$K' = K/\alpha_Y\alpha_M.$$

As a result, the conditional constant must be weaker than the thermodynamic one:

$$K' < K.$$

The complex yield we are seeking is less important than that expected according to the thermodynamic constant. The ratio K'/K may even be very weak. When the complex exhibits a basic or acidic character, α_{MY}, which is also superior to unity, tends to increase the value of the conditional constant. However, experience shows that conditional constants are always lower than the corresponding thermodynamic ones.

The change of K' with α_{MY}, α_Y, and α_M is difficult to predict, especially the change with α_M. This parasitic reaction coefficient is indeed composed of two polynomials, one taking the complexation with the buffer into account, the other the formation of hydroxo complexes. Changes in both polynomial values occur simultaneously and independently. This is the reason why the α_M coefficient changes and, consequently, why K' is difficult to predict.

26.3 Examples of Conditional Constants' Calculations

Exercise 1 Calculate the conditional constant of the nickel–EDTA complex in a solution buffered with a mixture of ammonia (0.1 mol/L) and ammonium ions (0.1 mol/L). The measured pH is 9.24 (pK_a value of ammonium ion). The stability constants are $\beta_1{}^{NH3} = 10^{2.80}$, $\beta_2{}^{NH3} = 10^{5.04}$, $\beta_3{}^{NH3} = 10^{6.77}$, $\beta_4{}^{NH3} = 10^{7.96}$, $\beta_5{}^{NH3} = 10^{8.71}$, $\beta_6{}^{NH3} = 10^{8.74}$, $\beta_1{}^{OH} = 10^{3.4}$, $\beta_2{}^{OH} = 10^{10.2}$, and $\beta_3{}^{OH} = 10^{13}$; $\alpha_{MY} = 1$, and $K(\text{Ni-EDTA}^{2-}) = 10^{18.56}$.

$[NH_3] \approx 0.1$ mol/L (pH $= pK_a$). Using the expression

$$\alpha_M = 1 + \beta_1{}^{NH3}[NH_3] + \cdots + \beta_1{}^{OH}[OH^-] + \cdots$$

gives $\alpha_M = 1.5 \times 10^4$. Through this calculation, we find that hydrolysis (formation of hydroxo complexes) has a very weak influence on the formation of the nickel–EDTA complex. It is the formation of ammine–nickel that exerts a strong such influence. From another standpoint,

$$\alpha_Y = 1 + 1.82 \times 10^{10}[H^+] + 2.62 \times 10^{16}[H^+]^2$$
$$+ 1.23 \times 10^{19}[H^+]^3 + 1.20 \times 10^{21}[H^+]^4;$$

$\alpha_Y = 11.5$. Since $\alpha_{MY} = 1$,

$$K' = 10^{18.56}/(1.51 \times 10^4 \times 11.5), \quad K' = 2.1 \times 10^{13}.$$

K' is considerably smaller than K (about five orders).

Exercise 2 Calculate the conditional constant of the Cu^{II}–EDTA complex at pH $=$ 13 in the absence of any buffer and of any other complexant. (We shall not take the acidic character of the complex or the formation of a binuclear complex into account.) We give $K = 10^{18.8}$ and $\beta_1{}^{OH} = 10^6$.

$$\alpha_Y = 1,$$
$$\alpha_M = 1 + 10^6 \times 10^{-14}/10^{-13},$$
$$K' = 10^{13.8}.$$

Exercise 3 Express the conditional constant of the EDTA–magnesium complex as a function of pH in the presence of the buffer ammonia at a 10^{-1} mol/L concentration. We give $\beta_1{}^{OH} = 10^{2.58}$, $K(\text{MHY}) = 10^{3.85}$, $pK_a(NH_4{}^+/NH_3) = 9.24$, β_n

Fig. 26.2 The coefficient α_{Mg} as a function of pH with and without the buffer NH_3 added

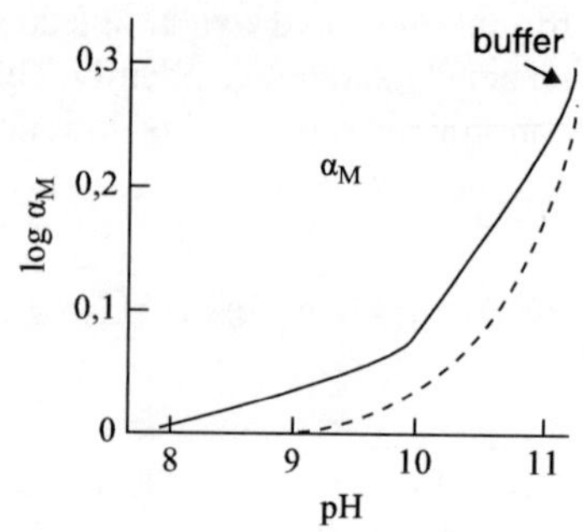

($Mg(NH_3)_n$) with $1 < n < 6$: 1.70, 1.20, 0.457, 0.091, 0.01, 5.0×10^{-4} (n going from 1–6). In order to solve the problem, we suppose that in the presence of EDTA, the metallic ion concentration is very weak (otherwise, complexation would present no interest!). The ammonia concentration obeys the following relations:

$$[NH_3] + [NH_4{}^+] = 0.10,$$

$$[NH_4{}^+] = 1.78 \times 10^9[H^+][NH_3],$$

$$[NH_3] = 0.10/(1 + 1.78 \times 10^9[H^+]).$$

(We must note that this expression is legitimate only if the ammine complexes' concentrations are negligible in the mass balance relation on the buffer.) Developing the stability constants' expressions as a function of the concentration of magnesium ions gives

$$\alpha_M = 1 + 1.70[NH_3] + 1.20[NH_3]^2 + \cdots + 3.98 \times 10^{-12}/[H^+].$$

Replacing $[NH_3]$ by its above expression as a function of $[H^+]$ permits us to draw α_M as a function of pH (Fig. 26.2). The figure also mentions the change in α_M with pH when no ammonia is added. In this case, at high pH values, α_M increases only because of the formation of the hydroxo complex $[Mg(OH)^+]$. The coefficient α_Y is given by the expression given in Exercise 1. In the pH range where the buffer is efficient,

$$\alpha_{MY} = 1;$$

i.e., $[MY'] = [MY]$.

As a result, we obtain

$$K' = K/\alpha_M\alpha_Y.$$

Evidently, K' is a complicated function of pH. At pH = 9, for example, the conditional constant is about 200 times weaker than the thermodynamic one.

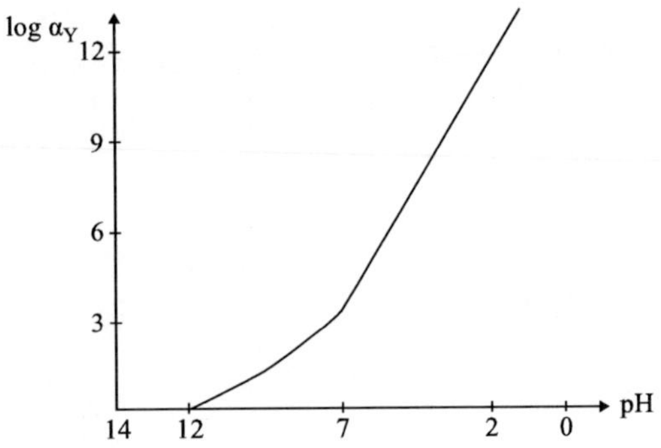

Fig. 26.3 Diagram α_Y/pH ($Y \equiv$ EDTA)

26.4 Quantitative Changes in Coefficients α

The important point to notice is that the α coefficients may change considerably with experimental conditions. This proves, in some manner, the vanity of the prediction of the extent of complexation reactions according to the only consideration of thermodynamic constants.

Concerning the α_L coefficients (devoted to the ligands), they usually are introduced to take their basic character into account. Hence, they vary with the pH value. EDTA is perhaps the most important example of this kind of behavior. The expression that permits the calculation of its α_Y coefficient is given in Exercise 1.The diagram α_Y/pH is calculated and drawn once and for all (Fig. 26.3).

Through this example, we see that the α_Y coefficients may considerably change with the pH value. For example, at pH $= 2.3$, $\log \alpha_Y = 12$, and at pH $= 12$, $\log \alpha_Y = 0$. Diagrams of this sort, but fitting some other basic ligands such as oxalate, acetate, fluoride, etc., have been drawn definitively. A particularly important case of such ligands for chemical analysis is that of metal ion indicators (see Sect. 26.7).

Concerning the metal cations, their coefficients, α_M, are usually introduced to take into account the formation of complexes with the buffer, in particular ammonia, and also to take their hydrolysis into account. In Fig. 26.4, we give $\log \alpha_M$ values for some metallic ions that form ammine complexes. Log α_M is plotted as a function of the analytical concentration of ammonia, the ammine complexes' concentrations being negligible compared to $[NH_3]$.

Again, we notice that changes in α_M may be considerable. Numerous tables giving α_M values for some metal cations and for several amines are published. Of course, when α_M coefficients are introduced to take into account the formation of hydroxo complexes, they depend on the pH value only.

We give α_M values as a function of pH for some metallic ions in Fig. 26.5.

Again, the changes in α_M values may be considerable.

Parasitic reactions such as those of the formation of ammine complexes and those of the formation of hydroxo complexes may occur simultaneously. In this case, a global parasitic reaction coefficient is introduced. It takes both phenomena into account. Actually, this was already done in Exercise 1, in which α_M was expressed

Fig. 26.4 Log α_M as a function of log(NH_3) for some metallic ions

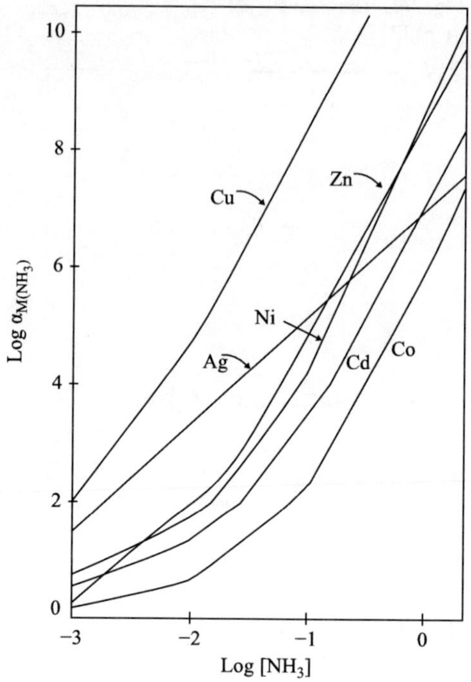

Fig. 26.5 Log α_M values concerning the formation of some hydroxo mononuclear complexes

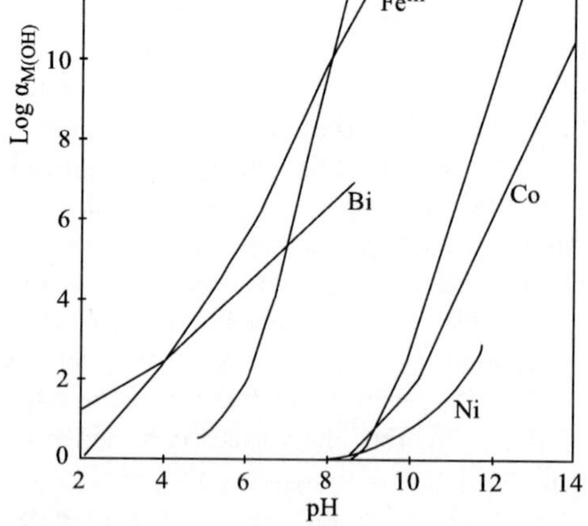

by the relation

$$\alpha_M = 1 + \beta_1{}^{NH3}[NH_3] + \cdots \beta_6{}^{NH3}[NH_3]^6 + \cdots + \beta_1{}^{OH}[OH^-] + \cdots + \beta_3{}^{OH}[OH^-]^3.$$

We can, of course, define two α_M coefficients, one relative to the ammine complexes and the other to hydroxo complexes; that is,

$$\alpha_M{}^{NH3} = 1 + \beta_1{}^{NH3}[NH_3] + \cdots + \beta_6{}^{NH3}[NH_3]^6,$$

$$\alpha_M{}^{OH} = 1 + \beta_1{}^{OH}[OH^-] + \cdots + \beta_3{}^{OH}[OH^-]^3.$$

To be rigorous from a mathematical standpoint, the global coefficient α_M is not equal to the sum of partial coefficients:

$$\alpha_M{}^{global} = \alpha_M{}^{OH} + \alpha_M{}^{NH3}$$

because of the fact that the number 1 occurs twice in the global coefficient. To be rigorous, we must write

$$\alpha_M{}^{global} = \alpha_M{}^{NH3} + \alpha_M{}^{OH} - 1,$$

and after generalization,

$$\alpha_M{}^{global} = \alpha_M{}^{L1} + \alpha_M{}^{L2} + \cdots + (1 - p),$$

where p is the number of types of parasitic reactions (in Exercise 2; formations of ammine and hydroxo complexes) and where L_1 and L_2 are the ligands giving rise to parasitic reactions. Actually, the difference $(1 - p)$ is often very weak compared to the terms $\alpha_M{}^{L1}$ and $\alpha_M{}^{L2}$ and is negligible. Thus, we can write

$$\alpha_M{}^{global} \approx \alpha_M{}^{L1} + \alpha_M{}^{L2} + \cdots.$$

In brief, we must bear in mind that conditional constants may differ considerably from thermodynamic ones. Moreover, they can also exhibit values that may differ considerably among experiments depending on the experimental conditions.

26.5 Conditional Constants, Masking, and Selective Complexations in the Presence of Several Metallic Ions

A parasitic reaction occurs because in the experimental conditions, its conditional stability constant value allows its development. The converse is very interesting in analysis. Indeed, one can conceive that in some experimental conditions, the conditional stability constant value is too weak for the reaction it governs to noticeably occur. For example, in very alkaline medium, zinc ions give zincate ions, which inhibit their reaction with EDTA. Hence, there is a possibility here of masking ions to their reagents. These ions are said to be *masked*. The selectivity that can be achieved in a titration with EDTA is based on the same principle. These notions will be extended in Chapters 27–29, which are devoted to complexometry.

26.6 Conditional Constants and Calculation of the Concentrations of the Different Species in Solution

Knowledge of the conditional constants permits the calculation of the apparent concentrations of species in solution once equilibria are reached. With these constants, the calculation is carried out as if only one reaction (that under study) were occurring. Conditional constants indeed take all the parasitic phenomena into account. An interesting point to notice is that once the apparent concentrations are known, it is easy to accede to the actual concentrations.

Exercise 4 A solution containing 1.0×10^{-2} mol/L magnesium ions and 2.0×10^{-2} mol/L H_4EDTA is buffered at pH $= 10$ by ammonia buffer at the analytical concentration 10^{-1} mol/L. Calculate the apparent concentration of uncomplexed Mg^{2+} and then its actual concentration. The conditional constant Mg-EDTA is $K' = 1.45 \times 10^8$.

According to the definition of the conditional constants, we can write

$$[MY']/[M'][Y'] = 1.45 \times 10^8,$$

$$[Y'] + [MY'] = 2.0 \times 10^{-2},$$

$$[M'] + [MY'] = 1.0 \times 10^{-2}.$$

The last two relations are the mass balance ones. (It is interesting to note that here we are faced with the simple scheme of concentration calculation with the system already encountered of three equations with three unknowns. This is the scheme encountered when the complexation under study is the only reaction occurring.)

Instead of solving the second-order equation resulting from the preceding system, we can consider that

$$[MY'] \approx 1.0 \times 10^{-2}$$

since the K' value is high and since EDTA is in excess. For the same reasons, we can consider that

$$[Y'] \approx 1.0 \times 10^{-2}.$$

Inserting both numerical values into the expression of K' gives $[M'] = 6.9 \times 10^{-9}$ mol/L. It is easy to check that $\alpha_M = 1.19$ at pH $= 10$ (see Exercise 3). As a result,

$$[Mg^{2+}] = [M']/1.19,$$

$$[Mg^{2+}] = 5.8 \times 10^{-9} \text{ mol/L}.$$

Exercise 5 The value of the conditional stability constant of the Mg–EDTA complex is 2.4×10^7 at pH $= 9$ in a solution buffered with the couple $NH_4{}^+/NH_3$. Calculate the fraction of complexed magnesium with EDTA when we mix 100 ml of a buffered solution of 10^{-2} mol/L magnesium ions with 100 ml of a 10^{-2} mol/L EDTA solution.

The first point to notice is that after the mixing of both solutions, the pH value has not changed since the pH of a buffered solution does not vary with the dilution (except in particular conditions of concentrations). When the equilibria are reached, we can write

$$[\mathrm{MY}] + [\mathrm{M}'] = 5 \times 10^{-3},$$

$$[\mathrm{MY}] + [\mathrm{Y}'] = 5 \times 10^{-3},$$

$$[\mathrm{MY}]/[\mathrm{M}'][\mathrm{Y}'] = 2.40 \times 10^{7},$$

$$[\mathrm{M}'] = [\mathrm{Y}'].$$

As a result,

$$[\mathrm{MY}] = 4.99 \times 10^{-3}\ \mathrm{mol/L}.$$

The fraction of magnesium complexed with EDTA is

$$(4.99 \times 10^{-3}/5.00 \times 10^{-3}) \times 100 = 99.7\ \text{percent}.$$

This result shows that the concentration of free magnesium is very low. Hence, it is possible to make the concentration of a metallic ion not only constant but also very weak by carrying out a mixture in judicious proportions of a metallic ion and a ligand that complexes it strongly, such as EDTA^{4-}. This is the origin of the concept of a metal buffer.

Exercise 6 A solution contains 1.0×10^{-2} mol/L of Mg^{2+} and 2.0×10^{-2} mol/L of H_4EDTA. It is buffered at pH = 10 by the buffer ammonia at the analytical concentration 10^{-1} mol/L. Calculate the concentration of free magnesium at equilibrium. In these conditions, $K' = 1.45 \times 10^{8}$ and $\alpha_M = 1.19$; repeat the same problem with a concentration of Mg^{2+} equal to 1.40×10^{-2} mol/L, all other parameters keeping the previous values.

By using the same equations as in the preceding exercise, we calculate in the first case

$$[\mathrm{MY}'] = 1.0 \times 10^{-2},\ [\mathrm{Y}'] = 1.0 \times 10^{-2},\ [\mathrm{M}'] = 6.9 \times 10^{-9}.$$

$$[\mathrm{Mg}^{2+}] = [\mathrm{M}']/\alpha_{\mathrm{M}},$$

$$[\mathrm{Mg}^{2+}] = 5.8 \times 10^{-9}\ \mathrm{mol/L}.$$

The remaining magnesium at the concentration ($10^{-2} - 5.8 \times 10^{-9}$ mol/L) is engaged in complexes formed with EDTA and with ammonia. In the second case, we calculate

$$[\mathrm{Mg}^{2+}] = 1.3 \times 10^{-10}\ \mathrm{mol/L}.$$

We see that the presence of an excess of EDTA makes the $[Mg^{2+}]$ concentration become a weak value that doesn't change much. Hence, the concept of a solution buffered in a metallic ion appears here. This sort of buffering phenomenon is quite comparable to that encountered in acid–base phenomena. Henderson's equation is

$$\mathrm{pH} = pK_a + \log([\mathrm{A}^-]/[\mathrm{HA}]).$$

In the case of complexes with EDTA, we can write

$$[\mathrm{M}'][\mathrm{Y}']/[\mathrm{MY}'] = 1/K'$$

or $\log\,[\mathrm{M}'] = pK' - \log([\mathrm{Y}']/[\mathrm{MY}'])$.

We see that for $[\mathrm{Y}']$ and $[\mathrm{MY}']$ concentrations of the same order, $\log(\mathrm{M}')$ differs little from the pK' value.

The concept of a metal buffer is important from a practical standpoint. It is used, in practice, to complex traces of $Fe^{(II)}$ in some solutions, $Fe^{(II)}$ being able to be the initiator of undesirable redox phenomena. This is the reason why EDTA is added in some buffers.

It is even possible to define a conditional solubility product by the expression

$$K_s' = [\mathrm{M}'][\mathrm{L}'].$$

It can easily be generalized to the case in which the metallic ion and the ligand do not bring the same electrical charge.

Exercise 7 Ni^{2+} gives a scarlet complex with dimethylglyoxime whose structure may be symbolized by the formula NiA_2, where A represents the dimethylglyoximate rest as its anionic form (see Chap. 30). The complex is precipitated from a solution buffered with the couple NH_4^+/NH_3. Express the conditional solubility product of the complex as a function of the pH of the buffer solution.

By definition,

$$K_s' = [\mathrm{Ni}'][\mathrm{A}']^2,$$

$$[\mathrm{Ni}'] = [\mathrm{Ni}^{2+}] + [\mathrm{Ni(NH_3)}^{2+}]$$

$$+\cdots[\mathrm{Ni(NH_3)_6}^{2+}] + \cdots[\mathrm{NiOH}^+] + \cdots + [\mathrm{Ni(OH)_3}^-]\ldots,$$

$$[\mathrm{A}^{-\prime}] = [\mathrm{A}^-] + [\mathrm{HA}].$$

HA is the conjugate acid of the dimethylglyoximate rest. By using the α_{Ni} coefficient, which is a function of the pH through the varied stability constants (see Exercise 1), and by introducing the coefficient α_{L} such that

$$\alpha_{\mathrm{L}} = 1 + [\mathrm{H}^+]/K_{\mathrm{a}} \quad \text{or} \quad \alpha_{\mathrm{L}} = ([\mathrm{A}^-] + [\mathrm{HA}])/[\mathrm{A}^-],$$

where K_a is the acid dissociation constant of dimethylglyoxime, it is easy to find the expression

$$K_s' = K_s\alpha_{\mathrm{M}}(\alpha_{\mathrm{L}})^2,$$

where K_s is the thermodynamic solubility product. K'_s is a complicated function of pH.

An interesting point developed in this exercise is the introduction of the particular coefficient α_{L}. This coefficient may be considered an extension of the other coefficients already encountered (see Sect. 26.8).

26.7 Case of Metal Indicators

Metal indicators are derivatives, more precisely dyestuffs, that exhibit a color change when the concentration of the metal cation they indicate changes within a certain concentration range. They form complexes with specific metal cations. In principle, they exhibit two colors depending on whether they are free or engaged in the complex with the metallic ion. They are used to detect endpoints in complexometric titrations.

Like numerous ligands, metal indicators can take up protons. The protons taken up also produce a color change; therefore, these derivatives are equally pH indicators. Of course, free metal indicators, once protonated, have lost a fraction of their complexing properties compared to their basic forms.

For analytical purposes, metal indicators can be studied from two standpoints, which, additionally, are interrelated. The first concerns their ability to complex metallic ions as a function of pH. The second standpoint concerns the stability of metal–indicator complexes compared to that of metal–titrant complexes. The latter standpoint will be studied in Chap. 29.

Using conditional stability constants is a good strategy to orient and control the use of metal indicators as a function of pH. As an example, we now consider the case of eriochrome black T in details. The structure of eriochrome black T in acidic medium is

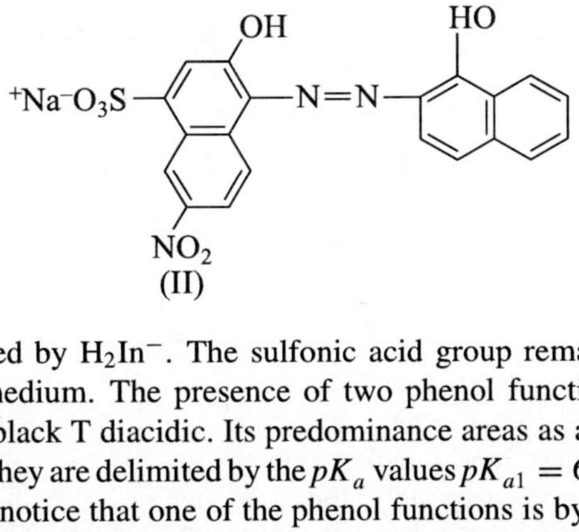

(II)

It can be symbolized by H_2In^-. The sulfonic acid group remains, at least partly, ionized in acidic medium. The presence of two phenol functions in its structure makes eriochrome black T diacidic. Its predominance areas as a function of pH are given in Fig. 26.6. They are delimited by the pK_a values $pK_{a1} = 6.3$ and $pK_{a2} = 11.6$.

Incidentally, we notice that one of the phenol functions is by far easier to ionize than the other, as indicated by the pK_a values. Eriochrome black T is an example of a metal indicator that exhibits properties of both a neutralization indicator and a metal cation indicator.

Drawing the diagrams $\log K'_{\text{MIn}} = p\text{M}$ as a function of the pH value of the solution where K'_{MIn} is the conditional stability constant of the [MIn] complex permits us to rationalize the use of metal indicators. The drawing of these diagrams directly follows from the mathematical expressions K'/pH. Let's investigate the indication of Mg^{2+} ions with eriochrome black T. The formation equilibrium for a metal ion indicator is

$$Mg^{2+} + In^{3-} \rightleftharpoons [MgIn^-].$$

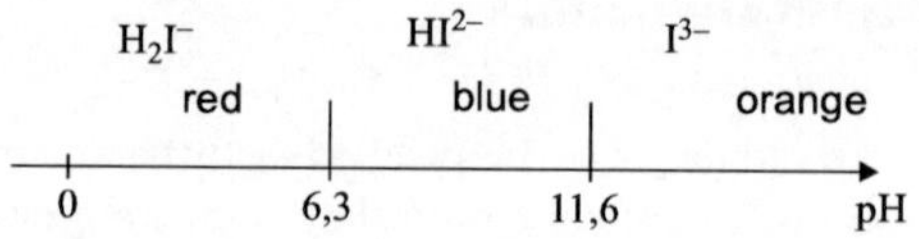

Fig. 26.6 Predominance areas of the protonated forms of uncomplexed eriochrome black T as a function of pH

The thermodynamic stability constant is defined by the expression

$$K = [\text{MgIn}^-]/([\text{Mg}^{2+}][\text{In}^{3-}]), \quad K = 1.0 \times 10^7.$$

According to what we said earlier,

$$[\text{In}'] = [\text{In}^{3-}] + [\text{HIn}^{2-}] + [\text{H}_2\text{In}^-].$$

or by using the expressions of K_{a1} and K_{a2}, we have

$$[\text{In}'] = [\text{In}^{3-}]\left\{1 + 3.5 \times 10^{11}[\text{H}^+] + 7 \times 10^{17}[\text{H}^+]^2\right\},$$

$$\alpha_L = 1 + 3.5 \times 10^{11}[\text{H}^+] + 7 \times 10^{17}[\text{H}^+]^2.$$

Concerning the magnesium ion, we know (see Exercise 3) that

$$[\text{Mg}'] = [\text{Mg}^{2+}] + [\text{Mg(OH)}^+],$$

$$\alpha_M = 1 + 10^{2.58} \times 10^{-14}/[\text{H}^+].$$

α_M plays a part only at high pH values. In every case,

$$K'_{\text{MIn}} = K/(\alpha_L \alpha_M).$$

The obtained diagram is mentioned in Fig. 26.7. It is interesting to notice that $\log K'_{\text{MIn}}$ is equal to—log M $= p$M when $[\text{In}^-] = [\text{MIn}]$, a condition that is required for the best detection of the equivalence point.

From the diagram, we deduce that the color change due to the complexation of the cation with the indicator (blue to red change) or the inverse can only be seen within the range $6.3 < \text{pH} < 11.6$. At pH < 6.3, both forms in equilibrium are H_2In^- and $MgIn^-$. They are both red. At pH > 11.6, magnesium hydroxide may precipitate. The diagram also mentions the precipitation area of magnesium hydroxide. It is delimitated by the straight line calculated from the solubility product expression $K_s = [\text{Mg}^{2+}][\text{OH}^-]^2$ with $K_s = 10^{-10.74}$. We notice that in this pH range, the K' value is located in the range $10^{1.9} < K' < 10^{6.5}$. We shall reconsider these values in Chap. 28 when we study the complexation of metallic ions with EDTA in the presence of metal indicators.

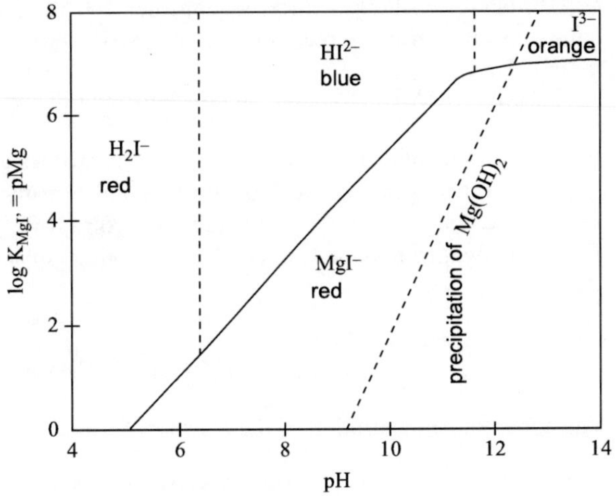

Fig. 26.7 Diagram log K'/pH of the Mg^{2+}–eriochrome black T complex (in dotted-line diagram *p*Mg/ pH) ($[In^-] = [InM]$; see text)

26.8 Extension of the Concept of Conditional Stability Constants

The notion of conditional solubility product is already an extension of the concept of conditional constants. Among other possible extensions, let's also mention the case for which in the studied complex, the metal and the ligand are combined in different ratios. For example, let's consider the case in which successive complexes ML and ML_2 are forming. We'll consider that the complex under study is ML; that is, it is considered the principal complex. After having adopted the same logic as that which prevailed in the adoption of conditional constants, we can write

$$[L'] = [L] + [ML_2].$$

$[L']$ is the ligand for any form except the principal complex. If K_1 and K_2 are the successive formation constants of the complexes ML and ML_2, we can write

$$[L'] = [L] + K_2[L][ML],$$
$$[L'] = [L](1 + K_2[ML]).$$

Hence, $\alpha_L = 1 + K_2[ML]$.

The introduction of this sort of coefficient is less interesting than that of the preceding coefficients. Indeed, we immediately see that α_L depends on the concentration [ML] of the complex under study, a concentration that is unknown. α_L cannot be known, and at first sight, its value appears extremely uncertain, as is, consequently, the value of K'. However, the order of the value [ML] may be sufficient, and it can be estimated as follows. Since we consider that the formation of the ML complex constitutes the principal reaction, we can use the approximation

$$[\mathrm{ML}] \approx \text{analytical concentration of the metallic ion.}$$

It is very often legitimate in the case of complexometric titrations of metal cations.

Another extension is possible in the case of the formation of mixed complexes. We know, for example, that mixed complexes such as [M(H)Y] and [M(OH)Y] in which EDTA is engaged may be formed. The following constants have been introduced:

$$K_{\mathrm{MHY}} = [\mathrm{MHY}]/([\mathrm{MY}][\mathrm{H}^+]),$$

$$K_{\mathrm{M(OH)Y}} = [\mathrm{M(OH)Y}]/([\mathrm{MY}][\mathrm{OH}^-]).$$

The apparent concentration of the complex formed is

$$[\mathrm{MY}'] = [\mathrm{MY}] + [\mathrm{MHY}]$$

or $[\mathrm{MY}'] = [\mathrm{MY}] + [\mathrm{M(OH)Y}]$.

We immediately find

$$[\mathrm{MY}'] = [\mathrm{MY}](1 + K_{\mathrm{MHY}}[\mathrm{H}^+]),$$

$$[\mathrm{MY}'] = [\mathrm{MY}](1 + K_{\mathrm{M(OH)Y}}[\mathrm{OH}^-]),$$

and the corresponding coefficients are

$$\alpha = 1 + K_{\mathrm{MHY}}[\mathrm{H}^+],$$

$$\alpha = 1 + K_{\mathrm{M(OH)Y}}[\mathrm{OH}^-].$$

Other types of mixed complexes also exist. For example, Hg(II) forms the mixed complex $[\mathrm{Hg(NH_3)Y}]^{2-}$ with EDTA. The following coefficient:

$$\alpha = 1 + K_{\mathrm{Hg(NH3)Y}}[\mathrm{NH_3}]$$

is introduced.

26.9 About the Interest in the Concept of Conditional Constants

In order to discuss the interest in the introduction of conditional constants, we shall compare the calculations, performed to calculate species concentrations, carried out with and without the help of conditional constants.

Let's again consider the example of the nickel–EDTA complex in a solution buffered with the couple $NH_4{}^+/NH_3$ (Exercise 1). In the case in which conditional constants are not used, the mass balance equations are

$$C_{Ni} = [NiY^{2-}] + [Ni^{2+}] + [Ni(NH_3)^{2+}] + \cdots + [Ni(OH)^+] + \cdots,$$

$$C_Y = [NiY^{2-}] + [Y^{4-}] + [HY^{3-}] + \cdots,$$

$$C_{NH3} = [NH_4{}^+] + [NH_3] + [Ni(NH_3)^{2+}] + 2[Ni(NH_3)_2{}^{2+}] + \cdots.$$

By introducing the expressions of the stability constants $\beta_1{}^{NH3}$, $\beta_2{}^{NH3}$, ..., $\beta_1{}^{OH}$, and so on, we find

$$C_{Ni} = [NiY^{2-}] + [Ni^{2+}]\left\{1 + \beta_1{}^{NH3}[NH_3] + \cdots + \beta_6{}^{NH3}[NH_3]^6\right\};$$

that is, $C_{Ni} = [NiY^{2-}] + [Ni^{2+}]\alpha_M$ and $C_Y = [NiY^{2-}] + [Y^{4-}]\alpha_Y$.

We observe that the parasitic reaction coefficients introduced themselves spontaneously in the previous calculations. Likewise, we again find the notion that when there is a great excess of the buffer $NH_4{}^+/NH_3$, the analytical concentration of ammonia decreases to

$$C_{NH3} = [NH_3](1 + K_a/[H^+]),$$

the concentrations of the ammine complexes being negligible. The concentration $[NH_3]$ is calculated immediately if we know the pH value of the solution (K_a being the acid dissociation constant of the ammonium ion). Hence, it turns out that conditional constants play the role of an intermediary variable and that the calculations carried out with them are strictly identical to those carried out without them.

As a result of these considerations, we can conclude that the concept of conditional stability constants brings nothing new from a theoretical standpoint. However, the advantage of their introduction is that in cases in which several equilibria occur simultaneously, we may follow reasoning quite analogous to that followed in the far more simple case in which only a single equilibrium is occurring. Their introduction focuses our attention on the reaction we want to study. Furthermore, they exhibit an undeniable practical interest, since numerous tables of conditional stability constants are published. Even if their values are, of course, legitimate only in particular experimental conditions, they let us appreciate the quantitative character of the studied reaction, quickly indeed, immediately in some favorable conditions.

In brief, using conditional stability constants appears to be an elegant way to study the formation of complexes and, moreover, is in accordance with chemical intuition.

Chapter 27
Complexometry I: Mercurimetry (Votocek–Dubsky's Method)

Complexometric titrations, namely, titrations based on the formation of complexes, occupied a minor place in the realm of titrations until the 1940s, except for the Liebig–Denigés and Votocek–Dubsky methods. Since 1940, the advent of complexones, particularly that of EDTA, has instigated very important progress in the field of titrimetric analysis based on the formation of complexes. This chapter is essentially devoted to Votocek–Dubsky's method, which is mercurimetric. (Titrations with EDTA and complexones are studied in Chap. 28. Liebig–Denigés' method is studied in the chapter devoted to argentimetric processes since it involves not only a complexation but also a precipitation; see Chaps. 36 and 37.)

We begin by discussing the most important reason why complexometric titrations were weakly developed before the advent of EDTA and various other complexones.

27.1 The Major Difficulty Encountered During Complexometric Titrations

The major difficulty encountered during complexometric titrations is the simultaneous formation of several complexes according to different stoichiometries with the same metallic ion and the same ligand. To be convinced of it, it is sufficient to examine Table 24.1 of Chap. 24, which mentions some cases of such complexations. The result of these cases is the existence of uncertain equivalence points that do not correspond to a well-defined stoichiometry.

Exercise 1 A titration method of iodide ions with mercuric nitrate has been proposed. According to it, the equivalence point might be indicated by the appearance of the red precipitate of mercuric iodide. Let's consider the titration of 100 ml of a 0.1 mol/L sodium iodide solution with a 0.1 mol/L mercuric nitrate solution. What is the volume added at the equivalence point in these conditions? What conclusion can we make?

Recall that mercuric nitrate is totally ionized in water. Let's also recall (see Chap. 2 and Part V) that when the precipitation of an ionizing product occurs, its solubility product is satisfied. In the present case, $K_s(HgI_2) = (Hg^{2+})(I^-)^2 = 10^{-27.7}$. Finally,

J.-L. Burgot, *Ionic Equilibria in Analytical Chemistry*,
DOI 10.1007/978-1-4419-8382-4_27, © Springer Science+Business Media, LLC 2012

recall that the complexes that may form in these conditions are $[HgI]^+$ ($K_1 = 10^{12.87}$), $[HgI_2]$ ($K_2 = 10^{10.95}$), $[HgI_3]^-$ ($K_3 = 10^{3.78}$), and $[HgI_4]^{2-}$ ($K_4 = 10^{2.23}$), where the K_j are the stepwise formation constants.

This exercise is solved by writing the mass balance relations on the Hg^{2+} and I^- ions (among others that are also satisfied—see below):

$$\frac{100 \times 0.1}{(100 + V)} = [I^-] + [HgI^+] + 2[HgI_2] + 3[HgI_3{}^-] + 4[HgI_4{}^{2-}],$$

$$\frac{0.1 \times V}{(100 + V)} = [Hg^{2+}] + [HgI^+] + [HgI_2] + [HgI_3{}^-] + [HgI_4{}^{2-}].$$

We notice that in these relations, no mention of the "precipitated concentration" $[HgI_2\downarrow]$ has been made since at the very beginning of the precipitation, it can be considered as being null even if, at that point, the solubility product is already satisfied. Because there is an excess of iodide ions until the final point, it appears reasonable to neglect the $[Hg^{2+}]$ and $[HgI^+]$ concentrations. The predominant species are the superior complexes. The previous equations become

$$\frac{100 \times 0.1}{(100 + V)} = [I^-] + 2[HgI_2] + 3[HgI_3{}^-] + 4[HgI_4{}^{2-}], \qquad (27.1)$$

$$\frac{V \times 0.1}{(100 + V)} = [HgI_2] + [HgI_3{}^-] + [HgI_4{}^{2-}]. \qquad (27.2)$$

By combining relations (27.1) and (27.2), we find

$$\frac{(100 - 2V)}{(100 + V)} = 10\left\{[I^-] + [HgI_3{}^-] + 2[HgI_4{}^{2-}]\right\}, \qquad (27.1')$$

$$\frac{V}{(100 + V)} = 10\left\{[HgI_2] + [HgI_3{}^-] + [HgI_4{}^{2-}]\right\}. \qquad (27.2')$$

The expressions K_1, K_2, K_3, and K_4 lead to the relations

$$[HgI^+] = 10^{12.87}[Hg^{2+}][I^-],$$

$$[HgI_2] = 10^{10.95} \times 10^{12.87}[Hg^{2+}][I^-]^2.$$

At the final point, the solubility product is satisfied:

$$[Hg^{2+}][I^-]^2 = 10^{-27.7};$$

as a result, we have

$$[HgI_2] = 10^{-3.88}\text{mol/L}.$$

(This is the concentration of dissolved HgI_2.) Therefore, we find

$$[HgI_3{}^-] = 10^{3.78} \times 10^{-3.88}[I^-]$$

and

$$[HgI_4{}^{2-}] = 10^{-2.13}[I^-]^2.$$

Equations (27.1′) and (27.2′) become

$$\frac{(100-2V)}{(100+V)} = 10\{[I^-] + 10^{-0.10}[I^-] + 2.10^{2.13}[I^-]^2\},$$

$$\frac{V}{(100+V)} = 10\{10^{-3.88} + 10^{-0.10}[I^-] + 10^{2.13}[I^-]^2\}.$$

Solving this system of two equations in two unknowns gives $V = 24$ ml and $[I^-] = 9.6 \times 10^{-3}$ mol/L. The concentrations of the other species are $[HgI_3{}^-] = 7.63 \times 10^{-3}$mol/L; $[HgI_4{}^{2-}] = 1.24 \times 10^{-2}$mol/L; $[HgI_2] = 1.32 \times 10^{-4}$mol/L; $[HgI^+] = 1.54 \times 10^{-13}$mol/L; $[Hg^{2+}] = 2.16 \times 10^{-2}$mol/L. The initial simplifying assumptions were good.

The important conclusion that results from this calculation is that if we adopt the beginning of the precipitation of HgI_2 as being the final point, the latter is by far too early. It occurs at $V = 24$ ml, whereas the equivalence point is located at $V = 50$ ml (see the reaction stoichiometry and the initial data). This method is absolutely not suitable for the titration of iodide ions. It appears that the premature endpoint indication is due to the fact that the following complexes, $[HgI_2]$, $[HgI_3]^-$, and $[HgI_4]^{2-}$, are forming quasi-simultaneously in nearly equal concentrations. The contrast existing between this proposed method and that of Votocek–Dubsky, which is also a mercuric titration (see below), is striking. However, it is important to already notice that the means to detect the final point used in both methods are different.

We'd like to point out that the possible hydrolysis of mercuric ions has not been taken into account in the above calculations. Additionally, this explains the fact that the charge balance relation has not been used. (Quite generally, the charge balance relation is not necessary to solve the mathematical systems governing the occurring phenomena if no acido-basic phenomenon occurs.) Taking the hydrolysis phenomenon in the above example into account might have no noticeable influence on the resulting values, but the calculations would have been seriously more complicated. Let's also notice, finally, that setting up the charge balance relation and then neglecting the $[H^+]$, $[OH^-]$, $[Hg^{2+}]$, and $[HgI^+]$ concentrations leads to relation (27.1′) already obtained in another way.

27.2 Mercurimetry: Votocek–Dubsky's Method

Mercurimetry groups all the titrimetric methods based on the use of titrant solutions consisting of ionized mercuric salts. The most important one is that due to Votocek and Dubsky. It is the sole method studied here.

The Votocek–Dubsky method is essentially a method of determining some halide ions by titration with a mercuric salt.

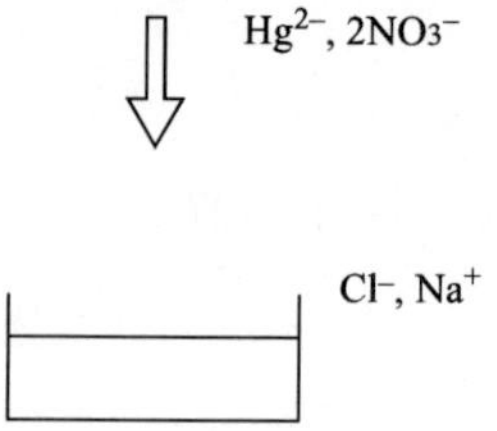

Fig. 27.1 Principle of Votocek's methode

27.2.1 Principle

Votocek–Dubsky's method is based on the property exhibited by some mercuric "salts" to be very poorly ionized in solution because they actually are complexes. This property is shared by halide, cyanide, and thiocyanate ions. Oxygenated salts, however, are normally ionized (nitrate, sulfate). From another standpoint, sodium nitroprusside forms a white precipitate of mercuric nitroprusside with Hg^{2+} ions, that is, with oxygenated mercuric salts but not with halides or parent compounds because of their property of being complexes. The principle of Votocek's method results from these considerations (see Fig. 27.1).

When we progressively add a solution of a normally ionized mercuric salt (nitrate or sulfate) to a solution of sodium chloride, for example, the mercuric ions disappear with equivalent quantities of chloride ions as they give the complexes $[HgCl_4]^{2-}$, $[HgCl_3]^-$, $[HgCl_2]$, and $[HgCl]^+$. Close to the equivalence point, it is essentially $HgCl_2$ that is formed (see below). After the equivalence point, free Hg^{2+} ions appear suddenly in the titrand solution. They are in sufficient quantity to be detected either

- by a turbidity indicator. They react with sodium nitroprusside according to the reaction

$$[Fe(CN)_5NO]^{2-} + Hg^{2+} \rightleftharpoons Hg[Fe(CN)_5NO]\downarrow$$

  This is Votocek's method;
- with colored indicators:

  - diphenylcarbazid

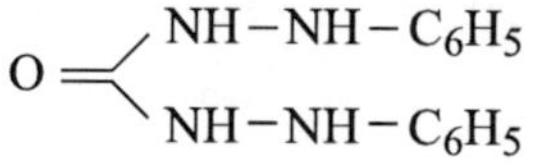

    This is Dubsky's method;
  - diphenylcarbazone

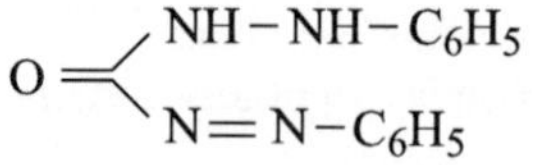

which, according to some authors, might give a violet chelate with Hg^{2+} ions, according to the reaction

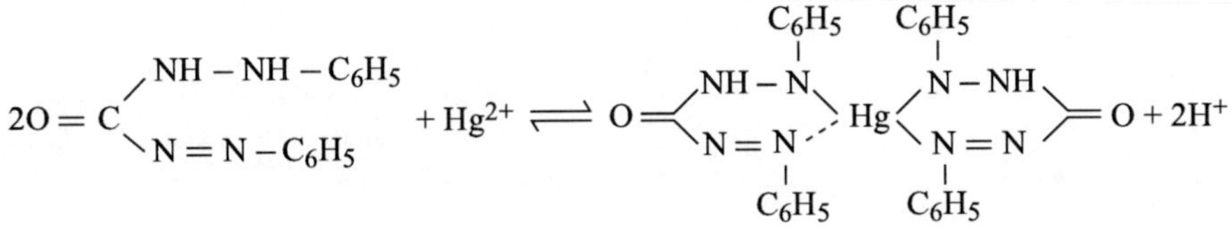

This is Schales' modification. Actually, in Dubsky's method, it would also be diphenylcarbazone that would give the same chelate as that obtained in Schales' modification. The same diphenylcarbazone would form *in situ* by oxidization of diphenylcarbazide by mercuric ions. For some other authors, diphenylcarbazide would simply be an acid–base indicator.

27.2.2 Equivalence Point

The stepwise formation constants of the complexes $[HgCl]^+$, $[HgCl_2]$, $[HgCl_3]^-$, and $[HgCl_4]^{2-}$ (respectively; $K_1 = 10^{6.74}$, $K_2 = 10^{6.48}$, $K_3 = 10^{0.85}$, $K_4 = 10^{1.00}$) and the value of the solubility product of mercuric nitroprusside ($K_s = 2.5 \times 10^{-9}$) are sufficient to theoretically justify Votocek's methode.

The difficulty is to understand why the only species to be quasi-exclusively formed near the equivalent point is $[HgCl_2]$.

Qualitatively speaking, we can say that

- at the beginning of the titration, due to the great excess of chloride ions, there is successively the formation of $[HgCl_4]^{2-}$, $[HgCl_3]^-$, and $[HgCl_2]$, which exhibit overall formation constants of nearly equal values: $\beta_4 = 10^{15.07}, \beta_3 = 10^{14.07}$, $\beta_2 = 10^{13.22}$;
- due to the high values of these overall formation constants and to the fact that $[HgCl_2]$ only exists in solution, few mercuric ions remain in solution. The solubility product of mercuric nitroprusside is not reached. It does not precipitate;
- after the formation of $[HgCl_2]$ and by a further addition of the ionized mercuric salt, there is formation of the complex $[HgCl]^+$, whose formation constant $\beta_1 = K_1 = 10^{6.7}$ is considerably lower than the three preceding ones. As a result, the $[Hg^{2+}]$ concentration becomes sufficient for mercuric nitroprusside to precipitate or for the formation of the colored chelate. Briefly, Hg^{2+} ions can be detected when $[HgCl]^+$ appears in solution;
- finally, there is only one equivalence point because of the fact that the β_4, β_3, and β_2 values are too close to each other to permit the detection of supplementary marked changes in the $[Hg^{2+}]$ concentration before the equivalence point.

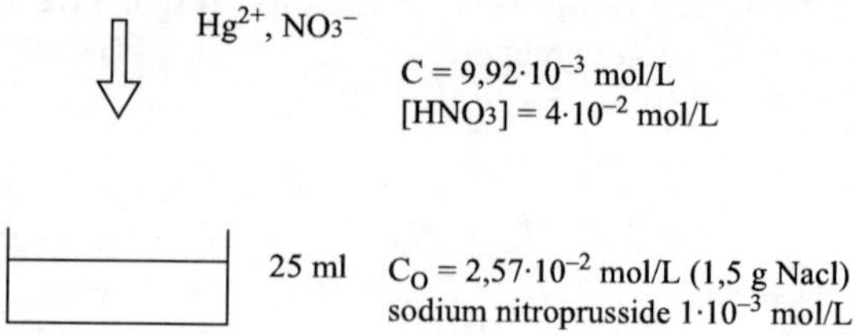

Fig. 27.2 Experimental parameters of the studied Votocek titration

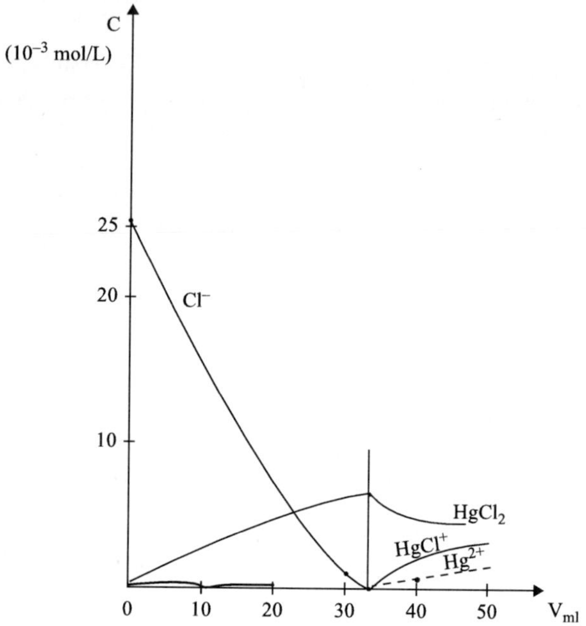

Fig. 27.3 Diagram of the concentrations of the different species as a function of the added volume of a 9.92×10^{-3} mol/L mercuric nitrate solution. The titrant solution contained 4×10^{-2} mol/L of nitric acid, and the titrand solution 2.57×10^{-2} mol/L NaCl and 1×10^{-3} mol/L sodium nitroprusside

In summary, the formation constant values β_1, β_2, β_3, and β_4 are such that they induce the presence of only one complex at the equivalence point, $[HgCl_2]$. Of course, this is purely accidental.

These qualitative considerations can be justified quantitatively after resolution of the mathematical system of the different equations that simultaneously govern the equilibria occurring in solution for each titration point. Figure 27.3 gives the profile of the calculated concentrations for a titration whose experimental parameters are mentioned in Fig. 27.2.

First, the titrant solution contains nitric acid (4×10^{-2} mol/L) in order to avoid the precipitation of mercuric ions by hydrolysis (see Chap. 25). The equivalence point must be located for the added volume of titrant $V_{eqp} = 32.33$ ml.

An examination of Fig. 27.3 shows that at the equivalence point, effectively

- the species $[HgCl_2]$ is the only one present. Other species, $[HgCl_3]^-$ and $[HgCl_4]^{2-}$, which were formed together with $[HgCl_2]$ just before, exhibit a quasi-null concentration;

- $[HgCl]^+$ appears only after the equivalence point, together with Hg^{2+}, whereas $[HgCl_2]$ begins to decrease.

At the equivalence point, $[Hg^{2+}] = 2.5 \times 10^{-6}$ mol/L. The distribution diagram of the complexes $[HgCl_4]^{2-}$, $[HgCl_3]^-$, $[HgCl_2]$, and $[HgCl]^+$ as a function of pCl is in accordance with these results (see Fig. 25.2 of Chap. 25). Indeed, the calculations evoked above give $[Cl^-] = 7.9 \times 10^{-6}$ mol/L at the equivalence point, or equivalently, $p\text{Cl} = 5.10$. For $p\text{Cl} = 5.10$, the distribution diagram gives $\alpha_{HgCl_2} \approx 1$. Calculation also shows that the final point, the point at which sodium nitroprusside precipitates, has been located for $V_{fp} = 33.10$ml. Hence, the relative titration error is

$$\frac{(V_{pf} - V_{eqp})}{V_{eqp}} \times 100 = +2.4\%.$$

The titrant solution exhibits the value pH = 1.40, whereas for the titrand solution, the value is 2.32 at the beginning of the titration. It decreases very rapidly to as low a value as 1.61, which remains constant after.

Exercise 2 Set up the principle of calculating the different species' concentrations during the previous titration.

The principle of the calculation is based on the following relations:

$$\frac{[HgOH^+]}{\{[Hg^{2+}][OH^-]\}} = 10^{10.30}; \qquad \frac{[Hg(OH)_2]}{\{[Hg(OH)^+][OH^-]\}} = 10^{11.40};$$

These constants are the hydrolysis constants. (Actually, hydrolysis is negligible because of the medium's acidity.) Other satisfied relations are

$$\frac{[HgCl^+]}{\{[Hg^{2+}][Cl^-]\}} = 10^{-6,74}; \qquad \frac{[HgCl_2]}{\{[HgCl^+][Cl^-]\}} = 10^{6.48};$$

$$\frac{[HgCl_3{}^-]}{\{[HgCl_2][Cl^-]\}} = 10^{0.85}; \qquad \frac{[HgCl_4{}^{2-}]}{\{[HgCl_3{}^-][Cl^-]\}} = 10^{1}.$$

The mass balances are (V: added volume of titrant solution at the titration point that is considered)

$$[Fe(CN)_5NO^{2-}] = \left(\frac{1 \times 10^{-3} \times 25}{25 + V}\right),$$

$$[Na^+] = \frac{(2.567 \times 10^{-2} \times 25)}{(25 + V)} + \frac{(2 \times 10^{-3} \times 25)}{(25 + V)}$$ (sodium nitroprusside brings two sodium ions),

$$[NO_3{}^-] = \left[\frac{(29.924 \times 10^{-3}V)}{(25 + V)}\right] + \left[\frac{(4 \times 10^{-2}V)}{(25 + V)}\right]$$ (do not forget nitric acid!).

The charge balance relation is

$$[Cl^-] + [HgCl_3^-] + 2[HgCl_4^{2-}] + [OH^-] + 2[\text{nitroprusside}] + [NO_3^-] = [H^+] + [Na^+] + 2[Hg^{2+}][HgCl^+] + [Hg(OH)^+]$$

The mass balance relations on chloride and mercuric ions are

$$[Cl^-] + 4[HgCl_4^{2-}] + 3[HgCl_3^-] + 2[HgCl_2] + [HgCl^+] = \frac{(2.567 \times 10^{-2} \times 25)}{(25 + V)} = X$$

$$[Hg^{2+}] + [HgCl^+] + [HgCl_2] + [HgCl_3^-] + [HgCl_4^{2-}] + [Hg(OH)^+] + [Hg(OH)_2] = \frac{(9.924 \times 10^{-3} V)}{(25 + V)} = Y.$$

This system of equations can be reduced to the fifth-order equation on $[Cl^-]$, which is solved by iterations for each titration point:

$$\beta_4[Cl^-]^5 + [\beta_3 + \beta_4(4Y - X)][Cl^-]^4 + [\beta_2 + \beta_3(3Y - X)][Cl^-]^3 + [\beta_1 + \beta_2\,(2Y - X)][Cl^-]^2 + [1 + \beta_1(Y - X)][Cl^-] - X = 0$$

This equation does not take the hydrolysis of Hg^{2+} into account, which is actually negligible. The titrant's pH value is calculated by using a charge balance relation in addition to those that govern the phenomena in solution. Its value is pH = 1.40.

27.2.3 Standard Solutions

Mercury-II nitrate and sulfate are used. They are prepared

- from mercury metal;
- from yellow mercury oxide HgO by dissolving it into concentrated nitric or sulfuric acid solutions. After dissolution, the obtained solution is diluted with distilled water and then titrated with standard solutions of alkaline halides.

(Mercuric nitrate salt is too difficult to handle because it is very hygroscopic.)

It is imperative to carry out titrations comparatively to a standard solution of sodium chloride for the reasons evoked above (simultaneous formation of several complexes, which makes the reaction not perfectly stoichiometric). As is the case in argentimetric processes (see Part V), the equivalent corresponds to the halide ion and, as a result, is half the molar weight of mercury, that is, Eq = 100.3 g of Hg^{2+} in 1 L of solution (Table 27.1).

Table 27.1 Calculated concentrations of some species during the course of a titration (unit: 10^{-3} mol/L)

	Cl^-	$HgCl_2$	$HgCl_3^-$	$HgCl_4^{2-}$	$HgCl^+$	Hg^{2+}
0 ml						
10 ml	12.38	2.58	0.23	0.028	–	–
20 ml	5.27	4.24	0.16	0.0083	–	–
30 ml	0.81	5.38	0.031	0.00025	–	–
32.33 ml	0.004	5.55	0.0016	0	0.043	0
40 ml	–	4.11	ε	0	1.6	0.36
50 ml	–	3.04	ε	0	2.47	1.1

27.2.4 Applications

Votocek–Dubsky's method is useful when argentometry can no longer be used. It permits us to determinate 200 μg of chloride ions with a ±5% precision. This is a very sensitive method. It has been systematically used for penurious reasons of different origins, for example, that of silver. Furthermore, mercurimetry has been called the "argentometry of poor people."

It permits the determination of Br^-, CN^-, and SCN^- ions. F^- ions cannot be determined in this way. It cannot be used to determine iodide ions because of the formation of the red precipitate of HgI_2 that begins too early (see the exercise at the beginning of the chapter). It gives good results in clinical chemistry for the determination of chloride ions in urine and in the cephalo-rachidian liquid, provided albumin and protides have been eliminated previously by defecation. Finally, let's recall that the titrant solution must be acidified to avoid the mercuric oxide precipitation after hydrolysis.

Later, we shall give further comments concerning mercurimetry when we speak of the formation of some complexes in organic analysis. Penicillins can indeed be titrated by mercurimetry.

Chapter 28
Complexometry II: Titrations with EDTA

In 1945, Schwarzenbach et al. initiated fundamental studies concerning the chelating properties of polyamino carboxylic acids in order to improve their interesting analytical properties. He named these new derivatives *complexones*. These studies followed previous ones, begun in 1939, which had led to the discovery that these compounds did form stable and water-soluble complexes with plenty of metallic ions. These derivatives were marketed during the 1930s by I. G. Farben Industie. Among other complexones, ethylenediaminetetraacetic acid (H_4EDTA) and its derivatives are probably the most commonly used reagents in complexometry. Their success originates from their chelating properties and also from the fact they give water-soluble complexes quasi-instantaneously. These attributes are required properties in titrimetry.

28.1 Some Properties of EDTA

(1,2-Ethanediyldinitrilo)tetraacetic acid or ethylenediaminetetraacetic acid of the structure

$$\begin{matrix} HOOC-CH_2 \\ HOOC-CH_2 \end{matrix} \!\!> N-CH_2-CH_2-N <\!\! \begin{matrix} CH_2COOH \\ CH_2COOH \end{matrix}$$

must be symbolized as H_4EDTA according to the IUPAC. The symbolism recalls that it is a tetraacid. Actually, in the literature, the symbolism EDTA involves either its more or less protonated forms or the unprotonated one. The literature also frequently uses the symbolisms H_4Y, H_3Y^-, H_2Y^{2-}, HY^{3-}, and Y^{4-}. We shall indiscriminately use the symbols EDTA, YH_4, $YH_3{}^-$, and so on.

J.-L. Burgot, *Ionic Equilibria in Analytical Chemistry*,
DOI 10.1007/978-1-4419-8382-4_28, © Springer Science+Business Media, LLC 2012

28.1.1 Acid Dissociation Constants of EDTA

H_4EDTA is a tetraacid (see also Chap. 25). Its pK_a values are $pK_{a1} = 1.99$, $pK_{a2} = 2.67$, $pK_{a3} = 6.16$, and $pK_{a4} = 10.26$. They are defined by the following expressions:

$$H_4Y \overset{K_{a_1}}{\rightleftharpoons} H_3Y^- + H^+ \qquad K_{a_1} = \frac{(H_3Y^-)(H^+)}{(H_4Y)} \qquad pK_{a_1} = 1{,}99$$

$$H_3Y^- \overset{K_{a_2}}{\rightleftharpoons} H_2Y^{2-} + H^+ \qquad K_{a_2} = \frac{(H_2Y^{2-})(H^+)}{(H_3Y^-)} \qquad pK_{a_2} = 2{,}67$$

$$H_2Y^{2-} \overset{K_{a_3}}{\rightleftharpoons} HY^{3-} + H^+ \qquad K_{a_3} = \frac{(HY^{3-})(H^+)}{(H_2Y^{2-})} \qquad pK_{a_3} = 6{,}16$$

$$HY^{3-} \overset{K_{a_4}}{\rightleftharpoons} HY^{3-} + H^+ \qquad K_{a_4} = \frac{(Y^{4-})(H^+)}{(HY^{3-})} \qquad pK_{a_4} = 10{,}26$$

These values are somewhat abnormal. We know from statistical data that aliphatic carboxylic acids exhibit pK_a values in the range 3.5–5.0, whereas for protonated aliphatic amines, they are located in the range 9.0–11.0. The abnormal third acidity value $pK_{a3} = 6.16$ is in favor of a mixture of the following two isomeric compounds, which are both acids:

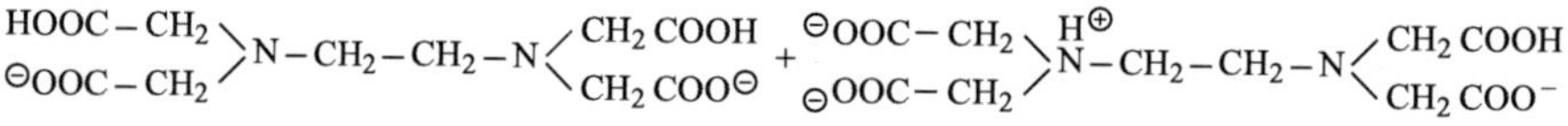

$: H_2Y^{2-}$

In one of the components of the mixture, the acidity is brought by a carboxylic rest, while in the other, it is one of the protonated amines that is acidic. This mixture is globally represented by the symbol H_2Y^{2-}. The composition of this mixture is constant regardless of the pH value (see the problem with the microconstants—Chap. 5). Moreover, $pK_{a1} = 1.99$ and even $pK_{a2} = 2.67$ are values that are too low to belong to normal carboxylic acids, but this point cannot be explained by the occurrence of several protonation sites. They probably should be ascribed to structural factors, such as, for example, the withdrawing inductive effect of the protonated amino group located at the proximity of the carboxylic groups of interest. $pK_{a4} = 10.26$ seems to be a normal value for a protonated amine. The structure of the HY^{3-} form should probably be written as

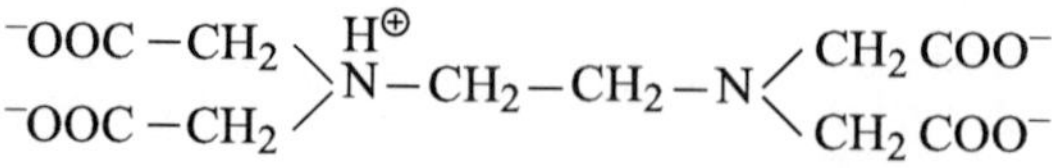 $:HY^{3-}$

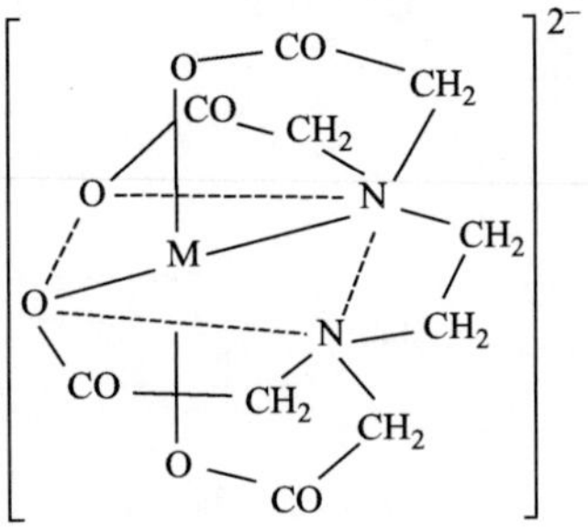

Fig. 28.1 Octahedral structure of a metal–EDTA chelate (example of a bivalent metal)

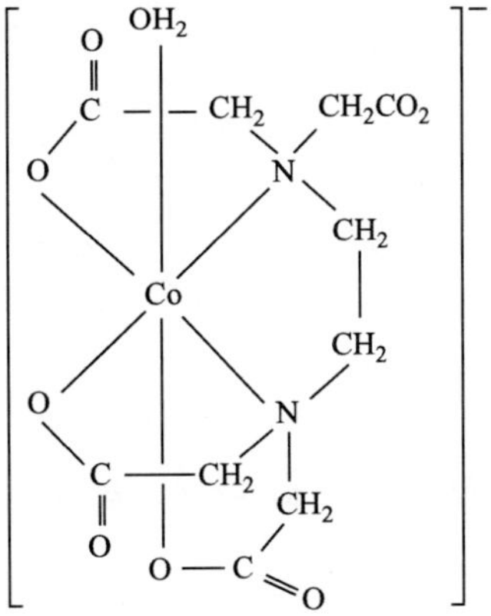

Fig. 28.2 Mixed complex aquo-cobalt(III)–EDTA: aqua [(1,2-ethanediyldinitrilo-κ^2N, N′) (tetraacetato-κ^3O, O″, O‴)] cobaltate (1^-)

28.1.2 EDTA: A Very Powerful Chelating Agent

EDTA is able to chelate numerous metal cations (about 50). The chelates it forms are octahedral. There are six sites of coordination per EDTA molecule for one metallic ion. These chelates are said to be one-to-one; that is, one molecule of chelate is formed from one EDTA molecule and from one of the metal cation. Figure 28.1 shows that the metallic ion in the chelate is tightly encircled, as it would be in a multiple–pincer device. In some rare cases of no interest in analysis, the stoichiometry is not one-to-one.

It is also possible that the formed chelates are mixed metal–EDTA—(or other ligand) chelates, such as, for example, in the cases of the following compounds: [M(OH)Y], [M(H)Y], and so forth. Another example of a mixed complex is provided by the aquo-cobalt(III)–EDTA **complex,** whose structure is seen in Fig. 28.2.

Table 28.1 gives some base-10 logarithms of stability constants at the ionic strength 0.1 mol/L. They take only the following equilibria into account:

$$M^{n+} + Y^{4-} \rightleftharpoons [MY]^{(n-4)+},$$

$$K = [MY]^{(n-4)+} / \{[M^{n+}][Y^{4-}]\}.$$

Table 28.1 Logarithms to base 10 of stability constants of metal–EDTA chelates

Mg^{2+}	8.7	Zn^{2+}	16.7	Ag^{+}	7.3
Ca^{2+}	10.7	Cd^{2+}	16.6	Li^{+}	2.8
Mn^{2+}	13.8	Hg^{2+}	21.9	Na^{+}	1.7
Fe^{2+}	14.3	Pb^{2+}	18		
Co^{2+}	16.3	Al^{3+}	16.3		
Ni^{2+}	18.6	Fe^{3+}	25.1		
Cu^{2+}	18.8	Th^{4+}	23.2		

Most values are high, even sometimes very high. This is not surprising. In these complexes we can see that all the factors that, according to Schwarzenbach, govern the stability of chelates are satisfied, that is,

- the ratio (number of chelating molecules/number of central ions) is equal to unity. It is the optimal value;
- the number of pentagonal cycles (within the chelate) in which the metal cation is engaged is high. It is equal to five (this number must be as high as possible);
- finally, the atoms of the chelating agent that share their electronic doublets with the metal are oxygen and nitrogen. Both atoms are known to give the most stable chelates.

28.1.3 Formation Reactions of Metal–EDTA Chelates

We have already seen that since H_4EDTA is a tetraacid, several reactions of formation of the same complex with the central metallic ion M^{n+} may occur. They are (see Chap. 25)

$$M^{n+} + H_4Y \overset{K_1}{\rightleftharpoons} [MY]^{(n-4)+} + 4H^+,$$

$$M^{n+} + H_3Y^- \overset{K_2}{\rightleftharpoons} [MY]^{(n-4)+} + 3H^+,$$

$$M^{n+} + H_2Y^{2-} \overset{K_3}{\rightleftharpoons} [MY]^{(n-4)+} + 2H^+,$$

$$M^{n+} + HY^{3-} \overset{K_4}{\rightleftharpoons} [MY]^{(n-4)+} + H^+,$$

$$M^{n+} + Y^{4-} \overset{K_5}{\rightleftharpoons} [MY]^{(n-4)+}.$$

According to the medium pH value, one of these reactions or even several ones occur to form the complex. This explains why the apparent stability constants of the metal–EDTA chelates may vary considerably with the pH value.

(Once again, here we find interest in the introduction of conditional constants since they take the influence of pH into account. Purely and simply, they permit us to investigate the reaction of formation of the complex according to the simpler scheme

$$Y' + M' \overset{K'}{\rightleftharpoons} [MY']$$

Table 28.2 Stability of some EDTA–metal chelates as a function of pH value

Minimal pH	Metallic ions
1–3	Zr^{4+}, Hg^{4+}, Th^{4+}, $Bi^{(III)}$, Fe^{3+}
4–6	Pb^{2+}, Cu^{2+}, Zn^{2+}, Co^{2+}, Ni^{2+}, Mn^{2+}, Fe^{2+}, Al^{3+}, Cd^{2+}, Sn^{2+}
8–10	Ca^{2+}, Sr^{2+}, Ba^{2+}, Mg^{2+}

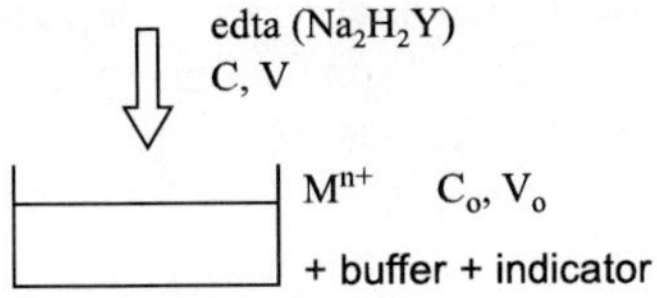

Fig. 28.3 Direct titration of the metallic ion M^{n+} with EDTA

without having to take the occurrence of parasitic chemical phenomena into account—see Chap. 26. Conditional constants are the only ones that are realistic since they govern the formation of the complex in the medium under study. This is not the case with the thermodynamic or even the formal constants—the latter being legitimate only for a given ionic strength).

The examination of the chelate formation reactions given above shows that the more acidic the medium is, the less the complex with EDTA tends to form. Inversely, the more stable the chelate is, the more acidic the medium in which it will be formed may be. For example, according to the stability of the chelate, whether or not the titration of a metallic ion is possible depends on the solution's pH. Table 28.2 mentions the minimal pH range values at which chelates may still form, in particular the pH values at which titrations with EDTA may still be carried out. A precise pH value cannot be given since the chelate formation is also conditioned by factors other than the pH value, such as the initial concentrations of the different species entering in the formation reaction, for example.

Roughly, complexes with the divalent ions are stable in slightly acidic or basic media, whereas complexes with tri- or tetravalent ions can be formed in more acidic media and, of course, in alkaline media.

28.2 Direct Titration Curve of a Metallic Ion with EDTA

Now, we'll study the direct titration of a metallic ion with EDTA. The reasoning followed in this section can be transposed to inverse titrations without any problem. Recall (Fig. 28.3) that the metallic ion is dissolved in a buffered solution. The pH value is imposed, at least in part, by the equilibrium constant K' value:

$$\mathrm{M'} + \mathrm{Y'} \overset{K'}{\rightleftharpoons} [\mathrm{MY'}].$$

We have already seen that most of the time titrations must be carried out in a rather alkaline medium. The required buffer that is added may induce the formation of parasitic complexes. A metal ion indicator is generally added to the titrand solution except in the case of a physicochemical indication. The titrant is usually a commercial solution of EDTA disodium salt.

The conditional equilibrium constant of the titration reactions takes all these experimental data into account. Knowing an approximate value of its value is sufficient to predict whether or not the titration will be satisfactory. The prediction is carried out by calculating the sharpness index of the titration (see below).

The titration curve is the diagram $p\mathrm{M}'$/added volume of titrant solution or $p\mathrm{M}'/\varphi$, with φ the fraction titrated:

$$\varphi = CV/C_oV_o$$

and $p\mathrm{M}' = -\log[\mathrm{M}']$.

The titration curve may be calculated point by point via the following three equations:

$$[\mathrm{M}'] + [\mathrm{MY}'] = C_oV_o/(V_o + V), \tag{28.1}$$

$$[\mathrm{Y}'] + [\mathrm{MY}'] = CV/(V_o + V), \tag{28.2}$$

$$[\mathrm{MY}'] = K'[\mathrm{M}'][\mathrm{Y}']. \tag{28.3}$$

By subtracting (28.1) from (28.2), we obtain

$$\varphi - 1 = [(V + V_o)/(C_oV_o)]\{[\mathrm{Y}'] - [\mathrm{M}']\}.$$

This is the equation of the titration curve. Indeed, φ is actually a function of $[\mathrm{M}']$ since $[\mathrm{Y}']$, which is still present in the last relation, may be expressed as a function of $[\mathrm{M}']$ after handling Eqs. (28.1)–(28.3). (In other words, $[\mathrm{Y}']$ is still present for brevity of writing.) Rather than solving the second-order equation issuing from Eqs. (28.1)–(28.3), we can simplify calculations by considering that the titration reaction is quasi-quantitative. If this were not the case, the titration would be devoid of any interest! With this condition,

- before the equivalence point, $[\mathrm{Y}']$ is negligible compared to $[\mathrm{MY}']$ in (28.2), and we immediately find

$$\varphi = 1 - [(V + V_o)/C_oV_o][\mathrm{M}'] \quad (\varphi < 1). \tag{28.4}$$

This simplification and this relation are legitimate only before the equivalence point.

Moreover, if the added volume V is negligible compared to V_o, we find

$$\varphi = 1 - [\mathrm{M}']/C_o \quad \text{or}$$

$$p\mathrm{M}' = \log[1/(1 - \varphi)] - \log C_o \quad (\varphi < 1 \quad \text{and} \quad V \ll V_o);$$

- after the equivalence point, $[\mathrm{M}']$ is negligible in (28.1). Therefore, we find

$$\varphi - 1 = [(V + V_o)/C_oV_o][\mathrm{Y}']$$

$$\text{and} \quad [\text{MY}'] = C_o V_o/(V_o + V),$$

from which we have $[C_o V_o/(V_o + V)]/\{[\text{M}'][\text{Y}']\} = K'$ and $[\text{Y}'] = [C_o V_o/(V_o + V)]/\{K'[\text{M}']\}$,

$$\varphi = 1 + 1/(K'[\text{M}']), \quad \varphi > 1, \tag{28.5}$$

$$\text{or} \quad p\text{M} = \log(\varphi - 1) - \log(1/K') \ (\varphi > 1 \text{ and } V \ll V_o);$$

- at the final point (very close to the equivalence point):

$$[\text{MY}'] \simeq C_o V_o/(V_o + V_{\text{pf}}).$$

Since $V_{\text{pf}} \approx V_{\text{eq p}}$, we can write

$$[\text{MY}'] \approx C_o V_o/(V_o + V_{\text{eq p}}),$$
$$[\text{MY}'] \approx C C_o V_o/(C V_o + C V_{\text{eq p}}),$$

and since $C V_{\text{eq p}} = C_o V_o$,

$$[\text{MY}'] = C_o C/(C_o + C) \quad \text{(equivalence and final points)}.$$

The general equation of the titration curve is at the final point:

$$\varphi_{\text{pf}} - 1 = [(C + C_o)/C_o C]\left\{[\text{Y}']_{\text{pf}} - [\text{M}']_{\text{pf}}\right\}$$

since at this point

$$\varphi_{\text{pf}} - 1 = [(V_{\text{pf}} + V_o)/(C_o V_o)]\left\{[\text{Y}']_{\text{pf}} - [\text{M}']_{\text{pf}}\right\}$$

Passing from the expression $(V_{\text{pf}} + V_o)/(C_o V_o)$ to $(C_o + C)/(C_o C)$ is carried out by following exactly the same reasoning as just before. In order to obtain $\varphi_{\text{pf}}/[\text{M}']$ at the equivalence point, we must express $[\text{Y}']$ as a function of $[\text{M}']$ close to the equivalence point. From a general standpoint,

$$[\text{MY}']/\{[\text{Y}'][\text{M}']\} = K',$$
$$[\text{Y}'] = [\text{MY}']/\{\text{K}'[\text{M}']\},$$
$$[\text{Y}']_{\text{pf}} = [C_o C/(C + C_o)]\left\{1/\left([K'][\text{M}']_{\text{pf}}\right)\right\},$$

$$\varphi_{\text{pf}} - 1 = 1/(K'[\text{M}']) - [(C_o + C)/C_o C][\text{M}']; \tag{28.6}$$

- at the equivalence point, $\varphi = 1$ and from (28.6),

$$[\text{M}']_{\text{eq p}} = \sqrt{\{C_o C/[K'(C_o + C)]\}} \quad \phi = 1 \quad \text{(equivalence point)}.$$

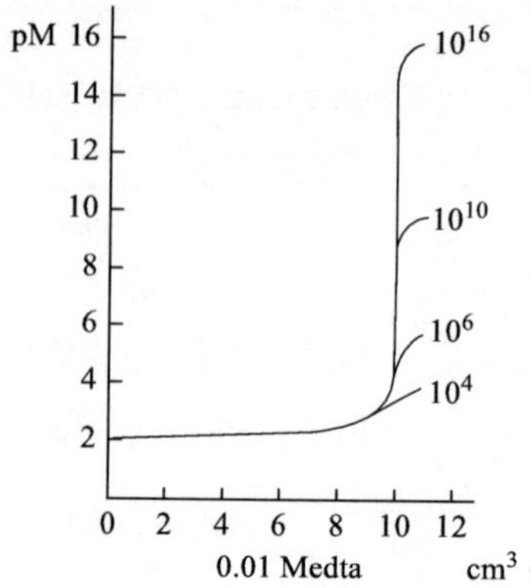

Fig. 28.4 Shapes of the curves obtained during the titration of 10 ml of a 10^{-2} mol/L metal cation with 10^{-2} mol/L EDTA

We immediately notice that the higher the equilibrium constant value is, the lower the remaining concentration of uncomplexed metallic ion becomes. From an intuitive standpoint, this result was expected, but the change in the remaining concentration $[M']$ with K' according to the inverse of a square root function was unexpected.

When the titrant solution is far more concentrated than the titrand solution, $V_{\text{eq p}} \ll V_o$ and according to what we saw earlier here,

$$(C + C_o)/C \approx 1 \quad C \gg C_o \quad \text{and}$$

$$[M']_{\text{eq p}} = \sqrt{(C_o/K')}, \quad \varphi = 1 \quad \text{(titrant solution far more concentrated than the titrand solution).}$$

The shapes of the titration curves are given in Fig. 28.4.

The higher the equilibrium constant K' is, the sharper the pM' change becomes close to the equivalence point.

- This is confirmed by the expression of the sharpness index η, which is nothing more than the slope of the curve at the equivalence point:

$$\eta = (dpM'/d\varphi)_{\text{eq p}}.$$

Remembering that

$$(dpM'/d\varphi) = -(1/2.3)(d\ln[M']/d\varphi),$$

$$(dpM'/d\varphi) = -(1/2.3)(d\ln[M']/d[M'])(d[M']/d\varphi),$$

$$(dpM'/d\varphi) = -1/(2.3[M'])(d[M']/d\varphi)$$

In order to calculate η, we still have to differentiate the approached expression that relates $[M']$ to φ close to the equivalence point. However, the differentiation must be carried out by taking into account the fact that $[M']$ is the dependent variable and

φ the independent one. In other words, [M′] is a function of φ. Starting from the equation of the curve at the equivalence point, we find

$$1 = -K'(d\,[\mathrm{M}']/d\varphi)/(K^2[\mathrm{M}']^2) - [(C_o + C)/(C_oC)]\,(d\,[\mathrm{M}']/d\varphi)\,;$$

as a result,

$$\left(d[\mathrm{M}']/d\varphi\left\{-1/(K'\,[\mathrm{M}'])^2 - (C_o + C)/C_oC\right\}\right) = 1,$$
$$d[\mathrm{M}']/d\varphi = -1/\left\{1/(K'\,[\mathrm{M}'])^2 + (C_o + C)/C_oC\right\},$$

and from the expression of $dp\mathrm{M}'/d\varphi$:

$$dp\mathrm{M}'/d\varphi = (1/2.3) \times 1/\left\{1/K'[\mathrm{M}'] + \frac{(C_o + C)}{C_oC}[\mathrm{M}']\right\}.$$

At the equivalence point,

$$[\mathrm{M}'] = \surd\{C_oC/K'(C_o + C)\};\text{ as a result,}$$
$$2.3(dp\mathrm{M}'/d\varphi) = 1/\mathrm{D},$$

$$\text{with}\quad \mathrm{D} = 1/\big(1/K'\surd\{C_oC/(C_o + C)K'\}$$
$$+ (C_o + C)\,C_oC\surd\{C_oC/[K'\,(C_o + C)]\}\big).$$

$$\text{Hence,}\quad (dp\mathrm{M}'/d\varphi)_{\mathrm{eq\,p}} = 0.217\surd\{K'C_oC/(C_o + C)\}.$$

The slope at the equivalence point is proportional to the square root of the (conditional) equilibrium constant K'. For the sharpness index to be superior to 10^3 when the concentrations C_o and C are both equal to 10^{-2} mol/L, the equilibrium must be at least higher than 5×10^5. With this value, the titration is difficult since the sharpness index is still lower than 10. In practice, some authors admit the principle that the titration will be satisfactory if $K' > 5 \times 10^7$.

The titration error is immediately given by the previous expression:

$$\varphi_{\mathrm{pf}} - 1 = 1/\left(K'[\mathrm{M}']_{\mathrm{pf}}\right) - \{(C_o+C)/C_oC\}\,[\mathrm{M}']_{\mathrm{pf}},$$

where [M′] is the concentration of the remaining M′ at the final point.

Exercise 1 Calculate the curve obtained during the titration of 50 ml of a solution containing 10^{-2} mol/L of magnesium cation in a 0.10 mol/L ammonia buffer at pH = 10 with a 10^{-2} mol/L EDTA solution. Calculate the sharpness index.

The calculation of the conditional constant (see Chap. 27) gives the value $K' = 1.4 \times 10^8$. Using Eqs. (28.1)–(28.3) permits us to calculate the curve given in Fig. 28.5.

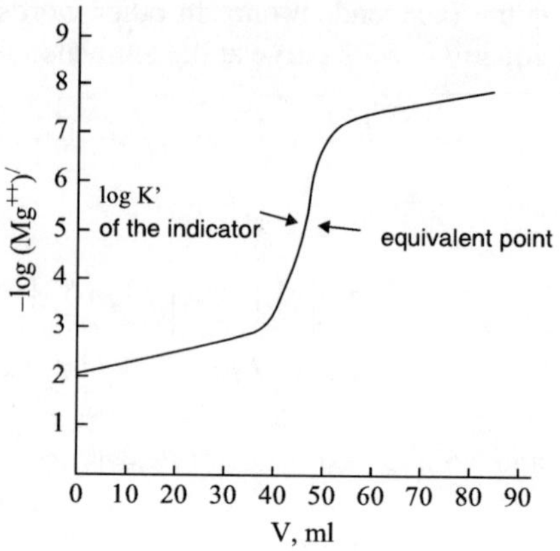

Fig. 28.5 Calculated curve resulting from the titration of a 10^{-2} mol/L of magnesium cation with a 10^{-2} mol/L EDTA solution in a buffered medium with ammonia at pH = 10

We immediately calculate $\eta = 180$. This value implicates that a change in pM' of 0.18 unity above or under the equivalence point corresponds to a 0.1% change in the fraction titrated. In other words, pMg must be determined (for example, with a metal indicator) in a range of 0.18 unity so that the error in determining the equivalence point might be less than 0.1%).

- The indication of the equivalence point is normally carried out with metal ion indicators (see Chaps. 26 and 29). We have already mentioned that not only do they exhibit the property to form chelates with metallic ions but also that they are acid–base indicators. Thus, their coloration depends on the presence of metal cations but also on the pH of the solution. We have already seen that their behavior may be described and explained by introducing their proper conditional constants. The point that remains to be investigated is their color change interval, which depends on several parameters.

In a direct titration of the metallic ion by EDTA, after the equivalence point, the metallic cation is totally complexed by the latter, in principle. Hence, the indicator is totally liberated; it is free. Hence, it exhibits the color of its free form at the pH solution. In order to obtain a color change at the equivalence point, the indicator must be complexed with the metal cation before the equivalence point. These considerations imply that the metal–indicator complex exhibits a sufficient stability but not too high for the metal–EDTA complex to be formed and the titration to be satisfactory.

In these conditions, in addition to the following reactions:

$$M' + Y' \overset{K'_{\mathrm{MEDTA}}}{\rightleftharpoons} MY'$$

$$\text{and} \quad M' + Ind' \overset{K'_{\mathrm{Mind}}}{\rightleftharpoons} Mind'$$

before the equivalence point, the following reaction also occurs:

$$Mind' + Y' \rightarrow MY' + Ind'.$$

We consider that the stability constant $K'_{\text{Mind}'}$ must be at least higher than 10^4. The ratio of constants $K'_{\text{MEDTA}}/K'_{\text{Mind}}$ must also be higher than 10^4. As a result, K'_{MEDTA} must be about 10^8. Finally, the analytical concentration of the indicator must not exceed 1% that of the metallic ion. With this ratio, the metal cation–indicator reaction does not disturb the location of the final part of the metal–EDTA reaction.

The smallest titration error is said to occur when the following relation is satisfied:

$$1/K'_{\text{ind}} = [\text{M}']_{\text{eq p}},$$

that is, when the remaining concentration of free ions at the equivalence point is equal to the inverse of the conditional equilibrium constant of the metal–indicator reaction. This means that the color change is the sharpest when the concentrations of both colored forms of the indicator are equal since

$$K'_{\text{ind}} = [\text{M} - \text{Ind}']/[\text{M}'][\text{Ind}'].$$

Exercise 2 Eriochrome black T is used for the titration of Mg^{2+} with EDTA in the conditions described in Exercise 1. Find the titration error if one considers that the color change occurs when the concentrations of both colored forms of the indicator are equal. The pH is 10. The complex–metal conditional constant of formation is 2.3×10^5.

In this case, the following relation is satisfied:

$$1/K'_{\text{Ind}} = [\text{M}']_{\text{eq p}},$$

where; $K'_{\text{Ind}} = [\text{MgInd}]/[\text{Mg}^{2+}][\text{Ind}]$.

Under the conditions of the exercise,

$$[Mg^{2+}]_{\text{pf}} = 1/K'_{\text{Ind}},$$

$$[Mg^{2+}]_{\text{pf}} = 4.35 \times 10^{-5}.$$

By reporting into the relation, we have

$$\varphi_{\text{pf}} - 1 = 1/K'[Mg']_{\text{pf}} - [(C_o + C)/C_oC][Mg']_{\text{pf}},$$

$$\varphi_{\text{pf}} - 1 = 7.3 \times 10^{-4}.$$

The relative error is about 0.07%, which is very weak.

Chapter 29
Complexometry III: Metal Cation Indicators and Types of EDTA Titrations

The success of EDTA titrations must undoubtedly be ascribed to the stability of the complexes it gives with metallic ions, to the versatility of the reaction, and also to the use of metal indicators to detect their equivalence point. In this chapter, we describe some of these indicators briefly. From another standpoint, there are other kinds of EDTA titrations than the direct ones. We examine these possibilities here. We are also interested in the possibility they offer to determine some anions. Finally, we'll examine the properties of other chelating agents whose structures are close to that of EDTA.

29.1 Some Metal Ion Indicators

The principal metal ion indicators are derived from several chemical families.

29.1.1 Azo Derivatives Possessing a Phenol Function

A first family is constituted by azo derivatives possessing one or several phenol functions in their structure. Eriochrome black T (solochrome black) is one of them (see Chaps. 26 and 28). It is used in the pH range 7–11. With a bivalent ion, for example, the reaction that is accompanied by a color change is

$$\underset{\text{(red)}}{\mathrm{MIn^-}} + \mathrm{H^+} + \mathrm{Y} \rightarrow \mathrm{M-Y} + \underset{\text{(blue)}}{\mathrm{HIn^{2-}}}$$

This is not a very selective indicator since it gives complexes with the following metallic ions:

Mg^{2+}, Mn^{2+}, Zn^{2+}, Cd^{2+}, Hg^{2+}, Pb^{2+}, Cu^{2+}, Al^{3+}, Fe^{3+}, Fe^{2+}, Co^{2+}, Ni^{2+}, and Ti^{IV}. However, the complexes formed by eriochrome black T with Cu^{2+}, Co^{2+}, Ni^{2+}, Al^{3+}, Fe^{3+}, and Ti^{IV} are too stable for direct titrations to reach a satisfactory extent, according to the equation:

$$\mathrm{MIn^-} + \mathrm{YH_2}^{2-} \rightarrow \mathrm{MY^{2-}} + \mathrm{H^+} + \mathrm{HIn^{2-}}.$$

J.-L. Burgot, *Ionic Equilibria in Analytical Chemistry*,
DOI 10.1007/978-1-4419-8382-4_29,

These metal cations are indeed masked by the indicator. The metallic ions are said to block the indicator. However, back titrations of Cu^{2+}, Co^{2+}, Ni^{2+}, and Al^{3+} can be carried out with this indicator. In such titrations, of course, it is the excess of EDTA that is determined. This is carried out by the addition of a standard solution of a metal cation, usually Zn^{2+} or Mg^{2+}. During the addition of the latter, the indicator is free as long as an excess of EDTA remains. The color change appears just after the equivalence point since then the indicator gives the complex Zn–black eriochrome T or Mg–black eriochrome T with Zn^{2+} or Mg^{2+}, which are then in excess.

There are other dyestuffs, named eriochrome black A, B, and R, whose structures are very close to that of black T. Another indicator of the same chemical family is Patton and Reeder's indicator, whose structure is as follows:

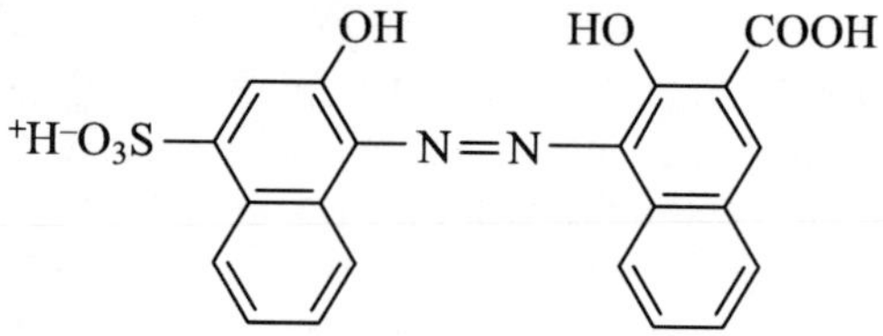

It permits the titration of Ca^{2+} in the presence of Mg^{2+} in the pH range 12–14. In a direct titration, it turns pure blue at the endpoint. Before, the solution is wine red.

In the same chemical family, we can also mention

- eriochrome blue black or calcon (solochrome dark blue);

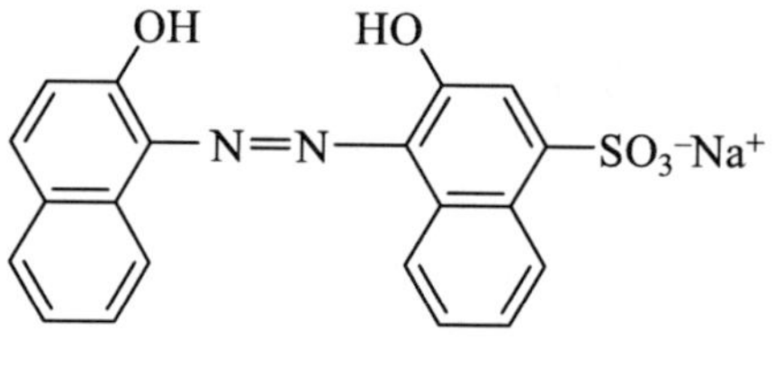

It is also used to titrate Ca^{2+} in the presence of Mg^{2+} at pH = 12.3;

- calmagite;

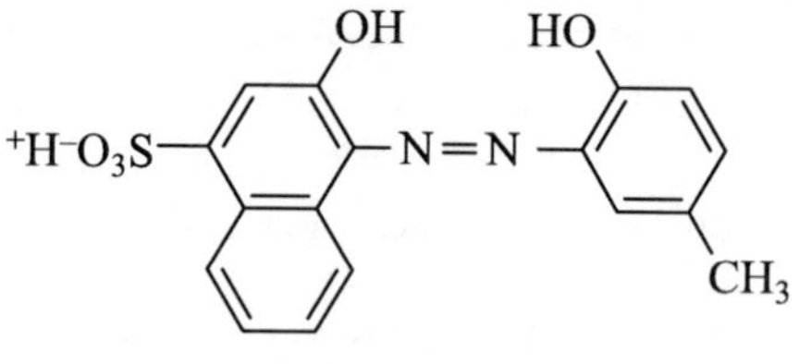

It exhibits the same color change as eriochrome black T, but the change is sharper and clearer. It gives rise to almost the same applications as eriochrome black T;

- Fast Sulphon Black F;

This is virtually specific indicator of Cu^{2+}. It can also form a complex with Ni^{2+}, but the color change is difficult to detect. The titration of copper is carried out in an ammonia solution or in pyridine. In a direct titration, the color change at the endpoint is from pale blue to bright green.

29.1.2 Triphenylmethane Derivatives

Among triphenylmethane derivatives, we can mention

- Catechol violet or pyrocatechol violet;

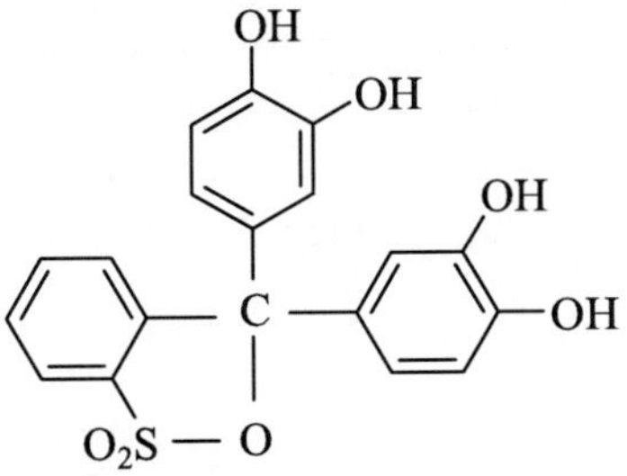

It forms blue or green-blue complexes with numerous metallic ions. It also exhibits acid–base indicator properties. Free, it is a tetraacid. At pH < 1.5, it is in solution as H_4In and the color is red. Between $2 \leq \mathrm{pH} \leq 6$, the yellow color is due to the anion H_3In^-. Between $7 \leq \mathrm{pH} \leq 10$, the violet color is due to the dianion H_2In^{2-}; after pH > 10, the color is blue. With some metal cations it gives blue-green stable complexes at $2 < \mathrm{pH} < 6$. As a result, there is a sharp color change from blue-green to yellow at the endpoint of a direct titration with EDTA. It is used for titrations of bismuth and thorium. With ions that can be quantitatively complexed only at high pH values, the endpoint is detected by a color change from blue-green to blue-violet. It is not as easy to detect;

- bromopyrogallol red;
  Its properties are analogous to those of the preceding indicator. It is used for the determination of bismuth at pH $= 2.3$.

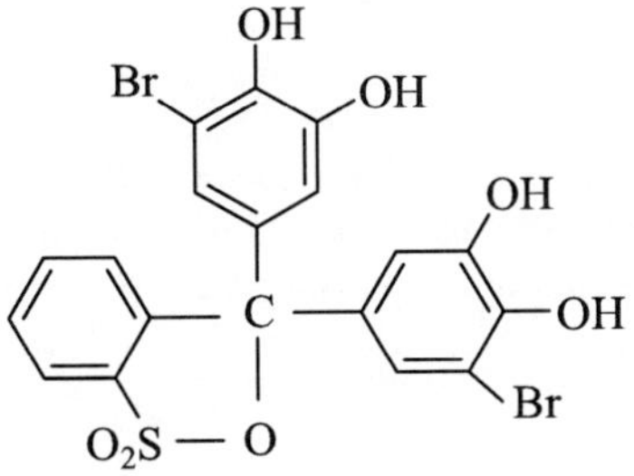

- xylenol orange;

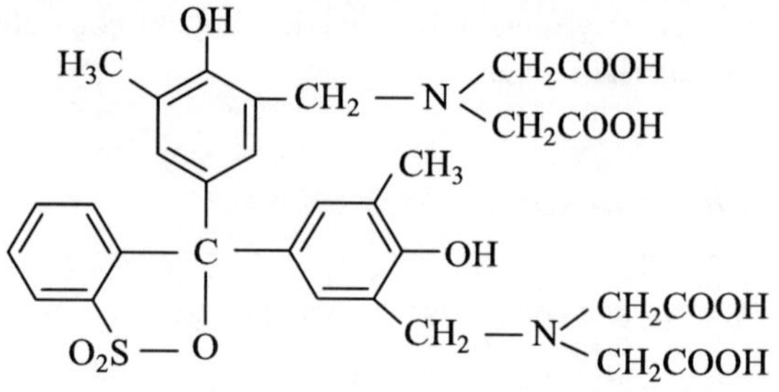

It also possesses acid–base indicator properties. It can form complexes with metal cations even in acidic solution $3 < pH < 5$. The free indicator in acidic solution exhibits a lemon yellow color. Its metal complexes are intensely red. It permits us to directly titrate Bi^{III}, Th^{4+}, Zn^{2+}, Pb^{2+}, and Co^{2+}. It is interesting to note that there are two imino-diacetic groups in its structure:

$$N{=}(CH_2COOH)_2,$$

which are also present in EDTA as well as in some cation-exchange resins;
- Thymolphthalein complexone (thymolphthalexone);
  It only reacts in alkaline medium. It is used for the titration of calcium.
- Methylthymol blue;

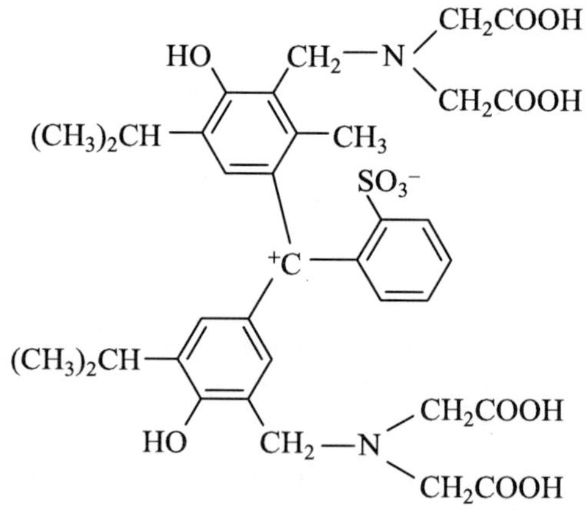

It functions in the pH range 0–12. Hence, it permits the titration of cations going from bismuth until alkaline earths (refer to the series of metallic ions that may be titrated by EDTA; see preceding chapters). So a wide variety of bivalent metal ions may be titrated. In particular, this is the case for Hg^{II}, for which few indicators are available.

29.1.3 Derivatives of Miscellaneous Structures

- Zincon

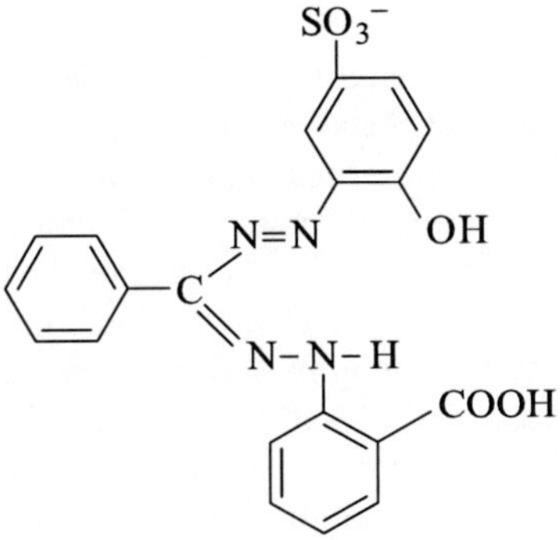

is a specific indicator of Zn^{2+}. At pH $= 9-10$, it gives a blue color.
- Murexide, which is an indicator of a particular structure. It is the ammonium salt of purpuric acid:

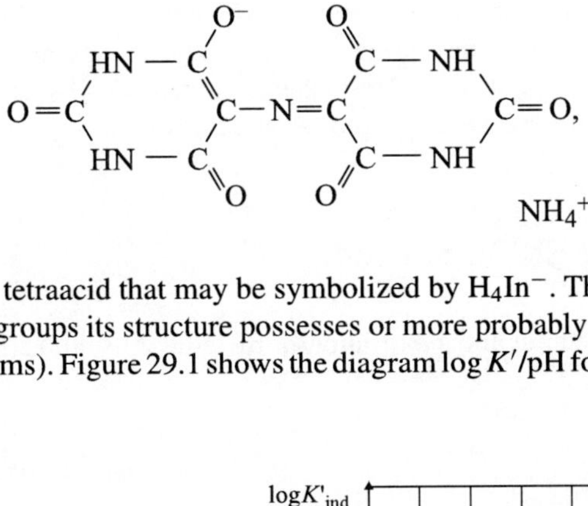

Purpuric acid is a tetraacid that may be symbolized by H_4In^-. The acidities are due to the four imido groups its structure possesses or more probably to their tautomeric forms (lactime forms). Figure 29.1 shows the diagram log K'/pH for the complexation with Ca^{2+}.

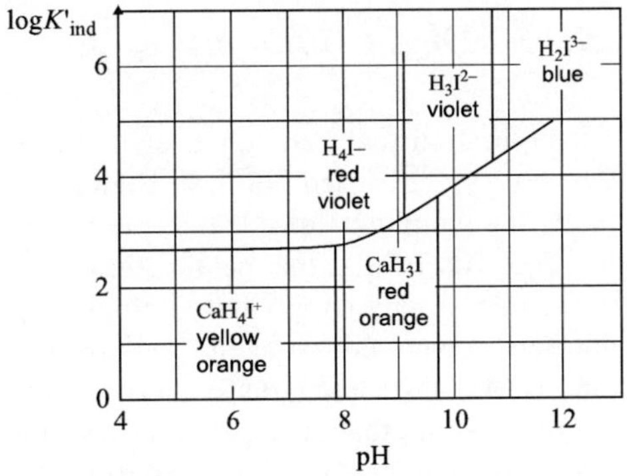

Fig. 29.1 Diagram log $K'_{in(Ca\text{-}murexide)}$/pH

Solutions of free murexide are red-violet until pH = 9, violet in the range 9–11, and blue beyond. Figure 29.1 shows that only two protons can be removed. The two pK_a values that must be considered are $pK_a(H_4In^-/H_3In^{2-}) = 9.2$ and $pK_a(H_3In^{2-}/H_2In^{3-}) = 10.5$.

Murexide forms complexes with numerous metal cations such as Cu^{2+}, Ni^{2+}, Co^{2+}, and Ca^{2+} and ions deriving from lanthanides. The complex's color depends on the pH and the nature of metallic ions. Murexide permits Ca^{2+} titration with EDTA at pH = 11. As the figure shows, the color turns red from violet blue.

There are also metal cation indicators that are unicolor.

Surprisingly, at first sight, redox indicators may also be used in some cases to detect the endpoint of a complexometric titration with EDTA. In fact, the endpoint of an EDTA titration may be accompanied by a change in the redox potential of the solution. When a mixture of Fe^{2+} and Fe^{3+} is titrated with EDTA, Fe^{3+} disappears before Fe^{2+} since Fe^{3+} gives more stable complexes with EDTA than Fe^{2+} does. A simple inspection of Nernst's equation shows that in these conditions, the solution's redox potential decreases markedly, in particular at the equivalence point. The sharp change may be detected by potentiometry with a platinum electrode or with a redox indicator such as Variamine blue.

We must also keep in mind that equivalence points of EDTA titrations can be detected by using several instrumental methods. A first method, potentiometry, was just mentioned. There are also other potentiometric methods, based on other principles than the previous one, that may be used. Amperometric and conductometric methods have been proposed equally (see electrochemical methods of analysis). Finally, we'll mention photometric and spectrophotometric indications.

29.2 Types of EDTA Titrations

The different procedures for the titration of metallic ions with EDTA are described in this section.

29.2.1 *Direct Titrations*

Direct titrations have already been discussed. The usual experimental conditions in which they are carried out have been seen. We must, however, stress the fact that under the conditions of alkaline pH values often required to obtain a satisfactory titration with EDTA, the metal cation may precipitate as oxide or hydroxide. This is the reason why an auxiliary complexing agent such as an acid-alcohol may be added to prevent this parasitic phenomenon. The acid-alcohols mostly used for this purpose are citric and tartaric acids (which at such a pH stand essentially as salts). There is also triethanolamine. Of course, the secondary complexation must be taken into account in the conditional stability constants of the metal–EDTA complexes. In

Table 29.1 Decimal logarithms of stability constants of triethanolamine–metal cations complexes

	β_1	β_2
Ag^+	2.3	3.6
CO^{2+}	1.7	
Cu^{2+}	4.4	
Ni^{2+}	2.7	
Zn^{2+}	2.0	

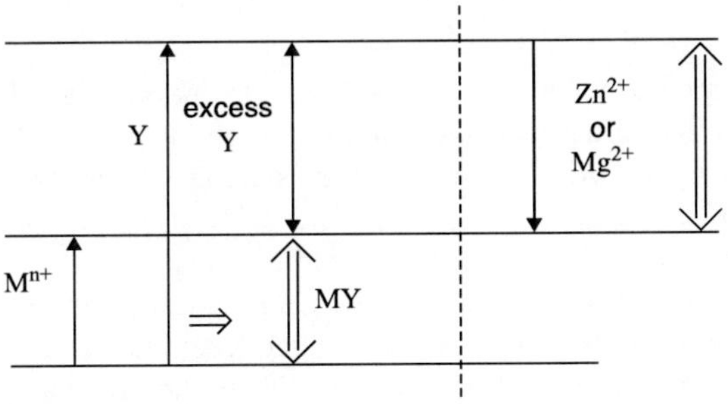

Fig. 29.2 Back titration of M^{n+} ion with EDTA

order to avoid the oxidization of some metallic ions, hydroxylamine hydrochloride, which is a reducing agent, may be added. Some masking agents such as sodium or potassium cyanide may also be added to avoid having some other cations interfere in the titration (see the end of this section). Finally, of course, buffers different from the couple $NH_4{}^+/NH_3$ may also be used. We mention in Table 29.1 some stability constants of metallic ions–triethanolamine complexes.

Triethanolamine gives a highly stable mixed complex with ferric ion in alkaline medium according to the equilibrium

$$Fe^{3+} + 4OH^- + L[Fe(OH)_4L]^- \quad (L: \text{triethanolamine}) \quad (\log K = 41.2).$$

29.2.2 Back Titrations

When a back titration is to be performed, a known volume of standard EDTA solution in excess is added to the solution of the metal cation to be titrated (Fig. 29.2).

Then the excess reagent is titrated with a standard metal ion solution. This procedure deserves further comments. First, the metal ion indicator must respond to the titrant ion, usually Zn^{2+} or Mg^{2+}. Second, the color change occurring at the endpoint is the inverse of that obtained in a direct titration. Before the endpoint of the back-titration reaction, the indicator is free; after it is complexed with the metallic ion (Zn^{2+} or Mg^{2+}), it is in excess. Third, a *sine qua non* condition for these titrations to be successful is that the MY complex must be more stable than the ZnY or MgY complex. If it were not the case, the following reaction might occur (charges omitted):

$$MY + Zn \rightarrow ZnY + M.$$

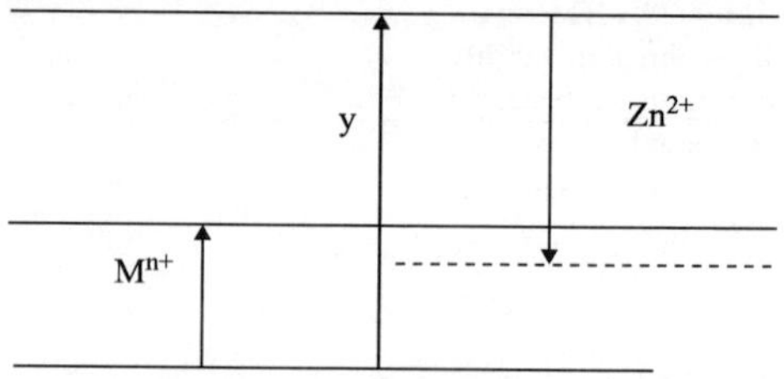

Fig. 29.3 Too great an addition of the back-titrating solution when the MY complex is not sufficiently stable compared to the ZnY or MgY complex

Too great a volume of back-titrating solution might be added, and consequently, the sought-after concentration M^{n+} might be appreciated by default (see Fig. 29.3). Actually, the M–Y complex must be about 10^6 times as stable as the ZnY or MgY complex.

This procedure is particularly interesting when the metal cation to titrate precipitates as oxide or hydroxide at the pH required to obtain a satisfactory direct titration. In the back titration, it can no longer precipitate since it is complexed with EDTA. Another interest in this procedure lies in the fact that it may turn out that no metal indicator properly functions in the conditions of the corresponding direct titration.

29.2.3 Titrations with Indirect Metal Indicators

When an indicator selectively reacts with a given metal cation, it can be used for the titration of other metallic ions with EDTA. This seems at first sight a little paradoxical, but the following example explains the principle. It concerns the determination of Ca^{2+} in the presence of eriochrome black T. According to this methodology, the titration is carried out in the presence of the complex $[MgY]^{2-}$. First, it must be known that a direct titration of Ca^{2+} with EDTA in the presence of eriochrome black T is not satisfactory. The reason for this fact is simple: The stability constant K_{CaIn} is too weak ($10^{5.4}$). The indicator is not sufficiently complexed with Ca^{2+} for a frank color change to occur at the endpoint. However, if we add a weak quantity of $[MgY]^{2-}$ to the titrand solution, we obtain a sharp color change. EDTA reacts with Ca^{2+} before Mg^{2+} since $[CaY]^{2-}$ is more stable than $[MgY]^{2-}$ [at pH = 10: $K'_{(CaY)} = 10^{10.2}$ and $K'_{(MgY)} = 10^{8.2}$ charges omitted]. However, the [MgIn] complex is more stable than the [CaIn] complex ($K'_{MgIn} = 10^{5.4}$ and $K'_{CaIn} = 10^{3.8}$; pH $\approx$ 10). The titration reaction is

$$Y^{4-} + Ca^{2+} \rightarrow [CaY]^{2-}.$$

The color change at the endpoint is due to the reaction

$$Y^{4-} + [MgIn] \rightarrow [MgY]^{2-} + In$$ (the charge of the indicator is omitted since it depends on the pH value).

The indicator becomes free. Before, the indicator was in the form of its complex with Mg^{2+}, the latter coming from the displacement reaction:

$$Ca^{2+} + [MgY]^{2-} \rightleftharpoons Mg^{2+} + [CaY]^{2-}.$$

For such a titration to be satisfactory, a judicious choice of the experimental conditions must be made. The variable to adjust is, for this example, the concentration $[MgY^{2-}]$.

In the general case, suppose that we want to titrate the metallic ion M_I with the complexing reagent L, which gives the complex M_IL, and that the colored indicator is sensitive to the metallic ion M_{II} with formation of the complex $M_{II}In$. M_{II} also forms the complex $M_{II}L$. (L is a common ligand to both metallic ions.) The problem is to calculate pM_I at the point at which the indicator color changes, that is, when $pM_{II} = \log K_{MIIIn}$. Expressing the concentration [L] at equilibrium from the expressions K_{MIL} and K_{MIIL} leads to the equality:

$$\frac{[M_IL]}{([M_I]K_{MIL})} = \frac{[M_{II}L]}{([M_{II}]K_{MIIL})}$$

$$\text{or} \quad pM_I = pM_{II} + \log\left(\frac{K_{MIL}}{K_{MIIL}}\right) + \log\left(\frac{[M_{II}L]}{[M_IL]}\right).$$

To illustrate the use of this relation, let's investigate the titration of a 10^{-3} mol/L Ca^{2+} solution with EDTA at pH = 10 in the presence of eriochrome black T. What must the value of the concentration of added $[MgY]^{2-}$be in order to obtain a satisfactory color change? (It is considered that at the color change, $pMg = 5.4$. A satisfactory titration of Ca^{2+} in these conditions is carried out if $pCa = 6.6$ at the equivalence point. The conditional stability constants are $K'_{CaY} = 10^{10.2}$, $K'_{MgY} = 10^{8.2}$.)

Identifying M_I with Ca^{2+} and M_{II} with Mg^{2+} and taking the numerical values already given into account lead to

$$6.6 = 5.4 + (10.2 - 8.5) + \log\left[MgY^{2-}\right] - \log\left(10^{-3}\right),$$

$$[MgY^{2-}] = 10^{-3.5},$$

$$[MgY^{2-}] \approx 3 \times 10^{-4}.$$

To obtain this result, we have used the assumption that $[Ca^{2+}] \approx 10^{-3}$ mol/L; that is, that the formation reaction of the $[CaY]^{2-}$ complex is quasi-complete. This is very reasonable given the value of the conditional stability constant.

The basis for titrations with indirect indicators of metallic ions is different from one that supports replacement titrations (see below).

29.2.4 Replacement or Substitution Titrations

According to the principle of these titrations, the metallic ion to be titrated M_I displaces a metallic ion M_{II} from its complex already formed with EDTA, namely, $[M_{II}Y]$:

$$M_I + M_{II}Y \rightarrow M_IY + M_{II}.$$

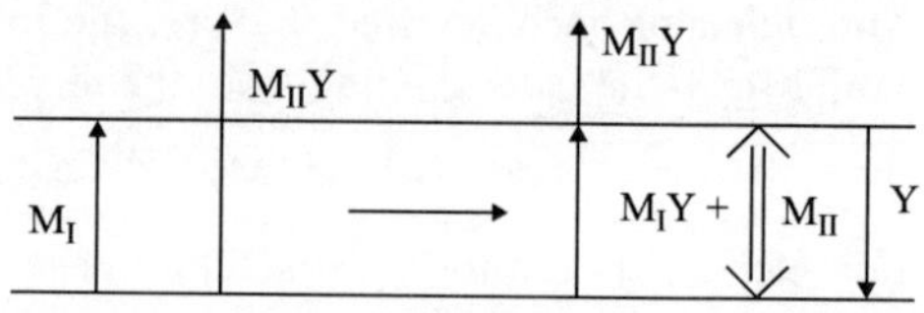

Fig. 29.4 Substitution titration of a metallic ion M_I by displacement with the $M_{II}Y$ complex

The displaced metallic ion M_{II} is titrated with EDTA. Figure 29.4 schematizes the principle.

For this sort of titration to occur satisfactorily, the M_IY complex must be more stable than the $M_{II}Y$ complex in the experimental conditions, that is,

$$K'_{\mathrm{MIY}} > K'_{\mathrm{MIIY}}.$$

These titrations may be used when no convenient indicator of M_I exists or when the direct reaction M_I/EDTA occurs too slowly. The replacement reaction may be carried out by warming, with $M_{(II)}$ then being titrated at room temperature with EDTA. According to the nature of the ion to be titrated and the experimental conditions, the most commonly used replacing ions are Zn^{2+}, La^{3+}, Mn^{2+}, and Mg^{2+}. Mn^{2+} has the disadvantage of being oxidized by dioxygen. Therefore, when it is used, it is necessary to add a reducing agent to the medium such as ascorbic acid or hydroxylamine. It is interesting to notice that this sort of titration may be generalized, since the free M_{II} ions liberated by $M_{II}Y$ may be titrated with a ligand L different from EDTA. In this case, the stability constants of the varied complexes playing a part in these titrations must be such that

$$K_{\mathrm{MIY}} > K_{\mathrm{MIIY}} > K_{\mathrm{MIIL}} \quad \text{and} \quad K_{\mathrm{MIY}} > K_{\mathrm{MIL}}$$

due to the two successive reactions

$$M_I + M_{II}Y \rightarrow M_IY + M_{II},$$

$$M_{II} + L \rightarrow M_{II}L.$$

29.2.5 Alkalimetric Titrations

If we add a solution of the disodium salt of EDTA to an accurately neutralized solution of the metal cation to be titrated, the anticipated complex is formed with the liberation of two protons:

$$M^{n+} + H_2Y^{2-} \rightleftharpoons [MY]^{(n-4)+} + 2H^+.$$

The liberated protons may be titrated with a solution of sodium hydroxide in the presence of an acid–base indicator or with a pH-metric indication. Alternatively, one may use the iodate–iodide reaction with profit. A solution of the iodate–iodide mixture is then added to the titrand solution. The limiting factor of iodine formation is the concentration of protons formed simultaneously with the complex. Actually, this sort of titration is no longer practiced. Its drawback is the hydrolysis of metallic ions in the initial solution, which must be absolutely neutralized.

29.2.6 Sequential Titrations of Mixtures, Selectivity, Masking, and Demasking

As for other sorts of titrations, the question of of carrying out sequential titrations with EDTA arises. It is admitted that the ratio $K'_{\mathrm{MIY}}/K'_{\mathrm{MIIY}}$ must be at least of about 10^8 for the sequential titration of the metal cations M_I and M_{II} to be satisfactory. Of course, the stability constants K_{MIY} and K_{MIIY} of the complexes of M_I and M_{II} ions with EDTA that pertain here must correspond to the experimental conditions, that is, must be the conditional ones. If the above value of the ratio is reached, the M_I ion is titrated first since it is designated conventionally as giving the more stable complex. As in the other sorts of titrations, the obtained precision in determining the first endpoint—in other words, the magnitude of the pM_I change—depends on the value of the ratio of both stability constants. This is not the case for the determination of M_{II}. In this case, the precision depends only on the K'_{MIIY} value and, of course, on the titrant's and titrand's concentrations.

If the ratio is less than 10^8, only the sum of the concentrations of both ions can be determined satisfactorily by titration with EDTA. Then, of course, we must use an indicator of the second ion M_{II}.

The influence of the presence of the M_{II} ion on the titration of M_I may also be taken into account by introducing an appropriate conditional constant. As usual in the definition of conditional constants, we begin with the recognition of the principal and parasitic reactions. Here we consider the parasitic reaction is that of the M_{II} ion with EDTA. If, in order to simplify, we suppose that M_I reacts only with EDTA to give M_IY and that EDTA reacts only with M_I and M_{II} (it gives rise to no further reaction, such as, for example, a protonation), we can then write

$$C_{\mathrm{MI}} = [\mathrm{M_I}] + [\mathrm{M_IY}],$$

$$C_{\mathrm{MII}} = [\mathrm{M_{II}}] + [\mathrm{M_{II}Y}],$$

$$C_{\mathrm{Y}} = [\mathrm{Y}] + [\mathrm{M_IY}] + [\mathrm{M_{II}Y}],$$

$$K_{\mathrm{MIY}} = \frac{[\mathrm{M_IY}]}{([\mathrm{M_I}][\mathrm{Y}])}, \quad K_{\mathrm{MIIY}} = \frac{[\mathrm{M_{II}Y}]}{([\mathrm{M_{II}}][\mathrm{Y}])}.$$

By introducing the following apparent concentrations:

$$C_{\mathrm{MI}} = [\mathrm{M_I}]' + [\mathrm{M_IY}] \quad \text{and} \quad [\mathrm{M_I}]' = [\mathrm{M_I}],$$

$$C_{\mathrm{Y}} = [\mathrm{Y}]' + [\mathrm{M_IY}] \quad \text{with} \quad [\mathrm{Y}]' = [\mathrm{Y}] + [\mathrm{M_{II}Y}],$$

or

$$[\mathrm{Y}]' = [\mathrm{Y}](1 + K_{\mathrm{MIIY}}[\mathrm{M_{II}}]),$$

we can now introduce the parasitic reaction coefficient. It is

$$\alpha_{\mathrm{YMII}} = 1 + K_{\mathrm{MIIY}}[\mathrm{M_{II}}],$$

and, hence,

$$[\mathrm{Y}]' = [\mathrm{Y}]\alpha_{\mathrm{YMII}}.$$

We see that the coefficient α_{YMII} indeed takes the influence of the presence of the $\mathrm{M_{II}}$ ion into account. It is evident that

$$K_{\mathrm{MIY}}' = \frac{K_{\mathrm{MIY}}}{\alpha_{\mathrm{YMII}}}.$$

The presence of the $\mathrm{M_{II}}$ ion exhibits the ability apparently to decrease the formation constant of the $\mathrm{M_IY}$ complex. When the apparent constant is endowed with a value less than 10^7, the titration of $\mathrm{M_I}$ alone is no longer possible. Of course, when EDTA is more or less protonated, we must introduce the coefficient $\alpha_{\mathrm{Y(H)}}$ in addition to α_{YMII}. That is to say, we must introduce the global coefficient:

$$\alpha_{\mathrm{Y}} = \alpha_{\mathrm{Y(H)}} + \alpha_{\mathrm{YMII}}.$$

In the same manner, when the metallic ion also gives parasitic reactions, we must introduce the corresponding coefficients α. The above expression α_{YMII} can be approximated to

$$\alpha_{\mathrm{YMII}} \approx [\mathrm{M_{II}}]K_{\mathrm{MIIY}}$$

$$\text{and} \quad K_{\mathrm{MIY}}' \approx \frac{K_{\mathrm{MIY}}}{K_{\mathrm{MIIY}}[\mathrm{M_{II}}]}.$$

The closer to K_{MIY} the conditional constant is relative to the interference of the $\mathrm{M_{II}}$ ion, the weaker the concentration of the parasitic ion $\mathrm{M_{II}}$ is. It is the same state of affairs with K_{MIIY}.

Actually, sequential titrations with EDTA are rather rare. It is sufficient to examine Table 28.1 in Chap. 28 to be convinced by the fact that EDTA is a poorly selective titrant. We can say that it is the drawback of its principal quality, that is, of the possibility it offers to titrate a great number of metal cations. However, its selectivity may be considerably enhanced by using the following strategies:

1. performing the titrations of mixtures at different successive judiciously chosen pH values. In this strategy, the fact that the complex's stabilities do vary with the pH is, of course, taken into account. For example, with a mixture of $\mathrm{Bi^{III}}$ and $\mathrm{Pb^{2+}}$, only $\mathrm{Bi^{III}}$ is titrated with EDTA at pH = 2. The titration is carried out in the presence of xylenol orange. It is easy to calculate that for this pH value, $K'_{\mathrm{BiY}} = 10^{8.6}$ and $K'_{\mathrm{PbY}} = 10^{5.2}$. The latter value is too weak for $\mathrm{Pb^{2+}}$ to be titrated. Hence, the direct titration of both ions is performed as follows. In a first reaction, the titrand solution pH is fixed at pH = 2 and $\mathrm{Bi^{III}}$ is titrated. The solution, initially

red (indicator complexed with the metal cation), turns yellow (free form). A little hexamine is then added to the titrand solution to buffer it at pH = 5. The solution again becomes red-violet and then turns yellow one more time at the new endpoint, that is, after the addition of an equivalent quantity of EDTA. The new endpoint permits us to determine Pb^{2+}. It is important to notice, concerning the mechanism of the color change, that the indicator combines with Pb^{2+} only for pH $\geq$ 4. This explains the color change from yellow to red-violet when hexamine is added;

2. using masking agents. Recall that masking is the process in which a substance is transformed into another that cannot enter a particular reaction. In other words, we can consider that the masking process is a means to decrease the conditional constant of formation of the complex of the substance (in the occurrence the metal–EDTA complex) by engaging it into another compound. Furthermore, forming complexes in acidic medium is a means to more or less mask EDTA toward metallic ions, and the preceding strategy might be inserted into the actual one.

A "good" masking agent is the cyanide ion. It forms stable "cyano complexes" with the cations Cd^{2+}, Zn^{2+}, Hg^{2+}, Cu^{2+}, Co^{2+}, Ni^{2+}, Ag^{+}, and Fe^{2+}. It does not form complexes with alkaline earths, Mn^{2+}, and Pb^{2+}. The formation reaction of these complexes is

$$M^{2+} + 4CN^- \rightarrow [M(CN)_4]^{2-}.$$

Ti^{IV}, Fe^{3+}, and Al^{3+} can be masked with triethanolamine ($N(CH_2CH_2OH)_3$: H_3tea). It usually gives the following kinds of complexes: $[M(tea)]^{n+}$ and $[M(tea)_2]^{n+}$. Fe^{III} gives the mixed complex $[Fe(OH)_4(tea)]^-$ ($\log K^\circ = 41.2$). Mercuric ions can be masked by iodide ions by the formation of tetraiodomercurate(II) $(HgI_4)^{2-}$ ($\log \beta_4 = 29.8$). Al^{3+}, Fe^{3+}, Ti^{IV}, and Sn^{2+} can be masked as "fluoro complexes," and so forth.

In some cases, the parasitic metallic ion can be masked by reduction or oxidization. For example, in some conditions, Cu^{2+} may be masked by reduction into Cu^{I}. This oxidation state is in principle not stable in aqueous solution, but it can be stabilized by complexation with ammonia (see Chaps. 14 and 15). Then the forming complexes are $[Cu(NH_3)]^+$ ($\log\beta_1 = 5.90$) and $[Cu(NH_3)_2]^+$ ($\log\beta_2 = 10.80$). The reduction can be carried out with hydroxylamine or ascorbic acid. Another example is provided by the reduction of Fe^{3+} into Fe^{2+}, which can be masked by the addition of cyanide ions.

After masking, *demasking* is possible in some cases. Demasking offers supplementary possibilities of selective titrations.

An example is provided by the titration of the Mg^{2+}, Zn^{2+}, and Cu^{2+} ions in admixture. The selective titration is carried out in three stages. In the first stage, a known quantity in excess of EDTA is added and after reaction, the excess is back-titrated by a standard solution of Mg^{2+} ions in the presence of eriochrome black T. This stage gives the sum of the concentrations of the three ions. (During this back titration, we notice that among the three ions, it is Mg^{2+} that gives the less stable complex with EDTA, whereby the back titration is theoretically justified. Recall that eriochrome black T gives excellent color changes with Mg^{2+} ions. Finally, still concerning this titration, there is no theoretical impossibility to back-titrate an Mg^{2+} solution with another Mg^{2+} solution.)

In the second stage, an excess of a cyanide solution is added to another sample of the initial solution. A known quantity in excess of EDTA is then added to the obtained solution. The EDTA in excess is finally back-titrated with a standard solution of Mg^{2+}. This stage gives the Mg^{2+} concentration.

In the third stage, Zn^{2+} is demasked. Hence, its complex with the Zn^{2+} ion, formed at the beginning of the second stage, is destroyed. The demasking process is carried out by adding an excess of an acetic solution of formaldehyde to the solution. The demasking reaction that occurs is

$$[Zn(CN)_4]^{2-} + 4H^+ + 4HCHO \rightarrow Zn^{2+} + \underset{\text{formaldehyde cyanohydrin}}{4HOCH_2CN}$$

(the same result may be obtained with chloral hydrate). The liberated Zn^{2+} is directly titrated with EDTA until the solution turns blue.

Another means to increase the selectivity is to carry out a preliminary separation of the ion to be titrated if, of course, it is easy to perform. For example, Ca^{2+}, Ni^{2+}, Mg^{2+}, and Cu^{2+} can be isolated as precipitates of, respectively, oxalate, dimethylglyoximate, ammonium magnesium phosphate, and Cu^{I} thiocyanate (see Part V of this book). After dissolution of these precipitates, the liberated metal cation is titrated with EDTA. In some rare cases, the metallic ion may be selectively extracted from the solution into an organic solvent and then determined with EDTA in convenient experimental conditions, for example, after dilution of the organic phase with water.

From a theoretical standpoint more than from an experimental one, the so-called kinetic masking is interesting. It is based on the fact that a parasitic metallic ion does not interfere in the titration of the ion under study because it is inert. It does not react sufficiently quickly with EDTA. A typical example of such a case is provided by the Cr^{3+} ion, which reacts very slowly with EDTA. Hence, a mixture of Cr^{3+} and Fe^{3+} may be titrated sequentially with EDTA. Fe^{3+} is titrated first by adding an excess of a standard solution of EDTA to the cooled sample solution. The EDTA in excess is titrated with a standard solution of lead nitrate in the presence of xylenol orange. Cr^{3+} does not react in these conditions if the solution remains cooled. In a second step, a known quantity in excess of EDTA is again added to the preceding solution. The solution is then acidified, and it is warmed until the red color of the Cr–EDTA complex appears. After cooling to room temperature, the EDTA in excess is titrated with mercuric nitrate at pH $= 6$. Another example of kinetic masking is provided by the back titration of Al^{3+}. The EDTA in excess is back-titrated with a standard solution of Zn^{2+} in neutral or slightly acidic medium. At first sight, this appears to be nonsense from a thermodynamic standpoint since Zn^{2+} forms a complex with EDTA that is slightly more stable than that formed with Al^{3+}. The displacement reaction governed by thermodynamics, which might be disturbing, is

$$[AlY]^- + Zn^{2+} \rightleftharpoons [ZnY]^{2-} + Al^{3+}.$$

It is ineluctable, but, actually, it occurs very slowly and does not disturb the titration.

29.3 Determination of Anions with EDTA

Using EDTA also permits us to quantitatively determine some anions. We can proceed in two ways.

The first approach consists of precipitating the anion with a judiciously chosen cation in order to precipitate it quantitatively, then collecting and dissolving it in an excess of a standard solution of EDTA, and, finally, titrating the EDTA excess with an appropriate cation. For example, the orthophosphate ion $PO_4{}^{3-}$ can be precipitated as ammonium magnesium phosphate $Mg(NH_4)PO_4$, $6H_2O$. The recuperated precipitate, once washed, is dissolved in excess EDTA. The EDTA excess is titrated at $pH = 10$ with Mg^{2+} in the presence of eriochrome black T. An analogous methodology is followed to titrate sulfate ions. They are precipitated with barium chloride. The $BaSO_4$ precipitate is dissolved in an excess of a standard solution of EDTA, which is titrated with Mg^{2+} in the presence of eriochrome black T.

The second approach is much like the preceding method. The precipitate in which the anion to be determined is engaged is treated with a reagent that stoichiometrically liberates a cation that is titrated with a standard solution of EDTA. Hence, a halide (except the fluoride) or the thiocyanate ion is precipitated as the corresponding silver salt. The latter is treated with a solution of potassium tetracyanonickelate $K_2[Ni(CN)_4]$. The silver salt dissolves into this solution with liberation of one nickel ion Ni^{2+} for two ions X^-. Ni^{2+} is titrated with EDTA in the presence of murexide. The reaction constituting the basis of the process is

$$2AgX\downarrow + [Ni(CN)_4]^{2-} \rightarrow Ni^{2+} + 2[Ag(CN)_2]^- + 2X^-.$$

The equilibrium constant of the reaction,

$$2Ag^+ + [Ni(CN)_4]^{2-} \rightleftharpoons Ni^{2+} + 2[Ag(CN)_2]^-,$$

is about 10^{15}. It is sufficiently high for the preceding reaction to be quasi-complete despite the very weak solubility of silver halides (or thiocyanate) (see the corresponding solubility products in Part V). Moreover, the reaction is carried out in an ammonia buffer. This may be considered to facilitate the dissolution of the precipitate AgX by the formation of "ammine-silver" complexes, but, inversely, such a formation precludes the displacement reaction of Ni^{2+}.

Some practical points concerning titrations with EDTA are sufficiently important to be briefly underlined:

- the EDTA solution mostly used is prepared with the disodium salt of EDTA Na_2 EDTA, $2H_2O$, also named trilon B, complexone III, sequestrene, versene, and chelaton 3. In addition to its two water molecules, it may contain a trace of moisture. Its formula agrees exactly with that given (two water molecules for one molecule of disodium salt) only after drying at 80°C during about 24 h. Usually, its solutions are 10^{-1} or 10^{-2} mol/L. It is not a primary standard and its solutions must be standardized. This may be done with a Ca^{2+} solution prepared by dissolution of $CaCO_3$ into a hydrochloric acid solution followed by an adjustment to a convenient

pH to carry out the titration. Its solutions may also be standardized with a Zn^{2+} solution obtained by dissolution of pure zinc pellets into hydrochloric or sulfuric acid or with a standardized magnesium chloride solution;
- the pH value is critical. Most of the time, only a narrow pH interval of about 1 unity is convenient. As has been seen, this constraint results from the fact that in some pH ranges, the completeness of the titration reaction may become insufficient;
- the indicator concentration must be weak and even very weak in order to favor the detection of the color change;
- the optimum metal cation concentration in such titrations seems to be in the 10^{-1}/ 10^{-2} mol/L range.

29.4 Other Complexones

Other complexones exist, but their use is more limited than that of EDTA:

- nitrilotriacetic acid;

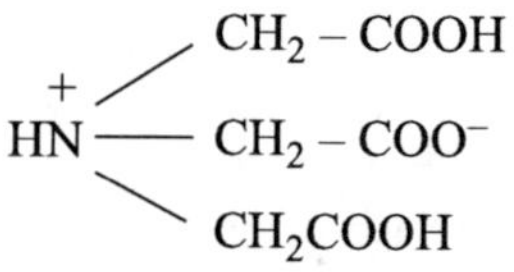

also named complexone I or H_3NTA. Its pK_a values are 1.9–2.5 and 9.7. This justifies the given formula:
- trans-1,2-diaminocyclohexane-N, N,N′,N′-tetraacetic acid (H_4CDTA):

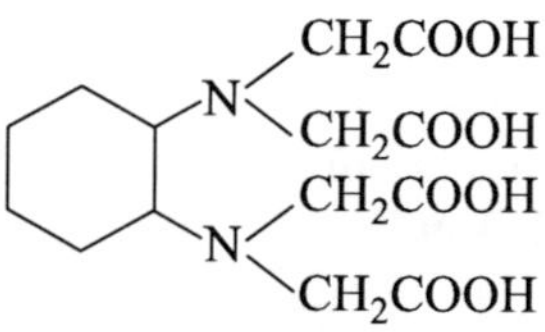

The given formula is probably not the only appropriate one to represent its true structure. There certainly exist other microforms, the existence of which is in connection with the occurrence of several protonation sites. Of course, all the microforms are in equilibria. This complexone often gives complexes more stable than those formed with EDTA, but it reacts more slowly than the latter;
- ethyleneglycolbis(2-aminoethyl ether)N, N, N′, N′-tetraa-cetic acid (EGTA) and triethylenetetramine-N, N, N′, N″, N‴, N‴-hexa-acetic acid (TTHA);

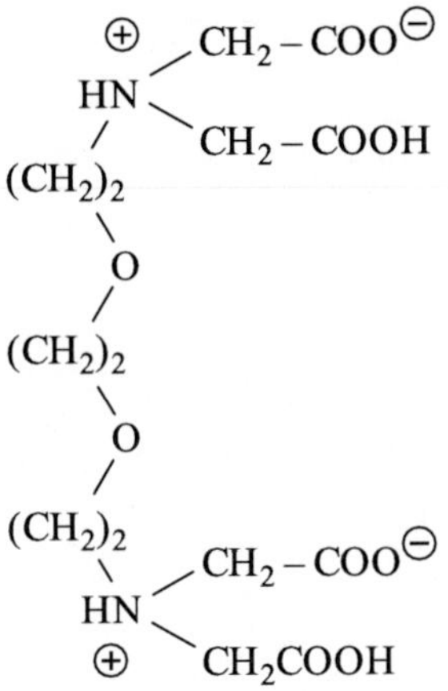
egta

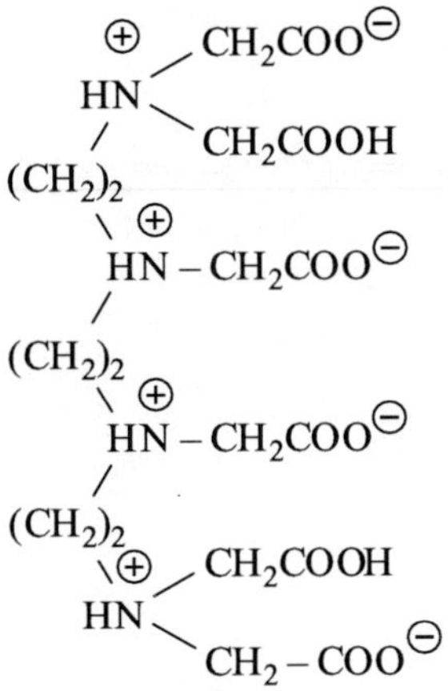
ttha

The principal application of EGTA is the determination of Ca^{2+} in admixture with Mg^{2+}. This is the reason why it is a useful reagent for the determination of the hardness of water (see Chap. 30). The origin of this property lies in the stability of the Ca–EGTA complex, whose stability constant is about 10^{11}, whereas that of the Mg–EGTA complex is only about 10^5. Hence, Mg^{2+} does not appreciably interfere during the titration of Ca^{2+}. The endpoint detection is carried out with zincon, which is the specific indicator of Zn^{2+} at pH 9–10. Ca^{2+} and Mg^{2+} do not give a colored complex with zincon. For the titration of $Ca^{2+}$with EGTA, a little quantity of the Zn–EGTA complex is added to the titrand solution. At pH $\approx$10, Ca^{2+} displaces Zn^{2+} from the Zn–EGTA complex according to the reaction

$$Ca^{2+} + [\text{Zn-EGTA}] \rightarrow [\text{Ca-EGTA}] + Zn^{2+}.$$

The liberated Zn^{2+} gives itself a complex with the indicator, making the solution blue:

$$Zn^{2+} + In^{4-} \rightarrow [ZnIn]^{2-}.$$

At the final point, the EGTA in excess displaces the indicator from its complex with Zn^{2+}. The liberated zincon gives the orange-red color of its free form:

$$[ZnIn]^{2-} + EGTA^{3-} \rightarrow [ZnEGTA]^{-} + In^{4-}.$$

Chapter 30
Applications of the Formation of Complexes in Inorganic Analysis

The formation of complexes is frequently used in qualitative and quantitative inorganic analysis.

Qualitative inorganic analysis carried out through the formation of complexes frequently involves other physicochemical phenomena in addition to that of complexation. Thus, there often exists a superimposition of several processes, among which complexation is present. The latter may indeed be followed, for example, by a precipitation. This is often the case with organic reagents such as oximes, dimethylglyoxime, cupferron, and so forth. As a result, gravimetric measurements might also be considered here. We will not consider them, however (see Part V, which is devoted to precipitation phenomena). The formed complexes may also be extracted in organic solvents that are very poorly miscible with water. Once extracted, and because of their characteristic UV spectra, these complexes can often be quantitatively analyzed by absorption measurements. All these cases are evoked in other parts of the book.

30.1 Qualitative Inorganic Analysis

In numerous cases, the formation of complexes permits the identification of inorganic ions after reaction with appropriate ligands. The identification may also be carried out with reagents that are already complexes. From another standpoint, it is possible that some ions disturb the preceding identifications. Hence, they may be masked to the identification reagents by complexing them. In the same connection, the complexation of an ion may allow it to avoid being precipitated in some experimental conditions and may even permit the dissolution of a precipitate to which it gives rise. Finally, it is sometimes possible to extract an ion into a nonpolar phase after complexation.

J.-L. Burgot, *Ionic Equilibria in Analytical Chemistry*,
DOI 10.1007/978-1-4419-8382-4_30,

30.1.1 Cations' Characterization

30.1.1.1 Fe^{3+} and Fe^{2+}

Fe^{3+} is often characterized with thiocyanate ions SCN^-. Potassium thiocyanate yields a bloody red color in even strongly acidic medium. It can be extracted into ether or benzylic alcohol. The color is attributed to the $[Fe(SCN)]^{2+}$ complex, which actually must be written as $[Fe(SCN)(H_2O)_5]^{2+}$. The simple salt $Fe(SCN)_3$ might also form. The $[Fe(SCN)_4]^-$ and $[Fe(SCN)_6]^{3-}$ complexes have also been isolated as salts. The thiocyanatofer(III) complex $[Fe(SCN)]^{2+}$ is decomposed by a solution of a mercury salt by exchange of the thiocyanate ligand:

$$[Fe\,(SCN)]^{2+} + Hg^{2+} \rightarrow \left[Hg\,(SCN)\right]^+ + Fe^{3+}$$

$$\text{and } [Fe\,(SCN)]^{2+} + \left[Hg\,(SCN)\right]^+ \rightarrow \left[Hg(SCN)_2\right] + Fe^{3+}$$

in accordance with the stability constant values of the different complexes. The $[Fe(SCN)]^{2+}$ complex is weakly stable ($K_1 = 10^{2.1}$). It decomposes upon dilution in water.

Fe^{3+} may also be characterized by the reaction it gives with the hexacyanoferrate(II) anion $[Fe(CN)_6]^{4-}$ formerly called the ferrocyanide ion. A deep blue precipitate called Berlin blue or Prussian blue forms. It is interesting to notice that the ferrocyanide ion is a complex itself. The structure of ferric hexacyanoferrate-(II), namely, $Fe_4[Fe(SCN)_6]_3$, has been ascribed to Prussian blue.

The soluble colloidal precipitate might have K^+ or $Na^+[Fe^{III}Fe^{II}(CN)_6]^-$, xH_2O as a formula and the insoluble form would have the structure $Fe^{3+}[Fe^{III}Fe^{II}(CN)_6]_3{}^-$, 14 to 16 H_2O. The intense color is obtained from the charge transfer phenomenon occurring between iron ions. It does not come from d-d transitions.

Otherwise, Fe-(III) may be characterized with o-phenanthroline after reduction to Fe(II) with sodium hydrogen sulfite or with ascorbic acid. Fe^{2+} may also be characterized by the complexes it gives with 2,2′-dipyrydile or with terpyridine, which yield a very stable red color with ferrous salts. The $[Fe(o\text{-}phen)_3]^{2+}$ complex exhibits the stability constant $\beta_3 = 10^{17}$ (see Chap. 23). Fe^{2+} can also be characterized with hexacyanoferrate(III), formerly called the ferricyanide ion. In this case, a blue color called Turnbull blue arises. For a long time, it was believed that this color was due to the formation of ferrous ferricyanide $Fe_3[Fe(CN)_6]_2$, but this was wrong. Actually, hexacyanoferrate(III) (which is an oxidant) oxidizes Fe^{2+} to Fe^{3+} by reducing itself to hexacyanoferrate(II):

$$Fe^{2+} + \left[Fe(CN)_6\right]^{3-} \rightleftharpoons Fe^{3+} + \left[Fe(CN)_6\right]^{4-}.$$

As a result, ferric hexacyanoferrate(II), that is, Prussian blue, is formed as in the characterization of Fe^{3+}. The analysis of both precipitates obtained in the preceding two characterizations proves that they are identical.

Fe^{3+} and Fe^{2+} are very interesting cations in inorganic and organic analysis (see also Chap. 31) because they form numerous complexes of analytical interest. This is the reason why we give some further information about them here. First, we recall that the aqua cation of a pale violet color $[Fe(H_2O)_6]^{3+}$ is easily hydrolyzed. It is an acidic cation. The hydrolysis phenomena to which it gives rise are complicated. For example, in some pH conditions, the following dimer forms (see Chap. 25):

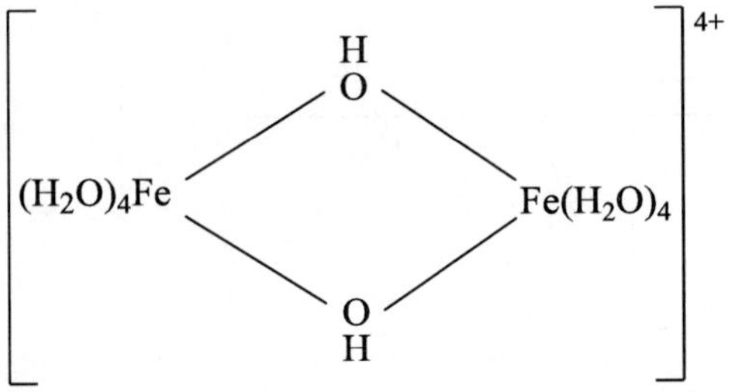

$[Fe(H_2O)_6]^{3+}$ only predominates at pH = 0. Fe^{3+} may also give other sorts of complexes in aqueous solutions according to the pH conditions. Hence, in acetic buffer (4 < pH < 5), Fe^{3+} precipitates as ferric hydroxide $Fe(OH)_3$, which forms due to the decomposition of the "basic acetate" $[Fe_3(OH)_2(CH_3COO)_6]^+$, CH_3COO^-, $3H_2O$, whose structure is

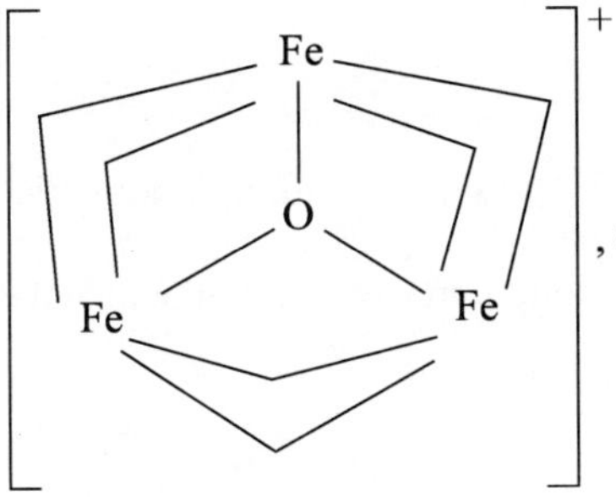

, $CH_3COO^- \cdot H_2O$

It is hexa-μ-acetato-μ_3-oxotrifer(III).

Fe^{3+} gives complexes easily despite its weak tendency to link with ligands that are only N-donors (Fe^{3+} does not give "amino" complexes, unlike Fe^{2+}). $[Fe(phen)_3]^{3+}$ also deteriorates over time with the formation of Fe^{3+} hydroxo complexes (phen: 1,10-phenanthroline). This is not the case with $[Fe(phen)_3]^{2+}$. However, Fe^{3+} gives complexes with ligands possessing both O and N donors. It is the case, for example, with EDTA, which forms the species $[Fe(EDTA)(H_2O)]^-$, whose structure is

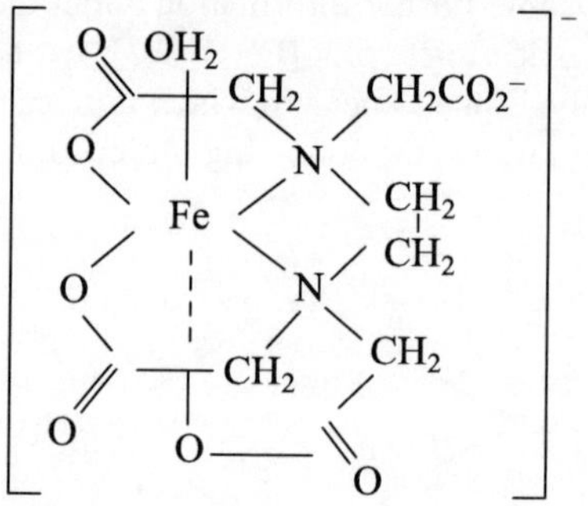

It is also the case with the ligand bis(salicylidene)ethylenediamine (H_2 salen) or 2,2′(-[1,2-ethanediylbis(nitromethylidyne)] diphenol (arrows indicate its anchoring points):

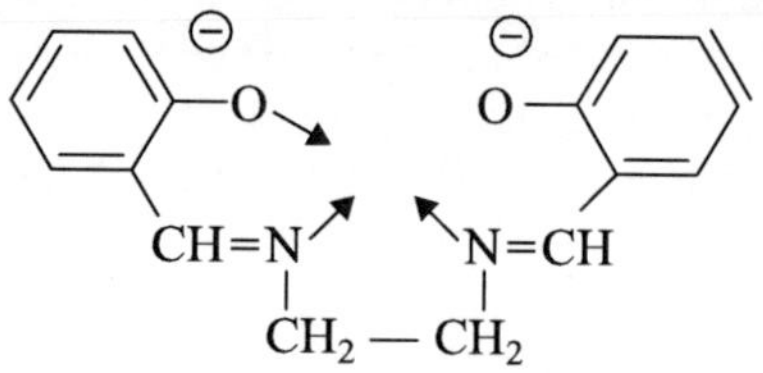

This ligand is a Schiff base. In the presence of chloride ions, the [(Fe(salen)Cl)] complex forms.

As some other ions, Fe^{3+} also yields complexes with the ligand diethyldithiocarbamato:

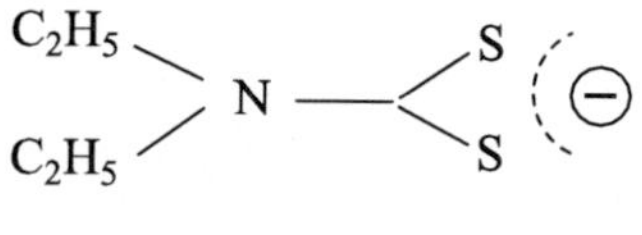

N, N-dialkyldithiocarbamatofer(III) $[Fe(S_2CNR_2)_3]$ exhibits the following structure:

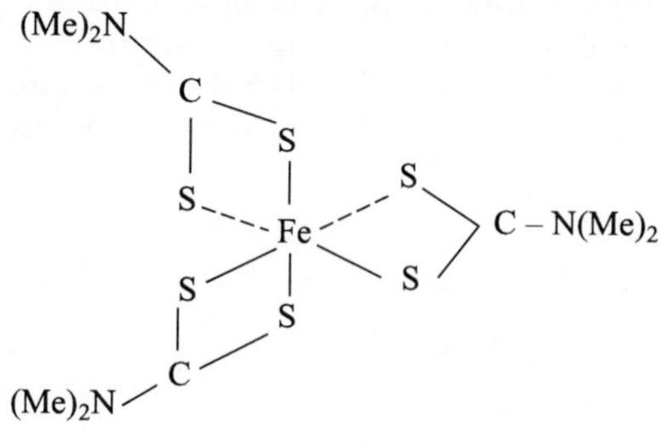

Finally, we recall that halide ions also yield complexes with Fe^{3+}. Their stability decreases with the nature of the halide.

The d^5 structure of Fe^{3+} does not particularly favor one geometry over another in the complexes formed. This is the reason why, with this ion, we find the following geometries:

- tetrahedral in $[FeCl_4]^-$,
- square-pyramidal in $[Fe(Hacac)_2Cl]$,where Hacac represents 2, 4-pentanedione,
- octahedral in $[Fe(CN)_6]^{3-}$ or in tris(oxalato)ferrate(III),
- bipyramidal-pentagonal in $[Fe(EDTA)(H_2O)]^-$,
- dodecahedral in $[Fe(NO_3)_4]^-$.

In aqueous solutions, Fe^{2+} is actually the pale green aqua complex $[Fe(H_2O)_6]^{2+}$. In the solid state, we find it as the Mohr salt $(NH_4)_2SO_4FeSO_4{,}6H_2O$ [diammonium hexaaquafer(II) bis sulfate], which is a double salt, weakly oxidizable, used in oxidoreductimetry (see Chap. 20). Fe^{2+} yields complexes with numerous ligands, but they are mostly less stable than those given by Fe^{3+} because of the cationic charge, which is weaker in the former cation. However, it is able to yield complexes with N-donor ligands. Most of the complexes given by Fe^{2+} are octahedral, but other geometries are known. The hexacyanoferrate-(II) complex is octahedral. The complexes given with bipyridine, terpyridine, and phenanthroline are also octahedral. Those resulting from the replacement of one ligand CN^- by H_2O, CO, NO_2, and NO^+ are known. In the case of NO^+, the complex obtained is nitroprusside $[Fe(CN)_5(NO)]^{2-}$ used as an indicator in Votocek's method (see Chap. 27). We would like to point out that an ambiguity exists for the description of this complex. It concerns the oxidation numbers of iron and of nitrogen oxide, which can be considered as being simultaneously either Fe^{2+} and NO^+ or Fe^{3+} and NO.

30.1.1.2 Cu^{2+}

Cu^{2+} may be characterized by several reagents with which it forms complexes:

- rubeanic acid (or dithioxamide). A black precipitate of copper-(II) rubeanate forms in ammonia. Actually, it is the bis-(ethanedithioamide)copper-(II) complex:

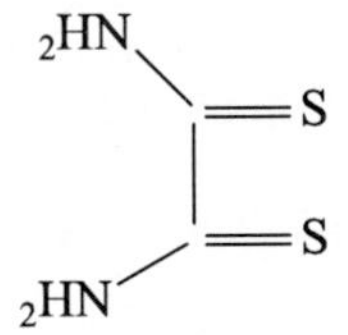

rubeanic acid

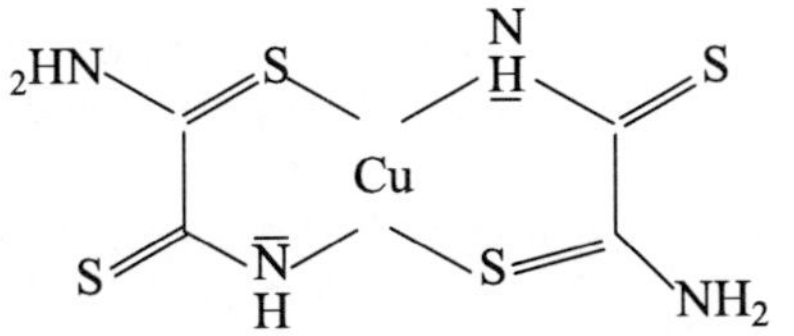

bis(ethanedithioamide)copper(II) complexe

- diphenylcarbazide $O{=}C(NH{-}NH{-}C_6H_5)_2$. A red-violet color is obtained in weakly acidic medium (pH > 2,2). Some other metallic ions may interfere.

- sodium N, N-diethyldithiocarbamate (NaEt$_2$dtc). In neutral medium or ammonia, this reagent yields a yellow color or a brown precipitate of diethyldithiocarbamato complex, sometimes wrongly called copper diethyldithiocarbamate. It is extractible into ethyl acetate. Fe^{3+} is disturbing. If it is present, EDTA is added in order to dissimulate it and more generally to dissimulate other disturbing cations. The dissimulation of parasitic cations by complexation with EDTA is justified by the stability constants of the involved complexes and by the extraction constant of diethyldithiocarbamato complex. We recall, incidentally, that NaEt$_2$dtc permits the extraction of about 20 "heavy" metallic cations into an organic phase.
- dithizone. This is diphenylthiocarbazone:

In solution, it may behave like the above mixture of two tautomeric forms. With Cu^{2+}, it yields a red-violet color in the pH range $0.5 < pH < 5$. It is also able to complex other metallic ions.

- ammonium tetrathiocyanatomercurate(II) (ammonium mercurithiocyanate) $[Hg(SCN)_4](NH_4)_2$. It yields a green precipitate with Cu^{2+}. ($[Hg(SCN)_4]^{2-}$ is a square complex.) Again, other cations may interfere.

Let's recall that with ammonia, Cu^{2+} gives the deep blue cation complex hexammine copper(II) (cupritetrammine). In analytical toxicology, Cu^{II} is searched for in the liquid resulting from the sulfonitric mineralization and is identified in it with ammonia. Because of its weak concentration in the mineralization liquid, Cu^{2+} is first displaced by metallic iron, on which it deposits in the form of a red coating of metallic copper. The deposit is purely and simply the result of the following redox reaction:

$$Cu^{2+} + Fe \rightarrow Cu + Fe^{2+}.$$

Depositing copper on the iron point permits its concentration. In the presence of ammonia, copper is oxidized to Cu^{II} and is simultaneously complexed. It is interesting to point out the fact that copper is oxidized by air in the presence of ammonia. The oxidization occurs because of the formation of the complex $[Cu(NH_3)_m]^{2+}$, which displaces the oxidization reaction toward the right.

The first step in the search for a mineral poison is the sample mineralization. Organic matters must indeed be destroyed since they may dissimulate the ion with respect to their reagents. After mineralization, the element is converted into an ion. Numerous methods have been proposed to carry out mineralizations. We'll only mention the mineralization with $Cl_{2(g)}$, which is insufficient since it does not destroy lipids, and the mineralization with the sulfonitric mixture alone or added with some adjuvants such as manganous salts, perhydrol, perchloric acid, and so on.

Recall also that copper(II) hydroxide yields deep blue anionic complexes with some acid-alcohols such as lactic and tartaric acids. They are the oxidizing reagents of Fehling's solution. The structure of such a reagent [the bis(tartrato)cuprate(II) complex]is given below:

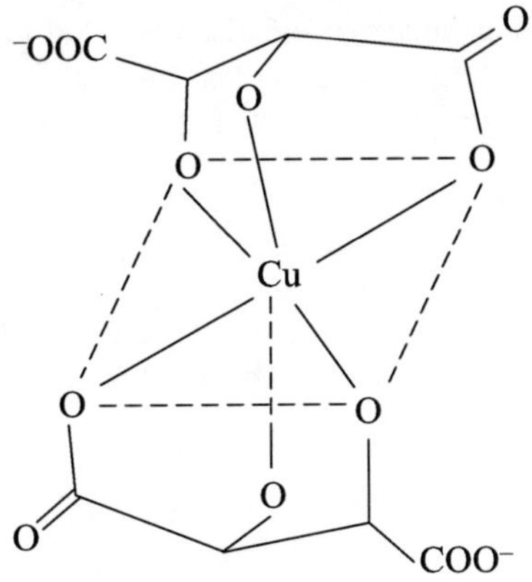

bis(tartrato)cuprate (II)

Fehling's solution is prepared by dissolution of Seignette's salt (still named Rochelle's salt) in an alkaline solution of copper(II) sulfate. Seignette's salt is the sodium and potassium double tartrate. It is a tetrahydrate.

Hydrogen sulfide H_2S yields a black precipitate of copper(II) sulfide CuS ($K_s = 10^{-35.1}$). Despite its very weak solubility product, it is soluble in solutions of alkaline cyanides by the formation of the very stable tetracyanocuprate(I) complex: $[Cu(CN)_4]^{3-}$, also named cuprocyanogen anion. We can schematize the dissolution of copper sulfide in the presence of alkaline cyanides by the following reaction:

$$2Cu_{(s)} + 10CN^- \rightleftharpoons 2S^{2-} + (CN)_2 + 2\left[Cu(CN)_4\right]^{3-}.$$

We point out the formation of cyanogen (ethanedinitrile) together with that of the complex. The overall reaction can be decomposed into the following three reactions:

$$2Cu_{(s)} + 4CN^- \rightleftharpoons 2Cu(CN)_2 + 2S^{2-},$$

$$2Cu(CN)_2 \rightleftharpoons 2CuCN + (CN)_2,$$

$$2CuCN + 6CN^- \rightleftharpoons 2\left[Cu(CN)_4\right]^{3-}.$$

The intermediary reaction that gives cyanogen is actually a redox reaction. The first redox couple is Cu^{2+}/Cu^+, and the second is $(CN)_2/CN^-$. We can remark that during the reaction, half of the cyanide ions play no part and remain at their initial oxidation state −I. A good argument in favor of the occurrence of these reactions is the fact that the unstable yellow copper(II) cyanide precipitates by yielding copper(I) cyanide and cyanogen. Copper(I) cyanide is soluble in an excess of alkaline cyanide, yielding the anion complex tetracyanocopper(I) (cuprocyanogen anion). Let's recall that copper cannot be precipitated by hydrogen sulfide from a solution of cuprocyanogen when a sufficient excess of cyanide ions exists in the solution.

Copper(II) yields a black precipitate of copper(II) thiocyanate, which is transformed into copper(I) thiocyanate, itself dissolving into an excess of thiocyanate ions with the formation of the anionic complex $Cu(SCN)_2{}^-$. The transformation copper(II) thiocyanate → copper(I) thiocyanate is favored by the presence of reductants such as sulfite ions:

$$2Cu(SCN)_2\downarrow + SO_3{}^{2-} + H_2O \rightarrow SO_4{}^{2-} + 2H^+ + 2SCN^- + 2CuSCN\downarrow$$

Finally, we'll mention that Cu^{2+} gives the biuret reaction with some organic structures endowed with some structural moieties. This is a means to characterize it.

30.1.1.3 Ag^+

With chloride ions, Ag^+ yields a white precipitate of silver chloride that blackens in bright light. It is easily solubilized

- in ammonia by forming the diamminesilver(I) complex: $[Ag(NH_3)_2]^+$;
- in potassium cyanide solutions by giving the dicyanosilver(I) $[Ag(CN)_2]^-$ complex according to the reaction

$$AgCl\downarrow + 2CN^- \rightarrow \left[Ag(CN)_2\right]^- + Cl^-;$$

- in sodium thiosulfate solutions by forming the di(thiosulfato)silver(I) complex:

$$AgCl\downarrow + 2S_2O_3{}^{2-} \rightarrow \left[Ag(S_2O_3)_2\right]^{3-} + Cl^-.$$

Again, the solubility product and the stability constant values are in accordance with these reactions, which are examples of the dissolution of a precipitate by complexation:

$$K_1\left(Ag(S_2O_3)^-\right) = 10^{8.82}, \quad K_2\left(Ag(S_2O_3)_2{}^{3-}\right) = 10^{4.64},$$
$$K_3\left(Ag(S_2O_3)_3{}^{5-}\right) = 10^{0.69}.$$

Recall that silver chloride is soluble in an excess of chloride ions by the formation of the ion complex chlorosilver(I) (see Chap. 25). Bromide ions give a clear yellow color of silver bromide that dissolves in the same reagents as the chloride ion but with more difficulty. Iodide ions give a silver iodide precipitate insoluble in ammonia but soluble in potassium cyanide and sodium thiosulfate. From another standpoint, Ag^+ ions give silver dithizonate with dithizone. The yellow precipitate, in acidic medium, is extractible into carbon tetrachloride. The reaction is nonspecific. Ag^+, in acidic medium, also yields a precipitate or a red color with para-dimethylaminobenzylidene rhodanine. It has the structure of the [5-[[2-hydroxy-5-(phenylazo)phenyl]methylene]2-(thioxo(κ)-S)-4-thiazolidinonato(κ)-N3]silver complex. It exhibits the following structure:

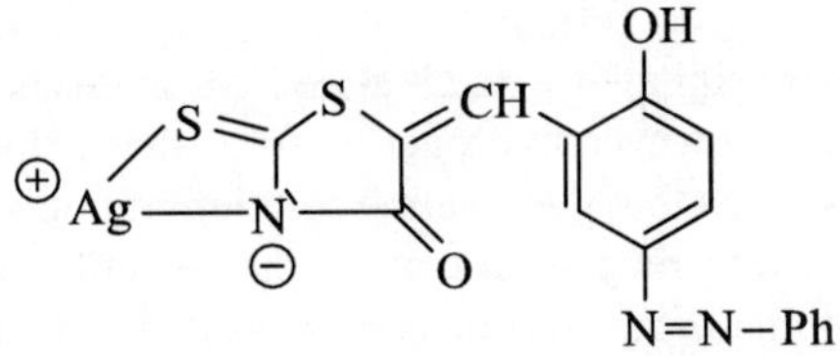

[The κ (kappa) symbol is used to signal the element through which the ligand is linked to the central atom. In this case, the atoms linked to Ag^+ are the sulfur of the thiocarbonyl group and the nitrogen of the thiazolidinone rest.]

30.1.1.4 Hg^{2+}

Hg^{2+} can be detected with dithizone. A yellow-orange precipitate forms that is extractible into carbon tetrachloride or chloroform in acidic medium ($1 < pH < 4$). The reaction is quasi-specific at $pH \leq 1.5$ in the presence of complexone(III) in order to complex other cations, in particular Cu^{2+}.

It can also be detected with diphenylcarbazide or diphenylcarbazone, which yields a blue-violet color in acidic medium, even in 0.2 mol/L nitric acid. The diphenylcarbazone/Hg^{2+} complex is extractible into benzene. Copper, iron, and cobalt are not disturbing in 0.2 mol/L nitric acid. However, $CrO_4{}^{2-}$, $Cr_2O_7{}^{2-}$, and molybdic anions are.

Alkaline hydroxides yield a yellow precipitate of mercuric oxide HgO as early as $pH >$ about 2.4. Ammonia transforms it into a yellow complex named Millon's base, with the formula $[Hg_2N(OH)(H_2O)_2]$. The latter, with dichloromercure(II) (wrongly called "mercuric chloride"), yields white precipitate with the formula $[Hg(NH_2)Cl]$, which is called an *infusible white precipitate* because it volatilizes without melting. The corresponding reactions are

$$HgCl_2 + 2NH_3 \rightleftharpoons \left[Hg(NH_3)_2Cl_2\right],$$

$$\left[Hg(NH_3)_2Cl_2\right] \rightleftharpoons \left[Hg\,(NH_2)\,Cl\right] + NH_4Cl.$$

The intermediary complex $[Hg(NH_3)_2Cl_2]$ is called a *fusible precipitate*.

Sodium carbonate yields a brown-red basic chloride $Hg_3O_2Cl_2$ in accordance with the reaction

$$3Hg^{2+} + 2Cl^- + 2CO_3{}^{2-} \rightarrow 2CO_2 + Hg_3O_2Cl_2.$$

Iodide ions yield a red precipitate of mercuric iodide soluble in excess of the reagent by formation of the colorless tetraiodomercurate(II) (mercuritetraiodide) complex:

$$Hg^{2+} + 2I^- \rightarrow HgI_2\downarrow$$

$$HgI_2\downarrow + 2I^- \rightarrow \left[HgI_4\right]^{2-}.$$

Hg^{II} can no longer be precipitated with alkaline hydroxides, sodium carbonate, and so on from this complex. However, hydrogen sulfide still precipitates Hg^{II} by forming its black sulfide. Let's recall that the alkaline solution of potassium tetraiodomercurate(II) constitutes Nessler's reagent. It permits the detection of ammonia and, from a more general standpoint, of nitrogenous bases. With ammonia, the $[Hg_2NI(H_2O)]$ complex forms, which is an ododerivative of Millon's base. Potassium iodide and strychnine sulfate also permit us to detect traces of mercury.

Cyanide ions do not give any precipitate with mercury(II) salts, $[Hg(CN)_2]$ being freely soluble in water. In the presence of an excess of reagent, higher complexes are formed according to the reactions

$$\left[Hg(CN)_2\right] + CN^- \rightleftharpoons \left[Hg(CN)_3\right]^-,$$

$$\left[Hg(CN)_3\right]^- + CN^- \rightleftharpoons \left[Hg(CN)_4\right]^{2-}.$$

Hg^{II} can no longer be precipitated from these complexes with alkaline hydroxides, ammonia, and alkaline carbonates. However, it can be precipitated with hydrogen sulfide and alkaline sulfides. The black sulfide HgS is formed.

As we have already said, chloride ions yield the successive complexes $[HgCl]^+$, $[HgCl_2]$, $[HgCl_3]^-$, and $[HgCl_4]^{2-}$. Likewise, Hg^{2+} gives the series of thiocyanato complexes $[Hg(SCN)]^+$ through $[Hg(SCN)_4]^{2-}$. The "mercuric thiocyanate" that precipitates is therefore soluble in excess of the reagent.

30.1.1.5 Pb^{2+}

Pb^{2+} can be detected with chloride ions. Lead chloride $PbCl_2$ forms and precipitates. It is soluble in concentrated solutions of sodium chloride or hydrochloric acid. The trichloroplombate(II) complex (chloro plombous anion) $[PbCl_3]^-$ forms. It is poorly stable. For example, it is dissociated upon simple dilution of its solutions. Then $PbCl_2$ crystallizes in white grains.

It is the same state of affairs with iodide ions. The yellow precipitate of lead iodide PbI_2 is soluble in the presence of an excess of iodide ions by formation of the triiodoplombate(II) complex: $[PbI_3]^-$, poorly stable. By cooling, lead iodide crystallizes as yellow grains. This reaction is carried out to detect Pb^{2+} in mineralization liquids used in analytical toxicology. The lead iodide precipitation is inhibited by the addition of complexone(III).

Cyanide ions yield a white precipitate of lead cyanide soluble in concentrated solutions of cyanide ions by formation of the tetracyanoplombate(II) complex, $[Pb(CN)_4]^{2-}$:

$$Pb^{2+} + 2CN^- \rightleftharpoons Pb(CN)_2\downarrow$$

$$Pb(CN)_2\downarrow + 2CN^- \rightleftharpoons \left[Pb(CN)_4\right]^{2-}.$$

Chromate ions $CrO_4{}^{2-}$ yield a yellow precipitate of lead chromate $PbCrO_4$:

$$Pb^{2+} + CrO_4{}^{2-} \rightarrow PbCrO_4\downarrow$$

The precipitate dissolves in alkaline hydroxide solutions by the formation of the tetrahydroxoplombate(II) complex, $[Pb(OH)_4]^{2-}$.

With Pb^{2+}, dithizone yields a red complex that is soluble in carbon tetrachloride and chloroform. This reaction is not specific. Cu^{2+}, Zn^{2+}, Cd^{2+}, and Bi(III) also yield a red color with dithizone.

In order to avoid these interferences, the reaction is carried out in the presence of potassium cyanide, which complexes Zn^{2+} and Cu^{2+} in ammonia. In these conditions, lead dithizonate is the only extractible species.

Let's recall that EDTA (trilon B) is used as a soluble antidote of some heavy metals, in particular lead. Lead reacts with it via a chemical mechanism. The principal drawback of its use is the complexation of Ca^{2+}, which is accompanied by a possible serious hypocalcic state with phenomena of tetany. This is the reason why the use of calcium complexonate $[Ca(EDTA)]^{2-}$ is recommended. Lead complexonate, by far more stable than the calcium one, forms at the expense of the latter. Actually, the administration of $[Ca(EDTA)]^{2-}$ quickly leads to the discharge of lead, eliminated as complexonate in urines.

30.1.1.6 Bi-III

Bi-III is detected at pH = 0 in hydrochloric acid medium. In these conditions, the cation Bi^{3+} seems to exist. By dilution with water, there is precipitation of basic salts or bismuthyl salts, which form through bismuth hydroxide $Bi(OH)_3$, which, by a subsequent loss of water, yields bismuthyl hydroxide BiO(OH):

$$Bi(OH)_3 \rightarrow BiO\,(OH) + H_2O.$$

Bismuth hydroxide is soluble in alkaline solutions of citric and tartaric acids and of mannitol and glycerol. The structure of these complexes would be similar to that of the complexes obtained with antimony in the same conditions. These complexes, treated by alkaline sulfides, generate the black precipitate of bismuth sulfide Bi_2S_3.

Iodide ions yield a brown-black precipitate of bismuth iodide BiI_3, which is soluble in excess of iodide ions by forming an orange-red solution due to the tetraiodobimuthate(III) complex, $[BiI_4]^-$ (iodobismuthic or iodobismuthate anion):

$$BiI_3\downarrow + I^- \rightarrow [BiI_4]^-.$$

If this solution is diluted, the orange bismuthyl iodide BiOI deposits. (Lead is disturbing because of the precipitation of yellow PbI_2. It is also the case with copper since it liberates iodine, which is brown. By adding a reductor such as sulfurous anhydride or sodium hypophosphite, iodine is reduced.) The iodobismuthic complex is extractible into varied solvents. The iodobismuthate ion is of particular importance. Like its potassium salt, it constitutes Draggendorf's reagent, which yields characteristic red precipitates with alkaloids together with varied organic nitrogenous molecules. Hence, with quinine hydrochloride, a red precipitate forms that is insoluble in water, having the formula

$$QH_2{}^{2+}, 2[BiI_4]^-,$$

where Q represents the quinine base molecule.

In toxicology, the bismuth identification is based on Leger's reaction, which involves the formation of red-orange precipitates with alkaloids that are alkaloid iodobismuthites. An example already mentioned is that given by quinine. Of course, the identification is carried out with the mineralization liquid.

Bi-III yields a yellow complex with thiourea in acidic medium (tu = thiourea):

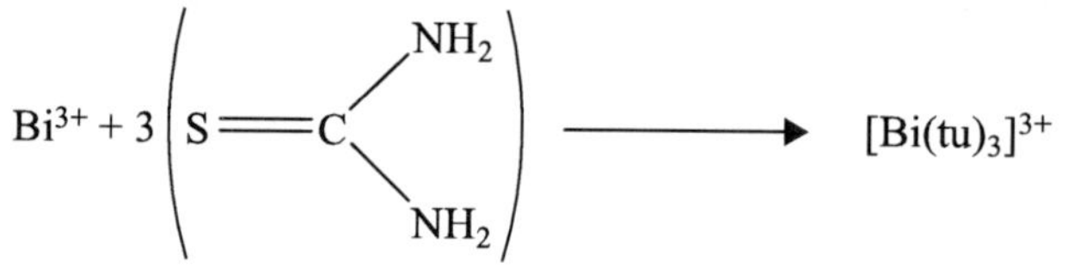

The tris(thiourea)Bi(III) complex is decomposed by fluoride ions with the formation of insoluble bismuth trifluoride BiF_3.

With sodium diethyldithiocarbamate, bismuth yields a yellow compound extractible into carbon tetrachloride. In the presence of complexone(III), of alkaline cyanide, the reaction is quasi-specific since only lead and mercury are disturbing.

30.1.1.7 Zn^{2+}

In mineral acidic medium, Zn^{2+} does not yield any precipitate with H_2S. In organic acidic medium (CH_3COOH, HCOOH), white zinc sulfide ZnS precipitates. ZnS is also precipitated by ammonium and alkaline sulfides. Alkaline hydroxides NaOH, KOH yield a white precipitate of zinc hydroxide $Zn(OH)_2$ for pH values such as $pH > 6.8$:

$$Zn^{2+} + 2OH^- \rightarrow Zn(OH)_2\downarrow$$

In the presence of the ammonium cation, zinc hydroxide cannot precipitate either with alkaline hydroxides or with ammonia because of the formation of soluble ammine zinc complexes such as the tetraamminezinc(II) cation $[Zn(NH_3)_4]^{2+}$. However, Zn^{2+} can be precipitated from these complexes by alkaline sulfides:

$$[Zn(NH_3)_4]^{2+} + S^{2-} \rightarrow ZnS\downarrow + 4NH_3.$$

Its hydroxide is soluble in an excess of alkaline hydroxide by the formation of zincates $[Zn(OH)_3]^-$ and $[Zn(OH)_4]^{2-}$, which are also complexes.

In analytical toxicology, the zinc identification is based on Montequi's reaction. It consists of the formation of a crystalline violet precipitate of copper and zinc mercurithiocyanate: $[Hg(SCN)_4]_2CuZn$.

Let's also recall that Zn^{2+} yields weakly stable amino, cyano, thiocyano, and chloro complexes.

30.1.1.8 Sb-III

Sb-III, as antimonous oxide Sb_2O_3, may be detected by its dissolution in tartaric acid or potassium acid tartrate solutions. With the latter, it forms the complex called *emetic*. Its formula is

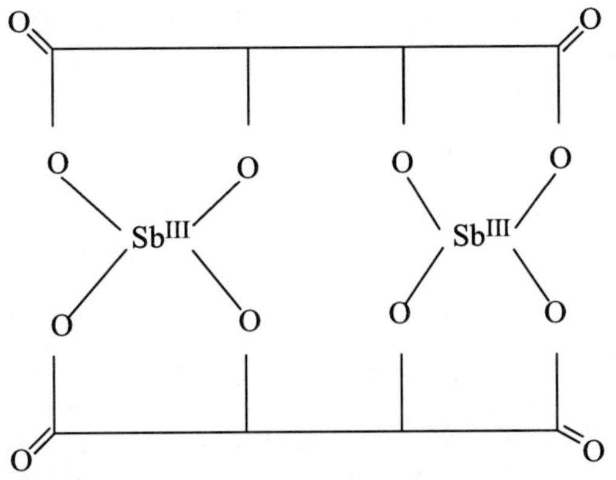

dipotassium bis[μ-2,3-dihydroxybutanedioato (4-)-0,1,02,O3,04] trihydrate

In this complex, antimony exhibits coordinence 4.

By reaction with hydrogen sulfide or with sodium sulfide in weakly acidic medium, the red-orange antimonous sulfide Sb_2S_3 precipitates from the solution of this complex. (It is soluble in 0.5 mol/L hydrochloric acid). The sulfide is soluble in alkaline hydroxides with the formation of antimonous anions $SbOS^-$, $SbOS_2^-$, and SbS_3^{3-}.

Sb-III also yields complexes with oxalic acid and with fluoride ions.

Sb(III) may also be identified after transformation into Sb(V), which gives a complex with methyl violet. This blue complex is soluble in toluene. This is Kuznetsov's reaction. Likewise, with rhodamine B, Sb(V) yields a violet color extractible into benzene. Note that in analytical toxicology, antimony has reached oxidation state +III after mineralization. Hence, Sb(III) must be converted firstly into Sb(V) before both preceding reactions being carried out. The oxidization can be carried out with permanganate or with perhydrol. In the first case, an excess of permanganate (which would be disturbing for the preceding reactions' developments) is eliminated by adding hydroxylamine chlorhydrate, which reduces it.

Finally, we'll mention that Sb(III) yields an orange-red precipitate due to the formation of a complex with fluorone.

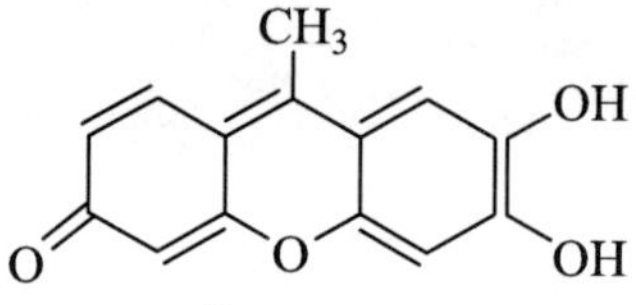

fluorone

30.1.1.9 Al^{3+}

With alkaline hydroxides, Al^{3+} yields hydroxide $Al(OH)_3$, which appears as a white precipitate at pH values about 3.9–4.0. It is soluble in an excess of alkaline hydroxide, as it gives the mononuclear aluminate ion $[Al(OH)_4]^-$:

$$Al(OH)_3\downarrow + OH^- \rightarrow \left[Al(OH)_4\right]^-,$$

which may be considered a complex. Adding an ammonium salt to the aluminate solution induces a new precipitation of $Al(OH)_3$. This phenomenon is explained by the fact that by adding the ammonium salt, the pH value, which was very high in order to obtain the aluminate ion (pH about 11–12), decreases until a value about 9–10 via the formation of the ammoniac buffer. It is also explained by the fact that aluminum hydroxide dissolves in the pH range 10.0–12.6, a value that varies depending on the polymerization of the precipitate as well as its age.

Al^{3+} also yields a colored complex with aluminon (ammonium aurinetricarboxylate):

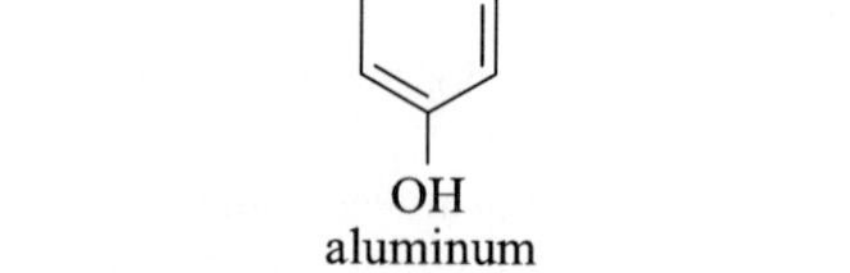

aluminum

The formed complex exhibits the structure Al(aluminon)$_3$.

Al^{3+} also gives a red-violet complex with

- a derivative of triphenylmethane, eriochrome cyanine R:

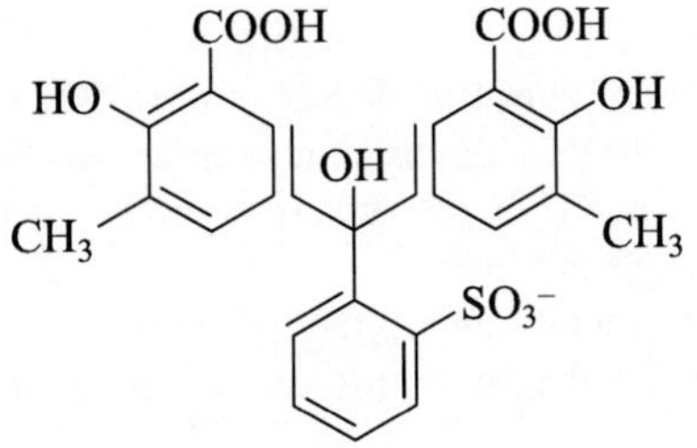

eriochromeacyanine[R]

- alizarine S (sodium 1,2-dihydroxyanthraquinone-3-sulfonate) in acetic medium:

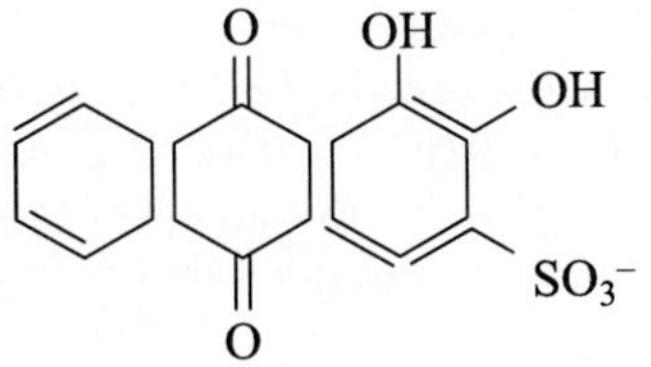

Alizarine S

The formula of the complex obtained is Al(alizarine)$_3$;

- acidic chrome blue 2R (2–2′-dihydroxy-4-sulfonaphthalene-azo-naphthalene):

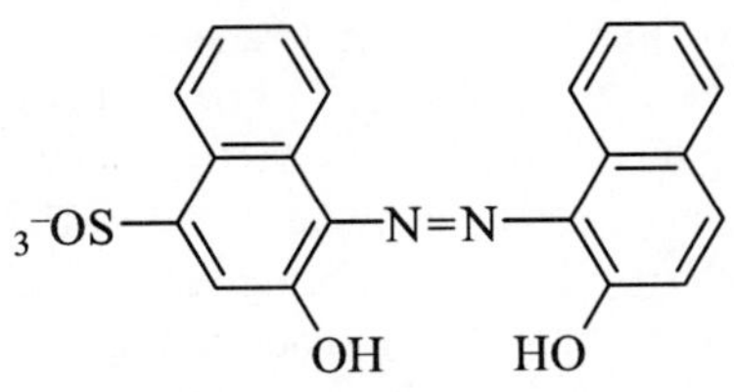

The complex obtained exhibits a very sensitive orange fluorescence.

30.1.1.10 Cd^{2+}

Cd^{2+} is very often identified by the precipitation of its yellow sulfide CdS in alkaline medium. The reaction is carried out in the presence of cyanide ions, which complex disturbing ions (Cu^{2+}, Fe^{3+}, etc.). Let's also recall that in alkaline medium, arsenic sulfide does not precipitate.

From another standpoint, with alkaline hydroxides, Cd^{2+} yields a white precipitate of cadmium hydroxide $Cd(OH)_2$ that is soluble in ammonia by the formation of the successive complexes $[Cd(NH_3)_m]^{2+}$, where m takes all the values up to 6. The stability of these complexes is weak.

Cyanide anion yields a white precipitate of cadmium cyanide $Cd(CN)_2$ soluble in an excess of reagent by the formation of the tetracyanocadmiate(II) complex: $[Cd(CN)_4]^{2-}$;

$$Cd(CN)_2\downarrow + 2CN^- \rightarrow \left[Cd(CN)_4\right]^{2-}.$$

In the latter, cadmium is masked to all of its reagents except to hydrogen sulfide, which yields cadmium sulfide. Therefore, concerning the reaction with cyanide ions, cadmium has a different behavior from that of Cu^{2+}.

In analytical toxicology, cadmium is detected by the formation of a white precipitate of brucine iodocadmiate: $(\text{brucineH})_2[CdI_4]$, where brucineH represents the conjugate acid of brucine and $[CdI_4]^{2-}$ the tetraiodocadmiate(II) complex. The precipitation must be carried out in acetic medium after evaporation until dryness of the mineralization solution. Cadmium complexes are generally weakly stable.

30.1.1.11 As(III) and As(V)

As(III) and As(V) exhibit a weak tendency to yield complexes. As(III) yields a compound with diethyldithiocarbamate, which, in 1–5 mol/L H_2SO_4, is extractible into chloroform. As(V) does not do so.

AsH_3 (hydrogen arsenide) reacts with a silver diethyldithiocarbamate $[AgSC(S)N(Et)_2]$ solution in pyridine to yield a soluble red complex.

Regarding arsenic(-V), it is interesting to notice that it can be reduced in two stages. In the first one, As(V) is reduced to As(III) by stannous ions in hydrochloric acid. In the second stage, As(III) is reduced to As(-III) (hydrogen arsenide) by zinc in hydrochloric acid. As(V) may also be detected by an ammonium molybdate solution, which is more exactly a polymolybdate solution. The polymolybdate anion might be a complex having formula $[Mo(MoO_4)_6]^{6-}$ and, hence, ammonium molybdate might have the formula $[Mo(MoO_4)_6](NH_4)_6$. Adding molybdic reagent (ammonium molybdate in nitric acid) into an arseniate solution yields a heteropolymolybdoarsenate (or an arsenomolybdate) of imprecise structure that is reduced by stannous chloride or by hydrazinium sulfate with the formation of blue complexes. Blue complexes would be mixtures of molybdenum oxides and hydroxides whose extreme oxidation numbers would be +VI (MoO_3) and +V ($MoO(OH)_3$). It seems that in these mixtures the $(MoO_3)_6$ species have been characterized.

It is interesting to mention 2,3-dimercaptopropanol (dimercaprol or B.A.L.—british anti-Lewisite) as a detoxicant for heavy metal poisoning. It is an antidote to arsenic, gold, and mercury. [Lewisite is the dichloro(2-chlorovinyl)arsine.] Dimercaprol is a chelating agent. Arsines were used during World War I, 1914–1918, as war toxins because of their extreme toxicity. Arsines belong to the group of thiolo-prive toxins because they are complexed by protein thiol rests with the formation of protein—As compounds. Dimercaprol complexes As and some heavy metals ions more strongly than protein thiol groups do. It is the reason why some heavy metals ions are eliminated as combinations with B.A.L.

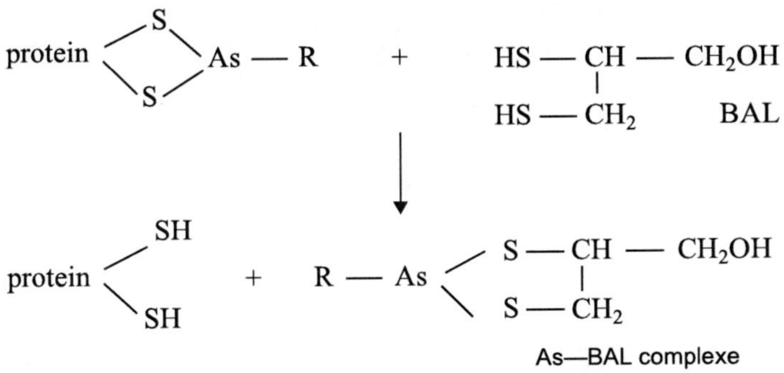

Excellent results have been obtained in the cases of mercuric intoxications. The detoxification mechanism is ultimately based on the ligand exchange:
protein $\rightleftharpoons$ BAL (see Chap. 18).

30.1.1.12 Ca^{2+}

Ca^{2+} can be detected by the formation of a red chelate with glyoxal-bis hydroxyanil:

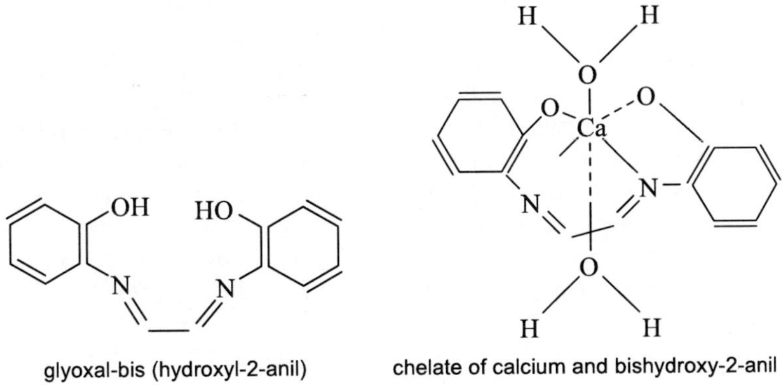

This reaction is not specific. Several cations respond positively to this reaction, such as uranyl cation: UO_2^{2+}. The chelate formed with Ca^{2+} is soluble in water. It is extractible into chloroform.

From another standpoint, Ca^{2+} gives a yellow complex in neutral medium with picrolonic acid. It may precipitate.

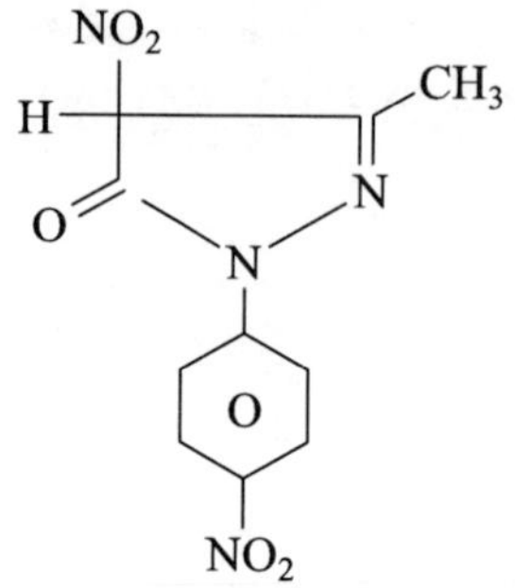

picrolonic acid

(Picrolonic acid is 1-p-nitrophenyl-3-methyl-4-nitro-5-pyrazolone.)

Ca^{2+} may also be detected via precipitation with oxalic acid as calcium oxalate, which is very poorly soluble. In a first reaction, barium and strontium must be eliminated as sulfates and other disturbing ions must be complexed by complexone(III). The latter, if added in excess, must not prevent the calcium precipitation. Its excess is eliminated by adding aluminum nitrate.

30.1.1.13 Mg^{2+}

Mg^{2+} gives weakly stable complexes with ammonia and also with HCO_3^- and NO_3^- ions. It gives more stable complexes with tartrate ion and EDTA. One means of detecting it is to use titane yellow. Properly speaking, this reaction is not a complexation one. Magnesium hydroxide indeed exhibits the property to adsorb on several dyes, in particular titane yellow, which passes from yellow to red; a red precipitate may possibly form. Titane yellow or thiazole yellow is 2,2′-[(diazoamino)di-p-phenylene]bis[6-methyl-7-benzothiazolesulfonic acid] disodium salt. The part played by the complexation is to allow the reaction to be selective. Most of the other ions that are able to precipitate together with magnesium may indeed interfere. They are maintained in solution by EDTA. An excess of a barium salt is added to eliminate the excess of EDTA, thus preventing the formation of the magnesium–EDTA complex. Of course, if it were the case, Mg^{2+} could not be evidenced.

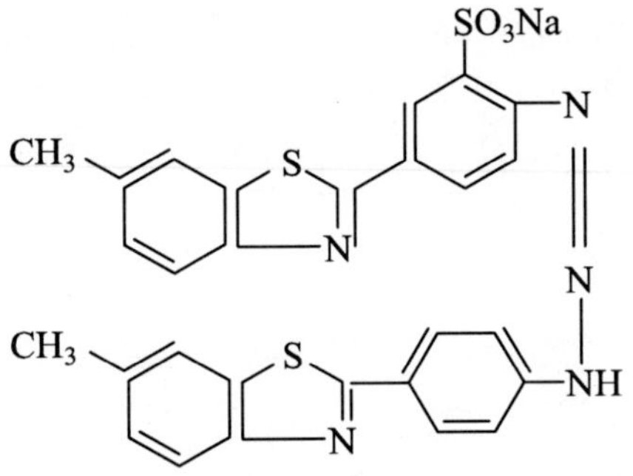

thiazole yellow

30.1.1.14 NH_4^+

NH_4^+ is detected by displacement of its conjugate base (ammoniac) via reaction with sodium hydroxide or magnesium oxide. Ammoniac yields a red-orange precipitate or a yellow color with Nessler's reagent [potassium tetraiodomercurate(II)]. Indeed, in alkaline medium, the liberated ammoniac gives the following reactions:

$$2[HgI_4]^{2-} + 2NH_3 \rightarrow 2\left[NH_3HgI_2\right] + 4I^-,$$

$$2\left[NH_3HgI_2\right] \rightarrow \left[NH_2Hg_2I_3\right] + I^- + NH_4{}^+.$$

These complexes easily dissociate. Well-defined conditions of dilution must be obeyed.

NH_4^+ also yields a yellow precipitate with sodium cobaltinitrite $Na_3[Co(NO_2)_6]$ in neutral medium. The precipitate obeys the formula of a double sodium and ammonium cobaltinitrite: $(NH_4)_2Na[Co(NO_2)_6]$. (In sodium cobaltinitrite, the ligand $NO_2{}^-$ is linked by the nitrogen atom. It is an N-donor ligand.)

30.1.1.15 Na^+

Na^+ yields a white precipitate with the anion complex antimoniate $[Sb(OH)_6]^-$ added as potassium salt. Sodium antimoniate forms according to

$$\left[Sb(OH)_6\right]^- + Na^+ \rightarrow \left[Sb(OH)_6\right]Na\downarrow$$

It also yields golden yellow precipitates with

- a uranyl magnesium acetate solution (this is Streng's reagent, which is a complex of uranyl and magnesium $[Mg(UO_2)_3(CH_3CO_2)_9]^-$):

$$H_2O + \left[Mg(UO_2)_3(CH_3CO_2)_9\right]^- + Na^+ \rightarrow Na\left[Mg(UO_2)_3(CH_3CO_2)_9\right], H_2O\downarrow$$

- uranyl zinc acetate reagent (Kolthoff's reagent, which is a complex of uranyl and zinc $[Zn(UO_2)_3(CH_3CO_2)_9]^-$):

$$nH_2O + Na^+ + \left[Zn(UO_2)_3(CH_3CO_2)_9\right]^- \rightarrow Na\left[Zn(UO_2)_3(CH_3CO_2)_9\right], nH_2O.$$

30.1.1.16 K^+

K^+ may be detected by the formation of several complexes:

- the octahedral anion complex hexachloroplatinate(II) $[PtCl_6]^{2-}$ (chloroplatinate). It yields a yellow crystalline precipitate of potassium chloroplatinate $K_2[PtCl_6]$, whereas the sodium salt is soluble. Ammonium ions give a similar precipitate. Hence, it must be eliminated as ammonia gas through boiling. Iodide ions must not be present in the solution because they form the octahedral complex hexaiodoplatinate-(II) $[PtI_6]^{2-}$, which is disturbing because of its intense brown-red color. Cyanide ions are also disturbing because they form the tetracyanoplatinate(II) complex $[Pt(CN)_4]^{2-}$;
- the anion complex hexanitratocobalt(III) (cobaltinitrite), which gives a yellow precipitate of sodium and potassium double cobaltinitrite $K_2Na[Co(NO_2)_6],H_2O$. As already seen, the ammonium ion gives the same reaction and must be eliminated as ammonia gas. Other disturbing ions must be complexed with complexone(III);
- the tris(thiosulfato)bismuthate(III) anion, $[Bi(S_2O_3)_3]^{3-}$ (bismuthothiosulfuric anion), which gives a white precipitate of potassium bismuthothiosulfate $K_3[Bi(S_2O_3)_3]$;
- the anion complex hexafluorosilicate(IV): $[SiF_6]^{2-}$ (fluosilicic anion or silicofluorhydric anion), which gives a white precipitate of potassium fluosilicate $K_2[SiF_6]$.

30.1.1.17 Ni^{2+}

Ni^{2+} yields a green precipitate of hydroxide $Ni(OH)_2$ once $pH > 7.2$, insoluble in excess of the reagent but soluble in ammonia through the formation of blue nickel ammine complexes, in particular of hexamminenickel(II): $[Ni(NH_3)_6]^{2+}$. Nickel hydroxide can again be separated from this solution with an excess of sodium or potassium hydroxide.

Cyanide ions yield a green precipitate of nickel cyanide $Ni(CN)_2$ soluble in excess of cyanide ions by the formation of the tetracyanonickelate(II) complex (or nickelocyanide anion) $[Ni(CN)_4]^{2-}$. The complex is moderately stable. It is decomposed by adding mineral acids, which liberate hydrocyanic acid and precipitate nickel cyanide. In a second stage, nickel cyanide itself decomposes into Ni^{2+} and hydrocyanic acid:

$$\left[Ni(CN)_4\right]^{2-} + 2H^+ \rightarrow 2HCN + Ni(CN)_2,$$

$$Ni(CN)_2 + 2H^+ \rightarrow Ni^{2+} + 2HCN.$$

Ni^{2+} reacts with sodium nitrite in concentrated solution in the presence of potassium ions. The precipitation of the hexanitritonickelate(II) complex as its potassium salt $K_4[Ni(NO_2)_6]$ is then observed. In the presence of alkaline earth ions, the following double salt is obtained: $CaK_2[Ni(NO_2)_6]$. It is scarcely soluble in water.

Actually, the detection of Ni^{2+} is carried out with dimethylglyoxime (Hdmg). A red complex forms and precipitates:

$$Ni^{2+} + 2Hdmg \rightleftharpoons \left[Ni\text{-}(dmg)_2\right]\downarrow + 2H^+.$$

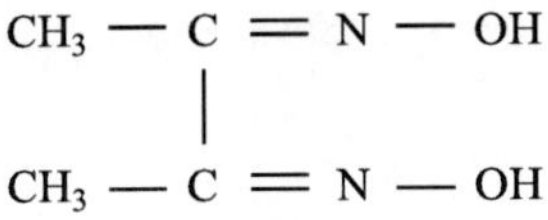

Hdmg

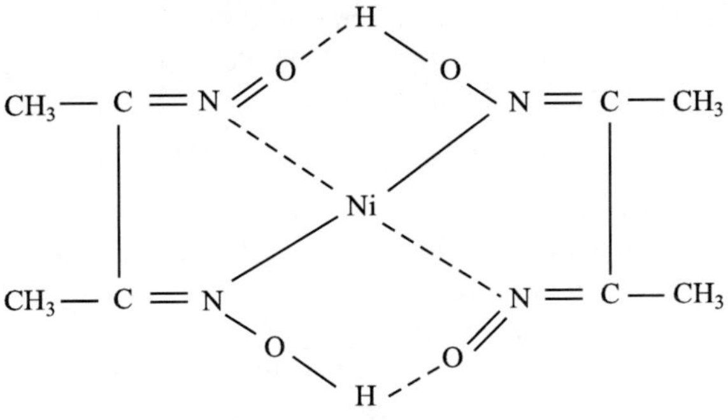

$[Ni(dmg)_2]$

30.1.2 *Anions' Characterization*

30.1.2.1 F^-

Fluoride anions are essentially characterized by the formation of a complex at the expense of another complex. Thus, zirconium Zr^{+IV} gives a red-colored complex with sodium alizarinesulfonate:

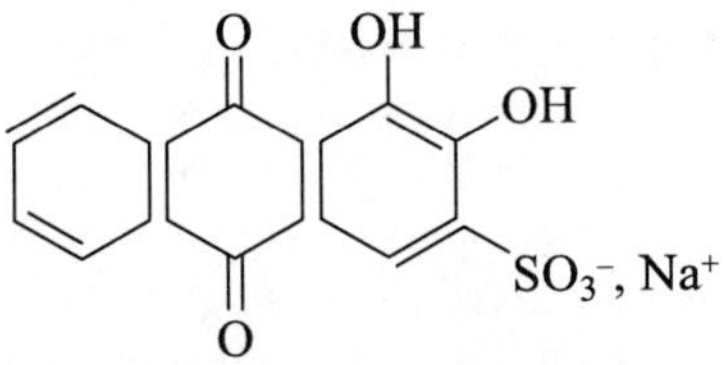

sodium alizarinesulfonate

Since fluoride ion complexes Zr^{+IV} more strongly than sodium alizarinesulfonate does, the latter is displaced and it turns yellow. Simultaneously, the new octahedral complex $[ZrF_6]^{2-}$ [hexafluorozirconate(IV)] forms. Complexes such as $[ZrF_7]^{3-}$ and $[ZrF_8]^{4-}$ could also be formed.

30.1.2.2 Cl^- and Br^-

Chloride ions yield a white precipitate of AgCl with Ag^+ cations. It is soluble in ammonia (see earlier), potassium cyanide, and potassium thiosulfate as it yields the following complexes: $[Ag(NH_3)]^+$, $[Ag(NH_3)]^{2+}$, $[Ag(CN)_2]^-$, $[Ag(CN)_3]^{2-}$, $[Ag(CN)_4]^{3-}$, $[Ag(S_2O_3)_2]^{3-}$, and $[Ag(S_2O_3)_3]^{5-}$, respectively. Let's also recall that silver chloride is soluble in concentrated hydrochloric solutions through the formation of chlorosilver(I) complexes: $[AgCl_2]^-$, $[AgCl_3]^{2-}$, and $[AgCl_4]^{3-}$.

Chloride ions are also characterized by their reaction with potassium dichromate in sulfuric acid in the presence of diphenylcarbazide. A violet color appears due to the Cr^{3+}–diphenylcarbazone complex. In a first stage, chromyl chloride CrO_2Cl_2 forms according to the reaction

$$4Cl^- + Cr_2O_7{}^{2-} + 6H^+ \rightleftharpoons 2CrO_2Cl_2\uparrow + 3H_2O.$$

The reaction is reversible. It is the reason why it must be carried out in concentrated sulfuric acid to fix water. Chromyl chloride, where chromium exhibits oxidation state +VI, oxidizes diphenylcarbazide into diphenylcarbazone and is simultaneously reduced to chromic ions Cr^{3+}, which react with the latter to yield the violet complex. It is interesting to point out that chromyl chloride is a readily volatile liquid that can be distilled. As a result, the formation of such a liquid characterizes the presence of chloride ions.

Bromide ions yield a precipitate of silver bromide with Ag^+ ions. However, in ammonia, they dissolve with more difficulty than chloride ions do. It is because silver bromide is markedly less soluble than silver chloride [$K_s(AgBr) = 10^{-12.2}$; $K_s(AgCl) = 10^{-9.75}$]. The following reaction:

$$AgBr\,(s) + 2NH_3 \rightleftharpoons \left[Ag(NH_3)_2\right]^+ + Br^-$$

is more equilibrated than that involving AgCl.

30.1.2.3 I^-

I^- ions also give a precipitate of silver iodide AgI, which is insoluble in ammonia excess ($K_s = 10^{-16.6}$). However, it is soluble in alkaline cyanides and in sodium thiosulfate solutions. Indeed, it forms complexes strictly analogous to those obtained with chloride ions. Let's recall the fact we've already mentioned several times that iodide ions give numerous complexes. This is the case, for example, with the tetraiodomercurate(II) cation, $[HgI_4]^{2-}$. They also react with some complexes such as the tetrachloropalladate(II) anion $[PdCl_4]^{2-}$ (palladohydrochloric anion). The deep brown palladous iodide separates out:

$$[PdCl_4]^{2-} + 2I^- \rightarrow 4Cl^- + PdI_2\downarrow$$

This precipitate is easily soluble in an excess of iodide ions by forming the tetraiodopalladate(II) complex $[PdI_4]^{2-}$:

$$PdI_2\downarrow + 2I^- \rightarrow [PdI_4]^{2-}.$$

30.1.2.4 CN^-

Dichlorodimercure(I) ("mercurous cation") yields a gray precipitate of metallic mercury formed by its disproportionation in the presence of cyanide ions:

$$Hg_2{}^{2+} + 2CN^- \rightarrow \left[Hg(CN)_2\right] + Hg.$$

The reaction is complete. The formed "mercuric cyanide" is not a salt. It is actually the bis-cyanomercury(II) complex. Quite generally, cyanide ions yield very stable complexes with ions of metals belonging to the columns VIII, Ib, and IIb, such as $[Ag(CN)_2]^-$, $[Ag(CN)_3]^{2-}$, $[Ag(CN)_4]^{3-}$, $[Fe(CN)_6]^{4-}$, and $[Co(CN)_6]^{3-}$.

Zn^{2+} yields a white precipitate of zinc cyanide soluble in an excess of alkaline cyanide by the formation of the tetracyanozincate(II) complex, $[Zn(CN)_4]^{2-}$:

$$Zn^{2+} + 2CN^- \rightarrow Zn(CN)_2\downarrow$$

$$Zn(CN)_2\downarrow + 2CN^- \rightarrow \left[Zn(CN)_4\right]^{2-}.$$

The Ag^+ cation yields a white precipitate of silver cyanide or of silver cyanosilver-(I) $Ag[Ag(CN)_2]$, whose structure consists of the silver salt of the anion complex cyanosilver(I) $[Ag(CN)_2]^-$. Silver cyanosilver(I) is easily soluble in an excess of cyanide ions through the formation of higher cyanosilver complexes. Likewise, it is soluble in ammonia and thiosulfate solutions through the formation of the silverammine and thiosulfatosilver complexes already encountered.

Cation Cu^{2+} (introduced as sulfate) yields a white precipitate of copper(I) cyanide soluble in an excess of alkaline cyanides through the formation of cyanocuprate(I) (cuprocyano) complexes, $[Cu(CN)_2]^-$ and $[Cu(CN)_3]^{2-}$. These complexes have very high stabilities. As a mark of this stability, let's note, for example, that hydrogen sulfide no longer precipitates copper from them, despite the extremely low value of its solubility product $K_s(CuS) = 10^{-35}$. A suspension of copper(II) sulfide becomes totally clear after the addition of a cyanide solution. It is interesting to remark that while Cu^{2+} is reacting with cyanide ions, two simultaneous reactions occur: a complexation reaction and a redox one. According to some authors, hydrocyanic acid could form a mixture of copper cyanides, called cuprosocupric cyanide, in which copper exhibits oxidation states +I and +II. This sort of cyanide would form according to the reaction

$$4HCN + 3SO_4Cu + H_2O \rightarrow \underset{\text{cuprosocopper-II cyanide}}{(CNCu)_2(CN)_2Cu} + 3H_2SO_4 + 1/20\ {}_2.$$

The formed dioxygen is detected by using oxidizable substances such as benzidine or gaiac resin, which yield intense colors in its presence. This reaction is used in toxicology in order to detect HCN in the atmosphere. It is named Schoenbein's or Gastaldi's reaction depending on the nature of the oxidizable substance.

The formation of Prussian blue also constitutes an excellent means for the characterization of cyanide ions. The latter combine easily to ferrous hydroxide to yield the hexacyanoferrate(II) (ferrocyanide) anion $[Fe(CN)_6]^{4-}$ easily detected with Fe^{3+} in acidic medium:

$$6CN^- + Fe(OH)_2 \rightarrow \left[Fe(CN)_6\right]^{4-} + 2OH^-,$$

$$3\left[Fe(CN)_6\right]^{4-} + 4Fe^{3+} \rightarrow Fe_4\left[Fe(CN)_6\right]_3\downarrow$$

30.1.2.5 SCN^-

SCN^- anions are most often detected by the formation of the ferrithiocyanate complex (see earlier). This reaction is very sensitive and is specific. It is also used to detect hydrocyanic acid. In the presence of hydrogen sulfide and sulfur, it is readily converted into thiocyanate anion. Let's recall that thiocanate anions yield numerous, often colored, complexes with diverse metallic ions: Fe^{3+}, Ca^{2+}, and Hg^{2+}. Mercuric cations, in the presence of thiocyanate anions, precipitate as "mercuric thiocyanate," which is not a salt but rather the dicyanomercury(II) complex $[Hg(CN)_2]$. It is soluble in an excess of thiocyanates by formation of the tri(thiocyano) and tetra(thiocyano) mercurate(II) complexes: $[Hg(SCN)_3]^-$ and $[Hg(SCN)_4]^{2-}$. The cation dichlorodimercury(I) ("mercurous") $Hg_2{}^{2+}$ yields a white precipitate of "mercurous" thiocyanate $Hg_2(SCN)_2$. The reaction is not simple. It involves the anion complex $[Hg(SCN)_3]^-$ as the intermediary. Hence, a disproportionation reaction occurs simultaneously with the complexation reaction

$$3SCN^- + Hg_2{}^{2+} \rightarrow \left[Hg(SCN)_3\right]^- + Hg\downarrow$$

When all of the thiocyanate ions have been transformed, a reaction of the "mercurous" cation with the intermediary occurs and there is formation of the "mercurous thiocyanate":

$$2\left[Hg(SCN)_3\right]^- + 3Hg_2{}^{2+} \rightarrow 3Hg_2(SCN)_2\downarrow + 2Hg^{2+}.$$

By adding thiocyanate anions, cobaltous cations Co^{2+} yield a blue complex, the tetrathiocyanocobaltate(II) complex, $[Co(SCN)_4]^{2-}$. Conversely, this reaction permits the detection of Co^{2+} by adding potassium or sodium thiocyanate. This reaction is disturbed by Fe^{3+} since the complex ($[Fe(SCN)]^{2+}$) it forms in the same conditions exhibits an intense red color that may mask that given by $[Co(CN)_4]^{2-}$. Hence, in order to detect Co^{2+} in the presence of Fe^{3+}, the latter must be complexed with fluoride ions (added as sodium fluoride) or with sodium and potassium tartrate. The stability constants of the complexes $[Fe(SCN)]^{2+}$ and $[FeF]^{2+}$ are, respectively, $K = 10^{2.1}$ and $K = 10^{5.5}$, and that of the tartrate complex is $K = 10^{7.5}$. This process is quite exemplary of the analytical possibilities offered by the formation of complexes.

30.1.2.6 $PO_4{}^{3-}$

Orthophosphate ions $PO_4{}^{3-}$ are detected with ammonium molybdate $(NH_4)_6[Mo_2O_{24}]4H_2O$ in the presence of an excess of ammonium nitrate. A yellow precipitate of ammonium phosphomolybdate $(NH_4)_3[PO_4(MoO_{13})_{12}]$ forms:

$$PO_4^{3-} + 12MoO_4^{2-} + 3NH_4^+ + 24H^+ \rightarrow (NH_4)_3\left[PO_4(MoO_{13})_{12}\right], 2H_2O\downarrow + 10H_2O.$$

The formula $[Mo(MoO_4)_6]^{6-}$ seems to be the best representation of the molybdate anion in ammonium molybdate. Phosphomolybdate anion may be considered a complex. The formula $[PMo_{12}O_{40}]^{3-}$ is sometimes attributed to it. (Molybdic anhydride MoO_3 is sparingly soluble in water. However, it dissolves in alkaline hydroxides as it gives "normal molybdates" M_2MoO_4. In addition to normal molybdates, there are other molybdates in which the ratio MoO_3/M_2O is superior to unity, a value found in normal molybdates.) Furthermore, molybdate ions MoO_4^{2-}, dissolved as sodium salt ($Na_2MoO_4,2H_2O$), condense with orthophosphate ions PO_4^{3-} in concentrated sulfuric acid medium. Molybdophosphoric acid (phosphomolybdic acid) is formed. It can be reduced by hydrazine sulfate to yield an intense blue color whose composition is uncertain. It permits the determination of PO_4^{3-} ions by absorption after the measurement of the absorbance at 820–830 nm.

Phosphate ions are also detected and determined as a phosphovanadomolybdate complex formed by reaction with ammonium vanadate NH_4VO_3 and ammonium molybdate. The yellow complex permits an absorption measurement in the range 460–480 nm. This complex, upon treatment with iodate ion in neutral or nitric acid medium, yields another complex of formula $[PV_2Mo_{10}O_{40}]^{5-}$.

Finally, PO_4^{3-} ions give a yellow precipitate of silver phosphate Ag_3PO_4 in neutral medium. It is easily dissolved in ammonia by the formation of ammine complexes:

$$Ag_3PO_4 + 6NH_3 \rightarrow PO_4^{3-} + 3\left[Ag(NH_3)_2\right]^+.$$

30.1.2.7 SiO_2

Silice SiO_2 is weakly reactive. It is attacked by hydrofluoric acid with the production of silicium fluoride:

$$SiO_2 + 4HF \rightarrow SiF_4\uparrow + 2H_2O.$$

The formed silicium fluoride is gaseous. In water, it decomposes with the precipitation of silice and with the formation of fluosilicic acid. Fluosilicic acid derives from the anion complex hexafluorosilicate(IV), $[SiF_6]^{2-}$ (fluosilicic ion). In alkaline medium, silice, whose formula is then SiO_3^{2-}, can be detected after precipitation with tetramminezinc(II) cation, $[Zn(NH_3)_4]^{2+}$, according to the reaction

$$SiO_3^{2-} + \left[Zn(NH_3)_4\right]^{2+} \rightarrow 4NH_3 + ZnSiO_3\downarrow$$

Ammonium molybdate $(NH_4)_6Mo_7O_{24},4H_2O$, which may also be represented by the formula $(NH_4)_2MoO_4$, gives a soluble acidic yellow silicomolybdic complex with silicates whose formula is $H_4[SiO_4(MoO_3)_{12}]$. It is extractible into ethyl acetate. By adding a reductor such as sodium stannite $Na[Sn(OH)_3]$ (stannous chloride in sodium hydroxide), molybdene blue appears. It is a complexe itself, having the following cumbersome formula:

$$[SiO_4(MoO_2)_n(MoO_3)_{(12-n)}]^{4-}.$$

It is interesting to point out incidentally that stannite ion is also a complex.

30.2 Quantitative Inorganic Analysis

In this section, we mention only some applications of titrations with EDTA.

30.2.1 Titration of Ca^{2+}

The titration of Ca^{2+} in the presence of Mg^{2+} is of practical importance. The best results are obtained with complexone(III) as the titrant and with calcein or calcon as the indicator. Murexide, as an indicator, is obsolete since both preceding indicators have been introduced. In any case, titrand solutions must be alkaline for the titration reaction to be sufficiently quantitative. Pb^{2+} and Zn^{2+} must be masked with 2,3-dimercaptopropanol. Some other heavy metal ions must also be masked with potassium cyanide. Fe^{3+} and Mn^{2+} can be masked by complexation with triethanolamine. Here, again, we find what we have already stressed several times about an interest of the complexation phenomenon that can be used to dissimulate disturbing ions. Ba^{2+}, Sr^{2+}, and Ca^{2+} are titrated simultaneously.

With calcein (a derivative of fluoresceine), the titration is very sensitive. Titrations with complexone(III) solutions of concentrations of about 10^{-2} mol/L can be carried out. At the endpoint, the yellow-green fluorescence disappears, whereas a violet color appears quasi-simultaneously.

With calcon, the color passes from red wine to blue. The reaction is somewhat less sensitive than the preceding one. With this indicator, however, the selective titration of Ca^{2+} in the presence of Mg^{2+} is possible since the pH value is sufficiently high for Mg^{2+} to be quantitatively precipitated as hydroxide $Mg(OH)_2$. EDTA does not react with the precipitated magnesium until free Ca^{2+} together with that fixed to the indicator are not yet themselves complexed by EDTA.

Ca^{2+} can also be directly titrated with EGTA in the presence of Mg^{2+} with zincon as the indicator. This selective titration is possible because the Ca^{2+}–EGTA stability constant is much higher than that of the Mg^{2+}–EGTA complex, respectively, 10^{11} and 10^5. Hence, this titration is well adapted for the titration of weak quantities of Ca^{2+} in the presence of Mg^{2+}. The final point is detected with zincon. This is an indirect detection and, for it to be effective, a weak quantity of the Zn^{2+}–EGTA complex must be added to the reaction vessel. In such a way, and because of the respective complexation constants,

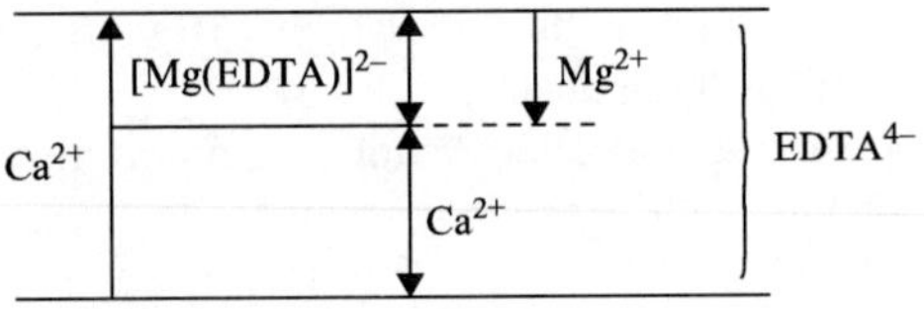

Fig. 30.1 Principle of the titration of Ca^{2+} with EDTA in the presence of eriochrome black T and $[MgEDTA]^{2-}$

- before the equivalence point, Zn^{2+} is displaced from its EGTA complex by Ca^{2+} according to the reaction

$$[ZnEGTA]^{2-} + Ca^{2+} \rightleftharpoons Zn^{2+} + [CaEGTA]^{2-}$$

but is immediately complexed by the indicator

$$Zn^{2+} + Ind^{4-} \rightleftharpoons [ZnInd]^{2-};$$

hence the blue color;

- after the equivalence point, Ca^{2+} is no longer present in the solution and the following ligand exchange reaction occurs:

$$[ZnInd]^{2-} + EGTA^{4-} \rightleftharpoons [ZnEGTA]^{2-} + Ind^{4-}$$

The indicator becomes free and an orange-red color appears.

It is interesting to notice that the titration with EDTA of Ca^{2+} alone is possible in the presence of eriochrome black T, but no sharp endpoint is obtained. A little quantity of the $Na_2[MgEDTA]$ complex must be added to the vessel solution containing Ca^{2+} in order for the indication to be possible. The latter complex is by far more stable than the Ca–indicator complex. In these conditions, the Mg^{2+} ions are liberated according to the reaction

$$[MgEDTA]^{2-} + Ca^{2+} \rightleftharpoons [CaEDTA]^{2-} + Mg^{2+}$$

since the EDTA–calcium complex is more stable than that of magnesium. The titration accuracy is not disturbed by the addition of $[MgEDTA]^{2-}$ since the liberated magnesium ions replace the calcium ions mole for mole (Fig. 30.1). The color change is sharp because Mg^{2+} is the last cation to be titrated.

It is easily conceivable that a Ca^{2+} and Mg^{2+} mixture may be globally titrated in such a way with EDTA. The last edition of the European pharmacopeia prescribed the Ca^{2+} titration with calcon as the indicator.

The EDTA method is sensitive enough to allow the determination of calcium in serum.

30.2.2 Titration of Mg^{2+}

Mg^{2+} can be directly titrated with $Na_2[H_2EDTA]$ with eriochrome black T as the indicator (see immediately above). The European pharmacopeia prescribes working with a solution buffered at pH = 10 with ammonia in the presence of "mordant noir 11R" as the indicator.

A practical determination of importance is that of the hardness of water, which is due to calcium and magnesium salts dissolved in it. Hence, this is the determination of Ca^{2+} and Mg^{2+} in admixture in water.

- A first methodology consists of carrying out two successive titrations, one in the presence of Patton and Reeder's indicator, the other in the presence of eriochrome black T. The former allows us to determine Ca^{2+} alone, the latter both globally. The global content in both ions is the total hardness. The difference between the added volumes at the two endpoints determines Mg^{2+}.
- A variation of this methodology consists of replacing Patton and Reeder's indicator by calcon as the indicator in the first titration and then in operating as in the above-mentioned conditions.

The hardness of water is a function of its Ca^{2+} and Mg^{2+} contents. The separated titrations allow us to determine the hardness due to Mg^{2+} and that due to Ca^{2+}. To clarify what has just been said, the total hardness is determined by a titration carried out at 40 °C (for kinetic reasons) with complexone(III) at pH 9–10 in the presence of eriochrome black T and of magnesium complexonate $[Mg(EDTA)]^{2-}$. In an alternative way to determine Mg^{2+}, it proceeds as before but after Ca^{2+} has been removed by precipitation as oxalate.

Ca^{2+} and Mg^{2+} may be in water as hydrogen carbonate, sulfate, chloride, or nitrate salts. Comparing the preceding results with those obtained during the determination of water's alkalinity permits distinguishing the hardness called the *permanent hardness* (obtained after elimination of CO_2) from that known as the *temporary hardness*. The latter is obtained by making the difference between the total and permanent hardnesses. A high value of the permanent hardness may be due to an overcharge in calcium sulfate. In this case, water may be unsuitable for domestic uses. From another standpoint, knowing the hardnesses and alkalinity of water permits us to estimate its CO_2 content, that is, its aggressiveness.

30.2.3 Titration of Lead

Pb^{2+} may be directly titrated with complexone(III) in hexamethylenetetramine medium with xylenol orange as indicator. The pH value is located about 5.6. A similar titration of Zn^{2+} can be carried out at pH about 6 with xylenol orange or methyl thymol blue as the indicator. Hexamine is then the buffer. This titration is also possible in ammonia buffer (pH about 10) in the presence of eriochrome black T.

30.2.4 Titration of Bismuth

Bi^{III} is also directly titrated by complexone(III) in the pH = 0–1 range with xylenol orange or methyl thymol blue as the indicator.

30.2.5 Titration of Aluminum

According to the European pharmacopeia, Al^{3+} must be back titrated in acetic buffer (pH = 4.6) with dithizone as the indicator. The titrant is a solution of zinc sulfate. The somewhat slowness of the EDTA–Al^{3+} reaction imposes the carrying out of the back titration. It is interesting to point out the fact that dithizone is used as an indicator for the complexometric titrations of metallic ions, which, precisely, are able to form complexes of formulas [MHDz] according to the reaction

$$M^{n+} + nH_2Dz \rightleftharpoons [M(HDz)_n] + nH^+,$$

where Dz is dithizone. It is the case of Zn^{2+} but also of Cd^{2+} and Pb^{2+}. The methodology is the same when using eriochrome black T as the indicator, but then the pH value solution must be about 7–8 (the lowest limit value obtained with ammonia buffer).

In adsorbed vaccines, Al^{3+} is back-titrated with a 2×10^{-2} mol/L copper sulfate solution at pH = 4.4. The indicator is the pyridylazonaphtol R or "PAN."

30.2.6 Miscellaneous Titrations

In this section, we'll mention the direct titration of Ba^{2+}, Cd^{2+}, Co^{2+}, Cu^{2+}, Fe^{3+}, Mn^{2+}, Hg^{2+},Ni^{2+}, Th^{4+}, and Sr^{2+}. Of course, the pH conditions vary depending on the stability of the formed complexes M^{n+}–EDTA.

Let's recall also that Ag^+, Pd^{II}, and Au^{III} can also be indirectly titrated with EDTA:

- For the Ag^+ determination, Ni^{2+} is displaced by it from the tetracyanonickelate(II) complex:

$$2Ag^+ + [Ni(CN)_4]^{2-} \rightleftharpoons 2[Ag(CN)_2]^- + Ni^{2+};$$

Ni^{2+} is determined with EDTA in the presence of murexide;

- Pd^{II} is determined according to the same principle. The liberated Ni^{2+} ions are back-titrated with an excess of EDTA in the presence of eriochrome black T. The back-titrant solution is a $MnSO_4$ solution.

Na^+ may also be titrated with the help of EDTA. At first sight, this is a surprising fact because EDTA gives very weak complexes with alkaline ions. (Actually, their formation is not even taken into account when complexone-III—which is a sodium salt—is used!) But the evoked titration is an indirect one. In a first stage, Na^+ is precipitated as sodium, zinc, and uranyl acetate: $Na(UO_2)_3[Zn(H_2O)_6](C_2H_3O_2)_9$. In a second stage, the precipitate is destroyed and the liberated Zn^{2+} ions are directly titrated with a 10^{-3} mol/L EDTA solution in the presence of eriochrome black T.

Finally, as we have already noticed, titrations of some anions with EDTA may also be carried out. The most often encountered case is that for which the anion to be determined gives a complex or a precipitate with a metallic ion added in excess in the titration vessel. Then the metallic ion in excess is titrated with EDTA. Hence,

- cyanide ions react with nickel salts to give the tetracyanonickelate(II) complex, $[Ni(CN)_4]^{2-}$;
- fluoride ions are precipitated by an excess of calcium chloride. Calcium ions in excess are back-titrated;
- sulfate ions are precipitated by an excess of barium chloride. The excess of barium is titrated with EDTA.

In another methodology, the precipitate is dissolved. The liberated metallic ions are then titrated with EDTA. For example,

- phosphate ions are precipitated as ammonium magnesium phosphate. The magnesium contained in the precipitate is liberated by acidification and titrated with EDTA. (From another standpoint, let's recall that it can be determined as ammonium phosphomolybdate, which quantitatively precipitates. The precipitate is separated, washed, and weighed. It can also be titrated. Then it is dissolved in an excess of sodium hydroxide and the excess titrated with hydrochloric acid in the presence of phenolphthalein according to the reaction:

$$(NH_4)_3\left[PO_4(MoO_3)_{12}\right] + 23OH^- \rightarrow HPO_4{}^{2-} + 12MoO_4{}^{2-} + 3NH_4{}^+ + 11H_2O.$$

The equivalence point is located at pH = 9, whence the chosen indicator.);

- tungstate ions $WO_4{}^{2-}$ are precipitated with a calcium chloride solution. The calcium tungstate precipitate is separated and decomposed in acidic medium. The liberated calcium ions are titrated with EDTA.

Part V
Precipitation Phenomena—Analytical Applications

Chapter 31
Applications of the Formation of Complexes in Organic Analysis

In this chapter, we give some examples of the use of the formation of complexes in qualitative and quantitative organic analysis. It is not surprising that the major part of these examples is in the realm of the characterization of organic compounds by the formation of colored complexes and in the realm of their quantitative analysis by absorption measurements. The formation of complexes permits not only the identification of an organic compound but also the detection of some structural moieties in unknown molecules. In brief, it permits the functional analysis.

31.1 Formation of Complexes with Fe^{3+} and Fe^{2+}

The cations Fe^{2+} and Fe^{3+} are very often used in organic analysis. In Chap. 30, we saw that the cation Fe^{3+} gives a brownish red color at $1.5 < pH < 5$ with acetate ion. At ebullition, a deposit of basic acetate occurs that is formed by hydrolysis. The basic acetate is a complex whose structure is $[Fe_3O(CH_3COO)_6(H_2O)]^+$, CH_3COO^-. This is hexa-μ-acetato-μ_3-oxotrifer(III). The reaction is characteristic of acetate ions. According to the fourth edition of the European pharmacopeia, an acetyl group present in an organic molecule can be detected as an acetate ion after hydrolysis. Let's also notice that at pH = 7 in the presence of Fe^{3+}, the benzoate ion $C_6H_5COO^-$ gives the complex ion hexa-μ-benzoato-μ_3-oxotrifer(III): $[Fe_3O(C_6H_5COO)_6(H_2O)]^+$, $C_6H_5COO^-$, which precipitates as benzoate.

Tartrates give soluble, more or less stable, complexes with miscellaneous cations such as copper, bismuth, and iron, with some metallic hydroxides, acids (such as boric acid), and antimony oxide. The cations become masked by complexation. The identity reaction is carried out, according to the European pharmacopeia, with Fenton's reagent (ferrous sulfate and perhydrol). In a first step, dihydroxyfumaric acid is formed:

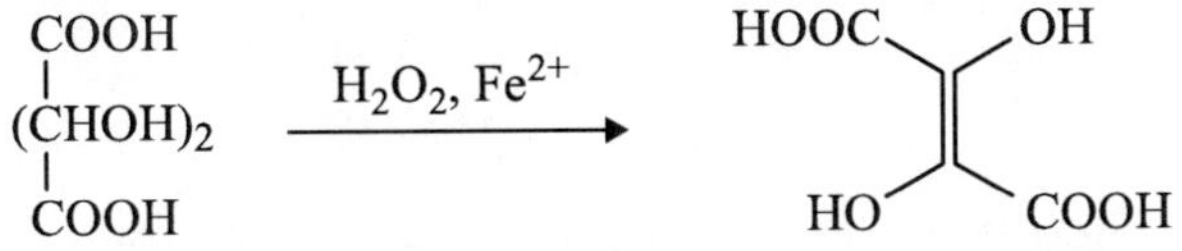

dihydroxyfumaric acid

J.-L. Burgot, *Ionic Equilibria in Analytical Chemistry*,
DOI 10.1007/978-1-4419-8382-4_31, © Springer Science+Business Media, LLC 2012

It yields a blue color in alkaline medium. The formation reaction of dihydroxyfumaric acid is complex. The development of a color is not surprising since a chelate of Fe^{3+} with the following structural moiety:

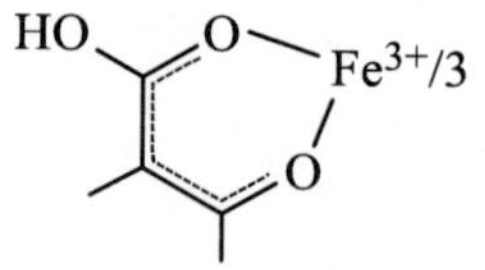

is probably formed. Tartrate ions (actually, tartaric acid) yield a stable yellow color with a solution of ferric chloride and hydrochloric acid whose pH value is located between 1 and 5 (Berg's reagent). Berg's reaction is characteristic of α-hydroxy acids. Uffelman's reaction exhibits similar properties except that phenol replaces hydrochloric acid. Rossi's reaction also involves tartaric acid by the formation of a complex with ammonium metavanadate. (Metavanadates correspond to the +V oxidation state of vanadium; their structure is $M_2O.V_2O_5$ or MVO_3.)

Citrate ions give soluble complexes with several cations, such as Cu^{2+}, Bi^{III}, Sb^{III}, Pb^{2+}, and above all, Fe^{2+} and Fe^{3+}. Generally, they are stable. Complexed cations cannot be precipitated from their solutions by some of their usual reagents.

Phenols yield characteristic colors in the presence of ferric chloride by forming complexes according to the reaction

$$6ArOH + Fe^{3+} + 3Cl^- \rightarrow \left[Fe(OAr)_6\right]^{3-} + 6H^+ + 3Cl^-,$$

which is poorly sensitive. Considering this reaction, we find the well-known tendency exhibited by Fe^{3+} to give very stable complexes with ligands possessing O-donor groups. Notice that the structure of these complexes is not perfectly known. The color obtained varies from blue to red, and the absorption intensity depends upon the structure of the phenol, that is, upon the substituents of the aromatic nucleus. This reaction is widely used in quantitative analysis by performing absorption measurements. It may also be considered a characteristic reaction of the group, although it is not general. For example, polynitrophenols yield no coloration. Moreover, the reaction is not selective. Numerous species that are not phenols or enols also give a color with Fe^{3+}. In the European pharmacopeia, phenol and chlorocresol are identified in this way. However, thymol cannot be so identified since it does not yield a color with $FeCl_3$. The explanation lies possibly in the fact that in this case, its weak solubility precludes the reaction.

The following reaction is probably connected to the preceding one. It is given by derivatives of 1,5-dimethyl-2-phenyl-3(2H)-pyrazolone (phenazone R = H and propylphenazone R = $_i$pr) with $FeCl_3$:

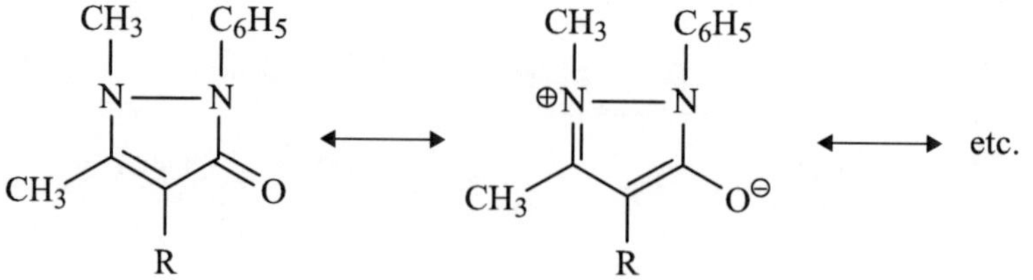

The weight of the mesoionic form in the true structure of these derivatives is important. This form exhibits the structure of a phenate or an enolate. The formation of a complex between this O donor and $FeCl_3$ probably explains the obtained red color. Vitamin B_6 (pyridoxine) can be

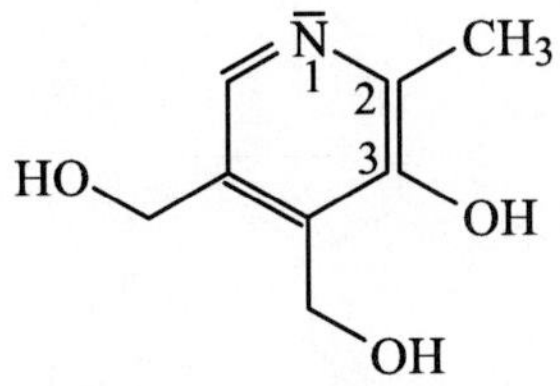

considered a phenol due to the presence of a hydroxy group on the pyridine aromatic nucleus. This explains the positive reaction obtained with $FeCl_3$. The tubocurarine structure,

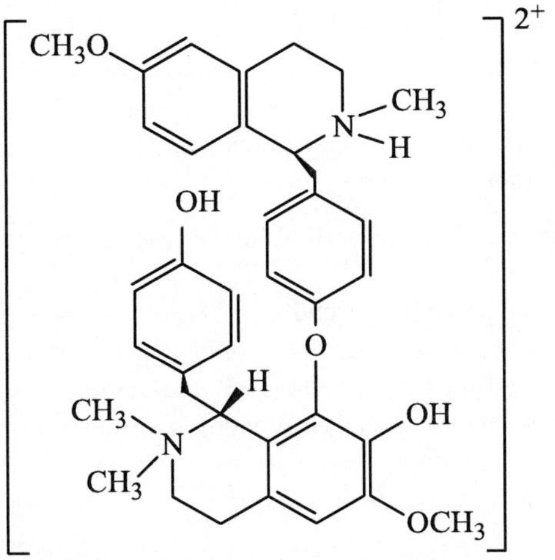

possesses two phenol functions brought by two different aromatic nuclei. They permit its identification through the green color obtained with $FeCl_3$.

The reaction developed by α-tocopherol seems to be connected to the preceding one. Its structure is

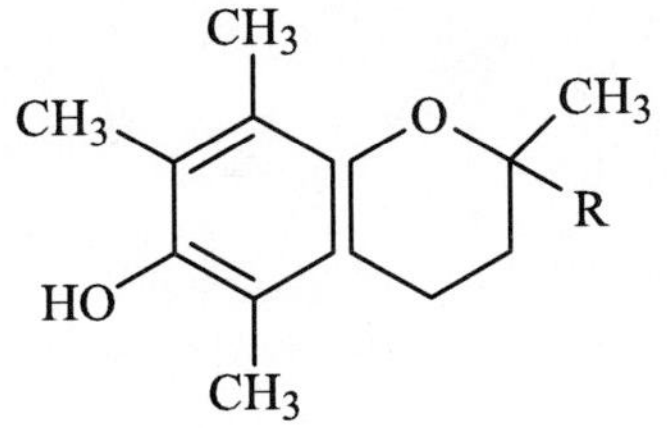

where R stands for an alkyl chain (a saturated alkyl chain with 16 carbon atoms: phytyl rest). Indeed, α-tocopherol possesses a phenol function, but in the presence of Fe^{3+}, the expected complexation reaction does not occur. A quinone forms. It results from the opening of the pyran cycle. Simultaneously, the reduction of Fe^{3+} into Fe^{2+} occurs. It is detected with o-phenanthroline:

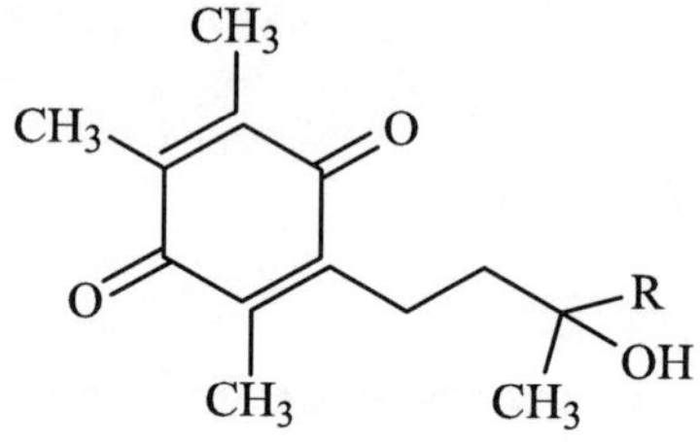

Actually, a redox reaction takes place. It is due to the reducing properties of para and ortho diphenols, structures that were masked in α-tocopherol. Hence, the driving force of this particular reaction is the formation of a quinone. It is sufficiently strong to induce the break of the ether–oxide bond of pyran. This is remarkable. The identification reaction of α-tocopherol is called *Emmeri–Engel's reaction.*

We point out, incidentally, that since phenols are reductants, they can be detected by the reduction of some complexes, such as

- ferric hexacyanoferrate(III) or Berlin blue (Prussian blue);
- phosphomolybdic acid;
- phosphotungstic and phosphomolybdic acids (Folin–Ciocalteu's reagent). In these two cases, there is formation of "molybdene blue." Initially, before reaction, molybdene exhibited oxidation state +VI (molybdate ion MoO_4^{2-}). Once the reaction is finished, the formed "molybdene blue" exhibits the behavior of colloidal compounds. It appears to be mixtures of oxides and hydroxides of different valences, such as $Mo^{VI}O_3$ and $Mo^{V}O(OH)^{3-}$.

Phenols also react with Millon's reagent, which is a complex solution containing NO_3^-, NO_2^-, Hg^{2+}, and $Hg\text{-}Hg^{2+}$ ions. In the presence of phenol, orthonitrosophenol might be formed in a first step. It might stand in equilibrium with its tautomeric form, o-quinonemonoxime:

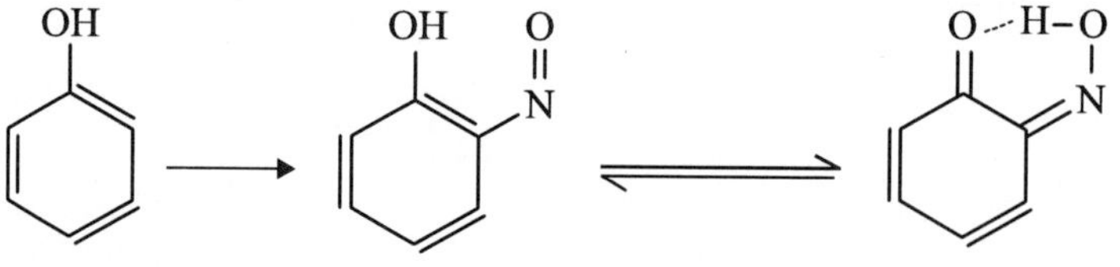

The latter might be stabilized by the formation of the following chelate in which the mercuric ion is engaged:

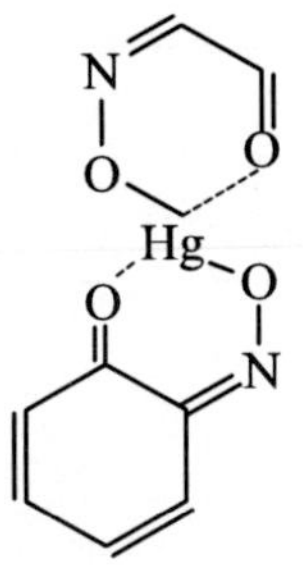

A yellow to intense red color is obtained.

Diphenols also yield colored reactions with $FeCl_3$. Different colors are obtained according to the ortho, meta, or para structure of the diphenol. Without any doubt, the formed complexes have different structures. Ortho diphenols such as levodopa, epinephrine (adrenaline), norepinephrine (noradrenaline), and isoprenaline:

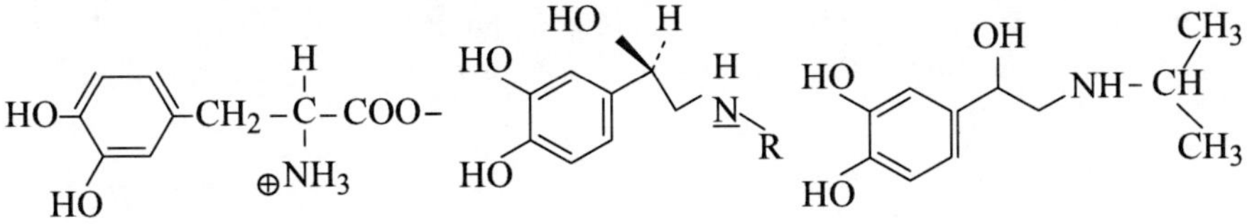

levodopa

R=CH_3 : adrenaline
R=H : noradrenaline

isoprenaline

yield a green color, characteristic of an ortho diphenol with $FeCl_3$. The color depends somewhat on the pH value and it may turn red upon the addition of ammonia. With adrenaline and noradrenaline, Fe^{3+} may behave as an oxidant in basic medium with the formation of adrenochrome and noradrenochrome, which are both orthoquinones:

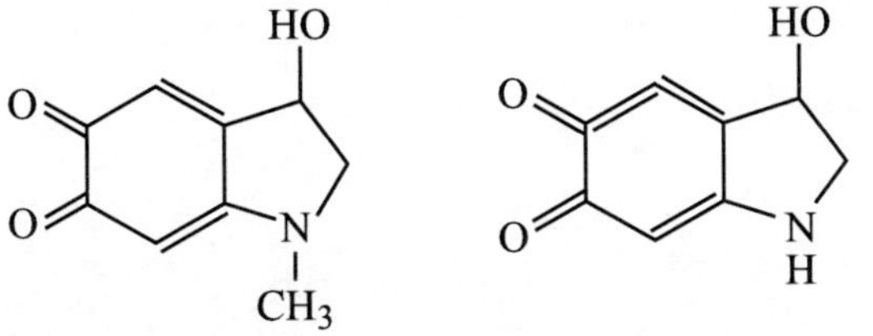

adrenochrome noradrenochrome

This is also the case for levodopa. The corresponding redox reaction is

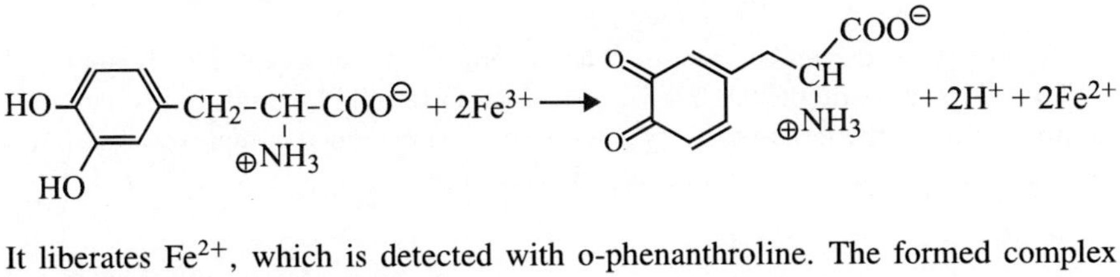

It liberates Fe^{2+}, which is detected with o-phenanthroline. The formed complex exhibits the following structure:

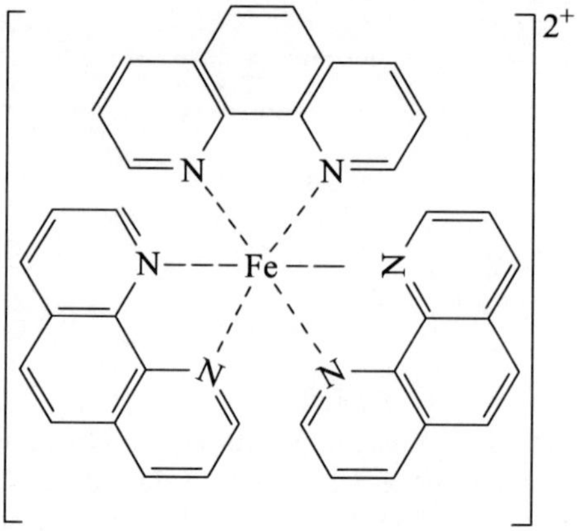

These complexes are octahedral ones.

Papaverine is neither a phenol nor a diphenol, but it is transformed into an ortho diphenol after heating with sulfuric acid. The transformation is accompanied by a demethylation:

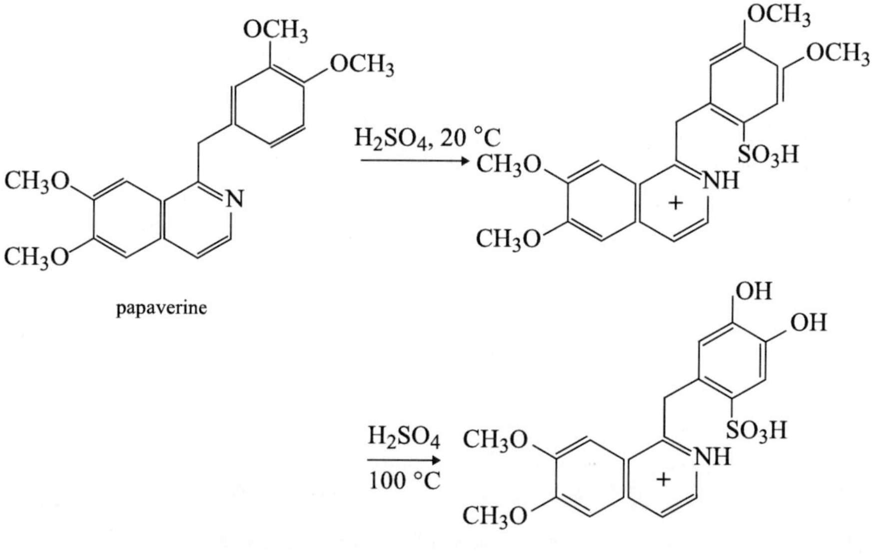

A deep green complex is formed with $FeCl_3$.

Carboxylic acids can be identified, and even sometimes quantitatively analyzed by absorption measurements, after transformation into hydroxamic acids and after complexation of the latter derivatives with Fe^{3+}. The formed complexes are violet-red. For example, the hydroxamic acid derived from acetic acid gives the reaction

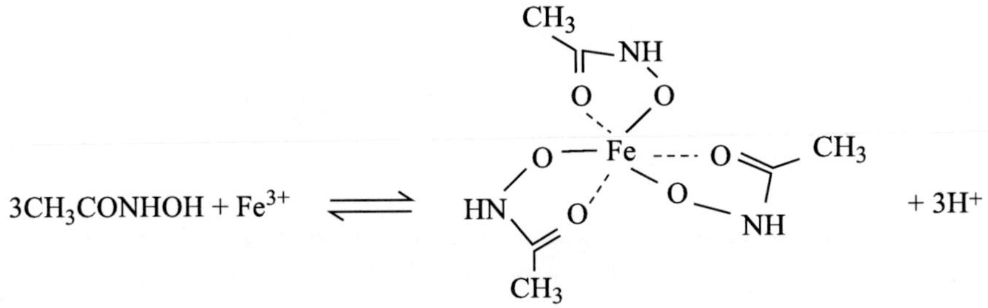

One hydroxamic acid molecule behaves as a bidentate ligand. The structure of the complex is probably very similar to that formed by Fe^{3+} and acetylacetone (acac): triacetylacetonefer(-III), that is, $[Fe(acac)_3]$.

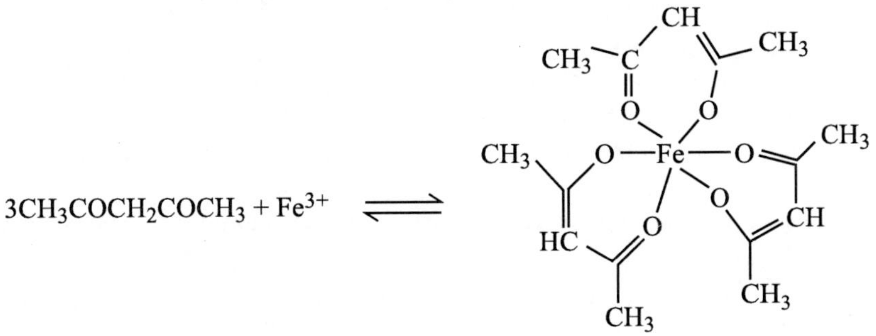

Both possess five unpaired electrons. The structure of the hydroxamate ligand is not wholly well known. It depends upon the structure of the substituent R. When R exhibits −I, −M effects, the proton loss takes place from the nitrogen atom, with, as a result, a stabilization by resonance of the ion:

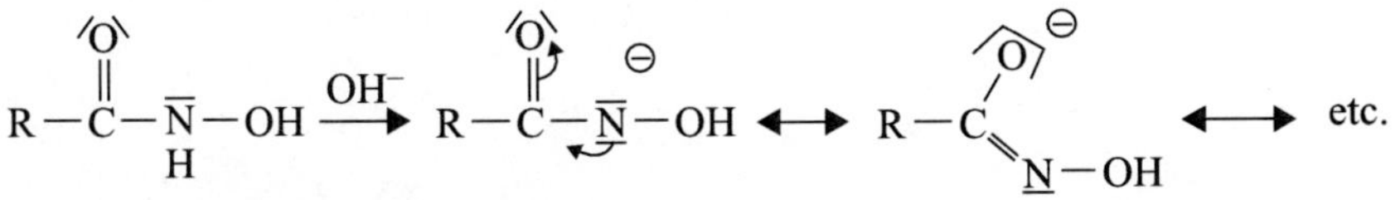

The color duration depends on the concentration $[Fe^{3+}]$ and the pH of the solution. The optimal value of the latter is located in the range 1.2–1.3. The characterization and quantitative analysis (through absorption measurements) of compounds that can be transformed into hydroxamic acids benefit from the formation of ferric hydroxamates. This is the case with

- esters, which are converted into the corresponding hydroxamic acids with hydroxylamine:

$$CH_3COOR \xrightarrow{NH_2OH} CH_3CONHOH;$$

- primary and secondary alcohols after transformation into esters:

$$>CH-OH \xrightarrow[\text{ou } (CH_3CO)_2O]{CH_3COCl} CH_3COO-CH< \xrightarrow{NH_2OH} CH_3CONHOH$$

- nitro and nitroso derivatives after reduction in substituted hydroxylamines, which are subsequently treated with benzoyl chloride:

$$RNO_2 \rightarrow RNHOH \xrightarrow{C_6H_5COCl} C_6H_5CON(R)OH;$$

- amides, which are converted into hydroxamic acids according to the general reaction

$$RCONH_2 + NH_2OH \rightarrow NH_3 + RCONHOH.$$

Usually, the transformation into hydroxamic acids is carried out with hydroxylamine sulfate. The hydroxylamine base is liberated by the addition of sodium hydroxide, but particular methods are also proposed.

As examples taken in the pharmaceutical field, we can mention the characterization of

- pantolactone formed by the treatment in alkaline medium of pantothenic acid:

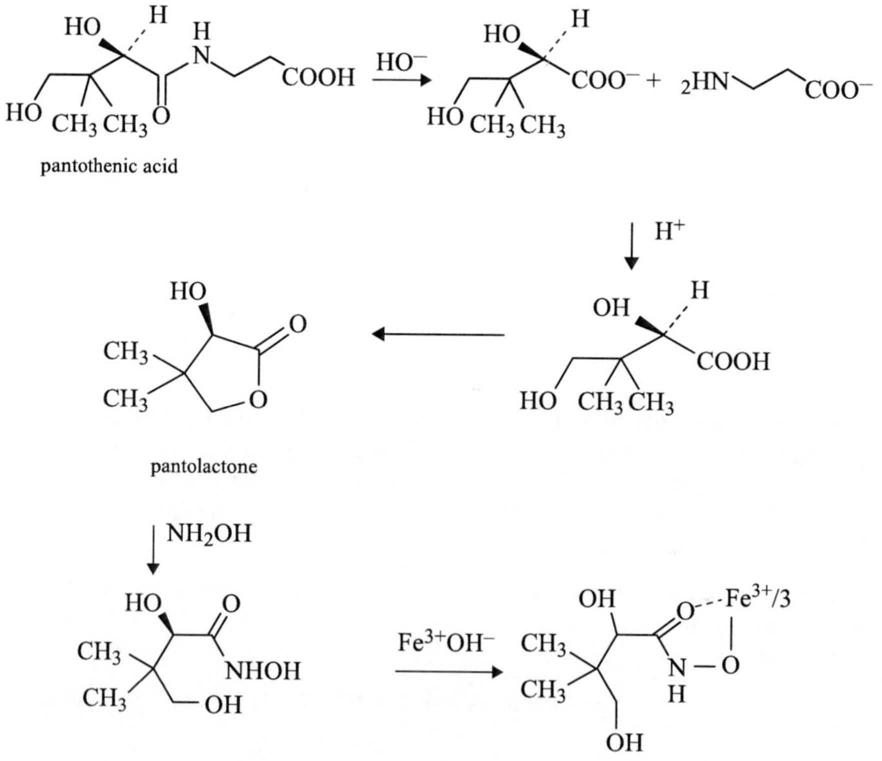

- cetyl palmitate:

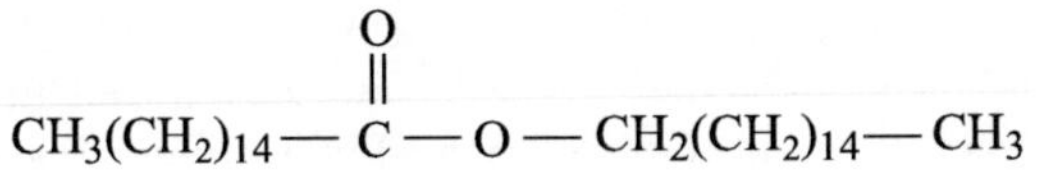

- clofibrate:

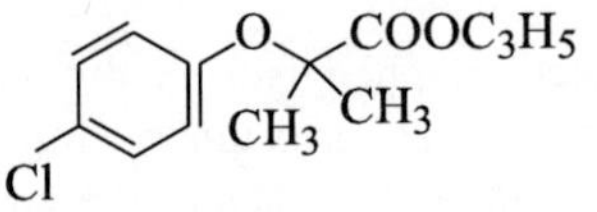

- glutethimide (whose corresponding hydroxamic acid formation reaction is slow):

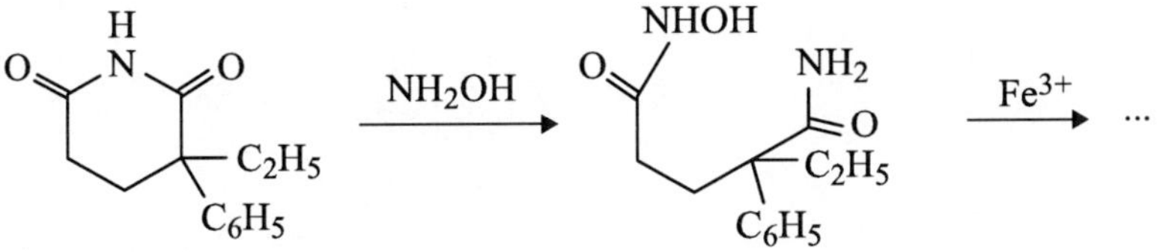

- benzylpenicillin (penicillin G):

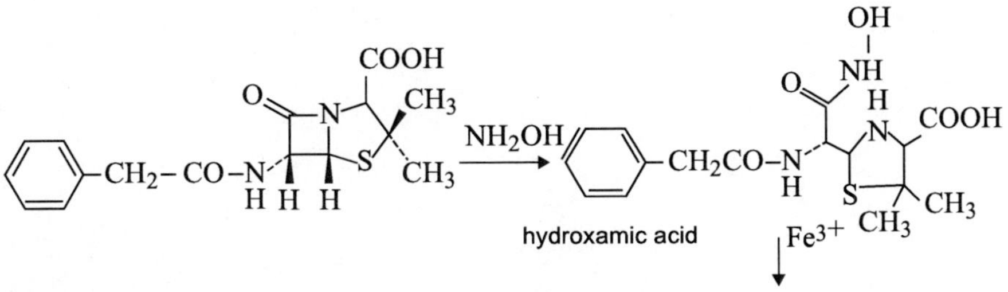

- 17-ketosteroids;

These derivatives are steroids whose cycle D possesses a keto chain (–$COCH_2OH$) on the carbon 17. First, they undergo the following reactions:

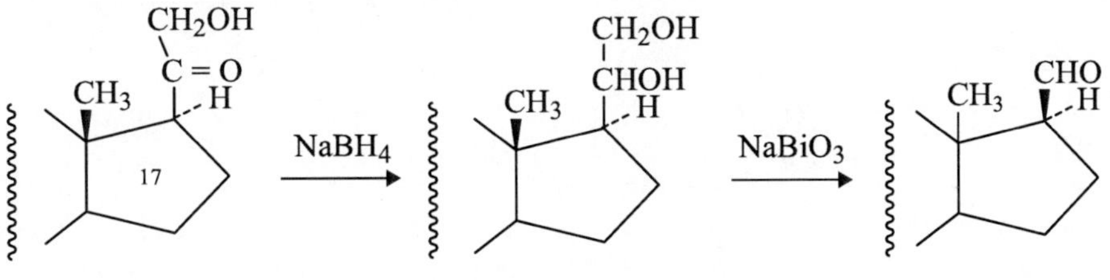

(Let's note that for the analytical reactions described ahead to be successful, carbon 17 must possess a hydrogen atom as the other substituent. This is the case for dehydrocorticosterone, corticosterone, desoxycorticosterone, and aldosterone. This is not the case for cortisone, hydrocortisone, and other synthetic glucocorticoids, which, most of the time, do possess a hydroxyl function in 17 in addition to the keto chain.) In order to be submitted to the hydroxamate reaction, the 17-ketosteroids (possessing a hydrogen atom on the carbon 17) must be reduced first into glycol with sodium borohydride. The yielded glycol, treated with sodium bismuthate, gives the corresponding aldehyde. (Bismuthic acid is a very strong oxidizing agent:

$$HBiO_3 + 5H^+ + 2e^- \rightleftharpoons Bi^{3+} + 3H_2O, \quad E°(HBiO_3/Bi^{3+}) = 1.70\ V)$$

The formed aldehyde, treated by phenylsulfohydroxamic acid in the presence of a ferric salt, offers the usual ferric hydroxamate complex. Phenylsulfohydroxamic acid gives sulfinic acid and nitroxyl (unstable intermediary of structure HNO) in basic medium. Nitroxyl is a nucleophilic species that attacks the carbonyl group of carboxaldehyde. A hydroxynitroso derivative forms. It is in tautomeric equilibrium with hydroxamic acid responsible for the color:

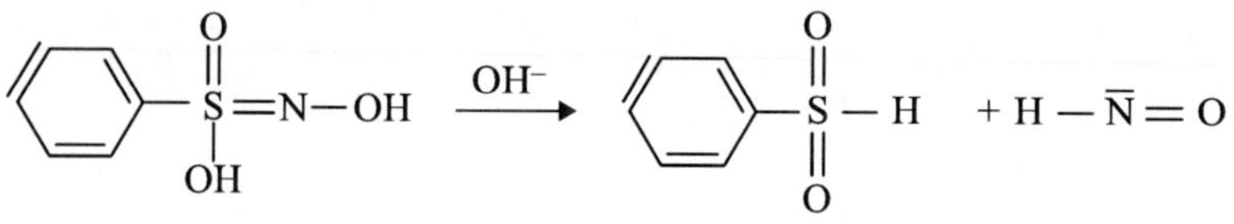

phenylsulfohydroxamic acid benzene sulfinic acid nitroxyle

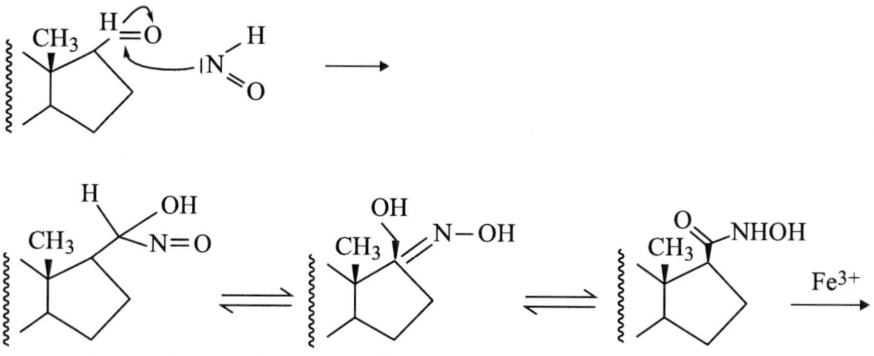

- indomethacine:

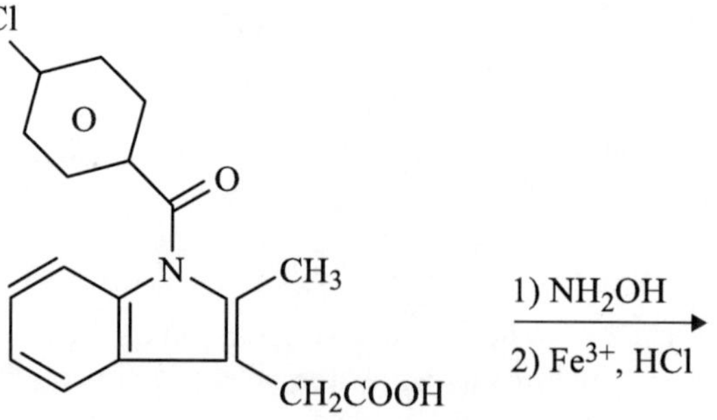

violet hydroxamate

However, we must point out that carboxylic acids do not directly react with hydroxylamine to give hydroxamic acids. Rather, a nucleophilic attack of hydroxylamine on the amide bond with the formation of 4-p-chlorophenylhydroxamic acid must be invoked in this case:

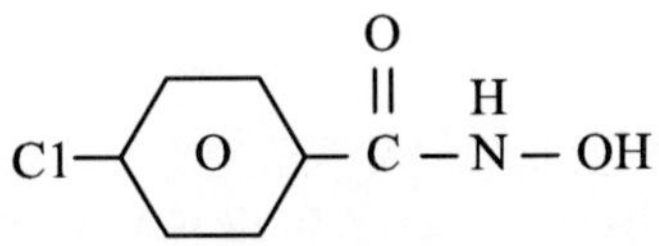

4-p-Chlorophenylhydroxamic acid is responsible for the color.

The formation of ferric hydroxamate permits the quantitative determination of pilocarpine:

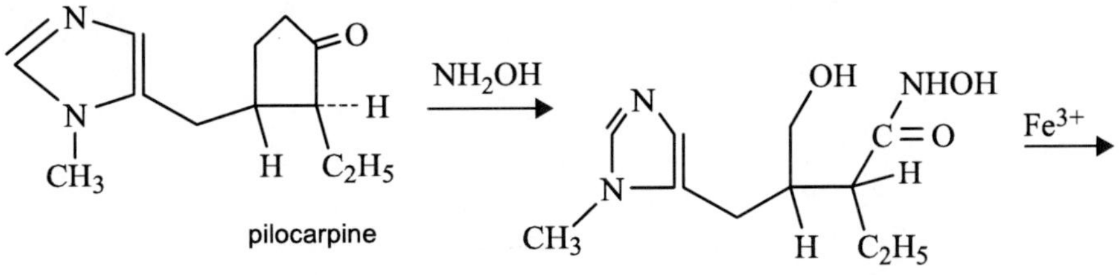

pilocarpine

Digitoxin, endowed with a lactonic cycle with five atoms (cardenolide),

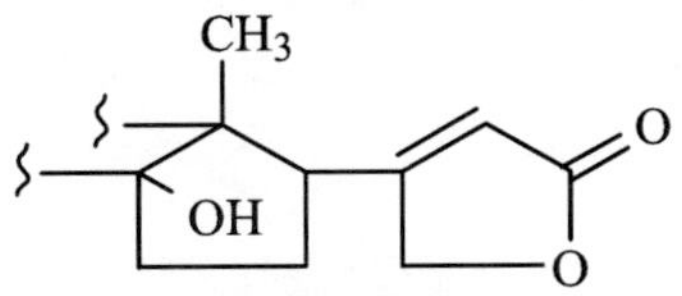

digitoxine

also gives the ferric hydroxamate assay according to a reaction analogous to that given by pantolactone (see above).

In addition to the already-mentioned ligands, Fe^{3+} reacts with some other complexants containing oxygen atoms. Salicylic acid, for example, exhibits a structure close to that of acetylacetone. Indeed, it possesses a carbonyl group (which is that of the carboxylic group) and a hydroxyl group brought by the aromatic nucleus, which may be considered a sort of enol function:

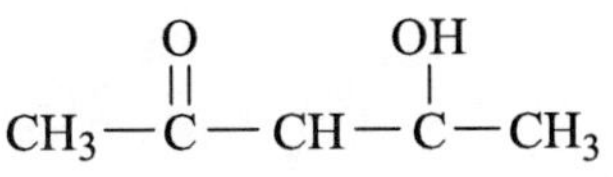

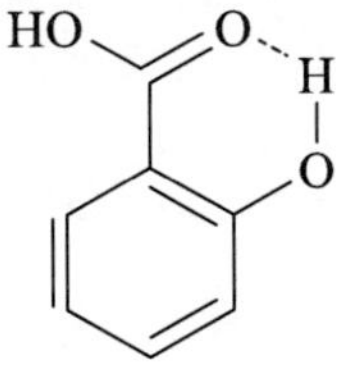

acetylacetone: tautomeric form salicylic acid

This structural sequence is responsible for the intense violet color obtained with ferric chloride. It may result from the following possible reaction:

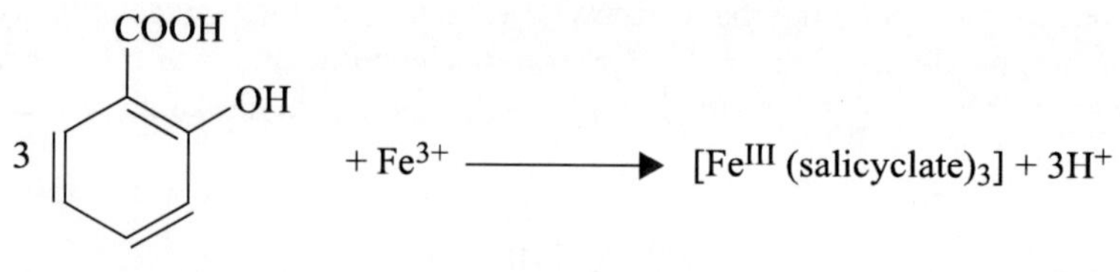

An analogous reaction is used to identify streptose in streptomycin. Streptose is an integral part of the streptomycin molecule. In alkaline medium, streptose decomposes into maltol in several steps:

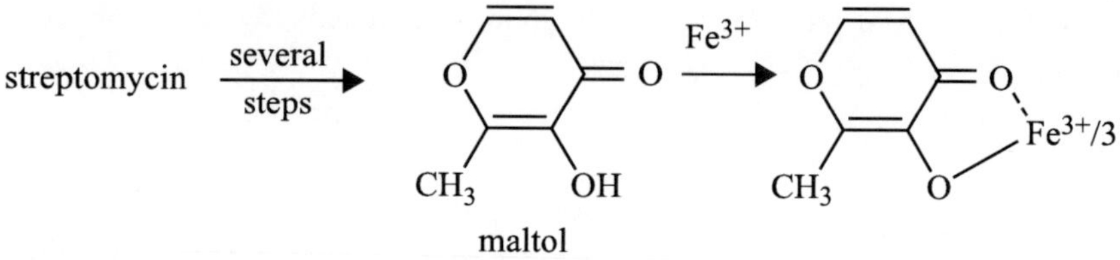

An identical reaction is given by dihydrostreptomycin. It is the same state of affairs for tropolones. This is not unexpected since they possess the same bond system as salicylic acid, with, in particular, the presence of an aromatic nucleus in the system:

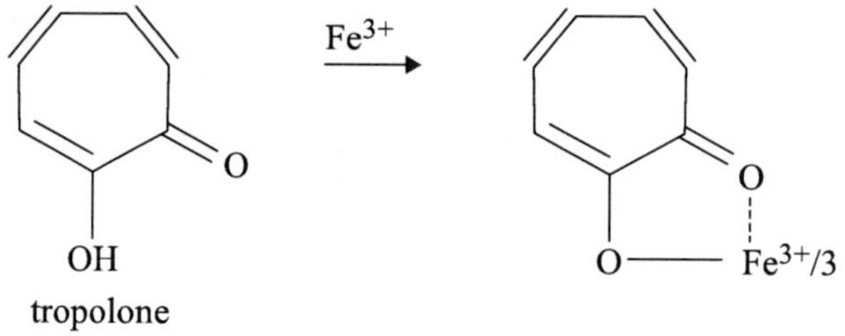

A purity assay of morphine is based on a similar reaction. Morphine base is contained in opium as a salt of meconic acid. Purified morphine in the presence of $FeCl_3$ in hydrochloric acid must not give a red color. Meconic acid indeed yields a red color with Fe^{3+}. The presence of residual meconic acid is a mark of insufficient purification. The reaction of meconic acid with Fe^{3+} is

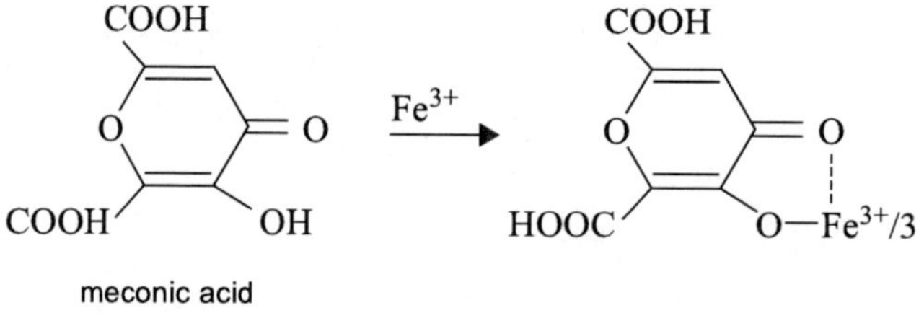

Now, concerning the morphine identification itself, we'll point out that one means to achieve it is to treat it with hexacyanoferrate(III) and Fe^{3+}. According to some works, a free radical reaction involving dioxygen may occur. Then a semiquinone radical may form, and the latter may duplicate into pseudomorphine, which may give a blue color with $FeCl_3$. Hydromorphone, whose structure is close to that of morphine and that also possesses a phenol function, gives a blue color with $FeCl_3$. The color turns yellow after the addition of hydrochloric acid. It is likely that a complexation reaction involving the phenol function occurs. At this stage, if there is an addition of hexacyanoferrate(-III), a deep blue precipitate of Prussian blue will appear. As in the morphine case, there is oxidization of the phenol function and a concomitant reduction of Fe^{3+} into Fe^{2+}.

Another kind of complex formed with $FeCl_3$ is obtained with α-hydroxycarboxylic acids. For example, trimethadione gives 2-hydroxyisobutyric acid after reaction with sodium hydroxide followed by an acidification. Hydroxyisobutyric acid gives a yellow complex with ferric chloride:

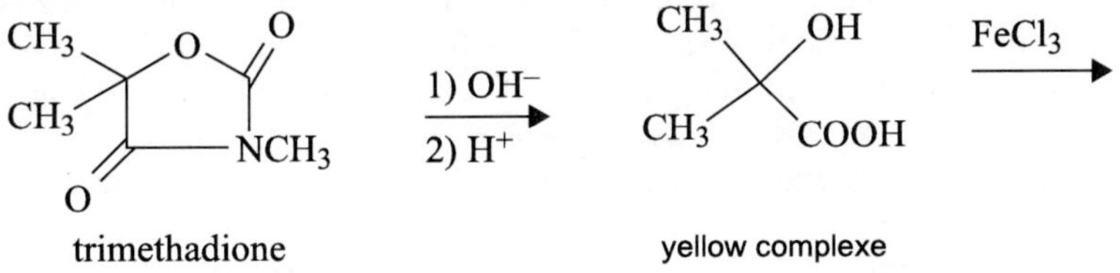

Finally, we'll recall that 8-hydroxyquinoleine gives chelates with some monocharged, bicharged, and tricharged cations. This reagent, in some favorable cases, may allow the selective separation of cations (see Chap. 19). It gives rise to several sorts of chelates. For example, with Mg^{2+}, two chelates are formed, one positively charged and the other neutral. With aluminum, a neutral one is formed.

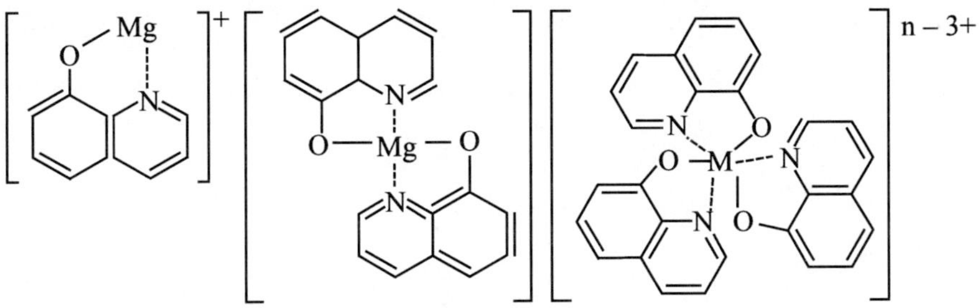

(Note that chelates formed by oxine and bivalent metallic ions may also complex water. Also note that magnesium "oxinate" is fluorescent and that the oxine ligand is electrooxidizable in some experimental conditions.)

The cation Fe^{2+} permits the identification of the bond system –N = C–C = N–. Because of its presence in its structure, bromazepam gives a violet complex with Fe^{2+} and hence is identified through this reaction:

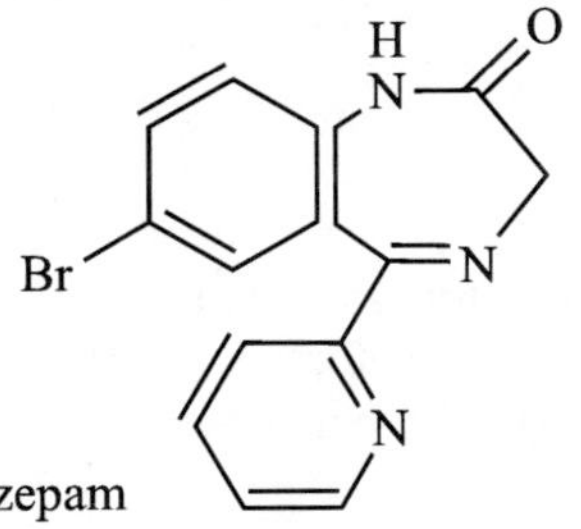

bromazepam

The same bond system also gives colored complexes with Cu^{2+} and Co^{2+}.

It is quite impossible to ignore Legal's reaction when we deal with the use of Fe^{2+} and Fe^{3+} in organic analysis. Legal's reaction is characteristic of α-methylene ketones or aldehydes. It involves nitroprusside ion as reagent. The reaction is carried out in alkaline medium, sodium hydroxide, ammoniac solution, ethylenediamine, piperidine, piperazine, sodium tetraborate, and sodium carbonate or hydrogen carbonate.

Nitroprusside is a complex of Fe^{III} of the structure $[Fe^{III}(NO)(CN)_5]^{2-}$. It is the pentacyanonitrosylferrate(III) anion. (An ambiguity exists about the oxidation state of iron in this complex. Indeed, nitric oxide as ligand can be considered neutral or positively charged. In the first case, iron is Fe^{III}, in the second Fe^{II}. In order to avoid any ambiguity, the nitroxyl ligand must always conventionally be considered as being neutral.) Nitric oxide reacts with numerous transition metals by forming nitrosyl complexes. Several geometries of the angle M–NO are encountered in these complexes. In nitroprusside, it is 178°; thus, its structure is said to be linear. Nucleophilic species such as OH^-, OR^-, SR^-, and NH_2R generally attack the linear complexes M–NO on the nitrogen atom.

With ketones, the structure of the obtained Legal's complexes depends on the successive reactions leading to them. In the case of acetone, for example, the reaction can be written as

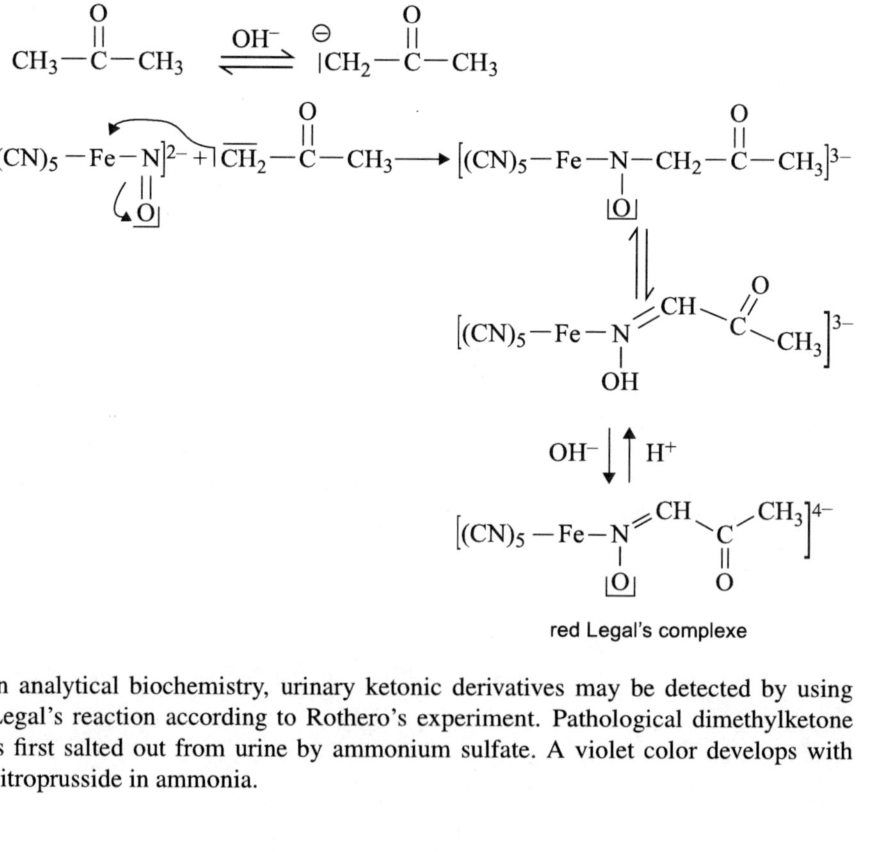

In analytical biochemistry, urinary ketonic derivatives may be detected by using Legal's reaction according to Rothero's experiment. Pathological dimethylketone is first salted out from urine by ammonium sulfate. A violet color develops with nitroprusside in ammonia.

Acetaldehyde, propionaldehyde, and acrolein yield a blue color. The reaction is positive only with ketones and aldehydes that can enolize. All such aldehydes do not necessarily react. Pyruvic acid and lactic acid (the latter probably after oxidization to the former) give the reaction. Likewise, citric acid, first transformed into dimethyl ketone-1,3-dicarboxylic acid and then into dimethyl ketone by reaction with potassium permanganate in acidic medium, gives a positive reaction.

In alkaline medium, sodium nitroprusside reacts with primary amines in the presence of dimethyl ketone and with secondary amines in the presence of acetaldehyde. For example, in addition to the cases already recalled, we can mention those of

- glycerol, which, according to the German pharmacopeia, gives acrolein after warming with potassium hydrogen sulfate. Acrolein gives Simon–Awe's reaction with piperidine and nitroprusside. The following complex forms:

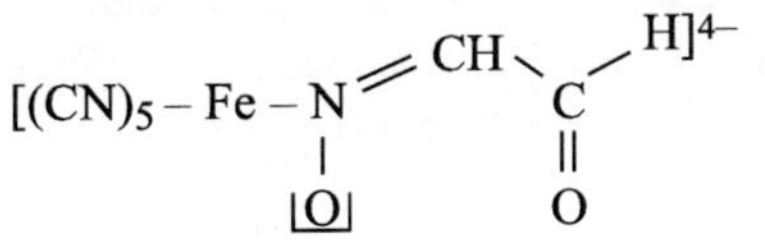

through, probably, the formation of acetaldehyde;

- the sodium salt of thiopental:

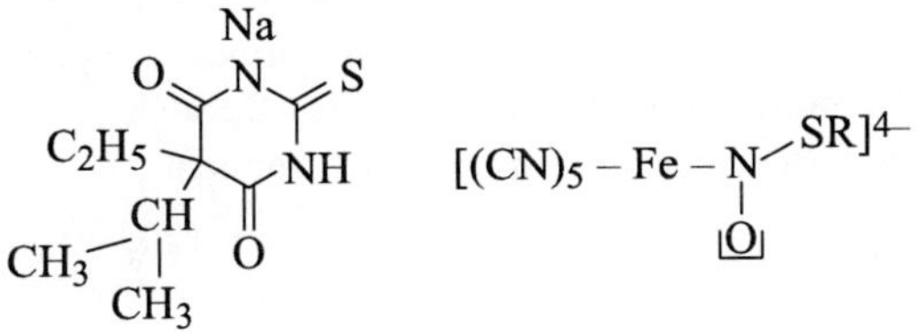

- piperazine, which reacts with nitroprusside and acetaldehyde. In this case, it is the basic agent necessary to carry out Legal's reaction, which is detected;
- paracetamol after an acidic hydrolysis into 4-aminophenol:

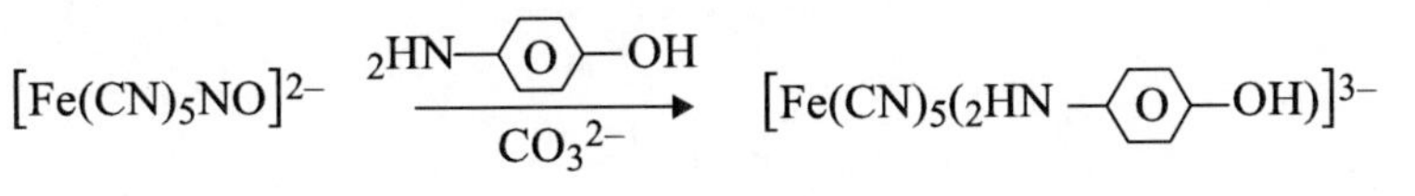

This is one of the purity assays to which it must be submitted. It is interesting to note that in this particular case, it is not Legal's reaction that occurs. It is rather an exchange reaction of nitroxyl ligand with the p-aminophenol rest;

- digitoxin, whose butenolide cycle possesses an active methylene:

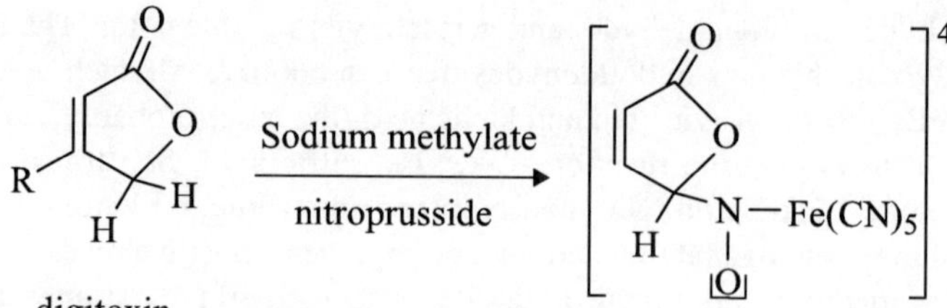

digitoxin

does react with sodium nitroprusside.

This is actually a general reaction of thiols that react as thiolates. Hence, it is the tautomeric form of thiopental that reacts.

Other molecules such as naphazoline and vitamin C also react, in alkaline medium, with sodium nitroprusside with the development of a blue color. The structures of the derivatives obtained have not yet been clearly established.

When it can be carried out, Legal's reaction is quasi-systematically used for quantitative analysis by absorption measurements.

31.2 Formation of Complexes with Cu^{2+}

Copper(II) salts are also of great use in organic analysis and the chemistry of Cu^{2+} is a very rich one. Cu^{2+} can indeed give complexes with N and O donors or with both simultaneously. Moreover, the geometries of the formed complexes are very varied. Finally, dimers sometimes appear that are formed through the copper atom. Their existence is related to the electronic structure of the metal.

- α-amino acids give blue derivatives with copper(II) salts. They are of the sort

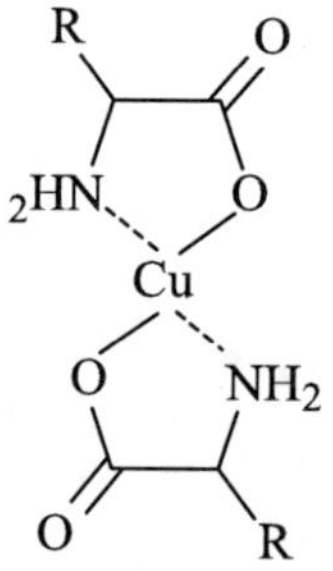

This reaction is an identification reaction of the group.

- in the presence of copper(-II) sulfate and concentrated sodium hydroxide, peptides yield a violet-red color. Actually, the following three bond systems give a positive reaction:

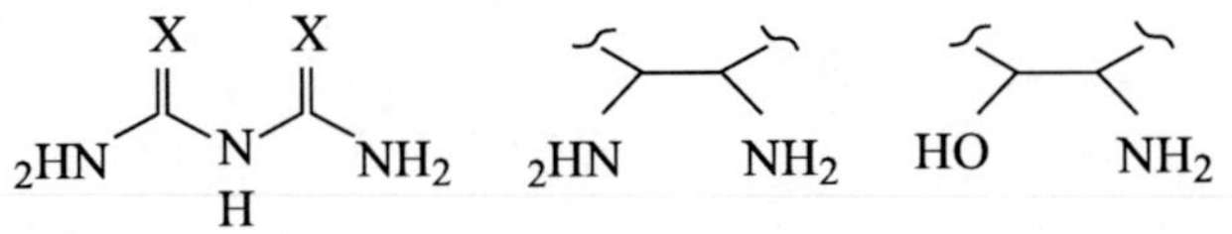

with
X = O, NH, S

It is the biuret reaction. (Recall that biuret forms by warming a mixture of urea, ammoniac solution, and cyanuric acid to dryness.)

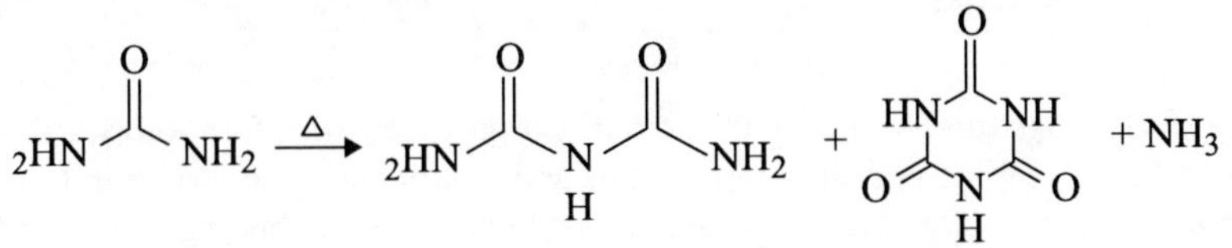

biuret cyanuric acid

Biuret itself may, depending on the experimental conditions and, in particular, on the concentrations, give three kinds of complexes with Cu^{2+} in alkaline medium. Recall also that globulins, albumins, peptones, polypeptides (provided the number of amino acids constituting them is suitable), and amino acids give the biuret reaction. Let's finally recall that this reaction, carried out in well-determined conditions, permits us to determine the quantity of proteins in the analyzed solution.

From the structural standpoint, the violet color of biuret is attributed to the following two complexes:

$$[Cu_2(\mu - OH)_2(NHCONHCONH)_4]^{2-},$$

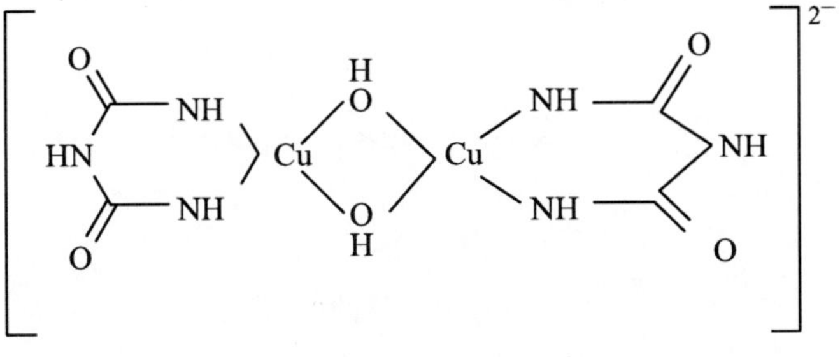

violet

and $[Cu(NHCONHCONH)_2]^{2-}$,

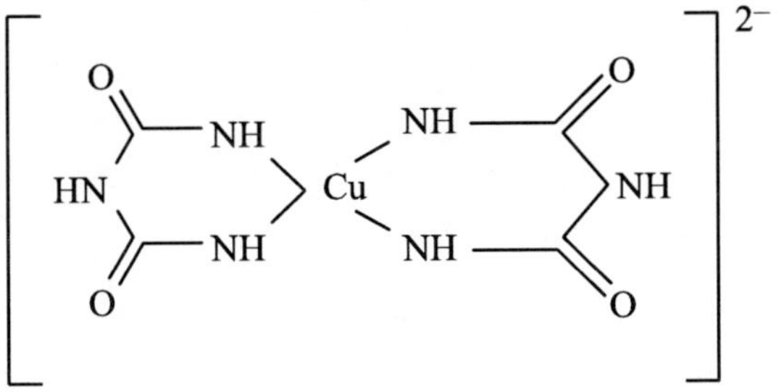

red

and

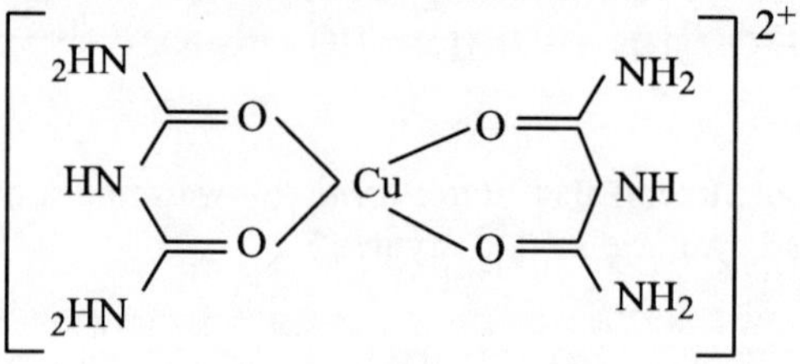

green

Not surprisingly, this is one of the identification reactions of antibiotics that are polypeptides, such as polymyxin, colistin, and bacitracin. Not surprisingly either, hydantoines, which are acylureas (a series to which phenytoin belongs, for example), react in such a manner:

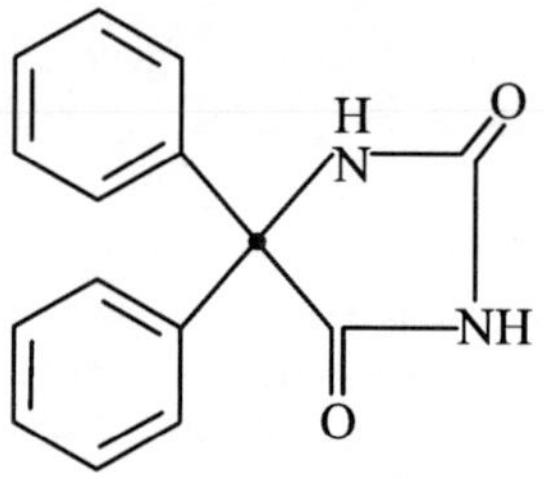

phenytoine (lactam form)

Several other compounds form complexes with Cu^{2+}. Thus, for example,

- ethylenediamine $NH_2–CH_2CH_2–NH_2$;
- ethambutol:

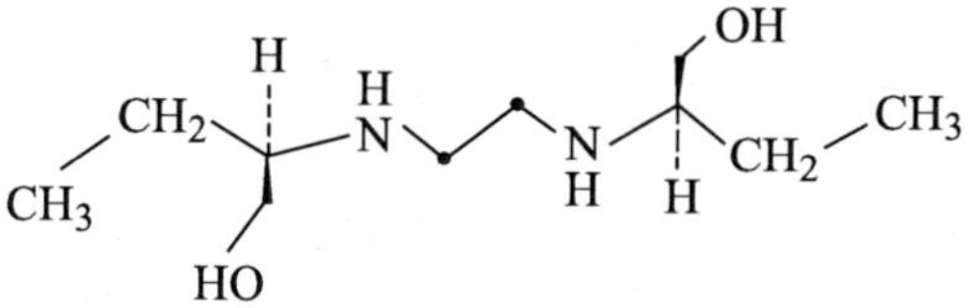

which is also a derivative of ethylenediamine. With Cu^{2+} in alkaline medium, the following complex is formed:

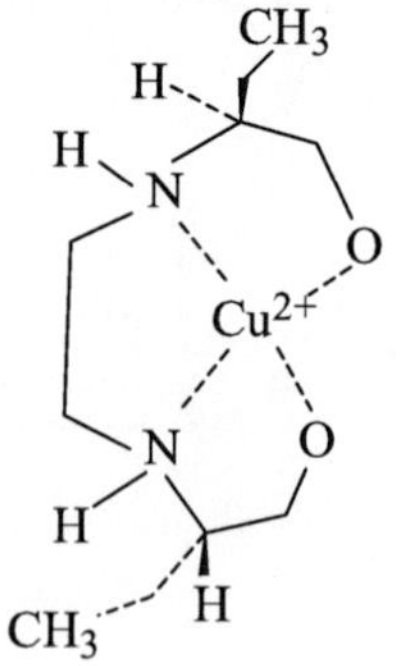

Its formation may be used for the quantitative analysis of the product. Indeed, a means to quantitatively determine it is to proceed through a polarimetric measurement. It is recommended to carry it out in an alkaline solution of copper(II) sulfate. There is an enhancement of the rotation and, hence, an enhancement of the determination precision by forming the copper complex. Of course, the rotation is compared to a reference in order to interpret the result.

- ephedrine (1R,2S)-2-methylamine-1-phenyl-1-propanol:

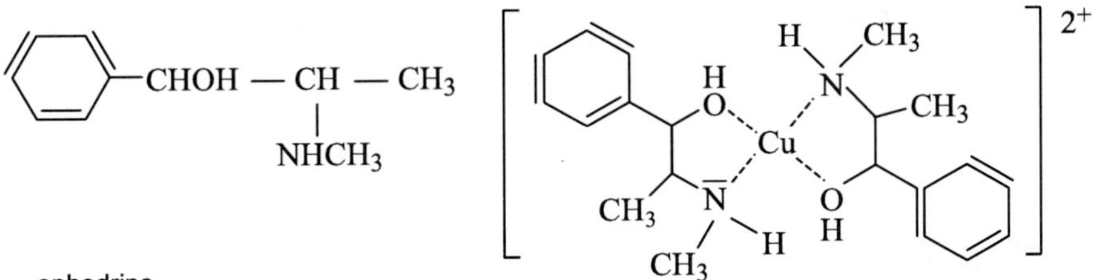

ephedrine

which yields a violet color with a concentrated sodium hydroxide solution and a copper(II) sulfate solution. If it is agitated with ether, the latter turns purple, whereas the aqueous solution is blue. The reaction, called *Chen–Kao's reaction*, is due to the formation of a chelate whose structure is given above. It is specific to alkylamines with vicinal amino and hydroxyl groups. We also notice that this bond system exists in the ethambutol molecule and that it is also one of the structural bond systems giving the biuret reaction.

- barbituric derivatives. They give Zwikker's reaction occurring in pyridine in the presence of copper(II) sulfate. A clear violet precipitate is produced whose structure is

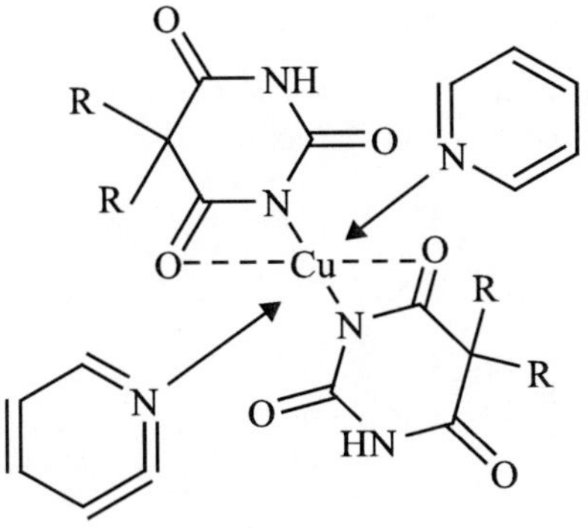

The copper coordinence is 6. The precipitation of the complex permits the elimination of the barbituric from the solution and, hence, favors its identification. An interesting point to note is the good solubility of this complex in chloroform, whereas the deep blue pyridine–Cu^{II} complex, which forms by the addition of an excess of reagent, is not soluble in this solvent. This fact favors the detection of the complex given by the barbituric whose color could be masked by that of the other complex. Furthermore, there is a means here to distinguish barbiturics from thiobarbiturics. The latter compounds color the chloroformic phase green.

- folic acid, which may be symbolized by the following structure:

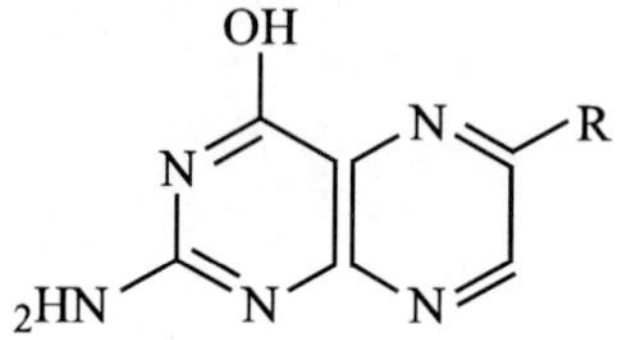

It yields the following copper(II) complex, which precipitates:

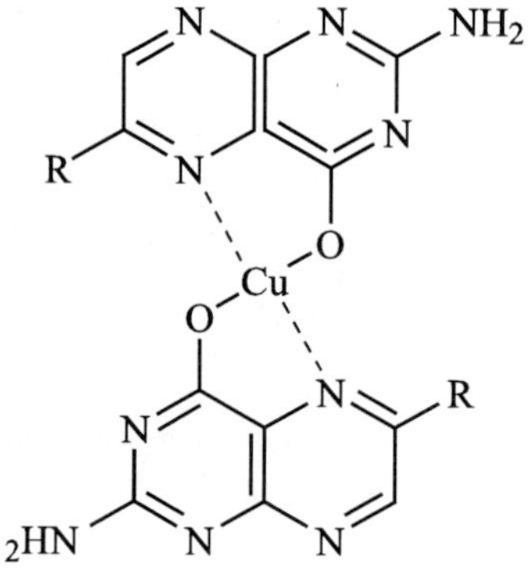

We notice that it looks like oxinates.

- isoniazid. In the presence of a copper(II) citrate solution, it yields a precipitate that is a complex of the following structure:

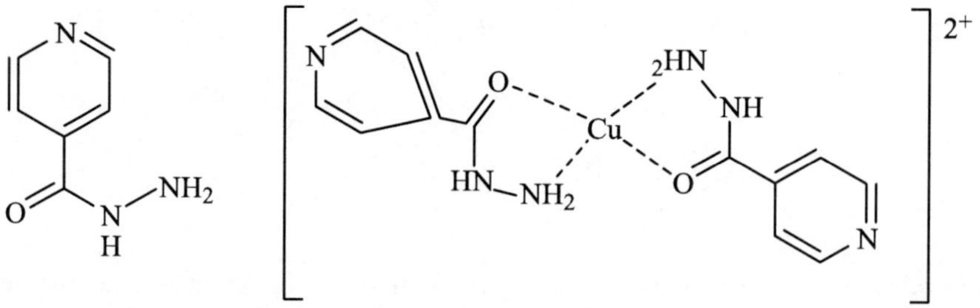

It belongs to the group of complexes built with ligands that are N and O donors. Let's recall incidentally that among the numerous methods of determination of isoniazid, one of them is based on the formation of the above complex in the presence of a known quantity in excess of Cu^{2+}. Its excess is titrated with EDTA. However, putting together $CuCl_2$ and isoniazid in aqueous solution yields a precipitate whose structure differs from that of the preceding complex. Its structure should be

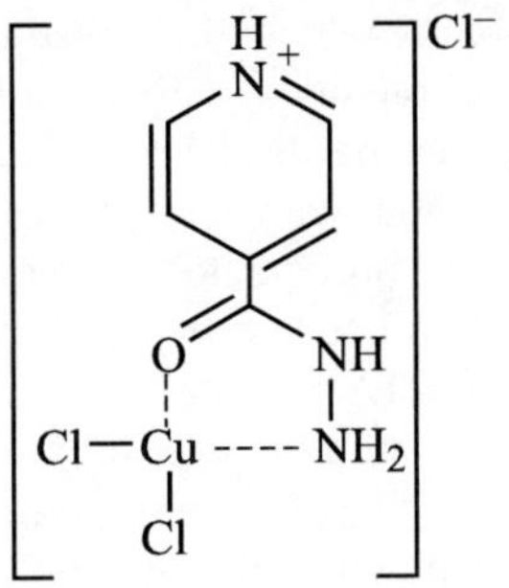

- pantothenic acid. Like pantothenate, it yields a blue color with copper(II) salts in alkaline medium. The β-alanin group does not participate in the complex's formation. According to recent studies, both alcohol functions and the carboxylic group of pantoic acid could be the ligand teeth;

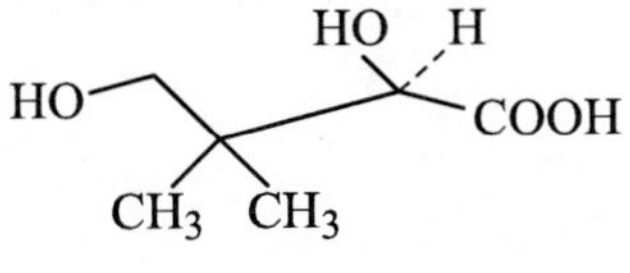

- sorbitol. The formed complex is not biuret-like but, rather, of the copper tartrate type;
- lidocaine. Not only can it be identified, but it can also be quantitatively determined through the formation of the following complex with copper(-II) salts in alkaline medium:

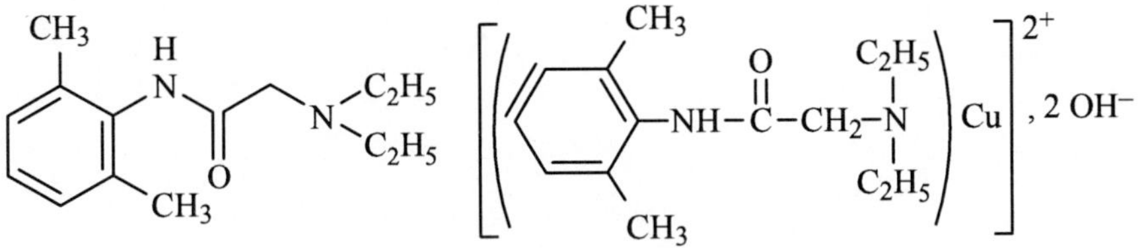

Again, we find the well-known tendency of Cu^{2+} ions to form complexes of coordinence 4 together with complexes of coordinences 5 and 6;

- sulfamides of the type

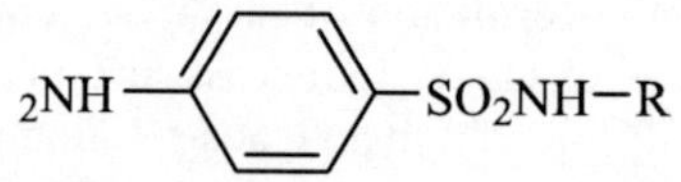

In alkaline medium, some sulfamides precipitate with copper(II) salts. Such sulfamides are principally those in which the substituent R brings a nitrogenous heterocycle. No precipitate is obtained with sulfalinamide, sulfacetamide, and sulfaguanidine. The reaction can be used for quantitative determinations. In order to avoid their coprecipitation with Cu^{II} hydroxide, pyridine is added. In these conditions, only the "good precipitate" is formed, whose overall formula is $(sulfonamide)_2CuPy_3$. The precipitate is filtered afterward and then destroyed. The liberated Cu^{2+} is titrated with EDTA.

Sulfisomidine, sulfadimidine, sulfathiazole, sulfadiazine, and sulfamerazine give this sort of complex. With some other sulfamides, other types are formed if pyridine is not added. Some of them are dimers of the type $[Cu_2(sulfonamide)_4]$, while others are polymers:

$[Cu(sulfonamide)_2]_n$. Acetazolamide, a sulfamide belonging to a pharmacological series different from the preceding one, also gives a blue-green precipitate with Cu^{2+}. This property constitutes the base of one of its identification reactions.

- indomethacine. It gives Büchi and Perlia's reaction. It forms a complex of the overall formula indometacine-copper-pyridine (2-1-2), extractible into chloroform, which then turns green;

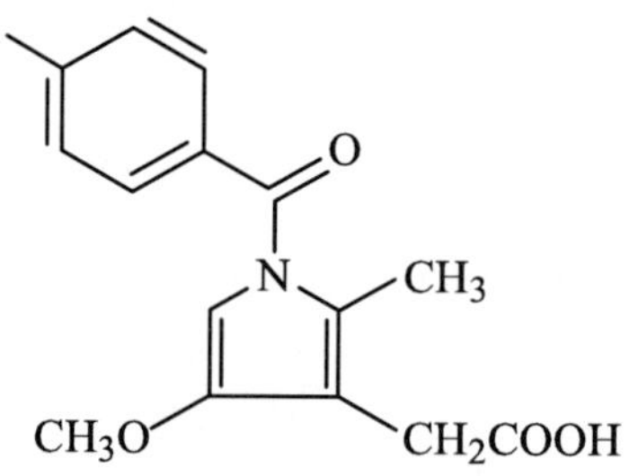

- nitrofural:

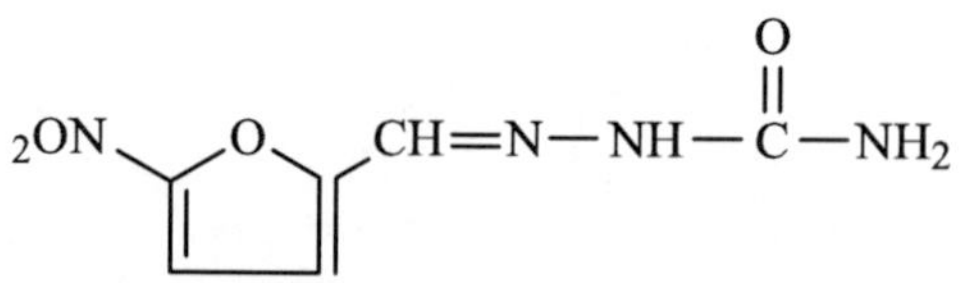

In the presence of copper(II) sulfate and pyridine, it yields a yellow color that is not extractible into chloroform. This reaction should probably be attributed to the ureid rest—$NHCONH_2$ Nitrofurantoin, whose structure is close to that of nitrofural, yields a green color:

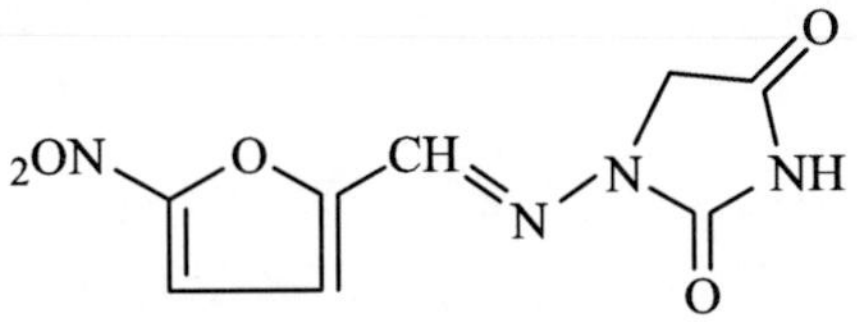

in the same conditions, but it is extractible into chloroform. It is, of course, interesting in order to distinguish both derivatives.

- histamine and disulfiram. They yield colored complexes with Cu^{2+}, but their structures are unknown.

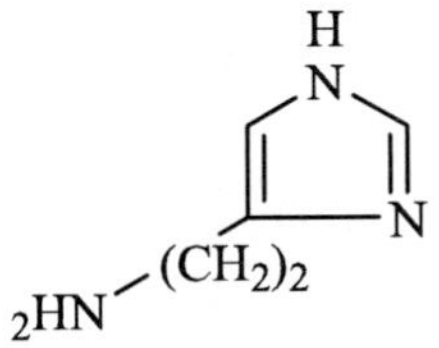

histamine

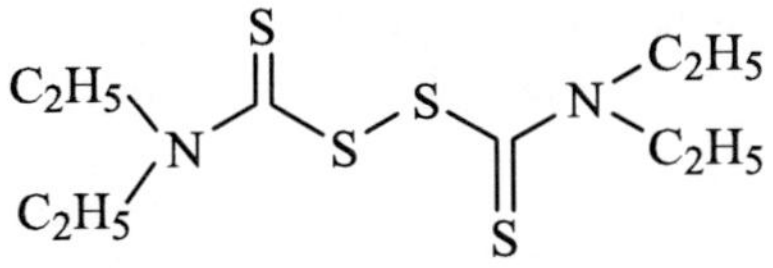

disulfiram

From the standpoint of analytical toxicology, it is interesting to notice that the following sulfoxide:

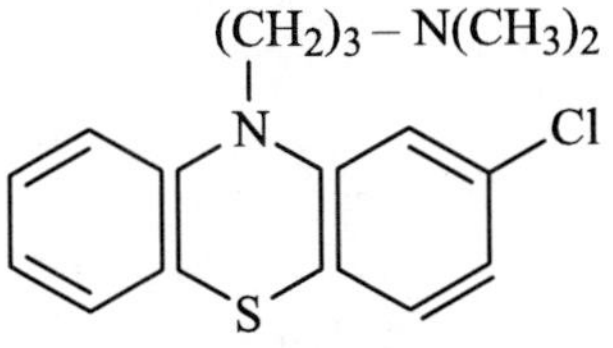

chlorpromazine

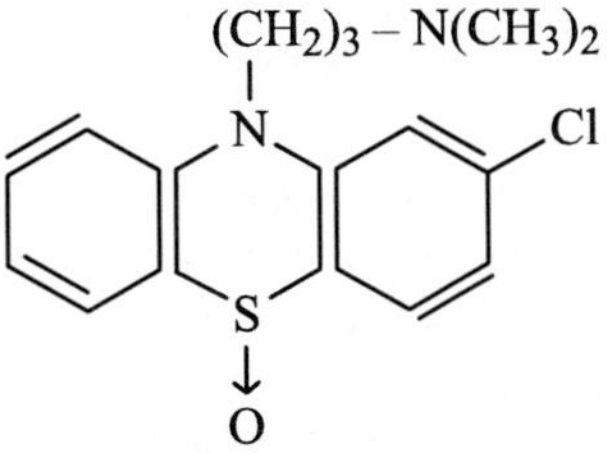

which is one of the principal metabolites of chlorpromazine, gives a complex with Cu^{2+} whose structure is unknown but that might be a dimer. Chlorpromazine does not undergo this reaction. Without any doubt, we can attribute the success of the reaction to the presence of the sulfoxide rest.

31.3 Formation of Complexes with Ag^+

The Ag^+ ion is, above all, used in qualitative analysis. However, in the pharmaceutical field, lets note the quantitative determinations of

- phenytoin in the presence of pyridine. The following complex is formed:

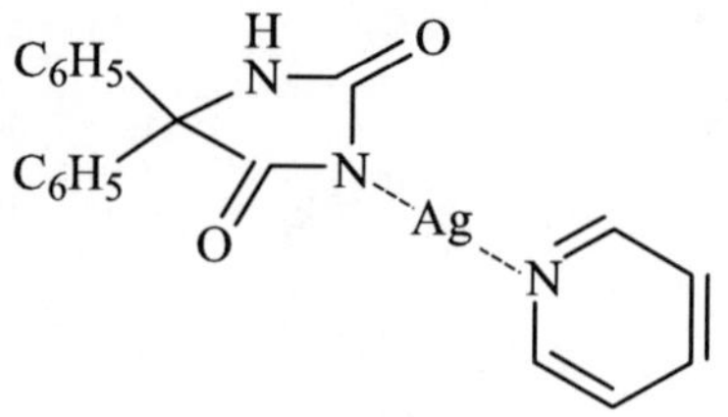

phenytoine

The addition of pyridine is doubly advantageous. On the one hand, it displaces the equilibrium of the reaction toward the right. Behaving in such a manner, it precludes the precipitation of silver oxide. On the other hand, the displaced protons become fixed on the nitrogen of pyridine, which is added in excess. The pyridinium ion, which is hence formed mole to mole with respect to phenytoin, can, as a result, be titrated with sodium hydroxide in the presence of phenolphthalein (titration of a weak acid with a strong base).

Let's recall at this point that the silver ion exhibits a strong tendency to form linear complexes with coordinence 2.

- disulfiram. It reacts with Ag^+ according to the scheme

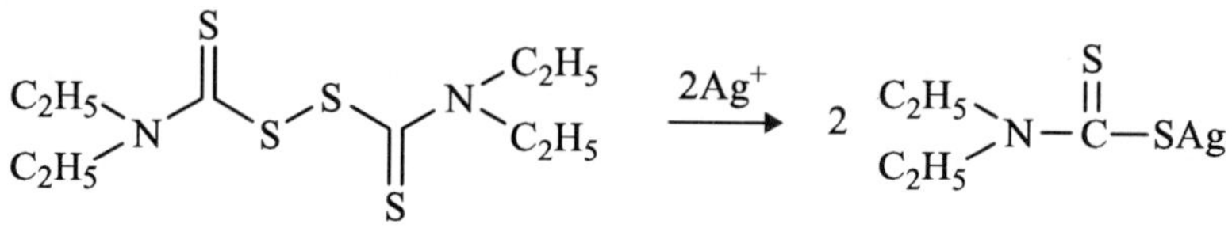

A question now comes to mind: Is the formed derivative a salt or a complex? (Let's note, from the standpoint of the oxidoreduction concept, that this reaction is puzzling because we may observe that two oxidized forms react upon each other.)

We shall see (Chap. 39) that it is possible to titrate other compounds after an addition of silver nitrate followed by a protometric titration of the pyridinium in aqueous or nonaqueous media. These determinations are called *titrations by transformation.*

From a qualitative standpoint, let's note the characterization of isionazid by the formation of a white complex that, by heating in a stream of nitrogen, leads to a deposit of metallic silver and to isonicotinic acid. Evidently, a redox reaction participates in the decomposition.

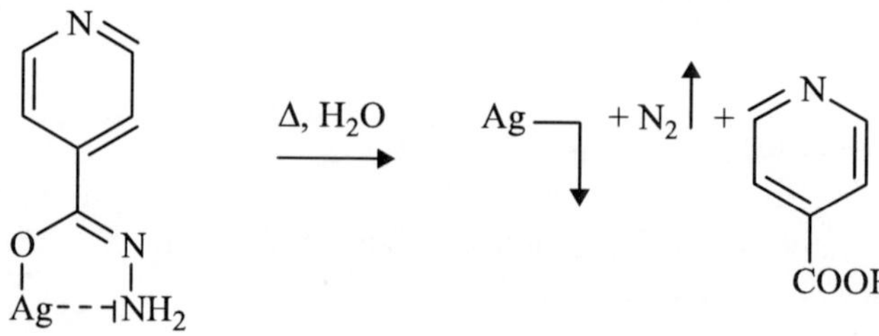

31.4 Formation of Complexes with Co^{2+}

Cobaltous ion Co^{2+} yield fewer complexes than cobaltic ions Co^{3+} do, but a large part of them can be isolated. They are endowed with a great variety of structures, and they are more labile than those given by Co^{3+}. Most of them are tetrahedral. The most striking difference between octahedral and tetrahedral complexes of Co^{II} lies in the fact that the former are pink, whereas the latter are blue. This criterion is not infallible.

An identification reaction of barbiturics consists of the formation of complexes with Co^{2+}. In alkaline medium, barbiturics, of which one nitrogen atom is not substituted, yield tetrahedral complexes of the formula $[Barb]_2CoX_2$ that are stabilized by the solvent or by neutral ligands that enter into the structure. In the following structure, X represents a primary amine:

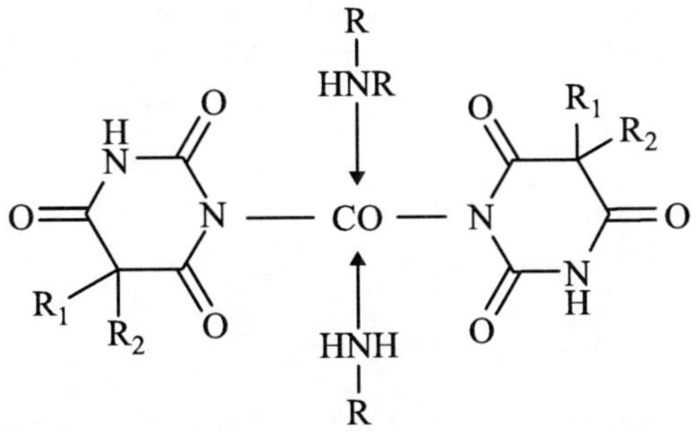

This reaction has been the matter of numerous studies and modifications. It seems to be most sensitive when X is an amine such as cyclohexylamine. The tetrahedral complex is intensively colored since the value of its molar absorptivity coefficient ε is high (100–1000). Probably, in the formation reaction of these complexes, amine plays the part not only of a ligand but also of a base, which favors the deprotonation of the barbituric. Therefore, this part favors the complex formation. When the reaction is carried out without an amine, but in the presence of methanol, the following octahedral complex forms:

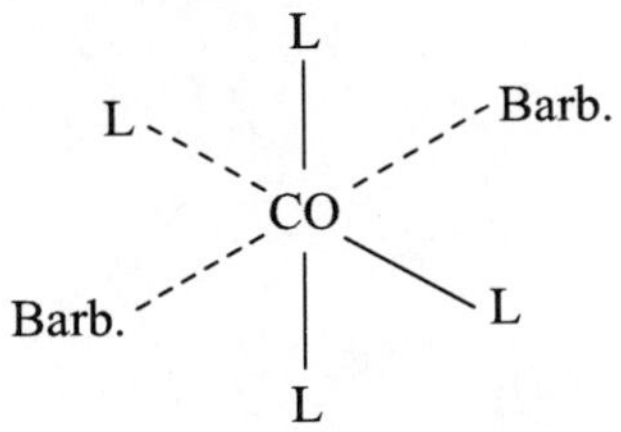

where L represents a methanol molecule. It is colored pink and the intensity is far weaker than that of the preceding complex ($\varepsilon < 10$). This is this sort of complex searched for in analytical toxicology in order to detect barbiturics according to Parri's method. In absolute methanol, barbiturics possessing a unsubstituted nitrogen atom yield a violet color in alkaline medium in the presence of cobalt nitrate. The alkaline medium is due to ammoniac vapors. This reaction is not specific to barbiturics. It is successful with all the compounds whose structure contains the bond system –CONH–CO– (succinimide, phthalimide, hydantoins, alloxane, cyanuric acid, allantoin, uric acid, theobromine, etc.).

Meprobamate can be identified and even quantitatively determined by absorption measurements after the formation of tetracyanato-cobaltate(II):

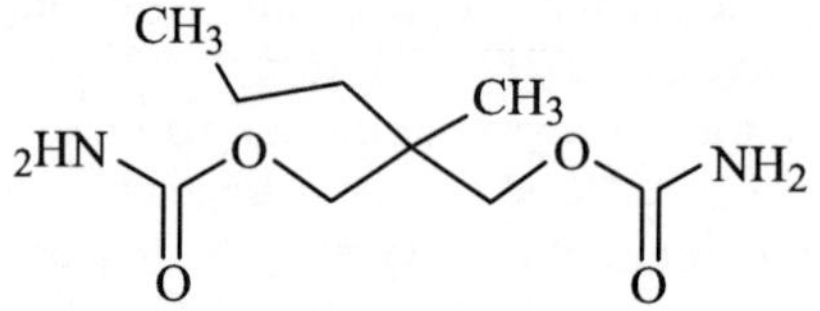

meprobamate

$[Co\,(OCN)_4]^{2-}$

tetracyanatocobaltate (II)

The complex's formation reaction is carried out by heating the reactional vessel in the presence of alcoholic potassium hydroxide. Cyanate ions form according to the reaction

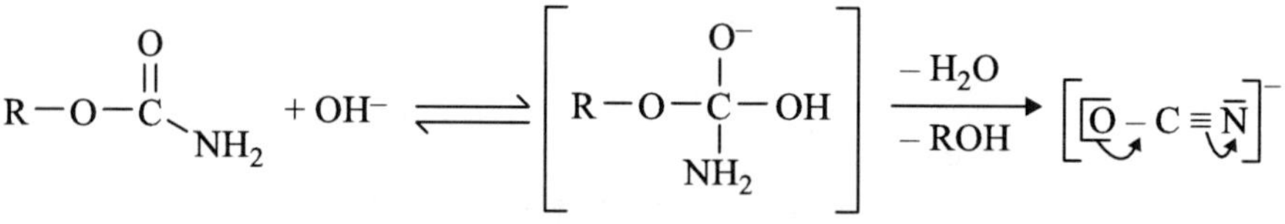

Histamine gives a stable complex with Co^{2+}, as it gives the same sort of complex with Cu^{2+} (see above). It is interesting to link this result with the fact that insulin dimerizes through the presence in its structure of histidine and through the participation of the zinc ion Zn^{2+}, which is divalent.

Recall that vitamin B_{12} may be considered a complex of Co^{2+}. For some authors, it is rather a complex of Co^{3+}. In any case, vitamin B_{12} does possess a true C–cobalt bond. Furthermore, it is the first organometallic structure found in nature. It can be schematized by the following formula:

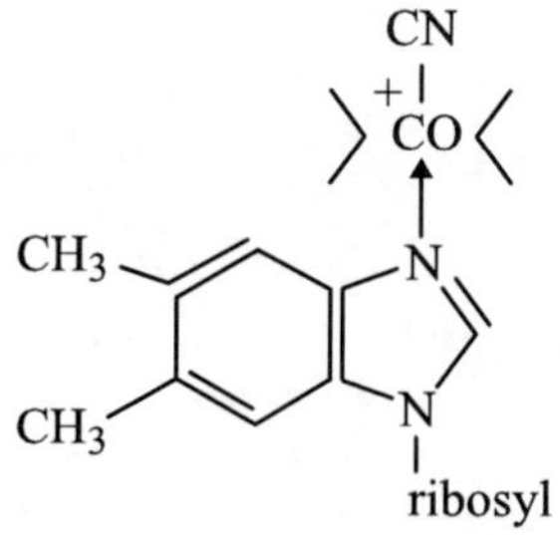

Cobalt is tightly encircled into the corrin cycle. Corrin is a more or less substituted cycle. It is coplanar and possesses a hydrogen atom that can be substituted. The last two bonds we must mention are the Co–CN and Co–benzimidazole ones.

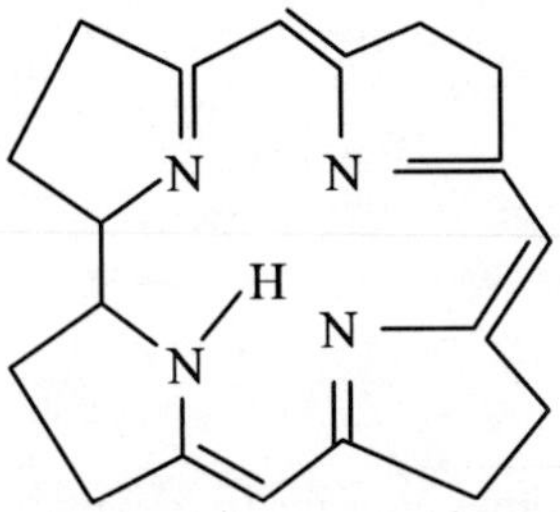

Finally, one of the group reactions permitting the identification of aliphatic amines and also their quantitative analysis through absorption measurements consists of making them react with potassium nitrate and hexahydrate cobaltous nitrate in solid phase. The blue tetrahedral complex is tetrathiocyanatocobaltate(II) $[Co(SCN)_4]^{2-}$. It is obtained in the form of the salt of the conjugate acid of the amine. The reaction may be schematized as follows:

$$2\text{ amine} + 4SCN^- + Co^{2+} + 2H_2O \rightarrow (\text{amineH})_2\left[Co(SCN)_4\right] + 2OH^-.$$

The salt of the complex is extractible into chloroform.

31.5 Formation of Complexes with Hg^{2+}

Mercuric ions permit the identification of

- barbiturics, which are not substituted on nitrogen atoms. With mercuric oxide in nitric acid medium, there is formation of the following complex:

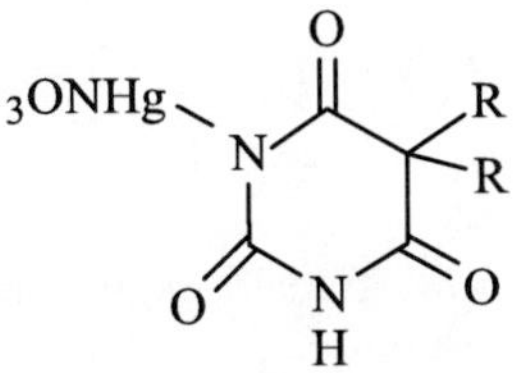

They give a white precipitate. According to the pH value and to the nature of the ions in the solution, the nitrate ion NO_3^- can be replaced by other ions, such as OH^-, Cl^-, and so forth.

If one or both nitrogen atoms is (are) substituted, the following compound forms, whose structure involves two molecules of barbituric. This derivative also precipitates but forms far more slowly than the preceding one.

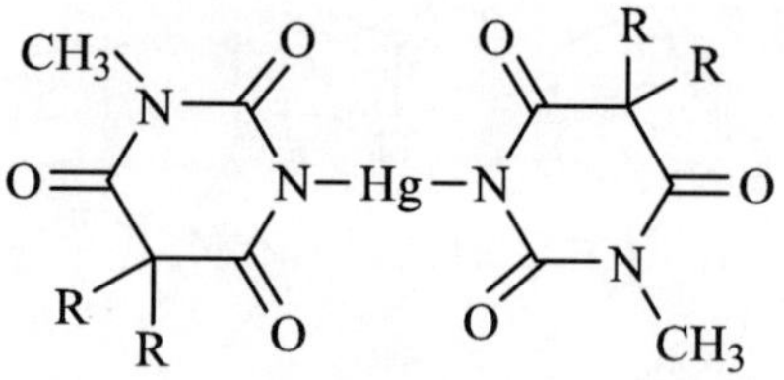

Therefore, these reactions constitute a means to distinguish both sorts of barbiturs. This result must be connected to Denigés' reaction used in analytical toxicology in order to search for barbiturics extracted in acidic medium. Denigés' reagent is mercuric sulfate in a mixture of sulfuric and acetic acids;

- mercaptopurine, under its tautomeric form. It may be identified through the complex it forms with Hg^{2+} (and also with Pb^{2+}) that precipitates:

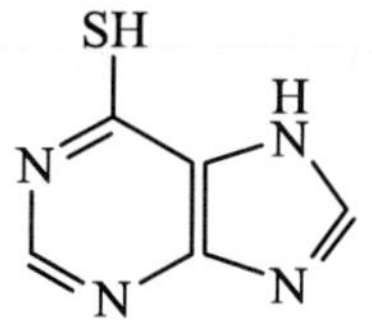

In both cases, the stoichiometry of the complex is mercaptopurine:bivalent ion: 2:1.

From a quantitative standpoint, let's mention the possible titration of penicillins by mercurimetry:

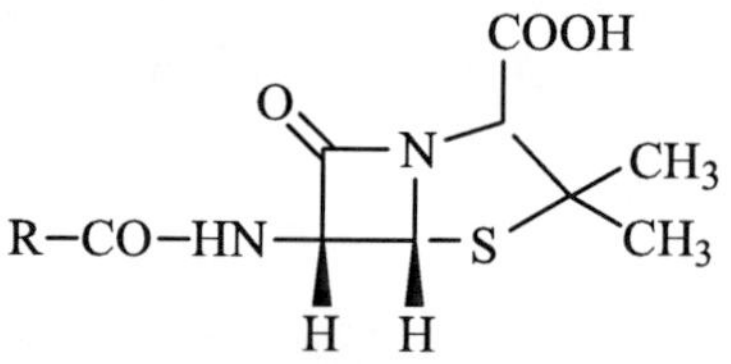

Penicillin is titrated with mercuric nitrate in alkaline medium. The equivalence point is detected by potentiometry. Two sharp potential changes exist, which correspond to the terms of the following two reactions:

$$Hg^{2+} + 2RSH \rightarrow Hg(SR)_2 + 2H^+,$$

$$Hg^{2+} + Hg(SR)_2 \rightarrow 2[Hg(SR)]^+,$$

The second change is the only one taken into account. Hence, the equivalence point corresponds to the ratio of one mercuric cation to two molecules of penicillin. The path of the titration reaction is still under study. In the opinion of some authors, it is possible that penicillamine should react with Hg^{2+} according to the above two reactions.

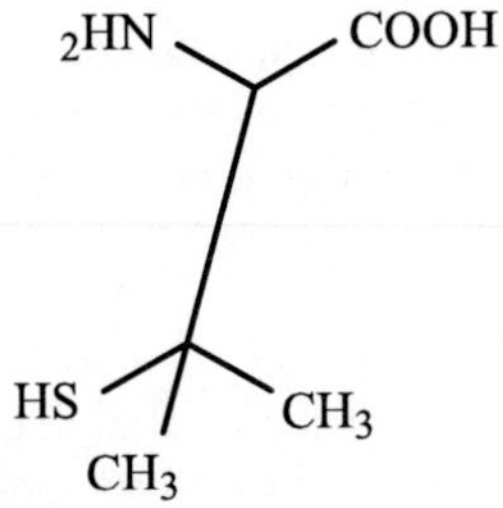

If this is the case, penicillamine reacts as classical thiols.

The β-lactam cycle of penicillins (and of cephalosporins) is very labile. The formation of (temporary) complexes with metallic ions, in particular with transition metals, favors the destruction of these molecules by breaking the β-lactam cycle. The formation of temporary complexes favors the nucleophilic attack on the lactam cycle. Therefore, in a first step, penicilloic acid should form probably as a reactional intermediate, and, after, it should evolve into penicillamine.

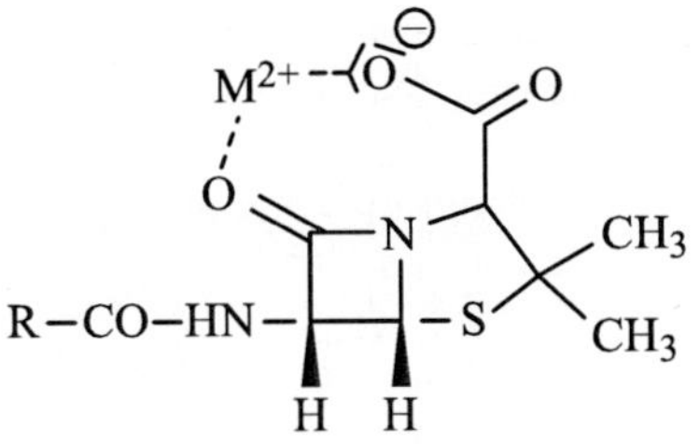

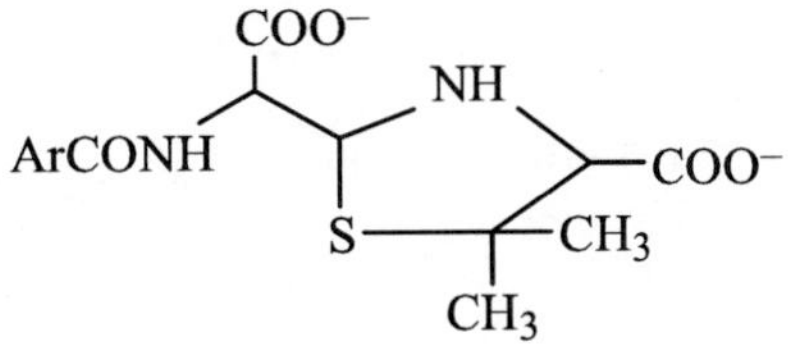

penicilloic acid

31.6 Formation of Miscellaneous Complexes and Analytical Applications

Pharmacologically interesting molecules such as tetracyclines (antibiotics) are interesting from the standpoint of complexation:

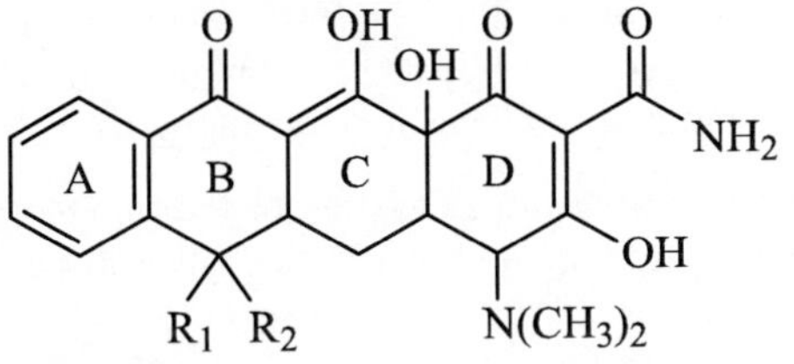

tetracyclines

Their structures possess an amid group, a phenol function, two enol functions, two tertiary alcoholic groups, an α,β-ethylenic ketone, and a tertiary amine group. Hence, it is not surprising that they are endowed with coordinating properties with respect to Fe^{2+}, Fe^{3+}, Al^{3+}, Cu^{2+}, Ni^{2+}, Mg^{2+}, Mn^{2+}, and Ca^{2+}. The formation of chelates of tetracyclines even occurs, whose structures consist of two tetracycline molecules, set head to tail with respect to the metallic ion through the cycles B and C or A and B. Tetracycline chelates are intensively fluorescent, as is tetracycline itself. From another standpoint, the presence of several functional groups allows us to carry out several colored reactions that permit tetracycline identification. Tetracycline itself has been used as a ligand in order to identify some metallic ions by using, for example, the ability exhibited by its complexes to fluoresce.

Now, concerning ceric ions Ce^{IV}, it is interesting to notice that alcohols react with them to yield the ion complex hexanitratocerate(IV), $[Ce(NO_3)_6]^{2-}$. A brownish-orange color appears. It is due to the formation of the alcoholate-pentanitratocerate(IV) complex, $[Ce(NO_3)_5(RO)]^{2-}$. This reaction can be used in quantitative analysis through absorption measurements. It is positive with the three types of alcohols. The reagent is constituted by ammonium hexanitratocerate(IV) in nitric acid medium. In the hexanitratocerate(IV) complex, cerium is dodecacoordinated since each nitrato substituent is bidentate through two oxygen atoms. By analogy with other complexes whose structure is already known, it is probable that in the complex having alcoholate groups as ligands, the latter are located in the *trans* position.

Otherwise, regardless of their class, alcohols react with di-8-quinolinolortho-vanadic acid:

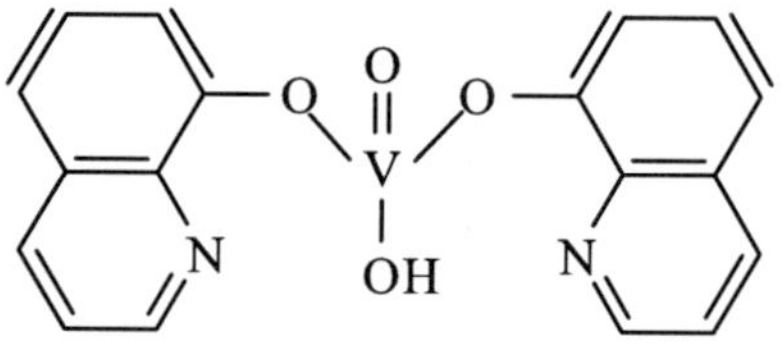

which is a reagent soluble in chloroform. A blue color forms. After the addition of an alcohol, the color turns red by the formation of a complex whose structure is still uncertain.

With acetate ions, cations La^{3+}, brought into solution as nitrate, yield a basic acetate poorly soluble in water. The latter has the property to adsorb free iodine; while doing this, it yields a blue color. The formation of a charge transfer complex between the basic acetate ion (which is itself a complex) and iodine is probable.

Concerning the quantitative analysis of active ingredients, let's point out those

- of dimercaprol (BAL, british anti-Lewisite). In addition to the already seen iodometric dosage, a back titration is proposed. Zn^{2+} gives a complex with BAL. Zn^{2+} in excess is titrated with EDTA;
- of calcium pantothenate and gluconate with EDTA. Notice, however, that in these cases, it is not the organic part of the drug that is quantitatively assayed.

The formation of Meisenheimer's complexes σ leads to interesting applications in quantitative organic analysis. Janovsky in the 1890s showed that *m*-dinitrobenzene or a substituted *m*-dinitrobenzene in solution in dimethyl ketone develops an intense color after the addition of potassium hydroxide. Globally, the reaction would be

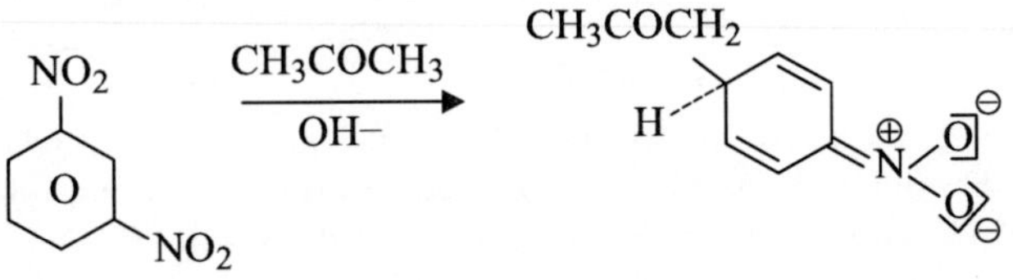

It has been extended to other ketones. Sometime later, it was discovered that this reaction is positive with molecules possessing an active methylene, such as α-methylene ketones. Zimmermann applied it to 17-ketosteroids of great pharmacological interest. Numerous experimental modifications have been proposed, including the use of organic bases instead of potassium hydroxide and the use of other nitro derivatives of benzene. It seems well established that when *m*-dinitro or 1,3,5-trinitrobenzene reacts with an excess of a compound possessing an active methylene in basic medium, a σ Meisenheimer complex forms:

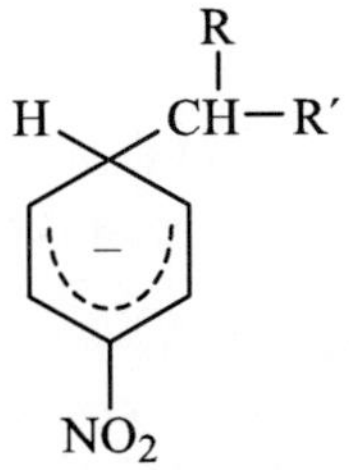

σ Meisenheimer's complexe

where R′ is an activating group. This complex is reversibly formed. (The most striking proof of the formation of this sort of complex is without any doubt furnished by NMR spectroscopy, with which it is easy to detect carbon atom hybridization changes.)

With an excess of the compound possessing an active methylene (Janovsky's conditions), the reaction is fast and stops at the stage-σ complex. In Zimmermann's conditions, namely, with an excess of the dinitro derivative, Meisenheimer's complex is irreversibly oxidized to yield the colored anion:

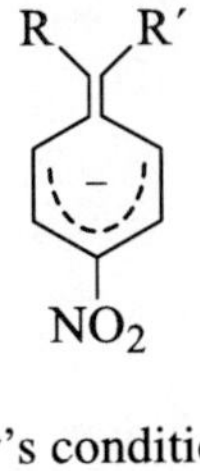

The reaction carried out in Janovsky's conditions can be applied to the quantitative analysis of aliphatic amines since they can play the part of the basic agent for the reaction development. Aromatic amines are not sufficiently basic to react. Numerous alkaloids react as aliphatic amines. Zimmermann's reaction can also be applied to 2,4-dinitrophenylhydrazones, which in the presence of nitromethane in alkaline medium, give a σ-complex. It can also be carried out with some benzene derivatives after nitration. In this case, nitromethane is the preferentially used compound among those possessing an active methylene. Reciprocally, nitrate ions can be detected in this way. Nitrobenzene is nitrated for a second time by nitrate ions in sulfuric acid. Dimethyl ketone is then the compound used with an active methylene. In Zimmermann's conditions, the reaction is, above all, applied to the quantitative analysis of derivatives with an active methylene by absorption measurements such as 3- and 17-ketosteroids.

In the pharmaceutical field, diazepam gives a σ-complex, the reaction terminating at this stage even with an excess of dinitrated derivative:

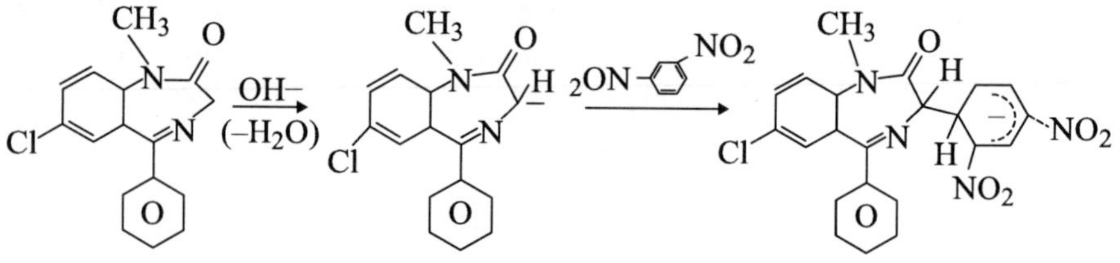

Zimmermann's reaction is positive with dihydrocodeinone, which possesses an active methylene in the α position of the keto group of the cyclohexanone rest. Finally, concerning this reaction, we also notice that digitoxin condenses with picric acid through the active methylene of the butenolide cycle. The absortion measurement at 495 nm allows its quantitative analysis and its identification (Baljet's reaction).

To end this chapter, we'll note that other kinds of complexes exist, whose formation may induce modifications of some physicochemical properties of the compounds giving them. Hence, their analysis may be perturbed. It is the case of ion pairs that we decided to include in the group of complexes and also of charge transfer derivatives. A good example is that of puric bases derived from 7H-purines: xanthine, caffeine, theophylline, and theobromine. Some abnormal physical properties are exhibited by these derivatives.

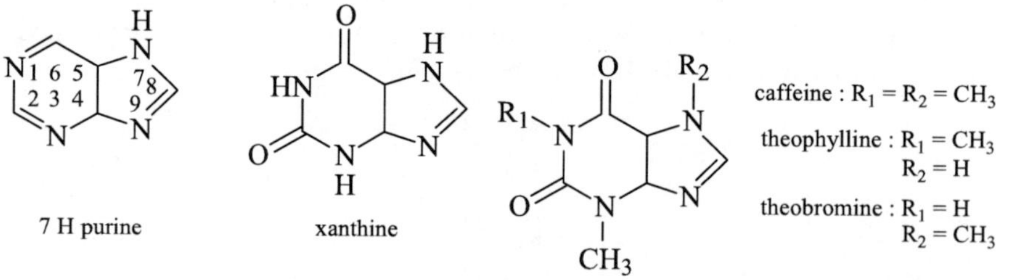

First, their water solubility clearly increases according to the series xanthine, theobromine, theophylline, and caffeine. Curiously, the most methylated derivative is the most soluble and the least methylated one the least soluble. Second, no regularity is found in the melting-point values of these derivatives. The explanation of these anomalies lies, at least in part, in the formation of hydrogen bonds. In caffeine, the three nitrogen atoms are methylated. Therefore, there are no = NH groups that would be able to give H bonds. Caffeine, hence, cannot give dimers, in any case, through the formation of H bonds. It is the most soluble. It sublimates at 178°C. Theophylline, which is twice less soluble than caffeine, gives a head-to-tail dimer by H-bond formation. Furthermore, there is insertion of water molecules in its crystalline network. Theobromine, for spatial reasons, forms larger aggregates than theophylline. As a result, it is about 20 times less soluble than the latter, and it melts at a temperature higher than 350°C, whereas theophylline melts at 274°C.

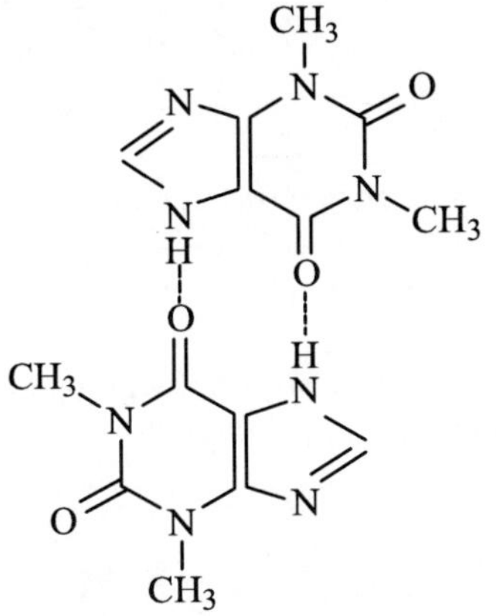

theophylline dimer by formation of H-bond

Nevertheless, caffeine is, abnormally, poorly soluble in water. The study of its NMR^{13C} spectra shows that it exists as a dimer form by the formation of a charge transfer complex. The same phenomenon occurs with theophylline. Moreover, caffeine also gives charge transfer complexes with some other compounds such as benzoic, salicylic, 4-aminobenzoic and acetylsalicylic acids, sulfamides, barbiturics, and local anesthetics. The charge transfer generic name results from the description of the molecular structure in terms of resonance. The charge transfer concept implicates a one-electron transfer from an electron-rich molecule called the donor to an electron-poor one called the acceptor molecule. These complexes are also called π-complexes since one of the halves of these complexes does possess π electrons. In the caffeine case, the central bond is a conjugated carbonyl. It can be considered a π-electron acceptor group, whereas bases must be considered donors. In the caffeine–benzocaine association, both molecules are located in two parallel planes and are superposed in the way presented below. In these conditions, the carbonyl oxygen atom of benzocaine is located near the nitrogen 7 of caffeine, and the benzocaine amino nitrogen near the carbonyl located in 2 in caffeine. These locations are those for which the contrast between electrical charges is the most marked. Additionally, other favorable positions exist in the spatial layout.

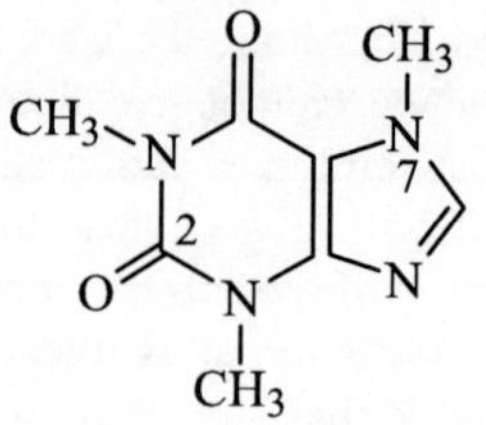

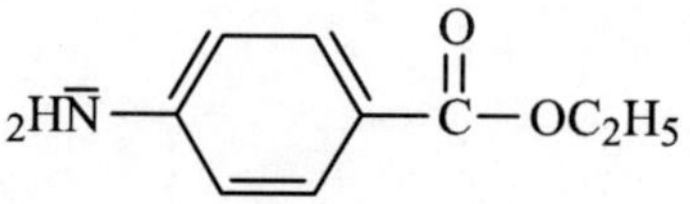

From the analytical standpoint, charge transfer complexes are usually more colored than their individual components. The shift toward longer wavelengths is explained by an enhanced resonance in the complex with respect to the resonances occurring in the individual components.

Another example of a complex formed through hydrogen bonds is given by the sulfamethoxazole–trimethoprime association:

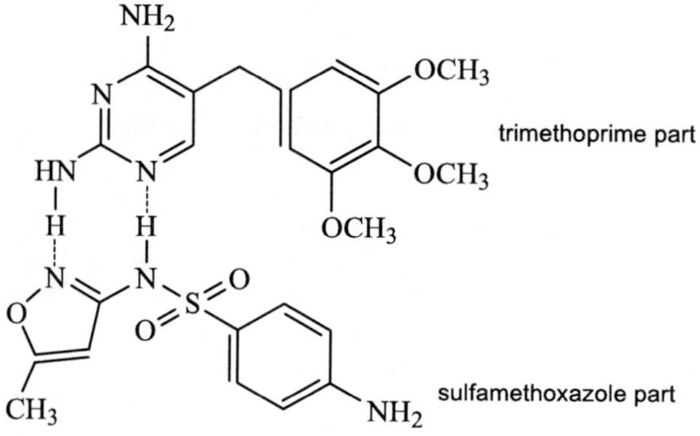

Chapter 32
Intrinsic, Ionic, and Total Solubilities; Solubility Product and Precipitation

The precipitation phenomenon is commonly used in chemical analysis. For example, in immediate analysis, it allows impurities to be eliminated from the studied medium by precipitating them as poorly soluble derivatives. Contrary to this methodology, in some other cases, the compound of interest may be isolated in a pure state by precipitation. In quantitative analysis, the precipitation phenomenon is the basis for gravimetry and for the titrations of some ions.

The precipitation of a product occurs when its solubility is reached. Its solubility is governed by its intrinsic solubility and also, when it is ionizable, by its solubility product.

32.1 Solubility Product and Intrinsic Solubility

Let's consider the substance MX and suppose that it can ionize into the cation M^+ and into the anion X^- according to the equilibrium

$$MX \rightleftharpoons M^+ + X^-.$$

The substance MX is said to belong to the group of 1,1-electrolytes since it gives two monocharged ions by ionization. The equilibrium is governed by the thermodynamic dissociation constant K_d:

$$K_d = a_{M^+} \times \frac{a_{X^-}}{a_{MX}}$$

or

$$K_d a_{MX} = a_{M^+} \times a_{X^-}$$

When the solution is saturated in MX, that is, when we are in the presence of the following equilibria:

$$MX_{(solid)} \rightleftharpoons MX_{(solution)} \rightleftharpoons M^+ + X^-,$$

we can demonstrate via thermodynamic reasoning (see Sect. 32.3) that the activity $a_{MX(solution)}$ is constant. Hence, we can write

$$K_d a_{MX} = \text{constant} \quad \text{(at saturation)}$$

J.-L. Burgot, *Ionic Equilibria in Analytical Chemistry*,
DOI 10.1007/978-1-4419-8382-4_32, © Springer Science+Business Media, LLC 2012

or
$$a_{M^+}a_{X^-} = \text{constant} \quad \text{(at saturation)}.$$

The product $a_{M^+} \times a_{X^-}$, at saturation of MX, is called the solubility product of MX. Its symbol is K_s:

$$K_s = a_{M^+} \times a_{X^-}$$

At a given temperature, K_s is a constant. This relation is satisfied only at saturation. In this case, it is said that the solubility product (of MX) is reached. When the saturation in MX is not reached, the product $a_{M^+} \times a_{X^-}$ is not constant. K_s is not reached.

The solubility product is defined in terms of activities. Hence, it is a dimensionless number, as are all the other equilibrium thermodynamic constants.

When the solution is saturated in MX, the concentration [MX] is constant at a given temperature, as is its activity a_{MX}.

$$\text{At saturation} \quad [\text{MX}\,] = \text{constant with}$$
$$\text{constant} = S_o$$

S_o is called the *intrinsic solubility* of M. It is also often called the solubility of MX. It is expressed in mol/L. It must not be confused with the solubility product. Note that in order to apply the notion of solubility, MX may be a dissociable molecule. It can also be an ion pair. When MX is nondissociable, of course, the concept of solubility product can no longer be applied.

32.2 Generalization of the Concept of Solubility Product

Let's consider the species M_zX_y, which dissociates according to the equilibrium

$$M_zX_y \rightleftharpoons zM^{+y} + yX^{-z}.$$

At saturation, both following equilibria are superimposing:

$$M_zX_y(\text{solid}) \rightleftharpoons M_zX_y(\text{solution}) \rightleftharpoons zM^{+y} + yX^{-z}.$$

The solubility product is:

$$K_s = (M^{+y})^z \times (X^{-z})^y$$

(in which the parentheses signify activities). This expression issues from the same considerations as the previous one. K_s is a constant at a given temperature. Of course, there also exists an intrinsic solubility S_o of M_zX_y. It is the number of moles in the species M_zX_y in the molecular state contained in 1 L of solution, at saturation, which are not dissociated.

Some poorly soluble compounds give more than two ions at saturation. For example, this occurs with zinc and potassium ferrocyanide $[Fe(CN)_6]_2Zn_3K_2$. It gives rise to the equilibrium

$$[Fe(CN)_6]_2Zn_3K_2 \downarrow \rightleftharpoons 2[Fe(CN)_6]^{4-} + 3Zn^{2+} + 2K^+.$$

The solubility product is

$$K_s = ([Fe(CN)_6)]^{4-})^2(Zn^{2+})^3(K^+)^2, \quad K_s = 10^{-95}.$$

The expression of a solubility product is not necessarily a product! It may be a ratio. Let's consider the dissociation of arsenic sulfide As_2S_3 in the presence of the sulfide ion S^{2-}. The reaction that occurs is

$$As_2S_3 \downarrow + S^{2-} \rightleftharpoons 2AsS_2{}^-.$$

We find:

$$\frac{(As_2S_2{}^-)^2}{(S^{2-})} = K_s.$$

Likewise, some hydroxides dissolve in alkaline media, such as aluminum hydroxide, which gives the aluminate ion $AlO_2{}^-$ according to the reaction

$$Al(OH)_3 \downarrow + OH^- \rightleftharpoons AlO_2{}^- + 2H_2O,$$

$$\frac{(AlO_2{}^-)}{(OH^-)} = K_s.$$

It is possible to transform these ratios into products. For example, by using the ionic product of water, the latter relation becomes

$$(AlO_2{}^-)(H^+) = K_s K_w.$$

In the case of arsenic sulfide, by using the acid dissociation constants of dihydrogen sulfide, we find the following relations:

$$H_2S \overset{K_{a1}}{\rightleftharpoons} HS^- + H^+$$

$$HS^- \overset{K_{a2}}{\rightleftharpoons} S^{2-} + H^+,$$

$$(S^{2-}) = K_{a1}K_{a2}\frac{(H_2S)}{(H^+)^2}$$

or

$$(AsS_2{}^-)^2(H^+)^2 = K_s K_{a1} K_{a2}(H_2S),$$

$$(AsS_2{}^-)^2(H^+)^2 = \text{constant}$$

for a constant activity of H_2S. (This is the case, for example, when the solution is saturated with H_2S.)

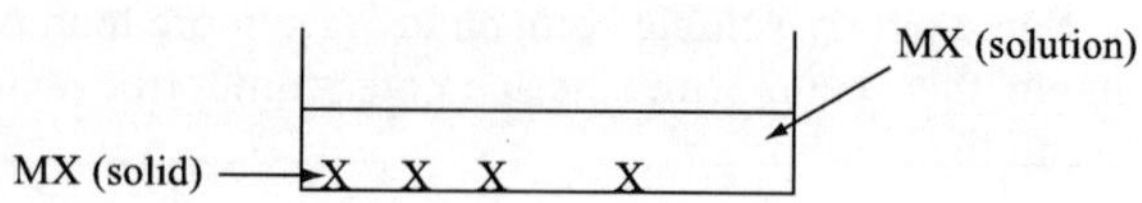

Fig. 32.1 Species MX at saturation

32.3 Thermodynamic Justification of the Concept of Solubility Product

The concept of solubility product, as we've already mentioned, is legitimate because of

- the existence of the dissociation constant K_d. We have already used it;
- the fact that, at saturation, we have considered the activity a_{MX} to be constant.

The existence of the dissociation constant is entirely justified by the thermodynamic developments of the second chapter. Hence, all that remains to do now is to justify the fact that the activity of a species at saturation in solution is constant.

When the species MX is at saturation, that is, when at equilibrium at least one minute particle of solid remains (see Fig. 32.1), the following relation must be satisfied (actually, it is the equilibrium condition between MX in solution and MX in the solid state—see Chap. 2):

$$\mu_{MXsolid} = \mu_{MXsolution}.$$

The chemical potential $\mu_{MXsolid}$ is given by the expression

$$\mu_{MXsolid} = \mu^\circ_{MXsolid} + RT\ln a_{MXsolid}$$

$\mu_{MXsolid}$ is the chemical potential of MXsolid in its standard state. It is a constant at a given temperature. From another standpoint, MXsolid constitutes a pure phase. As a result, due to the conventions concerning the activities,

$$a_{MXsolid} = 1$$

and

$$\mu_{MXsolid} = \mu^\circ_{MXsolid}.$$

Now, concerning MX in solution, its chemical potential is given by the expression

$$\mu^\circ_{MXsolution} = \mu^\circ_{MXsolution} + RT\ln a_{MXsolution}.$$

$\mu^\circ_{MXsolution}$ is the chemical potential of MX in its standard state in solution. It is a constant at a given temperature. According to the usual conventions concerning the activities, $\mu^\circ_{MXsolid}$ and $\mu^\circ_{MXsolution}$ are not equal since the standard states in the two phases are different. According to the equilibrium condition,

$$\mu^\circ_{MXsolid} = \mu^\circ_{MXsolution} + RT\ln a_{MXsolution}$$

and consequently,

$$a_{MXsolution} = \exp\left[\frac{(\mu^\circ_{MXsolid} - \mu^\circ_{MXsolution})}{RT}\right].$$

Hence, at saturation and at a given temperature, $a_{\mathrm{MXsolution}}$ is indeed constant since the standard chemical potentials in both phases are also constants.

32.4 Intrinsic Solubility, Total Solubility, and Ionic Product

Let's now consider the solubility of MX and suppose that it dissociates more or less in solution. What is usually called the solubility of MX is its total solubility, that is, the number of moles that can be dissolved in 1 L of solution, regardless of its state in solution.

In pure water and if M^+ and X^- do not exhibit any acidic or basic character, its solubility S is given by the expressions

$$S = [\mathrm{MX}]_{\mathrm{saturation}} + [\mathrm{M}^+]$$

or

$$S = [\mathrm{MX}]_{\mathrm{saturation}} + [\mathrm{X}^-]$$

since in these conditions, according to the dissociation equilibrium,

$$\mathrm{MX} \rightleftharpoons \mathrm{M}^+ + \mathrm{X}^-,$$

$$[\mathrm{M}^+] = [\mathrm{X}^-].$$

Indeed, by hypothesis, this equilibrium is not perturbed by supplementary equilibria, in particular by acid–base equilibria (see the following chapters).

We see that the (total) solubility is the sum of the intrinsic solubility S_o and of the ionic solubility. Hence, we can write

$$K_s = [\mathrm{M}^+][\mathrm{X}^-],$$

$$K_s = [\mathrm{M}^+]^2,$$

$$K_s = [\mathrm{X}^-]^2,$$

$$[\mathrm{M}^+] = \sqrt{K_s},$$

$$[\mathrm{X}^-] = \sqrt{K_s}.$$

In this case, $\sqrt{K_s}$ is the ionic solubility. As a result,

$$S = S_o + \sqrt{K_s}.$$

Usually, for mineral compounds in water,

$$S_o \ll \sqrt{K_s}$$

and

$$S \approx \sqrt{K_s},$$

the intrinsic solubility is negligible. For example, in the case of AgCl,

$$K_s = 10^{-9.70},$$

$$S_o = 2.2 \times 10^{-7}\ \text{mol/L},$$
$$S = 2.2 \times 10^{-7} + 10^{-4.88},$$
$$S \approx 10^{-4.88}\ \text{mol/L}.$$

It is interesting to notice that S_o is nothing more than K_{s1} already encountered (see Chap. 25) in the case of poorly soluble species that dissociate into two monocharged ions.

However, in the case of organic ions, the intrinsic solubility may be higher than the ionic solubility (when they dissociate). For example, 2,4,6-trichlorophenol $Cl_3C_6H_2OH$ exhibits the intrinsic solubility $S_o = 4 \times 10^{-3}$ mol/L and the dissociation constant $K_d = 1 \times 10^{-6}$ (it is merely the acid dissociation constant K_a):

$$K_d = K_a = \frac{\left[(H^+)(Cl_3C_6H_2O^-)\right]}{(Cl_3C_6H_2OH)}.$$

Its solubility product is

$$K_s = K_d S_o,$$
$$K_s = 4 \times 10^{-9}.$$

Its ionic solubility in pure water is $\sqrt{(4 \times 10^{-9})} = 6.3 \times 10^{-5}$ mol/L (by neglecting the hydrolysis of the trichlorophenate ion, which occurs according to the reaction:

$$Cl_3C_6H_2O^- + H_2O \rightleftharpoons Cl_3C_6H_2OH + OH^-\big).$$

Hence, we find

$$S = \left(4 \times 10^{-3} + 6.3 \times 10^{-5}\right)\ \text{mol/L},$$
$$S \approx 4 \times 10^{-3}\text{mol/L}.$$

Of course, the dissociated fraction depends on the total concentration of the product (Ostwald's law extension) and also of the dissociation constant K_d. Consequently, there is a limit to the dissociation when the saturation is reached.

In the case of a solid whose structure is M_zX_y, the solubility in pure water is (in the case in which the ions M^{+y} and X^{-z} do not participate in any supplementary equilibrium and, in particular, are not hydrolyzed) is given by the expressions

$$S = [M_zX_y]_{\text{saturation}} + z[M^{+y}]$$

or

$$S = [M_zX_y]_{\text{saturation}} + y[X^{-z}].$$

Exercise 1 Calculate the solubility product of barium sulfate given its solubility in pure water (determined by conductometry), which is 1.0×10^{-5} mol/L at 10 °C. The intrinsic solubility is negligible.

According to the equilibrium,

$$BaSO_4 \downarrow \rightleftharpoons Ba^{2+} + SO_4{}^{2-}.$$

One mole of $BaSO_4$ gives one mole of Ba^{2+} and one mole of $SO_4{}^{2-}$ by dissolution. As a result, at saturation,

$$\left[Ba^{2+}\right] = 1.0 \times 10^{-5} \text{ mol/L},$$
$$\left[SO_4{}^{2-}\right] = 1.0 \times 10^{-5} \text{ mol/L}.$$

By neglecting ionic strength effects (which is valid since the concentrations are weak), we can write

$$K_s = \left[Ba^{2+}\right]\left[SO_4{}^{2-}\right],$$
$$K_s = 1.0 \times 10^{-10}.$$

Exercise 2 The solubility product of calcium fluoride CaF_2 is $K_s = 4.0 \times 10^{-11}$. Calculate its solubility in pure water. Neglect its intrinsic solubility.

In these conditions, the equilibrium to be considered is

$$CaF_2 \downarrow \rightleftharpoons Ca^{2+} + 2F^-.$$

The solubility product is

$$[Ca^{2+}][F^-]^2 = 4.0 \times 10^{-11}.$$

With S the number of moles of calcium fluoride that can be dissolved in 1 L of solution, we have

$$\left[Ca^{2+}\right] = S,$$
$$[F^-] = 2S,$$
$$S(2S)^2 = 4.0 \times 10^{-11},$$
$$S = 2.15 \times 10^{-4} \text{ mol/L},$$
$$S = 1.68 \times 10^{-2} \text{ g/L} \quad \left(\text{experimentally } 1.60 \times 10^{-2} \text{ g/L}\right).$$

Exercise 3 Given the solubility products of dimercury(I) sulfate Hg_2SO_4 ("mercurous sulfate") and dimercury(I) dichloride Hg_2Cl_2 ("mercurous chloride-calomel"),

$$K_s(Hg_2SO_4) = \left[Hg_2{}^{2+}\right]\left[SO_4{}^{2-}\right] = 10^{-6.17},$$
$$K_s(Hg_2Cl_2) = \left[Hg_2{}^{2+}\right]\left[Cl^-\right]^2 = 10^{-17.88},$$

calculate the solubility of both compounds expressed in g/L. $M(Hg_2SO_4) = 497.24$ g/mol and $M(HgCl_2) = 472.09$ g/mol.

Let's first calculate in mol/L. For the sulfate, according to the equilibrium,

$$Hg_2SO_4 \downarrow \rightleftharpoons {Hg_2}^{2+} + {SO_4}^{2-},$$
$$S = [{Hg_2}^{2+}],$$
$$S = [{SO_4}^{2-}],$$
$$S^2 = 10^{-6.17},$$
$$S = 8.22 \times 10^{-4}\ \text{mol/L}, \quad S = 0.409\ \text{g/L}.$$

Now, for the dichloride, following the same reasoning but with $[Cl^-] = 2S$, we find

$$S = 6.91 \times 10^{-7}\ \text{mol/L}, \quad S = 3.26 \times 10^{-4}\ \text{g/L}.$$

Exercise 4 Calculate the weight of precipitate obtained when 20.0 ml of a 5.0×10^{-3} mol/L potassium iodate solution is added to 5.0 ml of a 1.0×10^{-2} mol/L lead nitrate solution. $K_s[Pb(IO_3)_2] = 10^{-12.59}$ and $M[Pb(IO_3)_2] = 556.98$ g/mol.
Let's first calculate the analytical concentrations of the IO_3^- and Pb^{2+} ions.

$$[{IO_3}^-] = 5 \times 10^{-3}\ \frac{20}{25} = 4 \times 10^{-3}\ \text{mol/L},$$
$$[Pb^{2+}] = 1.0 \times 10^{-2}\ \frac{5.0}{25} = 2 \times 10^{-3}\ \text{mol/L}.$$

Now, we'll verify that in these concentration conditions, the solubility product $K_s[Pb(IO_3)_2]$ has been reached. By forming the product of the analytical concentrations $[Pb^{2+}][{IO_3}^-]^2$, we find 3.2×10^{-8}. This value is well above that of the solubility product. The solubility product has been reached. There is precipitation. We'll call P the number of moles of lead iodate precipitated at equilibrium per liter of solution. The relations that are satisfied are

$$[{IO_3}^-] + 2P = 4.0 \times 10^{-3},$$
$$[Pb^{2+}] + P = 2.0 \times 10^{-3}.$$

The system leads to the following third-order equation in P:

$$\left(4 \times 10^{-3} - 2P\right)^2 \left(2 \times 10^{-3} - P\right) = 10^{-12.59}.$$

The root is $P = 2 \times 10^{-3}$ mol/L. We find that the precipitation is total. The result matches chemical intuition due to the weak value of the solubility product. For one liter of solution, the weight of precipitated lead iodate would be

$$2 \times 10^{-3} \times 556.98 = 1.11\ \text{g/L}$$

and for 25 ml, it is 27.85 mg, which is the desired value.

This exercise introduces the notion of precipitation extent. If the initial concentration of the ion to be precipitated is C_o, it is contained in a volume V_o of solution, and, after precipitation, the remaining concentration is C_f contained in the volume V_f, the fraction still dissolved is

$$\frac{C_f V_f}{C_o V_o},$$

and the precipitation extent α is

$$\alpha = \frac{(C_o V_o - C_f V_f)}{C_o V_o},$$

$$\alpha = 1 - \frac{C_f V_f}{C_o V_o}.$$

In the preceding exercise, it is 100%.

32.5 Difficulties Encountered in the Calculations of Solubilities

The solubility may differ considerably from the real value when it is calculated as we've shown through the solubility products. This is the case when supplementary equilibria involving the ionic species figuring in the solubility product are effective. These equilibria may be of several origins. For example, they may be acid–base, redox, or other precipitation equilibria. (In the enumeration of these supplementary equilibria, we have followed the systematic presentation of this book.) Common ion effects may also add themselves to the equilibrium under study. Of course, ionic strength effects may also occur. All these effects are controllable (see the following chapters).

Unfortunately, there are other effects that are far more difficult to control than the previous ones. They are due to the nature of the precipitates and to their evolution. They may concern either the ionic solubility or the molecular (intrinsic) solubility.

Chapter 33
Dependence of the Solubility on the Solution's Ionic Strength and on the Presence of Common Ions: Superimposition of Several Precipitation Equilibria

As we could have predicted, solubility depends on the solution's ionic strength. The ionic strength of the solution may be due to the presence of ions differing from those constituting the electrolyte whose solubility is under study. This influence is investigated in the first section of the chapter. The ionic strength may also be due to the addition of ions that constitute the electrolyte itself. In this case, an effect analogous to the preceding one occurs, but, in addition, another effect, which is far stronger than the preceding one, arises: the common ion effect. It is due to the shift of an equilibrium. This is the subject of Sect. 33.2. Finally, we investigate the superimposition of two precipitation equilibria.

33.1 Influence of the Ionic Strength on the Solubility

First, let's notice that the ionic strength of the solution influences the solubility of ions. We have indeed seen that the activity of a molecular species does not change with the ionic strength of the solution, provided the latter does not exceed about 0.1 mol/L. Therefore, in this section, we are interested only in its influence on the ionic solubility.

It is an experimental fact that the solubility of an electrolyte increases slightly when the ionic strength increases. Figure 33.1 gives the solubility changes of silver iodate as a function of the concentration of potassium nitrate present in the solution.

Quite evidently, the ionic strength of the solution increases with the concentration of potassium nitrate. [To be rigorous, we must notice, however, that in this example the ionic strength is not due only to K^+ and NO_3^-, but also to Ag^+ and IO_3^- ions resulting from the dissolution of silver iodate ($K_s = 10^{-7.52}$)].

We can easily explain the increase in solubility with the ionic strength if we remember that solubility products, defined in terms of activities, do not vary with the ionic strength as other thermodynamic constants do. Expressing solubility products in terms of concentrations and activity coefficients such as

$$K_s = [M^+][X^-]\gamma_\pm{}^2$$

J.-L. Burgot, *Ionic Equilibria in Analytical Chemistry*,
DOI 10.1007/978-1-4419-8382-4_33, © Springer Science+Business Media, LLC 2012

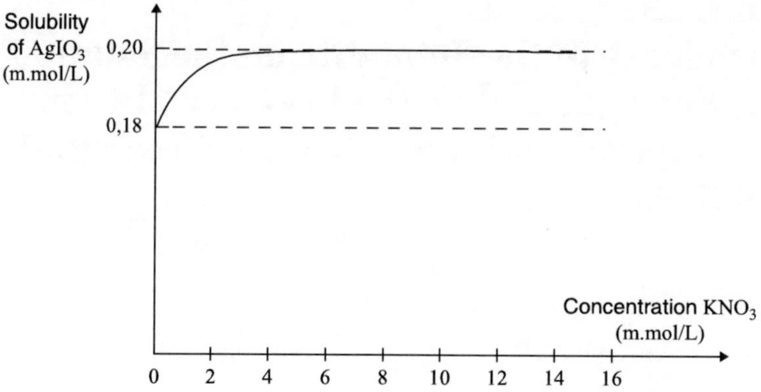

Fig. 33.1 Effect of the addition of potassium nitrate on the silver iodate's solubility

immediately shows that since activities decrease (at least in a first stage) when the ionic strength increases, the concentrations $[M^+]$ and $[X^-]$ must simultaneously increase for the K_s value to remain constant.

This example provides the occasion to recall that only the constants defined in terms of concentrations, that is, the formal or apparent constants, exhibit values that change with the ionic strength of the solution. Moreover, formal solubility products K_s defined as follows:

$$K_{s\,app} = [M^{+y}]^z[X^{-z}]^y$$

are seldom used for a simple reason: The values of the apparent and thermodynamic constants are identical to the concentrations at which the concept of solubility product is useful. Indeed, it is because these concentrations are very low.

From the standpoint of physical chemistry, the measurement of the solubility S at different ionic strengths I (whose values are imposed by the addition of a foreign electrolyte in judicious concentrations) permits us to determine both the thermodynamic solubility product and the mean activity coefficient. This determination can be carried out in a geometric or algebraic way. Let's consider the case of a one-to-one electrolyte 1.1. According to the preceding considerations, we can write

$$K_s/\gamma_\pm{}^2 = S^2,$$

$$\log K_s = 2\log\gamma_\pm + 2\log S.$$

At weak ionic strengths, according to Debye–Hückel's limiting law (see Chap. 3),

$$\log\gamma_\pm = -\left[Az^+z^-\sqrt{I}\right],$$

$$\log S = \frac{1}{2}\log K_s + \left[Az^+z^-\sqrt{I}\right].$$

Drawing the diagram of log S as a function of $\sqrt{I}$ permits us to determine K_s. Alternatively, combining two values S_1 and S_2, determined at two different ionic strengths I_1 and I_2, permits us to obtain K_s through the preceding relation. The better process is probably to draw the best straight line of log $S/\sqrt{I}$. It presents the advantage of taking all of the experimental data (i.e., all the couple values S_i/I_i) into account. For example, the solubility at null ionic strength of silver iodate and its solubility product value have been determined in this way. They are

$$S(\text{AgIO}_3) = 1.733 \times 10^{-4} \text{ mol/L},$$

$$K_s(\text{AgIO}_3) = S^2,$$

$$K_s(\text{AgIO}_3) = 3.00 \times 10^{-8}.$$

Then the calculation of the mean activity coefficient is immediate by using a couple of data S_i/I_i. Indeed, we can write

$$\gamma_{\pm}{}^2 = \frac{K_s}{S_i{}^2}.$$

It is interesting to notice that the graph of log $\gamma_{\pm}/\sqrt{I}$ is actually a straight line, in full accordance with Debye–Hückel's law. This fact signifies that the part of the ionic strength due to the concentrations $[IO_3^-]$ and $[Ag^+]$ is negligible compared to that due to the foreign electrolyte, since it is on this assumption that the straight line is drawn.

Exercise 1 The solubility at 25°C of thallium chloride in pure water is $S = 1.64 \times 10^{-2}$ mol/L. What are its solubility into a solution of potassium nitrate 2.5×10^{-2} mol/L and its solubility product value ($A = 0.509$)?

One can write for the solution in pure water

$$\log S = \frac{1}{2} \log K_s + \text{Az}^+\text{z}^- \sqrt{\left(1.64 \times 10^{-2}\right)}.$$

If S' is the desired solubility in the potassium nitrate solution, the ionic strength of the solution is

$$I = \left(S' + 2.50 \times 10^{-2}\right) \text{mol/L},$$

from which we find

$$\log S' = \frac{1}{2} \log K_s + \text{Az}^+\text{z}^- \sqrt{\left(2.5 \times 10^{-2} + S'\right)}.$$

Combining the relations giving log S and log S' gives

$$\log\left(\frac{S'}{S}\right) = \text{Az}^+\text{z}^- \left[\sqrt{\left(S' + 2.50 \times 10^{-2}\right)} - \sqrt{\left(1.64 \times 10^{-2}\right)}\right].$$

At 25°C

$$\log\left[\frac{S'}{(1.64\times10^{-2})}\right] = 0.509\left[\sqrt{(S'+2.50\times10^{-2})} - \sqrt{(1.64\times10^{-2})}\right]$$

$$\text{or}\quad S' = 1.80\times10^{-2}\ \text{mol/L}.$$

We notice that passing from an ionic strength equal to 1.64×10^{2} mol/L to another equal to 4.14×10^{-2} mol/L induces a change in solubility from 1.64×10^{-2} mol/L to 1.80×10^{-2} mol/L, that is, a change of $+9.8\%$. The calculation of the solubility product is immediate. For example, using the values found for the solubility of thallium chloride in pure water, we can write

$$\log\left(1.64\times10^{-2}\right) = \frac{1}{2}\log K_s + 0.509\sqrt{\left(1.64\times10^{-2}\right)},$$

$$K_s = 1.99\times10^{-4}.$$

33.2 The Common Ion Effect

In this section, we consider the influence of the presence of an ion identical to one of the ions constituting the electrolyte whose solubility is under study. For example, we study the influence of the presence of sodium sulfate in a solution upon the solubility of barium sulfate in the same solution. Of course, the added common ion influences the solubility through its contribution to the ionic strength, as the previous section described. This is not the phenomenon under study in this section. Here we study the *common ion effect*, which can be far more important than the preceding one.

For example, let's study the solubility of barium sulfate. Once its precipitation occurs, the solubility-product relation is satisfied:

$$\left[Ba^{2+}\right]\left[SO_4{}^{2-}\right] = K_s(\text{saturation}).$$

If the barium or sulfate ion concentration is increased by adding barium chloride or sodium sulfate (which are fairly soluble themselves), the barium sulfate precipitation is more pronounced since its solubility is decreased, as we can see now.

If S is the barium sulfate's solubility in a solution of barium chloride at concentration C, the relations that are satisfied once the saturation is attained are (by mixing activities and concentrations)

$$\left[Ba^{2+}\right]\left[SO_4{}^{2-}\right] = K_s,$$

$$\left[Ba^{2+}\right] = S + C,$$

$$\left[SO_4{}^{2-}\right] = S.$$

The barium sulfate's solubility is equal here to the sulfate ion concentration at saturation. In other words, the solubility S is the maximum number of moles of barium

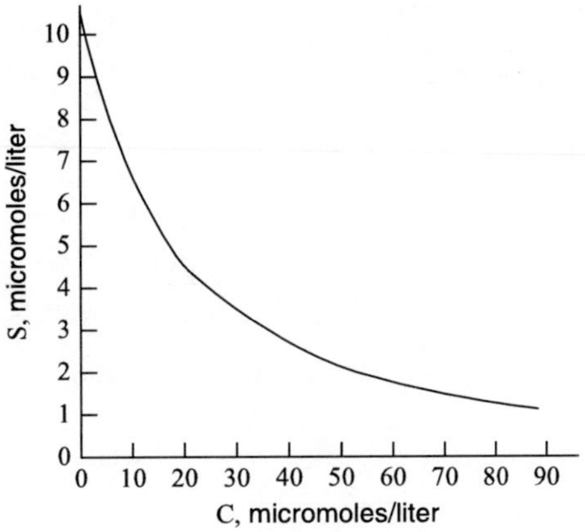

Fig. 33.2 Solubility S of barium sulfate in solutions of sodium sulfate at analytical concentrations C

sulfate that can be dissolved in 1 L of a barium chloride solution at concentration C. The above mathematical system is soluble since there are three equations for three unknowns, $[Ba^{2+}]$, $[SO_4{}^{2-}]$, and S. By mixing activities and concentrations, we've reduced the system to the second-order equation

$$S(S + C) = K_s.$$

If there is a large excess of common ions,

$$S \ll C$$

$$\text{and } S \approx \frac{K_s}{C}.$$

Exercise 2 Calculate the barium sulfate's solubility in a 1.0×10^{-2} mol/L barium chloride solution.

$$\left[Ba^{2+}\right]\left[SO_4{}^{2-}\right] = 1.1 \times 10^{-10},$$
$$\left[Ba^{2+}\right] = S + 1.0 \times 10^{-2},$$
$$\left[SO_4{}^{2-}\right] = S,$$
$$S\left(S + 1.0 \times 10^{-2}\right) = 1.1 \times 10^{-10},$$
$$S = 1.1 \times 10^{-8} \text{ mol/L}.$$

The solubility is considerably decreased compared to the case of pure water (in this case, $S = 1.05 \times 10^{-5}$ mol/L). In French, this phenomenon is called *recul de precipitation* ("solubility decrease"). From another standpoint, we can verify that, actually, $S \ll C$ in the term within brackets. Figure 33.2 shows the solubility changes of barium sulfate with the concentrations C of sodium sulfate.

These determinations may also be performed with logarithmic diagrams taking the different concentrations into account (activities and concentrations being mixed).

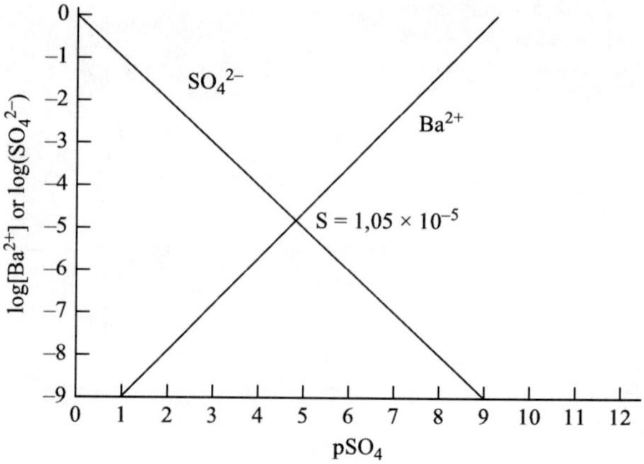

Fig. 33.3 Logarithmic representation of a divalent ion solubility product

For example, in the case of divalent metal sulfates, they are diagrams $\log(M^{2+})/pSO_4$ with $pSO_4 = -\log(SO_4{}^{2-})$. Starting from the solubility-product expression:

$$[M^{2+}][SO_4{}^{2-}] = K_s,$$

we find

$$\log[M^{2+}] = \log K_s + pSO_4.$$

The fact that the solubility product is obeyed results in a straight line with slope +1 and intercept $\log K_s$ (Fig. 33.3).

Now, if we draw the straight line $\log(SO_4{}^{2-})/pSO_4$, with slope −1, on the same diagram, it intersects the preceding for

$$\log(M^{2+}) = \log(SO_4{}^{2-}),$$

that is, for

$$[M^{2+}] = [SO_4{}^{2-}] = S.$$

S is the solubility in pure water. On the left side of the diagram and below, the following inequality holds: $[SO_4]^{2-} \gg [M^{2+}]$. It is the area of the weakest concentration of the metallic ion M^{2+}. This results from the common ion effect. It is the inverse for the upper area located on the right of the diagram in which $[M^{2+}] \gg [SO_4{}^{2-}]$. Such diagrams permit us to simplify solubility calculations immediately. For example, the solubility of $BaSO_4$ in a 10^{-2} mol/L solution of $BaCl_2$ is obtained by drawing the horizontal straight line $[Ba^{2+}] = 10^{-2}$ mol/L on the diagram. The horizontal line intersects the straight line $\log[Ba^{2+}]/pSO_4$ for $pSO_4 \approx 8$, namely, for $S = 10^{-8}$ mol/L. We can remark that this determination process implies that S can be neglected compared to 10^{-2} mol/L, as in the previous case in the algebraic calculation. Hence, this is confirmed in the present case by the graphical result

$$S = 10^{-8}\ \text{mol/L} \ll 1.0 \times 10^{-2}\ \text{mol/L}.$$

In the same manner, we could find the value 1.0×10^{-7} mol/L for the barium sulfate's solubility in a 1.0×10^{-3} mol/L sodium sulfate solution.

The graphical determination has been of little interest for the simple cases studied until now. However, the contrary is true for more complicated cases, such as those in which complex(es) as well as precipitates form, for example. Such diagrams permit us to immediately visualize species whose concentrations can be negligible in the calculations. As a result, the latter are simplified.

Exercise 3 Calculate the magnesium fluoride's solubility MgF_2 in a 10^{-4} mol/L magnesium nitrate solution. $K_s(MgF_2) = 10^{-8.18}$.

If S is the unknown solubility, we can write at saturation

$$[Mg^{2+}] = (S + 1.0 \times 10^{-4}) \text{ mol/L},$$
$$[F^-] = 2S.$$

By mixing activities and concentrations, we find

$$K_s = (4S^2)(S + 1.0 \times 10^{-4}) = 10^{-8.18},$$
$$S^3 + 10^{-4}S^2 - 1.65 \times 10^{-9} = 0,$$
$$S = 1.08 \times 10^{-3} \text{ mol/L}.$$

This is a case in which the solubility is higher than the added ion concentration.

Exercise 4 We want to precipitate thallium(I) as chloride. What is the volume of a 1 mol/L hydrochloric acid solution that must be added to a 10.0-ml solution containing 383 mg of thallium(I) for its precipitation to be total at the 99.5% extent?

$$K_s(TlCl) = 10^{-3.72} \quad \text{and} \quad M(Tl) = 204.37 \text{ g/mol}.$$

The number of moles of Tl(I) in the solution is $(383 \times 10^{-3})/(204.37) = 1.874 \times 10^{-3}$ mol. Let V be the unknown volume. The total concentration of Tl (i.e., precipitated and remaining in solution) is

$$[Tl^+]_o = \frac{(1.874 \times 10^{-3} \times 1000)}{(10 + V)} \text{ mol/L},$$
$$[Tl^+]_o = \frac{1.874}{(10 + V)} \text{ mol/L},$$
$$[Tl^+]_o = [Tl^+]\downarrow + [Tl^+].$$

The maximum concentration that must remain in solution is

$$[Tl^+] = \left(\frac{0.5}{100}\right)\left[\frac{1.874}{(10 + V)}\right] = \frac{9.37 \times 10^{-3}}{(10 + V)} \text{ mol/L}$$

$$\text{and} \quad [TlCl\,]\downarrow = \frac{(1.874 - 9.37 \times 10^{-3})}{(10 + V)} = \frac{1.865}{(10 + V)} \text{ mol/L}.$$

The total concentration of chloride ions is

$$[\mathrm{Cl}^-]_o = \frac{(V \times 1)}{(10 + V)}$$

and that remaining in solution after precipitation is

$$[\mathrm{Cl}^-] = \frac{(\mathrm{V} \times 1)}{(10 + \mathrm{V})} - \frac{1.865}{(10 + \mathrm{V})}.$$

Just at the end of the dissolution, the solubility product is still satisfied:

$$\left[\frac{9.37 \times 10^{-3}}{(10 + \mathrm{V})}\right] \times \left[\frac{(\mathrm{V} - 1.865)}{(10 + \mathrm{V})}\right] = 10^{-3.72},$$

$$V = 10\,\mathrm{ml}.$$

Exercise 5 Let's consider the ionic salt M_zX_y and let S_o be its solubility in pure water. Show that its solubility in a solution containing the salt Na_3X of analytical concentration C is always lower than S_o.

The dissolution equilibrium is

$$M_zX_y \downarrow \rightleftharpoons zM^{y+} + yX^{z-},$$

$$K_s = (zS_o)^z(yS_o)^y \quad \text{(pure water)}.$$

$$\text{Since} \quad [M^{y+}] = zS_o,$$

$$[X^{z-}] = yS_o.$$

In the presence of X^{2-},

$$K_s = (zS)^z(yS + C)^y.$$

The comparison of both expressions of K_s immediately shows that $S < S_o$ since the concentration C is positive.

The common ion effect is of great interest in gravimetry by precipitation (see Chap. 38). If we add a suitable excess of the precipitating reagent, the solubility of the precipitate becomes almost null and, as a result, losses by solubility also become null. For the same reason, the washing of the precipitates is very often carried out with a solution containing an ion that is common with one of the ions of the precipitate.

Exercise 6 In a gravimetric determination, 0.451 g of silver chloride is washed before ignition and weighed. Calculate the precipitate fraction lost during the washing:

1. with 2 ml of pure water,
2. with, successively, 150 ml of a 0.10 mol/L ammonium chloride solution and 50 ml of pure water.

$$\mathrm{M(AgCl)} = 143.323\,\mathrm{g/mol} \quad \text{and} \quad K_s(\mathrm{AgCl}) = 10^{-9.75}.$$

3. The solubility into pure water is $\sqrt{10^{-9.75}} \times 143.323$ g/L. The part dissolved in 200 ml of pure water is

$$\frac{(10^{-4.875} \times 143.323 \times 200)}{1000} = 3.82 \times 10^{-4} \text{ g}.$$

The answer to the first question is $(3.82 \times 10^{-4})/0.451 = 0.086\%$.

4. In the ammonium solution at saturation, we can write

$$S(S + 0.1) = 10^{-9.75},$$

$$S = 1.78 \times 10^{-9} \text{ mol/L},$$

$$\text{or equivalently,} \quad 1.78 \times 10^{-9} \times 143.323 = 2.549 \times 10^{-7} \text{ g/L},$$

and for 150 ml,

$$2.549 \times 10^{-7} \times \frac{150}{1000} = 3.823 \times 10^{-8} \text{ g}.$$

During the final washing with 50 ml of pure water, we find a loss of 9.556×10^{-5} g. (The principle of the calculation is the same as in question 1.) For the entirety of question 2, the loss is

$$9.556 \times 10^{-5} + 3.823 \times 10^{-8} = 9.56 \times 10^{-5} \text{ g},$$

$$\text{i.e.,} \quad \frac{(9.56 \times 10^{-5})}{0.451} = 0.021\%.$$

We notice that the loss occurring during the washing with 50 ml of pure water is far higher than that occurring during the washing with 150 ml of the ammonium chloride solution that contains the ion Cl^-, which is also engaged within the precipitate.

The process of mixing two solutions, each containing one ion that gives a very insoluble precipitate with the ion of the other solution, is a particular case of the preceding one. It is interesting because it is encountered in titrations by precipitation (see Chap. 36).

Let's take the example of the mixing of 50 ml of a 3×10^{-5} mol/L barium chloride solution with 100 ml of a 4.5×10^{-5} mol/L sodium sulfate solution. The problem is to determine the fraction of barium precipitated. Let's call P the number of moles of precipitated barium sulfate per liter of solution. The mass balance relations on Ba^{2+}, $SO_4{}^{2-}$ together with the solubility-product expression provide a system of three equations in three unknowns: $[Ba^{2+}]$, $[SO_4{}^{2-}]$, and P:

$$[Ba^{2+}] + P = 3.0 \times 10^{-5} \left[\frac{50}{(100 + 50)}\right] \text{ mol/L},$$

$$[SO_4{}^{2-}] + P = 4.5 \times 10^{-5} \left[\frac{100}{(100 + 50)}\right] \text{ mol/L},$$

$$[Ba^{2+}][SO_4{}^{2-}] = 1.1 \times 10^{-10}.$$

The second-order equation resulting from this system leads to

$$[Ba^{2+}] = 4.5 \times 10^{-6}\ \text{mol/L},$$
$$P = 5.5 \times 10^{-6}\ \text{mol/L},$$

and the precipitated fraction is

$$\left(\frac{5.5 \times 10^{-6}}{1.0 \times 10^{-5}}\right) 100 = 55\%.$$

Again, starting from these considerations, it is easy to show that the presence of a common ion decreases the solubility of the salt.

33.3 Superimposition of Two Precipitation Equilibria: Separation by Precipitation

The case of the superimposition of two precipitation equilibria is of practical importance in analysis since it occurs when the separation of two ions by precipitation is carried out. More precisely, the operation consists of preferentially precipitating one of the present ions in the admixture by adding an appropriate reagent. Of course, in this process, the markedly weaker solubility of one of the ions in the presence of the reagent is taken into account. Simple calculations based on the consideration of solubility products permit us to predict the maximum extent of the separation.

For example, let's consider a solution containing 10^{-2} mol/L of barium chloride and 10^{-2} mol/L of strontium chloride. A saturated solution of sodium sulfate is added to it. The problem is to calculate the concentration of the remaining ion (the most insoluble) when the second begins to precipitate. The solubility products are $K_s(BaSO_4) = 1.1 \times 10^{-10}$ and $K_s(SrSO_4) = 2.8 \times 10^{-7}$. An examination of the solubility products shows that barium ions precipitate first. Just at the beginning of the strontium precipitation, both solubility products are satisfied simultaneously and $[Sr^{2+}] = 10^{-2}$ mol/L. Therefore, we can write

$$[Ba^{2+}][SO_4{}^{2-}] = 1.1 \times 10^{-10},$$
$$[Sr^{2+}][SO_4{}^{2-}] = 2.8 \times 10^{-7},$$

from which we find $$\frac{[Sr^{2+}]}{[Ba^{2+}]} = 2.5 \times 10^{-3}$$

since $$[Sr^{2+}] = 10^{-2}\ \text{mol/L},$$
$$[Ba^{2+}] = 3.8 \times 10^{-6}\ \text{mol/L}.$$

This is the concentration of the remaining Ba^{2+}. Until this point, barium sulfate precipitated in the pure state, in principle. Its "precipitated concentration" is

$(10^{-2} - 3.8 \times 10^{-6})$ mol/L. However, this is an assertion based only on theoretical principles. In practice, in several examples, a phenomenon of coprecipitation also exists, which consists of an entrapment of the solution within the precipitate. Moreover, an adsorption of ions that are not those engaged in the precipitate may occur at the precipitate's surface. Eventually, the formation of a solid solution is also possible.

Exercise 7 A total of 200 ml of a 10^{-2} mol/L sodium sulfate is added to 100 ml of a solution containing 10^{-2} mol/L of barium chloride and 10^{-2} mol/L of strontium chloride. What is the fraction of each ion precipitated in the pure state?

$$K_s\,(BaSO_4) = 1.10 \times 10^{-10};\; K_s\,(SrSO_4) = 2.8 \times 10^{-7}.$$

Barium precipitates first as sulfate. At the beginning of the precipitation of strontium sulfate, we have

$$[Sr^{2+}] = \frac{(10^{-2} \times 100)}{(100 + 200)} = 3.33 \times 10^{-3}\ \text{mol/L},$$

$$[SO_4{}^{2-}] = \frac{2.80 \times 10^{-7}}{(3.33 \times 10^{-3})} = 8.40 \times 10^{-5}\ \text{mol/L}.$$

The concentration of the remaining barium is

$$[Ba^{2+}] = \frac{1.1 \times 10^{-10}}{(8.40 \times 10^{-5})} = 1.31 \times 10^{-6}\ \text{mol/L}.$$

The fraction of precipitated barium is:

$$\frac{(10^{-2} - 1.31 \times 10^{-6})}{(3.33 \times 10^{-3})} \approx 1.$$

Strontium cannot be obtained in a pure state.

Exercise 8 A solution contains 10^{-2} mol/L of sodium chloride and 10^{-3} mol/L of sodium bromide. A concentrated solution of silver nitrate is added to it. What is the maximum purity that can be reached by the solution of remaining chloride ions after the precipitation of silver bromide?

$$K_s(AgBr) = 10^{-12.28};\; K_s(AgCl) = 10^{-9.75}.$$

We'll begin by setting up the problem on qualitative grounds. Silver bromide precipitates first. As a result, the solution is purer in chloride ions when silver chloride begins to precipitate. However, once this point has been reached, silver bromide continues to precipitate together with silver chloride.

When silver chloride begins to precipitate, both solubility products are satisfied:

$$[Ag^+][Br^-] = 10^{-12.28},$$

$$[Ag^+][Cl^-] = 10^{-9.75},$$

from which we find $\dfrac{[Br^-]}{[Cl^-]} = 10^{-2.53}$.

Just at the beginning of the precipitation of chloride ions, $[Cl^-] = 10^{-2}$ mol/L. As a result,

$$[Br^-] = 10^{-2} \times 10^{-2.53} = 2.95 \times 10^{-5}\ \text{mol/L}.$$

For 10^{-2} mol/L of chloride ions, 2.95×10^{-5} mol/L of bromide ions remain, that is, 0.3%. The purity of chloride ions is $100 - 0.3 = 99.7\%$.

Exercise 9 A total of 1.15 mg of $PbSO_4$ is stirred with 100 ml of a 0.1 mol/L solution of potassium iodate KIO_3 until the equilibrium is reached. What is the fraction of $PbSO_4$ that has been converted into solid lead iodate, $Pb(IO_3)_2$?

$$K_s(PbSO_4) = 10^{-7.8},\ K_s\,(Pb(IO_3)_2) = 10^{-12.59},\ M\,(PbSO_4) = 303.25\ \text{g/mol},$$
$$M\,(Pb(IO_3)_2) = 556.99\ \text{g/mol}.$$

Let's set up the hypothesis that at equilibrium, both precipitates $PbSO_4\downarrow$ and $Pb(IO_3)_2\downarrow$ coexist. A total of 1.15 mg of $PbSO_4$ corresponds to 3.79×10^{-6} mol. The relations that are satisfied are

$$n(PbSO_4) + n'(Pb(IO_3)_2) + [Pb^{2+}]0.1 = 3.79 \times 10^{-6},$$
$$0.1[IO_3^-] + 2n'(Pb(IO_3)_2) = 0.1 \times 0.1,$$
$$n(PbSO_4) + [SO_4^{2-}]0.1 = 3.79 \times 10^{-6},$$
$$[Pb^{2+}][SO_4^{2-}] = 10^{-7.8},$$
$$[Pb^{2+}][IO_3^-]^2 = 10^{-12.59},$$

where $n(PbSO_4)$ and $n'(Pb(IO_3)_2)$ are, respectively, the number of moles of $PbSO_4$ and $Pb(IO_3)_2$ that have precipitated. The first three equations are expressed in mole numbers rather than in concentrations. They represent the mass balances on Pb^{2+}, IO_3^-, and SO_4^{2-}, respectively. The system leads to the equilibrium concentration $[Pb^{2+}] = 10^{-10.59}$ mol/L and consequently to the value $[SO_4^{2-}] = 10^{2.79}$ mol/L. This is an absurd value. The starting hypothesis was false.

Now we'll set up the hypothesis that lead exists only as lead iodate. At first sight, this hypothesis seems to be in accordance with intuition since lead iodate's solubility product is lower than that of lead sulfate. The system

$$n(Pb(IO_3)_2) + [Pb^{2+}]0.1 = 3.79 \times 10^{-6},$$
$$0.1[IO_3^-] + 2n(Pb(IO_3)_2) = 0.1 \times 0.1,$$
$$[Pb^{2+}][IO_3^-]^2 = 10^{-12.59}$$

leads to the concentration:

$$[Pb^{2+}] = 1.90 \times 10^{-7}\ \text{mol/L},$$
$$n(Pb(IO_3)_2) = 3.79 \times 10^{-6} - 1.90 \times 10^{-7} 10^{-1}$$
$$= 3.77 \times 10^{-6}\ \text{mol}.$$

These results are in accordance with the problem's data. Lead sulfate is totally converted into lead iodate.

Note that in some cases, separation by precipitation may be made selective by a judicious choice of the pH value (see Chap. 34).

Chapter 34
Solubility and pH

The medium's acidity may have a considerable influence on the solubility of numerous substances. It can indeed play different roles in a dissolution phenomenon. They are all exploited in chemical analysis. First, an acidic or basic reagent can be dissolved or precipitated as a function of the pH value. A second point to be considered is the ability to dissolve poorly soluble salts by changing the solution's pH value. The third point is connected to both preceding ones. It concerns the possible selective separation of species at controlled pH values.

34.1 Solubility of Acidic and Basic Solutes as a Function of pH

Numerous organic acids or bases are poorly soluble in water in their molecular states. Once they are ionized after a proton exchange, they become freely soluble. This property is evidently of the utmost practical importance. Examples are innumerable.

34.1.1 The Monoacid or Monobasic Case

The apparent solubility of a monoacid or a monobase is the sum of its intrinsic solubility So (solubility in its molecular state) and its ionic maximum concentration (solubility in its ionic state). The latter is a function of the pH value, a function that is easily stated.

Let's consider an acid HA that is poorly soluble in its molecular state. To the precipitation equilibrium

$$\mathrm{HA}\downarrow \rightleftharpoons \mathrm{HA}_{\mathrm{saturation}}$$

the following ionization equilibrium is superimposed:

$$\mathrm{HA} \rightleftharpoons \mathrm{H}^{+} + \mathrm{A}^{-}.$$

J.-L. Burgot, *Ionic Equilibria in Analytical Chemistry*,
DOI 10.1007/978-1-4419-8382-4_34, © Springer Science+Business Media, LLC 2012

Its acidic dissociation constant is

$$K_a = (\mathrm{H}^+)(\mathrm{A}^-)/(\mathrm{HA}).$$

At saturation, we can write

$$[\mathrm{HA}]_{\mathrm{saturation}} = \mathrm{So} \quad \text{(intrinsic solubility)}.$$

Hence (after mixing activities and concentrations),

$$[\mathrm{H}^+][\mathrm{A}^-] = K_a\mathrm{So} \quad \text{(saturation)}. \tag{34.1}$$

The product K_aSo is endowed with the meaning of the solubility product:

$$[\mathrm{H}^+][\mathrm{A}^-] = K_s \quad \text{(saturation)}. \tag{34.2}$$

By definition, the apparent (total) solubility S of HA is

$$S = \mathrm{So} + [\mathrm{A}^-],$$

$$S = \mathrm{So} + K_s/[\mathrm{H}^+],$$

$$S = \mathrm{So} + K_a\mathrm{So}/[\mathrm{H}^+],$$

$$S = \mathrm{So}(1 + K_a/[\mathrm{H}^+]), \tag{34.3}$$

$$\text{or} \quad \log S = \log \mathrm{So} + \log(1 + K_a/[\mathrm{H}^+]). \tag{34.4}$$

It is instructive to consider two limiting cases according to the values of the ratio $K_a/[\mathrm{H}^+]$ compared to unity.

- if $[\mathrm{H}^+] \ll K_a$, 1 is negligible in the logarithm argument. We can write

$$\log S = \log \mathrm{So} + \log K_a - \log[\mathrm{H}^+],$$

$$\log S = \log \mathrm{So} + \mathrm{pH} - pK_a.$$

The apparent solubility logarithm and hence the apparent solubility itself increases linearly with the pH value. This result is in accordance with chemical intuition since the higher the pH is, the more ionized the acid is;

- if $[\mathrm{H}^+] \gg K_a$,

  relation (34.4) reduces to

$$\log S = \log \mathrm{So};$$

- for pH values near pK_a, that is, for the transition between the two preceding limiting cases, no simplification can be done and relation (34.3) or (34.4) must be used. The limits of the transition range may be arbitrarily fixed about the values $pK_a - 2$ and $pK_a + 2$ pH units.

  In the monobase case, the following acid–base equilibrium:

$$\mathrm{B} + \mathrm{H_2O} \rightleftharpoons \mathrm{BH}^+ + \mathrm{OH}^-$$

is superimposed to that of precipitation:

$$B\downarrow \rightleftharpoons B_{saturation}.$$

The acid–base equilibrium is quantified by the ratio of constants K_w/K_a, where K_a is the acid ionization constant of the conjugate acid BH^+. The apparent solubility of B is defined as

$$S = [B]_{saturation} + [BH^+].$$

With the help of a similar reasoning as just above, we find

$$S = So(1 + [H^+]/K_a),$$

where $So = [B]_{saturation}$,

$$\log S = \log So + \log(1 + [H^+]/K_a);$$

- if $pH \ll pK_a$,

$$\log S = \log So + pK_a - pH.$$

The solubility increases when the pH value decreases. This is in accordance with chemical intuition since the base is increasingly protonated as the pH value decreases;

- for $pH \gg pK_a$,

$$\log S = \log So;$$

- for pH located in the range $pK_a - 2 < pH < pK_a + 2$ pH units, simplifications are no longer possible (Fig. 34.1b):

$$\log S = \log So + \log(1 + [H^+]/K_a).$$

A Hägg-type diagram (see Chap. 5) may be drawn (Fig. 34.1a, b).

Exercise 1 Consider an organic acid whose constants are $pK_a = 4.2$ and $So = 10^{-3}$ mol/L. Calculate the pH value of its saturated solution. Calculate its solubility in a buffered pH = 3 solution.

Let's write the equations that must be satisfied:

$$[H^+][A^-]/[HA]_{saturation} = 10^{-4.2},$$

$$[H^+] = [A^-] + [OH^-],$$

$$S = [A^-] + [HA]_{saturation},$$

$$[HA]_{saturation} = So.$$

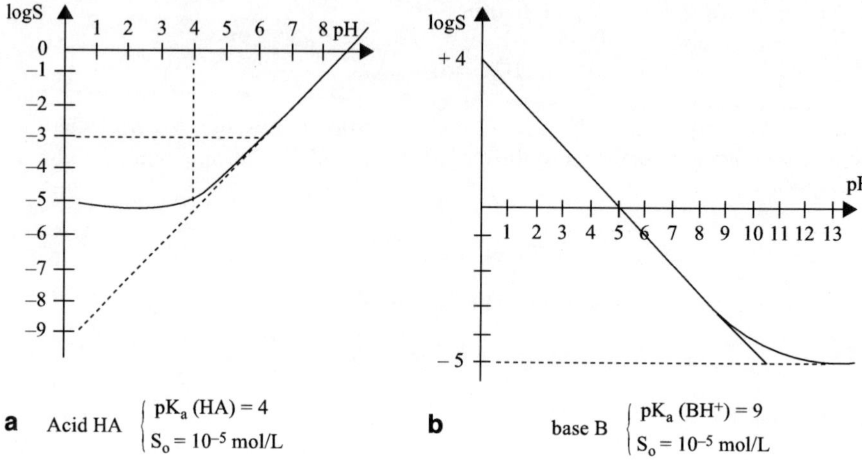

Fig. 34.1 Apparent solubilities of a monoacid and a monobase

Since we are in the presence of an acid, let's neglect $[OH^-]$. By replacing $[HA]_{saturation}$ with So, we find

$$[A^-][H^+] = 10^{-7.2}$$

and since $[A^-] = [H^+]$, we find

$$pH = 3.6.$$

The total solubility (which is not required) is

$$S = 10^{-3.6} + 10^{-3},$$

$$S = 1.25 \times 10^{-3} \text{ mol/L}.$$

At pH = 3,

$$S = 10^{-3}[1 + 10^{-4.2}/10^{-3}],$$

$$S = 1.06 \times 10^{-3} \text{ mol/L}.$$

The solubility has decreased, as we expected.

34.1.2 The Diacid or Dibase Case

The preceding considerations can be applied to polyacids and polybases. In the case of a diacid H_2A, the apparent solubility S is given by the expression

$$S = [H_2A] + [HA^-] + [A^{2-}].$$

Let K_{a1} and K_{a2} be the successive acid dissociation constants of H_2A:

$$K_{a1} = [H^+][HA^-]/[H_2A], \quad K_{a2} = [H^+][A^{2-}]/[HA^-].$$

Hence,

$$[HA^-] = K_{a1}So/[H^+] \text{ and } [A^{2-}] = K_{a1}K_{a2}So/[H^+]^2 \text{ with } [H_2A]_{saturation} = So.$$

As a result,

$$S = So(1 + K_{a1}/[H^+] + K_{a1}K_{a2}/[H^+]^2).$$

It is easy to conceive that

- for $[H^+] \ll K_{a2}$,

$$S \approx SoK_{a1}K_{a2}/[H^+]^2,$$
$$\log S = \log So - pK_{a1} - pK_{a2} + 2pH;$$

- for $K_{a2} \ll [H^+] \ll K_{a1}$,

$$S \approx SoK_{a1}/[H^+],$$
$$\log S = \log So - pK_{a1} + pH;$$

- for $[H^+] \gg K_{a1}$,

$$S \approx So,$$
$$\log S \approx \log So.$$

The diagram log *S*/pH is given in Fig. 34.2.

Analogous considerations may be given for a dibase.

34.1.3 The Ampholyte Case

For example, let's consider the 8-hydroxyquinoleine (oxine) case. It is a chelating and also precipitating reagent of several metallic ions (see Chap. 30).

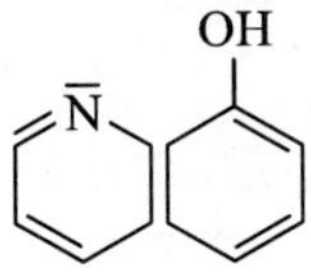

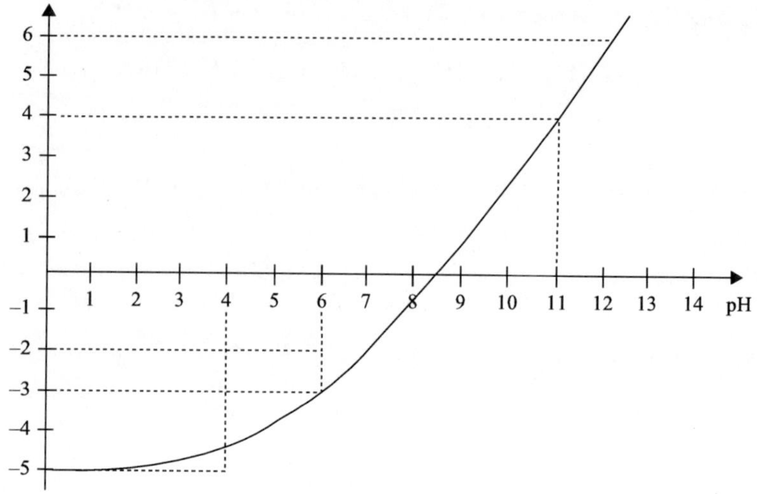

Fig. 34.2 Apparent solubility S of a diacid H_2A as a function of pH (So $= 10^{-5}$ mol/L, $K_{a1} = 10^{-4}$, $K_{a2} = 10^{-9}$)

Oxine is a base because of the presence in its structure of a basic nitrogen atom belonging to a quinoleine nucleus. It is also an acid because of the presence of a phenol function. Therefore, two prototropic equilibria exist:

$$HOx \rightleftharpoons Ox^- + H^+,$$

$$[Ox^-][H^+]/[HOx] = K_{a2} \text{ (phenol function} \quad K_{a2} = 10^{-9.9})$$

and

$$H_2Ox^+ \rightleftharpoons HOx + H^+,$$

$$[HOx][H^+]/[H_2Ox^+] = K_{a1} \text{(pyridinium ion} \quad K_{a1} = 10^{-5}).$$

(We recall incidentally that the nonionic form HOx is in equilibrium with the zwitterion form:

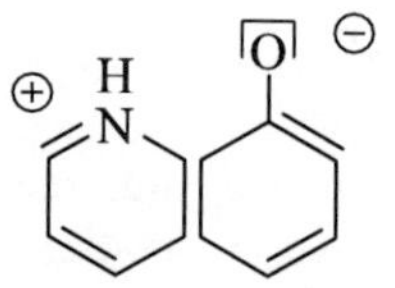

zwitterion form HOx±

and that the ratio $[HOx^{\pm}]/[HOx]$ is constant regardless of the pH solution (see Chap. 4). From now on, the symbol [HOx] will represent the sum of the concentrations of these two neutral species.)

The apparent solubility is defined by the expression

$$S = [HOx]_{saturation} + [Ox^-] + [H_2Ox^+].$$

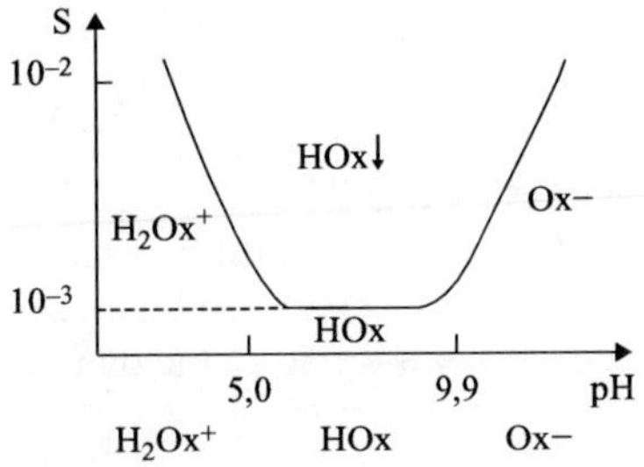

Fig. 34.3 Oxine's apparent solubility as a function of pH

With $[HOX]_{saturation}$ = So (intrinsic solubility) and by introducing the ionization constants, we find the relation

$$S = So(1 + K_{a2}/[H^+] + [H^+]/K_{a1}).$$

Figure 34.3 shows the apparent solubility changes with the pH value.

The apparent solubility is higher in acidic and basic ranges. The solubility change with pH is linear when each of the terms $K_{a2}/[H^+]$ and $[H^+]/K_{a1}$ dominates the two others that are present in the term included between brackets in the above relation giving S.

Analogous considerations may be applied to cases with other ampholytes.

34.2 Solubility of Poorly Soluble Salts as a Function of pH

34.2.1 Qualitative Aspect

In this section, we consider salts whose anion, bonded to the metallic ion in the precipitate, exhibits a weak basic character.

Consider the poorly soluble salt MA, which dissolves in water, giving ions:

$$MA\downarrow \rightleftharpoons M^+ + A^-. \quad (34.5)$$

The solubility product is

$$[M^+][A^-] = K_s. \quad (34.6)$$

According to the hypothesis, A^- is the base of the couple HA/A^-:

$$HA \rightleftharpoons H^+ + A^- \quad (K_a(HA/A^-)). \quad (34.7)$$

The overall equilibrium corresponding to both preceding simultaneous phenomena is

$$MA\downarrow + H^+ \rightleftharpoons M^+ + HA \quad K. \quad (34.8)$$

Its equilibrium constant K is given by the expression

$$K = [M^+][HA]/[MA\downarrow][H^+]$$

with $K = K_s/K_a$.

The more acidic the medium is, the more equilibrium (34.8) is displaced toward the right. Hence, the salt is more soluble in acidic medium. Moreover, we see that the weaker the acid is, the more sensitive the effect becomes.

34.2.2 Acid–Base Equilibria and Precipitation in a Buffered Medium: Quantitative Aspect

It is interesting to establish a quantitative relation between the solubility of the salt MA and the solution's pH value, which is fixed by the buffer.

The relations that must be satisfied are

$$[A^-][H^+] = K_a[HA], \qquad (34.9)$$

$$[M^+][A^-] = K_s, \qquad (34.6)$$

$$[M^+] = S, \qquad (34.10)$$

$$[HA] + [A^-] = S. \qquad (34.11)$$

We see that the solubility S is purely and simply the concentration of the dissolved metallic ion (at saturation) or equivalently of the dissolved anion regardless of its form, that is, protonated or not. Combining relations (34.9) and (34.11) gives

$$[A^-] = K_a S/(K_a + [H^+]).$$

Reporting this expression together with (34.10) into (34.6) leads to

$$S^2 = K_s(1 + [H^+]/K_a). \qquad (34.12)$$

The solubility increases once the concentration $[H^+]$ value approaches the K_a value. The phenomenon may be represented in a logarithmic diagram. Relation (34.12) may also be written as

$$-\log S = -1/2 \log K_s - 1/2\log(1 + [H^+]/K_a). \qquad (34.13)$$

For example, the diagram concerning silver acetate is represented in Fig. 34.4.

The curve obtained is sometimes called the *saturation line*. We see that the ionic solubility decreases when the pH increases. For the sake of discussion, the curve may be considered as resulting from three parts:

- a first part in which $[H^+] \gg K_a$, that is, pH $<$ about 2 (for the chosen example). The curve may be assimilated to the straight line whose equation is

$$-\log S = -1/2 \log K_s + 1/2\text{pH} - 1/2pK_a$$

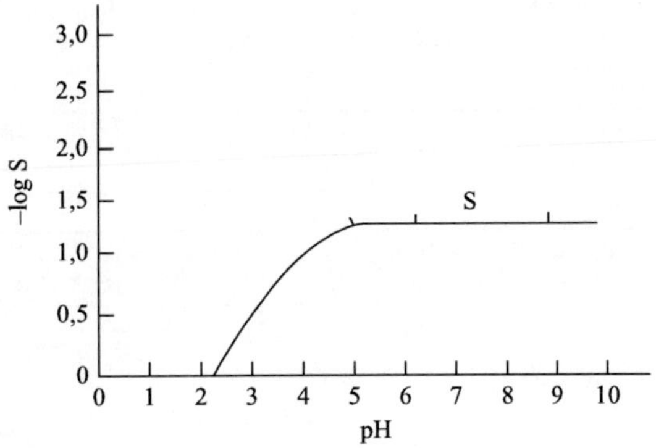

Fig. 34.4 Saturation line of silver acetate in solutions of known pH values: $K_a(CH_3COOH) = 10^{-4.75}$; $K_s(CH_3COOAg) = 4 \times 10^{-3}$

since in Eq. (34.13), $[H^+]/K_a \gg 1$;

- a second part, the middle part, for which the general Eq. (34.13) cannot be simplified in any way;
- the third part, for which $[H^+] \ll K_a$, that is, in (34.13), $[H^+]/K_a \ll 1$. This means that the saturation line may be reduced to the straight line

$$-\log S = -1/2 \log K_s,$$

that is, to $S = (K_s)^{1/2}$, a relation that hawse have already encountered and that is applicable when the anion is not basic.

Relation (34.12) may also be written as

$$S^2 = K_s\alpha,$$

with $\alpha = K_a/(K_a + [H^+])$. α is the fraction of the anion A^- remaining in solution after the prototropic and precipitation equilibria have been reached, that is,

$$\alpha = [A^-]/([A^-] + [HA]).$$

With the introduction α, a conditional solubility product K_s' may be considered. It is defined by the relation

$$K_s' = K_s/\alpha.$$

Using this symbolism is equivalent to considering that the solubility increase (when the fixed pH decreases) is apparently due to a change in the solubility product.

The salt under consideration may also be that of a diacid or a polyacid. The preceding reasoning is general, and so the solubility increases when the pH decreases. A suitable way to tackle the problem is to use conditional solubility products. In the case of an insoluble salt of the type M_2A, it is easy, starting from the equations,

$$K_{a1} = [HA^-][H^+]/[H_2A],$$
$$K_{a2} = [A^{2-}][H^+]/[HA^-],$$
$$S = [H_2A] + [HA^-] + [A^{2-}]$$

Table 34.1 Ionic solubility (mol/L) and apparent solubility products of silver acetate as a function of pH

pH	0	1	2	3	4	5
S (mol/L)	15.00	4.74	1.50	0.48	0.16	7.91×10^{-2}
K_s'	225	22.5	2.25	0.23	256×10^{-2}	6.26×10^{-3}

pH	6	7	8	9	10
S (mol/L)	6.50×10^{-2}	6.34×10^{-2}	6.33×10^{-2}	6.32×10^{-2}	6.32×10^{-2}
K_s'	4.22×10^{-3}	4.03×10^{-3}	4.00×10^{-3}	4.00×10^{-3}	4.00×10^{-3}

to show that

$$\alpha_{H_2A} = [H^+]^2/([H^+]^2 + K_{a1}[H^+] + K_{a1}K_{a2}),$$

$$\alpha_{HA^-} = K_{a1}[H^+]/([H^+]^2 + K_{a1}[H^+] + K_{a1}K_{a2}),$$

$$\alpha_{A2^-} = K_{a1}K_{a2}/([H^+]^2 + K_{a1}[H^+] + K_{a1}K_{a2}) \quad \text{(see Chap. 4).}$$

(Recall that $\alpha_{A2}{}^-$ is given by $\alpha_{A2}{}^- = [A^{2-}]/([A^{2-}] + [HA^-] + [H_2A])$ and so on for the other coefficients.)

The solubility product K_s' corresponding to the insoluble salt M_2A is

$$K_s{}' = K_s/\alpha_{A2^-}.$$

Table 34.1 mentions some values of the silver acetate ionic solubility together with its apparent solubility products.

At pH $=$ 1, the silver acetate solubility is 75 times higher than that found at pH $=$ 10.

It is quite evident that when the salt anion is the conjugate base of a strong acid, the pH value is devoid of any influence upon the solubility and, therefore, the ionic solubility remains equal to the square root of the "true" solubility product. Indeed, the base cannot be protonated.

34.2.3 Solubility of Poorly Soluble Salts in Unbuffered Media

The preceding considerations are only valid for buffered media. Indeed, if we begin by reasoning from the sole qualitative point of view, we can say that during the dissolution of the salt $MA_{(s)}$, two quasi-simultaneous processes occur: one that is purely physical (the dispersion of both ions, resulting from the solvolytic action of the solvent, among the solvent molecules), and the second is purely chemical: the hydrolysis of ion A^- according to the reaction

$$A^- + H_2O \rightleftharpoons HA + OH^-.$$

Of course, the hydrolysis results from the basic character of the anion. Therefore, the result of both processes is an increase in the pH value. In the previous section, it was not taken into account since the medium was buffered.

As a rule, we may consider that the dissolution of a salt whose anion is a base induces an increase in the solution's pH. This is the converse of the preceding phenomenon, which consisted of a solubility modification following changes in the solution's pH. To conclude with a more general statement, the dissolution or even the precipitation of a solute displaces the acid–base equilibria as far it is concerned.

From a quantitative standpoint, the relations that must be satisfied are:

$$[A^-][H^+]/[HA] = K_a, \tag{34.9}$$

$$[A^-] + [HA] = S, \tag{34.11}$$

$$[M^+][A^-] = K_s, \tag{34.6}$$

$$[H^+][OH^-] = K_w, \tag{34.14}$$

$$[M^+] = S, \tag{34.10}$$

$$[A^-] + [OH^-] = [H^+] + [M^+]. \tag{34.15}$$

The first four equations were used in the calculations concerning buffered media. They remain legitimate when the pH changes. However, they are not sufficient. This is the reason why we have added relations (34.14) and (34.15), which take pH changes into account. The system consisting of these six equations is difficult to solve. However, as we have seen, the first four equations can be reduced to relation (34.12), which remains valid:

$$S^2 = K_s(1 + [H^+]/K_a).$$

This is the saturation line equation. We still need to take relations (34.14) and (34.15) into account. Let's consider a solution that is not saturated in silver acetate. Let S' be its analytical concentration:

$$S' = [HA] + [A^-] \quad \text{with} \quad A^- \equiv CH_3COO^-,$$

$$[Ag^+] = S'.$$

According to calculations we've already seen concerning weak acids and bases, we can write

$$[A^-] = S'[K_a/(K_a + [H^+])].$$

Inserting this relation into the electroneutrality equation leads, after some rearrangements, to

$$S' = \{1 + K_a/[H^+]\}\{K_w/[H^+] - [H^+]\}. \tag{34.16}$$

Once the saturation is reached, and just at this point,

$$S' = S.$$

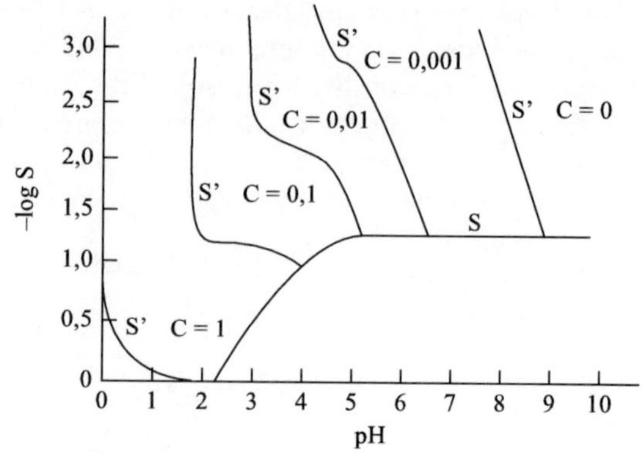

Fig. 34.5 Obtained solubility and pH values after dissolution of silver acetate until the very beginning of the precipitation (*C* is the strong acidic concentration added in the medium)

Therefore, $[H^+]$ may be expressed as a function of S from relation (34.12) and then introduced into (34.16). The obtained relation (34.17) permits us to calculate S and, hence, $[H^+]$ through (34.12) in a second stage:

$$(S^2 - K_s)(K_a/K_s + 1/S) - K_w K_s/K_a(S^2 - K_s) = 0. \tag{34.17}$$

Recall that this relation is valid only on the saturation line.

Solving (34.17) may be performed by iterations. A first alternative consists of drawing the curve S'/pH [Eq. (34.16)] and of finding its intersection with the saturation line (see Fig. 34.5). A second alternative would consist of simplifying it according to our chemical intuition. The chemical intuition, in a first step, consists of setting up

$$[HA] = [OH^-]$$

because of the hydrolysis reaction:

$$A^- + H_2O \rightleftharpoons HA + OH^-$$

and because, in the charge balance relation, $[H^+]$ and $[OH^-]$ are negligible. This amounts to saying that

$$[M^+] = [A^-] = K_s{}^{1/2}.$$

Inserting this value into the expression of the ionization constant of the conjugate acid leads to

$$[H^+]^2 K_s{}^{1/2}/K_w = K_a.$$

The value found for $[H^+]$ permits us to check the hypothesis that was made initially at the level of the charge balance equation. We can proceed through several iterations here also by starting from the preceding $[H^+]$ value and introducing it into the electroneutrality equation, and so on.

Exercise 2 Calculate the silver acetate's solubility in pure water. $K_s(CH_3COOAg) = 4.0 \times 10^{-3}$; $K_a(CH_3COOH/CH_3COO^-) = 1.75 \times 10^{-5}$.

Starting from the above hypothesis, which stipulates that $[H^+]$ and $[OH^-]$ are negligible in the electroneutrality, we find

$$[CH_3COO^-] = [Ag^+],$$

$$[CH_3COO^-] = K_s^{1/2},$$

$$[CH_3COO^-] = 6.32 \times 10^{-2}\ \text{mol/L}.$$

Likewise, with the hypothesis that

$$[CH_3COOH] = [OH^-],$$

the already-mentioned relation

$$[H^+]^2\, 6.3 \times 10^{-2}/10^{-14} = 1.75 \times 10^{-5}$$

gives the values

$$[H^+] = 1.60 \times 10^{-9} \quad \text{and} \quad S = 6.32 \times 10^{-2}\ \text{mol/L},$$

$$\text{pH} = 8.78, \quad [OH^-] = 6.00 \times 10^{-6}, \quad \text{and} \quad [CH_3COOH] = 6.00 \times 10^{-6}.$$

Inserting these values into the electroneutrality equation indicates that the starting hypothesis was satisfactory. It is interesting to notice that although the pH of the solution has changed markedly (from 7.00 until 8.78), the solubility does not differ significantly from $K_s^{1/2}$.

A slight complication in the above calculations occurs in the case of the dissolution of the salt in a strong acid of concentration C. Then the electroneutrality equation becomes

$$[M^+] + [H^+] = [A^-] + [OH^-] + [X^-],$$

with the other relations remaining valid. Equation (34.18), analogous to (34.17) and obtained in the same manner, permits us to calculate the salt's solubility and the solution's pH:

$$C = (S^2 - K_s)(K_a/K_s + 1/S) - K_w K_s/K_a\,(S^2 - K_s). \tag{34.18}$$

It can be solved by graphical means or, alternatively, by successive iterations.

Obviously, dissolving the salt MA in a strongly acidic solution induces an increase in the salt's solubility by equilibrium displacement due to the following equations:

$$A^- + H_2O \rightleftharpoons HA + OH^-,$$

$$OH^- + H^+ \rightleftharpoons H_2O.$$

Therefore, dissolving silver acetate in a 0.1 mol/L nitric acid solution leads to the values pH $= 4.24$ and $S = 0.13$ mol/L. The solubility is approximately two times that in pure water.

Exercise 3 Calculate the solubility of $Mg(OH)_2$ in a hydrochloric acid solution whose initial pH is 3.50. What is the pH of the saturated solution? $K_s(Mg(OH)_2) = 10^{-10.74}$.

At saturation, the relations that must be satisfied are

$$[H^+][OH^-] = K_w, \quad [Mg^{2+}] = S,$$

$$2[Mg^{2+}] + [H^+] = [OH^-] + [Cl^-], \quad [Mg^{2+}][OH^-]^2 = 10^{-10.74},$$

$$[Cl^-] = 10^{-3.5}.$$

As $Mg(OH)_2$ is a basic hydroxide, its dissolution induces a pH increase. Hypothesizing that $[H^+]$ is negligible in the electroneutrality relation, we find

$$2S = [OH^-] + 10^{-3.5},$$

$$S[OH^-]^2 = 10^{-10.74},$$

$$[OH^-] = 2.53 \times 10^{-4},$$

$$\text{pH} = 10.40 \quad \text{and} \quad S = 2.84 \times 10^{-4}\ \text{mol/L}.$$

Exercise 4 Calculate the quantity of 0.1 mol/L acetic acid that must be used in order to dissolve 5.83 mg of magnesium hydroxide, and find the pH of the solution obtained. $K_s(Mg(OH)_2) = 10^{-10.74}$ and $M(Mg(OH)_2) = 58.324$ g/mol.

We find that 5.83 mg of magnesium hydroxide corresponds to 10^{-4} mol. The relations satisfied at the saturation point are

$$[H^+][OH^-] = K_w, \quad [Mg^{2+}] = 10^{-4}/V,$$

$$[H^+][CH_3COO^-]/[CH_3COOH] = 10^{-4.75}, \quad [CH_3COOH] + [CH_3COO^-] = 0.1,$$

$$2[Mg^{2+}] + [H^+] = [OH^-] + [CH_3COO^-], \quad [Mg^{2+}][OH^-]^2 = 10^{-10.74},$$

where V(l) is the added volume at the total dissolution.

Globally, the addition of acetic acid is accompanied by the reaction

$$Mg(OH)_2\downarrow + 2CH_3COOH \rightarrow Mg^{2+} + 2CH_3COO^- + 2H_2O.$$

Hence, it seems logical to neglect $[CH_3COOH]$ in the mass balance on acetic acid and acetate ions. Therefore, $[CH_3COO^-] \approx 0.1$ mol/L. It also seems logical to neglect $[H^+]$ in the electroneutrality equation since two moles of the base CH_3COO^- are formed in accordance with the reaction above. Therefore, we can deduce the following two relations:

$$210^{-4}/V = 0.1 + [OH^-],$$

$$10^{-4}[OH^-]^2/V = 10^{-10.74}.$$

As a result, $V = 2 \times 10^{-3}$ L and pH = 9.72.

The starting hypothesis is satisfied.

Exercise 5 Calculate the volume of a buffer containing 0.05 mol/L of ammonia and 0.05 mol/L of ammonium chloride necessary to dissolve 5.83 mg of magnesium hydroxide; $K_s(Mg(OH)_2) = 10^{-10.74}$, $M(Mg(OH)_2) = 58.324$ g/mol, $pK_a(NH_4^+) = 9.24$.

According to the data, 10^{-4} mol of magnesium hydroxide must be dissolved. Just when the dissolution is achieved with all the process at equilibrium, the following relations are satisfied:

$$[H^+][OH^-] = K_w, \quad [NH_4^+] + [NH_3] = 10^{-1},$$
$$[Mg^{2+}][OH^-]^2 = 10^{-10.74}, \quad [Cl^-] = 5 \times 10^{-2},$$
$$[Mg^{2+}] = 10^{-4}/V, \quad [NH_3][H^+]/[NH_4^+] = 10^{-9.24},$$
$$2[Mg^{2+}] + [H^+] + [NH_4^+] = [OH^-] + [Cl^-],$$

where V is the data we are seeking.

It is logical to neglect $[H^+]$ in the electroneutrality equation (ammonia buffer) since, moreover, a strong base is added. From another standpoint, as for each weak monoacid, we can write

$$[NH_4^+] = 10^{-1}[H^+]/(10^{-9.24} + [H^+]).$$

The problem reduces to a system of two simultaneous equations in two unknowns:

$$2.10^{-4}/V + \left[H^+\right] 10^{-1}/(10^{-9.24} + \left[H^+\right]) = \left[OH^-\right] + 5 \times 10^{-2},$$
$$(10^{-4}/V)[OH^-]^2 = 10^{-10.74}.$$

The results are $V = 10$ ml and pH $= 9.63$.

Exercise 6 Five ml of a 1 mol/L ammonia solution are added to 10 ml of a solution of 0.2 mol/L $MgCl_2$ and 0.4 mol/L NH_4Cl. What is the fraction of magnesium precipitated as hydroxide? $K_s(Mg(OH)_2) = 10^{-10.74}$.

When there is saturation and the processes are at equilibrium, the following equations are satisfied:

$$[H^+][OH^-] = 10^{-14}, \quad [Cl^-] = \{(0.2 \times 2 \times 10) + (0.4 \times 10)\}/15 = 8/15,$$
$$[H^+][NH_3]/[NH_4^+] = 10^{-9.24}, \quad [Mg^{2+}][OH^-]^2 = 10^{-10.74},$$
$$[NH_3] + [NH_4^+] = (0.4 \times 10/15) + (5.0 \times 1.0)/15 = 9/15,$$
$$[Mg^{2+}] + Mg(OH)_2\downarrow] = 0.2 \times 10/15,$$
$$2[Mg^{2+}] + [H^+] + [NH_4^+] = [Cl^-] + [OH^-].$$

$[Mg(OH)_2\downarrow]$ is the precipitated magnesium's "concentration." By virtue of the expressions of the ionization constant of NH_4^+ and of its analytical concentration, we find

$$[NH_4^+] = (9/15)\left\{[H^+]/([H^+] + K_a)\right\}.$$

Because the medium is buffered with the ammonia–ammonium mixture, it is logical to neglect $[H^+]$ in the electroneutrality equation. Therefore, the preceding system of equations can be reduced to the following two relations:

$$2[Mg^{2+}] + (9/15)\left\{[H^+]/([H^+] + K_a)\right\} = 8/15 + [OH^-]$$

$$\text{and} \quad [Mg^{2+}] = 10^{-10.74}/[OH^-]^2,$$

where $[Mg^{2+}]$ is the concentration of Mg^{2+} still in solution. The values sought are $pH = 9.34$, $[Mg^{2+}] = 3.73 \times 10^{-2}$ mol/L, precipitated $Mg^{2+} = 9.62 \times 10^{-2}$ mol/L, and precipitated fraction: 72/100.

Exercise 7 Calculate the solubility of magnesium hydroxide in a 0.1 mol/L ammonia solution. $K_s(Mg(OH)_2) = 10^{-10.74}$.

The relations that must be satisfied are

$$[H^+][OH^-] = 10^{-14},$$

$$[NH_3][H^+]/[NH_4{}^+] = 10^{-9.24},$$

$$[NH_4{}^+] + [H^+] + 2[Mg^{2+}] = [OH^-],$$

$$[NH_3] + [NH_4{}^+] = 10^{-1},$$

$$[Mg^{2+}][OH^-]^2 = 10^{-10.74}.$$

$[NH_4{}^+]$ is given by the relation

$$[NH_4{}^+] = 10^{-1}\,[H^+]/(K_a + [H^+]);$$

hence, the system can be reduced to the two equations

$$\left\{[H^+]/(K_a + [H^+])\right\} 10^{-1} + [H^+] + 2[Mg^{2+}] = [OH^-],$$

$$[Mg^{2+}][OH^-]^2 = 10^{-10.74}.$$

It is legitimate to neglect $[H^+]$ (dissolution of a hydroxide in ammonia). Solving the system gives

$$S = 1.02 \times 10^{-5}\ \text{mol/L}\ (S = Mg^{2+}).$$

34.3 Fractional Precipitation of Ions as a Function of the Solution's pH Value

34.3.1 Qualitative Aspects

The possibility to fractionally precipitate ions as a function of the solution's pH value is important in qualitative as well as quantitative analysis. Indeed, by a judicious adjustment of the solution's pH (with the help of the theoretical considerations developed in this section, among other means), the ions present in a solution can be selectively and more or less simultaneously precipitated. The most commonly used precipitation agents are dihydrogen sulfide (more precisely, the sulfide ion S^{2-}), the hydroxide ion, and oxine. Of course, there are some other organic precipitating agents, such as cupferron and dimethylglyoxime, for example.

With hydrogen sulfide, the superimposing equilibria are

$$H_2S_{(g)} \rightleftharpoons H_2S_{(aq)} \quad K,$$

$$H_2S_{(aq)} \rightleftharpoons H^+ + HS^- \quad K_{a1},$$

$$HS^- \rightleftharpoons H^+ + S^{2-} \quad K_{a2},$$

and upon saturation of a bivalent metallic sulfide,

$$M^{2+} + S^{2-} \rightleftharpoons MS\downarrow \quad K_s.$$

It is quite evident that for a given concentration $[M^{2+}]$, the solubility product can only be attained by adjusting the concentration $[S^{2-}]$, that is, by adjusting the pH value, which indirectly plays a part through the constants K_{a1} and K_{a2}. In fact, the two parameters on which the precipitation and hence the separation are dependent are K_s, whose value is constant for a given ion, and the pH value, which can be easily controlled. Thus, one conceives that as a function of these parameters and also of the ions' concentrations, some of them can be precipitated and others can't.

Beyond this example, metallic ions can also be selectively precipitated:

- as hydroxides, by virtue of the reaction

$$M^{n+} + nOH^- \rightleftharpoons M(OH)_n\downarrow \quad K_s;$$

The precipitation depends on K_s and the pH since the water ionic product is obeyed;

- as oxinates (or as derivatives of other organic precipitant reagents) according to the reaction

$$M^{n+} + n \text{ oxinate} - H \rightleftharpoons M(\text{oxinate})_n + nH^+.$$

In this case, the precipitation depends on the solubility product of the oxinate K_s, its ionization constant K_a, and the pH value.

34.3.2 *Fractional Precipitation of Metallic Ions as Sulfides*

The dihydrogen sulfide acid dissociation equilibria are

$$H_2S \rightleftharpoons H^+ + SH^-, \quad K_{a1} = [H^+][SH^-]/[H_2S] = 10^{-7},$$

$$HS^- \rightleftharpoons H^+ + S^{2-}, \quad K_{a2} = [H^+][S^{2-}]/[SH^-] = 10^{-13}.$$

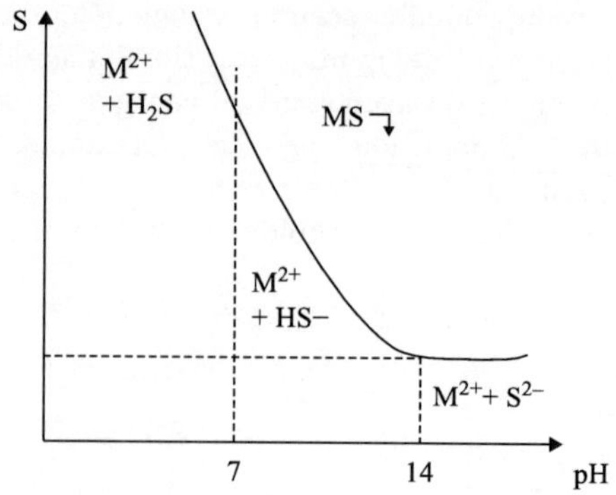

Fig. 34.6 Apparent solubility of metallic sulfides as a function of pH

It follows that

$$[S^{2-}][H^+]^2/[H_2S] = 10^{-20}. \tag{34.19}$$

From another standpoint, the solubility product is

$$K_s = [M^{2+}][S^{2-}]$$

for a bivalent ion (the generalization is immediate in the cases of other sorts of electrolytes). From a practical standpoint, an excess of H_2S or ammonium sulfide is used. If H_2S in excess is used, its saturated aqueous solution is such that $[H_2S] \approx 10^{-1}$ mol/L. Hence, according to (34.19),

$$[S^{2-}] = 10^{-21}/[H^+]^2,$$

and since $S = [M^{2+}] \quad [S^{2-}] = K_s/S$, we thus have

$$S = K_s[H^+]^2/10^{-21}.$$

The metallic sulfide's solubility increases when the pH value decreases.

The equation expressing the solubility$S$ of a metallic sulfide as a function of pH is

$$S = (K_s)^{1/2}(1 + [H^+]/K_{a2} + [H^+]^2/K_{a1}K_{a2})^{1/2},$$

with K_{a1} and K_{a2} defined above. It is established without any difficulty by settling

$$S = [M^{2+}] \quad \text{and simultaneously}$$
$$S = [S^{2-}] + [SH^-] + [H_2S].$$

Figure 34.6 represents the change in solubility with the pH.

The above relationship shows that the metallic sulfide's solubility changes with the K_s values. Figure 34.7 gives the changes of some of them as a function of pH. We

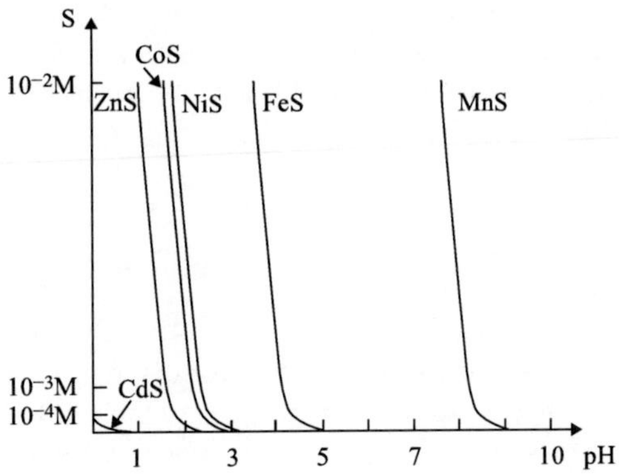

Fig. 34.7 Solubility of some metallic sulfides as a function of pH

can see that the solubility may remain very weak even in a very acidic solution. This is the case with CdS ($K_s = 10^{-27.8}$), PbS ($K_s = 10^{-27.9}$), CuS ($K_s = 8.5 \times 10^{-45}$), and HgS ($K_s = 10^{-51.8}$). This explains the origin of the following phrase used in qualitative analysis: "insoluble sulfide even in strongly acidic medium."

The sulfide ion concentration is the parameter that can be controlled for the separation. It is modulated through the pH value and determines whether or not we reach the solubility product. Let's now specify the range of exploitable concentrations $[S^{2-}]$. A saturated solution of dihydrogen sulfide exhibits a pH value of about 4 and a concentration $[S^{2-}]$ about 10^{-14} mol/L. In a 0.25 mol/L hydrochloric acid solution saturated in H_2S (required condition in order to precipitate sulfides of group 1 metals), the pH value is about 0.6 and the concentration $[S^{2-}]$ about 1.60×10^{-21} mol/L. We note that for a relatively narrow range of change in pH, that the change in concentration $[S^{2-}]$ is very large.

For example, let's take the separation of Fe^{2+} and Cu^{2+} by precipitation of their sulfides: $K_s(\text{CuS}) = 8.5 \times 10^{-45}$ and $K_s(\text{FeS}) = 1.5 \times 10^{-19}$. The precipitation is carried out from a 0.25 mol/L hydrochloric acid, where both ions are at a concentration of 10^{-2} mol/L. We are seeking the pH conditions in order to carry out the separation. We have seen that in these conditions, $[S^{2-}] = 1.60 \times 10^{-21}$ mol/L. The CuS solubility product is reached since

$$[Cu^{2+}][S^{2-}] = 10^{-2} \times 1.6 \times 10^{-21} = 1.6 \times 10^{-23},$$

a value far above $K_s(\text{CuS})$. The remaining concentration $[Cu^{2+}]$ at the end of precipitation is

$$[Cu^{2+}] = 8.5 \times 10^{-45}/1.6 \times 10^{-21},$$

$$[Cu^{2+}] = 5.10^{-24} \text{ mol/L} \quad \text{(end of precipitation)},$$

a ridiculously weak value. One may consider that all of the Cu^{2+} has been precipitated. Iron(II) cannot be precipitated in the same conditions since the product

$$[Fe^{2+}][S^{2-}] = 1.6 \times 10^{-23}$$

is weaker than K_s(FeS). For the precipitation of Fe^{2+}, the medium's acidity must be decreased; as a result, the concentration $[S^{2-}]$ is increased. The lower pH value beyond which it is impossible to work is calculated through the necessary concentration $[S^{2-}]$:

$$[S^{2-}] = 1.5 \times 10^{-19}/10^{-2}$$

$$\text{or} \quad [S^{2-}] = 1.5 \times 10^{-17} \text{ mol/L}.$$

In accordance with relation (34.19) (with a saturated solution in H_2S), the pH must be increased up to the value 2.10 in order to reach this limiting value $[S^{2-}]$. Furthermore, we must recall that this value is suitable only for the beginning of the precipitation. In order to totally precipitate iron(II), we would have to use the concentration $[S^{2-}]$ such as

$$[S^{2-}] = 1.5 \times 10^{-19}/10^{-5},$$

$$[S^{2-}] = 1.5 \times 10^{-14};$$

as a result,

$$[H^+]^2 = 10^{-21}/1.5 \times 10^{-14}, \quad pH = 3.60$$

if one admits that the precipitation is total when 10^{-5} mol/L (one thousandth of the initial concentration) of Fe(II) remains.

Exercise 8 Dihydrogen sulfide is bubbling until saturation through a solution initially containing $[Cd^{2+}] = 10^{-2}$ mol/L and $[Zn^{2+}] = 10^{-2}$ mol/L. Determine the pH limits between which its value must be located in order to totally precipitate cadmium sulfide (at the 0.1% level) without any precipitation of zinc sulfide [$K_s(CdS) = 10^{-26}$, $K_s(ZnS) = 10^{-21}$, $K_{a1}(H_2S) = 10^{-7}$, H_2S water solubility under 1 bar $= 10^{-1}$ mol/L].

The calculation principle is the same as above. Cadmium sulfide precipitates quantitatively at pH = 0. At pH = 1, zinc sulfide begins to precipitate.

Remarks

1. We must notice that the metallic sulfide precipitation is often accompanied by the formation of various complexes with some anions present in the medium, for example, with sulfide anion itself or with the anion of the added acid or with some species resulting from the hydrolysis of the metallic ions. The preceding calculations do not take into account these complications and, for this reason, may be considered as being weakly realistic (see Chap. 35).
2. Other inorganic reagents may be used in some cases in order to carry out the fractional precipitation of some metallic ions. Such is the case of chromate ion CrO_4^{2-}, which permits us to precipitate and separate barium and strontium ions from each other (see the following exercise).

Exercise 9 Let's consider a solution containing Ba^{2+} and Sr^{2+} ions, both at concentrations 10^{-2} mol/L. A 10^{-1} mol/L chromate solution is added. Determine the pH limits between which its value must be located for barium chromate to be precipitated at the 0.1% level without any precipitation of strontium chromate [$K_s(BaCrO_4) = 10^{-9.7}$, $K_s(SrCrO_4) = 10^{-4.4}$, $K_a(HCrO_4^-) = 10^{-6.4}$].

The two reactions involved are

$$Ba^{2+} + CrO_4{}^{2-} \rightleftharpoons BaCrO_4\downarrow$$

$$Sr^{2+} + CrO_4{}^{2-} \rightleftharpoons SrCrO_4\downarrow$$

From the qualitative standpoint, it must be realized that, depending on the solution's pH, the chromate ion $CrO_4{}^{2-}$ may be protonated (thus giving the acid chromate ion $HCrO_4{}^-$) and that only the former ion can precipitate both ions.

According to the solubility product values, Ba^{2+} precipitates first. When it is totally precipitated ($[Ba^{2+}] = 10^{-5}$ mol/L), the following relation is satisfied:

$$10^{-5}[CrO_4{}^{2-}] \geq 10^{-9.7}.$$

According to the usual relation applying to a weak acid,

$$[CrO_4{}^{2-}] = \left\{10^{-6.4}/([H^+] + 10^{-6.4})\right\} 0.1.$$

Due to the first relation,

$$[CrO_4{}^{2-}] \geq 10^{-4.7},$$

$$10^{-6.4} \times 0.1/([H^+] + 10^{-6.4}) \geq 10^{-4.7}, \quad \text{i.e.,} \quad pH \geq 2.7.$$

Strontium chromate begins to precipitate for

$$10^{-2}[CrO_4{}^{2-}] \geq 10^{-4.4},$$

$$[CrO_4{}^{2-}] = 10^{-6.4} \times 0.1/([H^+] + 10^{-6.4}).$$

As a result,

$$10^{-6.4} \times 0.1/([H^+] + 10^{-6.4}) \geq 10^{-2.4}, \quad \text{i.e.,} \quad pH \leq 5.02.$$

Hence, the pH range we are seeking is

$$2.70 \leq pH \leq 5.02.$$

At pH < 2.70, barium chromate does not precipitate quantitatively. At pH > 5.02, strontium chromate begins to precipitate.

34.3.3 Fractional Precipitation of Metallic Ions as Hydroxides

Poorly soluble metallic hydroxides give rise to the following equilibria:

$$\mathrm{MOH}\downarrow \rightleftharpoons \mathrm{MOH}_{\text{saturation}},$$

$$\mathrm{MOH}\downarrow \rightleftharpoons \mathrm{M}^+ + \mathrm{OH}^-.$$

The solubility S is given by the expression

$$S = [\mathrm{MOH}]_{\text{saturation}} + [\mathrm{M}^+].$$

By introducing the solubility product K_s,

$$K_s = [\mathrm{M}^+][\mathrm{OH}^-],$$

and the intrinsic solubility,

$$[\mathrm{MOH}]_{\text{saturation}} = \mathrm{So},$$

we find

$$S = \mathrm{So} + K_s[\mathrm{H}^+]/K_w.$$

Since the solubility product is related to the concentration $[\mathrm{MOH}]_{\text{saturation}} = \mathrm{So}$ (see Chap. 32), the apparent (total) solubility may be given by the expression

$$S = \mathrm{So}(1 + k[\mathrm{H}^+]/K_w),$$

$$S = \mathrm{So}(1 + K[\mathrm{H}^+]) \quad \text{with} \quad K = k/K_w \quad \text{and} \quad k = K_s/\mathrm{So}.$$

As a result, the solubility increases when the pH decreases. This was expected since the following two equilibria occur simultaneously:

$$\mathrm{MOH}\downarrow \rightleftharpoons \mathrm{M}^+ + \mathrm{OH}^-,$$

$$\mathrm{OH}^- + \mathrm{H}^+ \rightarrow \mathrm{H_2O}.$$

The dissociation equilibrium is displaced toward the right. At high pH values, the apparent solubility tends toward the intrinsic solubility So. The general form of the curve S/pH is mentioned in Fig. 34.8. We can notice that in acidic medium the apparent solubility increases linearly with $[\mathrm{H}^+]$, since the intrinsic one becomes negligible compared to $[\mathrm{M}^+]$ in these conditions. We likewise could demonstrate that with metallic hydroxides of the sort $\mathrm{M(OH)}_n$, the solubility increases linearly with $[\mathrm{H}^+]^n$.

In order to draw the solubility curves, the literature gives the pH value of precipitation for the usual 10^{-2} mol/L concentration of metallic ions. The intersection point of the straight line $S = 10^{-2}$ mol/L with the solubility curve gives the pH value at the beginning of the precipitation. The pH value at the end of the precipitation is often

Fig. 34.8 Solubility of a metallic hydroxide (nonamphoteric) as a function of the pH value

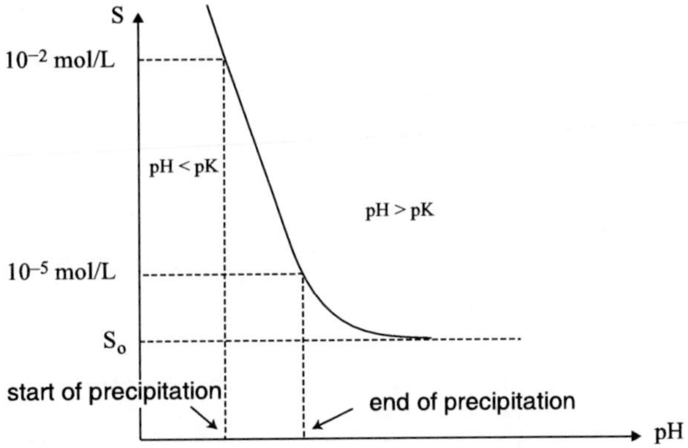

Table 34.2 Precipitation pH (pH_1) and dissolution pH (pH_2) of some amphoteric hydroxides ($[M^{n+}] = 10^{-2}$ mol/L)

Hydroxides	pH_1	pH_2	Hydroxides	pH_1	pH_2
$Be(OH)_2$	6.3	13.5	$Sn(OH)_2$	2	13
$Mg(OH)_2$	9.5	–	$Cr(OH)_3$	5	12
$Ca(OH)_2$	12.3	–	$Fe(OH)_3$	2.2	–
$Sr(OH)_2$	13.3	–	$Fe(OH)_2$	7.5	–
$Ba(OH)_3$	13.9	–	Ag_2O	8.5	15.7
$Al(OH)_3$	4.0	10	$Zn(OH)_2$	6.5	14
$Sn(OH)_4$	0.5	14	$Cd(OH)_2$	8	–

defined as being the pH at which 1/1000 of the initial concentration remains, that is, with the preceding convention for $S = 10^{-5}$ mol/L. Table 34.2 mentions some pH values at which some hydroxides begin to precipitate.

Some hydroxides are ampholytes since they exhibit an acidic character according to the equilibrium

$$MOH \rightleftharpoons MO^- + H^+,$$

which is quantified by the constant

$$K_2 = (MO^-)(H^+)/(MOH)$$

at saturation:

$$S = [M^+] + [MOH]_{saturation} + [MO^-],$$

$$S = So(1 + K_1[H^+] + K_2/[H^+]),$$

where K_1, at saturation, is given by the expression

$$K_1 = K_s So/K_w.$$

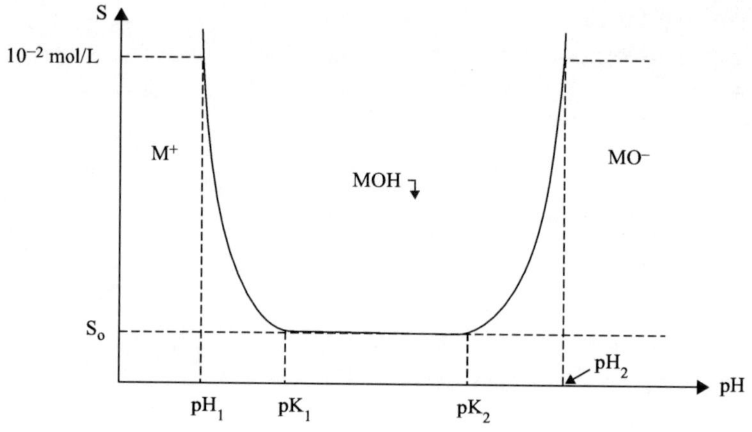

Fig. 34.9 Solubility of an amphoteric hydroxide as a function of the pH value

The curve S/pH looks like a U (see Fig. 34.9). More precisely, in alkaline medium for pH $> pK_2$, the solubility increases. The hydroxide precipitate dissolves by formation of the conjugate base MO^-. The value pH $=$ pH_2 corresponds to the dissolution of the initial 10^{-2} mol/L of M^{n+} as MO^-. The pH value pH $=$ pH_1 corresponds to the beginning of the precipitation.

We must be convinced here that we have given theoretical considerations. From a practical standpoint, a local excess of hydroxide ions in the solution may appear during the addition of a strong base (delivered in order to precipitate and separate the metallic hydroxide). The local excess may be the cause of the hydroxide precipitation, although, theoretically, it is soluble in these conditions. Some metallic ions may be trapped in it as a result. This is the reason why the alkali must be added slowly and with homogenization. It must also be known that the equilibrium between the hydroxide precipitate and the solution is often slowly attained because of the slow evolution of the precipitate. It is sometimes possible to quantitatively describe this parasitic phenomenon by handling several solubility products according to the state of the precipitate. As a result, several solubility curves exist.

34.3.4 Fractional Precipitation of Metallic Ions as Oxinates

The formation equilibrium of an oxinate with a bivalent metallic ion obeys the equilibrium

$$2Ox^- + M^{2+} \rightleftharpoons M(Ox)_2\downarrow$$

where Ox^- represents the oxinate rest.

Table 34.3 Solubility products of some oxinates

	K_s		K_s
Al(Ox)$_3$	$10^{-32.3}$	Mn(Ox)$_2$	$10^{-19.3}$
Cd(Ox)$_2$	$10^{-22.0}$	Ni(Ox)$_2$	$10^{-26.0}$
Co(Ox)$_2$	$10^{-24.8}$	Pb(Ox)$_2$	$10^{-22.0}$
Cu(Ox)$_2$	$10^{-29.6}$	Ti(Ox)$_3$	$10^{-37.0}$
Mg(Ox)$_2$	$10^{-15.2}$	Zn(Ox)$_2$	$10^{-24.5}$

It is governed by the solubility product:

$$K_s = [M^{2+}][Ox^-]^2$$

The two following equilibria are superimposed onto the preceding one:

$$H_2Ox^+ \rightleftharpoons HOx + H^+, \quad K_{a1} = 10^{-5},$$
$$HOX \rightleftharpoons Ox^- + H^+, \quad K_{a2} = 10^{-9.9}.$$

The ionic solubility is defined by the expressions

$$S = [M^{2+}],$$
$$S = 1/2\left\{[Ox^-] + [HOx] + [H_2Ox^+]\right\}.$$

The factor 1/2 stems from the fact that after having been dissolved, the oxinate M(Ox)$_2$ gives two molecules of oxinate ion for one molecule of metallic ion. After introducing the ionization constants into the above relation, we find

$$S = 1/2[Ox^-]\left\{1 + [H^+]/K_{a2} + [H^+]^2/K_{a1}K_{a2}\right\}.$$

Moreover, according to the solubility product,

$$[Ox^-]^2 = K_s/S,$$
$$[Ox^-] = (K_s/S)^{1/2}.$$

Inserting this relation into that above-expressed S, we get

$$S = \left\{K_s/4\,(1 + [H^+]/K_{a2} + [H^+]^2/K_{a1}K_{a2})^2\right\}.$$

With an oxinate of the sort M(Ox)$_3$, we would have obtained

$$S = \left\{K_s/27(1 + [H^+]/K_{a2} + [H^+]^2/K_{a1}K_{a2})^3\right\}.$$

Again, the solubility increases when the pH decreases. As indicated by the last two relations, the pH ranges in which the solution must be located in order to separate metallic ions from one another depend first and foremost on the corresponding solubility–product values. Table 34.3 mentions the solubility products of some oxinates. Hence, for example, the aluminum cation Al^{3+} can be separated from beryllium

Be^{2+} and magnesium Mg^{2+} ions. The experiment is carried out in acetic buffer. Only the aluminum oxinate precipitates in these conditions.

The selectivity may be improved by controlling some other parameters such as the analytical concentration of the precipitating reagent and the presence (or absence) of secondary masking agents.

Remarks

1. Metallic oxinates may also be extracted into organic solvents that are immiscible with water. This method involves another use of oxine and also of other precipitating organic reagents. It also permits the identification and even the quantitative determination of metallic ions.
2. The quantitative determination of metallic ions, which follows the precipitation as oxinates, may be carried out, depending on the case, by gravimetry (see Chap. 38) or by dissolution of the precipitate in hydrochloric acid and subsequent titration of oxine by bromometry (see Chap. 19). Moreover, in certain conditions, oxine is electroactive. As a result, it can be determined by amperometry.

Chapter 35
Precipitation and Complexation

In this chapter, we will investigate the case in which complexation and precipitation phenomena are superimposed. In particular, we study the dissolution of a precipitate by complexation and, inversely, the destruction of a complex by the precipitation of the metallic ion engaged within it. Furthermore, this chapter also concerns the case when the complexation is due to the ligand, which is the precipitant reagent. We also study the possibility of separating several metal cations in admixture by successively and even simultaneously using precipitation and complexation processes.

35.1 Dissolution of a Precipitate by Complexation of the Metal Cation Constituting it: Generalities

The solubility of a precipitate, within which a metallic ion is engaged, increases by complexation of the latter. In the best cases, we can arrive purely and simply at the total dissolution of the precipitate. Let us consider the precipitate $MA_{(s)}$ (s: solid). The equilibria that are superimposed are

$$MA_{(s)} \rightleftharpoons M^+ + A^-,$$

$$M^+ + L \rightleftharpoons ML,$$

$$ML + L \rightleftharpoons ML_2, \text{ etc.},$$

where L is the ligand (electrical charges are omitted for the sake of generality). It clearly appears that the dissolution equilibrium (the first above) is displaced toward the right by the following complexation equilibria. The apparent solubility S of the metal cation defined by

$$S = [M^+] + [ML] + [ML_2] \text{ etc.}$$

is increased. We can notice that this case may also be that in which L is identical to A^-. It is that of the dissolution of a precipitate with an excess of precipitating reagent. Of course, this case is particular to the phenomenon described in this section.

J.-L. Burgot, *Ionic Equilibria in Analytical Chemistry*,
DOI 10.1007/978-1-4419-8382-4_35,

Table 35.1 Successive formation constants of some ammine–metal complexes

Metal ion	Log K					
	K_1	K_2	K_3	K_4	K_5	K_6
Ag^+	3.20	3.83				
Hg^{2+}	8.80	8.70	1.00	0.78		
Zn^{2+}	2.37	2.44	2.50	2.15		
Mg^{2+}	0.23	−0.15	−0.42	−0.70	−0.95	−1.30
Cd^{2+}	2.65	2.10	1.44	0.93	−0.32	−1.66
Cu^{2+}	4.15	3.50	2.89	2.13	−0.50	−2.50
Ni^{2+}	2.80	2.24	1.73	1.19	0.75	0.03
Co^{2+}	2.11	1.63	1.05	0.76	0.18	−0.62
Co^{3+}	7.30	6.70	6.10	5.60	5.05	4.41

For example, silver iodide ($K_s = 10^{-16}$) is dissolved by the formation of complexes like $[Ag(CN)_2]^-$ $[Ag(S_2O_3)_2]^{3-}$ in the presence of ligands CN^- and $S_2O_3{}^{2-}$. As an example of dissolution in an excess of precipitating reagent, consider the precipitation of mercuric iodide:

$$Hg^{2+} + 2I^- \rightleftharpoons HgI_2\downarrow$$

An excess of iodide ions dissolves it according to the reaction

$$HgI_2\downarrow + 2I^- \rightleftharpoons [HgI_4]^{2-}.$$

Another example involves the reactions of nickel ion with cyanide ions:

$$Ni^{2+} + CN^- \rightleftharpoons [NiCN]^+,$$

$$[NiCN]^+ + CN^- \rightleftharpoons [Ni(CN)_2]\downarrow,$$

$$[Ni(CN)_2]\downarrow + CN^- \rightleftharpoons [Ni(CN)_3]^-,$$

$$[Ni(CN)_3]^- + CN^- \rightleftharpoons [Ni(CN)_4]^{2-}.$$

The precipitate $Ni(CN)_2$ is dissolved in an excess of cyanide ions. We'll also mention the complexes $[AgCl_2]^-$ and $[AgCl_3]^{2-}$, which are formed by dissolution of the AgCl precipitate into a solution containing a great excess of chloride ions.

35.2 Dissolution of a Precipitate by Complexation: Quantitative Aspects

Let's begin by recalling the successive formation constants of some ammine–metal complexes (Table 35.1). They govern the following equilibria:

$$M^{n+} + NH_3 \overset{K_1}{\rightleftharpoons} [M(NH_3)]^{n+}$$

$$[M(NH_3)]^{n+} + NH_3 \overset{K_2}{\rightleftharpoons} [M(NH3)_2]^{n+}, \text{ etc.}$$

Table 35.2 Formation constants of some complexes by reaction of insoluble salts with the same ligand

Ligand	Central ion	Logarithm of the equilibrium constant				
		K_{S0}	K_{S1}	K_{S2}	K_{S3}	K_{S4}
Cl^-	Cu^+	−6.73	−5.00	−1.12	−1.47	–
	Ag^+	−9.75	−6.70	−4.70	−4.70	−4.46
	Tl^+	−3.04	−3.15	−3.74	−4.70	–
	Hg^{2+}	−13.79	−7.05	−0.57	+0.28	+1.28
Br^-	Ag^+	−12.10	−7.96	−5.00	−4.15	−3.22
	Hg^{2+}	−19.10	−10.05	−1.77	+0.64	+1.90
	Tl^+	−4.81	−4.48	−4.62	−5.10	−5.80
I^-	Ag^+	−16.35	−8.22	−5.40	−2.60	−1.96
	Hg^{2+}	−27.70	−14.83	−3.88	−0.10	+2.13
	Pb^{2+}	−8.15	−6.23	−4.47	−4.65	−3.85
CN^-	Cu^+	−19.49	(−13)	−4.23	+0.36	+2.06
	Ag^+	−15.92	(−7)	+4.62	+5.32	+4.19
	Hg^{2+}	−35.10	(−17.10)	−0.40	+3.43	+6.41
SCN^-	Ag^+	−11.97	−7.22	−3.74	−2.52	−2.30
CH_3COO^-	Ag^+	−2.40	−1.67	−1.67	–	–

We notice that no precipitate appears during the development of these reactions.

We also recall the formation constants of some complexes by the reaction of insoluble salts with the same ligands as those that have prevailed in the formation of the precipitates (Table 35.2).

They govern equilibria of the following types:

$$AgCl\downarrow \rightleftharpoons Ag^+ + Cl^- \quad [Ag^+][Cl^-] = K_{s0},$$

$$AgCl\downarrow \rightleftharpoons AgCl_{saturation} \quad [AgCl_{saturation}] = K_{s1},$$

$$AgCl\downarrow + Cl^- \rightleftharpoons [AgCl_2]^- \quad [AgCl_2{}^-] = K_{s2}[Cl^-],$$

$$AgCl\downarrow + 2Cl^- \rightleftharpoons [AgCl_3]^{2-} \quad [AgCl_3{}^{2-}] = K_{s3}[Cl^-]^2,$$

$$AgCl\downarrow + 3Cl^- \rightleftharpoons [AgCl_4]^{3-} \quad [AgCl_4{}^{3-}] = K_{s4}[Cl^-]^3.$$

Let's consider a silver halide precipitate $AgX_{(s)}$ that undergoes a reaction with the ligand L. Let's also suppose, in order to simplify, that it gives the dicoordinated complex only, according to the reaction

$$AgX\downarrow + 2L \rightleftharpoons [AgL_2]^+ + X^-.$$

The constants playing a part in this reaction are those governing the following two equilibria:

$$AgX_{(s)} \rightleftharpoons Ag^+ + X^-, \quad K_s = [Ag^+][X^-],$$

$$[AgL_2]^+ \rightleftharpoons Ag^+ + 2L, \quad K_D = \frac{1}{\beta_2} = \frac{[Ag^+][L]^2}{[AgL_2{}^+]}.$$

The silver halide's solubility is

$$S = [X^-],$$

$$S = [Ag^+] + [AgL_2{}^+],$$

$$S = [Ag^+]\left[1 + \frac{[L]^2}{K_D}\right].$$

From these relations, we immediately obtain

$$S = \sqrt{K_s} \times \sqrt{1 + \frac{[L]^2}{K_D}}.$$

The solubility indeed increases with the solubility product and with the ligand's concentration. It also increases with the complex's stability. The relation permits us to easily calculate the concentrations of the different species once the equilibria have been reached. For example, with $[NH_3] = 1$ mol/L and in the case of AgCl $[K_s(AgCl) = 10^{-9.75}]$, we find

$$S = 4.4 \times 10^{-2}\,\text{mol/L};\ [Cl^-] = 4.4 \times 10^{-2}\,\text{mol/L};$$

$$[Ag^+] = \frac{10^{-9.75}}{(4.4 \times 10^{-2})} \approx 4 \times 10^{-9}\,\text{mol/L}.$$

The remaining concentration of silver ions is negligible. As a result,

$$[Ag(NH_3)_2{}^+] \approx 4.4 \times 10^{-2}\,\text{mol/L}.$$

In order to obtain this result, we must handle an ammonia solution with a concentration near 1.1 mol/L in the presence of the AgCl precipitate. The value 1.1 mol/L results from the fact that two ammonia molecules, together with one molecule of Ag^+, are necessary to form the complex (i.e., $4.4 \times 10^{-2} \times 2 = 8.8 \times 10^{-2} \approx 0.1$ mol/L) and from the fact that the concentration of free ammonia is equal to 1 mol/L in the calculation.

35.3 Dissolution of a Precipitate by Complexation: Further Calculations

While the above-described calculations are roughly correct, they are simplistic. To begin explaining their simplistic character, we must mention that they do not take the formation of successive complexes into account. For example, the formation of the diamminesilver(-I) complex $[Ag(NH_3)_2]^+$ necessarily implicates the formation of the monoamminesilver(-I) complex $[Ag(NH_3)]^+$. Moreover, other complexes may also be formed, such as hydroxo complexes. We did not take all these facts into account above.

From the quantitative standpoint, the superimposition of complexation and precipitation phenomena is difficult to tackle rigorously. Indeed, numerous relations must be satisfied and, as a result, there are numerous unknowns. However, the systems of equations may be considerably simplified after the adoption of judicious simplifications, based, of course, on chemical intuition.

35.3.1 Dissolution of Silver Bromide with a Sodium Thiosulfate Solution

Ag^+ and $S_2O_3{}^{2-}$ ions give rise to the following complexation equilibria:

$$Ag^+ + S_2O_3{}^{2-} \rightleftharpoons [Ag(S_2O_3)]^-, \quad K_1 = 6.6 \times 10^8,$$

$$[Ag(S_2O_3)]^- + S_2O_3{}^{2-} \rightleftharpoons [Ag(S_2O_3)_2]^{3-}, \quad K_2 = 4.4 \times 10^4,$$

$$[Ag(S_2O_3)_2]^{3-} + S_2O_3{}^{2-} \rightleftharpoons [Ag(S_2O_3)_3]^{5-} \quad K_3 = 4.9.$$

The problem consists of calculating the solubility of AgBr in a solution of 10^{-2} mol/L $Na_2S_2O_3$ [$K_s(AgBr) = 7.9 \times 10^{-13}$]. The relations that must be satisfied are

$$\frac{[Ag(S_2O_3)^-]}{[Ag^+][S_2O_3{}^{2-}]} = 6.6 \times 10^8, \quad \frac{[Ag(S_2O_3)_2{}^{3-}]}{[Ag(S_2O_3)^-][S_2O_3{}^{2-}]} = 4.4 \times 10^4,$$

$$\frac{[Ag(S_2O_3)_3{}^{5-}]}{[Ag(S_2O_3)_2{}^{3-}][S_2O_3{}^{2-}]} = 4.9,$$

$$[S_2O_3{}^{2-}] + [Ag(S_2O_3)^-] + 2[Ag(S_2O_3)_2{}^{3-}] + 3[Ag(S_2O_3)_3{}^{5-}] = 0.01,$$

$$S = [Ag^+] + [Ag(S_2O_3)^-] + [Ag(S_2O_3)_2{}^{3-}] + [Ag(S_2O_3)_3{}^{5-}],$$

$$S = [Br^-],$$

and since the solution is saturated,

$$K_s = [Ag^+][Br^-] \quad (K_s = 7.9 \times 10^{-13}).$$

It is interesting to note that the charge balance relation is not mentioned. Although it is satisfied, it does not bring anything that might help solve the problem since no prototropic equilibrium occurs. We also notice that we are faced with a system of seven equations with seven unknowns (six concentrations and the solubility S).

The simplifications that we can investigate are, in order of priority, based on the following assumptions:

- the $[Ag(S_2O_3)_2]^{3-}$ complex probably forms in the greatest quantity. This appears through the examination of the equilibrium constants of the following reactions:

$$AgBr\downarrow + S_2O_3{}^{2-} \rightleftharpoons [Ag(S_2O_3)]^- + Br^-,$$

$$AgBr\downarrow + 2S_2O_3{}^{2-} \rightleftharpoons [Ag(S_2O_3)_2]^{3-} + Br^-;$$

- the concentration $[Ag^+]$ is probably negligible with respect to $[Ag(S_2O_3)_2{}^{3-}]$.

Therefore, by setting into the global equilibrium constant

$$S \approx [Ag(S_2O_3)_2{}^{3-}] \quad \text{and hence} \quad [S_2O_3{}^{2-}] \approx 0.01 - 2S,$$

the system of the preceding equations becomes considerably simpler and can be reduced to a system in two unknowns, $[S_2O_3{}^{2-}]$ and S. Finally, we find

$$S = 4.53 \times 10^{-3}\,\text{mol/L},$$

$$[Ag(S_2O_3)_2{}^{3-}] = 4.53 \times 10^{-3}\,\text{mol/L},$$

$$[Ag^+] = 1.74 \times 10^{-10}\,\text{mol/L},$$

$$[Br^-] = 4.53 \times 10^{-3}\,\text{mol/L},$$

$$[Ag(S_2O_3)^-] = 1.09 \times 10^{-4}\,\text{mol/L},$$

$$[S_2O_3{}^{2-}] = 9.48 \times 10^{-4}\,\text{mol/L},$$

$$[Ag(S_2O_3)_3{}^{5-}] = 2.10 \times 10^{-5}\,\text{mol/L}.$$

These results are in accordance with the starting hypothesis. For them to be fully rigorous, supplementary equilibria must be taken into account. For example, hydrogen thiosulfuric acid $HS_2O_3{}^-$ exhibits the pK_a value 1.92×10^{-2}. Its eventual protonation has not been taken into account. From another standpoint, the formation of molecular silver hydroxide according to the reaction

$$Ag^+ + OH^- \rightleftharpoons AgOH(\text{solution}), \quad K = 200,$$

has not been considered either, and we also have not considered the formation of the superior complexes $[AgBr_2]^-$, $[AgBr_3]^{2-}$, and $[AgBr_4]^{3-}$, whose overall formation constants are, respectively, $10^{7.12}$, $10^{7.96}$, and $10^{8.90}$ when $[AgBr]_{\text{solution}} = 1.1 \times 10^{-8}$ mol/L. Taking these new reactions and these new data into account necessitates modifying the above system of equations and, in particular, adding the charge balance relation and the ionic product of water. Carrying out the rigorous calculations should be done by starting with the same hypothesis as above. The results would be identical. Thus, the supplementary equilibria do not play an appreciable role. It is interesting to notice that rigorous calculations lead to the value pH = 7.01. The solution remains neutral during the dissolution process.

35.3.2 *Dissolution of a Precipitate with an Excess of Precipitating Reagent*

The phenomenon of dissolving a precipitate with an excess of precipitating reagent is not rare (see Table 35.2).

35.3.2.1 Dissolution of Copper(I) Chloride in the Presence of Chloride Ions

In order to calculate the solubility increase of Cu^{I} due to its complexation with chloride ions, we must first calculate the cuprous chloride's solubility in water. The equilibria that must be taken into account are

$$CuCl_{(s)} \rightleftharpoons Cu^{+} + Cl^{-} \quad K_s = 1.85 \times 10^{-7},$$

$$CuCl_{(s)} + Cl^{-} \rightleftharpoons [CuCl_2]^{-} \quad K_{s2} = 7.60 \times 10^{-2},$$

$$[CuCl_2]^{-} + Cl^{-} \rightleftharpoons [CuCl_3]^{2-} \quad K_{s3} = 3.40 \times 10^{-2},$$

and the mass balance relations are

$$S = [Cl^{-}] + 2[CuCl_2{}^{-}] + 3[CuCl_3{}^{2-}],$$

$$S = [Cu^{+}] + [CuCl_2{}^{-}] + [CuCl_3{}^{2-}].$$

Solving this system is carried out by successive approximations. In a first stage, the complexes' formation is neglected. This leads to the provisory value $S = 4.30 \times 10^{-4}$ mol/L. With this value in mind, we calculate $[CuCl_2{}^{-}]$ and $[CuCl_3{}^{2-}]$ through K_{s2} and K_{s3}. The values are, respectively, 3.3×10^{-5} mol/L and 6.30×10^{-9} mol/L. It appears that $[CuCl_2{}^{-}]$ is not negligible and that the approximation adopted was not good. Thus, in a second stage, the calculation is recommenced with the following new set of approximations:

$$S = [Cu^{+}] + [CuCl_2{}^{-}],$$

$$S = [Cl^{-}] + 2[CuCl_2{}^{-}].$$

The definitive values are

$$[Cu^{+}] = 4.45 \times 10^{-4}\ \text{mol/L}, \quad [Cl^{-}] = 4.15 \times 10^{-4}\ \text{mol/L},$$

$$[CuCl_2{}^{-}] = 3.15 \times 10^{-5}\ \text{mol/L}, \quad [CuCl_3{}^{2-}] = 5.85 \times 10^{-9}\ \text{mol/L},$$

$$S = 4.77 \times 10^{-4}\ \text{mol/L}.$$

Let's now consider the solubility of CuCl in a solution of hydrochloric acid at the concentration C mol/L. The relations satisfied are the same as those given above, except for the mass balance on chloride ions, which is now

$$C + S = [Cl^{-}] + 2[CuCl_2{}^{-}] + 3[CuCl_3{}^{2-}].$$

Handling these relations leads to the following two:

$$C = 1.076[Cl^{-}] + 6.8 \times 10^{-2}[Cl^{-}]^2 - \frac{1.85 \times 10^{-7}}{[Cl^{-}]}$$

$$\text{and} \quad S = \frac{1.85 \times 10^{-7}}{[Cl^{-}]} + 7.60 \times 10^{-2}[Cl^{-}] + 3.40 \times 10^{-2}[Cl^{-}]^2.$$

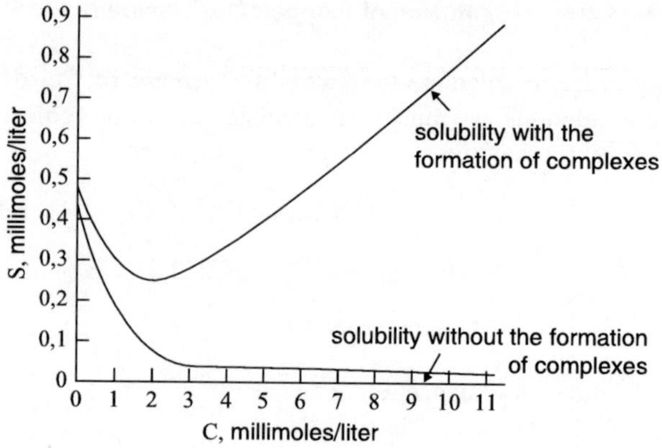

Fig. 35.1 : Solubility of Cu^I chloride into hydrochloric acid solutions of concentrations C

Introducing varied values of $[Cl^-]$ into the last relations allows us to determine the coordinates of the points C_i, S_i and to draw the corresponding curve S/C (Fig. 35.1).

In the case of the curve qualified on the diagram as "formation of complexes neglected," the solubility S has been calculated using the relation $S = [Cu^+]$. The diagram shows very clearly that the solubility can be doubled when complexes are formed.

Figure 35.1 shows that the solubility exhibits a well-defined minimum. The curve decrease corresponds to the common ion effect, which is a consequence of the solubility-product expression. Indeed, $[Cu^+]$ decreases when $[Cl^-]$ increases. The increasing portion of the curve corresponds to the formation of complexes. The solubility minimum occurs for

$$\frac{dS}{dC} = 0 \quad \text{or} \quad \left(\frac{dS}{d[Cl^-]}\right) \times \left(\frac{d[Cl^-]}{dC}\right) = 0.$$

Solving the last relation gives

$$\frac{dS}{d[Cl^-]} = -\frac{1.85 \times 10^{-7}}{[Cl^-]^2} + 7.6 \times 10^{-2} + 6.8 \times 10^{-2}[Cl^-],$$

which vanishes for $[Cl^-] = 1.56 \times 10^{-3}$ mol/L.

35.3.2.2 Formation of Silver Oxide and Its Dissolution in the Presence of Ammonia

The formation of silver oxide and its dissolution in the presence of ammonia is slightly more complicated than the preceding case. The ligand ammonia NH_3 is basic. Not only does the superimposition of precipitation and complexation phenomena occur, but there also are prototropic equilibria. Janik Bjerrum[1] in the 1950s studied the formation of these complexes by pH-metry.

The addition of ammonia to a solution of Ag^+ ions induces the precipitation of silver oxide, which is forming from the corresponding hydroxide according to the reaction

$$2AgOH \rightarrow Ag_2O\downarrow + H_2O.$$

But when an excess of ammonia is added, the precipitate dissolves essentially by formation of the diamminesilver(I) $[Ag(NH_3)_2]^+$ complex. (Ammine–metal complexes are numerous and have a great analytical importance—see Table 35.1. They are also of historical interest.)

It is possible to calculate the minimum concentration of ammonia N that must be added to induce the precipitation of silver oxide and also the concentration necessary in order to totally dissolve it. The relations that must be satisfied in order to provide the answer to the first question are

$$\frac{[Ag(NH_3)^+]}{([Ag^+][NH_3])} = K_1, \quad \frac{[Ag(NH_3)_2{}^+]}{([Ag(NH_3)^+][NH_3])} = K_2,$$

$$\frac{[AgOH]}{([Ag^+][OH^-])} = K', \quad [AgOH\,] = S \quad \text{(at saturation of } Ag_2O\text{);}$$

$$\frac{[Ag(OH)_2{}^-]}{([AgOH][OH^-])} = \kappa, \quad \frac{[NH_3][H^+]}{[NH_4{}^+]} = K_a, \quad [Ag^+][OH^-] = K_s.$$

The constant values are $K_1 = 1.58 \times 10^3$, $K_2 = 6.75 \times 10^3$, $K' = 200$, $S = 3.7 \times 10^{-6}$ mol/L, $\kappa = 52.5$, $K_a(NH_4{}^+/NH_3) = 10^{-9.24}$, and we'll suppose that the concentration of silver nitrate is 10^{-1} mol/L.

First, we notice that the calculation involves dissolved silver hydroxide and the formation of the unusual complex $[Ag(OH)_2]^-$. The above system of equations seems to be incredibly difficult to solve. This is not true if we make the following assumptions:

- hypothesis 1: the monoammine–silver $[Ag(NH_3)]^+$ concentration is negligible. This assertion is based on the fact that the distribution diagram $\alpha/\log(NH_3)$, calculated once and for all, shows that this complex is always in minority with respect to $[Ag^+]$ or to $[Ag(NH_3)_2]^+$;
- hypothesis 2: $[Ag^+]$ is nearly equal to 0.1 mol/L when the precipitation begins. In order to be convinced, it is sufficient to consider the mass balance on silver at the time when the precipitation exactly begins. We can write

$$0.1 = [Ag^+] + [Ag(NH_3)^+] + [Ag(NH_3)_2{}^+] + [AgOH] + [Ag(OH)_2{}^-].$$

[1] Janik Bjerrum, Danish chemist. Among other discoveries, he demonstrated that monodentate ligands invariably become fixed to the central metallic ion through successive steps. He also devoted much work to the electrostatic forces that operate within solutions.

Indeed, nothing has yet precipitated and ammonia has not been sufficiently added for the complex's concentration to be significant.

Setting up these two hypotheses is sufficient to solve the mathematical system easily. In these conditions, we find

$N = 9.14 \times 10^{-6}$ mol/L (concentration of ammonia added in order to obtain the beginning of precipitation),

$[NH_3]_{free} = 3.33 \times 10^{-8}\,mol/L, \quad [NH_4^+] = 3.85 \times 10^{-6}\,mol/L;$

$[Ag(NH_3)^+] = 5.26 \times 10^{-6}\,mol/L, \quad [Ag(NH_3)_2^+] = 1.18 \times 10^{-9}\,mol/L;$

$[Ag(OH)_2^-] = 3.6 \times 10^{-11}\,mol/L.$

The ammonia concentration must be very weak (about $\approx 10^{-5}$ mol/L) in order to obtain the beginning of the precipitation.

The calculation of the concentration of ammonia necessary to dissolve the last trace of silver oxide having precipitated in the preceding solution is carried out with the same relations, but with the assumptions that

- hypothesis 3: $[Ag(NH_3)_2^+] = 10^{-1}$ mol/L;
- hypothesis 4: $[H^+]$ is negligible (alkaline medium due to the addition of ammonia);
- hypothesis 5: $[Ag(OH)_2^-] \ll [Ag(NH_3)_2^+]$.

With these conditions, we obtain

$$[NH_3]_{free} = 1.80 \times 10^{-2}\,mol/L, \quad [NH_4^+] = 6.43 \times 10^{-4}\,mol/L,$$

$$[OH^-] = 6.39 \times 10^{-4}\,mol/L, \quad [Ag^+] = 2.90 \times 10^{-5}\,mol/L,$$

$$[AgOH] = 3.70 \times 10^{-4}\,mol/L, \quad [Ag(NH_3)^+] = 8.25 \times 10^{-4}\,mol/L,$$

$$[Ag(OH)_2^-] = 1.24 \times 10^{-7}\,mol/L, \quad [Ag(NH_3)_2^+] = 0.10\,mol/L,$$

$$[H^+] = 1.60 \times 10^{-11}\,mol/L.$$

The number of moles of ammonia that must be added in order to dissolve Ag_2O totally is $N = 0.22$ mol. This number is a little more than two times the initial concentration of Ag^+ ions. This is easily explained by the fact that there is the quasi-exclusive formation of the $[Ag(NH_3)_2^+]$ complex.

35.3.3 Precipitation of Metallic Sulfides

The precipitation of metallic sulfides is often accompanied by the formation of varied complexes. Thus, their solubility may differ, even considerably, from that predicted by only taking the solubility products into account.

Let's consider the solubility of mercuric sulfide in a 10^{-1} mol/L hydrochloric solution. Concerning this example, we recall that the systematic research of cations in a solution or a powder requires a precipitation by H_2S in hydrochloric acid solution. This is the case in order to detect Hg^{2+} and also Sn^{2+} and Sb^{III}.

Mercuric sulfide (black sulfide) gives a complex with sulfide ion S^{2-}. It is the dithiomercurate(II) ion (thiomercuric anion) $[HgS_2]^{2-}$, which forms according to the reaction

$$HgS_{(s)} + S^{2-} \rightleftharpoons [HgS_2]^{2-}.$$

This complex may be protonated:

$$[HgS_2]^{2-} + 2H^+ \rightleftharpoons [Hg(SH)_2]$$

Mercuric ions also undergo hydrolysis by giving hydroxo complexes $[Hg(OH)]^+$ and $[Hg(OH)_2]$ (see Chap. 25). They also form chlorinated complexes: $[HgCl]^+$, $[HgCl_2]$, $[HgCl_3]^-$, and $[HgCl_4]^{2-}$.

The relations that must be satisfied are

$$\frac{([H^+][HS^-])}{[H_2S]} = 1.0 \times 10^{-7}, \tag{35.1}$$

$$\frac{([H^+][S^{2-}])}{[HS^-]} = 1.2 \times 10^{-13}, \tag{35.2}$$

$$\frac{[HgS_2{}^{2-}]}{[S^{2-}]} = 0.58, \tag{35.3}$$

$$\frac{[Hg(SH)_2]}{([HgS_2{}^{2-}][H^+]^2)} = 1.0 \times 10^{-7}, \tag{35.4}$$

$$\frac{[Hg(OH)^+]}{([Hg^{2+}][OH^-])} = 2.0 \times 10^{-10}, \tag{35.5}$$

$$\frac{[Hg(OH)_2]}{([Hg(OH)^+][OH^-])} = 2.5 \times 10^{11}, \tag{35.6}$$

$$\frac{[HgCl^+]}{([Hg^{2+}][Cl^-])} = 5.5 \times 10^{6}, \tag{35.7}$$

$$\frac{[HgCl_2]}{([HgCl^+][Cl^-])} = 3.0 \times 10^{6}, \tag{35.8}$$

$$\frac{[HgCl_3{}^-]}{([HgCl_2][Cl^-])} = 7.08, \tag{35.9}$$

$$\frac{[HgCl_4{}^{2-}]}{[HgCl_3{}^-][Cl^-]} = 10.0, \tag{35.10}$$

$$[Hg^{2+}][S^{2-}] = 1.6 \times 10^{-52}. \tag{35.11}$$

The mass balance equations are

- on mercury:

$$\begin{aligned} S = {} & [Hg^{2+}] + [HgS_2{}^{2-}] + [Hg(SH)_2] + [Hg(OH)^+] + [Hg(OH)_2] \\ & + [HgCl^+] + [HgCl_2] + [HgCl_3{}^-] + [HgCl_4{}^{2-}]; \end{aligned} \tag{35.12}$$

- on sulfur:

$$S = [S^{2-}] + [SH^-] + [H_2S] + 2[HgS_2{}^{2-}] + 2[Hg(SH)_2]; \tag{35.13}$$

- on chloride ions:

$$0.10 = [Cl^-] + [HgCl^+] + 2[HgCl_2] + 3[HgCl_3{}^-] + 4[HgCl_4{}^{2-}]. \tag{35.14}$$

The charge balance equation is

$$\begin{aligned} & [Cl^-] + [OH^-] + [SH^-] + 2[S^{2-}] + 2[HgS_2{}^{2-}] + [HgCl_3{}^-] + 2[HgCl_4{}^{2-}] \\ & \quad = [H^+] + [HgCl^+] + [Hg(OH)^+] + 2[Hg^{2+}]. \end{aligned} \tag{35.15}$$

Combined with the three preceding relations, Eq. (35.15) leads to a simpler one:

$$\begin{aligned} & 0.10 + [HgOH^+] + 2[Hg(OH)_2] + [OH^-] \\ & \quad = [H^+] + [SH^-] + 2[H_2S] + 2[Hg(SH)_2]. \end{aligned} \tag{35.16}$$

At first sight, solving this system of equations seems to be an arduous task. However, it is not the case if the following assumptions are made:

- hypothesis 1: the concentrations of all species containing Hg and S are negligible compared to 10^{-1} mol/L. If it is the case, according to relations (35.14) and (35.16):

$$[Cl^-] \approx 0.10\,\text{mol/L},$$

$$[H^+] \approx 0.10\,\text{mol/L};$$

- hypothesis 2: it is also the case for $[Hg^{2+}]$ and $[HgCl^+]$ with respect to $[HgCl_2]$, $[HgCl_3{}^-]$, and $[HgCl_4{}^{2-}]$ since the solution is very rich in chloride ions, so that, according to (35.12), we can write
  $S \approx [HgCl_2] + [HgCl_3{}^-] + [HgCl_4{}^{2-}]$. ($[Hg(OH)^+]$ and $[Hg(OH)_2]$ can be also neglected in (35.12); see just below);
- hypothesis 3: $[Hg(OH)^+]$, $[Hg(OH)_2]$, $[S^{2-}]$, and $[SH^-]$ are negligible due to the acidity of the solution. As a result, according to (35.3), $[HgS_2{}^{2-}]$ is also negligible.

Combining Eqs. (35.1)–(35.4) lets us easily find

$$[Hg(SH)_2] = 7.0 \times 10^{-14}[H_2S].$$

This concentration is also negligible. Finally, according to (35.13), $S = [H_2S]$;

In these conditions, handling the remaining equations leads to the value

$$S = 7.15 \times 10^{-12}\ \text{mol/L}$$

The solubility is very weak. However, if the solubility product $K_s(HgS)$ and the prototropic equilibria of hydrogen sulfide had been taken into account, the calculation would have given $S = 1.15 \times 10^{-17}$ mol/L, which is exceedingly weaker than the preceding value. The difference between both results is due essentially to the formation of the $[HgCl_2]$, $[HgCl_3{}^-]$, and $[HgCl_4{}^{2-}]$ complexes. Thus, chloride ions may be considered the "disturbing ligand."

Qualitatively, according to these calculations, we can assert that the occurring reactions are

$$Hg^{2+} + S^{2-} \rightleftharpoons HgS_{(s)},$$

$$Hg^{2+} + 2Cl^- \rightleftharpoons [HgCl_2],$$

$$[HgCl_2] + Cl^- \rightleftharpoons [HgCl_3{}^-],$$

$$[HgCl_3{}^-] + Cl^- \rightleftharpoons [HgCl_4{}^{2-}],$$

$$S^{2-} + 2H^+ \rightleftharpoons H_2S.$$

The concentration values of all the solution species are

$[Cl^-] = 0.10$ mol/L

$[H^+] = 0.10$ mol/L

$[H_2S] = 7.15 \times 10^{-12}$ mol/L

$[HS^-] = 7.15 \times 10^{-18}$ mol/L

$[S^{2-}] = 8.60 \times 10^{-30}$ mol/L

$[Hg^{2+}] = 1.86 \times 10^{-23}$ mol/L

$[HgS_2{}^{2-}] = 5.00 \times 10^{-30}$ mol/L

$[Hg(SH)_2] = 5.00 \times 10^{-25}$ mol/L,

$[OH^-] = 1.00 \times 10^{-13}$ mol/L,

$[Hg(OH)^+] = 3.70 \times 10^{-26}$ mol/L,

$[Hg(OH)_2] = 9.30 \times 10^{-28}$ mol/L,

$[HgCl^+] = 1.00 \times 10^{-17}$ mol/L,

$[HgCl_2] = 3.00 \times 10^{-12}$ mol/L,

$[HgCl_3{}^-] = 2.10 \times 10^{-12}$ mol/L,

$[HgCl_4{}^{2-}] = 2.10 \times 10^{-12}$ mol/L.

These values have been calculated with $S = 7.15 \times 10^{-12}$ mol/L, a value obtained above. They are in accordance with the starting assumptions.

Another interesting calculation is that of the solubility S of HgS in a solution of 10^{-1} mol/L sodium sulfide Na_2S. From a qualitative standpoint, mercuric sulfide dissolves into a sulfide solution by formation of the $[HgS_2]^{2-}$ complex according to the reaction

$$HgS\downarrow + S^{2-} \rightleftharpoons [HgS_2]^{2-}.$$

It is one of the characteristic reactions of Hg^{2+}.

The relations satisfied are the same as those earlier except for the mass balance on mercury, which leads to the relation

$$S = [Hg^{2+}] + [HgS_2{}^{2-}] + [Hg(SH)_2] + [Hg(OH)^+] + [Hg(OH)_2]$$

since there are no longer chloride ions in the solution. The charge balance becomes

$$[Na^+] + [H^+] + 2[Hg^{2+}] + [Hg(OH)^+] = [OH^-] + 2[S^{2-}] + [SH^-] + 2[HgS_2{}^{2-}].$$

The simplifying assumptions are based on the following reasoning: $[S^{2-}]$ is probably around 10^{-2}–10^{-1} mol/L since the solution is very basic. As an immediate result, $[Hg^{2+}]$ is unbelievably weak according to the solubility–product value $K_s(HgS) = 1.6 \times 10^{-52}$, as is also the case for $[Hg(OH)^+]$, $[Hg(OH)_2]$, and $[Hg(SH)_2]$ for the same reason. The value found by starting with these assumptions is $S = 1.80 \times 10^{-2}$ mol/L. From a practical standpoint, this value is sufficiently high to separate HgS from other metallic sulfides by dissolving it into an Na_2S solution.

Other calculated concentrations are

$$[HgS_2{}^{2-}] = 1.80 \times 10^{-2} \text{ mol/L} \qquad [H_2S] = 1.00 \times 10^{-7} \text{ mol/L},$$

$$[S^{2-}] = 3.10 \times 10^{-2} \text{ mol/L} \qquad [Hg(SH)_2] = 6.90 \times 10^{-21} \text{ mol/L},$$

$$[SH^-] = 5.10 \times 10^{-2} \text{ mol/L} \qquad [Hg^{2+}] = 5.15 \times 10^{-51} \text{ mol/L},$$

$$[OH^-] = 5.10 \times 10^{-2} \text{ mol/L} \qquad [Hg(OH)^+] = 5.30 \times 10^{-43} \text{ mol/L},$$

$$[H^+] = 1.96 \times 10^{-13} \text{ mol/L} \qquad [Hg(OH)_2] = 6.70 \times 10^{-34} \text{ mol/L}.$$

They are in accordance with the starting hypothesis. We must notice the inconceivably weak value

$$[Hg^{2+}] = 5.15 \times 10^{-51} \text{ mol/L}.$$

35.4 Destruction of a Complex by Formation of a Precipitate

Let's consider the complex ML_n and investigate the case in which the metal cation engaged within it precipitates with some reagent by formation of the species $MY_{(s)}$ according to the reaction

$$ML_n + Y \overset{K}{\rightleftharpoons} MY_{(s)} + nL$$

The equilibria of the dissociation of the complex and of the formation of the precipitate are

$$ML_n \rightleftharpoons M^{n+} + nL \quad K_D = \frac{1}{\beta_n} = \frac{\{[M^{n+}][L]^n\}}{[ML_n]},$$

and

$$M + Y \rightleftharpoons MY_{(s)} \quad \frac{1}{K_s} = \frac{[MY_{(s)}]}{\{[M][Y]\}} = \frac{1}{\{[M][Y]\}}.$$

The global constant K of the destruction reaction of the complex is

$$K = \frac{\{[MY_{(s)}][L]^n\}}{\{[ML_n][Y]\}}.$$

Since $[MY(s)] = 1$ (in terms of activities, $MY_{(s)}$ constitutes a pure solid phase),

$$K = \frac{[L]^n}{\{[ML_n][Y]\}},$$

$$\text{or} \quad K = \frac{K_D}{K_s},$$

$$K = \frac{1}{(\beta_n K_s)}.$$

It appears that, according to the global constant K value and the ligand and precipitating reagent concentrations, the complex ML_n may be dissociated by precipitation of the species MY. This is the inverse of the dissolution of a precipitate by complexation.

An example is provided by the diamminesilver(I) complex from which the precipitation of silver chloride is possible.

Both equilibria that are superimposing are

- $[Ag(NH_3)_2]^+ \rightleftharpoons Ag^+ + 2NH_3$,

$$\text{with} \quad \frac{1}{\beta_2} = \frac{\{[Ag^+][NH_3]_2\}}{[Ag(NH_3)_2{}^+]} = 10^{-7.03},$$

- $Ag^+ + Cl^- \rightleftharpoons AgCl_{(s)}$,

$$\frac{1}{K_{(s)}} = 10^{9.75}.$$

The destruction reaction of the complex is

$$[Ag(NH_3)_2]^+ + Cl^- \overset{K}{\rightleftharpoons} AgCl_{(s)} + 2NH_3.$$

It gives the equilibrium constant $K = 10^{2.72}$, which is not too high. It allows the reaction to occur in both directions, which actually takes place. The part played by the ligand L's concentration to determine the direction of the reaction is illustrated by the following two experiences:

- let's consider a solution where the diamminesilver(I) complex is dissolved at a 10^{-2} mol/L concentration. (In the following study, we shall systematically neglect the monoamminesilver(I) complex, which is in the minority in any case—see above.) The dissociation constant of the former complex is $1/\beta_2 = 10^{-7.03}$. The following question arises: What chloride ion concentration induces the beginning of the silver chloride precipitation? A simplified reasoning provides the answer, as follows. Due to the $1/\beta_2$ value, the dissociation of the complex is very weak. We can also consider that the precipitation occurs as soon as we add chloride ions, because of the very weak value of K_s(AgCl). This hypothesis remains to be checked. In these conditions, we can write

$$[Ag(NH_3)_2{}^+] \approx 10^{-2}\ \text{mol/L} \quad \text{(at the beginning of the precipitation).}$$

As a result, $[Ag^+][NH_3]^2 = 10^{-9.03}$. By dissociation, the complex gives very weak concentrations of Ag^+ and NH_3, ε and 2ε. Therefore,

$$\varepsilon(2\varepsilon)^2 = 10^{-9.03},$$

$$[Ag^+] = 10^{-3.20}\ \text{mol/L}.$$

With this result in mind, we find

$$[Cl^-] = \frac{K_s}{[Ag^+]}$$

$$[Cl^-] = 10^{-6.5}\ \text{mol/L}.$$

Ag^+ precipitates immediately.

From the rigorous calculation based on the following relations:

$$2[Ag(NH_3)_2{}^+] + [NH_3] = 2 \times 10^{-2}\ \text{mol/L}$$

$$[Ag(NH_3)_2{}^+] + [Ag^+] = 10^{-2}\ \text{mol/L}$$

$$[Ag^+][Cl^-] = K_s$$

$$\frac{[Ag(NH_3)_2{}^+]}{\{[Ag^+][NH_3]^2\}} = \beta_2$$

we immediately verify that $[Ag(NH_3)_2^+] \approx 10^{-2}$ mol/L. It is important to notice that this calculation strictly concerns the beginning of the precipitation, because on the one hand, it implicates the solubility product, and on the other hand, it does not take into account any precipitate in the mass balance equations.

Of course, the presence of ammonia favors complex formation. The destruction of the latter by precipitation becomes more difficult than without the addition of ammonia. Let's add ammonia to the preceding solution in such a way that its concentration is 1 mol/L. What concentration of chloride ions must be added in order to attain the beginning of the silver chloride precipitation? By following the previous reasoning, we find

$$[Ag^+] \approx 10^{-9.03}\,\text{mol/L},$$

$$[Cl^-] = \frac{10^{-9.75}}{10^{-9.03}},$$

$$[Cl^-] = 0.19\,\text{mol/L}.$$

The concentration of added chloride ions must be considerably higher than in the previous case.

- It is interesting to investigate the inverse of the phenomenon of destruction of a complex by precipitation by taking the same chemical species as we did previously. Suppose that we want to dissolve 10^{-2} mol of silver chloride into 1 L of an ammonia solution. What must the ammonia concentration be in order to reach this goal? We can write

$[Ag^+][Cl^-] = 10^{-9.75}$ (relation still legitimate exactly at the end , of the dissolution)

$[Cl^-] = 10^{-2}$ mol/L (requirement of the problem),

$[Ag(NH_3)_2{}^+] \approx 10^{-2}$ mol/L (requirement of the problem),

$$\frac{[Ag(NH_3)_2{}^+]}{\{[Ag^+][NH_3]^2\}} = \beta_2.$$

We find

$$[Ag^+] = 10^{-7.75}\ \text{mol/L},$$

$$[NH_3] = 0.23\ \text{mol/L}.$$

We might find a strictly identical result by using the relation given in Sect. 35.2 devoted to the dissolution of a precipitate by complexation.

From a general standpoint and considering the expression giving the global constant:

$$K = \frac{1}{(\beta_n K_s)},$$

it immediately appears that

- the more stable the complex is, the more difficult its dissolution by complexation is;
- the more poorly soluble the forming product is (K_s small), the more easily the complex is destroyed, and inversely. This is the reason why the dissolution of AgBr ($K_s = 10^{-12.3}$) in an ammonia solution is more difficult than that of AgCl ($K_s = 10^{-9.75}$), whereas silver iodide ($K_s = 10^{-16}$) is only very poorly soluble into ammonia.

35.5 Separation by Complexation and Precipitation

The superposition of complexation and precipitation phenomena permits us to proceed to separations.

A simple example is provided by the separation of Cd^{2+} from Cu^{I}. Both ions give a soluble complex with cyanide ions. They are tetracyanocadmiate(II) $[Cd(CN)_4]^{2-}$ and tetracyanocuprate(I) $[Cu(CN)_4]^{3-}$;

$$Cd^{2+} + 4CN^- \rightleftharpoons [Cd(CN)_4]^{2-} \quad \beta_4 = 10^{16.9},$$

$$Cu^+ + 4CN^- \rightleftharpoons [Cu(CN)_4]^{3-} \quad \beta_4 = 10^{27.3}.$$

Let's first recall that the formation of cyanocomplexes of metal cations constitutes a means to separate some of them from others (see Chap. 29). This is the case for $Cd^{2+}$and Cu^{I}. Disulfide ions may precipitate Cd^{2+} from tetracyanocadmiate, but this is not the case for Cu^{I}, as explained by the constant values of the following two equilibria:

$$[Cd(CN)_4]^{2-} + S^{2-} \rightleftharpoons CdS\downarrow + 4CN^- \quad K = 10^{11.5},$$

$$2[Cu(CN)_4]^{3-} + S^{2-} \rightleftharpoons Cu_2S\downarrow + 8CN^- \quad K = 10^{-7.9}.$$

A second example is provided by the separation of the elements of the Al^{3+} group from those of the Zn^{2+} group. At pH = 9, hydroxides of the ions Al^{3+}, Fe^{3+}, Be^{2+}, Cr^{3+}, Ti^{IV}, Zr^{IV}, Ce^{III}, Th^{IV}, No^{V}, Ta^{V}, and Ga^{3+} are poorly soluble. It is the same state of affairs for the hydroxides of the following ions: Zn^{2+}, Mn^{2+}, Ni^{2+}, Co^{2+}, and Cd^{2+}. In the presence of ammonia, only the elements of the first series precipitate. Those of the second series do not since they are complexed essentially as ammine complexes $[M(NH_3)_2]^{n+}$. Parameters governing the separations are the solubility products, the pH value, the complex formation constants, and, of course, the metallic ion and ammonia concentrations.

A third example concerns the separation of Ti^{IV}, Be^{2+}, U^{VI}, Nb^{V}, and Ta^{V} from Hg^{2+}, Pb^{2+}, Bi^{III}, Cu^{2+}, Cd^{2+}, Fe^{3+}, Al^{3+}, Cr^{3+}, Ni^{2+}, Co^{2+}, Mn^{2+}, Zn^{2+}, W^{VI}, Th^{IV}, Ce^{III}, Ca^{2+}, Sr^{2+}, Ba^{2+}, and Mg^{2+}. The first elements precipitate as hydroxides in ammonia in the presence of EDTA. Of course, the hydroxide ions necessary for the precipitation are due to the basic character of ammonia. We can infer from these experimental facts that the presence of EDTA does not preclude their precipitation.

In other words, EDTA is not a sufficiently powerful complexing agent to dissolve the hydroxides of these metal cations according to the reaction (charges omitted)

$$M(OH)_n + Y \longrightarrow MY + nOH \quad \text{(charges omitted for simplicity)}.$$

It is the inverse phenomenon for the second series of metal cations.

It is quite evident that in the latter example, the fact that a reaction may occur with EDTA brings a supplementary selectivity factor.

The last example is that given by Mo^{VI}, W^{VI}, V^{V}, and Ti^{V}, which precipitate with oxine in the acetic buffer. It is not the case with Fe^{3+}, Al^{3+}, Be^{2+}, Zn^{2+}, Ni^{2+}, Co^{2+}, Mn^{2+}, Bi^{III}, Pb^{2+}, Cd^{2+},Cu^{2+}, Hg^{2+}, and V^{IV}.

Chapter 36
Theoretical Study of Some Precipitation Titration Curves

In this chapter, we study some titration curves involving a precipitation from a theoretical standpoint. This chapter is an introduction to the so-called Charpentier–Volhard, Mohr, and Fajans methods, which all involve standard solutions of silver nitrate. Liebig-Denigés' method, which also involves such silver nitrate solutions, will be considered in the next chapter.

As in the studies of other sorts of titration curves, those investigated here are carried out to determine the curve's shape in the vicinity of the endpoint in order to choose the best experimental conditions to detect it.

36.1 Case of a Symmetric Titration: Titration of A Halide by Silver Ions and Inversely

We call *symmetric titrations* those putting ions that bring the same (but opposite) electrical charge face to face, such as Ag^+ and Cl^- ions, for example. We also want to make clear that in this chapter, fluoride ions are excluded when we speak of halides.

Let's consider the titration of the halide X^- by a solution of silver nitrate. The formalism used is given in Fig. 36.1.

Let us call C_o and V_o the initial concentration (mol/L) and volume (ml) of the halide solution, respectively, C the concentration of the titrant solution, and v its added volume at each stage of the titration.

36.1.1 Titration Reaction

The titration reaction is

$$X^- + Ag^+ \rightleftharpoons AgX\downarrow \quad 1/K_s.$$

Its equilibrium constant is the inverse of the solubility product, once the precipitate does appear.

J.-L. Burgot, *Ionic Equilibria in Analytical Chemistry*,
DOI 10.1007/978-1-4419-8382-4_36, © Springer Science+Business Media, LLC 2012

Fig. 36.1 Titration of halide ion X^- by silver nitrate: symbolism

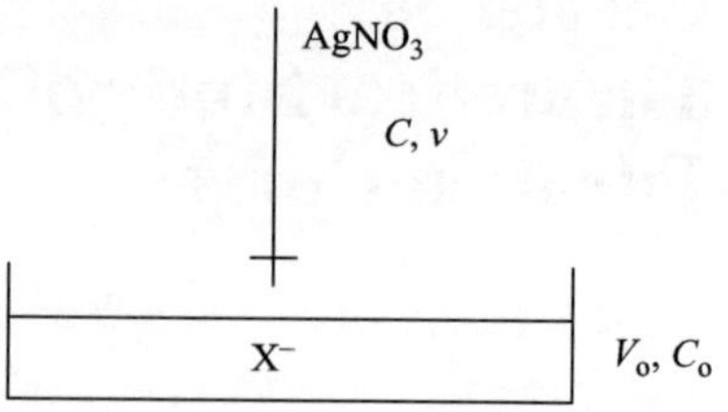

First, we must demonstrate that there is precipitation of AgX once the first drop of silver nitrate solution has been added. In order to do that, we can seek the added volume v for which the precipitation just begins. Let's set $C_o = 10^{-1}$ mol/L, $C = 10^{-1}$ mol/L, and $V_o = 10^{-1}$ L with the example of chloride ions.

At the very beginning of the precipitation, the following relations must be satisfied:

$$[\mathrm{AgCl}]_{\mathrm{solution}} = K_{s1} = 10^{-6.66}\ \mathrm{mol/L}$$

and

$$[\mathrm{AgCl}]_{\mathrm{solution}} = Cv/(V_o + v).$$

The product Cv is the mole number of Ag^+ ions added at saturation. By solving these relations, we find $v \approx 10^{-3.7}$ ml. The volume of a drop is higher than $10^{-3.7}$ ml. Hence, the titration reaction is indeed, from the beginning until the end, that of the silver halide precipitation.

Due to the very weak values of silver halide's solubility products, titration reactions are quasi-totally displaced toward the right. This condition, which is required to carry out a good titration, is obeyed.

36.1.2 General Equation of the Titration Curve

At any stage of the titration curve, the mass balance relations are

$$[\mathrm{Ag}^+] + P = CV/\,(V_o + v)\,, \tag{36.1}$$

$$[\mathrm{X}^-] + P = C_oV_o/\,(V_o + v)\,, \tag{36.2}$$

where P is the "concentration" in mol/L of precipitated silver halide. [In Eqs. (36.1) and (36.2), we have deliberately neglected $[AgX]_{solution}$, which is actually negligible. Taking this concentration into account would not preclude solving the system from a mathematical standpoint since then the new relation $[AgX]_{solution}$ = constant would be at our disposal.] The solubility product is reached:

$$[\mathrm{Ag}^+][\mathrm{X}^-] = K_s \tag{36.3}$$

From a mathematical standpoint, the system is soluble since it is composed of three simultaneous equations in three unknowns: $[Ag^+]$, $[X^-]$, and P. It is worth noting

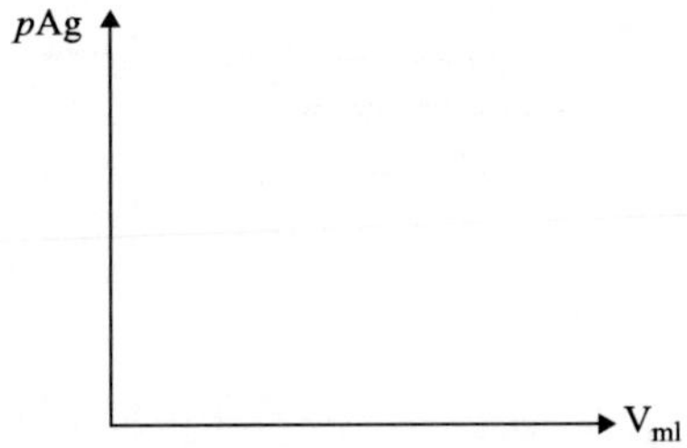

Fig. 36.2 Diagram drawn during the titration of X^- by Ag^+

that we did not use the ionic product of water or the charge balance. Although they are always satisfied, they are of no use if no acid–base phenomenon occurs. This is the case here: No pH variation occurs during the titration since X^- is a null base in water.

Usually, in these kinds of titrations, the quantity that is experimentally measured as a function of the volume v of titrant solution added is $p\text{Ag} = -\log[\text{Ag}^+]$ (Fig. 36.2). A good reason to choose pAg as the dependent variable lies in the fact that it is directly given by a silver electrode. [To be rigorous from a physical point of view, pAg (whose value is given by a silver electrode) is actually $-\log(\text{Ag}^+)$ and not $-\log[\text{Ag}^+]$ (furthermore, from the mathematical standpoint, the argument of a logarithm must be a dimensionless number).]

The general equation of the titration curve is found by expressing P from Eq. (36.2) and by replacing it in (36.1) with its expression. This gives

$$[\text{Ag}^+] - [\text{X}^-] = (Cv - C_oV_o)/(V_o + v).$$

By replacing $[\text{X}^-]$ with its expression coming from the solubility product, we find the general expression

$$[\text{Ag}^+] - K_s/[\text{Ag}^+] = (Cv - C_oV_o)/(V_o + v). \tag{36.4}$$

It is absolutely general. It is obeyed at each stage of the titration. It is strongly similar to that obtained for the titration of a strong base with a strong acid (see Chap. 9). In this case, the titration curve was

$$[\text{H}^+] - K_w/[\text{H}^+] = (Cv - C_oV_o)/(V_o + v).$$

Therefore, it will be not surprising to find analogous developments and results concerning these two kinds of titrations in the following developments.

By introducing the fraction titrated $\varphi = Cv/C_oV_o$, we find

$$[\text{Ag}^+] - K_s/[\text{Ag}^+] = (\varphi - 1)C_oV_o/(V_o + v). \tag{36.5}$$

At the equivalence point, by definition, $Cv = C_oV_o$ ($\varphi = 1$):

$$[\text{Ag}^+] = K_s/[\text{Ag}^+], \quad \varphi = 1 \quad \text{(equivalence point)},$$

$$[\text{Ag}^+] = K_s^{1/2},$$

$$[\text{Ag}^+] = 1.33 \times 10^{-5}\ \text{mol/L},$$

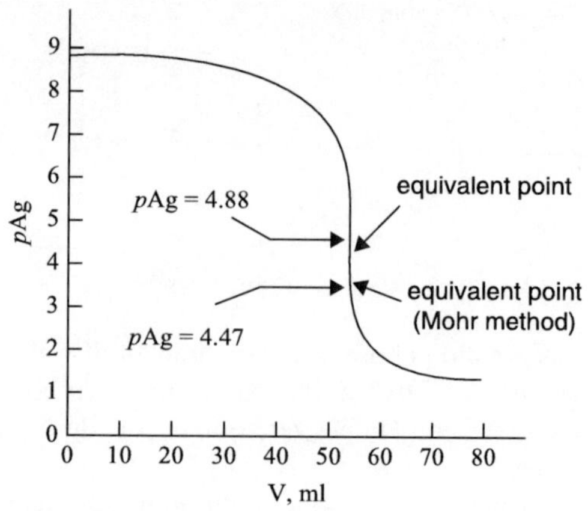

Fig. 36.3 Titration curve of a sodium chloride solution by a silver nitrate solution ($V_o = 50$ ml, $C_o = C = 10^{-1}$ mol/L)

pAg = 4.88 (equivalence point for the titration of chloride ions). This is the value already found for the solubility of silver chloride in pure water after having neglected the molecular solubility $[AgCl]_{solution}$.

36.1.3 *Shape of the Titration Curve*

The titration curve obtained in the titration of 50 ml of 10^{-1} mol/L sodium chloride with a 10^{-1} mol/L silver nitrate solution is reproduced in Fig. 36.3.

The curve is experimentally obtained with a silver electrode that directly gives pAg, or it can be calculated using Eqs. (36.4) and (36.5) by mixing activities and concentrations. The important point to be stressed here is the abrupt change in pAg close to the equivalence point.

36.1.4 *Simplified Equations of the Titration Curve*

As with other kinds of titration curves, the general titration equation may be simplified in certain conditions. The curve may be divided into three parts: before, at, and after the equivalence point.

- Before the equivalence point, we can consider that there is a large excess of halide ions compared to silver ions and hence that $[Ag^+]$ is negligible with respect to P in Eq. (36.1). As a result,

  $$P = Cv/(V_o + v),$$

  and according to Eq. (36.3),

  $$[X^-] = (C_oV_o - Cv)/(V_o + v),$$

$$K_s/[Ag^+] = (C_oV_o - Cv)/(V_o + v), \tag{36.6}$$

$$K_s/[Ag^+] = (1 - \varphi)C_oV_o/(V_o + v), \quad \varphi < 1, \tag{36.7}$$

$$\text{or} \quad pAg = \log(1 - \varphi) + \log[(C_oV_o)/(V_o + v)] + pKs, \quad \varphi < 1. \tag{36.8}$$

Some authors go further into approximations: They assume that $v \ll V_o$. This implies that the titrant solution is very concentrated and the titrand solution very diluted. In these conditions, the titration curve equation before the equivalence point becomes

$$pAg = \log(1 - \varphi) + \log C_o + pKs, \quad \varphi < 1 \quad \text{and} \quad v \ll V_o. \tag{36.8'}$$

- at the equivalence point, $[Ag^+] = [X^-]$. No decent simplification can be done. The general equation must be used.
- after the equivalence point, we can consider that $[X^-]$ is negligible compared to P in Eq. (36.2). As a result,

$$P = C_oV_o/(V_o + v),$$

$$[Ag^+] = (Cv - C_oV_o)(V_o + v),$$

$$pAg = -\log(\varphi - 1) - \log[C_oV_o/(V_o + v)], \tag{36.9}$$

$$\text{and if} \quad v \ll V_o,$$

$$pAg = -\log(\varphi - 1) - \log C_o, \quad \varphi > 1 \quad \text{and} \quad v \ll \text{Vo}. \tag{36.9'}$$

The drawback of successively neglecting $[Ag^+]$ compared to P and after $[X^-]$ compared to P lies in the fact that such simplifications are no longer possible close to the equivalence point. Indeed, at this stage of precipitation, both concentrations $[Ag^+]$ and $[X^-]$ are of the same order of magnitude. Equations (36.8) and (36.8′) can no longer be applied just before the equivalence point, and it is the same for Eqs. (36.9) and (36.9′) just after. This is a serious handicap because the most interesting points of a titration curve are precisely those located close to the equivalence point.

36.1.5 Titration Error

Let $\varphi_{p.f.}$ be the fraction titrated at the final point. By inserting it into Eq. (36.5), which is legitimate for all points on the curve (and hence for the final point), we find

$$C_oV_o(\varphi_{p.f.} - 1)/(V_o + v_{p.f.}) = [Ag^+]_{p.f.} - K_s/[Ag^+]_{p.f.} \tag{36.10}$$

$(\varphi_{p.f.} - 1)$ is actually the absolute titration error expressed in fraction titrated, since at the equivalence point, $\varphi_{eq.p.} = 1$. $(\varphi_{p.f.} - 1)$ is also the relative titration error since

$$\varphi_{p.f.} - 1 = (\varphi_{p.f.} - 1)/\varphi_{eq.p.} = (\varphi_{p.f.} - 1)/1 \quad \text{(relative titration error)}.$$

For the sake of easy handling, Eq. (36.10) can be modified as usual. One may consider that the titration error is weak and as a consequence, from a numerical standpoint, we can consider that

$$Cv_{\text{p.f.}} \approx C_o V_o,$$

from which $v_{\text{p.f.}} \approx C_o V_o / C$. By inserting this expression into (36.10), we find

$$\varphi_{\text{p.f.}} - 1 \approx [(C_o + C)/C_o C]\left\{[\text{Ag}^+]_{\text{p.f.}} - K_s/[\text{Ag}^+]_{\text{p.f.}}\right\}. \qquad (36.11)$$

The titration error is easily calculated once the concentration $[\text{Ag}^+]_{\text{p.f.}}$ is known. The result is absolutely analogous to that obtained for the titration of a strong base with a strong acid.

Equations (36.10) and (36.11) can be used for titrations in which the two antagonistic ions bring the same electrical charge, for example, $Hg_2{}^{2+}$ with $SO_4{}^{2-}$, Pb^{2+} with $SO_4{}^{2-}$, Ag^+ with SCN^-, and so on. It is because the titration curve equation is the same for all these titrations.

36.1.6 Inflection Point of the Titration Curve

Very close to the equivalence point, the titration curve equation can be written

$$\varphi - 1 \approx [(C_o + C)/C_o C]\left\{[\text{Ag}^+] - K_s/[\text{Ag}^+]\right\}. \qquad (36.12)$$

Comparatively to Eq. (36.5) (which is absolutely general), the simplification carried out to obtain (36.12) consisted of setting

$$v \approx C_o V_o / C.$$

This relation can only be applied to points very close to the equivalence point ($\varphi \approx 1$).

The slope of the titration curve ($dp\text{Ag}/d\varphi$) is

$$dp\text{Ag}/d\varphi = -d(\log[\text{Ag}^+])/d\varphi = -(1/2.303)d(\ln[\text{Ag}^+])/d\varphi,$$

and after applying the chain derivation rule, we have

$$dp\text{Ag}/d\varphi = -(1/2.303)(d(\ln[Ag^+])/d[Ag^+])(d[\text{Ag}^+]/d\varphi),$$

$$dp\text{Ag}/d\varphi = -0.434(1/[\text{Ag}^+])(d[\text{Ag}^+]/d\varphi).$$

The derivation of both members of (36.12) with respect to φ permits us to express $d[\text{Ag}^+]/d\varphi$ and finally leads to

$$dp\text{Ag}/d\varphi = -0.434 C_o C/(C + C_o)\left\{[\text{Ag}^+] + K_s/[\text{Ag}^+]\right\}^{-1}. \qquad (36.13)$$

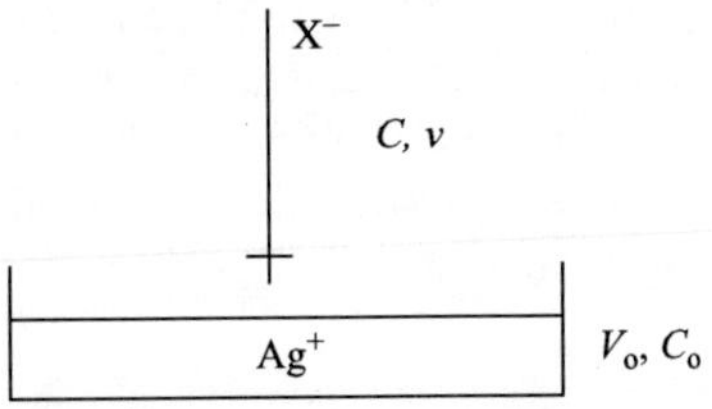

Fig. 36.4 Parameters of a titration experiment of a silver ion by a halide solution

The curve's inflection point must be such that the slope should be maximal; that is,

$$d\{(dp\text{Ag}/d\varphi)/d\varphi\} = 0.$$

The derivation of Eq. (36.13) with respect to φ gives the expression

$$d^2p\text{Ag}/d\varphi^2 = 0.434[C_oC/(C + C_o)]\left\{1 - K_s/[\text{Ag}^+]^2\right\}\{d[\text{Ag}^+]/d\varphi\}.$$

It vanishes for $[\text{Ag}^+]^2 = K_s$, that is, exactly at the equivalence point.

Therefore, we can assert that in a symmetric precipitation titration, the equivalence point is located exactly at the inflection point of the titration curve.

So far, the titration that we have just studied corresponds exactly to Mohr's titration (see Chap. 37).

36.1.7 *Inverse Titration*

Now, we will briefly investigate the titration of silver ion by a halide solution. The parameters of the titration are given in Fig. 36.4.

It is easy to demonstrate that the general equation of the titration curve is

$$[\text{Ag}^+] - K_s/[\text{Ag}^+] = (C_oV_o - Cv)/(V_o + v)$$

$$[\text{Ag}^+] - K_s/[\text{Ag}^+] = (1 - \varphi)C_oV_o/(V_o + v).$$

We can notice that compared to Eq. (36.5), we find the term $(1 - \varphi)$ instead of $(\varphi - 1)$. Equivalently, these two relations can be written as

$$K_s/[\text{Ag}^+] - [\text{Ag}^+] = (Cv - C_oV_o)/(V_o + v),$$

$$K_s/[\text{Ag}^+] - [\text{Ag}^+] = (\varphi - 1)C_oV_o/(V_o + v).$$

The titration error is given by the relation

$$(\varphi_{\text{p.f.}} - 1) = [(C_o + C)/C_oC]\left\{K_s/[\text{Ag}^+]_{\text{p.f.}} - [\text{Ag}^+]_{\text{p.f.}}\right\}. \qquad (36.12)$$

It is found by using the approximate relation

$$V_{\text{p.f.}} \approx C_oV_o/C,$$

which implies, actually, that the error is weak. It is easy to demonstrate that the inflection and equivalence points are exactly superimposed. As previously, this is only true when we admit the hypothesis that $v_{\text{p.f.}} \approx C_o V_o / C$, that is, for $\varphi \approx 1$.

These theoretical considerations correspond to the last stage of Charpentier–Volhard's method with $X^- = SCN^-$.

36.2 Dissymmetric Titrations

Dissymmetric titrations lead to the formation of dissymmetic poorly soluble salts, such as Ag_2SO_4. Let's consider the titration of sulfate ions with Ag^+ ions according to the reaction

$$SO_4{}^{2-} + 2Ag^+ \rightarrow Ag_2SO_4\downarrow$$

Silver sulfate exhibits the solubility product $K_s(Ag_2SO_4) = 10^{-4.80}$. The rigorous general equation of the titration curve is obtained from the following three relations:

$$[SO_4{}^{2-}] + P = C_o V_o/(V_o + v),$$

$$[Ag^+] + 2P = CV/(V_o + v),$$

$$[Ag^+]^2\,[SO_4{}^{2-}] = K_s,$$

with the same symbolism as earlier. P is the "concentration" (mol/L) of the precipitated sulfate. Since for one mole of precipitated sulfate two moles of silver are precipitated, the factor 2 intervenes in the mass balance of silver equations. The following equation comes immediately:

$$[Ag^+] - 2K_s/[Ag^+]^2 = (Cv - 2C_o V_o)/(V_o + v),$$

$$[Ag^+] - 2K_s/[Ag^+]^2 = (\varphi - 1)2C_o V_o/(V_o + v),$$

where φ (fraction titrated) is given by the relation

$$\varphi = Cv/2C_o V_o.$$

Indeed, at the equivalence point, $\varphi = 1$, we must have added a silver ion concentration that is twice as important that of the $SO_4{}^{2-}$ ions.

Close to the equivalence point, that is, at the final point,

$$v_{\text{p.f.}} \approx 2C_o V_o/C.$$

As a result, the relation giving the titration error is

$$\varphi_{\text{p.f.}} - 1 = \{(C + 2C_o)/2C_o C\}\left\{[Ag^+]_{\text{p.f.}} - 2K_s/[Ag^+]_{\text{p.f.}}\right\}.$$

A calculation quite analogous to that already carried out above leads to the following relation, which expresses the titration curve slope close to the equivalence point ($\varphi \approx 1$):

$$dp\mathrm{Ag}/d\varphi = -0.434[2C_oC/(2C_o + C)]\left\{[\mathrm{Ag}^+] + 4K_s/[\mathrm{Ag}^+]^2\right\}^{-1}.$$

The second derivative with respect to φ is given by the following relation:

$$d^2p\mathrm{Ag}/d\varphi^2 = 0.434\,[2C_oC/(2C_o + C)]\left\{1 - 8K_s/[\mathrm{Ag}^+]^3\right\}(dp\mathrm{Ag}/d\varphi).$$

This relation is only valid close to the equivalence point, that is, for $\varphi \approx 1$ or $v_{\mathrm{p.f.}} \approx 2C_oV_o/C$. From a strict analytical standpoint, it is interesting since it shows that the inflection point is located for

$$[\mathrm{Ag}^+]^3 = 8K_s, \tag{36.14}$$

whereas the equivalence point is located for $[\mathrm{Ag}^+]^3 = 2K_s$ (see the general equation with $\varphi = 1$). The result is that if we choose the final point at the inflection point (this is frequently done when the equivalence point is detected by potentiometry), a titration error will be committed. It is, of course, a systematic error. For example, when a solution of silver ions at a 1 mol/L concentration is titrated, the inflection point is located 7.5% later than the equivalence point. In this case, taking the endpoint at the point of maximum slope introduces a titration error of 7.5%. This value is immediately calculated by inserting Eq. (36.14) into Eq. (36.13). We find

$$\varphi_{\mathrm{p.f.}} - 1 = \pm(3/2)[(2C_o + C)/2C_oC](K_s)^{1/3}.$$

The plus sign applies to the titration of a divalent ion by a monovalent one. The minus sign would correspond to the inverse titration.

The above relations apply to the case of a dissymmetric titration in which it is the titrant's concentration (more precisely, *p*Ag through a silver electrode) that is measured. Sometimes, in the same sort of titration, we might be interested in the titrand's concentration. This is the case, for example, with the cation dimercury(I) Hg–Hg^{2+} titrated with chloride ions according to the equation

$$\mathrm{Hg_2}^{2+} + 2\mathrm{Cl}^- \rightarrow \mathrm{Hg_2Cl_2}, \quad K_s(\mathrm{Hg_2Cl_2}) = 10^{-17.88}.$$

The titration curve equation is

$$[\mathrm{Cl}^-] - 2K_s/[\mathrm{Cl}^-]^2 = (CV - 2C_oV_o)/(V_o + v)$$

or

$$[\mathrm{Cl}^-] - 2K_s/[\mathrm{Cl}^-]^2 = (\varphi - 1)\,[2C_oV_o/(V_o + v)].$$

The titration error close to the equivalence point is given by the expression

$$\varphi_{\mathrm{p.f.}} - 1 = [(2C_o + C)/2C_oC]\left\{[\mathrm{Cl}^-]_{\mathrm{p.f.}} - 2K_s/[\mathrm{Cl}^-]^2{}_{\mathrm{p.f.}}\right\} \tag{36.13'}$$

Since, by hypothesis, we follow the changes in the titrand's concentration (Hg_2^{2+}) with the volume of added titrant (for example, with an appropriate electrode), we find the relation

$$\varphi_{\text{p.f.}} - 1 = [(2C_o + C)/2C_oC]\left\{K_s{}^{1/2}/[\mathrm{Hg_2}^{2+}]_{\text{p.f.}}{}^{1/2} - 2[\mathrm{Hg_2}^{2+}]_{\text{p.f.}}\right\} \quad (36.15)$$

by using the solubility product

$$K_s = [\mathrm{Hg_2}^{2+}]_{\text{p.f.}}[\mathrm{Cl}^-]^2{}_{\text{p.f.}}.$$

For the inverse dissymmetric titrations (for example, the titration of silver ions with SO_4^{2-} ions or that of chloride ions with Hg_2^{2+} ions), it is sufficient to permute both terms of the differences in Eqs. (36.13), (36.13′), and (36.15) in order to calculate the titration errors.

Chapter 37
Titrimetric Methods Involving a Precipitation

This chapter is essentially devoted to argentometry, in which we find the Charpentier–Volhard, Mohr, Fajans, and Liebig–Denigés methods. They are the most important methods involving a precipitation phenomenon. We shall also say a few words about ferrocyanimetry and uranometry.

37.1 Argentometry

37.1.1 Definitions

This term "argentometry" designates all titrimetric methods that use standard solutions of silver ions Ag^+. The term "argentimetry" designates determination techniques of Ag^+ ions and its salts. Argentometry groups titration methods involving a precipitation as well as those involving the formation of complexes within which Ag^+ is engaged.

37.1.2 Generalities

For our purposes, let's recall that silver forms poorly soluble compounds in water, except for silver nitrate, perchlorate, fluoride, acetate, and chlorate. It forms very insoluble compounds such as silver chloride ($K_s = 10^{-9.75}$), bromide ($K_s = 10^{-12}$), iodide ($K_s = 10^{-16.3}$), and thiocyanate ($K_s = 10^{-11.4}$). It also forms very insoluble derivatives with the cyanide ion CN^-, i.e., AgCN ($K_s = 10^{-14.92}$), and silver dicyanosilverate(I), "silver argentocyanide" ($K_s = 10^{-12}$). It gives weakly soluble salts with some anions such as arsenite, phosphate, and chromate anions.

Argentometry encompasses four great titration methods:

- argentometry in acidic medium: Charpentier–Volhard's method;
- argentometry in weakly acidic or neutral medium: Fajans' method;
- argentometry in neutral or weakly alkaline medium: Mohr's method;
- cyanoargentimetry, or Liebig–Denigés' method.

J.-L. Burgot, *Ionic Equilibria in Analytical Chemistry*,
DOI 10.1007/978-1-4419-8382-4_37, © Springer Science+Business Media, LLC 2012

Standard solutions are prepared by taking into account the fact that the silver ion is monovalent. Thus, the equivalent is 1 mole of Ag^+ ions. The salt used most often to prepare standard solutions is silver nitrate $AgNO_3$. Its solutions can also be prepared in a state of high purity (99.9%) by dissolving silver metal in a nitric acid solution. They are easily stored provided they are protected from light. Usually, they are standardized against

- a thiocyanate solution in the presence of a ferric salt according to Volhard's method (see Sect. 37.2);
- a sodium or potassium chloride solution of warranted purity according to Mohr's method (see Sect. 37.3).

37.2 Argentometry in Acidic Medium: Charpentier–Volhard's Method

37.2.1 *Principle*

Following Charpentier–Volhard's method (1874),[1] a nitric acid solution and a known quantity in excess of silver nitrate solution are added to the medium under study. Silver ions in excess are determined by titration with a standard solution of an alkaline thiocyanate according to the reaction

$$Ag^+ + SCN^- \rightarrow AgSCN\downarrow$$

The indicator is iron alum $Fe_2(SO_4)_3$, $(NH_4)_2SO_4$, $24H_2O$, which, evidently, is a ferric salt. The equivalence point is detected by the appearance of a red color due to the formation of the thiocyanatoiron(III) complex ("thiocyanoferric" complex) $[Fe(SCN)^{2+}]$.

Hence, it is a back titration. The titration reaction is

$$X^- + Ag^+ \rightarrow AgX\downarrow \quad \text{with} \quad X^- = Cl^-, Br^-, I^-.$$

37.2.2 *Theoretical Justification of the Method*

Concerning their very strong trend toward precipitation, both reactions given above are justified by the solubility-product values of the precipitates formed. The important question that arises is the following: Does the formation of the colored thiocyanatoiron(III) complex close to the equivalence point distort the titration (for example,

[1] Jakob Volhard, who was Liebig's assistant in Munich (1834–1910). Charpentier was a french "civil engineer" in the 1960's.

by precluding the silver thiocyanate precipitation or by dissolving the latter)? This phenomenon might occur through the possible parasitic reaction

$$\mathrm{AgSCN}\downarrow + \mathrm{Fe}^{3+} \overset{K}{\rightleftharpoons} [\mathrm{Fe(SCN)}]^{2+} + \mathrm{Ag}^{+}.$$

The answer to the question is negative, as the following reasoning proves. Indeed, the formation constant of the colored ferric complex, according to the reaction

$$\mathrm{Fe}^{3+} + \mathrm{SCN}^{-} \rightleftharpoons [\mathrm{Fe(SCN)}]^{2+},$$

is too weak ($K' \approx 126$). For this reason, the equilibrium constant K of the possible K' parasitic reaction is itself far too weak. Its value is about $10^{-9.9}$.

$$K = \{[\mathrm{Fe(SCN)}^{2+}][\mathrm{Ag}^{+}]\}/\{[\mathrm{Fe}^{3+}][\mathrm{Ag(SCN)}]_{(s)}\},$$
$$K = \{[\mathrm{Fe(SCN)}^{2+}][\mathrm{Ag}^{+}]\}/[\mathrm{Fe}^{3+}]$$

since the activity of a pure substance constituting a solid phase is equal to unity. We can also write

$$K = \{[\mathrm{Fe(SCN)}^{2+}][\mathrm{SCN}^{-}][\mathrm{Ag}^{+}]\}/\{[\mathrm{Fe}^{3+}][\mathrm{SCN}^{-}]\},$$
$$K = 10^{-9.9}.$$

The equilibrium of the parasitic reaction is quasi-totally displaced toward the left. It is the reason why it does not disturb the titration of thiocyanate ions in excess.

Another point concerning this titration deserves a further comment. It concerns the possible formation of the superior complexes $[\mathrm{Ag(SCN)_2}]^{-}$, $[\mathrm{Ag(SCN)_3}]^{2-}$, and $[\mathrm{Ag(SCN)_4}]^{3-}$. Due to their formation constants, which are given ahead, they do not form $\{K_1[\mathrm{AgSCN}] = 10^{4.59},\ K_2[\mathrm{Ag(SCN)_2}]^{-} = 10^{3.70},\ K_3[\mathrm{Ag(SCN)_3}]^{2-} = 10^{1.77}$, and $K_4[\mathrm{Ag(SCN)_4}]^{3-} = 10^{1.20}\}$.

37.2.3 Conditions in Which the Titration Must Be Carried Out

In nitric acid solution, Cl^{-}, Br^{-}, I^{-}, CN^{-}, and SCN^{-} are exclusively precipitated. According to some authors, this specificity may be relative.

In order to determine chloride ions, it is necessary to remove the silver chloride precipitate, for example, by filtration or by centrifugation, before carrying out the back titration with thiocyanate ions. The reason is the following: after the addition of an excess of silver ions, the precipitation of chloride ions as silver chloride may be considered complete because of the common ion effect. During the second stage of the determination, the excess of silver ions disappears with the addition of thiocyanate ions. Close to the equivalence point of the back titration, when both precipitates coexist, the following solubility products are satisfied simultaneously:

$$[\mathrm{Ag}^{+}][\mathrm{Cl}^{-}] = 10^{-9.75},$$
$$[\mathrm{Ag}^{+}][\mathrm{SCN}^{-}] = 10^{-12}.$$

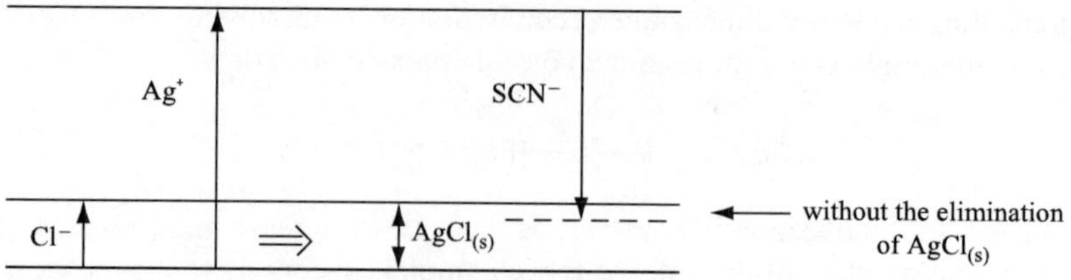

Fig. 37.1 Error made in the determination of chloride ions by Charpentier–Volhard's method when the silver chloride precipitate is not removed before the addition of thiocyanate ions

The ratio of the two expressions gives

$$[Cl^-]/[SCN^-] \approx 177.$$

Both concentrations must be in this ratio. At the end of the back titration, the thiocyanate ions' concentration is higher than that of chloride ions ($[SCN^-] \gg [Cl^-]$) since the former are added. For the ratio to be satisfied, the following reaction must occur:

$$AgCl\downarrow + SCN^- \rightarrow AgSCN + Cl^-.$$

Hence, we see that if we do not remove the silver chloride precipitate, an excess of thiocyanate ions is added and the chloride ions' determination is endowed with a negative error (Fig. 37.1).

From a practical standpoint, the silver chloride precipitate is agglomerated by ebullition. It is isolated from the solution by the addition of some nitrobenzene, which facilitates its separation from the aqueous phase. There is no problem of this sort with bromide and iodide ions due to their solubility products compared with that of silver thiocyanate.

Washing the precipitates and a vigorous stirring are necessary in order to avoid the adsorption of the silver ions in excess on the precipitate. This might induce a positive error.

37.2.4 Titration Error

Now, we're interested in the error arising in the back titration. In a typical determination, the encountered concentrations are as given in Fig. 37.2. One considers that at the final point,

$$[SCN^-]_{fp} = 10^{-5}\,\text{mol/L},$$
$$[Ag^+]_{fp} = 10^{-7}\,\text{mol/L}.$$

By using the formula giving the titration error (see Chap. 36),

$$\varphi_{fp} - 1 = [(2 \times 10^{-2} + 10^{-2})/(2 \times 10^{-2} \times 10^{-2})][(10^{-12}/10^{-7}) - 10^{-7}],$$
$$\varphi_{fp} - 1 = +1.5 \times 10^{-5}, \text{that is, } +0.15\%.$$

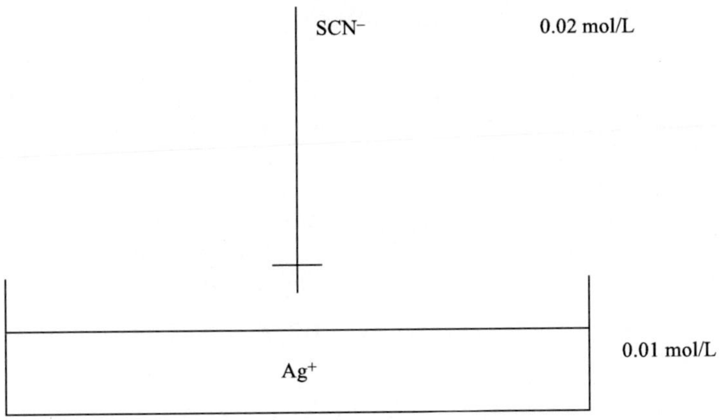

Fig. 37.2 Typical concentrations encountered during the second stage of a Charpentier–Volhard determination

Let's not focus too much on this value, which is the result of arbitrary choices. We must remember only that thiocyanate is added in a slight excess and that, consequently, the determination of chloride ions by Charpentier–Volhard's method has a slight negative error.

37.3 Argentometry in Neutral or Weakly Alkaline Medium: Mohr's Method

37.3.1 Principle

Mohr's method[2] is a direct titration. It consists of making a silver nitrate solution to react with the solution containing the halide to be determined in neutral or weakly alkaline medium according to, for example,

$$Cl^- + Ag^+ \rightarrow AgCl\downarrow$$

By definition, once the equivalence point is reached, an excess of silver ions appears. It is detected with potassium chromate K_2CrO_4, which gives a red precipitate of silver chromate at the equivalence point, according to the reaction

$$CrO_4{}^{2-} + 2Ag^+ \rightarrow Ag_2CrO_4\downarrow$$

In this context, potassium chromate is sometimes called a *precipitation indicator*.

[2] Karl Friedrich Mohr: a german chemist (1806–1879), manufacturer of chemicals in Munich.

37.3.2 Mechanism of the Endpoint Indication: Titration Error

The processes governing the endpoint indication are the successive precipitations of silver halide and silver chromate. The sensibility and accuracy of the endpoint indication depend on the concentration of chromate ions, which, in turn, depends on the total concentration of chromate ions and the pH value. Examining the solubility products permits us to determine the best experimental conditions.

According to the principle behind Mohr's method, at the equivalence point, two precipitation phenomena coexist. As a result, we can write

$$[Ag^+][Cl^-] = 1.1 \times 10^{-10},$$

$$[Ag^+]^2[CrO_4{}^{2-}] = 2.0 \times 10^{-12}.$$

We know that for the titration to be accurate, the following equality must be obeyed:

$$[Ag^+] = [Cl^-] = \sqrt{K_s(\text{AgCl})},$$

$$[Ag^+] = 1.05 \times 10^{-5}\ \text{mol/L}.$$

As a result,

$$[CrO_4{}^{2-}] = 2.0 \times 10^{-12}/[Ag^+]^2,$$

$$[CrO_4{}^{2-}] = 1.8 \times 10^{-2}\ \text{mol/L}.$$

It is the theoretical concentration that must exist in order to obtain a perfect titration. However, with this concentration, an experimental difficulty occurs: The intensive yellow color of chromate ions may mask the color change to red. This is the reason why a lower chromate concentration is chosen. It is about 5×10^{-3} mol/L. This value induces a titration error, which is easily calculated (see Chap. 36). Indeed,

$$[CrO_4{}^{2-}]_{fp} = 5 \times 10^{-3}\ \text{mol/L},$$

$$[Ag^+]_{fp}{}^2 = 2 \times 10^{-12}/(5 \times 10^{-3}),$$

$$[Ag^+]_{fp} = 2 \times 10^{-5}\ \text{mol/L}.$$

In a typical titration, the concentrations C_o and C (titrand and titrant concentrations) are about 10^{-1} mol/L, from which we find

$$\varphi_{fp} - 1 = (0.2/10^{-2})[2 \times 10^{-5} - (1.1 \times 10^{-10})/(2 \times 10^{-5})],$$

$$\varphi_{fp} - 1 = 3 \times 10^{-4}.$$

The relative titration error is about +0.03%, which is very weak.

37.3.3 pH Conditions

The pH value must be contained within a narrow range: $6.5 < pH < 10.5$. At values higher than 10.5, the poorly soluble silver oxide Ag_2O begins to precipitate before the equivalence point. For $pH < 6.5$, the following supplementary equilibria occur:

$$2CrO_4^{2-} + 2H^+ \rightleftharpoons 2HCrO_4^- \rightleftharpoons Cr_2O_7^{2-} + H_2O.$$

There is formation of hydrogen chromate ions (or acidic chromate ions) and also of dichromate ions (see Chap. 20). Due to the former equilibrium, if the pH value is too low, the concentration $[CrO_4^{2-}]$ is considerably lowered and the solubility product $K_s(Ag_2CrO_4) = [Ag^+]^2[CrO_4^{2-}]$ is never reached. The change at the final point is not seen.

Exercise 1 Calculate the concentrations of the different present species when potassium chromate K_2CrO_4 is dissolved at concentration 2×10^{-2} mol/L in a 10^{-1} mol/L nitric acid solution. The data are

$$[HCrO_4^-]/([H^+][CrO_4^{2-}]) = 3.16 \times 10^6, \quad [Cr_2O_7^{2-}]/[HCrO_4^-] = 33,$$
$$[H_2CrO_4]/([H^+][HCrO_4^-]) = 0.83, \quad [HCr_2O_7^-]/([H^+][Cr_2O_7^{2-}]) = 1.17.$$

In addition to these relations, we must write the following equations in order to solve the problem:

$$[CrO_4^{2-}] + [HCrO_4^-] + [H_2CrO_4] + 2[Cr_2O_7^{2-}] + 2[HCr_2O_7^-] = 2 \times 10^{-2}\ \text{mol/L},$$

$$[OH^-] + [NO_3^-] + 2[CrO_4^{2-}] + [HCrO_4^-] + 2[Cr_2O_7^{2-}] + [HCr_2O_7^-] = [K^+] + [H^+].$$

Because of the value $pK_a(HCrO_4^-)$, we can set up that $[CrO_4^{2-}]$ is negligible and that, on the contrary, the concentrations $[HCrO_4^-]$, $[Cr_2O_7^{2-}]$, $[HCr_2O_7^-]$, and $[H_2CrO_4]$ are predominant. The first reaction to occur is

$$CrO_4^{2-} + H^+ \rightarrow HCrO_4^-.$$

Since the reaction is considered complete (we are using the concept of the principal reaction here—see Chap. 5), we can set $[H^+] \approx 8 \times 10^{-2}$ mol/L. Using the relation that expresses the mass balance on chromium and following the above hypothesis, we obtain a first set of values that gives the value $[H^+] = 7.8 \times 10^{-2}$ mol/L. The calculation reiterated with this value taken as the initial value leads to values identical to those obtained before. Finally, we find

$$[H^+] = 7.80 \times 10^{-2}\ \text{mol/L}, \quad [HCr_2O_7^-] = 3.54 \times 10^{-4}\ \text{mol/L},$$
$$[HCrO_4^-] = 1.08 \times 10^{-2}\ \text{mol/L}, \quad [H_2CrO_4] = 7.01 \times 10^{-4}\ \text{mol/L},$$
$$[Cr_2O_7^{2-}] = 3.88 \times 10^{-3}\ \text{mol/L}, \quad [CrO_4^{2-}] = 4.38 \times 10^{-8}\ \text{mol/L}.$$

In practice, if the solution is too acidic, sodium hydrogen carbonate is added to buffer the medium.

37.3.4 Applications

Mohr's method is essentially used for the titration of chloride and bromide ions. It is not satisfactory for iodide and thiocyanate ions because they exhibit a strong tendency to be adsorbed on the silver iodide and silver thiocyanate precipitates. Finally, some ions that give poorly soluble compounds with silver (see Sect. 37.1.2) may disturb the titration.

37.4 Argentometry in Weakly Acidic or Neutral Medium: Fajans' Method

37.4.1 Principle

Like Mohr's method, Fajans' method[3] is a direct titration method. It is based on the use of adsorption indicators. A small volume of an acetic acid solution and an adsorption indicator are added to the solution containing the halide to be determined. Then a silver nitrate solution is added. At the equivalence point, the precipitate suddenly turns colored and separates completely from the solution.

37.4.2 Indication Mechanism

At the beginning of the addition of silver nitrate, a colloidal precipitate of halide chloride forms. This first fraction of precipitate tends to preferentially adsorb bromide ions (for example) and not the anionic anions A^- of the indicator. This phenomenon occurs until the silver ions are completely exhausted.

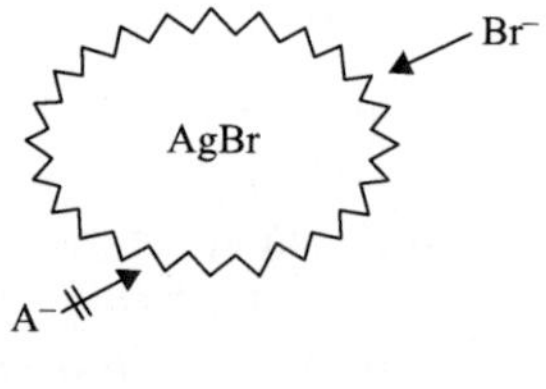

Once the equivalence point has been reached, halide ions are no longer available since they all precipitated. The ions A^- of the anionic indicator take the place of halide ions on the crystalline lattice. This induces the sudden color change and the separation of the precipitate.

[3] Kasimir Fajans, hungarian physicist (1887–1975).

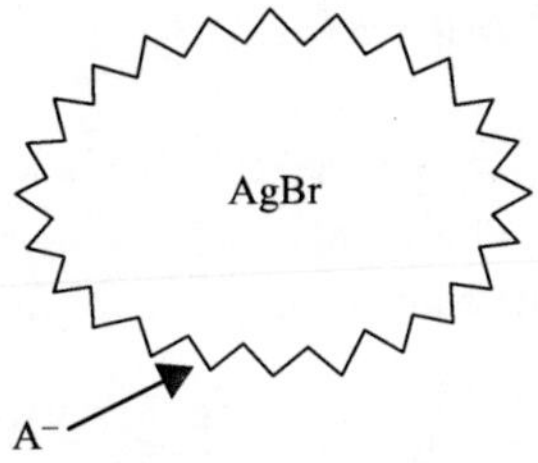

For the determination of silver salts by halides, all the preceding considerations are inverted.

37.4.3 Experimental Conditions

37.4.3.1 Indicators

In order to determine halides (anions), anionic dyestuffs must be used (eosine, fluorescein, dibromo and dichlorofluorescein). For the determination of silver ion with halides, cationic dyestuffs are used (rhodamine G, p-ethoxychrysoidine).

37.4.3.2 pH

The pH value must not be too far from neutrality for the indicator to remain ionized. For example, for the determination of iodide ions, too acidic a medium might protonate the anionic indicator.

37.4.3.3 Titration with Exposure to Diffuse Light

This requirement comes from the fact that silver halides are sensitive to the action of light by a layer of adsorbed dyestuff.

37.5 Liebig–Denigés' Method

37.5.1 Definitions

Liebig[4]–Denigés'[5] method determines cyanide ions with silver ions.

[4] Justus von Liebig (1803–1873): german analyst and organic chemist, most well-known by his works concerning organic elementary analysis. He was a student of Gay-Lussac.

[5] Georges Denigés: Professor of medical and biological chemistry at the University of Bordeaux. His most abundant publications are in the realm of the analysis.

37.5.2 *Principle of Liebig's Method*

The principle of Liebig's method (1851) is as follows: when a silver nitrate solution is added to a slightly alkaline solution of an alkaline cyanide, the soluble ion complex dicyanosilverate(I) ("argentocyanide") forms according to the reaction

$$2CN^- + Ag^+ \rightarrow [Ag(CN)_2]^-. \tag{37.1}$$

This is the titration reaction. Once the equivalence point has been reached, that is, once there is an excess of silver ions, the solution becomes turbid because of the formation of insoluble silver dicyanosilverate(I) ("silver argentocyanide"):

$$[Ag(CN)_2]^- + Ag^+ \rightarrow Ag[Ag(CN)_2]\downarrow$$

The equivalence point is defined by the appearance of a persistent turbidity.

The value of the solubility product of silver cyanide is $K_s(AgCN) = 10^{-15.92}$. It is very insoluble. Once the first drop of silver nitrate solution is added, the silver cyanide precipitates according to the reaction

$$Ag^+ + CN^- \rightarrow AgCN\downarrow$$

but it dissolves by complexation with cyanide ions according to the reaction

$$AgCN\downarrow + CN^- \rightarrow [Ag(CN)_2]^-, \quad K = 10^{4.62}.$$

Indeed, the "argentocyanide ion" is extremely stable (overall stability constant $\beta_2 = 10^{20.54}$). Its formation induces the dissolution of silver cyanide. Once the equivalence point has been reached, that is, after one equivalent of Ag^+ has been added for two equivalents of silver cyanide, silver dicyanosilverate(I) forms and induces the turbidity of the solution since it is poorly soluble $K_s(Ag[Ag(CN)_2]) = 10^{-11.3}$.

The obtained titration curve is given in Fig. 37.3.

Remark This last value of the solubility product ($10^{-11.3}$) is usually noted in the literature as being that of silver cyanide. Actually, as we have seen, it is that of silver "argentocyanide."

37.5.3 *Equations of the Titration Curve*

The titration curve, which is the diagram pAg/φ, comprises two portions, since the relations that are satisfied before and after the equivalence point are not the same. In the second portion of the curve, the solubility products must be taken into account since a precipitation is occurring. This is not the case, of course, with the first portion. Hence, at the equivalence point, there is a discontinuity of the titration curve.

Fig. 37.3 Liebig's titration curve $C = 0.1$ mol/L and $C_o = 0.2$ mol/L

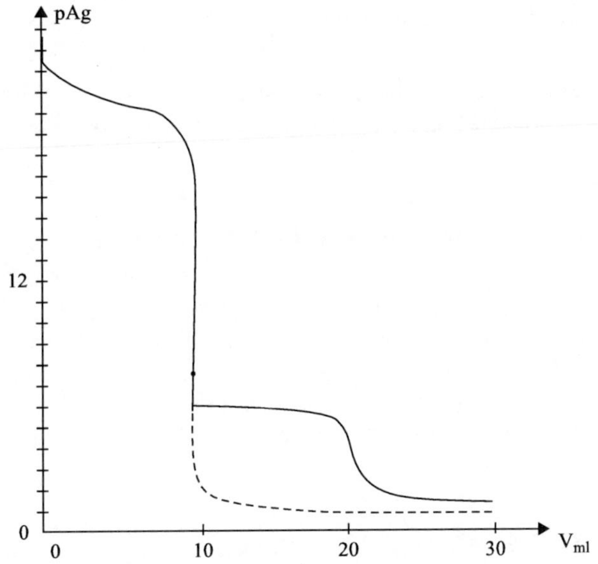

1. An expression $\varphi/[Ag^+]$ of the first portion cannot be expressed explicitly. However, a calculation of the curve point by point up to the point where a precipitation exists can be carried out through the following three relations (see Appendix F):

$$\varphi = 1 - [CN^-](1 + [H^+]/K_a)/C_o,$$

$$[H^+]^2(1 + [CN^-]/K_a) = K_w,$$

$$[Ag^+] = \{C_o C\varphi/(2C + C_o\varphi)\}\,(1/\beta_2[CN^-]^2).$$

   The parameters have the same meaning as usual. The point-by-point calculation is carried out as follows. A value of $[CN^-]$ is chosen first. It leads to a value $[H^+]$, and then to φ, and finally to pAg.
2. An expression of the second portion is (Appendix F)

$$\varphi = \left\{2 + 2[Ag^+]/C_o - 2K_{s0}K_{s2}/[Ag^+]C_o\right\},$$

   where the product $K_{s0}K_{s2}$ is purely and simply the solubility product of silver "argentocyanide." Assuming values for $[Ag^+]$, we can calculate φ.
3. Close to the endpoint, the following relation is satisfied (Appendix F):

$$\varphi_{fp} - 1 = \{(C_o + 2C)/2C_oC\}\left\{2K_{s0}/[CN^-]_{fp} - [CN^-]_{fp}(1 + [H^+]/K_a)\right\},$$

   where fp means final point.

37.5.4 Titration Error

The titration error is very small in principle. With concentrations $C = C_o = 10^{-2}$ mol/L, it is estimated to be approximately −0.02% (see Appendix F).

37.5.5 Drawback to Liebig's Method

Liebig's method suffers from the following drawback: the delay necessary for the dissolution of silver cyanide by cyanide ions still in excess (before the equivalence point). The persistence of the insoluble silver cyanide not yet dissolved may induce an error concerning the final point detection, which may be detected prematurely. This is the reason why it is recommended to carry out the titration very slowly.

37.5.6 Denigés' Modification

Denigés developed a very nice amelioration to Liebig's method. He proposed using an ammonia solution instead of "pure" water as solvent and, moreover, adding iodide ions as potassium iodide to the solution. The equivalence point is detected by the precipitation of silver iodide, which is very insoluble [$K_s(\mathrm{AgI}) = 10^{-16.35}$].

The phenomena that occur are the following: Once the addition of silver ions to the solution under study is carried out, they are converted into amminesilver ions, essentially diamminesilver(I) ions $[Ag(NH_3)_2]^+$ ($\beta_2 = 10^{7.03}$). Before the equivalence point, diamminesilver(I) ions react with two cyanide ions to give the dicyanosilverate(I) complex, "argentocyanide," according to

$$[Ag(NH_3)_2]^+ + 2CN^- \rightarrow [Ag(CN)_2]^- + 2NH_3. \qquad (37.1')$$

This reaction is, properly speaking, that of titration. We can note that its equilibrium constant is markedly weaker than previously ($10^{13.5}$ instead of $10^{20.55}$ in Liebig's method). However, it remains sufficiently high for the ion complex "argentocyanide" to form. After the equivalence point, silver ions in excess (always as diamminesilver complex) react with "argentocyanide" ions according to the global reaction

$$[Ag(NH_3)_2]^+ + [Ag(CN)_2]^- \rightleftharpoons Ag[Ag(CN)_2]\downarrow + 2NH_3,$$

but contrary to Liebig's method, this reaction is too equilibrated to induce a precipitation occurring close to the equivalence point. Indeed,

$$K = \{[Ag\,[Ag(CN)_2]][NH_3]^2\}/\{[Ag(NH_3)_2{}^+][Ag(CN)_2{}^-]\},$$

$$K = 10^{4.27}.$$

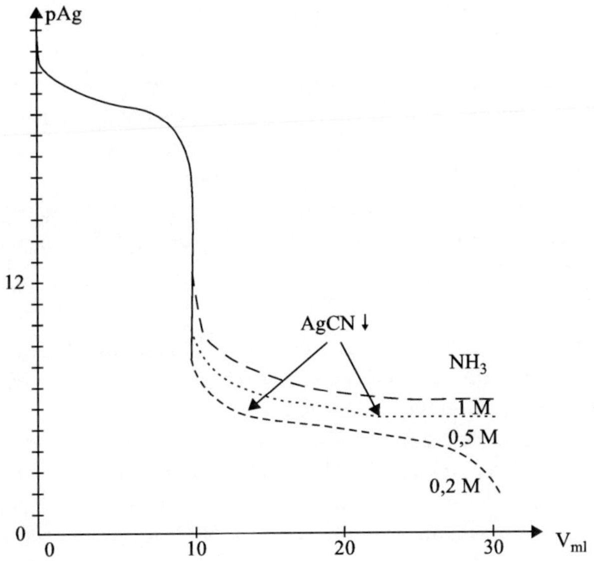

Fig. 37.4 Titration curve according to Denigés' method of 0.2 mol/L cyanide ions ($V_o = 10$ ml) with a solution of 0.1 mol/L silver nitrate in the presence of several ammonia concentrations

We can also note that the reaction is also governed by the total concentration of the ammonia. Iodide ions are added to play the part of an indicator of the reaction's end. The solubility product $K_s(\text{AgI})$ is sufficiently low for the precipitation reaction

$$[\text{Ag(NH}_3)_2]^+ + \text{I}^- \rightleftharpoons \text{AgI}\downarrow + 2\text{NH}_3, \quad K = 10^{9.32}.$$

to occur just once the complex ion $[\text{Ag(NH}_3)_2]^+$ is in excess, that is, when the cyanide ions no longer remain in solution. From another standpoint, it is interesting to notice that insoluble silver iodide, which might possibly form before the equivalence point, is not sufficiently insoluble not to be dissolved by the action of the remaining cyanide ions according to the reaction

$$\text{AgI}\downarrow + 2\text{CN}^- \rightleftharpoons [\text{Ag(CN)}_2]^- + \text{I}^-, \quad K = 10^{4.20}.$$

Hence, before the equivalence point, there are no Ag^+ ions out of their normal reactional pathway, a pathway in which "argentocyanide" ions are forming.

Figure 37.4 shows titrations curve obtained according to the principle of Liebig–Denigés' method, by using several ammonia concentrations.

A rough calculation permits us to know the lowest limit of the iodine concentration that must be added in order to obtain an accurate titration (see Appendix F).

One must adhere to the following condition:

$$[\text{I}^-] > 10^{-8.35}\,\text{mol/L}.$$

The values chosen in this example are typical.

37.5.7 Applications and Extensions of Liebig–Denigés' Method

The principal application of Liebig–Denigés' method is the determination of alkaline cyanide and hydrocyanic acid in laurel cherry water, for example. Pharmacopeias recommend the determination of total hydrocyanic acid according to Liebig–Denigés' method, whereas free HCN must be determined according to Charpentier–Volhard's method. Liebig–Denigés' method seems to be the better one to determine very small quantities of cyanide ions due to the fact that the "argentocyanide" complex is very stable ($\beta_2 = 10^{20.54}$).

Liebig's method gives rise to two extensions:

- the first one consists of determining a precipitated quantity of silver halide $AgX\downarrow$. The latter is dissolved in an excess of alkaline cyanide, and cyanide ions in excess are titrated with silver nitrate;
- the second extension concerns the indirect determination of some metallic ions such as $Ni^{2+}$and Zn^{2+}, which form complexes of well-known composition with cyanide ions. For example, the nickel salt in ammonia is treated with an excess of potassium cyanide solution. Ni^{2+} forms the tetracyanonickelate(II) complex quantitatively according to

$$[Ni(NH_3)_6]^{2+} + 4CN^- \rightarrow [Ni(CN)_4]^{2-} + 6NH_3.$$

The excess of cyanide ions is titrated with a standard silver nitrate solution according to a principle close to that prevailing in Liebig–Denigés' method. If a small quantity of silver iodide is added to the initial ammonia solution containing Ni^{2+} ions, the turbidity it induces because of its insolubility disappears after an addition of cyanide ions because of the occurrence of the reaction

$$AgI\downarrow + 2CN^- \rightarrow [Ag(CN)_2]^- + I^-.$$

But once there is no longer an excess of cyanide ions, that is, after the equivalence point of the cyanide ions' back titration, the precipitation of the iodide ions as silver iodide occurs according to the reaction

$$I^- + [Ag(NH_3)_2]^+ \rightarrow AgI\downarrow + 2NH_3.$$

37.5.8 Standard Solutions

Equation (37.1′) shows that the cyanide equivalent is equal to 2 mol/L. Standard solutions are $AgNO_3$, NaCN, or KCN solutions. They are prepared by weighing.

37.6 Some Other Precipitation Methods

Other precipitation methods are no longer used except in some particular cases. We can mention, however, the following methods:

- ferrocyanimetry, which is based on the fact that several metallic ferrocyanides are poorly soluble. They give precipitates of a well-known composition;
- uranometry, which is based on the precipitation of uranyl cations UO_2^{2+} with phosphate ions. This method has been totally supplanted today by colorimetric methods;
- that using Ba^{2+} ions in order to determine sulfate ions. The precision of this method is very poor.

Chapter 38
Gravimetry by Precipitation

Gravimetry is a method of quantitative chemical analysis. It qualifies as a macroscopic quantitative method of analysis because it involves relatively important quantities of a substance to be determined compared to more recent methods, such as electrochemical, spectroscopic and chromatographic means. From this standpoint, it should instead be compared to titrimetric methods. However, it has remained a method of choice for the analysis of standard compounds, those compounds with which the more recent instrumental methods of analysis listed above are calibrated.

38.1 Principle and Some Definitions

In gravimetric analysis, the substance to be determined (the *analyte*) is transformed quantitatively into an insoluble precipitate that is isolated and weighed.

Most of the time, the substance to be analyzed is transformed quantitatively into another that yields the precipitate that will be weighed. The weight of the analyte, and through it its mole number, can easily be calculated from that of the precipitate if the precipitate's composition is known. (Here we shall confuse weights and masses. This has no effect on the accuracy of the different results, provided calculations do not mix both quantities.)

One distinguishes among

- gravimetry by precipitation, which is characterized by the fact that the separation of the substance to be determined (or of one of its derivatives) is carried out by precipitation. This is the sole sort of gravimetry that will be studied here;
- electro-gravimetry, in which the element or the compound to be determined is deposited quantitatively upon a suitable electrode. Weighing the electrode before and after the process permits the determination;
- gravimetry by volatilization, in which the loss of weight by volatilization is determined by weighing. The volatilization is due to a gain in energy by the sample. Most often, it is thermal energy. An important example of such a process is provided by the determination of moisture in a sample;

J.-L. Burgot, *Ionic Equilibria in Analytical Chemistry*,
DOI 10.1007/978-1-4419-8382-4_38,

- particulate gravimetry, in which the analyte is already under a physical or chemical form that can be easily separated from a liquid, that is, as a gas or a solid matrix. Briefly, in this process, no prior chemical transformation of the compound to be determined is needed.

Some authors distinguish two sorts of gravimetry by precipitation:

- direct gravimetry, in which the analyte itself is weighed. For example, when nickel is determined by the weighing of its dimethylglyoximate, a direct gravimetry is carried out;
- indirect gravimetry, in which the analyte is not weighed. For example, phosphite ions PO_3^{3-} may be determined by weighing dichlorodimercury(I) ("calomel") that has previously been formed from it and that has precipitated. Phosphite ions do indeed quantitatively reduce Hg(II) into dichlorodimercury(I) while simultaneously being oxidized into phosphate ions $[E°(H_3PO_4/H_3PO_3) = -0.276\ V]$. The reaction is

$$2HgCl_2 + PO_3^{3-} + 3H_2O \rightarrow Hg_2Cl_2 + 2H_3O^+ + 2Cl^- + PO_4^{3-}.$$

Exercise 1 A sample of impure sodium chloride is dissolved in water and chloride ions are precipitated with silver nitrate added in sufficient quantity. A total of 1 g of silver chloride is obtained. What is the weight of chlorine in the sample? [M(AgCl) = 143.34 g/mol, M(Cl) = 35.46 g/mol.]

The precipitation reaction is

$$Ag^+ + Cl^- \rightarrow AgCl\downarrow$$

The reasoning is the following:

143.34 g of AgCl contains 35.46 g of chlorine,
1 g of AgCl contains 143.34 less chlorine,

whereby the weight of chlorine in the sample is

$$35.46/143.34 = 0.247\ g.$$

38.2 Conditions for the Success of a Gravimetry by Precipitation Determination

The conditions for the success of a gravimetry by precipitation determination depend on some characteristics of the precipitation reaction and on the precipitate's properties.

The precipitate must

- be very insoluble in the medium in which it is formed. This is, of course, a *sine qua non* condition;
- possess a well-defined composition or be converted into a compound of a well-defined composition. This is also a *sine qua non* condition. The stoichiometric

connection between the precipitate and the analyte formula must indeed be known. This imposes that the precipitate be pure;

- separate easily from the medium where it is formed;
- resist the successive washing operations;
- possess an analyte percentage in its composition that is as weak as possible.

The precipitation reaction must be as quantitative and specific as possible. From a practical standpoint, fewer prior separations must be carried out if the latter condition is obeyed. (We specify some of these conditions in the following sections.)

38.3 Insolubility of the Precipitate in the Medium

The solubility of the precipitate must be very weak. Losses by solubilization may be controlled by a judicious adjustment of the chemical conditions that govern the precipitation. For example, silver chloride is formed according to

$$Ag^+ + Cl^- \rightarrow AgCl\downarrow.$$

If we only consider this reaction, we can write (see Chap. 32)

$$S_{AgCl} = [Ag^+] \quad \text{(saturation)},$$
$$S_{AgCl} = K_s/[Cl^-].$$

From the latter relation, we immediately deduce that the losses incurred during the dissolution of the precipitate are reduced by adding an excess of chloride ions. Actually, adding too great an excess of chloride ions, contrary to what was said just above, induces an increase in S_{AgCl} by forming the "superior" complexes $[AgCl_2]^-$ and $[AgCl_3]^{2-}$ (see Chap. 35). Indeed, the solubility must be defined by the expression

$$S = [Ag^+] + [AgCl_{solution}] + [AgCl_2{}^-] + [AgCl_3{}^{2-}].$$

We have seen that solubility exhibits a minimum value for $pCl \approx 3$. Then it is about $10^{-5.8}$ mol/L. For $pCl > 3$, the solubility increases when pCl increases. For $pCl < 3$, the solubility increases when pCl decreases. In this example, we are faced with the superimposition of two sorts of phenomena: precipitation and complexation. From studying this example, we see that perfect knowledge of the precipitation reaction and of all the chemical phenomena occurring simultaneously with it is absolutely necessary to control the losses by making the precipitate soluble. A parameter that in some cases may be of great interest to limit losses is the pH value. For example, let's investigate the precipitation of calcium phosphate $Ca_3(PO_4)_2$. Its solubility depends on the following four chemical reactions:

$$Ca_3(PO_4)_2\downarrow \rightleftharpoons 3Ca^{2+} + 2PO_4{}^{3-},$$
$$PO_4{}^{3-} + H_2O \rightleftharpoons HPO_4{}^{2-} + OH^-,$$
$$HPO_4{}^{2-} + H_2O \rightleftharpoons H_2PO_4{}^- + OH^-,$$
$$H_2PO_4{}^- + H_2O \rightleftharpoons H_3PO_4 + OH^-.$$

The ions PO_4^{3-}, HPO_4^{2-}, and $H_2PO_4^-$ are indeed bases. The solubility of calcium phosphate is defined by the expression

$$S_{Ca_3(PO_4)_2} = [Ca^{2+}]/3.$$

By virtue of the solubility product,

$$K_s = [Ca^{2+}]^3[PO_4^{3-}]^2.$$

in order for the solubility to be at its minimum, the concentration $[PO_4^{3-}]$ must be at its maximum. This implies that the pH must be as high as possible $[pK_a(HPO_4^{2-}) = 12.3]$. The solubility is at its minimum for pH = 14.

38.4 Composition of the Precipitate; Impurities of the Precipitate

In gravimetry by precipitation, the final precipitate must have a well-defined composition. It must not contain any impurity. Indeed, the proportion of an impurity is generally poorly reproducible from one experiment to another.

There are several sorts of impurities. They can be classified into two groups:

- impurities of coprecipitation, which group impurities by inclusion and by occlusion and the adsorbates. They have in common the ability to be formed in the solid state simultaneously with the precipitate of interest;
- impurities formed by the precipitation of any other solution species in the conditions of the analysis. They may be qualified as being impurities resulting from a lack of selectivity of the precipitation reaction.

38.4.1 Impurities by Coprecipitation

Coprecipitation impurities are due to surface and occlusion phenomena. The presence of coprecipitation impurities whose origin results from surface phenomena is due to the existence of the precipitate as a crystalline lattice, even if the latter exists only on a microscopic scale. This signifies that the ions constituting the precipitate appear regularly within the lattice. For example, in the precipitate of silver chloride, one silver ion is surrounded by six chloride ions within the lattice. However, at the surface of the lattice, silver ions are not surrounded by more than five chloride ions. As a result, they bear a partial positive charge. Chloride ions exhibit an analogous phenomenon and bear a partial negative charge at the surface of the lattice. Two consequences arise:

- first, there is attraction of the counterion. As a result, the counterion fixes on the lattice and contributes to the growth of the latter. This phenomenon is not a drawback. It is called *chemical adsorption*;

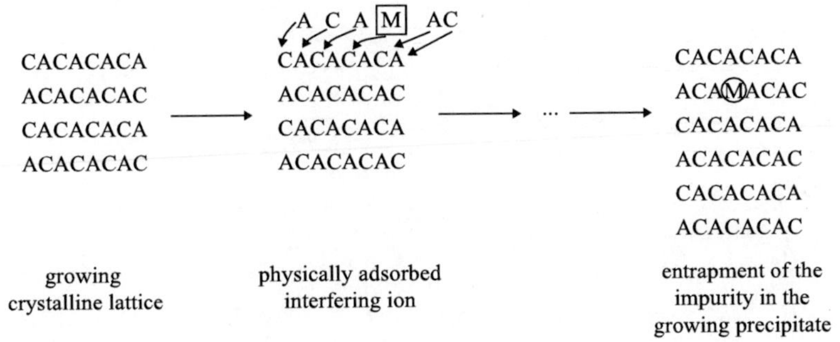

Fig. 38.1 Inclusion process of a foreign ion into a precipitate

- second, the surface charge may attract a counterion that is not that of the analyte. In such a case, we speak of *physical adsorption*. After a further growth of the lattice, the foreign ion becomes totally included within it. In that case, we speak of *impurity by inclusion* (Fig. 38.1).

Of course, the more the foreign and expected ions exhibit similar charges and volumes, the more pronounced the inclusion is.

When an impurity is physically adsorbed without being included, it can be displaced by chemically adsorbed impurities. It is indeed an experimental fact that the first sort of adsorbed impurities are less strongly adsorbed than the second. It is not surprising that the probability of obtaining an impurity by inclusion is higher when the foreign ion concentration is higher than that of the ion of interest. Impurities by inclusion are difficult to eliminate because they enter as a true component in the structure of the crystal. The only way to eliminate them is to alternatively carry out precipitation, dissolution, "reprecipitation," and so forth. To be more precise, the process is the following: After isolation of the precipitate from the mother solution, it is dissolved into a minimum quantity of the pure solvent at an elevated temperature. Then the solution is cooled at room temperature or even lower. The precipitate forms again. At this stage, the percentage in mass of the impurity is less than that in the initial liquid since a first dilution has already occurred. This process may be repeated several times until the impurity totally disappears. Such a process increases the risk, of course, of losses by making the precipitate soluble. Of course, the precipitate must exhibit a lower solubility at room temperature than at higher temperatures.

When the precipitation is achieved, the lattice surface continues to attract the solution ions. Another sort of impurity, sometimes called *adsorbates*, is formed in such a way. Their adsorption is minimized by a decrease in the crystalline surface. This is in agreement with the fact that the probability of a particle to dissolve is inversely proportional to its diameter. By using the process called *digestion* (see ahead), one makes the size of the particules of the precipitate increase. Adsorbates can be eliminated by washing the precipitate. Of course, there is a risk here of making the precipitate soluble. Let's note that some impurities constituting adsorbates may also be the ions that get into the precipitate (see Fig. 38.2 for the case of the silver

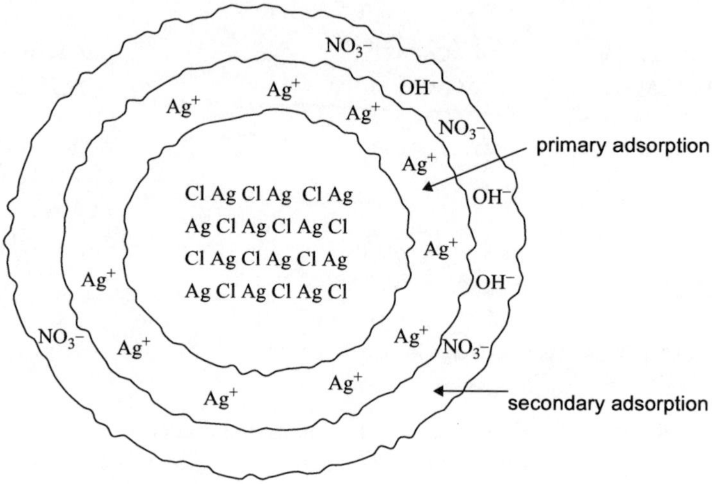

Fig. 38.2 Representation of the precipitate of AgCl in a solution of $AgNO_3$

chloride precipitate in the presence of a solution of silver nitrate). According to Fig. 38.2, we observe that the precipitate is surrounded by two adsorption sheaths: the primary and secondary ones.

Among the impurities by coprecipitation, one can also distinguish those by occlusion. This is the case when a small volume of solution containing some solutes is trapped within the precipitate. As a result, the mass of the precipitate is higher than expected. The formation of impurities by occlusion is minimized by maintaining the precipitate in equilibrium with the solution from which it was formed for as long as possible. This is the process of digestion. It can be carried out at any temperature.

38.4.2 Impurities by a Lack of Selectivity of the Precipitation Reaction

It is possible that the experimental conditions necessary to minimize the solubility of the anticipated precipitate induce a new precipitation in addition to the one of interest. For example, the precipitation of Ni^{2+} as nickel dimethylglyoximate necessitates working at a slightly basic pH. In these conditions, Fe^{3+}, if present, precipitates as ferric hydroxide. These supplementary precipitations may be, in principle, avoided (or at least minimized) by a judicious adjustment of some parameters, in particular the pH value. The supplementary precipitates, which are more insoluble than the anticipated precipitate, may be eliminated by filtration, with the analyte remaining in the filtrate. Another means to overcome this problem is, in a first stage, to mask the parasitic impurity by complexation. The inverse process may also be carried out: The analyte may be masked in a first step and the impurity precipitated and filtered off. For example, with a solution containing Cu^{2+}, Pb^{2+}, Fe^{3+}, Co^{2+}, and Ni^{2+} and in order to determine the two latter ions by gravimetry, Pb^{2+} is first precipitated and eliminated by filtration as $PbSO_4$ after the addition of H_2SO_4, Cu^{2+} by precipitation with H_2S as CuS, and Fe^{3+} as $Fe(OH)_3$ after the addition of ammonia. Ni^{2+} and Co^{2+} remain in solution as ammine complexes and can be precipitated as dimethylglyoximate for Ni^{2+} or by reaction with 1-nitroso-2-naphthol or with $[Hg(SCN)_4]^{2-}$ for Co^{2+}. Finally, we mention that in some cases, the rate according to which a precipitate forms permits us to avoid the interference due to the formation of another precipitate. For example, the formation of magnesium and calcium hydroxides may occur in the same experimental conditions, but it is not necessarily a difficulty to determine the concentration of both ions because it is possible to filtrate $Ca(OH)_2$, which precipitates before $Mg(OH)_2$ begins to do it. The gravimetric determination of Ca^{2+} is therefore possible in these conditions.

An interesting point to notice, connected with the preceding one, is that it may be possible, in some cases, to carry out the gravimetric determination of an analyte through the weighing of a precipitate that is a mixture of two precipitates. In such a case, the determination of the compound giving rise to the interfering precipitate is carried out simultaneously. It is easily conceivable that since there is one supplementary unknown (the concentration of the interfering compound), we must have supplementary experimental data at our disposal. For example, the concentrations $[Mg^{2+}]$ and $[Ca^{2+}]$ in a mixture can be determined first by isolating the precipitated mixture of their carbonates $CaCO_3$ and $MgCO_3$ and by weighing it. Other necessary data are obtained, in a second stage, by heating the preceding precipitate. It is converted into a mixture of calcium and magnesium oxides. The following mathematical relations are satisfied:

$$\text{Mass of precipitate 1} = \text{mass}(CaCO_3) + \text{mass}(MgCO_3),$$

$$\text{Mass of precipitate 2} = \text{mass}(CaO) + \text{mass}(MgO).$$

At this point, it seems that there are only two equations for four unknowns, $m(CaCO_3)$, $m(MgCO_3)$, $m(CaO)$, and $m(MgO)$. However, this is only one appearance, since there are two additional relations. They are connected directly to the stoichiometries of the reactions that are occurring, which are

$$MgCO_3 \rightarrow MgO\downarrow + CO_2,$$

$$CaCO_3 \rightarrow CaO\downarrow + CO_2.$$

It appears that the numbers of moles of magnesium oxide and of calcium oxide formed are the same as the numbers of moles of the corresponding carbonates present in the first precipitate. The mathematical system becomes determined. Finally, it can be reduced to the following two equations with two unknowns:

$$\text{Mass of precipitate 1} = n_1 M(CaCO_3) + n_2 M(MgCO_3),$$

$$\text{Mass of precipitate 2} = n_1 M(CaO) + n_2 M(MgO),$$

where n_1 and n_2 are the numbers of moles of calcium carbonate and magnesium carbonate in precipitate 1. $M(CaCO_3)$ is the molar weight of calcium carbonate, and so on.

38.5 Obtaining a Suitable Precipitate

It is obvious that the precipitate must be filtered easily. The basic principle in order to satisfy this condition is that the precipitate's particules must be as big as possible. In this case, they are easily filtered and, moreover, they are also the purest.

38.5.1 Purity of Precipitates and Size of Particules

The more extended the specific surfaces of precipitates are, the more impurities they tend to adsorb. The specific surface is defined as the surface of the precipitate of unit mass. Usually, it is expressed in cm^2/g. For a given mass of solid, the specific surface increases considerably when the particule sizes decrease. For example, a 1-cm-long cube exhibits a specific surface of $6 \times 1 \times 1 = 6\,cm^2$. If its mass is 2 g, its specific surface is $6/2 = 3\,cm^2/g$. If the same mass is distributed over 1000 cubes that are 0.1 cm long, the total surface is $0.1 \times 0.1 \times 6 \times 1000 = 60\,cm^2$ and its specific surface is $30\,cm^2/g$. In order to clarify this point, let's consider 2 g of a colloid containing about 10^{18} particules, whose length is about 10^{-6} cm. They exhibit a specific surface of $30\,m^2/g$.

38.5.2 Size of Particules

The sizes of particules formed by precipitation may vary within a large range. Those giving rise to colloidal suspensions exhibit a length in the range 0.1–1.0 mm. From the standpoint of the filtration, these suspensions behave as true solutions since a filtering paper only retains particules whose diameters are located in the range 10^{-2}–10 mm (concerning this point, recall that the limit of vision under the ordinary microscope is 0.2 μm. Tyndall's effect decreases the limit of vision down to 10 nm by using an ultra-microscope, which permits us to see with scattered light. Using X-rays permits a further decrease to this limit). At the opposite end of the scale from particule sizes are crystalline suspensions. Their particule dimensions are some tenths of millimeters in size, and even far bigger. These particules are easily filtered.

The mechanism of the formation of particules is not perfectly known. It is, however, acknowledged that particule sizes depend on the solubility of the precipitate's constituents, the temperature, the reagents' concentrations, and also the rate according to which the reagents have been mixed. A relation given by von Weymarn[1] (1925) connects the size of particules to only one property of the system, called the *relative supersaturation* (RSS), defined by the expression

$$RSS = (Q - S)/S.$$

[1] Peter von Weymarn (1879–1935), Russian chemist; colloids were the center of his interests.

Q is the instantaneous concentration of the solute and S its solubility at equilibrium. We may be surprised at first sight that Q can be greater than S. This indeed seems to contradict equilibrium laws. But, to be precise, the system is out of equilibrium since precipitation rates are slow and, even, supersaturation does exist when the precipitating reagent is added very slowly. The important experimental point to stress here is that particule sizes are inversely related to the average relative supersaturation during the time interval spent to add the reagent. More precisely, when the relative supersaturation is high, obtaining colloidal suspensions is probable. For example, with RSS = 25,000, the precipitate is colloidal. With RSS = 125, statistically, we can affirm that we will obtain crystals.

The effect of the supersaturation upon the particule sizes may be explained if we start from the hypothesis that the precipitate forms in one of two ways, that is, by nucleation or by growth of the particules. The size, in fact, is determined by the faster way.

During the nucleation phenomenon, a very weak number of ions, atoms, or molecules agglomerate to give a stable solid. The precipitation may then be considered the result of the competition between a new nucleation and a further growth of the particule issuing from the first nucleation. The nucleation rate is considered an increasing exponential function of the relative supersaturation RSS, whereas the growth rate seems to be weakly dependent on this one. This explains the fact that when supersaturation is weak, the tendency for crystalline suspensions to form is high.

The experimental control of the smallness of the size of particules is founded on the information given by von Weymarn's relation. Quite evidently, it results from the latter that the relative supersaturation must be minimized. Consequently,

- the solute solubility must be as high as possible. The precipitation temperature must be increased so that this may be the case;
- the solutions used must be diluted in order to decrease Q;
- the precipitant must be added slowly under efficient stirring in order to decrease the instantaneous concentration Q.

One method used to increase the size of the particules is to induce their coagulation or flocculation. The coagulation is the tendency exhibited by particules, when they are neutral, to agglomerate, that is, to reduce their surface area. But the adsorption of varied ions at the surface of precipitates causes them to be positively or negatively charged and induces an electrostatic repulsion that precludes their coagulation. More precisely, the adsorption induces the presence of two adsorption sheaths (Fig. 38.2), and coagulation cannot take place if the size of the second sheath is too great. Coagulation may be induced in two manners:

- by an increase in the concentration of the ions constituting the secondary adsorption sheath. In this case, there is a decrease in the size of the sheath when there is an increase in the neutral electrolyte concentration. The minimum amount of electrolyte to cause coagulation is called the *critical concentration of coagulation,* or *flocculation value,* or *coagulation value*;

- by an increase in the temperature. When this is the case, the kinetic energy of the varied species of the solution, including ions, is increased. As a result, ions exhibit a tendency to go away from the central particule, facilitating the coagulation.

The inverse phenomenon of coagulation is peptization. It occurs when the precipitate passes into a colloidal solution. (Recall that the colloidal state of matter is characterized, among other properties, by a certain range of particule sizes: 1 nm–0.1 μm.) This may occur if the electrolyte concentration in the supernatant liquid goes below the coagulation value. These considerations may be summarized by the scheme

$$\text{Colloid} \underset{\text{Peptization}}{\overset{\text{Coagulation}}{\rightleftharpoons}} \text{crystalline precipitate.}$$

Another point that must not be forgotten is that the composition of the precipitate, which finally is weighed, must be well defined. In addition to the impurities already mentioned, some precipitates may contain volatile impurities and also substantial quantities of water. Hence, they must be dried at a sufficient temperature. For example, a standard method of magnesium determination is to precipitate it as ammonium magnesium phosphate $MgNH_4PO_4, 6H_2O$. Unfortunately, this precipitate is difficult to dry at relatively low temperatures without its losing hydration water and ammoniac in controllable and reproducible manners. As a result, the precipitate must be dried by ignition at temperatures higher than 1000°C, at which it decomposes into magnesium pyrophosphate $Mg_2P_2O_7$.

Concerning the presence of some quantities of water in precipitates, we'll mention that there are several types of water accompanying precipitates. In addition to superficially adherent water, the following other types must be encountered:

- adsorbed water, which is present on all solid surfaces. Its amount depends on the atmospheric humidity;
- sorbed water. It is encountered with several colloidal substances, such as starch, proteins, coal, zeolites, and silica. It differs from adsorbed water by the fact that its quantity may be very high (about 20% of the total weight of the solid) although the powder still has the aspect of a perfectly dry powder;
- occluded water, which is present within the precipitate;
- essential water, such as that of crystallization (for example, as in calcium oxalate CaC_2O_4,H_2O) or that of constitution. In the latter case, it is not present as molecules. Essential water forms extemporaneously by heating. For example, calcium hydroxide liberates water by heating according to the reaction

$$Ca(OH)_2 \longrightarrow CaO\downarrow + H_2O$$

38.6 Precipitation from Homogeneous Solution

The technique of precipitation from homogeneous solutions is practiced in order to carry out a very slow addition of the precipitating reagent. It consists of extemporaneously and slowly generating the precipitant in the solution containing the analyte.

Table 38.1 Some examples of reactions generating precipitating reagents used in homogeneous precipitation

Precipitating agent	Generating reactant	Generated reaction	Precipitated elements
$PO_4{}^{3-}$	Trimethyl phosphate	$(CH_3O)_3PO + 3H_2O \rightarrow 3CH_3OH + H_3PO_4$	Zr, Hg
$C_2O_4{}^{2-}$	Diethyl oxalate	$(C_2H_5)_2C_2O_4 + 2H_2O \rightarrow 2C_2H_5OH + H_2C_2O_4$	Mg, Zn, Ca
$SO_4{}^{2-}$	Dimethyl sulfate	$(CH_3O)_2SO_2 + 4H_2O \rightarrow 2CH_3OH + SO_4{}^{2-} + 2H_3O^+$	Ba, Ca, Sr, Pb
$CO_3{}^{2-}$	Trichloroacetic acid	$CCl_3COOH + 2OH^- \rightarrow CHCl_3 + CO_3{}^{2-} + H_2O$	La, Ba, Ra
H_2S	Thioacetamide	$CH_3CSNH_2 + H_2O \rightarrow CH_3CONH_2 + H_2S$	Sb, Mo, Cu, Cd

This is performed by a very slow chemical reaction whose product is the precipitant. The local excess of precipitant is maintained constantly in a weak excess, and, as a result, the RSS is constantly low. An example is provided by urea, which generates hydroxide ions. In these conditions, they satisfactorily precipitate varied metallic ions as hydroxides or oxides. Hydroxide ions result from the reaction

$$\begin{array}{l} {}_2HN \\ \quad \diagdown \\ \quad\;\, \rangle\!= O + 3H_2O \longrightarrow CO_2 + 2NH_4{}^+ + 2OH^- \\ \quad \diagup \\ {}_2HN \end{array}$$

The reaction occurs slowly up to a temperature of 100°C. Table 38.1 gives some other examples of such reactions.

Hydroxide ions formed by the decomposition of urea permit us to slowly and regularly increase the pH of a solution of sodium hydrogen oxalate COOH–COO$^-$. This methodology can be considered a process to obtain convenient crystals of calcium oxalate. The reactions that occur are

$$\begin{array}{l} {}_2HN \\ \quad \diagdown \\ \quad\;\, \rangle\!= O + 2HC_2O_4{}^- + H_2O \rightleftharpoons 2NH_4{}^+ + CO_2 + 2C_2O_4{}^{2-} \\ \quad \diagup \\ {}_2HN \end{array}$$

$$C_2O_4{}^{2-} + Ca^{2+} \rightarrow C_2O_4Ca\downarrow$$

Magnesium and zinc ions may be precipitated satisfactorily using this process. Oxalate ions may also be obtained by hydrolysis of diethyl oxalate. In the following example, the reagent may be slowly generated by modifying the pH value at will. It is the case with Ca^{2+}, which is displaced from its EDTA-complex by decreasing the pH. Then Ca^{2+} precipitates with the oxalate ion:

$$[Ca - EDTA]^{2-} + 2H^+ \rightleftharpoons [H_2 - EDTA]^{2-} + Ca^{2+},$$

$$Ca^{2+} + C_2O_4{}^{2-} \rightarrow CaC_2O_4\downarrow$$

Another example is provided by the precipitation of barium ions with sulfate ions generated by the hydrolysis of sulfamic acid:

$$NH_2SO_3H + H_2O \rightleftharpoons NH_4{}^+ + H^+ + SO_4{}^{2-}.$$

Despite all these processes, numerous precipitates cannot form crystals. A colloidal solid is generally encountered if its solubility S is very low. This is the case with heavy metal sulfides. A very weak value of S confers a large value to the RSS.

38.7 The Gravimetric Factor

In order to introduce this central notion of gravimetry, we'll investigate the following examples.

Exercise 2 An impure sample of sodium chloride is dissolved in water. Chloride ions are precipitated with silver nitrate. A total of 1.000 g of silver chloride is formed. What is the weight of chlorine in the sample? [M(AgCl) = 143.337 g/mol, Cl = 35.457 g/mol^{-1}.]

143.337 g of AgCl contains 35.457 g of chlorine,
1 g contains 143.337 less chlorine.

The chlorine mass of the sample is 35.457/143.337 = 0.2474 g.

In this calculation, we notice that the ratio (mass of the substance to be determined/ mass of the weighed substance) intervenes. We also notice that the numerator and denominator of the ratio are both affected by the coefficient 1.

Exercise 3 Iron of an impure sample of iron carbonate $FeCO_3$ is converted into oxide Fe_2O_3 after successive dissolution, oxidization, precipitation, and ignition. The mass of oxide obtained is 1.000 g. What is the mass of the iron, expressed in iron carbonate, in the initial sample? The impurities contained in the sample are inert. [M(Fe_2O_3) = 159.70 g/mol, M($FeCO_3$) = 115.86 g/mol, M(Fe) = 55.85 g/mol.]

159.70 g of oxide contains 2 × 55.85 g of iron, i.e., 111.7 g.
1 g of Fe_2O_3 contains 159.70 less iron, that is, 111.7/159.7 = 0.6994 g.

Since 55.85 g of iron is contained in 115.86 g of iron carbonate, 1 g of iron is contained in 55.85 lesser mass of carbonate and 0.6994 g in 0.6994 higher carbonate mass. Hence, the mass of iron in the sample, expressed in carbonate, is 1.4509 g.

The first part of this exercise involves the ratio of the molar masses of the analyte and the precipitate in which it is engaged, i.e., 55.85/159.70. It was the same case in the preceding exercise. However, in the last one, it is multiplied by the stoichiometric factor 2 since the Fe_2O_3 oxide contains two iron atoms, whereas the carbonate contains only one atom.

These two exercises introduce the notion of gravimetric factor or analytical factor. It is defined as the ratio of the analyte mass and the mass of the compound that is weighed. In the first example, the analytical factor was 1 [M(Cl)/M(AgCl)], and in the second, it was 2[M(Fe)/M(Fe_2O_3)]. In the expression of the analytical factor, the analyte mass is always located in the numerator.

Exercise 4 Calculate the gravimetric factors relative to the conversions of $BaSO_4$ into Ba, $Mg_2P_2O_7$ into MgO, $KClO_4$ into K_2O, and Fe_3O_4 into Fe_2O_3.

Answers 0.5885; 0.3621; 0.3399; 1.035.

Exercise 5 Calculate the gravimetric factors relative to the following transformations:

$(NH_4)_2PtCl_6 \rightarrow NH_3$; $MoS_3 \rightarrow MoO_3$; $U_3O_8 \rightarrow U$; $B_2O_3 \rightarrow Na_2B_4O_7, 10H_2O$; $(NH_4)_3PO_4, 12MoO_3 \rightarrow P_2O_5$.

Answers 0.07671; 0.7492; 0.8480; 2.738; 0.03783. The introduction of the gravimetric factor permits us to immediately calculate the analyte's percentage in the initial sample.

Exercise 6 An iron ore is analyzed by dissolving 1.1324 g of sample in concentrated hydrochloric acid. The resulting solution is diluted with water and Fe(-III) is precipitated as ferric hydroxide Fe_2O_3,xH_2O by the addition of ammonia. After filtration and washing, the remaining is ignited at high temperature. Finally, we obtain 0.5394 g of Fe_2O_3. Calculate the percentage of iron in the initial sample. [M(Fe_2O_3) = 159.690 g/mol, M(Fe) = 55.847 g/mol.]

159.690 g of Fe_2O_3 contains 2×55.847 g of iron,
1 g contains 159.690 less iron,
0.5394 g contains 0.5394 more.

The sample iron content is

$$(2 \times 55.847/159.69) \times 0.5394 = 0.3773 \text{ g}.$$

This is the value of the gravimetric factor (GF) multiplied by the obtained mass. The result in percentage is accessible through the following reasoning:

1.1324 g of sample contains 0.3773 g of iron,
1 g contains 1.1324 less iron,
100 g contain 100 more, i.e.:

$$0.3773 \times 100/1.1324 = 33.32\%.$$

In this case,

$$\text{Percentage of iron} = (\text{gravimetric factor} \times \text{weight of precipitate/weight of sample}) \times 100,$$

and generalizing,

$$\text{analyte percentage} = (\text{weight of precipitate/weight of sample}) \times \text{GF} \times 100.$$

We give some gravimetric factor values in Table 38.2.

Table 38.2 Values of some gravimetric factors (HQ: hydroxyquinoline; dmg: dimethylglyoxime)

Analyte	Precipitated species	Weighed species	Gravimetric factor
MgO		$Mg_2P_2O_7$	$\frac{2M(MgO)}{M(Mg_2P_2O_7)} = 0.3622$
Fe_3O_4		Fe_2O_3	$\frac{2M(Fe_3O_4)}{3M(Fe_2O_3)} = 0.9666$
Fe^{3+}	$Fe(OH)_3$	Fe_2O_3	$\frac{2M(Fe)}{M(Fe_2O_3)} = 0.6994$
Al^{3+}	$Al(O)_3$	$Al(O)_3$	$\frac{2M(Al)}{M[Al(O_3)]} = 0.0587$
Ni^{2+}	$Ni(dmg)_2$	$Ni(dmg)_2$	$\frac{M(N_1)}{M[N_1(dmg)_2]} = 0.2032$

38.8 Sensitivity of Gravimetry

According to the preceding considerations, it is legitimate to set up the following relation:

$$\text{weight of the precipitate} = k.\ \text{weight of the analyte.}$$

k appears to be the sensitivity factor of the method. It also appears as an amplification factor. For example, let's consider the determination of iron as Fe_2O_3. We can write

$$2 \times \text{moles}(Fe_2O_3) = \text{moles}(Fe).$$

$$m(Fe_2O_3)/M(Fe_2O_3) = {}^1/_2[m(Fe)/M(Fe)].$$

The ratios $m(Fe_2O_3)/M(Fe_2O_3)$ and $m(Fe)/M(Fe)$ are the mole numbers of Fe_2O_3 and Fe. We can also write

$$m(Fe_2O_3) = {}^1/_2[M(Fe_2O_3)/M(Fe)]m(Fe),$$

where $m(Fe_2O_3)$ and $m(Fe)$ are the obtained masses. From this relation, we deduce two ways to increase the sensitivity of the method:

- the first involves increasing the ratio of the molar masses of the precipitate and the analyte. This implies that the molar mass of the precipitate must be as high as possible. This is, of course, interesting when traces are to be analyzed. For example, phosphate ions may be analyzed as oxinium phosphomolybdate of the global formula $[P(MoO_7)_6]H_4$, $(C_9H_7NO)_3$, $2H_2O$. In this precipitate, the percentage of phosphor, estimated as phosphoric anhydride (P_2O_5), intervenes only for 3% of the weight;
- the second way is less evident than the preceding one. It takes the stoichiometric factor into account. It indeed appears that the precipitate must possess the fewest possible number of motifs. In the above example, the motif is the iron atom.

38.9 Some Experimental Details

38.9.1 *Quantitative Filtration*

In gravimetry by precipitation, the following devices are used for the filtration process:

- filter papers. First, they are used for the filtration of spongy precipitates (such as some hydroxides) or gelatinous and voluminous ones and coarse particules that do not deteriorate by ignition of the paper. The latter is ignited in a platinum crucible that is weighed ahead of time;
- sintered glass filtering crucibles. They are essentially used to retain thermally sensitive precipitates such as those given by organic complexes of metal cations;
- porous porcelain filtering crucibles used to filtrate very thin precipitates, such as those of $BaSO_4$, CaC_2O_4, $MgNH_4PO_4$, and so on.

Filter papers and glass filtering crucibles are used at temperatures up to 500°C and porcelain crucibles at higher temperatures.

There are other microanalytical techniques that enable us to work with samples whose masses are about 1 mg. In such cases, filtering sticks are used.

38.9.2 *Drying of Precipitates*

Precipitates may be dried in a dessicator in order, in a first stage, to obtain the so-called equilibrium weight (limits of agreement with previous weights of the same precipitate: ± 0.5 mg). The filter paper must be dried ahead of time, as the filtering crucibles must also be. For the latter, this is done in furnaces.

38.9.3 *Precision Balances*

We'll only mention the existence of electronic balances here. They are classified in three groups:

- macrobalances, which can accept loads up to 200 g with an accuracy of 0.1 mg;
- microbalances, loads up to 20 g with an accuracy of 0.1 mg;
- ultramicrobalances, loads up to 5 g with an accuracy of 0.1 mg.

38.9.4 *Thermobalances*

Thermobalances permit us to register the weight change(s) of a precipitate as a function of temperature. Hence, by a simple study of thermogravimetric curves, they permit us to fix the optimal temperature conditions of a gravimetric determination. For example, the study of the curve obtained with calcium oxalate shows that up to 200°C, the precipitate formula is CaC_2O_4, H_2O. In the temperature interval 200–400°C, the formula is CaC_2O_4; between 400–700°C, it is $CaCO_3$; and, finally, it is CaO after 900°C.

38.10 Some Characteristics of Gravimetry by Precipitation

The range of applications is governed by the balance sensitivity. If we want to obtain an accuracy of ±0.1% together with a sensitivity of ± 0.1 mg by using an analytical balance, the precipitate must weigh at least 100 mg.

The accuracy of "macromeasurements" is about 0.1–0.2%. Actually, it is limited by losses due to an excessive solubility, losses occurring during the different handlings, and losses due to impurities. One of the characteristics of gravimetry lies in the fact that, in principle, the origin of errors may easily be found since the filtrate may be analyzed for an eventual presence of the analyte. The precipitate itself may be analyzed in order to detect the presence of impurities within it.

The relative precision of gravimetry by precipitation depends on the sample and the precipitate weights. A precision of 1–2% may be obtained with small samples. With more important quantities, the relative precision may attain some parts per million. Few quantitative methods exhibit such a precision.

The specificity of the method is connected to the precipitation reaction. Generally, precipitating reagents are more selective of a group of ions than they are specific of only one metallic ion.

Gravimetry by precipitation does not require a prior standardization or a prior drawing of a calibration curve. In this sense, it is qualified as being an absolute method, as is, for example, coulometry (see electrochemistry).

Its principal drawback is the fact that it is a time-consuming method. This may be ascribed to the several successive operations, such as drying, ignition, digestion, and evaporation, that are carried out. However, this criticism must be moderated by the fact that it does not mobilize the operator permanently.

Finally, we'll recall that gravimetry by precipitation, because it is an absolute method, is used to check and calibrate instrumental methods.

Chapter 39
Some Applications of the Precipitation Phenomenon in Inorganic and Organic Qualitative and Quantitative Analysis

In this chapter, we mention some analytical applications of the precipitation phenomenon. We group them arbitrarily under the following headings: precipitations as silver salts; other titrimetric methods involving the precipitation phenomenon; gravimetry; determination of inorganic ions after precipitation with organic precipitant reagents; and some aspects of organic and inorganic qualitative analysis.

In our opinion, these headings are sufficient to provide good insight into the importance of the precipitation phenomenon.

39.1 Titrations Involving the Precipitation of Insoluble Silver Salts

In the French literature, these titrations are called *argentometric* titrations.

39.1.1 Determination of Organic Halogens

Organic halides are principally designated as chlorine and bromine and (more rarely) iodine, one or several atoms of which are linked to a carbon atom of an organic molecule through a true covalent bond. In them, halogen atoms are absolutely not free and hence do not at all exhibit the reactivity of a halide ion. For the determination of organic halides, we must first break the C–X bond by mineralization. Once the break has been carried out, it becomes possible to determine the halogen as the halide ion with silver ions, as usual.

The breakdown of the bond C–X with the formation of the X^- ion is relatively easy. It depends on the structure of the initial halogenated derivative. Distinctions between the different organic halides may be made. They are based on their reactivity with silver nitrate in ethanol. Three groups are distinguished:

J.-L. Burgot, *Ionic Equilibria in Analytical Chemistry*,
DOI 10.1007/978-1-4419-8382-4_39, © Springer Science+Business Media, LLC 2012

- the first one is composed of derivatives that quickly give a precipitate at room temperature. They are allylic halides, benzylic halides, and tertiary alkyl halides. The high reactivity of the first two must be ascribed to the stabilization by resonance of the carbonium ion resulting from the ionization of the initial halide:

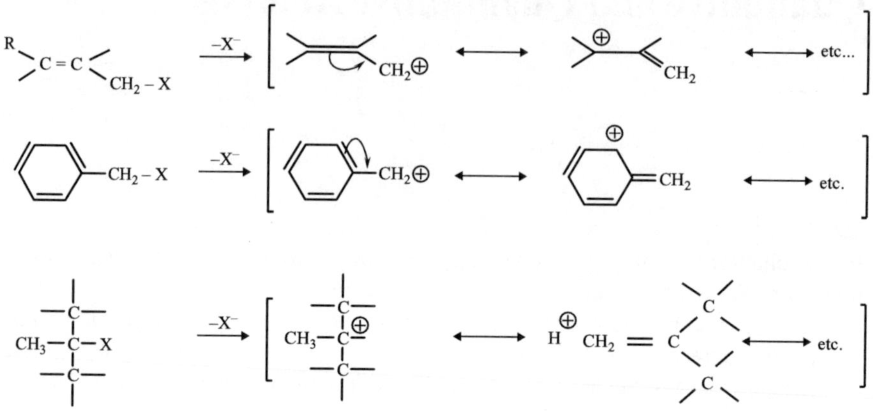

The last kind of halides (last scheme) involves the phenomenon known as *hyperconjugation*;

- the second group is composed of derivatives that exhibit a reaction slower than the previous ones. Heating the system is often necessary for the reaction to occur. Primary and secondary alkyl halides and vicinal alkyl dihalides are located in this group. These compounds do not give a stabilized carbonium ion by expulsion of the halide ion;
- the third group is composed of vinylic and aryl halides that do not give any reaction, even by heating. Contrarily to both preceding cases, the compounds that are stabilized by resonance are the initial halides. For example, for vinylic derivatives, we see that the carbon–halogen bond possesses some double–bond character. As a result, the ionization ability of the halogen atom to form the corresponding halide ion decreases drastically from the former group to the latter:

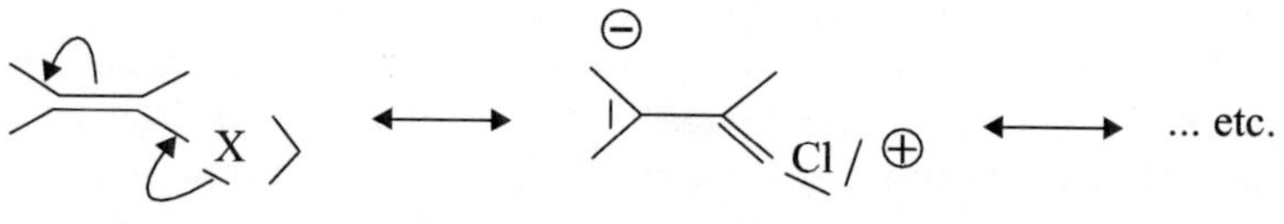

Mineralization of the initial halide may be carried out by a reductive or an oxidative process. The reductive mineralization is performed either with metallic sodium (Lassaigne's method) or with zinc in mineral acid medium or by catalytic hydrogenation with Raney's nickel. The oxidative mineralization may be carried out in the presence of dioxygen according to Schöniger's method. It consists of burning 5–10 mg of substance in air.

39.1.2 *Determination of Hydrochlorides, Hydrobromides, and Hydroiodides*

All the previous considerations do not apply, of course, to hydrochlorides, hydrobromides, and hydroiodides of varied organic compounds since, in water, the halide ions of these derivatives are already totally free. As a result, their behavior is totally identical to that of an inorganic halide. We recall that quaternary ammonium ions as well as sulfonium and oxonium ions may have halide ions as counterions.

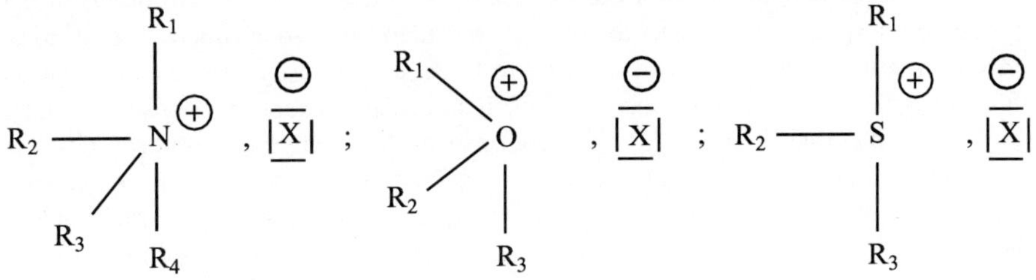

39.1.3 *Some Examples*

We begin with some examples taken from the fourth edition of the European pharmacopeia.

39.1.3.1 Titrations Involving a Prior Mineralization or, More Generally, a Prior Transformation of the Compound to Be Determined

We can mention the examples of

- chlorobutanol:

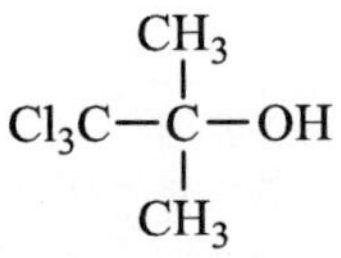

In a first stage, it must undergo the action of a hot concentrated sodium hydroxide solution to be mineralized. After acidification of the solution by nitric acid, it is titrated according to Charpentier–Volhard's method.

- chloral hydrate:

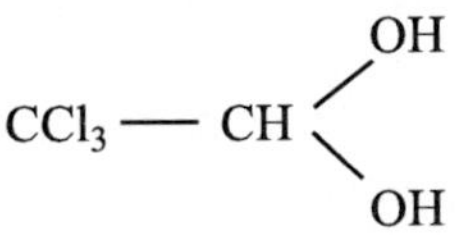

In a first stage, it is hydrolyzed by an excess of concentrated sodium hydroxide. Then the solution is neutralized by a standard sulfuric acid solution in the presence of phenolphthalein. The liberated chloride ions are titrated according to Mohr's method. The reactions occurring during the hydrolysis are

$$CCl_3CH(OH)_2 + OH^- \rightarrow CHCl_3 + HCOO^- + H_2O,$$
$$CHCl_3 + 4OH^- \rightarrow HCOO^- + 3Cl^- + 2H_2O.$$

The first comment that must be made is to assert that it is a prototropic back titration with sulfuric acid, and not only an "argentometric" titration, although, in the procedure, a quantitative determination of chloride ions is mentioned. The titration of chlorine ions with silver ions is used to be sure that an excessive consumption of hydroxide ions does not exist. In fact, it is just this excessive consumption of hydroxide ions that liberates the chloride ions from chloroform. In other words, the consumed hydroxide ions that correspond exactly to the concentration of chloral hydrate are those given in the first chemical equation. The second comment is that Mohr's method is particularly well adapted to determine the liberated chloride ions, due to the basic pH of the solution imposed by the use of phenolphthalein as indicator;

- thiamphenicol, cyclophosphamid, and chlorambucil:

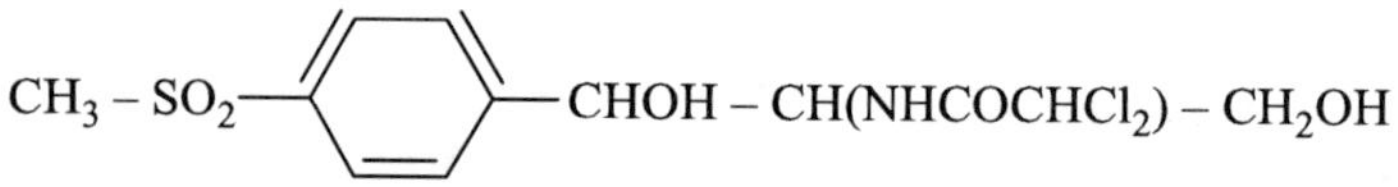

thiamphenicol

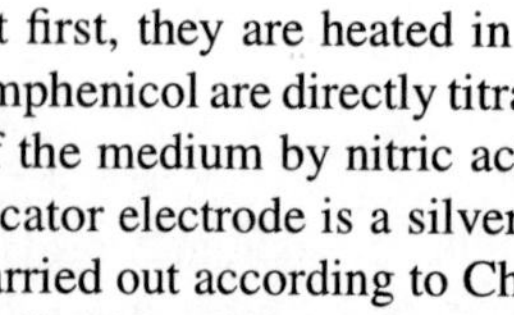

cyclophosphamide

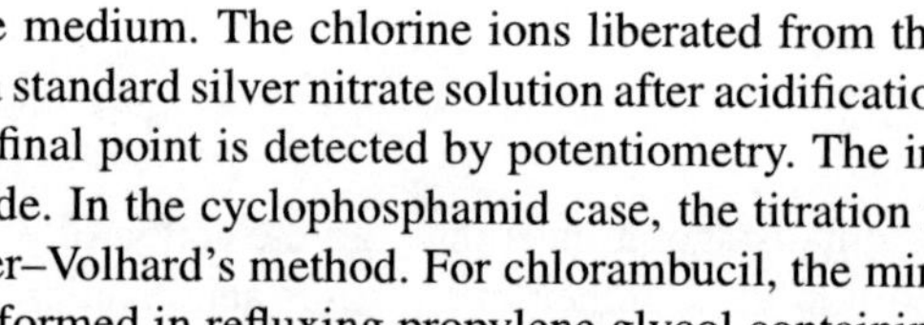

chlorambucil

At first, they are heated in alkaline medium. The chlorine ions liberated from thiamphenicol are directly titrated by a standard silver nitrate solution after acidification of the medium by nitric acid. The final point is detected by potentiometry. The indicator electrode is a silver electrode. In the cyclophosphamid case, the titration is carried out according to Charpentier–Volhard's method. For chlorambucil, the mineralization of chlorine atoms is performed in refluxing propylene glycol containing potassium hydroxide.

Numerous determinations of the same kind as those given here have been recommended in previous pharmacopeias. Generally, they consist of titrating halide ions, once they are formed by mineralization, by an argentometric method. Usually, it is Charpentier–Volhard's method. We can also cite the examples of lindane (γ-benzenehexachloride, DTT, iodoform, eosine, chlorotrianisene, etc.). The preliminary mineralization is carried out in alkaline medium, that is, with sodium or potassium hydroxide in alcohol over a relatively important time interval. Mohr's method has been recommended to titrate iodide ions resulting from the mineralization of iodinated contrast products. The preliminary mineralization is carried out with potassium permanganate in alkaline medium followed by the reduction of the formed iodine by a bisulfite solution. Fajans' method is also recommended sometimes.

39.1.3.2 Argentometric Titrations Requiring No Preliminary Mineralization

First, we recall that several direct titrations with silver nitrate have been recommended in formularies and pharmacopeias older than the present European pharmacopeia. These include the titrations of

- ammonium and sodium chlorides, especially in Ringer's liquid,
- silver nitrate in ophthalmic solutions,
- calcium chloride,
- chloride ions in lauryl sulfate and potassium perchlorate solutions,
- iodhydric acid,
- tetramethylammonium bromide and chloride,
- thiourea,
- zinc chloride,
- gallamine triiodoethylate,
- quaternary ammoniums formed by protonation of amines or not, etc.

Most of these titrations are recommended according to Charpentier–Volhard's method. In this context, its great advantage lies in the fact that during the titration, the medium remains acidic. In these conditions, the quaternary ammoniums to be titrated, which were formed by the protonation of amines and which are usually hydrochlorides or hydrobromides, do not precipitate because they remain protonated. The reaction that occurs is

$$BH^+, X^- + Ag^+ \rightarrow BH^+ + AgX\downarrow$$

In alkaline medium, as with Mohr's method, the following reaction might occur:

$$-\overset{|}{\underset{|}{N}}H^+ + OH^- \rightarrow -\overset{|}{\underset{|}{N}}\downarrow + H_2O$$

with the result that the conjugate amine precipitates. This precipitation would disturb the titration and even distort it. Recall that amines, regardless of their class and alkaloids, protonated or not, constitute the most important class of drugs.

In the list of the above-mentioned compounds, some of them form silver salts completely only in the presence of ammonia. This is the case for thiourea and also theophylline. The precipitation of thiourea by silver ions might be the result of the following two reactions:

$$2AgNO_3 + 2NH_4OH \rightarrow Ag_2O + 2NH_4NO_3 + H_2O,$$

$$Ag_2O + (NH_2)_2C = S \rightarrow Ag_2S\downarrow + (NH_2)_2C = O.$$

A black precipitate of silver sulfide appears in these conditions. In acidic medium, a white precipitate of the silver salt of the tautomeric form of thiourea appears:

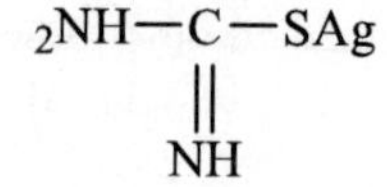

The titration is a back titration according to Charpentier–Volhard's method after filtration of silver sulfide as well as acidification of the solution. Theophylline gives a silver theophillinate precipitate only totally in ammonia and after heating. Silver theophillinate does not form in diluted acid medium, and its precipitation is incomplete in pure water.

Of course, in ammonia, silver ions stand as ammine complexes.

39.1.4 Zeisel's Method: Determination of Methoxy and Ethoxy Groups

Zeisel's method is used to determinate methoxy and ethoxy groups. It is a method of both qualitative and quantitative organic analysis. It is mentioned under the heading "General Methods of Analysis" of the 10th edition of the french pharmacopeia. It ends with an argentometric titration. Its principle is based on the reaction of hydroiodic acid (Berthelot's universal reductor) with ether oxides, according to the reactions

$$ROCH_3 + HI \rightarrow ROH + CH_3I,$$

$$ROC_2H_5 + HI \rightarrow ROH + C_2H_5I.$$

Methyl and ethyl iodides are the only sufficiently volatile alkyl iodides to be expelled from the initial solution by a flow of carbonic gas (propyl iodide can be slightly expelled away and butyl iodide not at all). Herein lies the specificity of the method. The formed alkyl iodides are collected in a solution containing a known concentration in excess of silver nitrate. Silver iodide is formed and precipitates:

$$RI + AgNO_3 \rightarrow AgI\downarrow + RNO_3.$$

(According to some authors, silver iodide might be formed through an intermediary organic complex.) The silver ions in excess are titrated with ammonium thiocyanate according to the last stage of Charpentier–Volhard's procedure.

Zeisel's method is now used less than it was previously since the advent of nuclear magnetic resonance (NMR). In ^1H NMR, the proton chemical shifts of the groups $-OCH_3$ and $-OCH_2CH_3$ are indeed located in a characteristic range and, moreover, offer a characteristic multiplicity. However, NMR remains a poorly sensitive method. An overview of the previous importance of Zeisel's method may be immediately gained if we recall the fact that numerous natural substances, among which are several alkaloids, do possess one or several methoxy groups.

The primary method Zeisel conceived has undergone several modifications since it was originally published.

39.1.5 *Prototropic Titrations in the Presence of Silver Ions*

A very interesting method is that consisting of carrying out prototropic titrations of the compounds, which are endowed with an insufficient acidic character, in the presence of silver nitrate and pyridine. The titrant solution may be an alkaline aqueous solution or a basic ethanolic solution.

The silver ions become fixed both on the conjugate base and on the nitrogen atom of pyridine. Hence, there is the formation of a kind of complex. Simultaneously, the liberated protons are fixed on the nitrogen atom of pyridine, giving the pyridinium ion ($pK_a = 5.19$). As a global result, the titration reaction is displaced toward the right. It becomes complete. Actually, it is probably the pyridinium ion that is titrated with the basic titrant. The pyridine's presence offers another interest. It complexes silver ions in water and in pyridine as well. In water, the silver pyridine(I) complexes $[AgPy]^+$ and $[Ag(Py)_2]^+$ exhibit the following stability constants, respectively: $\log \beta_1 = 2.05$ and $\log \beta_2 = 4.10$. This formation of complexes precludes the precipitation of silver oxide.

For example, barbital is a weak diacid ($pK_{a1} = 7.9$ and $pK_{a2} = 12.3$).

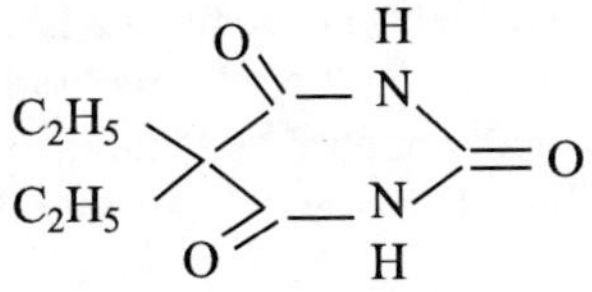

barbital

Without any particular condition, the second acidity cannot be totally titrated by a sodium hydroxide solution in water. When the titration is carried out in pyridine with a sodium hydroxide solution in ethanol in the presence of thymolphthalein as indicator, the difficulty no longer remains. It is the same state of affairs with phenobarbital and amobarbital. In the presence of pyridine, the following overall reaction occurs:

$$\text{barbH}_2 + 2\text{Ag}^+ + 4\text{Py} \rightarrow \text{barbAg}_2(\text{Py})_2 + 2\text{HPy}^+,$$

where barb denotes the barbital molecule as its dibasic form. According to some authors, the characteristic atom chain in the silver complex might be

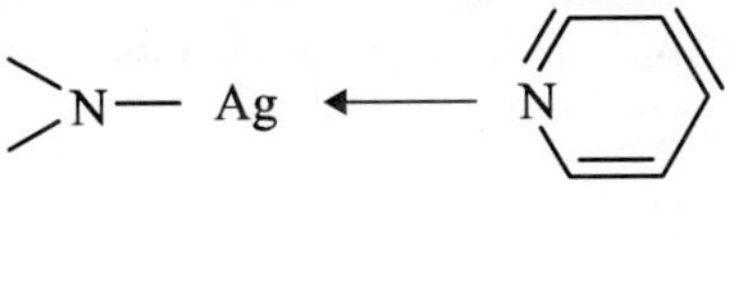

Again, we may observe the tendency exhibited by silver ions to form linear complexes of coordinence 2.

This kind of titration is not an argentometric one, strictly speaking. It would be classified either in the group of titrations involving formation of complexes or in the group of prototropic titrations after transformation (see Chap. 11). The classification in the first group is justified by the fact that it seems that the bound between the nitrogen atom (of barbital) and the silver ion does possess some characteristics of a covalent bond.

Also titrated in a similar way are the sodium salts of phenytoin and phenobarbital:

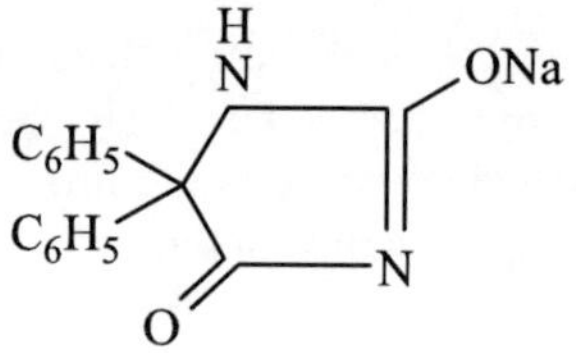

phenytoin (sodium salt)

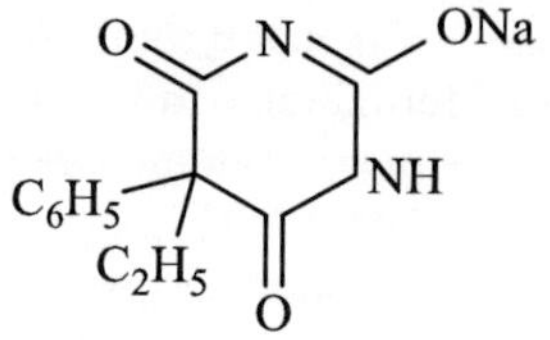

phenobarbital (sodium salt)

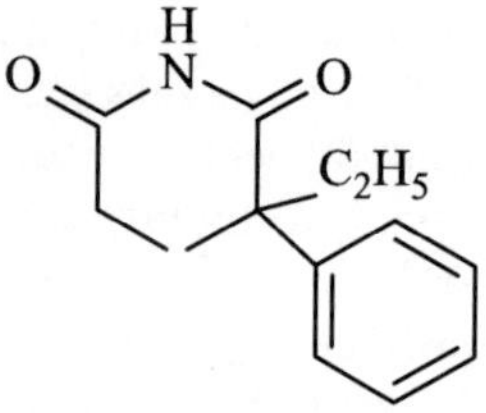

glutethimide

Once the first acidity is neutralized by adding sodium hydroxide without the presence of pyridine, silver nitrate in pyridine solution is then added. The titration is followed by potentiometry with a silver electrode as indicator electrode. Methyl and propylthiouracils may exist as several tautomeric forms. The lactime and thiolactime forms cause their two acidities to appear. The value of the former is known ($pK_a = 8.2$). By titration in aqueous solution with sodium hydroxide, without any particular condition, the second acidity cannot be totally neutralized, but its neutralization becomes possible in the presence of silver ions. Thiouracils may complex two silver ions:

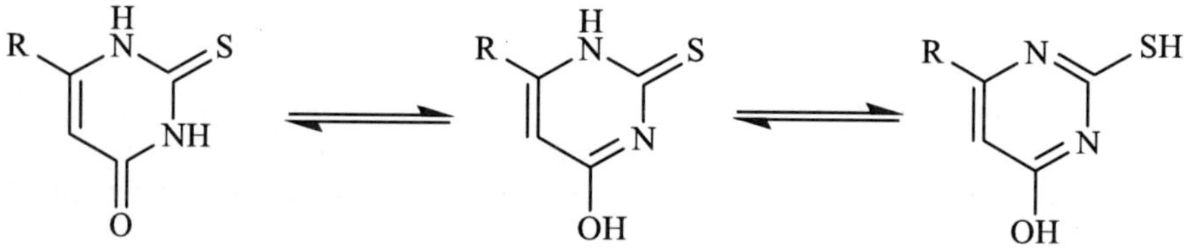

$R = CH_3$
and
$R = -CH_2 - CH_2 - CH_3$

lactam form

lactim-thiolactim form

All the molecules just mentioned that are subject to this titration method do actually exhibit structures of ureides or thioureides. They possess the structural pattern

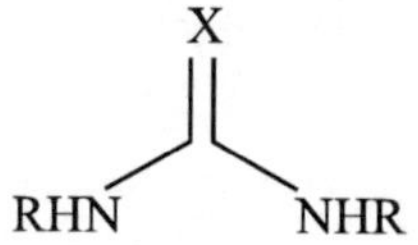

or

$X = 0$

$X = S$

Other derivatives that are not ureides may, however, also be subject to this reaction. This is the case, for example, of theophylline, which is endowed with a weak acidic character due to its mobile hydrogen brought by the nitrogen in position 7. Theophylline's pK_a value is 8.6. To perform the titration, an excess of silver nitrate is added and the titration is carried out by sodium hydroxide. The indicator is bromothymol blue.

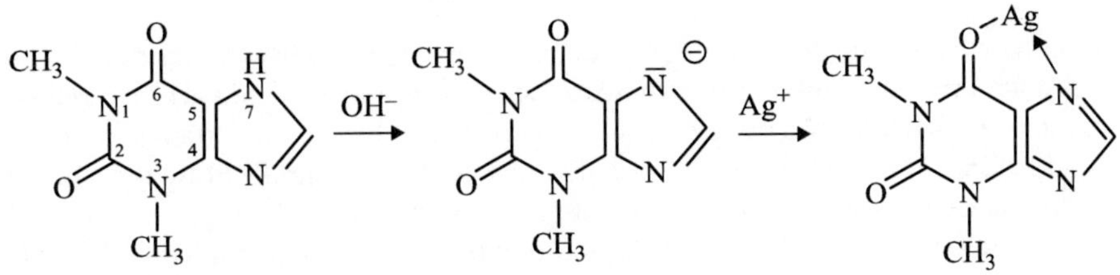

Let's also recall that 1-alkynes ("true alkynes") give silver salts. These derivatives exhibit an acidic character (pK_a of acetylene ≈ 25 to compare to the pK_a of methane ≈ 42). It is ascribed to the sp^1 hybridization of carbon atoms bringing the mobile hydrogen. Let's also note that, according to the European pharmacopeia, 17-ethynylsteroids (which are also 1-alkynes) are titrated by sodium hydroxide in tetrahydrofuran as solvent in the presence of silver nitrate (see Chap. 11).

39.2 Other Titrimetric Methods Involving a Precipitation Phenomenon

We are strictly content with studying quantitative determinations with tetraphenylboron ion (TPB): $(C_6H_5)_4B^-$. Its sodium and lithium salts are soluble in water. Its rubidium, cesium, potassium, and ammonium salts are very poorly soluble. The solubility product of potassium tetraphenylboron is $K_s(K(C_6H_5)_4B) = 2.25 \times 10^{-8}$. Likewise, numerous quaternary ammoniums and protonated amines form slightly soluble salts with tetraphenylboron ion according to the reaction

$$(R_4)N^+ + (C_6H_5)_4B^- \rightarrow (R_4)N, B(C_6H_5)_4\downarrow$$

This property gives rise to numerous qualitative and quantitative applications. Among the quantitative ones are some gravimetric determinations (see Sect. 39.3) and some titrations we are studying now.

A first procedure is based on the fact that the salts of amines and of TPB are more soluble in a mixture of water–acetone than is the silver salt of TPB. Therefore, the protonated amine is precipitated with TPB. The obtained salt is dissolved in acetone and is titrated with a silver nitrate solution according to the reaction

$$(R_4)N, B(C_6H_5)_4 + Ag^+ \rightarrow Ag, B(C_6H_5)_4\downarrow + (R_4)N^+ \quad \text{(water-acetone).}$$

A back titration may also be carried out by the addition of an excess of silver nitrate and titration of the excess with thiocyanate ions. This alternative may be classified in the group of argentometric titrations.

A second procedure consists of allowing mercuric chloride to react with the ammonium salt of TPB to be titrated. Hydrochloric and boric acids are liberated according to the reaction

$$(R_4)NB(C_6H_5)_4 + 4HgCl_2 + 3H_2O \rightarrow (R_4)NCl + 4C_6H_5HgCl + 3HCl + H_3BO_3.$$

The liberated hydrochloric acid can be determined in the presence of boric acid. The reaction of the ammonium salt with mercuric chloride added in excess is carried out in acetone. Then a known quantity of a standard sodium hydroxide certainly in excess is added in water. The excess of base that has not been neutralized by the formed hydrochloric acid is titrated with a standard acid solution. During the course of the titration—more precisely after the addition of the sodium hydroxide—an iodide solution is added. The complex tetraiodomercurate(II) forms, which precludes any interference with the dichloromercury(II) in excess. Due to the stability constants of the possibly existing complexes, it is indeed the following reaction that takes place:

$$[HgCl_2] + 4I^- \rightleftharpoons \left[HgI_4\right]^{2-} + 2Cl^-.$$

Additionally, the ammonium salts of TPB can also be determined by using electrochemical methods such as potentiometry and amperometry. TPB is, indeed, oxidizable on a graphite electrode (see the electrochemistry course). Finally, the ammonium salts of TPB may also be titrated in nonaqueous media and can be determined by spectrophotometry.

We will now very briefly mention some other titrations by precipitation. They are the precipitations of

- sulfate ions as barium sulfate. A standard solution of barium chloride is added to the solution of sulfate ions. The indicator is either the tetrahydroxyquinone or rhodisonic acid.

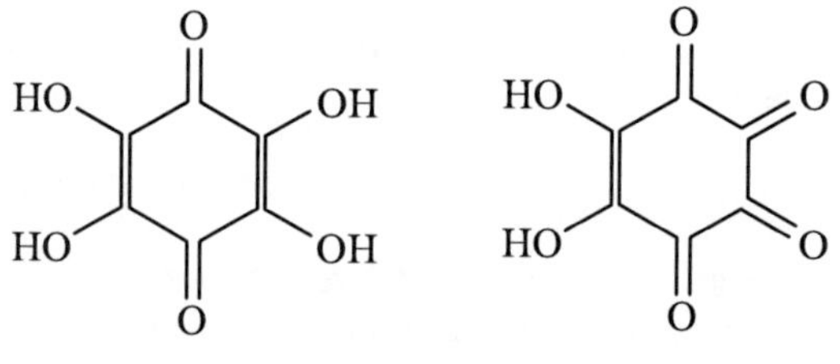

tetrahydroxyquinone rhodisonic acid

At the equivalence point, the indicator turns red. The standard solutions contain either barium nitrate or sodium or ammonium sulfate. In both cases, the equivalent is half the molar mass. These titrations are rarely performed because the detection of their equivalence point is unreliable. Most of all, it is used to quickly determine Ba^{2+} concentrations. In the presence of barium ions, the indicator is red. After the addition of an excess of sulfate ions, the color disappears;

- Zn^{2+}, Cd^{2+}, $UO_2{}^{2+}$, Pb^{2+}, Ni^{2+}, and Mn^{2+} ions with ferrocyanide ion. Numerous poorly soluble ferrocyanides indeed have a well-defined composition in some given conditions. At the equivalence point, the ferrocyanide ion in excess is detected by an external indication. For example, Zn^{2+} is determined in diluted strongly acidic medium by the following precipitation reaction:

$$2\left[Fe(CN)_6\right]^{4-} + 2K^+ + 3Zn^{2+} \rightarrow K_2Zn_3\left[Fe(CN)_6\right]_2\downarrow$$

(the precipitate formed is a mixed one). The ferrocyanide in excess is revealed with the uranyl cation [as nitrate $UO_2(NO_3)_2$], which gives a brown precipitate of uranyl ferrocyanide. This method is superseded by titrations with EDTA;
- phosphate ions by the formation of uranyl monoacid phosphate according to the reaction

$$UO_2(NO_3)_2 + NaH_2PO_4 \rightarrow H^+ + 2NO_3{}^- + Na^+ + UO_2HPO_4\downarrow$$

In nitric acid medium, the precipitation of uranyl phosphate is incomplete. Uranyl monohydrogen phosphate is insoluble in water and in an acetic acid solution. It is soluble in mineral acid medium. These points explain the use of the acetic buffer to carry out the precipitation. One may also use uranyl acetate instead of nitrate, but the former is less stable than the latter. The titration is performed at about $t = 80°C$. The indicators are either cochineal tincture as internal indicator or potassium ferrocyanide as an external one. Cochineal tincture is an acid–base indicator. It is obtained by the extraction of cochineal by alcohol. The principal application of this precipitation reaction occurs in the field of analytical biochemistry. More precisely, it is the determination of phosphates in urine. Although this procedure is superseded by colorimetric methods, it remains an interesting replacement method.

39.3 Gravimetry

Today few gravimetric determinations remain in the European pharmacopeia. The reason is simple: gravimetric determinations are time-consuming and generally are laborious. However, we'll mention the determination of nitrofural after hydrolysis and after formation of the 2,4-dinitrophenylhydrazone of the obtained 5-nitrofurfural:

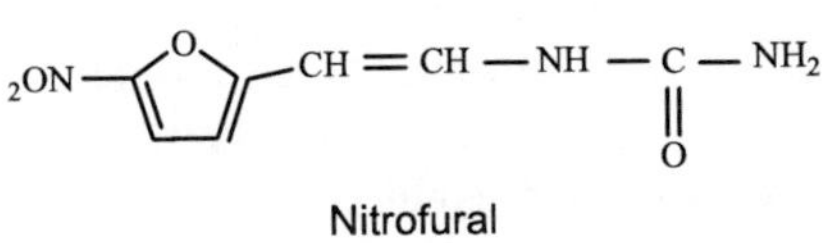

Nitrofural

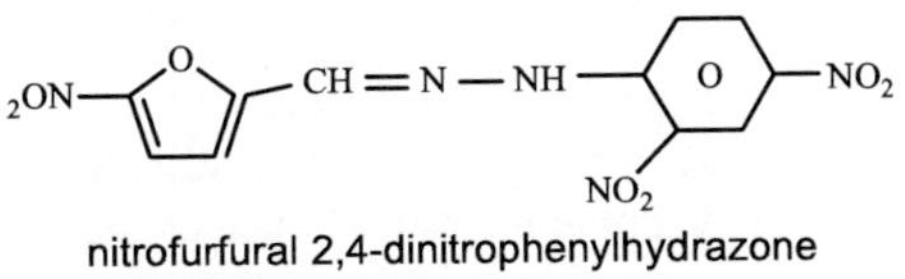

nitrofurfural 2,4-dinitrophenylhydrazone

Evidently, it is the phenylhydrazone mass that is determined. Otherwise, the determination of thiols is also carried out by gravimetry in some cases. Thiols are good nucleophilic agents and moreover react easily as thiolates with 1-chloro-2,4-dinitrobenzene. Thioethers form according to the reaction

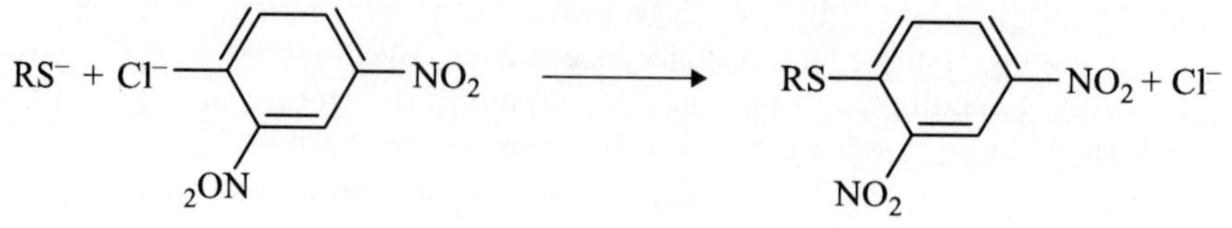

These derivatives crystallize very easily. Of course, numerous other methods exist to determine thiols, such as colorimetric ones and procedures by argentometry (see Sect. 39.1).

However, numerous gravimetric determinations of active ingredients have been recommended in old pharmacopeias and formularies. Without entering into the details of these determinations, we'll give some examples of such determinations.

39.3.1 Gravimetric Assays Involving Ignition

Sometimes the sample is directly ignited. This is especially the case with inorganic compounds. For example, this is the case with

- bismuth derivatives such as the basic subsalicylate, which is transformed, after ignition, into the oxide Bi_2O_3,
- magnesium trisilicate, which gives SiO_2 as residue,
- zinc derivatives, which yield the oxide ZnO.

In some other cases, the ignition follows a preliminary precipitation. We can then mention derivatives of aluminum that are successively transformed into aluminum hydroxide and then, after ignition, into alumine.

39.3.2 Gravimetric Assays Involving a Prior Solvent Extraction

According to the technique of gravimetric assays involving a prior solvent extraction, the compound to be determined is extracted with an appropriate solvent. The removal of the solvent yields a residue whose weight is determined. In some cases, the extraction can be carried out after a preliminary chemical treatment of the initial sample. Hence, we can distinguish the following two procedures:

- extraction without preliminary treatment. Some barbiturics have thus been determined. In the group of the derivatives treated with this procedure, we can also include some natural substances that, additionally, are often mixtures;
- extraction after a preliminary treatment with acid or alkali, which liberates acid or basic materials from initial salts, which, as a result, can then be extracted. Sodium salts of some barbiturics have been determined in such a way after acidification. Inversely, amine hydrochlorides and sulfates such as those of quinine, papaverine, quinidine, and tetracaine have been extracted and determined after alkalinization.

39.3.3 Assay Involving Solvent Extraction and Drying to Constant Weight

The successive operations of solvent extraction and drying to a constant weight may be carried out under different conditions of temperature, pressure, and time intervals. Concerning the moisture of a sample, a drying agent may be used as a supplementary parameter. The specificity of such a process is evidently weak since every substance extracted and not volatile in the experimental conditions is finally weighed. Under this heading, we can mention the notion of sulfuric ashes. They consist of the residue obtained after extraction of 100 g of sample in the presence of sulfuric acid. They are impurities that leave a fixed residue in the presence of sulfuric acid. Likewise, the notion exists of residue after extraction of essential oils. A classical example is provided by the notion of dry residue, whose determination is a procedure recommended for the analysis of a wine. It consists of the totality of substances that are not volatile in the experimental conditions of the assay.

39.3.4 Gravimetric Determinations Involving the Formation of a Precipitate that Is Weighed

Numerous determinations are located in this group, which consists of gravimetric determinations involving the formation of a precipitate that is weighed. In the pharmaceutical field, we find the determinations of aneurine (vitamin B1) as its silicotungstate, of hexamethonium tartrate as its semicarbazone, of penicillin G, and of procain after precipitation with ethylpiperidinium ion. Reinecke's salt is also used as a precipitant.

The formula of silicotungstic acid is SiO_2, $12WO_3$, $26H_2O$. It is used as a precipitating reagent of alkaloids. Reinecke's salt exhibits the formula $NH_4[Cr(NH_3)_2(SCN)_4]H_2$. It is the ammonium diamminetetracyanatochromate(III). Actually, the potassium salt is known as Reinecke's salt. It is a precipitant of primary and secondary amines and of some amino acids.

We can also mention the gravimetric determinations of protonated amines and quaternary ammoniums as tetraphenylboron salts (see above). Furthermore, alkaloids seem to have been the subject of the greatest number of this type of gravimetric determination. Usually, the precipitation of tetraphenylboron salts is carried out in the range $3 < pH < 5$. Gravimetric determinations of alkaloids as phosphotungstates and reineckates have also been proposed.

Phosphotungstic acid exhibits the approximate formula $24WO_3$, $2H_3PO_4$, $48H_2O$. It is also a precipitant of numerous nitrogenous bases, alkaloids, phenols, amino acids, and so forth.

39.4 Determination of Inorganic Ions After Precipitation with Organic Precipitants

Very often today, gravimetric determinations are carried out with organic precipitants. The precipitation of numerous metallic ions with organic precipitants provides an example of this procedure. The most important precipitants are those able to give chelates with metallic ions that are of a great stability. The chelate precipitates, which form in water, are soluble in organic phases. This property is advantageous because it permits their extraction in these media. Their extraction also opens a mode of analysis other than gravimetry or different from any other method of metal ion analysis. In addition, organic precipitants present another advantage: The intense color of the chelates they form permits their colorimetric determination, which can be highly sensitive in some experimental conditions. Moreover, the precipitates they form with metallic ions exhibit high molar masses very often for one metallic ion. The result is the high sensitivity of their gravimetric determination. Finally, according to the nature of the complex, the precipitate may be directly weighed after judicious drying, or alternatively, if its composition is not perfectly known, it may be transformed into the metal oxide by ignition.

The drawback of these precipitants is their lack of specificity. They are only selective of different groups of ions, the limits of which are ill defined. However, it is often possible to precipitate only one ion of a group by controlling the experimental conditions very sharply, in particular, the pH value.

Let's mention some organic precipitants:

- oxine or 8-hydroxyquinoleine (C_9H_7ON). We have already given its structure and some of its properties (see Chap. 35). From the standpoint of their structures, formed chelates are of the form $M(C_9H_6ON)_2$ when the ion coordinence is 4 (Mg^{2+}, Zn^{2+}, Cu^{2+}, Cd^{2+}, Pb^{2+}), of the form $M(C_9H_6ON)_3$ when the ion coordinence is 6 (Al^{3+}, Fe^{2+}, Bi^{III}, Ga^{3+}), and of the form $M(C_9H_6ON)_4$ when the coordinence is 8 (Th^{+IV}, Zr^{+IV}). Table 39.1 gives the pH ranges in which varied oxinates precipitate (their solubility products are given in Chap. 34);
- dimethylglyoxime H_2dmg. It precipitates Ni^{2+} and Pd^{2+} ions to give the complexes $M(C_4H_7M_2O_2)_2$. The complex given by Pd^{2+} has the same structure as that obtained with Ni^{2+} (see Chap. 30);
- cupferron, which is the ammonium salt of N-nitroso-N-phenylhydroxylamine. It precipitates Fe^{3+}, $VO_2{}^+$, Ti^{4+}, Zr^{4+}, Ce^{IV}, Nb^V, Ta^V, W^{VI}, Ga^{3+}, and Sn^{4+} and hence separates them from Al^{3+}, Be^{2+}, Cr^{3+}, Mn^{2+}, Ca^{2+}, Zn^{2+}, U^{VI}, Sr^{2+}, and Ba^{2+} in acidic medium.
- benzoin-α-oxime is above all used to precipitate copper(II) by forming the $Cu(C_{14}H_{11}O_2N)$ complex in ammonia:

$$\begin{array}{l} C_6H_5-CH-OH \\ \qquad\quad | \\ C_6H_5-C{=}N-OH \end{array}$$

Table 39.1 pH ranges for precipitation of some metal oxinates (see Vogel's 1978 in References)

	pH Initial precipitation	Complete precipitation
Thorium	3.9	4.4–8.8
Titanium	3.6	4.8–8.6
Tungsten	3.5	5.0–5.7
Uranium	3.7	4.9–9.3
Vanadium	1.4	2.7–6.1
Zinc	3.3	4.4...
Aluminum	2.9	4.7–9.8
Bismuth	3.7	5.2–9.4
Cadmium	4.5	5.5–13.2
Calcium	6.8	9.2–12.7
Cobalt	3.6	4.9–11.6
Copper	3.0	3.3...
Iron(III)	2.5	4.1–11.2
Lead	4.8	8.4–12.3
Magnesium	7.0	8.7...
Manganese	4.3	5.9–9.5
Molybdenum	2.0	3.6–7.3
Nickel	3.5	4.6–10

Cu^{2+} may therefore be separated from Cd^{2+}, Pb^{2+}, Ni^{2+}, Co^{2+}, Zn^{2+}, Al^{3+}, and Fe^{3+};

- nitron:

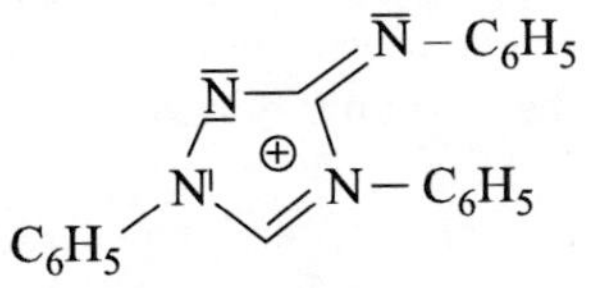

which precipitates the following anions: NO_3^-, ClO_4^-, BF_4^-, and WO_4^{2-} in acidic medium;

- tetraphenylarsonium chloride $(C_6H_5)_4As^+$, Cl^-, which precipitates the anions $Cr_2O_7^{2-}$, MnO_4^-, ReO_4^-, MoO_4^{2-}, WO_4^{2-}, ClO_4^-, and I_3^-;
- salicylaldehyde oxime:

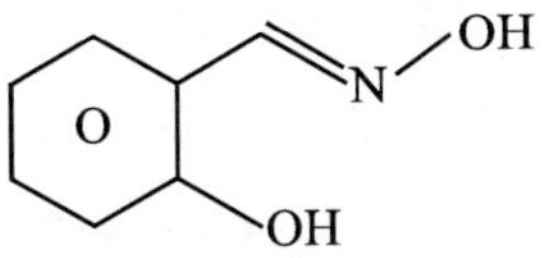

which precipitates Cu^{2+}, Pb^{2+}, Bi^{III}, Zn^{2+}, Ni^{2+}, and Pb^{2+};

- anthranilic acid:

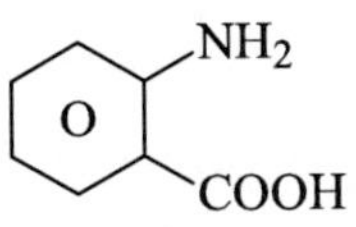

whose sodium salt, in weakly acidic or neutal medium, precipitates Cd^{2+}, Zn^{2+}, Ni^{2+}, Co^{2+}, and Cu^{2+}.

39.5 Qualitative Organic Analysis

We use the term "qualitative organic analysis" for the identification of organic compounds as well as for that of some atoms or of some groups of atoms present in an organic molecule. Both identifications are actually located in two different fields of organic analysis:

- the immediate analysis. Very often this branch of analysis uses precipitation phenomena, for example, to separate the product under study from a mixture. For some authors, the identification of organic compounds constitutes one part of the immediate analysis;
- the functional and elementary analysis.

An inexpensive means of identification of a pure crystallized organic compound is to determine its melting point. If the melting point of the product under study is identical to that of the expected one, it is very probable that both products are identical. By no means is this assertion an absolute certitude, for several reasons. First, it is possible that two different products fortuitously exhibit the same melting point. Second, the compound under identification may decompose by heating. This is the reason why two determinations are often recommended, one instantaneous and the other slow. We must also keep in mind that an impurity may considerably decrease the melting point.

In this book, we only give some examples of recommended identifications of pharmaceutical products through measurements of melting points. There are two possibilities: either the measured melting point is that of the substance to be identified or it is that of one of its immediate derivatives. The second possibility is the more frequent one. We'll mention the identifications of

- methanol and menthol both as 3,5-dinitrobenzoates. It is interesting to note in passing that the melting points of 3,5-dinitrobenzoates of both enantiomers of the latter differ greatly from that of the racemic (respectively, 154–157°C and 130–131°C);
- propylenglycol as di-p-nitrobenzoate;
- sorbitol as hexaacetate;
- camphor as oxime. Again, the melting point of enantiomers differs markedly from that of the racemic (118°C and 121°C);
- lactic acid as 2,4-dinitrophenylhydrazone after its oxidization into pyruvic acid;
- amphetamine as the benzoylated derivative;

- benzalkonium as the tetraphenylboron derivative (benzalkonium is a chloride of quaternary ammonium). Hence, an anion exchange occurs before the measurement of the melting point;

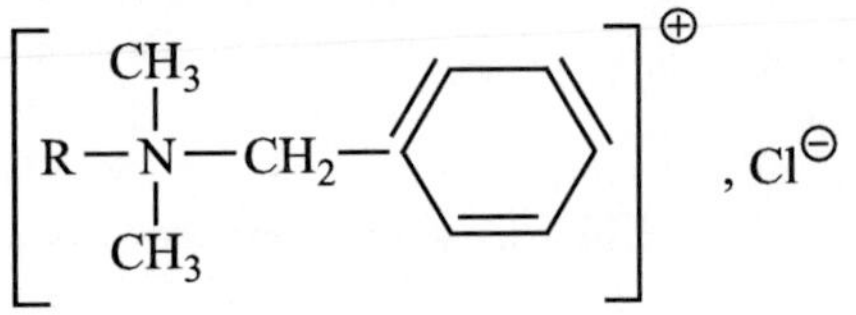

- chlorprothixene:

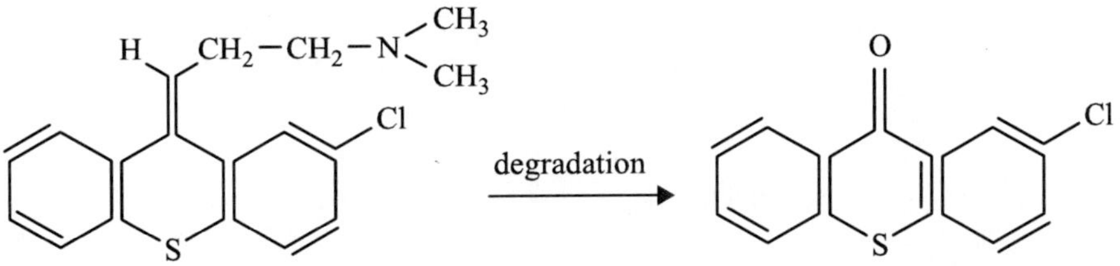

after degradation in the corresponding ketone. The latter is identified through its melting point without any transformation. Often, ketones are identified through the melting points of their oximes or of their 2,4-dinitrophenylhydrazones. Camphor provides a good example of this assertion. Benzyl alcohol is also identified by the formation of a 2,4-dinitrophenylhydrazone. In the occurrence, it is that formed from benzaldehyde. The latter derivative results from the oxidization of benzyl alcohol.

Now, in terms of identifying some general classes of organic derivatives, we know that it requires reagents that must be at once sufficiently general to detect the whole of the derivatives of the class and, however, not too general for the products of the group to be identified. Let's recall the identification of the following classes by a precipitation process:

1. amines and alkaloids. With Draggendorf's reagent (tetraiodobismuthate(III) or potassium "iodobismuthite" $K[BiI_4]$) in acidic media, they give red or orange amorphous precipitates that can be decomposed in alkaline media. (Draggendorf's reagent, more precisely, is used in particular to identify papaverine and morphine. It is also used as revealing agent in surface chromatography.) The same class of compounds yields white or yellow precipitates with Valser–Mayer's reagent (potassium tetraiodomercurate(II) $K_2[HgI_4]$). Amines and alkaloids also yield precipitates whose colors may vary from yellow to brown-black with Bouchardat's reagent, which is a mixture of iodine and iodide ions. Precipitates may dissolve in an excess of iodide ions. This is the reason why the reagent must contain a well-defined quantity of iodide ions. The precipitate is decomposed by treatment with sulfurous anhydride (sodium acid sulfite+hydrochloric acid). Iodine oxidizes sulfurous anhydride and then alkaloids give soluble hydrochlorides. Inversely,

alkaloids (or amines) may be extracted into organic solvents after alkalinization. As their precipitant reagents, we also mention chloroplatinic acid $H_2[PtCl_6]$, phosphotungstates and phosphomolybdates, Reinecke's salt already mentioned, and black Roussin's salt $K[Fe_4(NO)_7S_3]$. Finally, the same class of compounds yields poorly soluble addition products with picric acid and with some other nitrated organic acids. Their melting points are often sharp. It is the case, for example, with alkaloids derived from tropane and also with naphazoline;

2. urea derivatives such as amides, semicarbazones, hydrazines, imides, sulfoamides, amines, mercaptans, and compounds possessing an active methylene that precipitate with xanthydrol.

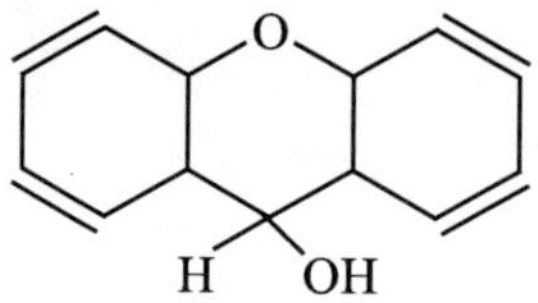

This reaction involves an intermediary carbonium ion whose formation is accompanied by an energy gain due to an increase in resonance. Later, the carbonium ion undergoes an attack from a nucleophilic species, with a proton loss. Xanthydrol permits us to obtain characteristic precipitates with bromisoval and sulfalinamide. The xanthydrol reaction may be considered to be relatively selective of amides. With barbiturics, which may be considered diamides, the reaction may occur twice.

The scheme is as follows:

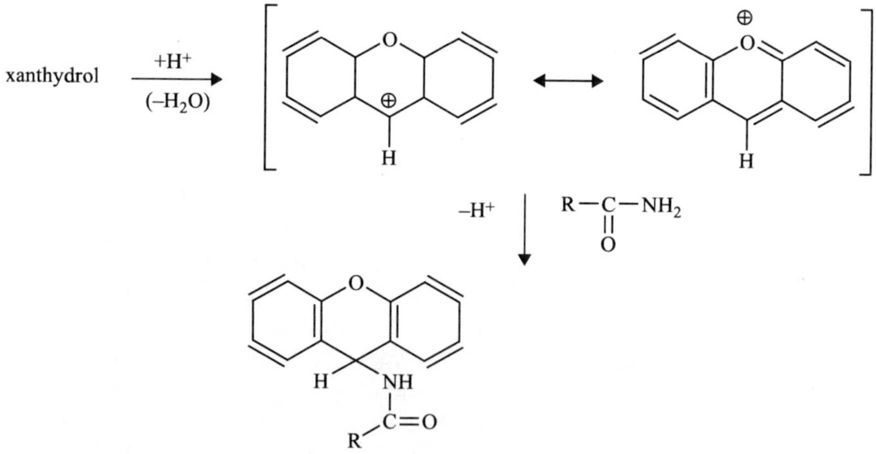

3. oses which react with Fehling's solution. They give a precipitate of Cu^{I} oxide (cuprous oxide) due to their reducing power. Let's also recall the formation of osazones for the identification of oses. Oses react with phenylhydrazine in well-defined conditions to yield crystallized yellow phenylosazones poorly soluble in

water. By heating the ose with an excess of phenylhydrazine, three molecules of the latter react with the sugar to give the previous phenylosazone in which two supplementary phenylhydrazine rests are linked. The linear structure of oses, although not totally satisfactory for all their chemical properties to be taken into account, is, however, sufficient to write a plausible reactional scheme explaining this reaction:

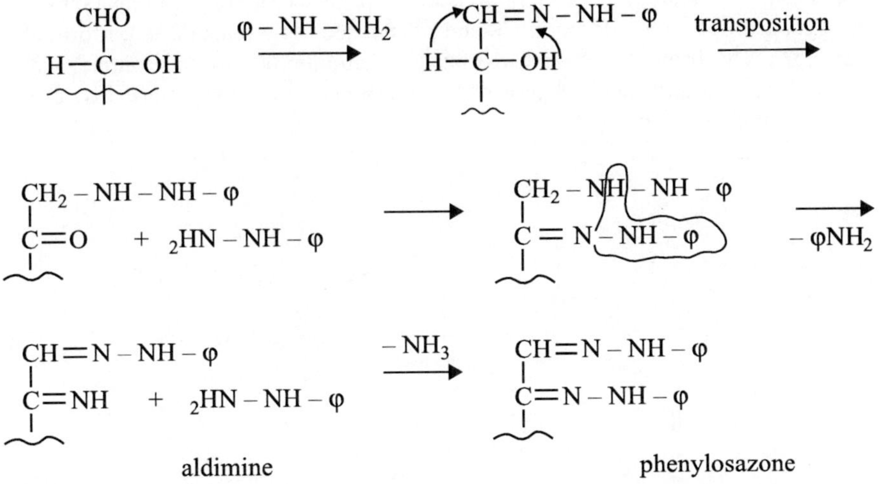

We can ask about the stoppage of the reaction after the introduction of the last last phenylhydrazine rests. The given explanation is the occurrence of intramolecular hydrogen bonds in the phenylosazone. However, a point remains puzzling in the scheme. It is the fact that one molecule of phenylhydrazine behaves like an oxidant since it is reduced into aniline and the oxidizing behavior of phenylhydrazine is unknown so far in other reactions!

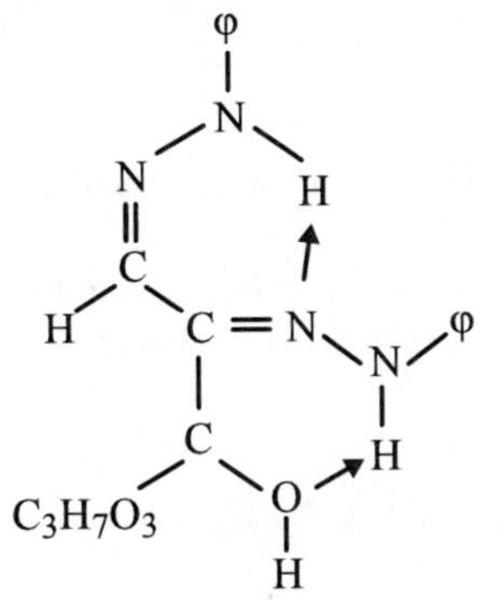

Finally, in order to stress the interest in precipitation reactions in order to study the structure of some molecules, we choose the case of halothane:

$$CF_3 - CH(Br) - Cl,$$

whose structure possesses three fluorine, one bromine, and one chlorine atoms. It is decomposed by hydrogen peroxide in ammonia with heating at 50°C. An aqueous solution containing the three halides F^-, Br^-, Cl^- is obtained. The problem is to distinguish them from the others, that is, to identify them without identifying the others.

Fluorine ions are evidenced by the destruction of the zirconium–alizarin complex (see Chap. 30). Revealing bromide ions in the presence of chloride ions is trickier. It is carried out with chloramine T and phenol red. The reaction is performed at pH = 5.2. The bromide ions are oxidized by chloramine T into bromine, which reacts by substitution with phenol red. A purple color appears. Normally, a pure solution of phenol red is yellow.

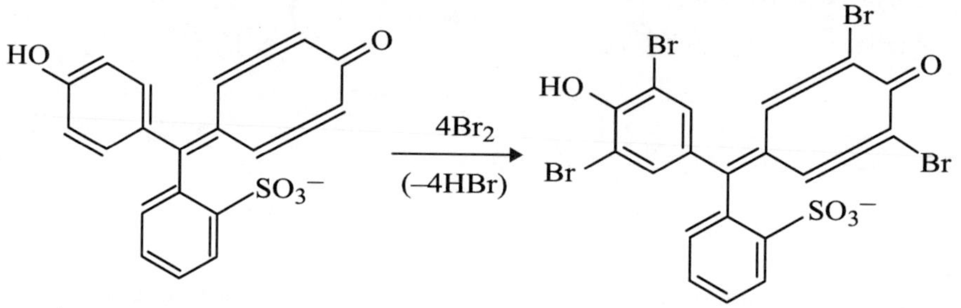

Seeking chloride ions in the mineralization solution containing bromide and iodide ions is carried out as follows. Sulfuric acid, acetone, and potassium bromate are first added to the solution. Bromine is formed by retrodismutation from the bromide ions already present and from bromate ions added. Bromine reacts with acetone to yield pentabromoacetone:

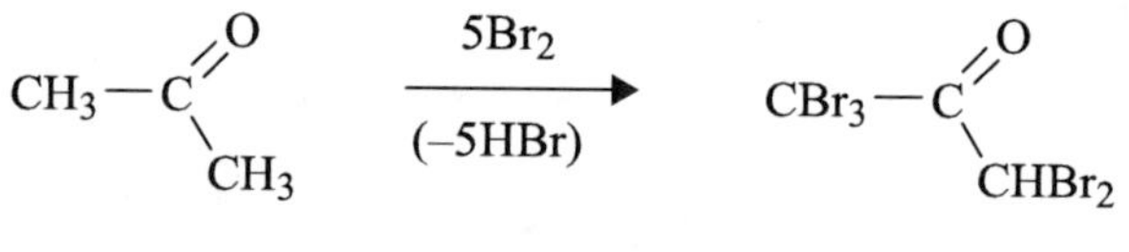

In such a way, the bromide ions are eliminated and the chloride ions may be precipitated as silver chloride.

Numerous halogenated derivatives belonging to the pharmacopeias must be analyzed so that the presence of halogen atoms in their structures may be checked.

39.6 Inorganic Qualitative Analysis

Innumerable examples prove the great interest that the precipitation phenomenon in qualitative inorganic analysis presents. Some have been mentioned in Chapter 30 in connection with complexation phenomena. Often both phenomena do indeed coexist. For example, a formed complex may precipitate.

Perhaps the most emblematic example of the interest in the precipitation phenomenon is provided by the method of systematic research of cations and anions in a solution or powder. We limit ourselves to recalling the following strategy, which is sufficiently illustrative for our purpose. It is due to Fresenius and has been improved by Villiers.[1]

Concerning the cations, the principle of the method consists of using general reagents that classify metallic ions in six different groups. For the first five groups, the discrimination is based on precipitation phenomena. These phenomena are checked in a chronologic manner. Therefore, one can detect

- metals of group I, which precipitate in the presence of hydrochloric acid: Ag, Hg^{I}, Pb;
- metals precipitating with dihydrogen sulfide in hydrochloric acid medium: As, Sb, Sn, Hg^{II}, Pb, Bi, Cu, Cd;
- metals precipitating with ammonia: Fe, Cr, Al;
- metals precipitating with dihydrogen sulfide in ammonia or with ammonium sulfide: Co, Ni, Mn, Zn;
- metals precipitating with ammonium carbonate: Ba, Ca, Sr and those precipitating with disodium phosphate in ammonia;

Metals of group VI are sought in the primitive solution. In some cases, they are revealed by the formation of a precipitate. Moreover, we must know that in each group, more acute and more numerous distinctions are performed on the basis of new precipitation reactions.

[1] Villiers, professor at the École Supérieure de Pharmacie in Paris.

Appendix A
The Chain Rule or Differentiating a Function of a Function

A function of a function may be simply represented under the form

$$y = f(u),$$
$$\text{with } u = \varphi(x),$$

where u is itself a function of x. The sole independent variable is x. $u(x)$ is an intermediary function. The problem is to derive y with respect to x. If Δx represents a change in x and Δy the corresponding change in y, we know by definition that

$$\frac{\mathrm{d}y}{\mathrm{d}x} = \lim_{(\Delta x \to 0)} \left(\frac{\Delta y}{\Delta x}\right)$$

When the change in x is Δx, that of u is Δu and that of y is Δy. It is perfectly legitimate to write

$$\frac{\Delta y}{\Delta x} = \left(\frac{\Delta y}{\Delta x}\right)\left(\frac{\Delta u}{\Delta u}\right) = \left(\frac{\Delta y}{\Delta u}\right) / \left(\frac{\Delta u}{\Delta x}\right).$$

When Δx approaches zero, Δu and Δy also approach zero. As a consequence,

$$\lim_{(\Delta x \to 0)} \left(\frac{\Delta y}{\Delta x}\right) = \lim_{(\Delta u \to 0)} \left(\frac{\Delta y}{\Delta u}\right) \cdot \lim_{(\Delta x \to 0)} \left(\frac{\Delta u}{\Delta x}\right)$$

By definition,

$$\operatorname*{limit}_{(\Delta u \to 0)} \left(\frac{\Delta y}{\Delta u}\right) = \frac{\mathrm{d}y}{\mathrm{d}u},$$
$$\operatorname*{limit}_{(\Delta x \to 0)} \left(\frac{\Delta u}{\Delta x}\right) = \frac{\mathrm{d}u}{\mathrm{d}x}.$$

Hence, we have the chain rule formula:

$$\frac{\mathrm{d}y}{\mathrm{d}x} = \left(\frac{\mathrm{d}y}{\mathrm{d}u}\right)\left(\frac{\mathrm{d}u}{\mathrm{d}x}\right).$$

J.-L. Burgot, *Ionic Equilibria in Analytical Chemistry*,
DOI 10.1007/978-1-4419-8382-4, © Springer Science+Business Media, LLC 2012

Of course, using the chain rule efficiently requires that we find a judicious choice of the intermediary function u, that is, a function that permits an easy calculation of the derivative dy/dx.

As an example, we'll set up the formula giving the buffer index of a weak monoprotic acid (Chap. 6).

We know (Chap. 6) that in a solution containing C_b mol/L of sodium hydroxide, C_a mol/L of hydrochloric acid, and C mol/L of a weak acid more and less dissociated as HA and A^- the charge balance can be written as

$$C_b = C_a + K_w/[H_3O^+] - [H_3O^+] + \{K_a/(K_a + [H_3O^+])\}C.$$

By definition, the buffer index β is

$$\beta = \frac{dC_b}{dpH}.$$

According to the chain rule, we can write

$$\beta = \left(\frac{dC_b}{d[H_3O^+]}\right)\left(\frac{d[H_3O^+]}{dpH}\right).$$

The continuation of the calculation is as follows:

$$pH = -\log[H_3O^+] \quad \text{(mixing up activity and concentration)},$$

$$pH = -\left(\frac{1}{2.303}\right)\ln[H_3O^+],$$

$$\frac{dpH}{d[H_3O^+]} = -\left(\frac{1}{2.303}\right)d\ln\frac{[H_3O^+]}{d[H_3O^+]},$$

$$\frac{dpH}{d[H_3O^+]} = -\left(\frac{1}{2.303}\right)\left(\frac{1}{[H_3O^+]}\right),$$

$$\frac{d[H_3O^+]}{dpH} = 1/dpH/d[H_3O^+],$$

$$\frac{d[H_3O^+]}{dpH} = -2.303[H_3O^+],$$

and so on.

Appendix B
Sharpness Index for the Titration of a Strong Acid with a Strong Base

Let's recall (Chap. 10) that the exact equation of the titration curve is

$$\frac{C_o V_o}{(V_o + V)}(\varphi - 1) = K_w/\mathrm{H} - \mathrm{H}.$$

At the final point (f.p.), near the equivalence point (e.p.), we assume, from a numerical standpoint, that the number of moles of added titrant is very close to that of the titrand:

$$C V_{\mathrm{f.p.}} \approx C_o V_o.$$

With this hypothesis, the exact equation becomes

$$\varphi - 1 = (C_o + C)/C_o C(K_w/\mathrm{H} - \mathrm{H}) \quad \text{near the equivalence point.}$$

By definition,

$$\eta = \frac{\mathrm{dpH}}{\mathrm{d}\varphi}.$$

According to the chain rule,

$$\eta = \left(\frac{\mathrm{dpH}}{\mathrm{dH}}\right)\left(\frac{\mathrm{dH}}{\mathrm{d}\varphi}\right) \quad \text{with} \quad \frac{\mathrm{dH}}{\mathrm{d}\varphi} = \frac{1}{(\mathrm{d}\varphi/\mathrm{dH})}.$$

Near the equivalence point,

$$\left(\frac{\mathrm{d}\varphi}{\mathrm{dH}}\right)_{\mathrm{eq.p.}} = [(C_o + C)/C_o C][-K_w/\mathrm{H}_{\mathrm{eq.p.}^2} - 1],$$

$$\left(\frac{\mathrm{d}\varphi}{\mathrm{dH}}\right)_{\mathrm{eq.p.}} = -\left[\frac{(C_o + C)}{C_o C}\right]\left[\frac{(K_w + \mathrm{H}_{\mathrm{eq.p}}{}^2)}{\mathrm{H}_{\mathrm{eq.p.}}{}^2}\right],$$

$$\left(\frac{\mathrm{dH}}{\mathrm{d}\varphi}\right)_{\mathrm{eq.p.}} = -\left[\frac{C_o C}{(C + C_o)}\right]\left[\frac{\mathrm{H}_{\mathrm{eq.p.}}{}^2}{(K_w + \mathrm{H}_{\mathrm{eq.p.}}{}^2)}\right].$$

[1] From now on, we symbolize $[H_3O^+]$ by H for simplicity of writing.

J.-L. Burgot, *Ionic Equilibria in Analytical Chemistry*,
DOI 10.1007/978-1-4419-8382-4, © Springer Science+Business Media, LLC 2012

From another standpoint, we know that (see Appendix A)

$$\frac{\text{dpH}}{\text{dH}} = -\frac{1}{(2.303\text{H})},$$

from which we have

$$\left(\frac{d\eta}{d\varphi}\right)_{\text{f.p.}} \approx \left[\frac{C_{\text{o}}C}{(C + C_{\text{o}})}\right]\left(\frac{1}{2.303}\right)\left[\frac{\text{H}_{\text{eq.p.}}}{(K_{\text{w}} + \text{H}_{\text{eq.p.}}{}^2)}\right].$$

At the equivalence point,

$$\text{H}_{\text{eq.p.}} = 10^{-7}\ \text{mol/L},$$

$$\frac{\text{H}_{\text{eq.p.}}}{(K_{\text{w}} + \text{H}_{\text{eq.p.}}{}^2)} = \left(\frac{1}{2}\right) \times 10^7,$$

$$\left(\frac{d\eta}{d\varphi}\right)_{\text{f.p.}} \approx 0.217 \times 10^7 \left[\frac{C_{\text{o}}C}{(C + C_{\text{o}})}\right].$$

Appendix C
Sharpness Index for the Titration of a Weak Acid with a Strong Base and Conversely

- Titration of a weak acid with a strong base

Recall (Chap. 10) that the exact equation of the titration curve is

$$\varphi = K_a/(K_a + \mathrm{H}) + [(V_o + V)/C_o V_o][K_w/\mathrm{H} - \mathrm{H}].$$

Near the equivalence point (see Appendix B),

$$\varphi = K_a/(K_a + \mathrm{H}) + [(C + C_o)/CC_o][K_w/\mathrm{H} - \mathrm{H}].$$

Due to the fact that the equivalence point is located in the basic range, we have

$$\mathrm{H} \ll K_a \quad \text{and} \quad \mathrm{H} \ll \frac{K_w}{\mathrm{H}},$$

from which we have

$$\varphi = 1 + \left[\frac{(C_o + C)}{C_o C}\right]\frac{K_w}{\mathrm{H}}.$$

As a result,

$$\frac{d\varphi}{d\mathrm{H}} = -\left[\frac{(C_o + C)}{C_o C}\right]\frac{K_w}{\mathrm{H}^2},$$

$$\frac{d\mathrm{H}}{d\varphi} = -\left[\frac{C_o C}{(C + C_o)}\right]\frac{\mathrm{H}^2}{K_w},$$

$$\eta = \left(\frac{d\mathrm{pH}}{d\mathrm{H}}\right)\left(\frac{d\mathrm{H}}{d\varphi}\right),$$

$$\eta = \left(\frac{1}{2.303}\right)\left[\frac{C_o C}{(C + C_o)}\right]\frac{\mathrm{H}}{K_w},$$

$$\eta_{\mathrm{f.p.}} = 0.436\left[\frac{C_o C}{(C + C_o)}\right]\frac{\mathrm{H}_{\mathrm{f.p.}}}{K_w}.$$

J.-L. Burgot, *Ionic Equilibria in Analytical Chemistry*,
DOI 10.1007/978-1-4419-8382-4, © Springer Science+Business Media, LLC 2012

A convenient explicit expression may be obtained in terms of the equilibrium constants alone. Starting again from the expression

$$\varphi = K_a/(K_a + \mathrm{H}) + [(C + C_o)/CC_o][K_w/\mathrm{H} - \mathrm{H}].$$

and, at this point in our reasoning, neglecting only H vs. K_a, we find

$$\frac{d\varphi}{d\mathrm{H}} = -[(C + C_o)/CC_o](K_w/\mathrm{H}^2 + 1),$$

$$\frac{d\mathrm{H}}{d\varphi} = -\left[\frac{C_oC}{(C + C_o)}\right]\left[\frac{\mathrm{H}^2}{(K_w + \mathrm{H}^2)}\right],$$

$$\eta = \left(\frac{1}{2.303}\right)\left[\frac{CC_o}{(C + C_o)}\right]\left[\frac{\mathrm{H}}{(K_w + \mathrm{H}^2)}\right].$$

At the equivalence point, $\varphi = 1$. The titration curve equation reduces to

$$[(C + C_o)/C_oC][K_w/\mathrm{H} - \mathrm{H}] = 1 - K_a/(K_a + \mathrm{H})$$

$$\text{or} \quad \mathrm{H}/(K_a + \mathrm{H}) = [(C_o + C)/CC_o](K_w/\mathrm{H} - \mathrm{H}).$$

By neglecting H vs. K_a in the term on the left, we have

$$\mathrm{H}/K_a = [(C_o + C)/CC_o](K_w/\mathrm{H} - \mathrm{H}),$$

and moreover, neglecting H vs. K_w/H in the term on the right leads to

$$\mathrm{H}^2 = K_wK_a\left[\frac{(C_o + C)}{CC_o}\right].$$

Introducing this expression of H^2 into that of η gives

$$\eta = 0.434\left[\frac{C_oC}{(C + C_o)}\right]^{1/2}\left(\frac{K_a}{K_w}\right)^{1/2}.$$

- Titration of a weak base B with a strong acid

The reasoning is quite analogous to that followed in the preceding case. The only difference lies in the fact that the equivalence point is located in the acidic range. As a result, the hypotheses that must be made are

$$[\mathrm{H}^+]_{\mathrm{eq.p.}} \gg K_{a2}$$

(where K_{a2} is the ionization constant acid of the conjugated acid BH^+) and

$$[\mathrm{H}^+]_{\mathrm{eq.p.}} \gg [\mathrm{OH}^-]_{\mathrm{eq.p.}}.$$

The formula obtained is

$$|\eta| = 0.434\left[\frac{C_oC}{(C + C_o)}\right]^{1/2}\left(\frac{K_{b2}}{K_w}\right)^{1/2}$$

$$\text{with} \quad K_{b2} = \frac{K_w}{K_{a2}}.$$

Appendix D
Sharpness Index for the Titration of a Weak Acid with a Weak Base

It is not a difficult task to find from the charge balance that the exact equation of the titration curve of a weak acid with a weak base is

$$\mathrm{H} + \frac{\mathrm{H}}{(K_{a2} + \mathrm{H})}\left[\frac{CV}{(V_o + V)}\right] = \frac{K_w}{\mathrm{H}} + \left[\frac{K_{a1}}{(K_{a1} + \mathrm{H})}\right]\left[\frac{C_o V_o}{(V_o + V)}\right],$$

where K_{a1} is the acid ionization constant of the titrated acid HA and K_{a2} that of the conjugate acid of the titrant base B. We know that the pH at the equivalence point must be located between pK_{a1} and pK_{a2}. Near the equivalence point, we can assume that simultaneously

- $[\mathrm{BH}^+] \gg \mathrm{H}$ i.e., $\left[\frac{\mathrm{H}}{(K_{a2} + \mathrm{H})}\right]\left[\frac{CV}{(V_o + V)}\right] \gg \mathrm{H}$ or

$$\left[\frac{1}{(K_{a2} + \mathrm{H})}\right]\left[\frac{CV}{(V_o + V)}\right] \gg 1;$$

- $[\mathrm{A}^-] \gg [\mathrm{OH}^-]$ i.e., $\left[\frac{K_{a1}}{(K_{a1} + \mathrm{H})}\right]\left[\frac{C_o V_o}{(V_o + V)}\right] \gg \frac{K_w}{\mathrm{H}}.$

Indeed, the $\mathrm{H}_{\mathrm{eq.p.}}$ and $[\mathrm{OH}^-]_{\mathrm{eq.p.}}$ values are not very far from 10^{-7} mol/L. According to these hypotheses, we find that near the equivalence point,

$$\varphi = \left[\frac{K_{a1}}{(K_{a1} + \mathrm{H})}\right] \times \left[\frac{(K_{a2} + \mathrm{H})}{\mathrm{H}}\right].$$

After derivatization,

$$\frac{d\varphi}{d\mathrm{H}} = \frac{[-K_{a1}\mathrm{H}^2 - K_{a1}K_{a2}(K_{a1} + 2\mathrm{H})]}{(K_{a1}\mathrm{H} + \mathrm{H}^2)^2}.$$

At the equivalence point, $\varphi = 1$ and

J.-L. Burgot, *Ionic Equilibria in Analytical Chemistry*,
DOI 10.1007/978-1-4419-8382-4, © Springer Science+Business Media, LLC 2012

$$K_{a1}\mathrm{H} + \mathrm{H}^2 = K_{a1}K_{a2} + K_{a1}\mathrm{H},$$
$$\mathrm{H}^2 = K_{a1}K_{a2}.$$

Replacing H^2 and H by this expression and by $(K_{a1}K_{a2})^{1/2}$ in the derivative, we find the following relations after straightforward but somewhat lengthy calculations:

$$\frac{d\varphi}{d\mathrm{H}} = -\frac{2}{[K_{a1} + (K_{a1}K_{a2})^{1/2}]},$$

$$\frac{d\mathrm{H}}{d\varphi} = -\frac{[K_{a1} + (K_{a1}K_{a2})^{1/2}]}{2},$$

$$\eta = \left(\frac{1}{(2.303\mathrm{H})}\right)\left\{\frac{[K_{a1} + (K_{a1}K_{a2})^{1/2}]}{2}\right\},$$

$$\eta = \left(\frac{1}{4.606}\right)\left\{\frac{[K_{a1} + (K_{a1}K_{a2})^{1/2}]}{(K_{a1}K_{a2})^{1/2}}\right\},$$

$$\eta = 0.217\left\{\left[\frac{K_{a1}K_{b2}}{K_{\mathrm{w}}}\right]\frac{1}{2} + 1\right\}.$$

Appendix E
Finding an Approximate Expression of the Fraction α of the Added Titrant That Has Reacted

The exact expression that links the parameter α to the fraction titrated φ and to the parameter β is

$$\alpha^2\varphi - \alpha(\varphi + \beta + 1) + 1 = 0.$$

As we have said, a good approximation of the root is

$$\alpha \approx \frac{1}{(\varphi + \beta + 1)}.$$

It is found by using the binomial theorem, which is

$$(a+b)^m = a^m + \left(\frac{m}{1!}\right) a^{m-1} b + \left[\frac{m(m-1)}{2!}\right] a^{m-2} b^2 + \left[\frac{m(m-1)(m-2)}{3!}\right] a^{m-3} b^3 + \dots,$$

which itself results from the MacLaurin series formula of the function $y = (a + x)^m$. Replacing x by b gives the binomial theorem.

The above equation to be solved has the following roots:

$$\alpha = \frac{\{(\varphi + \beta + 1) \pm [(\varphi + \beta + 1)^{2-} \times 4\varphi]^{1/2}\}}{2\varphi}.$$

The square root $[(\varphi + \beta + 1)^2 - 4\varphi]^{1/2}$ is developed according to Newton's theorem by setting $a \equiv (\varphi + \beta + 1)^2$, $b \equiv -4\varphi$, and $m \equiv ½$. Retaining the negative root and the first two terms of the series brings the result.

J.-L. Burgot, *Ionic Equilibria in Analytical Chemistry*,
DOI 10.1007/978-1-4419-8382-4, © Springer Science+Business Media, LLC 2012

Appendix F
A Study of Liebig–Denigés's Titration Curves

1. Liebig's titration curve

The titration curve is the diagram pAg/φ.

- the relations used to set up the equations of the different portions of the titration curves are

– mass balance on silver:

$$\frac{CV}{(V_o + V)} = [Ag^+] + P + [Ag(CN)_2^-] + [Ag(CN)_3^{2-}] + [Ag(CN)_4^{3-}]. \quad (F.1)$$

The higher complexes are mentioned in order to be rigorous, but their concentrations may be considered negligible, in particular close to the equivalence point, where an excess of cyanide ions does not exist due to the values of the overall constants ($\beta_3 = 10^{21.8}$, $\beta_4 = 10^{20.4}$, $\beta_2 = 10^{20.54}$). Henceforth, they will be neglected. P is the number of moles of AgCN precipitate per liter of solution;

– mass balance on cyanide:

$$\frac{C_o V_o}{(V_o + \mathrm{v})} = [CN^-] + [HCN] + P + 2[Ag(CN)_2^-] + 3[Ag(CN)_3^{2-}] + 4[Ag(CN)_4^{3-}]. \quad (F.2)$$

Here, too, both of the last complexes will be neglected.

– mass balances on K^+ and NO_3^- (we suppose that cyanide ions have been added as potassium salt and that silver ions are added as nitrate):

$$[K^+] = \frac{C_o V_o}{(V_o + V)} \quad (F.3)$$

and

$$[NO_3^-] = \frac{CV}{(V_o + V)}; \quad (F.4)$$

J.-L. Burgot, *Ionic Equilibria in Analytical Chemistry*,
DOI 10.1007/978-1-4419-8382-4, © Springer Science+Business Media, LLC 2012

– solubility product of silver cyanide:

$$[\mathrm{Ag}^+][\mathrm{CN}^-] = K_{\mathrm{so}} \quad K_{\mathrm{so}} = 10^{-15.92}; \tag{F.5}$$

– overall stability constant of the "argentocyanide ion":

$$\beta_2 = \frac{[\mathrm{Ag(CN)_2}^-]}{\{[\mathrm{Ag}^+][\mathrm{CN}^-]^2\}} \quad \beta_2 = 10^{20.5}; \tag{F.6}$$

– complex formation constant of "argentocyanide ion" from solid silver cyanide:

$$K_{\mathrm{s2}} = 10^{4.62}.$$

(K_{s2} is the equilibrium constant of the following reaction:

$$\mathrm{AgCN}\downarrow + \mathrm{CN}^- \rightarrow [\mathrm{Ag(CN)_2}^-].)$$

– solubility product of silver "argentocyanide":

$$[\mathrm{Ag}^+][\mathrm{Ag(CN)_2}^-] = K_{\mathrm{so}}K_{\mathrm{s2}} \quad K_{\mathrm{so}}K_{\mathrm{s2}} = 10^{-11.3};$$

– acid dissociation constant of hydrocyanic acid:

$$K_a(\mathrm{HCN}) = \frac{[\mathrm{H}^+][\mathrm{CN}^-]}{[\mathrm{HCN}]} \quad pK_a(\mathrm{HCN}) = 9.32; \tag{F.7}$$

– the charge balance equation:

$$[\mathrm{OH}^-] + [\mathrm{CN}^-] + [\mathrm{Ag(CN)_2}^-] + [\mathrm{NO_3}^-] = [\mathrm{Ag}^+] + [K^+] + [\mathrm{H}^+]; \tag{F.8}$$

– the fraction titrated:

$$\varphi = \frac{2CV}{C_{\mathrm{o}}V_{\mathrm{o}}};$$

- first portion of the curve (before the equivalence point).

Actually, the equation $\varphi/[\mathrm{Ag}^+]$ cannot be expressed explicitly. However, the curve may be calculated point by point with the help of the following three relations:

$$\varphi = 1 - [\mathrm{CN}^-]\left(\frac{1 + [\mathrm{H}^+]}{K_a}\right) \Big/ C_{\mathrm{o}}, \tag{F.9}$$

$$[\mathrm{H}^+]^2\left(1 + \frac{[\mathrm{CN}^-]}{K_a}\right) = K_{\mathrm{w}}, \tag{F.10}$$

$$[Ag^+] = \left\{\frac{C_oC\varphi}{(2C + C_o\varphi)}\right\}\left(\frac{1}{\beta_2[CN^-]^2}\right). \tag{F.11}$$

The calculation is carried out as follows:

- a value of $[CN^-]$ is chosen first. (Notice that this value is still noticeable, since we are before the equivalence point);
- with this value, $[H^+]$ is calculated through (F.10);
- with the preceding values $[H^+]$ and $[CN^-]$, φ is calculated through (F.9);
- finally, $[Ag^+]$ is calculated through (F.11).

Equations (F.9), (F.10), and (F.11) are found as follows.

Before the equivalence point, the concentrations $[Ag^+]$ and P together with those of the higher complexes may be neglected. As a result, Eqs. (F.1) and (F.2) give

$$\frac{Cv}{(V_o + V)} \approx [Ag(CN)_2{}^-], \tag{F.1$'$}$$

$$\frac{C_oV_o}{(V_o + V)} \approx [CN^-] + [HCN] + 2[Ag(CN)_2{}^-]; \tag{F.2$'$}$$

- Concerning Eq. (F.9):

from Eq. (F.7), we find

$$[HCN] = \frac{[CN^-][H^+]}{K_a}.$$

Introducing this equation and (F.1′) into (F.2′) gives

$$\frac{C_oV_o}{(V_o + V)} \approx [CN^-]\left(\frac{1 + [H^+]}{K_a}\right) + \frac{2CV}{(V_o + V)}.$$

Dividing both members by the term on the left permits us to introduce φ. After rearranging, we find

$$\varphi = 1 - [CN^-]\left(\frac{1 + [H^+]}{K_a}\right)\left[\frac{(V_o + V)}{C_oV_o}\right].$$

Neglecting V vs. V_o affords (F.9);

- Concerning Eq. (F.10):

It is found from the charge balance equation (F.8) by taking into account mass balance relations involving the K^+ and $NO_3{}^-$ ions (F.3) and (F.4) and neglecting P, $[Ag^+]$, and the higher complexes, as before. We find

$$[H^+] + [HCN] = [OH^-]$$

(note that this relation is purely and simply the proton condition). Due to (F.7), the last equation immediately results in (F.10);

- Concerning Eq. (F.11):

It is found from the expression of β_2 [Eq. (F.6)]. It is equivalent to

$$[Ag^+] = \frac{[Ag(CN)_2^-]}{(\beta_2[CN^-]^2)}$$

and from (F.1′):

$$[Ag^+] = \frac{\{CV/(V_o + V)\}}{(\beta_2[CN^-]^2)},$$

or, since $2CV = \varphi C_o V_o$,

$$[Ag^+] = \left\{\frac{C_o C\varphi}{(2C + C_o\varphi)}\right\}\left(\frac{1}{\beta_2[CN^-]^2}\right) \quad \text{relation. (F.11).}$$

(In order to demonstrate that

$$\frac{CV}{(V_o + V)} = \frac{C_o C\varphi}{(2C + C_o\varphi)},$$

we must multiply the numerator and the denominator of the left-hand term by the factor $2C_oC$ and introduce the definition of φ into the expression obtained.)

- after the equivalence point, the only important terms in both mass balance equations (F.1) and (F.2) are P, $[Ag^+]$, and $[Ag(CN)_2^-]$:

$$\frac{CV}{(V_o + V)} = [Ag^+] + P + [Ag(CN)_2^-], \qquad (F.1'')$$

$$\frac{C_oV_o}{(V_o + V)} = P + 2[Ag(CN)_2^-].$$

We deduce that

$$P = \frac{C_oV_o}{(V_o + V)} - 2[Ag(CN)_2^-].$$

Inserting this into (F.1″) gives

$$\frac{CV}{(V_o + V)} = \frac{C_oV_o}{(V_o + V)} + [Ag^+] - [Ag(CN)_2^-].$$

Dividing both members by $C_oV_o/ (V_o + V)$ permits us to introduce φ:

$$\frac{\varphi}{2} = 1 + [Ag^+]\frac{(V_o + V)}{C_oV_o} - [Ag(CN)_2^-]\frac{(V_o + V)}{C_oV_o}.$$

Neglecting V vs. V_o gives

$$\frac{\varphi}{2} = 1 + \frac{[Ag^+]}{C_o} - \frac{[Ag(CN)_2^-]}{C_o}.$$

Now we need to express $[Ag(CN)_2^-]$ in relation to $[Ag^+]$. This is done by the introduction of K_{s2} and subsequently by that of K_{so}. Since

$$K_{s2} = \frac{[Ag(CN)_2^-]}{[CN^-]} \quad \text{and} \quad K_{so} = [Ag^+][CN^-],$$

we find

$$\varphi = 2 + \frac{2[Ag^+]}{C_o} - \frac{2K_{s2}K_{so}}{[Ag^+]C_o}.$$

Actually, the product $K_{s2}K_{so}$ is purely and simply the solubility product K_s of silver "argentocyanide." The relation becomes

$$\varphi = 2 + \frac{2[Ag^+]}{C_o} - \frac{2K_s}{[Ag^+]C_o} \quad \text{relation given in Chap. 37;}$$

- close to the equivalence point, both mass balance relations are satisfied as well as the solubility product $K_s(Ag[Ag(CN)_2])$, although we must still consider that $P = 0$. After handling the mass balance equations and taking into account the definition of φ, we find

$$\frac{\varphi}{2} = \frac{\{[Ag^+] + [Ag(CN)_2^-]\}}{\{[CN^-] + [HCN] + 2[Ag(CN)_2^-]\}}.$$

(We must turn our attention to the fact that, henceforth, all the concentrations that appear in the above equation are those at the equivalence point, such as $[Ag^+]_{peq}$ and so on. The subscript eq.p. is systematically omitted for the simplicity of writing.)

At this point, it is convenient to replace $[Ag^+]$ by an equivalent relation involving $[CN^-]$. [HCN] can also be expressed as a function of $[CN^-]$. Both operations are carried out using Eqs. (F.5) and (F.7). Moreover, $[Ag(CN)_2^-]$ can also be expressed in relation with $[CN^-]$ through K_{s2}. We obtain

$$\frac{\varphi}{2} = \left\{\frac{K_{so}}{[CN^-]} + K_{s2}[CN^-]\right\} \Big/ \left\{[CN^-]\left(1 + \frac{[H^+]}{K_a}\right) + 2K_{s2}[CN^-]\right\}$$

or

$$2K_{s2}[CN^-](\varphi - 1) = \frac{2K_{so}}{[CN^-]} - \varphi[CN^-]\left(1 + \frac{[H^+]}{K_a}\right).$$

Close to the equivalence point, $\varphi \approx 1$; as a result,

$$2K_{s2}[CN^-](\varphi - 1) \approx \frac{2K_{so}}{[CN^-]} - [CN^-]\left(1 + \frac{[H^+]}{K_a}\right).$$

It remains to express $[CN^-]$. According to the definition of K_{s2}:

$$[CN^-] = \frac{[Ag(CN)_2{}^-]}{K_{s2}},$$

and since the reactions can be considered complete,

$$2[Ag(CN)_2{}^-] = \frac{C_o V_o}{(V + V_o)}.$$

Hence,

$$[CN^-] = \frac{C_o V_o}{[(V_o + V)2K_{s2}]}$$

and

$$\varphi_{f.p.} - 1 = \left[\frac{(V_o + V)}{C_o V_o}\right]\left\{\frac{2K_{so}}{[CN^-]_{f.p.}} - [CN^-]_{f.p.}\left(1 + \frac{[H^+]}{K_a}\right)\right\}.$$

It is not difficult to demonstrate that $[(V_o + V)/C_o V_o] = (C_o + 2C)/2C_o C$. To do so, it is sufficient to multiply the numerator and denominator of the term on the left-hand side and to replace CV by $C_o V_o/2$ since close to the equivalence point, $\varphi \approx 1$:

$$\varphi_{f.p.} - 1 = \left\{\frac{(C_o + 2C)}{2C_o C}\right\}\left\{\frac{2K_{so}}{[CN^-]_{f.p.}} - [CN^-]_{f.p.}\left(1 + \frac{[H^+]}{K_a}\right)\right\} \quad \text{(F.12)}$$

(an equation given in Chap. 37);

- titration error

The equation just above is very interesting since it permits us to calculate the titration error once the concentration $[CN^-]_{f.p.}$ is known. The latter is accessible because of the reasoning below. Since the precipitation occurs, the constant K_{s2} is satisfied:

$$K_{s2} = \frac{[Ag(CN)_2{}^-]}{[CN^-]}.$$

From another standpoint, since the formation reaction of "argentocyanide" may be considered complete, we can write

$$[Ag(CN)_2{}^-] = \frac{C_o V_o}{(V_{e.p.} + V_o)}/2.$$

Just above, we have seen that

$$[CN^-] = \frac{C_o V_o}{[(V_o + V)2K_{s2}]}$$

and that

$$\frac{(V_o + V)}{C_o V_o} = \frac{(C_o + 2C)}{2C_o C}.$$

As a result, we find

$$[CN^-]_{f.p.} = \frac{CC_o}{[K_{s2}(C_o + 2C)]}.$$

Therefore, in order to calculate the titration error, $[CN^-]_{f\cdot p.}$ is computed with the last equation, $[H^+]$ is still calculated through (F.10), and the titration error through (F.12). With $C = C_o = 10^{-2}$ mol/L, with $K_{s2} = 10^{4.62}$, we find

$$[CN^-]_{f.p.} = 8.0 \times 10^{-8},$$

$[H^+] = 7.7 \times 10$ $\quad \varphi_{f.p.} - 1 = -2.05 \times 10^{-4}$ result given in Chap. 37.

2. Denigés' modification

The essential difference between Liebig's and Denigés' methods lies in the fact that the constants that govern the phenomena occurring in Denigés' modification exhibit values different from those found in Liebig's titration since silver ions stand as ammine complexes, above all as the diamminesilver(I) complex.

a. Before the equivalence point, the mass balance equations on Ag^+ and CN^- are, respectively,

$$\frac{Cv}{(V_o + v)} = [Ag^+] + [Ag(NH_3)^+] + [Ag(NH_3)_2{}^+] + [Ag(CN)_2{}^-], \quad (F.13)$$

$$\frac{C_o V_o}{(V_o + v)} = [CN^-]\left(1 + \frac{[H^+]}{K_a}\right) + 2[Ag(CN)_2{}^-]. \quad (F.14)$$

There is a great excess of ammonia relative to silver ions; due to the previously known distribution diagram of the ammine complexes of silver, it is legitimate to consider that

$$[Ag(NH_3)_2{}^+] \gg [Ag(NH_3)^+].$$

Due to the constant β values [β overall formation constants of the "amminosilver" complexes], diamminesilver-I predominates ($\beta_2 = 10^{7.03}$), and Eq. (F.13) reduces to (F.13′):

$$\frac{Cv}{(V_o + v)} = [Ag(NH_3)_2{}^+] + [Ag(CN)_2{}^-]. \quad (F.13')$$

Before proceeding with calculations, let's notice that with the assumptions just adopted, the reaction that is considered as occurring before the equivalence point is

$$[Ag(NH_3)_2]^+ + 2CN^- \rightleftharpoons [Ag(CN)_2]^- + 2NH_3 \qquad \beta'_2.$$

(It results from the superimposition of the following two equations:

$$[Ag(NH_3)_2]^+ \rightleftharpoons Ag^+ + 2NH_3 \qquad 1/\beta_2{}^*$$

$$Ag^+ + 2CN^- \rightleftharpoons [Ag(CN)_2]^- \qquad \beta_2$$

$$\text{with} \quad \beta'_2 = \frac{\beta_2}{\beta_2{}^*} = 10^{13.51}.)$$

The titration reaction is still quantitative, and since we are presently studying the titration curve before the equivalence point, we can consider that the concentration $[Ag(NH_3)_2{}^+]$ is negligible. Equation (F.13′) becomes Eq. (F.13″):

$$\frac{Cv}{(V_o + v)} = [Ag(CN)_2{}^-]. \tag{F.13″}$$

Now, the calculations are quite analogous to those encountered in Liebig's method. The three relations describing the titration before the equivalence point are

$$\varphi = 1 - [CN^-]\left\{1 + \frac{[H^+]}{K_a}\right\}\Big/C_o. \tag{F.15}$$

It is found by handling (F.13″), (F.14), and the expression of φ, and by neglecting v compared to V_o;

$$[H^+]^2\left\{1 + \frac{[CN^-]}{K_a}\right\} = K_w. \tag{F.16}$$

This relation remains valid if one acknowledges that the proton condition itself remains valid (this is the case if the concentration $[NH_4+]$ can be considered as being neglected);

$$[Ag(NH_3)^{2+}] = \left[\frac{C_oC\varphi}{(2C + C_o\varphi)}\right]\left\{\frac{[NH_3]^2}{(\beta'_2[CN^-]^2)}\right\}, \tag{F.16′}$$

$$\text{or} \quad [Ag^+] = \left[\frac{C_oC\varphi}{(2C + C_o\varphi)}\right]\left\{\frac{1}{(\beta'_2\beta_2{}^*[CN^-]^2)}\right\}; \tag{F.16″}$$

b. After the equivalence point, as we have already seen, the following reaction occurs:

$$[Ag(CN)_2]^- + [Ag(NH_3)_2]^+ \rightleftharpoons Ag[Ag(CN)_2]\!\downarrow + 2NH_3.$$

The only important terms in the mass balance on the CN^- and Ag^+ equations are P, $[Ag(NH_3)_2{}^+]$, and $[Ag(CN)_2{}^-]$. As a result, the equations become

$$\frac{Cv}{(V_o + v)} = [Ag(NH_3)_2{}^+] + P + [Ag(CN)_2{}^-],$$

$$\frac{C_oV_o}{(V_o + v)} = 2[Ag(CN)_2{}^-] + P.$$

The reasoning is analogous to that concerning the analogous part of Liebig's titration curve, but now we use the constant K_{s2}, which governs the equilibrium:

$$AgCN\downarrow + CN^- \rightleftharpoons [Ag(CN)_2]^-.$$

It leads to the relation

$$\varphi = 2 + \frac{2[Ag(NH_3)_2{}^+]}{C_o} - 2K_{s2}\frac{[CN^-]}{C_o}.$$

Until now, $[CN^-]$ has remained in this equation. It is more convenient to express φ totally as a function of $[Ag(NH_3)_2{}^+]$, which is in excess. To do so, it is convenient to introduce the apparent (in the given experimental conditions) solubility product of silver argentocyanide $K_s{}^*$ defined by

$$K_s{}^* = [Ag(NH_3)_2{}^+][Ag(CN)_2{}^-],$$

Taking account of K_{s2} and $K_s{}^*$, we find

$$[CN^-] = \frac{K_s{}^*}{(K_{s2}[Ag(NH_3)_2{}^+])} \quad \text{and}$$

$$\varphi = 2 + \frac{2[Ag(NH_3)_2{}^+]}{C_o} - \frac{2K_{s2}K_s{}^*}{([Ag(NH_3)_2{}^+]C_o)}.$$

Now we need to calculate $K_s{}^*$. It can be expressed from the "true" solubility product K_{so} as follows:

$$K_{so} = [Ag^+][Ag(CN)_2{}^-].$$

According to the expression of $\beta_2{}^*$ (see above),

$$[Ag^+] = \frac{[Ag(NH_3)_2{}^+]}{\beta_2{}^*[NH_3]^2} \quad \text{and,}$$

$$K_{so} = [Ag(CN)_2{}^-]\left\{\frac{[Ag(NH_3)_2{}^+]}{(\beta_2{}^*[NH_3]^2)}\right\}.$$

As a result,

$$K_s{}^* = K_{so}\beta_2{}^*[NH_3]^2.$$

$K_s{}^*$ is easily calculated. To do so, it is sufficient to add a large excess of ammonia (this is the case—see above) and then to assimilate the concentration of free ammonia to its analytical concentration. This is legitimate, since a great excess of ammonia is added.

c. It is useless to consider final and equivalence points from a theoretical standpoint as in Liebig's method because silver "argentocyanide" precipitates too late. Indeed, its formation reaction is too equilibrated because of the interference from ammonia. This is the reason for the addition of iodide ions to detect the equivalence point (see Chap. 37).

Bibliography

J. N. Butler—Ionic equilibria: A mathematical approach. Addison-Wesley, Reading, MA,1964. (A true masterpiece. In this book, I have no other ambition than applying Butler's principles of calculations or to analytical chemistry. In my opinion, this book was and is still, purely and simply, the key to understanding and handling equilibria in aqueous solutions without intuitive approximations.)

J. N. Butler—Ionic equilibrium: Solubility and pH calculations. Wiley-Interscience, New York, 1998. (An update of the preceding book with the last chapter written by D. R. Cogley devoted to general computer programs capable of performing equilibrium calculations.)

J. Bassett, R. C. Denney, G. H. Jeffery, and J. Mendham (Rev. Eds.) —Vogel's textbook of quantitative inorganic analysis, 4th ed. Longman, London and New York, 1978. (A very lucid and valuable book. It contains all that is necessary to practice inorganic analysis.)

G. Charlot—Cours dechimie analytique générale, tome I. Masson ed., Paris,1967. (A typical book from the French school in analytical chemistry that was driven by G. Charlot. In my book, I have followed the presentation of the different points that this author has systematically stressed.) On more of an ad hoc basis, I have consulted the following books:

U. R. Kunze—Grundlagen der quantitativen Analyse, 3rd ed. G. Thieme Verlag, Stuttgart, 1990 (An excellent summary.)

A. Ringbom—Les complexes en chimie analytique. Dunod, Paris, 1967 (French translation of "Complexation in analytical chemistry." John Wiley and Sons, New York, 1963). (No need to present the Scandinavian school in the chemistry of aqueous solutions any further!)

K. Eger, R. Troschütz, and H. J. Roth—Arzneistoffanalyse: Reaktivität, Stabilität, Analytik, 4th ed. Deutscher Apotheker Verlag, Stuttgart, 1999. (An excellent compilation of the analysis methods of the drugs most commonly used at the time of publication.)

R. Rosset, D. Bauer, and J. Desbarres—Chimie analytique des solutions et microinformatique. Masson ed., Paris, 1979. (This book concisely and generally presents a new way, based on the use of computers, to tackle equilibria in aqueous solutions.)

M. Pesez and J. Bartos—Colorimetric and fluorimetric analysis of organic compounds and drugs. Marcel Dekker, New York, 1974. (A very useful book for the quantitative organic analysis by colorimetry.)

J. Burgess—Ions in solution: Basic principles of chemical interactions. Horwood Publishing, Westergate, Chichester, UK, 1999. (The state of the art concerning ions in solution.)

J.-L. Burgot, *Ionic Equilibria in Analytical Chemistry*,
DOI 10.1007/978-1-4419-8382-4,

Index

J.-L. Burgot, *Ionic Equilibria in Analytical Chemistry*,
DOI 10.1007/978-1-4419-8382-4, © Springer Science+Business Media, LLC 2012

D

E